Telecommunications
Protocols and Design

John D. Spragins
Clemson University
with Joseph L. Hammond
Clemson University
and Krzysztof Pawlikowski
University of Canterbury

 Addison-Wesley Publishing Company

Reading, Massachusetts ▪ Menlo Park, California ▪ New York
Don Mills, Ontario ▪ Wokingham, England ▪ Amsterdam ▪ Bonn
Sydney ▪ Singapore ▪ Tokyo ▪ Madrid ▪ San Juan ▪ Milan ▪ Paris

Sponsoring Editor: Tom Robbins
Production Supervisor: Helen M. Wythe
Production Coordinator: Amy Willcutt
Art Consultant: Joseph K. Vetere
Copy Editor: Stephanie Argeros-Magean
Illustrators: Scientific Illustrators
Text Designer: Hunter Graphics
Cover Designer: Marshall Henrichs
Manufacturing Manager: Hugh Crawford

Library of Congress Cataloging-in-Publication Data

Spragins, John D.
    Telecommunications : protocols and design / John D.
    Spragins, Joseph L. Hammond, Krzysztof Pawlikowski.
        p.   cm.
    Includes bibliographical references.
    ISBN 0-201-09290-5
    1. Computer networks.   2. Computer network protocols.
I. Hammond, Joseph L.   II. Pawlikowski, Krzysztof.   III. Title.
TK5105.5.S67   1991
621.382—dc20   89-27327
                                                                                  CIP

Reprinted with corrections August, 1994

8 9 10 DO 9796

To Cathy

Contents

Chapter 1   **Introduction to Telecommunications**   *1*

1.1   Introduction   *1*
    1.1.1   Approaches to Telecommunications   *2*
    1.1.2   Future of Telecommunications   *2*

1.2   Motivations for Growth   *4*

1.3   Evolution of Telecommunication Systems   *6*
    1.3.1   Telecommunications Milestones   *7*
    1.3.2   Additional Milestones   *12*

1.4   Major Design Problems   *18*
    1.4.1   Resolving Incompatibilities of Equipment   *18*
    1.4.2   Coordination of Sender and Receiver   *18*
    1.4.3   Maximizing Reliability and
           Freedom from Errors   *21*
    1.4.4   Optimizing Performance   *22*
    1.4.5   Minimizing Cost   *23*
    1.4.6   Network Management   *24*

1.5   The Telecommunications Environment   *25*
    1.5.1   Technological Trends   *25*
    1.5.2   Impact of Installed Equipment Base   *26*
    1.5.3   Legal and Regulatory Factors   *27*
    1.5.4   Standards Activities   *27*

1.6   Summary   *28*

1.7   Preview of Later Chapters   *29*
    Problems   *30*

Chapter 2   **Fundamental Limits**   *31*

2.1   Introduction   *31*

2.2   Physical Constraints   *32*

2.3   Common Communication Channels   *33*
    2.3.1   Telephone Networks   *34*
    2.3.2   Radio Links   *38*
    2.3.3   Communication Satellites   *42*
    2.3.4   Copper Conductors   *43*
    2.3.5   Fiber Optics   *44*
    2.3.6   Locally Installed Facilities   *45*
    2.3.7   Communication Circuit Types and Operation   *46*

2.4   Accommodating Signals to Channels   *47*
    2.4.1   Time and Frequency Domain Representations
           of Signals   *47*
    2.4.2   Amplitude, Frequency, and Phase Modulation   *48*

☐ 2.4.3   Typical Modem Signal Constellations   *52*
☐ 2.4.4   Binary Signal Encoding Techniques   *57*
   2.4.5   Multiplexing and Concentration   *60*
   2.4.6   Relative Capacities of Multiplexers
           and Concentrators   *63*

2.5   Fundamental Limits on Transmission Rates   *66*
   2.5.1   Intersymbol Interference and Nyquist
           Transmission Rate   *66*
   2.5.2   Equalization Techniques   *72*
   2.5.3   Shannon Channel Capacity Theorem   *73*
   2.5.4   Attempting to Reach Shannon Bounds   *75*

2.6   Digital Communications   *77*
   2.6.1   Nyquist Sampling Theorem   *77*
   2.6.2   Digital Transmission of Analog Signals   *77*
   2.6.3   Characteristics of Major Types of
           Information Sources   *79*

2.7   Computing Power Limitations   *81*
2.8   Summary   *83*
      Problems   *85*

Chapter 3   **Introduction to Networks**   *95*

3.1   Introduction   *95*
3.2   Switching Techniques   *95*
   3.2.1   Circuit Switching   *95*
   3.2.2   Message Switching   *97*
   3.2.3   Packet Switching   *98*
   3.2.4   Datagrams Versus Virtual Circuits   *99*
   3.2.5   Performance Tradeoffs   *101*
   3.2.6   Optimal Packet Lengths to
           Maximize Pipelining   *103*

3.3   Classes of Networks   *105*
   3.3.1   Local Area Networks   *106*
   3.3.2   Wide Area Networks   *106*
   3.3.3   Metropolitan Area Networks   *107*
   3.3.4   Radio and Satellite Networks   *107*
   3.3.5   Internetworks   *108*
   3.3.6   Special Purpose Networks   *109*
   3.3.7   Major Types of Geometries   *109*

3.4   Naming and Addressing Conventions   *112*
3.5   The Structure of Telecommunication Networks   *113*
   3.5.1   Access Path Requirements   *114*
   3.5.2   Characteristics of Protocols   *117*
   3.5.3   Layered Protocol Architectures   *117*

3.6   The OSI Reference Model   *118*
   3.6.1   OSI Architecture Layers   *118*
   3.6.2   Peer-to-Peer Communication   *121*
   3.6.3   Interfaces Between Layers   *123*
   3.6.4   Typical Communication Scenario   *125*
   3.6.5   Limitations of Reference Model   *126*

3.7   The DoD Reference Model   *127*
       3.7.1   Differences in Emphases   *127*
       3.7.2   DoD Reference Model Architecture   *129*
3.8   Vendor Architectures   *133*
       3.8.1   Digital Network Architecture   *133*
       3.8.2   Systems Network Architecture   *138*
3.9   Other Important Standards   *144*
       3.9.1   Local Area Network Protocol Architectures   *144*
       3.9.2   MAP and TOP   *146*
       3.9.3   Integrated Services Digital
               Network Architecture   *146*
3.10   Comparison of Architectures   *147*
3.11   Summary   *149*
       Problems   *149*

Chapter 4   **Computer Equipment—Communications
             Equipment Interface Design**   *153*

4.1   Introduction   *153*
4.2   Overview of Physical Layer Protocols   *155*
4.3   EIA-232D/RS-232C/V.24   *156*
       4.3.1   Overview of RS-232C Interface   *156*
       4.3.2   Major Signals   *157*
       4.3.3   Typical States and Timing Diagrams   *157*
       4.3.4   Null Modems   *163*
       4.3.5   Limitations of RS-232C   *163*
4.4   X.21   *165*
       4.4.1   Overview of X.21   *165*
       4.4.2   Major Signals   *165*
       4.4.3   Typical States and Timing Diagrams   *167*
       4.4.4   Limitations of X.21 Interfaces   *171*
4.5   I.430 and I.431 ISDN Interfaces   *172*
       4.5.1   Overview of I.430   *172*
       4.5.2   Signal Formats   *173*
       4.5.3   I.430 Bidding for Access to D Channel   *175*
       4.5.4   I.431 Interface   *178*
       4.5.5   Limitations of I.430 and I.431   *179*
4.6   Physical Layer in DNA and SNA   *180*
4.7   Physical Layer in IEEE 802 and MAP/TOP   *180*
4.8   Invoking Physical Layer Services   *181*
4.9   Computer Equipment Interface   *183*
       4.9.1   A Typical Computer Equipment Interface   *184*
       4.9.2   USARTs   *186*
       4.9.3   More Powerful Computer
               Equipment Interfaces   *187*
4.10   Interface to Communications Channel   *188*
4.11   Impact of Interfaces on Performance   *189*
4.12   Summary   *192*
       Appendix 4A   Details of EIA-232D/RS-232C/V.24 Interface   *192*
       Problems   *196*

Chapter 5   **Medium Access Control** *205*

5.1   Introduction *205*
5.2   Modes of Accessing Communications Media *207*
    5.2.1   TDMA/Reservations *208*
    5.2.2   Polling *213*
    5.2.3   Token Passing *216*
    5.2.4   Random Access with no Sensing *224*
    5.2.5   Random Access with Sensing
           Before Transmission *226*
    5.2.6   Random Access with Sensing Before
           and During Transmission *229*
5.3   Performance Modeling *231*
    5.3.1   Performance Parameters and Environment *231*
    5.3.2   Basic Performance Relations *234*
    5.3.3   Performance of Scheduling Methods *235*
    5.3.4   Performance of Random Access Methods *242*
5.4   Assessment of Medium Access Protocols *254*
5.5   Summary *257*
    Problems *257*

Chapter 6   **Synchronization and Error Control** *261*

6.1   Introduction *261*
6.2   Types of Synchronization *261*
6.3   Control of Transmission Errors *263*
    6.3.1   Sources of Transmission Errors *263*
    6.3.2   Approaches to Error Control *264*
    6.3.3   Early DLC Error Control Techniques *264*
    6.3.4   Two-Dimensional Parity Checks *266*
    6.3.5   Simple Block Codes *267*
    6.3.6   Error Correcting Capabilities of
           Hamming (7,4) Code *269*
    6.3.7   Error Detecting Capabilities of
           Hamming (7,4) Code *270*
    6.3.8   More Powerful Codes *271*
    6.3.9   Polynomial Codes *272*
☐ 6.3.10   Implementation of Polynomial Codes *274*
☐ 6.3.11   Error Detecting Capabilities of
           Polynomial Codes *277*
    6.3.12   Polynomials in Standards *279*
    6.3.13   FEC versus ARQ *279*
6.4   ARQ Protocols *280*
    6.4.1   Stop-and-Wait ARQ *281*
    6.4.2   Continuous ARQ *284*
    6.4.3   Piggy-backed Acknowledgments and
           Sequence Counts *285*
☐ 6.4.4   Communications Efficiency Calculations *286*
☐ 6.5   Optimum Frame or Packet Sizes *293*
6.6   Flow Control for DLCs *295*
6.7   Summary *296*
    Problems *296*

Chapter 7   **Data Link Layer Protocols**   *303*

     7.1   Introduction   *303*
     7.2   Asynchronous or Start-Stop Data Link Controls   *304*
         7.2.1   Reception of Start-Stop Data   *305*
         7.2.2   Error Control   *306*
         7.2.3   DLC Overhead   *306*
         7.2.4   Limitations of Start-Stop DLC Protocols   *306*
     7.3   Character-Oriented DLC Protocols—Bisync   *307*
         7.3.1   Normal Text Transmission in Bisync   *307*
         7.3.2   Transmission of Transparent Text   *311*
         7.3.3   Control of Access to Data Link   *311*
         7.3.4   Limitations of Bisync   *314*
     7.4   ARPANET DLC   *315*
     7.5   Byte-Count–Oriented Protocols—DDCMP   *316*
         7.5.1   Interpretation of Fields in DDCMP Frame   *317*
         7.5.2   Typical Frame Sequences   *319*
         7.5.3   Limitations of DDCMP   *321*
     7.6   Bit-Oriented Protocols—HDLC   *321*
         7.6.1   HDLC Frame Format   *323*
         7.6.2   HDLC Operational Modes   *327*
         7.6.3   Typical Data Exchanges   *328*
         7.6.4   Initialization Sequences   *334*
         7.6.5   HDLC Subsets Used in ISO (X.25) and ISDN   *334*
         7.6.6   Limitations of HDLC   *335*
     7.7   Local Area Network and MAP/TOP DLC Protocols   *337*
     7.8   Overhead in Synchronous DLC Protocols   *341*
     7.9   Invoking Data Link Layer Services   *344*
     7.10   Summary   *347*
     Appendix 7A   Listing of Bisync Control State Frames   *348*
     Problems   *349*

Chapter 8   **Routing and Flow Control**   *355*

     8.1   Introduction   *355*
     8.2   Classification of Routing Algorithms   *357*
     8.3   Routing Tables   *359*
     8.4   Shortest-Path Routing   *362*
         8.4.1   The Bellman-Ford-Moore Algorithm   *363*
         8.4.2   Dijkstra's Algorithm   *369*
       □   8.4.3   The Floyd-Warshall Algorithm   *374*
         8.4.4   Distributed Asynchronous Bellman-Ford-Moore Algorithm   *376*
     8.5   Stability Problems in Shortest-Path Routing   *379*
   □   8.6   Optimal Routing   *385*
     8.7   Other Routing Algorithms   *391*
     8.8   Introduction to Flow Control   *394*
         8.8.1   Functions of Flow Control   *395*
         8.8.2   Levels for Exercising Flow Control   *396*
     8.9   Sliding Window Flow Control   *398*
         8.9.1   Permits Issued at End of Windows   *400*

8.9.2   Permit Issued After Receipt of First
        Packet in Window   *401*
8.9.3   Windows Advanced by Acknowledgments
        After Packets   *404*
8.9.4   Credit Mechanisms Allowing Variable
        Size Windows   *407*
8.9.5   End-to-End Versus Hop-by-Hop
        Window Flow Control   *407*
8.9.6   Use of Windows at Receivers   *410*
8.9.7   Impact of Window Flow Control on Delay   *411*
8.10   Buffer Allocation Approaches to Flow Control   *412*
8.10.1 Buffering Strategies for
        Deadlock Prevention   *413*
8.10.2 Buffering Strategies for
        Maximizing Throughput   *416*
8.11   Other Approaches to Flow Control   *418*
□ 8.12   Combined Optimal Routing and Flow Control   *419*
8.13   Summary   *421*
        Problems   *422*

Chapter 9   **Network Layer Protocols**   *429*

9.1   Introduction   *429*
9.2   X.25   *431*
9.2.1   X.25 Layers   *432*
9.2.2   Virtual Circuit Services   *432*
9.2.3   X.25 Packet Formats   *435*
9.2.4   Virtual Calls   *435*
9.2.5   Data Transfer   *439*
9.2.6   Error Handling   *441*
9.2.7   Interconnection of Simple Terminals—
        X.3, X.28, and X.29   *442*
9.2.8   X.25-Based Packet Data Networks   *443*
9.2.9   Limitations of X.25   *445*
9.3   Circuit-Switched Network Layer   *447*
9.4   OSI Network Layer   *447*
9.4.1   Internal Organization of the Network Layer   *448*
9.4.2   Network Layer Services   *449*
9.5   ARPANET CSNP-CSNP Network Layer Protocols   *454*
9.5.1   ARPANET Routing   *456*
9.5.2   ARPANET Flow Control   *461*
9.6   DNA Routing Layer   *462*
9.6.1   DNA Routing   *462*
9.6.2   DNA Routing Layer Flow Control   *464*
9.7   SNA Path Control Layer   *464*
9.7.1   SNA Routing   *467*
9.7.2   Transmission Group Control   *470*
9.7.3   SNA Path Control Layer Flow Control   *470*
9.8   The Network Layer in IEEE 802 and MAP/TOP   *472*
9.9   ISDN Network Layer   *473*
9.10   Summary   *475*

Appendix 9A   Descriptions of Four Packet-
Switching Networks   *475*
9A.1   DATAPAC   *475*
9A.2   TELENET   *477*
9A.3   TRANSPAC   *478*
9A.4   TYMNET   *479*
Problems   *481*

Chapter 10 — **Internetworking**   *489*

10.1   Introduction   *489*
10.1.1   Motivations for Internetworking   *489*
10.1.2   Requirements for Network Interconnection   *491*
10.1.3   Relays, Bridges, Routers, and Gateways   *491*
10.1.4   Network Interconnection Issues   *493*
10.1.5   Common Network Differences   *493*

10.2   Bridges   *496*
10.2.1   Definition of Bridges   *496*
10.2.2   Source Routing Bridges   *499*
10.2.3   Spanning Tree Bridges   *503*
10.2.4   Comparison of IEEE
802 Bridge Approaches   *506*

10.3   Routers and Gateways   *509*
10.3.1   X.75   *510*
10.3.2   DoD's Internet Protocol (IP)   *512*
10.3.3   ISO's Internet Protocol   *520*
10.3.4   DNA/SNA Gateway   *522*

10.4   Problems in Large Internetworks   *528*
10.4.1   Disagreement over Broadcast Addresses   *528*
10.4.2   Chernobyl Packets and Network Meltdown   *529*
10.4.3   Synchronization Loading Peaks   *529*
10.4.4   Black Holes and Other Routing Problems   *530*
10.4.5   General Comments on Large Internetworks   *531*

10.5   Summary   *532*
Appendix 10A   The ISO Checksum Algorithm   *532*
Problems   *533*

Chapter 11 — **Transport Layer Protocols**   *539*

11.1   Introduction   *539*
11.1.1   End-to-End Error Control   *541*
11.1.2   Transport Layer Connection Management   *542*
11.1.3   Transport Layer Timeouts and
Congestion Control   *546*

11.2   DoD Transport Layer Protocols   *548*
11.2.1   Transmission Control Protocol   *548*
11.2.2   TCP Primitives   *549*
11.2.3   TCP Data Unit Format   *549*
11.2.4   TCP Initial Connection Establishment   *550*
11.2.5   TCP Data Transmission   *555*
11.2.6   Closing a TCP Connection   *557*
11.2.7   User Datagram Protocol   *559*

11.2.8 Limitations of DoD Transport
        Layer Protocols *560*

11.3   OSI Transport Layer Protocols *561*
        11.3.1 OSI Network Service Types *561*
        11.3.2 Classes of OSI Transport Layer Protocols *562*
        11.3.3 OSI Transport Layer Primitives and TPDUs *563*
        11.3.4 OSI Transport Layer
                Connection Establishment *569*
        11.3.5 OSI Transport Layer Data Transfer *571*
        11.3.6 OSI Transport Connection Termination *573*
        11.3.7 OSI Transport Layer Error Recovery *574*
        11.3.8 Limitations of OSI Transport
                Layer Protocols *580*

11.4   The End Communications Layer in DNA *581*
        11.4.1 End Communications Layer Primitives
                and Data Units *582*
        11.4.2 End Communications
                Connection Establishment *583*
        11.4.3 End Communications Layer Data Transfer *586*
        11.4.4 End Communications Layer
                Connection Termination *587*
        11.4.5 Critique of DNA End
                Communications Layer *587*

11.5   Transmission Control Layer in SNA *588*
        11.5.1 SNA Transmission Control Data
                Units and Primitives *591*
        11.5.2 Transmission Control LU-LU Connection
                Establishment and Termination *591*
        11.5.3 Transmission Control Data Transfer *596*
        11.5.4 Critique of SNA Transmission
                Control Layer *596*

11.6   Transport Layer in IEEE 802 Networks and MAP/TOP *597*
11.7   Summary *597*
        Problems *598*

Chapter 12   **Integrated Services Digital Networks** *603*

12.1   Introduction *603*
12.2   Evolution of Telephone Networks *606*
12.3   ISDN Overview *611*
12.4   ISDN Reference Model *615*
12.5   Common Channel Signaling *621*
12.6   Evolution Toward ISDN *625*
12.7   Growth in ISDN Applications *626*
12.8   Evolution Toward Broadband ISDN *629*
12.9   Summary *637*
        Problems *638*

Glossary *641*
Bibliography *657*
Index *703*

Preface

Although telecommunications is one of the most important and rapidly growing fields of current technology, there does not seem to be a fully adequate text for a first course on telecommunication networks at the upper level undergraduate or first year graduate level. This book has been written to fill the need for such a text for students in electrical engineering, computer engineering, computer science, and related disciplines. At Clemson University the material is taught at the senior or graduate level. In addition to serving as a university text, the book is suitable for a course taken by practicing engineers or computer scientists, and as a reference book.[1] The primary background required is a level of technical expertise that is typical of the students indicated—that is, a general knowledge of computer systems, plus mathematics through calculus, and a first course in probability or statistics.[2]

The book discusses a wide variety of problems encountered in designing telecommunication networks and presents common techniques to solve them. A good understanding of telecommunication protocols is essential for anyone working on the design of telecommunication networks, and their study is a major emphasis of this text. Nontechnical factors, such as legal, regulatory, and economic factors, plus standards activities, are also discussed since they have major impacts on design of telecommunication networks. The emphasis is on basic principles, and on motivations for designs, rather than on encyclopedic coverage of the state of the art. When practical, within constraints imposed by the background assumed for students, a quantitative approach is used. The field cannot be fully covered in a quantitative manner, however, since aspects that do not lend themselves to quantitative treatment are important. For example, only limited parts of the design of protocols can be treated quantitatively.

The basic approach to presenting this material is to describe problems that must be solved during the design of telecommunication networks and then to discuss typical approaches used to solve these problems. Chapters that concentrate on typical problem-solving techniques alternate with chap-

---

[1] Footnotes have been used liberally to present precise details useful for those using the book as a reference. Many of these can be ignored during a first reading, or in a first course based on the text.

[2] Students who do not have the first course in probability or statistics should be able to understand most of the material. Subsections that need to be omitted for classes with such students are listed here; their omission will not significantly impact coverage of other topics.

ters presenting standard telecommunications protocols employing these techniques. In some instances, a chronological sequence of approaches to a problem is given to illustrate the manner in which new designs evolve from earlier ones in order to solve problems with the earlier versions or extend their capabilities. More than 240 problems are included; a few are design problems for which a variety of reasonable solutions are possible. A detailed solutions manual is available from the publisher to instructors adopting the text for classroom use.

The field of telecommunications is evolving so rapidly that writing a text to cover the field has proved to be an extremely challenging task. Despite extensive efforts at standardization, there are major differences between different telecommunication architectures, aggravated by inconsistencies in documentation, terminology, and so forth. We have attempted to be more consistent in terminology than the standard literature, at times deviating from terminology in manuals to achieve consistency and make it easier to appreciate similarities and differences between architectures. This may make it slightly more difficult to learn the accepted terminology in the manuals for a particular architecture, but we hope it may prove to be an initial step in achieving more uniform terminology to be used in the field.[3]

The book covers communications aspects of the eight telecommunication architectures we feel are most important: the OSI Reference Model; U. S. Department of Defense architecture; IBM's Systems Network Architecture (SNA); Digital Equipment Corporation's Digital Network Architecture (DNA); the architecture for local area networks (LANs) being developed by the Institute of Electrical and Electronics Engineers; General Motors' Manufacturing Automation Protocol (MAP); Boeing Computer Services' Technical and Office Products System (TOP); and the Integrated Services Digital Network (ISDN) architecture being developed by telephone companies throughout the world.

To the extent possible, our discussions of telecommunication architectures have been based on original sources, rather than on secondary sources. Some secondary sources we have consulted contain disturbing numbers of inconsistencies. Most of the figures and tables from other sources have been adapted to be in a format consistent with the rest of the text, and sometimes to make corrections to the originals. The text also includes corrections to some common errors in the literature, such as statements of Shannon's channel capacity theorem that do not mention accuracy of data transmission (the key point of the theorem).

---

[3]Despite our attempt to use more uniform terminology than the literature, we faced problems in trying to create a consistent index, since there are still numerous points where different terminology has been used for essentially the same concept. In some cases we have resolved this by including cross references (e.g. references from fragmentation to segmentation and vice versa), while in other instances we have included index references for pages where basic concepts are discussed, regardless of whether the exact words in the index are used on the pages or not.

Table 1   Sections to Be Included in Protocols Course.

3.4– 3.10	
4.1– 4.10	
7.1– 7.7	7.9
9.1– 9.9	
10.1–10.4	
11.1–11.6	
12.3–12.5	

Table 2   Sections to Be Included in Design Problems Course.

2.1–2.7
3.1–3.5
5.1–5.4
6.1–6.6
8.1–8.12

Essentially all of the material in the book can be covered in a two-semester course. For a one-semester course, some selectivity in material covered is needed. Suggested course outlines and guidelines for selecting material from individual chapters to be covered are given below. An excellent course sequence would start with a course based on this text followed by a course explicitly oriented to telecommunication network performance modeling, using texts such as [SCHW87] or [BERT87]. Our current program at Clemson University follows this sequence.

The major single challenge in using this book as a text for a course lasting less than a full year is selecting an appropriate subset of the material to be covered. There are at least two possible one-semester courses that could be based on this material: one primarily devoted to telecommunication network protocols and one primarily devoted to standard approaches to solving the telecommunication network design problems. Each course would start with a general outline, such as that in Chapter 1; additional sections to be included are listed in Tables 1 and 2. Some additional selectivity may be necessary to limit the number of protocol architectures covered in the protocols course or the number of design approaches treated in the design problems course and guidelines for this are given below. The sections and subsections listed for the two courses are virtually disjoint, but skimming material in the complementary course would help students appreciate the significance of topics in the selected course. Instructors who wish to emphasize mathematical modeling techniques should also consider including the sections and subsections in Table 3. Advanced sections and subsections, indicated by □ in the margin,

Table 3   Material Emphasizing Mathematical Modeling Techniques.

2.4–2.7
3.25–3.2.6
4.11
5.3–5.4
6.3–6.5
8.4–8.6
11.1

Table 4   Material Requiring Mathematical Sophistication.

2.4.3–2.4.4	2.4.6	2.5.4
3.2.6		
4.11		
5.3–5.4		
6.3.6–6.3.7	6.3.11	6.4.4 6.5
7.8		
8.6	8.12	

may be omitted for courses with students who have limited mathematical sophistication. These sections and subsections are listed in Table 4.

Some instructors teaching the "protocols course" may want to limit the number of different protocol architectures covered. If so, we have the following suggestions.

- The first five sections of Chapter 3 contain material essential for a good understanding of telecommunication networks, though parts of Section 3.5 on the structure of networks can be skimmed. Treatment of Sections 3.6–3.10, which introduce the various network architectures discussed in the text, can be limited to cover only those architectures selected for coverage in a particular course.

- A detailed presentation of only one of the three physical layer interfaces treated in Chapter 4, with brief sketches of the other two, is adequate for a first course. The interface chosen can be a matter of personal preference: EIA-232D/RS-232C/V.24 is by far the most common, but it is based on obsolescent technology; X.21 is the standard for the OSI Reference Model architecture but is less common; I.430/I.431 represent the state of the art but are just coming in. A treatment of all three is an excellent introduction to how telecommunications has evolved as designers learn from experience.

- The computer interfaces and communications channel interfaces treated in Sections 4.9 and 4.10 are essential, but are not often considered to be part of networking architectures; neither is officially part of the standard architectures we treat. Instructors wishing to limit their treatment to these networking architectures may omit this material, but we have found that it clarifies essential steps in the communications process.

- A choice of DLC protocols in Chapter 7 should be based on the following considerations. Start-stop DLCs are the oldest and most rudimentary, but may still be the most common. Bisync was the first really successful synchronous protocol and is still widely used. ARPANET DLC involves interesting and useful modifications of Bisync but it is not widely used. DDCMP and the bit-oriented protocols represent the current generation of DLC protocols, with HDLC and variants most prominent in standards. A survey of all protocols treated gives a useful picture of how protocols evolve with experience.

- A choice of network layer protocols in Chapter 9 should be consistent with coverage of protocol architectures treated throughout a course. Portions of the discussion of X.25, such as the treatment of important X.25 based vendor networks, have been relegated to the appendix since their coverage is not essential.

- If it is not practical to cover all of the material on internetworking in Chapter 10, we suggest that at least one of the two types of bridges in Section 10.2 be discussed, along with at least an introduction to both types of Internet Protocols in Section 10.3. Brief treatments of the other topics would be worthwhile.

- The most important transport layer protocols, from Chapter 11, are TCP and the TP4 version of the OSI transport layer. The corresponding DNA and SNA layers should be covered if DNA and SNA are examined throughout a course.

- Changes to the ISDN architecture in Chapter 12 are currently being made more rapidly than changes in any other architecture treated. Hence, we recommend that treatment of this material be supplemented with up-to-date material from the current literature. On the other hand, this architecture to a large extent represents the direction in which future development of the telecommunications field is moving.

Corresponding comments about material to cover in a "design problems" course are as follows.

- If students already have a strong communications background, the communications material in Chapter 2 (Sections 2.4–2.6) may be briefly skimmed or even completely eliminated. Alternatively, instructors (especially those who are not engineers) may prefer to skim over this material as not being of major importance to the course they are teaching.

- We suggest that at least an overview of the major classes of medium access control algorithms in Section 5.2 be treated. Only a brief summary of the performance modeling results in Section 5.3 needs to be included, however, especially if students have limited mathematical backgrounds.

- We suggest that the basic fundamentals of error control coding from Chapter 6 be treated, but a detailed understanding of the various types of codes discussed in Subsections 6.3.6–6.3.12 is not necessary.

- It may be impossible to cover all of the routing algorithms in Chapter 8 in a short period of time. Either the Bellman-Ford-Moore algorithm in Subsection 8.4.1 or the Dijkstra algorithm in Subsection 8.4.2 should be covered in reasonable detail, but the other could be treated more briefly. The Floyd-Warshall algorithm in Subsection 8.4.3 is more complex and less common, so it could be omitted. The sections on optimal routing and combined optimal routing and flow control are the most mathematically sophisticated sections of the entire text and are not necessary for an understanding of later material.

- At least the essentials of the sliding window and buffer allocation approaches to flow control in Sections 8.9 and 8.10 should be covered, but not all variants need to be examined. The variants to be covered may be selected by instructors according to their personal preferences.

A number of people have helped in the preparation of this book. Sections of the text have been written by Joe Hammond and Krys Pawlikowski, with the major portion written by John Spragins. All three have edited various sections of the text. Most of the book has been written at or around Clemson University, where the material has been taught in a course taken by both senior and graduate students. A major part of John's material was written during a semester's leave of absence from Clemson plus summers devoted to the book, with a minor part written during a summer (winter there) at the University of Canterbury in New Zealand and final details completed during a sabbatical at Hewlett Packard Laboratories, Bristol, England. Much of Joe's material was written during summers at his mountain cabin, and a large part of Krys' contributions were completed during a sabbatical (from the University of Canterbury) spent at Clemson University.

We would like to express our appreciation to the following referees for their careful reviews: Mostaffa H. Ammar, Georgia Institute of Technology; Robert E. Bradley, Digital Equipment Corp.; Mario Gerla, UCLA; Tony Hsiao, Purdue University; Thomas Robertazzi, SUNY at Stony Brook; and B. W. Stuck, Business Strategies. The manuscript has benefitted greatly from their reviews, plus reviews by others who remain anonymous.

Typing and other secretarial work by B. Rathz and R. Watkins are also appreciated. Others who have helped in preparation of the manuscript and in other ways include A. Aletty, S. S. Bijjahalli, M. Den, N. El-Khoury, H. Jafari, S. Kamat-Timble, W. P. Lyui, P. Moghe, M. Raman, J. Song, D. Stevens, H. R. Wang, J. Ward and W. P. Yeh. A number of persons at Addison-Wesley gave us valuable assistance, including Tom Robbins and Barbara Rifkind—editorial assistance, Stephanie Argeros-Magean—copyediting, Joseph Vetere—art coordination, and Helen Wythe—production supervision. Finally, I would like to thank my wife, Cathy, and my children, Katrina, Jennifer and Matthew for their encouragement and to apologize for the way manuscript preparation impacted the time I had available for them. Cathy's support has been especially valuable, including serving as primary breadwinner for our family during the half year when I took leave of absence from my job to work on the book. Joe and Krys also wish to thank their wives, Edith and Barbara, respectively, for their support.

J. D. S.
Clemson, S. C.

1 Introduction to Telecommunications

1.1 Introduction

We are surrounded by telecommunication networks. From their beginnings in the 1950s in military networks interconnecting sensors, weapon systems, and command-and-control centers, telecommunication networks have evolved to find applications in virtually all areas of modern society: in banking and finance (remote teller banking systems, electronic funds transfer networks, stock brokerage networks, and credit verification systems); in reservation systems (the airline and travel industry, hotels and motels, sporting events, concerts, etc.); in grocery and retail store checkout systems and for home shopping; and in offices and factories (automation, time-shared computing in corporate networks, and electronic mail systems). Personal computers with modems now access a wide variety of networks and provide a major impetus for network growth.

LANs, MANs, and WANs Networks are classified by geographical coverage as local area networks (*LANS*), spanning areas with diameters not more than a few kilometers, metropolitan area networks (*MANs*), spanning areas with diameters up to a few dozen kilometers, and wide area networks (*WANs*), some with world-wide extent.

A few examples will illustrate the scale of today's networks: General Motors currently operates a network that links together more than 500,000 computing devices and telephones and connects 18,000 locations world-wide. American Airlines' SABRE reservations network links more than 60,000 video terminals all over the globe to six large-scale computers (processing about 1800 messages per second, or 60 million messages per day [SEMI89]). Sometimes the SABRE network posts larger annual revenues than the airline itself. DARPA Internet, the world's largest internetwork (interconnecting other networks), links several thousand networks together and had a growth rate estimated, in late 1987, at 15 percent per month! The number of installed local area networks was estimated at 423,000 in 1987, with the number predicted to double again by the end of 1988. By 1992 there are expected to be over 80 million personal computers as well as tens of millions of other kinds of computers, terminals, and peripherals installed.

As [CIO88, p. 10] states, "Today communications systems that send voice, data, and video signals span the earth. Information is bounced off satellites and shot through cables that cross entire oceans. Computer-based systems operate in the dead of night receiving messages from places where it is high noon. And the organization that turns a blind side to how communications and computer technologies together are transforming the world will be out-performed and outmaneuvered by its competitors—competitors that can now come from any country on earth."

1.1.1   Approaches to Telecommunications

Telecommunications has evolved from two main technological fields: the computer field and the communications field. There are also two primary approaches to telecommunications, one evolving from each field. One approach is described as communications for computers and the other as computers for communications. We use the general term, telecommmmunica-tions, to encompass both.

We are surrounded by computers, ranging from small personal com-puters to large mainframes—computers that we are unaware of that handle such things as switching our telephone calls or processing our financial trans-actions. There are computers purchased as computers and computers that are part of other things we purchase—such as our televisions, our automo-biles, or our washing machines. A large percentage of these computers do more than computing—they communicate with other devices also. The pro-liferation of computers using communications is a major incentive to devel-opment and leads to what we call *communications for computers*.

communications for computers

A parallel incentive to progress is *computers for communications*. Much current work in communications is based on computer techniques. One view-point [STAL88b] is that, "There is no fundamental difference between data processing (computers) and data communications (transmission and switch-ing equipment)." Current communication facilities use digital transmission techniques developed in the computer field. Digital switching offices now in use and being installed throughout the world are special purpose computers and include some of the most sophisticated and powerful computers ever built. The U.S. telecommunications network has been characterized as the largest collection of computers in the world [PHIL82]. It would be more accu-rate, though, to so characterize the world-wide telecommunications network as over 98 percent of the world's telephones are interconnected in the sense that it is physically possible to establish a connection between any two.

computers for communications

1.1.2   Future of Telecommunications

A utopian view of future progress is given in [MART77, p. 327]. The majority of the telecommunications technologies in this utopia are either already available or do not require major advances. Martin envisions "a city with parks and flowers and lakes, where the air is crystal clear and cars are kept in large parking lots on the outskirts. . . . The high-rise buildings are not too close, so they all have good views, and all persons living in the city can walk through the gardens or rain-free pedestrian malls to shops, restaurants, and pubs. . . . There is almost no robbery in the streets because most persons carry little cash."

Specific facilities or services in Martin's utopia include:

- Wall-sized television screens and computer terminals in houses and apartments
- Banking and shopping from home
- Good delivery service and fast public transportation
- Videophones enabling users to work from home
- Telephone and video channel switching to allow conferences
- Computer-assisted instruction
- Educational films accessible from homes
- Information retrieval for accessing news, sports information, stock market figures, weather reports, encyclopedias, and reports and documents
- Browsing in libraries via home computer terminals
- Use of bank-card terminals for financial transactions
- Radio devices worn by individuals for calling police or ambulances
- Home burglar and fire alarms wired to police and fire stations
- Television cameras wired to centralized guard stations for security
- Automatic utility billing with wired meters
- Telemedical facilities available from homes, with computerized or video connections to nurses and doctors
- Emergency communications for ambulances
- Home printers or facsimile machines to obtain business documents
- Custom-selected news items and financial or stock market reports
- Electronic mail, bank statements, airline schedules, and so forth
- Computer editing plus automatic transmission of documents to people who need them

Realization of this utopia is more likely to be limited by legal, regulatory, and economic factors, and by basic human nature, than by technologies available. Many technologies needed are already available. In most other cases they represent advances that can be expected within the next few years if adequate markets develop.

1.2   Motivations for Growth

An excellent example of the need for the type of automation exemplified by telecommunication networks was the realization by the Bank of America in the 1950s that manual handling of checks would require, in the foreseeable future, a work force equivalent to the total adult population of California [FANO72, p. 1249].[1] As Fano states (emphasis added), "We hear many complaints about inadequacy of medical care, of education, and of many other services. These inadequacies are usually blamed on a variety of factors, such as lack of funds, incompetence, poor planning, or resistance to change. It may well be, however, that these factors are merely symptoms of a more fundamental limitation, namely, that *society does not possess the human resources necessary to perform adequately all the necessary tasks,* particularly in view of its growing complexity and the rising expectations on the part of the population." Many services that we take for granted would not be available without the type of automation related to telecommunications, and other services that we expect in the future cannot be made available without further automation.

If information production and dissemination, and the means to manage this information, were viewed as a single industry, it would account for more than half of the U.S. gross national product (GNP) [KOST84], and a comparable percentage of the GNP in other developed countries. Not all such information is amenable to telecommunication techniques, but an appreciable fraction is. Important applications of telecommunications include the following:

- *Systems to automate corporate operations.* These include networks for order entry, centralized purchasing, distributed inventory control, and factory automation systems. Manufacturers are moving from batch-oriented systems for functions such as inventory control and requirements planning to interactive on-line systems to manage their entire resource flow.

- *Automated checkout systems.* Retail stores and supermarkets often use local networks for functions such as automated price lookup, with input from optical or magnetic scanners for product identification, and access to remote locations for inventory control, automated ordering systems, credit verification and similar functions (or for automatic backup if the local network at one site fails).

- *Corporate information networks.* These contain information about marketing, customers, products, payroll, employees, etc. and may be accessed from a variety of locations by persons with adequate authorization.

---

[1] At that time, the Bank of America was legally restricted to operate only in California.

- *Reservation systems.* Airlines and other transportation systems, car rental agencies, hotels, theaters, sports arenas, and so forth use telecommunications as the standard approach to business. Currently more than 100 million airline reservation transactions are handled each year through INTELSAT alone [DAM88].[2]

- *Electronic mail and message systems.* A tremendous potential market exists here. It has been estimated that in the United States alone a potential 50 billion pieces of mail per year could be delivered electronically in the near future [MART77]. The major use of important networks now in existence, originally designed for other purposes, has been for electronic mail and file transfers between network users.[3]

- *Electronic transfer of funds.* For banks and check cashing houses, this is now the standard approach for handling financial transactions, with billions of dollars regularly shunted between banks by means of electronic signals. About $9 trillion in electronic funds transfers are completed annually through international communication facilities, an amount equal to two thirds of the global GNP [DAM88]. Other banking applications include remote teller machine banking and automated access to customer account information. Recent deregulation of banking in the United States has led to such applications as handling brokerage and investment accounts by banks.

- *Stock brokerage networks and related networks.* A wide variety of financial transactions are handled through these networks, including automated price quotation and order entry systems. Significant numbers of stock purchases and sales are currently made automatically, via programs that buy and sell stocks based on information from automated information systems. There is some controversy regarding the degree to which the stock market crash of October 19, 1987—when the Dow-Jones average in the United States dropped over 20 percent in one day—was due to computers buying and selling stock without human intervention.

- *Consumer check and credit card verification via computer networks.* This is becoming standard and is being extended to the use of automatic debit cards; when these cards are used, the amount of a purchase is automatically (and immediately) deducted from the customer's bank account. Some projections of the future envision a "cashless" society where all financial transactions are handled electronically.

- *Intercorporate networks.* Tying together various corporations, these networks can allow a computer in one firm to transmit orders or invoices to a computer in another firm. This type of electronic data interchange (EDI)

---

[2] INTELSAT refers to the International Telecommunications Satellite Organization, the dominant carrier of international traffic.

[3] For example, see discussion of the ARPANET in the next section.

is seeing increasing use, and standards are now being developed. Similar transactions occur regularly in automatic reservation systems, with orders for airline tickets, hotel reservations, and so forth placed by travel agents and invoices returned from the airlines and hotels. Current and potential applications of such networks are much broader, however; they could handle a large fraction of the ordering and invoicing transactions of many businesses.

- *Access to remote data bases.* This is a primary function of many networks. At one time access to computing power was equally important, but the proliferation of personal computers and other equipment with computing power has reduced the need for this type of access, except in cases where access to powerful computers is required. Over 3000 different data bases in the United States are now accessible via personal computers and the number is growing. Numerous other data base applications can be found in various corporations and other organizations.

This list could be extended greatly, and the applications listed will continue to grow rapidly. A survey by International Data Corporation [IDC87a] indicates that, among the largest companies in the United States, the fraction of personal computers connected into networks went from 35 percent in 1985 to 65 percent in 1986. A related article, [IDC87c], estimates that expenditures on data processing and voice communications during 1987 by U.S. businesses would total approximately $300 billion. The former article also states, "The 1980s saw the incredible outward migration of computing technology, primarily through the personal computer revolution, and the buildup of ever larger islands of automation. The 1990s will see an equally incredible cross-hatching of those islands with networking and systems integration. By the turn of the century, one hopes, the islands will form a single continent."

1.3   Evolution of Telecommunication Systems

The current telecommunications state of the art has been reached via thousands of incremental steps. Most have been small, though a few have been major "breakthroughs." Some progress has been unpredictable, but important progress has resulted from alleviating shortcomings of previous equipment or techniques. We trace evolutionary steps throughout this book, indicating how designs evolve as technology evolves and techniques are developed. To illustrate evolution, we present the most important developments in roughly chronological order. It is important to understand how new techniques are built on earlier techniques. An overview of telecommunications milestones will illustrate this type of progress.

1.3.1   Telecommunication Milestones[4]

A dozen telecommunication milestones are discusssed by Green [GREE84]. We summarize his list (with minor modifications) plus others that are similarly important. Some milestones have resulted from individual inventions, while others resulted from large projects. Still others are of a nontechnical nature. Extremely important milestones that have impacted numerous fields—such as the invention of the transistor and integrated circuits, or the invention of the computer and its evolution—are excluded in favor of milestones specifically associated with telecommunications.

Green's first milestone is the cathode ray tube, invented by K. F. Braun in 1897 and now a standard input/output device in telecommunications. The technology has evolved over an extremely long learning interval to a state where CRT terminals are the most common terminals for fast, inexpensive computer to human communications. In addition, virtually all TV sets use CRT technology. Television has, in fact, been the greatest single motivator for refinement of CRT technology.

The second milestone is the teletype machine, dating back to World War I [KAHN73]. This was a mechanical marvel for its time, with hundreds of precisely machined and high-speed moving parts. Although no fundamental changes have been made in these devices for over half a century, they are still in widespread use. The number of computer-connected mechanical keyboard-printers was not surpassed by the number of CRT terminals until 1978.

automatic repeat request (ARQ)

The *automatic repeat request (ARQ)* technique was devised by van Duuren during World War II [STUM84]. The principle is to ensure successful transmission of a block of characters by relying on error detection, followed by a signal from the receiver asking for retransmission if an error is detected. Sophisticated techniques for error correction at the receiver (forward error correction, or FEC) without retransmission have been evolving since Shannon's development of information theory and proof of his channel capacity theorem.[5] Nevertheless, the simple ARQ approach has been almost universally adopted for telecommunication networks (see Chapters 6 and 7).

The SAGE (Semi-Automatic Ground Environment) system for air defense [EVER83], designed and developed by M.I.T.'s Lincoln Laboratory in the early 1950s, was the first real computer network and was amazingly extensive. It involved 23 computer networks, each connecting approximately 100 radar sites, ground-to-air data link sites, and other locations to what was for that time a huge computer. Each duplexed computer system, with air conditioning

---

[4] Adapted from P.E. Green, Jr., "Computer Communications, Milestones and Prophecies," *IEEE Communications Magazine,* © 1984 IEEE. Reprinted by permission.

[5] This theorem proves it is possible to transmit information over a channel, with arbitrarily low error probability, at a positive rate known as channel capacity, if adequate error correction codes are used. See Chapter 2 for more details.

Figure 1.1   Communication Links for One of the SAGE Direction Centers. (Figure originally published in [EVER83].) © 1983 AFIPS Press. Reprinted by permission.

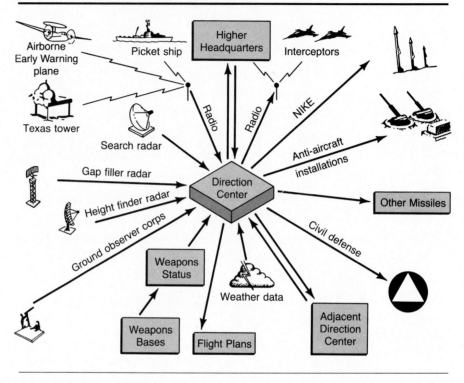

and power supply, required one-and-a-half stories of a building but was roughly equivalent in power to some current day desktop personal computers. Figure 1.1 illustrates the sensors, weapons systems, and other locations connected to one of the 23 SAGE Direction Centers.

The SAGE computers did not talk directly to each other but used communications technology to talk to peripheral sites. Voice-grade phone lines were used to send data at approximately 1300 bits per second (bps), with erroneous frames discarded. Innovations included keyboard/CRT terminals, light pens, a real-time multiuser operating system, ground-to-air data transmission, cycle stealing for fast memory access, computer duplexing, and the magnetic core memory that made reliable computers possible. The first large commercial computer network—the SABRE airline reservation system [JARE81], built by IBM for American Airlines in 1964—owed a great deal to SAGE and involved many SAGE alumni.

The first time-sharing computer operating system was the Compatible Time-Sharing System (CTSS) developed by Corbato and others at M.I.T. in

1961 [MORE84]. CTSS introduced the idea of time-slicing execution of the processor and passing slices around only to jobs awaiting service. This eliminated the previous limitation of a computer network being a truly one-user network. Today, this type of time-sliced operation is standard for multiuser systems.

The idea of synchronous communication satellites originated with science fiction writer Arthur C. Clarke in 1945 [CLAR45] and was realized by H. A. Rosen of Hughes in the SYNCOM satellite of 1963. By placing a communication satellite in orbit approximately 36,000 km above the equator, it is possible to have it remain in a fixed position relative to ground stations; in this "geostationary" orbit, the rotational period of the satellite is the same as that of the earth. Close to a quarter of the earth's surface may be within the coverage area of a satellite, with communications cost within this area independent of terrestrial distance. Such satellites are ideal where wide geographical coverage and high bandwidth are needed, but they may not be satisfactory where extremely fast response times are needed because of the long propagation delays they incur. At the speed of light, it requires roughly a quarter of a second for information to propagate to a satellite and back down, or half a second to send a message and get a reply, so applications requiring response times of a fraction of a second must use terrestrial transmission facilities.

Automatic equalization techniques for telephone transmission links [QURE82] are making it possible to achieve rates of transmission of bit streams across telephone links far above the rates achievable over the links without such equalization. Such techniques were developed around 1965 by Lucky [LUCK65] and others. During communication or when a new connection is made, they adaptively adjust characteristics of the transmitter to compensate for those of the communication channel. Without such equalization, error rates may be unacceptably high for speeds higher than about 2400 bps on ordinary telephone lines. However, because of the use of automatic equalization, plus recent advances in modulation and coding [HOLS87], some modems for such lines now operate at speeds as high as 19,200 bps.

The *RS-232 physical layer interface standard,* first published in 1960 and stabilized to its most common current form (*RS-232C*) nine years later, established a standard plug and protocol convention between modems and the machines to which they attach. The 25-pin plug used for this interface (and the equivalent V.24 interface in other countries) has become almost as universal as the standard cord used for electrical power.

RS-232C

The Carterfone FCC decision of 1968 [MATH72] was a legal and regulatory event that has had more impact than many important technological events. Prior to this decision, telephone companies in the United States were able to enforce a blanket prohibition against interconnecting equipment they did not manufacture to the telephone network. The decision marked the beginning of a series of regulatory decisions opening up communications in the United States to competition. Without these decisions, the competition between manufacturers that has resulted in much rapid progress would have been outlawed. Although the Carterfone decision applied only in the United States, increased competition in many other countries has developed since the decision.

The Carterfone decision was a key step initiating a chain of events leading up to January 1, 1984, when AT&T, the dominant telephone company in the United States, divested its operating companies [SODO85], [NEWM87b]. The 22 Bell Operating Companies were split into seven regional companies, with court-imposed constraints on their entering other lines of business. (Recent regulatory decisions have relaxed some of these constraints.) In turn, the parent company, AT&T, was allowed to enter lines of business, including computers, previously prohibited to it. AT&T's reasons for accepting divestiture of the major part of its previous business were largely based on potential growth in fields the company had not been allowed to enter. Many aspects of divestiture are very controversial (see [SODO85] and [NEWM87b]), but divestiture has resulted in more competition in telecommmunications.

ARPANET

The *ARPA Network (ARPANET),* which began service in 1971, greatly accelerated development. L. A. Roberts, then in the Advanced Research Projects Agency (ARPA) of the U.S. Department of Defense, pushed through the development of the ARPANET. It connected heterogeneous machines at universities and military installations using a then new technique known as "packet switching" [KAHN78b]. Configurations of the ARPANET as it existed in 1970 and in 1979 are illustrated in Fig. 1.2.

A wide variety of "firsts" came out of the ARPANET project. These included layered protocols, mesh network—host backbone topologies, flow control, and fail-soft or fault-tolerant performance exemplified by the ability of a node or link to disappear without bringing the network down or requiring operator intervention.[6] In addition the ARPANET developers popularized use of analytical and simulation models for predicting and understanding performance. L. Kleinrock and his colleagues at the University of California in Los Angeles were especially productive and influential in these efforts [KLEI75b, 76b]. Much material in this text can be traced to work on the ARPANET.

The ARPANET has had a strong impact on commercial network evolution. Before this, it was not unusual to find commercial networks with several identical terminals in an office, each connected via a different phone line to a different CPU and incapable of being used to access the other CPU.

The ARPANET was split into two networks in October 1983. The first is the new Defense Data Network (DDN), also called MILNET, which consists of approximately 160 nodes, including 24 in Europe and 11 in the Pacific and Far East. The second is ARPANET, which consisted of approximately 50 nodes scattered across the continental United States and Western Europe. The ARPANET itself is now being phased out, with hosts on the network being moved into other domains interconnected by the DARPA Internet (an out-

---

[6] Precise descriptions of these terms are given in later chapters.

Figure 1.2 Configurations of Two Versions of the ARPANET. (a) The network as it existed in 1970. (b) ARPANET in 1979 (from [ROSN82b]).

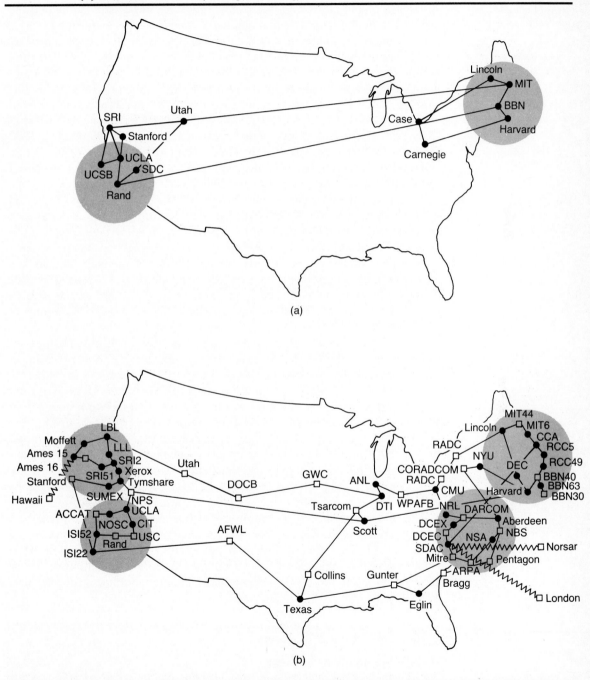

(a)

(b)

growth of the ARPANET, which links together a wide variety of other networks).

layered network architectures

Development of *layered network architectures* has greatly simplified the design and evolution of telecommunication networks by establishing hierarchical structures of network functions and protocols. To date, the most ambitious commercially developed telecommunication network architecture is

Systems Network Architecture

Systems Network Architecture (SNA) IBM's layered architecture for telecommunications. It is currently the most widely implemented commercial network architecture, rivaled only by Digital Equipment Corporation's *Digital*

Digital Network Architecture

Network Architecture, also known as DECNET. The number of current installations of DECNET is comparable to the number of installations of SNA, though SNA installations tend to be larger. The most prominent example of a layered network architecture, however, is the OSI Reference Model discussed in the next subsection as one of our additions to Green's list.

Ethernet

Ethernet, which originated at Xerox in 1974 as a laboratory project by R. M. Metcalfe and others [METC76], has become the best known local area network (LAN). As of late 1988 approximately 75 percent of the installed LANs used the Ethernet protocol [CAMP88]. The original objective was to devise a cheap wideband way of sending information between office machines such as terminals, files, and printers. Ethernet has been successful in both office situations and universities. Its basic protocol is a refinement of the ALOHA random-access protocol invented for a radio application by N. M. Abramson [ABRA70].

Videotex can be used to make data-processing terminals out of color television receivers [MART82]. It has evolved from older teletext services, in which a constantly updated library of 100 to 200 screens of information is cycled through and broadcast during vertical blanking intervals of TV broadcast. Videotex operates in a two-way interactive mode using a connection with a central processor provided by telephone lines or two-way CATV.

Teletext and videotex services are presently more common in Europe than in the United States, with teletext especially popular in Great Britain where it has millions of users. The most successful videotex system to date is the French Teletel, offered by the PTT (Postal Telephone and Telegraph) and developed originally as an electronic telephone directory. A "dumb" Teletel terminal is given free to any home or business interested in using the system. The number of French Minitel (terminal) owners is approximately 3.4 million, with 4.7 million connect hours per month [JUDI88]. In addition to the electronic telephone directory, users can access more than a thousand other services, each offered by an independent company and linked into the network by gateways.

1.3.2   Additional Milestones

We give five additions to Green's list.

T-1 carrier system

The first is the development by Bell Labs of the *T-1 Carrier System* for telephone transmission [AARO62], [CRAV63]. The first T-1 system was installed in the United States in 1962, with similar systems installed in Europe in the late 1960s. The T-1 system represented a distinct break from earlier telephone carrier systems, which transmitted multiple voice signals over one communication facility by stacking them in different frequency bands, an approach known as frequency division multiplexing. The T-1 system relies on digital transmission of voice signals. Different time slots are allotted to different voice channels, with 24 different voice channels, each using 64,000 bps, transmitted in purely digital fashion in a 1,544,000 bps pulse stream—often over facilities previously used for the earlier frequency-based carrier systems. This use of different time slots for different sources is called time division multiplexing. The purely digital type of carrier system first exemplified by T-1 has become the preferred type of carrier system for new installations, with millions of circuit miles now installed. Such carrier systems have definite cost and performance advantages over older analog systems.

Telephone companies have recently extended access to digital carrier systems to user locations rather than terminating them in switching offices. This is having a revolutionary impact on telecommunications since it will eventually provide data transmission at close to 64,000 bps, or even 1.544 million bps, at costs comparable to those for much slower transmission (say, 4800 bps) over analog channels with modems. Still higher speed digital transmission systems are used widely. Digital transmission systems are now being combined with digital switching systems in the Integrated Services Digital Networks discussed below.

Our second addition to Green's list is related to the first since it also involves digitization of telephone facilities; it is development of *digital switching offices*. The first electronic switching systems were actually computer controlled electromechanical switching systems, known as Stored Program Controlled Systems (SPCs). The first prominent system was the U.S. Bell System's No. 1 ESS, first installed in the early 1960s. Its most remarkable design goal, maintained in later generations of switching systems, was a reliability goal of not more than two hours down time over a 40-year design lifetime. Major advances in design of reliable computers and associated software have been necessary to approach this goal.

The newer electronic switching systems, first installed in the mid-1970s and typified by the Bell System's No. 4 ESS, operate in a purely digital manner. The signals switched are digital signals, such as those for T-1, and they are switched using digital techniques developed in the computer field.[7] These switching techniques are those motivating the comment by Stallings, quoted in Section 1.1, that there is no fundamental difference between data processing and data communications.

Our third addition is development of *fiber optic communication systems*. The invention of the laser in 1960 by T. H. Maiman first provided a narrow-

---

[7] Any incoming signals encoded in older analog formats are converted to digital signals before switching.

band source of optical radiation suitable for use as a carrier of information. Kao and Hockham of S.T.L. (Standard Telecommunication Laboratories) proposed a clad glass fiber as a suitable dielectric waveguide in 1966 [KAO66], and Kapron and others of Corning Glass Works reported on the first low-loss fiber in 1970 [KAPR70]. Since that time there has been rapid development in all phases of fiber optic technology. Early field trials of fiber optic communication systems began in the mid-1970s [SCHW84], with expansion to the point where total fiber installed in the United States alone was about 3.2 million Km by the end of 1987 [KAIS87].

The major impact fiber optics is having on telecommunications is a result of the tremendous gains in digital transmission speeds it provides. Early fiber optic systems operated at 45 and then 90 million bps; since 1984, though, there has been a rapid changeover to single-mode fiber transmission systems operating in the 405 to 565 million bps range, with systems operating at 1.6 to 1.7 billion bps now being installed [FISH86]. Furthermore, the cost differential between fiber and copper already favors fiber in many situations, and it is expected that fiber will be more economical for installations in individual living units by the early 1990s.[8] In 1988 the first transatlantic fiber optic cable, TAT8, went into commercial operation between the United States and Great Britain. It can carry 40,000 conversations at once, more than all other existing transatlantic cables and satellite links [BELL89]. Thus tremendous increases in digital transmission speeds are rapidly occurring. The possibility of providing circuits operating at terabit per second (10^{12} bps) rates to communication system users is being discussed at technical workshops and conferences. It is too early to assess the impact of this on telecommunications, but the higher speeds will make many additional applications feasible.

OSI Reference Model

Our fourth addition to Green's list is the development of the Reference Model for Open Systems Interconnection by the International Organization for Standardization (ISO). Work on the *OSI Reference Model* was motivated by development of noncompatible network architectures. It is intended to provide an international standard architecture for telecommunication networks. Work on the model began in 1978, with recommendations adopted by an ISO technical committee at the end of 1979, a draft proposal in 1980, a draft international standard in 1982, and adoption of the basic reference model as an international standard (ISO 7498) in 1983 [DAY83]. This model defines the seven layer architecture for telecommunication networks depicted in Fig. 1.3. The basic organization of this text is largely motivated by the reference model.

A standards committee normally takes commercial practices and research results and codifies them into a standard for commercial products. The approach to OSI Reference Model development was to establish standards for emerging products *before* commercial products were in place and before some fundamental research problems had been solved. This has not

---

[8] This projection was included in a presentation by BellSouth at Clemson University in 1987.

Figure 1.3 Layered Architecture of the OSI Reference Model.

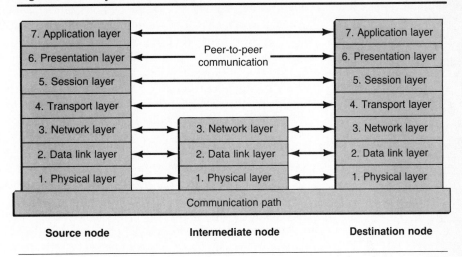

7. Application layer		7. Application layer
6. Presentation layer	Peer-to-peer communication	6. Presentation layer
5. Session layer		5. Session layer
4. Transport layer		4. Transport layer
3. Network layer	3. Network layer	3. Network layer
2. Data link layer	2. Data link layer	2. Data link layer
1. Physical layer	1. Physical layer	1. Physical layer

Communication path

Source node   **Intermediate node**   **Destination node**

been fully successful, and valid criticisms of the OSI Reference Model can be, and have been, made. (See [PADL85] for a spirited discussion of some of the most important problems by a proponent of alternative approaches.)[9] Nevertheless, work on the model has greatly increased our understanding of networking architectures, and basic principles developed during this work will be influential for years to come.

General concern about compatibility of networks is pushing more and more users and vendors towards recognition of the OSI Reference Model as the international standard. In 1988 the U.S. federal government decided all governmental networks purchased after 1990 should conform to OSI standards. The model is also influencing directions taken by developers of other networking standards who try to make new networks reasonably compatible with the OSI standard or to modify architectures for greater compatibility. Thus large computer manufacturers, including IBM and DEC, have now begun to trumpet their support for international standards, and the OSI Reference Model in particular [STIX89].

The OSI Reference Model has been heralded as the final solution to network compatibility problems, but significant weaknesses are becoming apparent as the model matures. Since use of the model has tremendous momentum, we expect it to be a dominant architecture in the near future. Its long term future will be jeopardized, however, if it is not modified to incorporate some features that can yield significant improvements and are present in other architectures we discuss in this book.

---

[9] One problem is the number of options available with the OSI Reference Model; when several approaches to a problem with reasonably equal merits and support have been proposed, committees developing the model have tended to allow all as options.

Integrated Services
Digital Network

    Our final addition to Green's list is development of the *Integrated Services Digital Network* (ISDN), with demonstration networks currently being installed in several countries. Initial efforts date back to a report by a special study group of the CCITT (Consultative Committee for International Telegraph and Telephone) in 1972 [IRME86]. The first recommendation was approved in 1980 with initial standards set in 1984. Converting telecommunication networks to digital transmission and switching techniques is well underway. Although the primary incentive has been the need to provide economical voice communications, the result has been development of facilities well adapted to handle data. This has led to the ISDN effort.

    Concisely described, "An ISDN is a network, in general evolving from a telephony integrated digital network, that provides end-to-end digital connectivity to support a wide range of services, including voice and nonvoice services, to which users have access by a limited set of standard multipurpose user-network interfaces" [DECI86a, p. 320]. The primary goal of ISDN is to develop one worldwide public telecommunications network that will meet all user needs. Another goal is "to be synergistically responsive to the evolving needs of the information society, and in addition, to spur the demand for information services" [KOST84, p. 11].

    Applications of the ISDN can range from meter readings, security or alarm systems, telecontrol and opinion polling systems (each with data rates in the 10 to 100 bps range), through services such as teletex, videotex, home computer, facsimile and low-speed switched services (in the 1000 to 10,000 bps range), digitized voice (commonly transmitted at 64,000 bps), high-speed data, digital facsimile and slow-scan video (typically in the 10,000 to 100,000 bps speed range), high-speed facsimile, high-quality music and video conferencing (requiring 100,000 to 2 million bps capacity) to broadcast quality TV (20 million to 100 million bps) and high-definition TV (over 100 million bps) [FEHE87b]. Figure 1.4 illustrates the basic integration concept of the ISDN. All services and devices in the figure, plus others, should be accessible by the user via the same interfaces and with essentially the same protocols.

    Regardless of whether the ISDN effort ever meets its goal of providing what appears to the user to be one unified network satisfying all user needs, it is already having a major impact. Essentially all major telephone companies world-wide, as well as a wide variety of manufacturers and other groups, are involved. Work includes development of digital data transmission techniques providing major performance improvements, using both conventional transmission channels and newer technologies such as fiber optics; similar improvements in digital switching and in integration of transmission and switching; new protocols and interface standards; and new applications of digital transmission systems, and expanded use of systems previously available.

    The milestones include several categories. A few can be attributed to a single inventor or a small research group (for example, invention of the cathode ray tube or of ARQ), others are the results of major development efforts (sometimes, as in the OSI Reference Model and ISDN efforts, involving thou-

Figure 1.4 ISDN Network Integration Concept (from [KOST84]). © 1984
IEEE. Reprinted by permission.

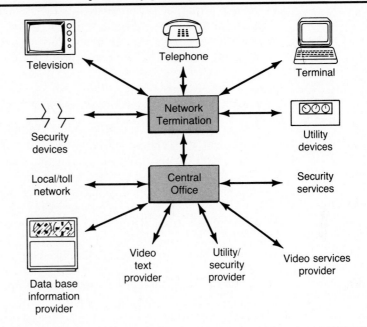

sands of people), while others are nontechnological events with important
impacts (for example, the Carterfone decision and succeeding events culmi-
nating in divestiture of AT&T). These plus thousands of other incremental
steps have combined to produce the state of the art in telecommunications.

An important point is that the field is evolving as designers learn how to
design better networks by eliminating the weaknesses of earlier techniques.
Many (if not all) of the major techniques and protocols that we discuss here
represent early phases in this learning process and will be supplanted by new
and better approaches to the same problems in the not-too-distant future. As
[PADL85, p. 40] states, "We went through two or more passes over every protocol . . . as we were learning our craft, and there's no reason to suppose we
were much dumber than anybody else who might have gotten into the field
subsequently, so unless they went to the trouble of learning from our mis-
takes they ought to expect to have to do a couple of passes themselves."[10] To
illustrate this learning process, we point out weaknesses of many of the protocols and techniques we discuss, and discuss evolutionary versions of
several.

---

[10] The most important of these protocols are discussed in later chapters. We use the term DoD
Reference Model for what Padlipsky calls the Arpanet Reference Model, or ARM.

1.4   Major Design Problems

Despite variations in applications, technologies used, and so forth, a few major problems must be solved in the design of any telecommunication system. These problems recur in different guises at different levels of design. We summarize here some important problems that continue to recur.

1.4.1   Resolving Incompatibilities of Equipment

A basic problem is to resolve, or if possible prevent, incompatibilities of equipment. For example, physical attributes of the different pieces of equipment must be compatible with each other and able to handle the application. Plugs, connectors, and so forth must be matched so it is physically possible to interconnect equipment. Fundamental constraints imposed by electromagnetic theory dictate general dimensions of equipment as a function of signals used for transmission and the signalling speed. Similar considerations limit factors such as speeds and energy or power dissipation.

In order for communication of useful information to take place, communicating devices must use compatible signals. Compatibility of voltage or current levels (optical pulse levels for fiber optic systems), frequency bands occupied, and time durations of pulses or other signal elements are required. In addition, techniques to encode information must be uniform. This includes interpretation of basic signal elements, grouping signal elements into characters (and choice of character set), and formats for messages. Also, device speeds must match up for transmitter, communication channel, and receiver. Compatible buffering schemes are needed and adequate processing capability to handle the workload must be present at nodes.

A higher level of compatibility is imposed by the requirement that compatible protocols must be used. This is analogous to stipulating that a discussion must be carried on in a common language or with translators provided. It is a problem of great significance because of the large variety and range of protocols used in a telecommunication network. Some protocols communicate individual bits, while others organize bits into characters and messages and transmit them across communication channels. Some minimize the impact of errors, govern conventions for addressing and naming, route messages across networks, or set up and terminate sessions or periods of communication. Protocols are also involved in many applications.

1.4.2   Coordination of Sender and Receiver

If two devices are to successfully communicate with each other, their communication must be coordinated so that one knows (or can learn) when to talk (transmit) and the other knows (or can learn) when to listen (receive). No useful communication takes place if both are only listening or only talking.

(Data processors may usefully transmit and receive at the same time,[11] but techniques for doing so require coordination.)

There must be adequate synchronization of basic signal elements—for example, sinusoidal-like waveforms for some types of communication and pulse strings for others—so the receiver knows when to look for data. This involves clocking at the transmitter and clock recovery at the receiver. Signal elements must be processed to yield bit patterns, with the transmitter knowing when to transmit and the receiver knowing when to look for them.

Next, bits must be assembled into characters, with the receiver able to determine when complete characters are available. Characters must be assembled into frames or messages, with techniques to determine beginning and end of frames. Furthermore, it is common for different types of information, such as control, address, data, and error protection overhead, to be included in one frame so methods for locating different types are needed.

Determining which device has access to, or control over, a communications medium at a given time is another type of coordination. Initiation, use and termination of a connection, or of a dialogue using a connection, must be synchronized. Connection establishment and termination can be required at virtually any level of an architecture; it may include establishment and termination of connections between data sources/sinks and communication facilities, of connections over physical links, and of connections at the data transfer level or at higher levels.

Another category of related problems occurs in error recovery, which often involves putting both ends of a communication link back into a known state. Synchronization points established in some of the higher layers of protocol architectures are points at which the states of both ends of a link are known (at least with high probability), so the system can be restarted from these points if problems occur.

Major problems for coordinating sender and receiver arise from the fact that they must communicate over communication links that are not perfectly reliable. A classic problem in distributed communications illustrates the difficulties. Analogous situations occur in computer networks.

The problem is known as the three army problem. There are three armies, two orange and one black, positioned as shown in Fig. 1.5. The black army is larger than either orange army but not as large as both combined. In our simplistic model of a battlefield, the larger army always wins a battle. Hence, the orange armies can be assured of victory if both attack the black army simultaneously, but either orange army is assured of defeat if only one attacks the black army.

The only method of communication between the orange armies available to them is via messengers who must go across black army lines to reach the

---

[11] Although the data processors can perform only one function at a time, computation speeds are often so much greater than communication speeds that the processors can keep both the transmitting and receiving communications equipment simultaneously busy.

Figure 1.5   Three Army Problem.

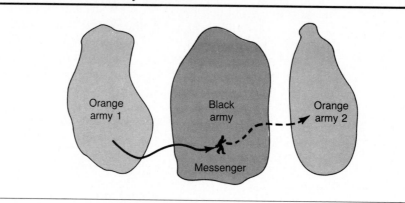

other army. This type of communication is far from being perfectly reliable, since the black army will try to capture any messengers passing through.

Assume that the commander of the two orange armies, with headquarters in the area occupied by Orange Army 1, wants both orange armies to attack at 2:00 P.M. on Saturday. No attack will take place, however, unless each orange army is certain that the other will attack at the same time. To ensure this, Orange Army 1 sends a messenger through the lines to Orange Army 2. There is no guarantee this messenger will get through, however, so neither army will attack until receipt is verified. Hence, after receiving the message, Orange Army 2 sends back a messenger to notify Orange Army 1 of receipt.

If the confirmation message gets through, both orange armies will know the schedule for an attack, but neither attacks until it is absolutely sure the other will attack simultaneously. The exchange of messages so far is not adequate, since Orange Army 2 is not sure its confirmation message got through, and knows Orange Army 1 will not attack if it did not. Orange Army 1 can, of course, return still another message to indicate the confirmation got through, but it has no way of guaranteeing the confirmation of the confirmation will get through. Without this, Orange Army 2 will not attack on schedule.

Although the two armies can continue indefinitely to send messengers back and forth, there is no way to absolutely guarantee a simultaneous attack with this approach. This can be deduced from the following reasoning. Assume a finite number of messages is adequate and that the most recently sent (or received) message is the last essential one—that is, a simultaneous attack is guaranteed if and only if it has been received. If it is known to have been received, no more messages need to be exchanged. However, the army sending this last message will not attack unless it knows the message got through. This implies another (acknowledgement) message is also essential. This gives a contradiction, so the assumption that a finite sequence of messages is adequate must be false.

There are techniques for guaranteeing a very high probability of a simultaneous attack (see Problem 1.5), but no way of gaining an absolute guarantee. Similarly, there is often no way to absolutely guarantee coordination of sender and receiver in computer networks. Under virtually all circumstances where such coordination is needed, however, it is not the life or death situation depicted in the three army problem, and techniques for guaranteeing a high probability of coordination are satisfactory.

1.4.3 Maximizing Reliability and Freedom from Errors

If a network is to be successful and provide good service to users, it must be reliable and minimize or eliminate erroneous results. Design for reliability and freedom from errors spans all levels of network design. Current technological trends leading to availability of ever more powerful types of equipment appear to make it likely that reasonable response time and throughput goals may be easy to meet in the near future, with achievement of adequate reliability a more and more important design focus. High reliability may in the future be a stronger selling point than good response time or throughput performance.

A variety of reliability measures are in common use. *Availability* is commonly defined to be the probability that a system or piece of equipment is operational at a randomly chosen instant of time (for example, the instant when a user wants to use it). *Reliability* is defined in most discussions of reliability theory as a function of time; it is the probability that a system will remain operational for a specified interval of time, assuming that it was operational at the beginning of the interval. *Maintainability* refers to ease of maintenance, including ease of diagnosis and repair. *Survivability* refers to the ability of a network to continue operation despite failures of significant amounts of equipment; for commercial networks such failures would normally be the result of natural disasters such as earthquakes or floods; for military networks, though, they could result from enemy attacks. Finally, *performability* refers to the network's ability to perform specified tasks. Optimizing any such reliability measure requires careful design, ranging from careful design of equipment through careful manufacturing and maintenance to use of appropriate protocols, such as fault diagnosis, checkpointing, and other recovery techniques.

Maximizing freedom from errors is closely related to design techniques emphasized here. Errors in bits sent across communication channels are inevitable because of physical phenomena such as thermal noise resulting from random motion of electrons in all dissipative electrical elements. The number of errors can be reduced by appropriately designing the communication facilities and transmitting and receiving equipment used, but techniques for recovering from their effects are also needed.

Specific types of errors that need to be handled are bit transmission errors, inserted or deleted bits, loss of synchronization, lost messages, invalid messages or message sequences, addressing errors, and protocol errors. Some problems can lead to deadlocks, when communication comes to a screeching halt because neither transmitter nor receiver can proceed further. Excessive delays can, for some applications, be as damaging as errors, so techniques for preventing excessive delay may be classified in this category. Standard techniques include techniques to regulate user traffic allowed into the network so delays do not become excessive.

A final category of problems we include here is preventing unauthorized access, although this could be listed as a separate category. Unauthorized access can be access to data, to transmission facilities, to computational facilities, or to users. It may be unintentional or intentional but not malicious (for example, access by computer hackers who are just experimenting), or it may be malicious. A variety of techniques for ensuring network security have been developed.

1.4.4   Optimizing Performance

A large part of the telecommunication literature is devoted to techniques for optimizing performance. Most of this work emphasizes optimizing quantitative performance measures. However, other important criteria, such as user friendliness, network expandability, and compatibility of new equipment with existing equipment may be equally important, or even more important, in deciding on network architectures or equipment.

The reliability measures discussed in the previous subsection are important quantitative performance measures, though we treat them separately. The other most common measures are *response time* and *throughput*. Response time is commonly defined as a measure of time from submission of work to a system until its completion. Throughput is defined as a measure of the amount of work done per unit time. Unfortunately, it is not normally possible to simultaneously optimize both response time and throughput since increased (improved) throughput leads to increased (degraded) response time due to queuing effects.

Other performance criteria considered include congestion, equipment utilization, fairness, and combinations of measures—for example, power is defined as throughput divided by response time. These other measures are less commonly studied, however.

Performance criteria that are important to network managers and those important to network users only partially overlap. A user is most commonly interested in his or her experience with a network, not in its technical performance measures, while a manager is more likely to be interested in technical performance measures—especially those related to maximizing network revenue. For example, the criteria most important to a network manager may well be total network throughput (total number of jobs handled in a given period of time), and network down time (that is, non–revenue-

Figure 1.6   Relationship Between User-oriented and Manager-oriented Network Performance Criteria. Network designers should consider both classes of criteria (adapted from [ILYA85]). © 1985 IEEE. Reprinted by permission.

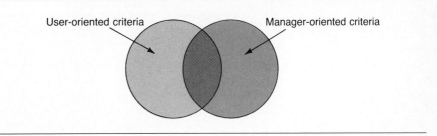

producing time), while the criteria most important to the user are more likely to be response time (how rapidly a user's jobs are processed) and user friendliness. The criteria used do overlap to some extent, however. For example, the network manager is interested in satisfied users, so must consider user-oriented criteria in addition to the other criteria mentioned (see Fig. 1.6). Both categories should be considered by a network designer.

Design decisions that have an impact on performance include selection of computer equipment and communication facilities. Performance modeling techniques normally verify that more powerful computers or higher speed communication lines reduce response time or increase potential throughput, although there are occasions where the opposite may occur.[12] The network topology, the communications approach, and a variety of algorithms used in the network can also have an impact on performance.

1.4.5   Minimizing Cost

Minimizing cost is a criterion that permeates all levels of design. It is often combined with other criteria, since optimizing cost/performance, or at least achieving acceptable performance while minimizing cost, is important in virtually all applications.

Cost is a standard design criterion for any component of a network, but we will not always go into fine enough detail on design of computers to illustrate how cost is factored into their design. Detailed studies are highly device specific and outside the scope of this text.

---

[12] For example, in situations involving frequent changes in the direction of transmission some high-speed modems may take so long to reverse their direction of transmission that the response time of a network using these modems is worse than that of a network the same but using lower-speed modems that reverse direction more rapidly.

Figure 1.7   Evolution of Freedom to Make Changes and the Cost of Changes During the Network Planning, Implementation, and Operation Stages (from [ILYA85]). © 1985 IEEE. Reprinted by permission.

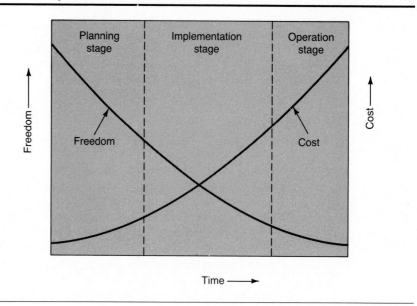

The network designer's freedom to make changes in network configuration, parameters, and so forth is maximum at the start of network design, especially during the planning stage, and drops rapidly as time advances, and the design evolves through planning and implementation stages to the final operation stage. Conversely, the cost of changes is minimum during the initial planning stage and maximum during the operation stage. This is illustrated by Fig. 1.7, which graphically shows why a thorough design job during the initial stages can pay major dividends later.

1.4.6   Network Management

Network management functions may not be visible to users under normal circumstances, but they are essential for the network operator. They include such functions as network configuration and, when necessary, reconfiguration, network status monitoring, reacting to conditions such as failures or overloads, and various accounting and billing functions. System management functions are needed at most levels of network design, though the most critical management functions may be present only at higher protocol levels.

1.5   The Telecommunication Environment

The design, development, and use of telecommunication networks are strongly influenced by the telecommunication environment. We use the term "environment" to denote trends such as legal and regulatory trends, and technological and economic trends, which are difficult to quantify but have a major impact on the field. Specific trends are the following.

1.5.1   Technological Trends

Dramatic advances in price/performance ratios for data processing and switching system hardware, in contrast with slower advances for transmission components, have a strong influence on network design. A common projection for data processing hardware cost is reduction by a factor of ten every five years for equivalent capability. This corresponds to approximately a 37 percent reduction each year. Corresponding transmission cost reductions are occurring, but at a considerably slower rate. A value of 8 percent per year reduction can be estimated from a curve in [ROSN82b]. The quoted numbers yield the projections in Fig. 1.8.

Since computing costs are dropping far more rapidly than transmission costs, a major trend is to distribute intelligence (logic, memory, and so forth)

Figure 1.8   Projected Variations in Relative Computing and Transmission Hardware Costs, for Constant Capabilities, from 1980 to 2000.

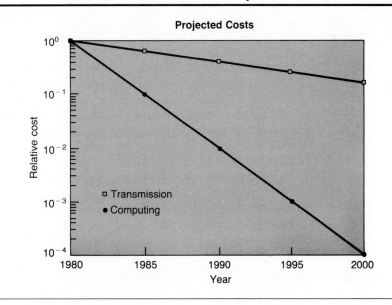

throughout a network rather than to rely on large amounts of computing power at a few locations. This can substantially reduce the communication necessary by doing computations, or at least data reduction, locally. It is interesting to note, however, that expenditures for computing hardware for networks have not tended to drop. Instead, the trend has been to provide greatly increased computational power at costs comparable to those for less power in earlier networks. Profits of computer hardware manufacturers have not been dropping!

Other cost trends have had similar impact. For example, costs of digital transmission and switching equipment have been dropping rapidly in comparison with costs of older analog equipment for the same functions. This has greatly hastened the changeover to digital equipment, which also tends to have considerably enhanced performance. The cost of optical fibers for transmission is dropping rapidly while the cost of copper is increasing slowly (largely because of increases in labor costs). This is hastening installation of new equipment and services with far greater capability than older equipment.

1.5.2   Impact of Installed Equipment Base

The installed base of many types of equipment, such as personal computers, terminals, and work stations, has now reached critical mass, so costs for such equipment are rapidly decreasing and systems using them have a relatively low revenue threshold to be profitable. Networks requiring equipment types that have not reached critical mass require much more investment to become profitable and are less likely to be developed.

An excellent illustration of the impact of the installed base of equipment on the development of networks is given by the videotex systems. As was indicated in Subsection 1.3.1, the French Teletel videotex system has by far the largest installed base of terminals of any such network, thanks to the French PTT's decision to offer a simple Teletel terminal free to any telephone subscriber wishing to have one. The result has been far faster development of videotex offerings in France than in other countries. The number of French Minitel (terminal) owners is approximately 3.4 million, with 4.7 million connect hours per month [JUDI88].

A different aspect of an installed base of equipment can at times slow innovation. A large investment in installed equipment may make replacing the equipment with new and better equipment economically unsound, especially if the new equipment is not compatible enough with older equipment to allow gradual migration to the new technology. An example is the impact of the telephone companies' installed base of equipment (hundreds of billions of dollars). The installed base of twisted pair local loops (pairs of copper wires running from local switching offices to subscriber locations) in the United States alone is estimated at over $130 billion. If the telephone companies were to render obsolete this installed equipment base, it would be difficult to avoid bankruptcy, especially since only a small percentage of the investment may have been written off.[13]

1.5.3   Legal and Regulatory Factors

Some legal and regulatory factors were discussed in Subsection 1.3.1 in the description of the Carterfone decision and subsequent events leading up to the divestiture of AT&T. During the past 20 years, from the Carterfone decision in 1968 until now, the trend of legal and regulatory decisions in the United States has been to open up more and more of the communications field to competition, with consequent major advances in speed of innovation. Major businesses, such as those manufacturing equipment to be interconnected with the telephone network or developing new networks to compete with older established ones, would have been illegal two decades ago.

There are substantial differences between the legal and regulatory environments in the United States and those in most other countries. In the United States, telephone service is offered by private, profit-making companies that are regulated by government agencies. A similar environment exists in Canada and a few other countries, but in the majority of countries around the world telephone service is offered by a government agency, often called a PTT (Postal Telephone and Telegraph) agency. This implies that the government both offers and regulates communications service, which often leads to slow rates of innovation. On occasion, however, the PTT model has led to rapid innovations, as in the French PTT's Teletel offering. Furthermore, there are fewer constraints on obsoleting old equipment with the PTT model since the profit motivation that governs the U.S. model is not present.

The status of computer industry regulation also varies from country to country, though in virtually all countries the computer industry has had fewer regulatory constraints. However, a number of countries subsidize nationally based computer firms and put limits on competition.

Although legal and regulatory factors are sometimes even more important than technical factors, it is not practical to discuss the subject in a few pages, and we will thus not attempt comprehensive coverage here.

1.5.4   Standards Activities

Adherence to recognized standards for various aspects of network design has become absolutely essential if networks are to have broad markets. A major part of this book is devoted to discussing important standards for telecom-

---

[13]The speed at which the investment is written off is often limited by regulatory agencies. AT&T, however, announced a $6.8 billion writeoff at the end of 1988 (causing the first annual net loss in the company's history) in order to "clear the decks" for more purely digital facilities. This indicates the importance such companies place on participating in the new technologies described here.

munications. Standards adhered to by equipment manufacturers make it possible to have equipment from various vendors work together.

A commonly cited problem with standards is that they stifle innovation because new and better ways of doing things may not have a chance to be used if other techniques have been standardized. This is a valid criticism, but the benefits of standardization greatly outweigh losses from stifling innovation. Standards do evolve with time, so major improvements in techniques can still win out eventually.

Another weakness of standards is that there are a wide variety of them. We discuss incompatible standards for accomplishing essentially the same tasks numerous times in the book. Furthermore, a number of prominent standards contain so many options that there is no guarantee that different pieces of equipment supposedly adhering to the same standard will work together. This complicates life for equipment manufacturers and network designers, but things would be far more complex without the standards.

1.6   Summary

In this chapter we have discussed the history of telecommunications and outlined the factors causing the current wave of explosive growth in the field. Some major milestones indicating progress in the field have been unpredictable, resulting from work by individuals or changes in the legal and regulatory climate, but other milestones have been reasonably predictable results of development efforts. Other important factors determining approaches to solving problems have been changes in capabilities and relative costs of different systems or subsystems. Another important factor has been progress in developing standards to ensure compatibility of different pieces of equipment.

A number of general problems are common to various areas of telecommunication network design, and they appear in various guises throughout this book. Such problems include the following:

- Resolving incompatibilities of equipment
- Coordination of sender and receiver
- Maximizing reliability and freedom from errors
- Optimizing performance
- Minimizing costs
- Network management

Major parts of this book are devoted to discussion of techniques for solving these problems.

1.7   Preview of Later Chapters

In this book we develop basic principles of telecommunication network architecture, concentrating on those levels of the architecture primarily concerned with providing transparent connections between users.

Chapter 2 demonstrates that what is achievable in telecommunications is determined by fundamental factors. These range from limits determined by physical dimensions of equipment used, through limits imposed by the communication facilities available, through limits resulting from intrinsic computability of basic quantities needed in algorithms for telecommunication networks.

Chapter 3 introduces prominent approaches to telecommunication network design, along with major problems that must be solved by the designer. Brief introductions to the major communications architectures discussed in the rest of the text are also given.

Chapter 4 discusses standard techniques for interconnecting various types of equipment and resolving incompatibilities of equipment. The interfaces include those between data communications equipment (for example, modems) and data processing equipment, computer ports or similar interfaces, and interfaces with communications lines.

Chapter 5 focuses on contention among multiple users wishing to access a communications medium. Algorithms range from those with preallocation of portions of the medium to each potential user to techniques with largely free contention for facilities—each user transmitting when ready and recovering from conflicts with other transmissions in a prescribed manner.

Chapter 6 discusses common techniques for synchronization and error control. Synchronization is a major aspect of coordination of sender and receiver. Error control is used to help maximize reliability and freedom from errors.

Chapter 7 discusses the handling of information transfer over single data links without relays or intermediate nodes. The data link layer protocols discussed assemble characters into messages or other data units and apply techniques for medium access control, synchronization, and error control to guarantee reliable transmission of data.

Chapter 8 is concerned with techniques for ensuring that data gets to the correct destination and that facilities are not so overloaded that they cannot function well.

Chapter 9 discusses protocols that provide end-to-end communications capability. They mask peculiarities of the data transfer technology, establish, maintain, and terminate network layer connections, and handle routing and flow control plus related functions. Some of these functions may be handled in a cooperative manner by all nodes in the network.

Chapter 10, Internetworking, extends the earlier material to include the case where different networks are interconnected despite numerous incompatibilites among them.

Chapter 11 presents the highest level of protocols that we discuss. These protocols interface the more communications-oriented lower level protocols with data processing–oriented protocols and are sometimes considered to be the keystone of a protocol architecture.

Chapter 12, on Integrated Services Digital Networks, concludes the book. Although many details of ISDNs are treated in earlier chapters, this chapter ties the material together and presents current plans for what many consider to be the future of telecommunications.

Problems

1.1   List and briefly describe six telecommunication networks you have come in contact with.

1.2   Briefly discuss recent developments in telecommunications, such as major technological changes, new applications and services, and new products.

Hint: In addressing Problems 1.1 and 1.2 you may wish to look into such technical journals as *IEEE Spectrum, IEEE Computer, IEEE Network, IEEE Communications, Datamation, Data Communications,* or *Telecommunications.*

1.3   At the end of section 1.1, factors likely to limit achievement of Martin's utopian vision of the future were listed as legal, regulatory and economic factors, and basic human nature. Briefly discuss each of these factors and indicate your impressions of how each is likely to limit progress.

1.4   The bits transmitted in the T-1 Carrier System (described in Subsection 1.3.2) include bits used to maintain frame synchronization as well as bits used to encode voice signals. How many bits per second are used to maintain frame synchronization?

1.5   The simplest approach to modifying the three army problem discussed in Subsection 1.4.2 is for the commander of Orange Army 1 to send a number of messengers carrying the attack schedule to Orange Army 2. The commander assumes at least one messenger will get through and has Orange Army 1 go ahead with the scheduled attack. Orange Army 2 attacks if at least one message gets through.

a.   Assuming the probability of any individual messenger getting through is p, and that N messengers are sent, what is the probability that the two orange armies will attack simultaneously?

b.   Assume that $p = 0.5$, that is each messenger is equally likely to get through or be captured. How large does N need to be for the probability of a simultaneous attack to exceed 0.999?

1.6   Compute and plot a modified version of Fig. 1.8 with the rate of decrease of communications costs increased to 12 percent per year. Comment on the significance of this change.

1.7   The formula for response time (time in queue plus service) in the simplest queuing system is $t_s/(1 - \lambda t_s)$, with t_s service time and λ throughput. Give an explicit formula for power, as defined in Subsection 1.4.4, and compute the value of $\rho = \lambda t_s$ for which power is maximized. Assume t_s is a constant.

2   Fundamental Limits

2.1   Introduction

The designer of a telecommunication network should be aware of the fundamental limits of what can be achieved, since it is folly to try to violate such limits. These limits include capabilities of equipment: electronic circuitry, communication channels, and data processing equipment. Properties of electronic circuitry largely determine speeds at which such circuitry can operate as well as voltages and currents it can tolerate. Different communication channels are best adapted for operation at different speeds and are characterized by different noise statistics, error rates, and other perturbations of signals. Some important fundamental theorems determine the maximum rates at which it is reasonable to transmit data in terms of bandwidth and noise statistics of channels.

In this chapter we first give a brief discussion of physical constraints on equipment. Next we give an overview of the most important types of communications systems, including an introduction to modulation and digital transmission techniques. Two fundamental theorems describing theoretical limits to achievable performance are given. Another fundamental theorem that indicates it is feasible to transmit analog signals digitally with no loss of information is given next. This is followed by a presentation of data transmission techniques typically employed in today's systems and a discussion of the characteristics of major information sources. The chapter concludes with a discussion of limitations on computing power needed to perform computations for network operation.

The entire chapter should be of interest to specialists in telecommunications, but some specialized topics may be omitted by those wishing an overview of the field. Subsections 2.4.3 and 2.4.4, covering modem signal constellations and binary signal encoding techniques, are the most specialized of the subsections in the chapter (and are hence marked with a special symbol ($\square$)); they may be omitted (or skimmed) with no loss of continuity. A brief overview of the different types of communication channels described in Subsections 2.3.1 to 2.3.6 should also be adequate (but Subsection 2.3.7 should be studied in reasonable detail).

2.2　Physical Constraints

During the early days of telecommunication networks, most electronic circuits were implemented using discrete components—resistors, capacitors, inductors, and electron tubes of various types. Today, most are implemented using integrated circuits. This has resulted in tremendous advances in operating speeds, reliability, and so forth, but it has also placed new constraints on system designers. Only a limited number of types of components can be readily fabricated as integrated circuits—primarily transistors, diodes, resistors and capacitors. Furthermore, the range of values of these components is limited, so designers try to avoid the use of resistors with resistance less than about 50 Ω or greater than about 100 kΩ; when capacitors must be used, total capacitance should not exceed 100 pF, though a slightly broader range of values can be fabricated [GLAS77], [HAMI75].[1]

natural operating speeds of circuits

A very rough, and simplistic, estimate of the *"natural operating speed"* of logic circuitry is given by its "time constant" RC, since deviations of simple RC circuit outputs from their final values are proportional to $e^{-t/RC}$. Using the R and C values above, we find typical time constants in the range 10^{-5} to 10^{-9}. Time constants for transients in integrated circuits obtained through more complex analyses are also within this range [GLAS77]. Taking a time equal to several time constants, say three to ten, as an adequate "settling time," we find the natural operating speeds of integrated circuits are on the order of tens of thousands to hundreds of millions of operations per second. As circuit sizes decrease for very large scale integration, both R and C also decrease, which means that natural operating speeds increase still further. In addition, size decreases reduce distances between components, decreasing delays caused by electromagnetic propagation. These effects further increase natural rates.

It is not always easy to use newer technologies to implement older standards. Operating speeds of the discrete circuits used in earlier days, and still incorporated in widely used standards such as RS-232C, were relatively slow. Such speeds are achievable with integrated circuits but do not take advantage of their capabilities. Furthermore, voltages used for the older discrete circuits are not compatible with integrated circuits, which causes problems in trying to implement older standards with newer technology.

Physical constraints on operating speeds are especially obvious for systems involving radio links. An efficient transmitting or receiving antenna for electromagnetic energy needs to be an appreciable fraction of the wavelength (preferably half or more) of the signals used. The wavelength, in meters, of a signal at frequency *f* hertz (Hz) is given by

---

[1]The symbol Ω denotes ohms, the standard unit for resistance values. One pF represents one picofarad, that is, 10^{-12} farad, a farad being the standard unit for capacitance values.

$$\lambda = \frac{3 \times 10^8}{f}, \tag{2.1}$$

since the speed of light in free space (or the atmosphere) is 3×10^8 meters per second. The most important frequencies for speech communication are around 300 to 3000 Hz, implying antenna sizes on the order of 50 to 500 km would be needed for direct electromagnetic propagation of speech—very impressive, and unrealistic, figures! This has motivated use of modulation to shift frequency bands occupied by speech, or other signals, to bands for which reasonable antenna sizes can be used. By shifting different signals to different frequency bands, a technique known as *frequency division multiplexing,* it is also possible to transmit a large number of signals over the same links.

Another constraint encountered by heavy volume users of communication facilities has been a space problem. In some cases, companies have displaced one or more elevators to run communication cables in elevator shafts. Telephone companies sometimes run into major problems routing all necessary cables into and out of switching offices.[2] Techniques for sharing facilities, including frequency division multiplexing (mentioned above), *time division multiplexing* (giving different users access to facilities during different time intervals), and to some extent *space division multiplexing* (a term telephone companies use for grouping different conductors into single cables) can be used to alleviate such problems. An alternative to multiplexing, known as *concentration,* involves demand assignment of facilities and is especially applicable for bursty information sources, that is, sources that generate information only during a small part of the time. Information sources for telecommunication networks usually fall in this class.

Both multiplexing and concentration techniques have another important advantage. By allowing sharing of communication facilities, they can often reduce communication costs. Trade-offs between use of multiplexers and concentrators are analyzed in a later subsection.

2.3   Common Communication Channels

A variety of communication channels are used to provide data transmission facilities for telecommunication networks. The most common are telephone channels, used because of their almost universal availability. Other channels are largely characterized by the media they use. Prominent media include radio links, twisted pair cables, coaxial cables, and fiber optic cables. Com-

---

[2]It is standard for a separate pair of copper wires or local loop to run from each telephone subscriber's location to the closest telephone company switching office.

munication satellite channels are treated as a separate category, although they use radio communication, since they have important characteristics that distinguish them. It is also useful to distinguish locally installed facilities from those installed by telephone companies or similar vendors, since local installation allows users to take advantage of the full capabilities of media rather than being restricted to vendor standards.

2.3.1   Telephone Networks

The *world-wide telephone network,* viewed as a single entity, is arguably the most complex system ever designed and installed by mankind. There are over 300 million telephones in the world. At least 98 percent of them are interconnected in the sense that it is physically possible to interconnect any pair for a telephone conversation, so it is meaningful to speak of the worldwide network as a single entity.

An important characteristic of the world-wide telephone network is its diversity of equipment. Design lifetimes for equipment are 30 to 40 years, and limited amounts of equipment twice this age are in use. The tremendous strides made by technology in the past 30 to 40 years would alone be sufficient to produce immense variability, but diverse standards for equipment around the world also lead to variability.[3] Environments for equipment are also extremely diverse, ranging from some comparable to typical computer rooms to situations where extreme temperatures (either hot or cold), high or low humidity, noxious gases, and so forth are regularly encountered. These *variability of communi-cation channels* extremes produce corresponding variability in parameters characterizing communication channels, which means that "typical" parameter values should be treated with caution. Three or four decimal orders of magnitude variation in important parameters for different telephone calls is common, especially for dialed calls. A guarantee that important parameters fall within certain ranges can be obtained for leased (private) lines, but at extra cost.

Media used for telephone circuits include copper wires, coaxial cables, microwave radio, waveguides, optical fibers, communication satellite links, and undersea cables. Until recently the majority of the circuit kilometers in the United States[4] have been handled by microwave radio links, each transmitting thousands of voice channels plus other signals, but the more than 3 million km of fiber optic links recently installed have more than doubled the capacity of the networks, though they comprise only a small percentage of the mileage of copper circuits [SHUM89]; this implies the majority of circuit kilometers are now optical fiber circuits.

---

[3]There are major efforts to introduce world-wide standards, but they are still far from universal.

[4]A circuit kilometer is defined here to be one voice-grade channel transported one km; thus a microwave link handling thousands of voice-grade channels has its length multiplied by a factor of several thousand in computing circuit kilometers.

Figure 2.1 North American Telephone Switching Office Hierarchy.

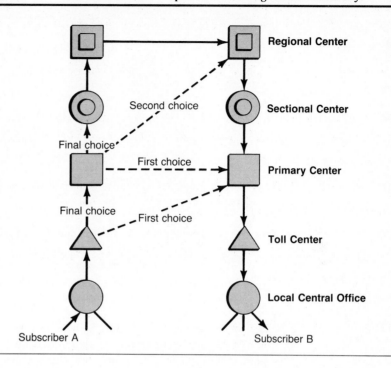

Figure 2.1 illustrates the telephone switching office hierarchy in North America[5] along with the manner in which alternate routing is available between subscribers such as the Subscriber A and Subscriber B illustrated. The five levels of switching offices may be supplemented by one or two more for international calls. Similar hierarchies are used in other areas.

Solid lines connecting switching offices indicate circuits that are always available, provided calling volume is not too great.[6] Dotted lines indicate circuits installed only when the amount of traffic between different points in the network merits installation. Calls are routed by the most direct route when appropriate circuits are available, with fallback to alternate routes when preferred routes are unavailable. For the case illustrated, with Subscriber A and Subscriber B in regions served by different regional centers, a maximum of ten switching offices and nine interoffice trunks can be used for one call, but the probability of this is very low. However, alternate routing means different calls between two locations may go through radically different equipment.

---

[5] Figure 2.1 illustrates the hierarchy that has been used in North America for a number of years. It is gradually being replaced by a nonhierarchical structure; see [SCHW87] for details.

[6] In the United States such circuits are almost always available except on Mother's Day and Christmas, the two days of the year when the telephone network is most likely to saturate.

Figure 2.2   Typical Frequency Response for Analog Voice
Grade Telephone Channel.

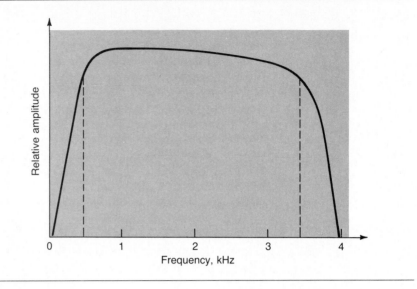

Numerous types of telephone channels are available, but the most common is a *voice-grade channel* with bandwidth restricted to 4 kHz of which approximately 3 kHz is usable for data transmission. Such channels, built originally to carry voice traffic, have a bandwidth that is adequate to convey speech even though the higher frequencies are not transmitted. A rough sketch of the frequency response of a typical voice-grade channel is given in Fig. 2.2, which gives relative amplitudes with which signals of different frequencies would be received if transmitted input amplitudes were equal. The limited bandwidth results from filtering (that is, band limiting) equipment in switching offices, and it is motivated by the fact that telephone companies stack voice bands 4 kHz apart for transmission. As the figure shows, the frequency range from approximately 300 Hz to 3300 Hz is transmitted with adequate amplitude to be useful for data transmission. It is not an inherent characteristic of the transmission media used.

voice-grade channel

Modems (*mo*dulator/*dem*odulator*s*) are needed to send digital data over analog voice grade channels (or any analog channels). The basic function of modems for voice-grade lines is to modulate incoming digital signals, transforming them into a form that is enough like voice signals to be transmitted over voice-grade lines, and to demodulate received signals to transform them back to digital form. Although most modems for voice-grade lines still operate at speeds not exceeding 1200 bps to 2400 bps, major advances in modem technology in recent years (such as automatic equalizers to adjust modem characteristics to line characteristics and track such characteristics during

modems

transmission, plus trellis coding to protect transmitted signals against errors with minimal impact on effective transmission rates) now allow transmission over analog voice grade lines at rates as high as 19.2 kbps—if one is willing to pay for expensive modems.

multiplexing

Signals sent between switching offices, on interoffice trunks, share communication facilities through the telephone companies' frequency division and time division *multiplexing hierarchies,* sometimes passing back and forth between analog and digital formats. At times, they may share communication facilities with thousands of other signals. The older frequency division hierarchy is based on channel groups, a total of 12 voice-grade channels frequency division multiplexed together in a total 48 kHz bandwidth, plus still larger groupings known as supergroups, mastergroups, and so forth. The newer time division hierarchy is based on the 24 voice-grade channel T-1 Carrier System used in North America and a few other countries,[7] plus a similar 30 (or 32) CCITT channel carrier system used in other countries; higher levels of time division multiplexing are also used.

Demodulation and/or modulation, analog to digital conversion, or digital to analog conversion may occur at each switching office that a signal passes through en route from source to destination. The quality of received signals is usually remarkably good after all these transformations, especially if one considers the variability in types of equipment and of environments en route.

The great majority of circuits in telephone networks are analog voice-grade lines (either switched or private lines),[8] but direct access to purely digital facilities is available in many areas. This allows transmission at rates up to 56 kbps (in areas following Bell System standards) or 64 kbps (in areas following CCITT standards),[9] with much better quality of service (error rates, and so forth) than can be obtained using analog lines with modems.

Although subscriber costs for purely digital service are currently higher than costs for analog service, the telephone companies are moving toward wider and wider use of digital transmission and switching facilities, with a major motivation being that such facilities are more economical to install and operate. This is also a major factor motivating development of ISDNs, which will be purely digital. Since it is cheaper for telephone companies to provide purely digital 56 kbps or 64 kbps channels than to provide voice-grade analog channels, the subscriber cost of these digital channels should eventually be at least as low as the cost of voice-grade analog channels.

In addition to the amplitude distortion (characterized by distorting relative amplitudes of different frequencies) of signals illustrated by Fig. 2.2, analog channels are characterized by delay distortion (different frequencies

---

[7] See the list of telecommunication milestones in Section 1.3.

[8] A wide variety of offerings are actually available, ranging from very narrow bandwidths to full group (48 kHz), or even wider, bandwidths. See [DOLL78] for a reasonably detailed listing of the offerings available.

[9] Still higher rates, such as the full 1.544 Mbps T-1 rate, may sometimes be available.

delayed by different amounts). The human ear is not sensitive to delay distortion, but it can be very detrimental for transmission of digital signals. Noise is another major limiting factor for data transmission, with major noise types including thermal noise caused by thermal motion of electrons in dissipative elements (that is, resistors), shot noise from quantization of electrical charge flows (individual electrons or holes) in active circuits, intermodulation noise from nonlinearities in circuits shared by signals at different frequencies, and impulse noise or short noise spikes resulting from lightning, switching system transients, maintenance work, and so forth. Each can cause transmission errors, with impulse noise often the most difficult to counteract.

The impairments mentioned above are critically important to modem designers, but the data communications user is primarily interested in their impact on data communications quality of service. The most relevant quality parameters are usually error rates. Error rates for analog channels vary widely, by three or four decimal orders of magnitude for different channels, with type of equipment, transmission speeds, type of modem, and so forth strongly affecting them. A typical bit error rate quoted is 10^{-5}, implying an average of one bit error for each 100,000 bits transmitted, but bit error rates anywhere in the range of 10^{-3} to 10^{-7} are common for transmission over analog channels. Bit error rates over purely digital facilities and newer analog facilities are usually closer to 10^{-7}. Older facilities in a communication path, say at either end of a communication satellite link (which normally has very low error rate) tend to dominate error statistics, however, so the advantages of newer facilities are lost. Errors also tend to occur in bursts, on either analog or digital channels, so the conditional probability of additional errors in a block of bits increases significantly after the first error occurs.[10]

2.3.2   Radio Links

As we indicated above, until recently the majority of the circuit miles (or circuit kilometers) in the U.S. telephone network were handled by *microwave radio links.* Direct use of radio links for communication channels is also common for telecommunication networks. In some situations radio links may be the only feasible channels to use. Situations where it may be necessary to use radio links include communication with remote areas where no telephone or other alternative channels exist or situations where nodes are mobile (for example, in cars, planes, or other vehicles). Simply communicating between two locations separated by a public right of way such as a street or highway may require use of radio links, since permission to lay cables over or under public rights of way may be impossible to obtain. Appropriate permits for using radio channels are normally needed, but these are usually easier to obtain if the application is justifiable and suitable radio frequency bands have not all been previously assigned.

---

[10]This also means that fewer characters or blocks are affected by errors, since those that do have errors are likely to have multiple errors.

A wide variety of radio links can be used for communication channels, since the term "radio" encompasses use of essentially any frequencies in the electromagnetic spectrum. A brief summary of the characteristics of various frequency bands is given in Table 2.1. National and international agreements regulate allocation of specific bands for different uses.

As the table indicates, different frequency bands have radically different propagation characteristics and are best adapted for different applications. Although a variety of bands have been used for telecommunication networks, the most common are HF, VHF, and UHF bands. These include commercial FM and television bands, plus frequencies normally used for microwave links. Most transmission in the VHF and UHF bands uses line-of-sight or nearly line-of-sight propagation, since radio waves at these frequencies experience relatively little refraction or reflection from ionospheric layers.[11] Their line-of-sight propagation normally restricts transmission ranges, for reasonable antenna heights, to no more than about 20 km, though this depends on terrain and other obstacles.[12] On the other hand, HF transmission is strongly influenced by ground wave propagation (following the curvature of the earth) and sky wave propagation (reflected off ionized atmospheric layers). This has, at times, allowed communication half way around the earth by radio amateurs with very limited radiated power. Such HF links are not as reliable as VHF or UHF links, and HF bandwidths tend to be limited, so VHF and UHF bands are more commonly used for telecommunication networks.

Because of the wide variations in propagation characteristics, it is difficult to treat concisely the transmission limitations likely to be encountered. Most types of noise sources listed for telephone channels are applicable here also. Amplifier noise can be especially severe for cases where the received signal is weak. Deep fades of the received signals, often frequency selective, are a problem with most radio channels. Such fades are often caused by multi-path propagation, with the received waveform the sum of contributions propagated by different paths; the contributions may reinforce or cancel each other, depending on frequency and relative propagation delays. Atmospheric conditions, such as heavy rainfall, also contribute to fades. Fades of as much as 40 dB are not unusual and can effectively disable most communication links.[13] Digital radio links must also be carefully designed to avoid interfer-

---

[11]There are some exceptions, such as meteor-burst data communications links [KOKJ86], which rely on reflections from ionization trails left by the estimated 100 billion meteors, of mass greater than the order of a microgram, bombarding the earth during a typical day. Such links have been used to provide reliable transmission (with delays not normally exceeding a few minutes to find and use suitable meteor trails) for distances up to 2000 km or so.

[12]Tall antennas can also increase the distance; for example, a 500 m antenna would give coverage over a radius of approximately 60 km, using calculations based on the earth's curvature.

[13]A power ratio in dB is given by $10 \log_{10} (P_1/P_2)$, with P_1 and P_2 the absolute power levels. Thus 40 dB of fading corresponds to a multiplicative factor of 10^{-4} in power caused by fading.

Table 2.1   Radio Frequency Bands.

Frequency Band	Designation	Propagation Characteristics	Typical Uses
3–30 kHz	Very low frequency (VLF)	Ground wave, low attenuation day or night, high noise level	Long-range navigation, submarine communication
30–300 kHz	Low frequency (LF)	Similar to VLF but not quite as reliable	Long-range navigation, marine radio beacons
300–3000 kHz	Medium frequency (MF)	Ground wave and night skywave, low attenuation at night and high in day, atmospheric noise	AM broadcasting, maritime radio, direction finding, emergency frequencies
3–30 MHz	High frequency (HF)	Ionospheric reflection varying with time of day, season, and frequency	Amateur radio, military communication, international broadcasting, long-distance plane and ship communication
30–300 MHz	Very high frequency (VHF)	Nearly line-of-sight (LOS) propagation, scattering due to temperature inversions, cosmic noise	VHF television, FM broadcasting, FM two-way radio, AM plane communication, aircraft navigational aids
0.3–3 GHz	Ultra high frequency (UHF)	LOS propagation, cosmic noise	UHF television, radar, microwave links, navigational aids.
3–30 GHz	Super high frequency (SHF)[a]	LOS propagation, rainfall attenuation above 10 GHz, atmospheric attenuation due to oxygen and water vapor	Satellite communication, microwave links, radar.
30–300 GHz	Extremely high frequency (EHF)[b]	Same as above	Radar, experimental satellite uses
10^3–10^7 GHz	Infrared, visible light, ultraviolet	LOS propagation, atmospheric attenuation due to water vapor (fog) for some wavelengths (for example, visible)	Optical communication

[a]Smaller bands within this range are also designated by letters, for example, S, C, X, or K.
[b]The upper end of this band is called the mm (millimeter) band because of wavelengths used.

ence with other communication links since digital radio is notorious for generating interfering emissions.

Data rates vary from a few hundred or few thousand bps (especially for HF links) to millions of bps for other links. The rates 34 million bps (Mbps), 45 Mbps, 90 Mbps and 140 Mbps figure prominently in digital microwave standards in the United States, Europe, and other areas.

Figure 2.3 Configuration of Cellular Mobile Radio Network.

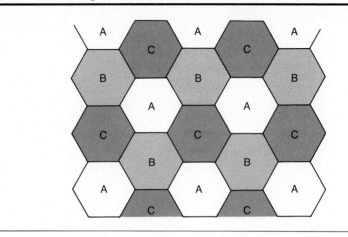

Most radio transmissions can be overheard by any receivers in the geographic vicinity of the transmitter. Directional antennas can reduce the number of directions in which significant electromagnetic energy is radiated, but multiple receivers for each transmitted signal are the norm. This significantly affects the algorithms that should be used to determine locations allowed to transmit at a given time, how information should be routed from source to destination, which node should acknowledge transmissions when acknowledgment is desirable, and so forth.

Cellular mobile radio is a special type of radio network. Such systems use fixed transmitters located in each cell of a grid pattern to communicate with mobile stations that can move around within the grid. The cells are often represented as being hexagonal, as in Fig. 2.3, but actual shapes may vary substantially.[14]

The fixed transmitters within different cells transmit at different frequencies, with frequencies chosen so the same frequency is never used in adjacent cells. Assignment of frequencies to cells is an example of the well known map coloring problem in mathematics. One suitable assignment, with frequencies labeled A, B, and C is indicated in the figure; other assignments with more than three frequencies are often recommended, however, since they can increase the minimum distance between cells using the same frequency.

The fixed transmitters are used both to originate messages and to relay messages from vehicles moving about within the grid. Each vehicle uses a "frequency agile" transmitter and receiver, which monitor the frequencies received at any location and automatically switch to transmit and receive in the frequency band giving the best reception (the strongest signal) at any

---

[14]A square grid of cells is often preferred to the hexagonal grid illustrated [KURI87].

point. As a vehicle moves about within the grid, it switches frequency bands whenever a different band gives better reception. This automatically adjusts the grid shape to compensate for obstructions to radio wave propagation.

2.3.3   Communication Satellites

Communication satellites could be grouped under the radio transmission category, since they use radio propagation; because of special characteristics, we treat them separately, however. The characteristic most obviously distinguishing satellite links, from the point of view of the typical user, is long propagation delay. The great majority of the communication satellites in use are in geosynchronous orbits approximately 36,000 km above the equator. At this elevation, the rotational period of a satellite is 24 hours, which means that it can be positioned so it appears to be stationary from an earth station, eliminating the need for repositioning transmitting and receiving antennas. This elevation means that radio transmissions up to the satellite and back down travel on the order of 72,000 km, giving propagation delay, at the speed of light, of approximately 0.25 sec. Sending a message and getting a reply incurs total propagation delay of around 0.5 sec. Even transcontinental terrestrial radio links seldom incur more than 10 to 20 ms propagation delay.

The "footprint" of a satellite antenna can be approximately one third of the earth's surface, so a large number of receiving stations may be able to hear each transmission. Directional satellite antennas can significantly reduce the size of the footprint, but it is inevitably large. Thus the modifications in algorithms due to multiple receivers for transmissions mentioned above are especially relevant for satellite transmission. The cost of communication between different earth stations is largely independent of terrestrial distance between the stations.

A number of frequency bands are, by international agreement, available for satellite transmission. The major bands are in the 4 to 6 GHz, 12 to 14 GHz and 20 to 30 GHz ranges. The lower frequency bands, which have been used the longest, are becoming congested;[15] in addition the better "parking spaces" for satellites are being filled up. This has forced a move toward using higher frequencies in newer satellites. Although there are advantages to higher frequencies, such as smaller antenna sizes to obtain a given beamwidth, attenuation due to rain is much more severe at higher frequencies.

During the past few years, two areas of satellite communications receiving emphasis have been very small aperture terminal (VSAT) systems and ultra small aperture terminal mobile satellite (MSAT) systems [CHAK88], [MURT88]. Tens of thousands of such systems are expected to be deployed during the next few years, allowing satellite communications to individual buildings or vehicles.

---

[15]The 4 GHz to 6 GHz band is also commonly used for terrestrial microwave links.

A typical communication satellite contains one or more transponders. Each listens to a frequency band, amplifies it, and then rebroadcasts the information at another frequency to avoid interference with the upward transmissions. Approximately a dozen transponders, each capable of encoding a 50 Mbps data stream or various lower speed combinations is typical. At the 50 Mbps rate, a total of approximately 25 million bits can be transmitted before an acknowledgment could be obtained. Modifications to common communication protocols are needed to cope with this.

Error rates for communication satellite links are usually much lower than error rates on telephone networks. Even VSAT links, with antenna diameters in the range 1.2 m to 1.8 m and power amplifiers in the range 1 to 3 watts, are designed to operate at bit error rates less than 10^{-7}, and links using larger antennas and higher power may have error rates orders of magnitude lower. This advantage is quickly lost if telephone links are used at either end of a satellite link, though, since telephone error rates dominate.

2.3.4   Copper Conductors

The majority of the communication circuits in use are based on use of copper conductors. The two major categories are twisted pair cables and coaxial cables. Large numbers of either or both are often bound together in multiconductor cables to reduce the number of cables installed.[16]

twisted pairs

Twisted pairs are pairs of insulated copper wires, typically 1 mm or so in diameter. They are used for almost all local loops, extending from telephone switching offices to subscriber locations, in telephone networks.[17] Although their frequency response is normally limited to less than 4 KHz in the analog portions of telephone networks, they are capable of transmitting frequencies up to the MHz range over short distances.[18] For purely digital communications, approximately 1.5 Mbps is common (twisted pairs are regularly used to transmit T-1 signals), and still higher rates are occasionally transmitted. Twisted pairs can be noisy, especially during electrical storms or when they pass near electrical machinery or other sources of strong electrical interference. Although optical fibers should soon reach a point where they are preferred for new installations, twisted pairs will be ubiquitous for many years since they represent a large part of the investments made by telephone companies.

---

[16]Use of multipair cables also significantly reduces visual clutter in areas with many circuits.

[17]Twisting the wires in a pair greatly reduces the electrical interference it may pick up. During the early days of telephony, single wires were used instead of pairs, with the earth as a current return. Far too much noise was encountered, though, so pairs of wires, later twisted, soon replaced these single wire circuits.

[18]Only a small part of this bandwidth is normally used in telephone networks because of filtering out frequencies outside the standard telephone voiceband at switching offices.

Figure 2.4   Coaxial Cable.

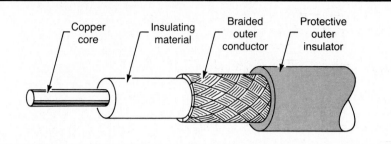

coaxial cables

A *coaxial cable* consists of inner and outer conductors, separated by insulating material, with the entire cable also insulated (see Fig. 2.4). This construction gives a broad effective bandwidth, up to hundreds of MHz for analog transmission. Digital rates up to 50 Mbps for single channels are used on some coaxial cables (baseband transmission), with cumulative transmission rates up to a few hundred Mbps when the bandwidth is split up among multiple channels (broadband transmission). Error rates for data transmission in carefully designed systems can be excellent, possibly three orders of magnitude or more below error rates in twisted pairs.

Coaxial cables have been installed in many regions, especially metropolitan areas, as distribution networks for cable tv (CATV) systems, commonly with potential bandwidths on the order of 300 MHz. This gives a large installed base of cables that could be used for high-speed data transmission. In some areas there are almost as many locations served by CATV as by telephones.

2.3.5   Fiber Optics

Optical fibers are being installed widely. Their main advantage over copper cables is the tremendous gains in digital transmission speeds they provide. Early fiber optic systems operated at 45 to 90 Mbps; but single-mode fiber transmission systems operating in the 405 to 565 Mbps range are now common, with systems operating at 1.6 to 1.7 Gbps being installed. Furthermore, the cost differential between fiber and copper already favors fiber in many situations. Providing circuits operating at terabit-per-second (10^{12} bps) rates is being discussed.

Three basic types of optical fibers are illustrated in Fig. 2.5. Early installations used multimode stepped index fibers, with more recent installations using multimode graded index and (for higher speeds) single mode fibers.

The fibers rely on total reflection of light waves from an outer (cladding) layer to confine waves within the fiber. With appropriate differences in the refractive indices of cladding and core materials, all waves striking the cladding layer at angles less than a critical angle are totally reflected. Those

Figure 2.5 Characteristics of Three Basic Types of
Optical Fibers (from [NAGE87]). © 1987 IEEE.
Reprinted by permission.

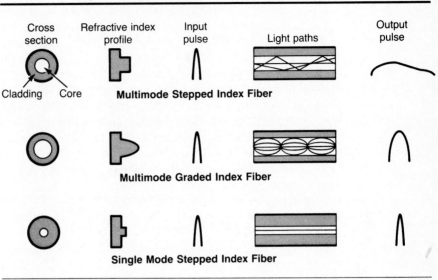

waves traveling parallel to the center axis of the fiber travel shorter distances
and arrive at the destination earlier than waves reflecting off of the cladding
layer. This causes the dispersion of output pulses illustrated for the multimode stepped index fibers. This dispersion is considerably reduced for multimode graded index fibers for which the refractive index varies gradually in
a manner that causes the velocity of propagation to be greater near the outer
edges of the core, partially equalizing delays for different waves. Single mode
fibers, on the other hand, use a core that is so thin that only a single mode
can propagate along the fiber, and essentially no pulse broadening occurs.[19]

Optical fiber systems in the U.S. telephone network are normally designed for cumulative bit error rates $< 10^{-8}$, with repeater spacings on the
order of 40 km [JACO86]. This requires design of individual repeater sections
with much lower bit error rates, typically $< 10^{-11}$.

2.3.6 Locally Installed Facilities

Locally installed facilities can include virtually any of the types of facilities
discussed above. The reason for making a distinction is to emphasize the fact
that locally installed facilities are not subject to the restrictions on band-

---

[19]These are not rigorous descriptions of how light propagates through fibers, but they are
useful for descriptive purposes. For more rigorous descriptions see [ADAM81] or [SNYD83].

width, data rate, and so forth imposed by communications common carrier (telephone company) standards. Locally installed facilities can be used in any situation where the entire span of a network is privately owned; this includes networks in a building, university campus, or industrial park.

The most common locally installed communication circuits are twisted pairs, coaxial cables, and optical fibers. Some radio links are also used, especially when it is necessary to cross a public right of way. Twisted pairs are often used for data transmission rates of 1 to 3 Mbps, with even higher speeds over short distances or with closely spaced repeaters. Coaxial cables are most often used for bit rates of 3 to 50 Mbps, or even a few hundred Mbps in broadband networks. Optical fibers extend bit rates even higher; rates of Gbps (billions of bps) are feasible though most optical fiber used in telecommunication networks is for rates no higher than 100 Mbps or so. Bit error rates for locally installed facilities can be kept within the 10^{-7} to 10^{-11} range with careful design.

2.3.7   Communication Circuit Types and Operation

simplex, half duplex, and full duplex

Communication circuits may be *simplex, half duplex,* or *full duplex.*[20] A simplex circuit transmits information in only one direction; it is analogous to a one-way street for traffic flow. A half duplex circuit can transmit in either direction, but only one direction at a time; it is analogous to a single lane bridge. A full duplex circuit can transmit in both directions simultaneously; it is analogous to a two-way street.

The same three terms can also be used to describe the way in which a communication circuit is operated. Thus simplex operation involves transmission in only one direction. It is possible to operate this way over half duplex or full duplex circuits, as well as simplex circuits. Since simplex telephone circuits are not available, use of the term "simplex" for telephone circuits always refers to simplex operation. Similarly, half duplex operation involves transmitting in only one direction at a time; it is possible to operate over either half duplex or full duplex circuits in this manner. As we discuss in Chapter 4, it is sometimes desirable to use full duplex circuits in this manner to reduce the time to reverse the direction of transmission. Finally, full duplex operation involves transmitting in both directions simultaneously and is only possible when using full duplex circuits.

asynchronous and synchronous transmission

Another important distinction is between *asynchronous and synchronous* transmission. Asynchronous transmission involves transmitting individual characters, with overhead added to enable the receiver to locate the begin-

---

[20] We use these terms as they are normally defined by telecommunication engineers. Unfortunately, others (including some standards committees) use "simplex" to mean what we call "half duplex" and "duplex" for what we call "full duplex" (plus "channel" for what we call "simplex"). Hence, caution must be used when interpreting these terms in published material.

ning and end of characters but no prescribed clocking between times of transmission of different characters. Synchronous communication, on the other hand, involves transmission of character streams in such a way that characters follow previous ones at precisely clocked times. Descriptions of protocols using these techniques are given in Chapter 7. Asynchronous communication is the standard mode at low speeds (up to approximately 1200 bps), while synchronous communication is standard at higher speeds.

2.4 Accommodating Signals to Channels

The signals sent over telecommunication networks are not usually generated in a form appropriate for direct transmission. For example, they may not occupy frequency bands appropriate for the transmission medium or the pulse shapes used or the encoding of information may not be appropriate. Techniques for accommodating signals to channels include *modulation,* use of appropriate *codes, signal sets,* and *pulse shaping.* The first three techniques are discussed briefly here; pulse shaping is deferred to Sections 2.5 and 2.6 since a fundamental theorem presented there indicates the best pulse shapes to use.

2.4.1   Time and Frequency Domain Representations of Signals

Fourier transform pair

Some techniques we discuss are best described in terms of their effect on the frequency content of signals, while others are best described in the time domain. Frequency and time domain representations of signals are described in terms of direct and inverse Fourier transforms. If $s(t)$ is a signal, its *Fourier transform, $S(f)$,* is given by

$$S(f) = \int_{-\infty}^{\infty} s(t)e^{-j2\pi ft}dt, \tag{2.2}$$

where $j^2 = -1$. The inverse Fourier transform

$$s(t) = \int_{-\infty}^{\infty} S(f)e^{j2\pi ft}df, \tag{2.3}$$

can be used to obtain $s(t)$ from $S(f)$.

These transform relationships are valid for general types of signals, including all those we may be concerned with.[21] Note that the transform pair allows either $S(f)$ or $s(t)$ to be obtained from the other, so signals can be

---

[21] See a standard text, such as [OPPE83], for precise conditions for validity of the transforms.

Figure 2.6   Time and Frequency Domain Representation of
Linear System or Channel.

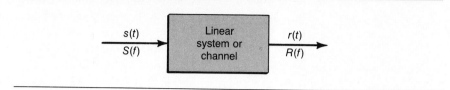

represented with equal validity in either the *time domain* or the *frequency domain*.

Although both representations are equally valid, there are times when one is simpler to use, especially when considering the impact of signal transformations. An important example is illustrated in Fig. 2.6. If $H(f)$ represents the frequency response of the linear system or channel illustrated, then

$$R(f) = S(f)H(f), \tag{2.4}$$

so the frequency domain representation of the output can be computed by multiplication. Calculations performed entirely in the time domain for this situation are often considerably more complex.[22] The frequency domain representation is more convenient, under most circumstances, for representing effects of modulation and demodulation of signals. The next subsection gives a brief introduction to modulation and demodulation.

bandwidth/time duration tradeoffs

The bandwidth of a signal can be defined to be the frequency interval within which its frequency domain representation is significantly nonzero. Similarly, the time duration is the time interval within which it is significant.[23] The time duration and bandwidth of a signal can be shown to be inversely proportional to each other, that is, if B represents the bandwidth of a signal and τ its time duration, then

$$B\tau = C, \tag{2.5}$$

with C a constant (depending on definitions of bandwidth and time duration). Thus, short duration signals (that is, short pulses) need wide bandwidth channels for accurate transmission.

2.4.2   Amplitude, Frequency, and Phase Modulation

A major reason for modulation is to shift the frequency band occupied by a signal to a band appropriate for transmission over communication facilities. This is especially obvious for radio communication, since (as we indicated in

---

[22] It is possible to compute $r(t)$ directly from $s(t)$ by a convolution integral. See a text such as [OPPE83] for details.

[23] A variety of precise definitions for each of these two terms are used in the technical literature.

Section 2.2) efficient transmission of radio signals is only possible at frequencies greatly exceeding those normally occupied by many common signals. Similar frequency band shifts are also frequently needed with other types of channels. Three types of modulation are common: amplitude, frequency, and phase modulation. Each involves modulation of a carrier

$$C(t) = A \cos (2\pi f_c t + \Theta) \tag{2.6}$$

by an information-bearing signal. For amplitude modulation, the amplitude (A) is varied, or modulated, by the signal, $s(t)$. For frequency modulation, the frequency (f_c) is modulated, and for phase modulation, the phase (Θ) is modulated. In each case the result is a new modulated signal with frequency band shifted to be centered around the carrier frequency, f_c.

Figure 2.7 illustrates these types of modulation for a simple case where the information signal, $s(t)$, is a binary signal taking on the values 0 or 1 for T second intervals. For the version of amplitude modulation illustrated, the amplitude of the carrier waveform is multiplied by 0 for a zero signal and by 1 for a signal of one; for frequency modulation, the frequency is f_0 for a zero and f_1 for a 1; and for phase modulation, the value of Θ is $0°$ for a one and $180°$ (the maximum possible phase difference) for a zero.

Part (a) of the figure shows the unmodulated carrier. Multiplication by the binary digits indicated gives the amplitude shift keying (ASK) waveform in (b). (This particular variant of ASK is also called on-off keying (OOK) for obvious reasons.) The third waveform illustrates frequency shift keying (FSK) with different frequencies for one and zero signals. The final waveform, phase shift keying (PSK), shows a $180°$ phase shift between waveforms transmitted for one and zero. (This particular form of PSK is also an alternative form of ASK, with multipliers of 1 and -1 instead of 1 and 0.)

Variations on the types of modulation illustrated are common. It is not necessary to restrict the number of signals to two. If four signals are used, each can encode two bits of information; if eight are used, each can encode three bits, and so forth. The number of signal changes per second (amplitude, frequency or phase changes, or changes in more than one parameter[24]) is defined to be the *baud rate*. The *bit rate* is the number of bits per second transmitted. The two are related by

bit rate and baud rate

$$\text{Bit rate} = \log_2 M \cdot \text{baud rate}, \tag{2.7}$$

with M representing the number of different signals used.[25]

---

[24]Signal sets using combined amplitude and phase changes are common. Several are illustrated in the next subsection.

[25]Although this distinction between bit rate and baud rate is given in essentially any text on data communication, maintaining the distinction is a losing battle since the term "bauds" has been misused to mean "bits per second" so often, especially in computer science literature, that this is becoming accepted usage. We will soon need a new term to denote signal changes per second.

Figure 2.7   Digital Modulation Techniques. (a) Unmodulated carrier.
(b) Amplitude shift keying (ASK). (c) Frequency shift keying
(FSK). (d) Phase shift keying (PSK).

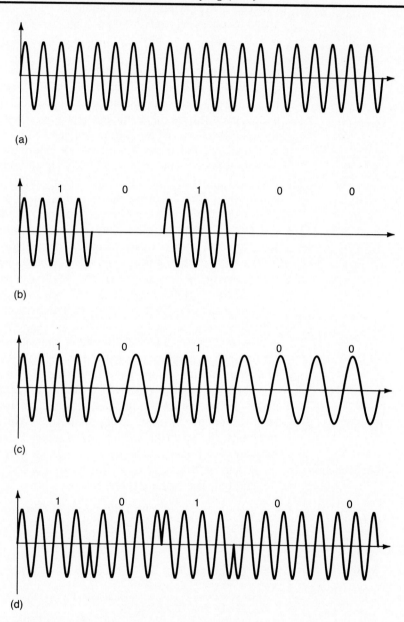

differential modulation   Another important variant is *differential modulation.* This is especially common with phase shift keying. In differential phase shift keying, phase differences, rather than absolute phase values, convey the information. For example, a 1 could be encoded by using the same phase as was used in the previous signaling interval and a 0 into a phase reversal. This is easier to implement than plain PSK, since an absolute phase reference at the receiver is difficult to maintain, but tends to increase the burstiness of errors. This is because an error during one signalling interval implies the phase reference used during the following interval is incorrect.

Frequency translations introduced by modulation are especially easy to illustrate for amplitude modulation. Consider a simple sinusoidal information signal, $\cos(2\pi f_i t)$ multiplied by the carrier in Eq. (2.6). Using the trigonometric identity

$$\cos A \cdot \cos B = \frac{1}{2} \{\cos(A + B) + \cos(A - B)\}, \qquad \textbf{(2.8)}$$

we can write the product as

$$s_m(t) = \frac{1}{2} \{\cos[2\pi(f_c + f_i)t + \Theta] + \cos[2\pi(f_c - f_i)t + \Theta]\}, \qquad \textbf{(2.9)}$$

which contains frequency components at the carrier frequency, f_c, plus or minus the information frequency, f_i. This can be extended to show that, in general, amplitude modulation produces the type of frequency translation illustrated in Fig. 2.8.[26]

The original signal spectrum illustrated is a simple triangular spectrum for ease of representation and interpretation. Each signal component in the original spectrum is mapped into two signal components in the modulated spectrum, one on each side of the carrier frequency. This causes the modulated signal to occupy twice the bandwidth, W, of the unmodulated signal. Signal components at frequencies above the carrier frequency are called the upper sideband and those below the carrier frequency the lower sideband. Either sideband can be shown to contain full information about the original signal; *single sideband (SSB)* amplitude modulation transmits only one sideband (either is satisfactory) and avoids bandwidth doubling.

Frequency translations with frequency or phase modulation are complex and do not preserve the bandwidth or shape of the unmodulated spectrum. Wideband frequency or phase modulation is often preferred because of improved noise immunity but can use many times the original bandwidth.

---

[26] In figures showing frequency spectra, we only draw spectra for positive frequencies, although the Fourier transform math is simplified by considering positive and negative frequencies. The positive and negative frequency portions of spectra are always symmetric (if the time functions are real), though, so the positive frequency portions are adequate to describe spectra.

Figure 2.8   Frequency Translations in Amplitude Modulation.
(a) Spectrum of unmodulated signal. (b) Spectrum
of modulated signal.

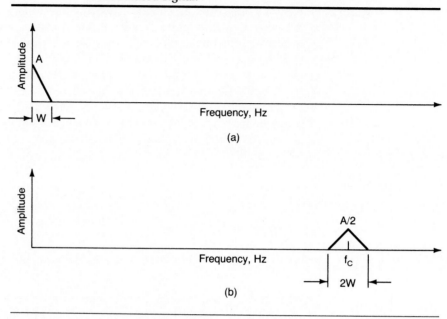

Both shift the original frequency band to one about a carrier frequency, and thus allow shifting frequencies to some appropriate band for transmission.

This has been a bare introduction to modulation. Any of a variety of standard texts (for example, [SCHW80] or [ZIEM85]) can be consulted for details on the modulation techniques discussed.

□ 2.4.3 Typical Modem Signal Constellations

Although Fig. 2.7 portrays basic signal sets for the types of modulation illustrated, other *signal constellations* are often used. Complex constellations are used in high-performance modems and other systems such as digital microwave systems. Signal constellation selection has been especially highly developed for modem design.

We concentrate on constellations for variants of amplitude shift keying and phase shift keying, or combined amplitude/phase modulation since these are commonly used for high-performance modems. Frequency shift keying is excellent for inexpensive low-speed modems because of its simplicity, but it is less amenable to high-speed modems.

Figure 2.9 illustrates typical modem signal constellations. Recall that bit rate is equal to baud rate multiplied by the log to the base two of the number of different signals, so each of the constellations, with the exception of that in (c), can be used to send several bps per baud.

Figure 2.9 Possible Modem Signal Constellations. (a) Four-signal constellation for ASK. (b) Eight-signal constellation for ASK. (c) Two-signal constellation for PSK. (d) Four-signal constellation for PSK. (e) Sixteen-signal constellation for two-dimensional modulation. (f) Thirty-two–signal constellation for two-dimensional modulation.

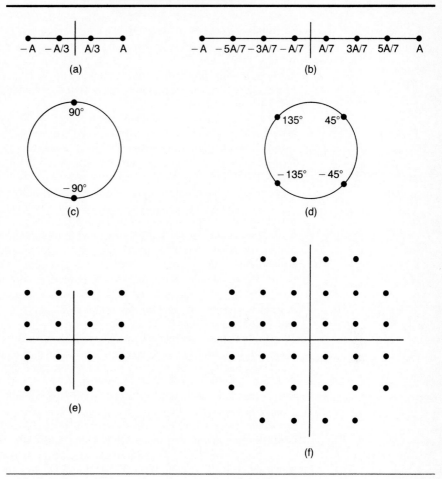

Parts (a) and (b) illustrate constellations for ASK, (c) and (d) constellations for PSK, and (e) and (f) constellations for two-dimensional modulation, achievable by combining ASK and PSK or through quadrature amplitude modulation (QAM), a technique based on amplitude modulation of "quadrature" carriers, 90° out of phase with each other. If carriers $\cos(2\pi f_c t + \Theta)$ and $\sin(2\pi f_c t + \Theta)$ are multiplied by independently chosen amplitudes, the two ASK waveforms can be recovered independently at the receiver. Thus parts (e) and (f) represent either amplitude and phase (in polar coordinates) for combined ASK/PSK or in-phase and quadrature components (in rectangular coordinates) for QAM.

The ASK signal amplitudes are symmetrical about zero and evenly spaced. For K amplitudes in the interval $-A$ to A, there are $K-1$ subintervals within a total interval of length $2A$, so the amplitudes are spaced $2A/(K-1)$ apart, with one interval centered about zero.[27] The phase angles for PSK are also spaced as far apart as possible; alternative choices, equally good for standard PSK, are $0°$ and $180°$ for (c) and $0°$, $\pm 90°$, and $180°$ for (d). For differential PSK, however, the phase angles shown are preferred since they guarantee phase transitions between all pairs of signalling intervals. The number of possible absolute phases is doubled, however, to $0°$, $\pm 90°$, and $180°$ for binary DPSK and $0°$, $\pm 45°$, $\pm 90°$, $\pm 135°$, and $180°$ for quaternary DPSK. To see this, assume an initial phase of $0°$ and quaternary DPSK with $+45°$ encoding input bits 00; a continuous string of input zeros would then be encoded by phases $+45°$, $+90°$, ..., $-45°$, $0°$, and so forth.

The two-dimensional signal constellations in (e) and (f) represent rectangular arrays of signals and have been widely used. The set in (e) could be realized by QAM with four levels each for in-phase and quadrature components while that in (f) could be realized with six levels for each (and the four combinations giving corner values prohibited to keep the number of signals a power of two).[28] These constellations can be used to transmit four and five bps/baud, respectively.

Although rectangular two-dimensional constellations are easy to implement, they are suboptimum. Alternative constellations are given in Fig. 2.10. Part (a) illustrates a 16-signal constellation that is insensitive to a common type of noise (phase jitter) and has been adopted by some modem manufacturers and accepted by the CCITT as a standard; parts (b) and (c) illustrate constellations based on equilateral triangles (or interlaced equilateral hexagons), which are optimum (for fixed power) in the presence of the most common type of noise (Gaussian noise). The 64-signal constellation is used in at least one modem operating at 14.4 kbps on voice-grade lines (at 2400 baud).

Until recently, choosing constellations, such as those in Fig. 2.10, to maximize signal distance subject to a power limitation, or to have good performance in handling types of distortion such as phase jitter, was felt to be optimum. For telephone channels, the signal set in Fig. 2.10(c) appears to be close to ultimate for increasing bit rate in this way, since spacing signals even closer together than the spacing in this signal set, to give more signals for the same signal power, would lead to unacceptable error rates. Recent advances have been made by realizing that distance between signal points is less critical than distance between sequences.

trellis coding

Trellis coding [UNGE87a,87b] uses dense signal sets but restricts sequences with which signals can be used. When this is done properly, the error rate can be reduced below that for transmission at the same rate without coding. Signal constellations with hundreds of signals can sometimes be used without excessive error rates.

We illustrate trellis coding for a case with eight signals used to send two bps/

---

[27] To minimize transmitted power for a given probability of transmission error, the center of gravity of the signal set should always be at the origin.

[28] The four signals prohibited are the best ones to eliminate since they would require more power (proportional to the sum of squared amplitudes) than any of the remaining signals.

Figure 2.10   Nonrectangular Signal Sets. (a) Sixteen-signal set insensitive to phase jitter and used in a number of 9600 bps modems. (b) Optimum 16-signal set for use in the presence of Gaussian noise. (c) Optimized 64-signal set.

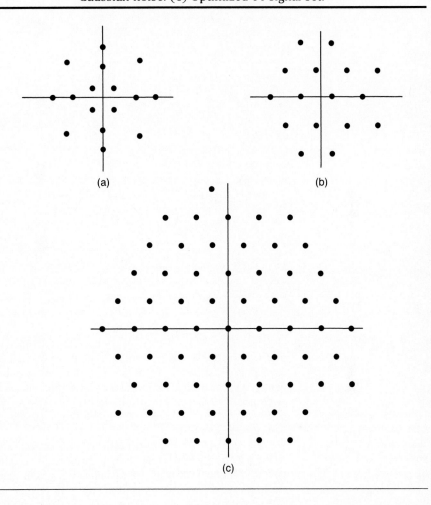

(a)

(b)

(c)

baud with better error rate than a comparable four-signal set (also sending two bps/baud) without coding; eight-phase PSK is illustrated for simplicity, but other code sets can be used. Figure 2.11 illustrates the eight signals plus a coding trellis to restrict code sequences.

The coding trellis has four states and is used to encode two bits at a time. The second bit of each pair determines state transitions while the first bit determines which of two possible outputs accompanies each transition. Different outputs accompanying the same state transition are always as far apart as possible, for example 0

Figure 2.11   Trellis Coding Example. (a) Eight-phase PSK signals used. (b) Coding trellis.

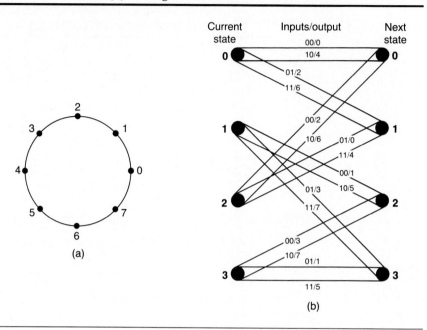

(a)

(b)

and 4, 1 and 5, 2 and 6, or 3 and 7. Thus these parallel transitions always have maximum possible distance. Any other transitions involve state changes. The code is chosen so this causes code sequences to have total distances greater than that between sequences differing only by parallel transitions. We use ordinary digits for input values (for example, 0 or 1) italics for outputs (for example, *0, 1, . . . 7*) and boldface for states (for example, **0, . . . 3**).

An example of a minimum distance pair of code sequences, other than one differing only by a parallel transition, is given by the sequences generated by inputs 00 00 00 00 00 and 00 01 00 00 00, assuming the system starts in state **0** in each case. In the first case, the system remains in state **0** throughout and has output symbols *0 0 0 0 0,* while in the second case the state sequence is **0 0 1 2 0** (the first representing the initial state) and the output sequence is *0 2 1 2 0.* Assuming that the illustrated circle has unit radius, distances between outputs can be computed as 2.0 for symbols on opposite sides of the circle (for example, *0* and *4*), $2 \sin (67.5°) = 1.848$ for the next most distant pairs (for example, *0* and *3*), $\sqrt{2} = 1.414$ for the next pairs (for example, *0* and *2*), and $2 \sin (22.5°) = 0.765$ for the closest pairs (for example, *0* and *1*). Thus the distance between output symbol sequences for the input sequences given is

$$\sqrt{0^2 + (\sqrt{2})^2 + [2 \sin (22.5°)]^2 + (\sqrt{2})^2 + 0^2} \approx 2.141 > 2.$$

Since this is greater than the distance of 2 between output sequences differing only by parallel transitions, the minimum distance between output sequences is 2. This is greater than the comparable minimum distance of $\sqrt{2}$ for four-phase PSK without encoding, so the error rate obtained for the same transmission rate is lower.

Although trellis codes have helped greatly in increasing transmission rates with modems, other improvements have also helped. These include improvements in timing recovery at receivers, in adaptive filtering and equalization techniques (see Subsection 2.5.2), in techniques for cancelling effects of transmission line echoes, and in techniques for pulse shaping. Today's modems are approaching theoretical limits for transmission rates over voice-grade lines.

□ 2.4.4 Binary Signal Encoding Techniques

Binary signal encoding is used to modify digital data to improve data communications performance. Techniques in this subsection are normally used for purely digital transmission systems, though at times the resulting waveforms may be used as inputs to standard types of modulators.

The two primary criteria for choosing binary signal encoding techniques are spectrum shaping and clock extraction at the receiver. Other criteria include error detection capability (available with some techniques without adding extra error-detection bits) and noise immunity.

The major consideration for spectrum shaping is minimizing low-frequency content of encoded signals, since most communication channels do not do a good job of transmitting low frequencies. Minimizing spectrum width is also desirable, especially when transmission bandwidth is limited.

Figure 2.12 shows common encoding techniques. Each is illustrated for rectangular pulse shapes for clarity, though implementations normally use rounded pulse shapes of the types discussed in Sections 2.5 and 2.6, since these are better adapted to transmission over communication channels.

Polar NRZ (for *nonreturn to zero*) encoding is the most common technique within computers or terminals, though it is not well adapted for transmission across communication channels. A full width positive pulse represents a one and a negative pulse a zero. The "nonreturn to zero" designation reflects the fact that the pulse does not return to zero during a bit interval.

Unipolar RZ (*return to zero*) encoding uses a half width positive pulse to represent a one and no pulse to represent a zero. It is the only encoding technique illustrated that produces signals with a nonzero *dc* (average) value.[29]

Bipolar RZ (also called *AMI, alternate mark inversion*)[30] encoding uses either positive or negative RZ pulses to represent ones and no pulses to represent zeros. Positive and negative pulses alternate in sign to eliminate *dc* components. It is one of the most

---

[29] This assumes ones and zeros are equally likely. Otherwise nonzero *dc* values can occur with polar NRZ encoding also.

[30] Terminology in the literature for this encoding technique is far from uniform. Bipolar RZ is sometimes used to denote using a positive pulse of the unipolar RZ form to represent one and a negative version of this same pulse for zero, but the terminology given here is also common. Another term for the encoding illustrated is pseudoternary encoding, since three levels are used to represent binary information.

Figure 2.12   Common Digital Signal Encoding Techniques.

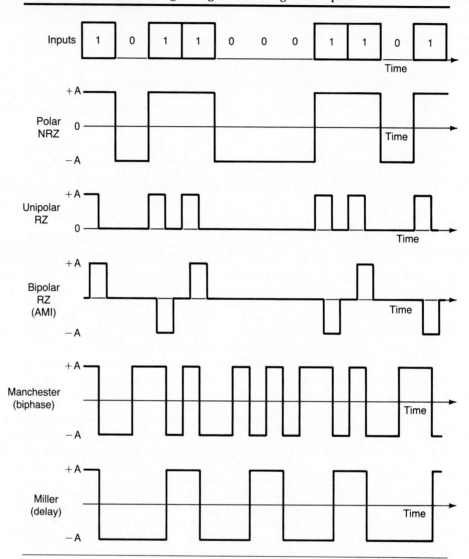

common encoding techniques, including application in the T-1 carrier system discussed in Subsection 1.3.2.[31]

Manchester (also called *biphase*) encoding uses a pulse shape consisting of a

[31] T-1 and some other systems using this encoding modify it to avoid long strings of zero signals (which make clocking difficult). If the number of zeros in data exceeds a predetermined maximum, a special signal with some nonzero pulses is sent to represent this string of zeros. The special signal is usually identified as such by means of a "bipolar violation" consisting of two successive positive or two successive negative pulses.

Figure 2.13 Spectra for Signal Encoding Schemes shown in Fig. 2.12. Vertical arrows indicate relative magnitudes of discrete frequency components, present *only* for the unipolar RZ code among those illustrated.

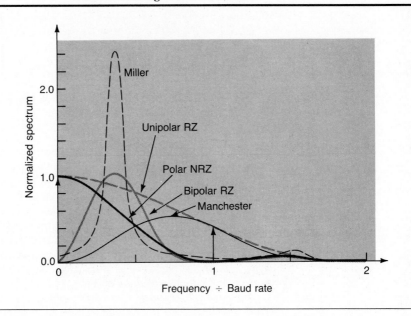

positive pulse in the first half bit time followed by a negative pulse in the second half bit time to represent a one and the negative of this shape for a zero. It is also common, with application in a variety of local area networks.

Miller (sometimes called *delay*) encoding represents a one by a signal transition in the middle of a bit interval. A zero is represented by no transition if it is followed by a one or a transition at the end of an interval if it is followed by another zero.

Spectra for the encoding techniques in Fig. 2.12 are given in Fig. 2.13. As in Fig. 2.12, rectangular pulses have again been assumed.

A disadvantage of polar NRZ and unipolar RZ is that their spectra peak near zero (*dc*). Only unipolar RZ encoding produces a true *dc* component, or finite power at precisely zero frequency,[32] but both have strong components near *dc*, which means that they are unsatisfactory for transmission over channels that do not transmit low frequencies well. On the other hand, it can be shown that polar NRZ has the least sensitivity to noise of any encoding technique illustrated;[33] so when frequencies near *dc* can be handled and noise sensitivity is important, it may be preferred. Neither polar

---

[32] This assumes ones and zeros are equally likely. The nonzero *dc* component otherwise present with polar NRZ also corresponds to finite power at zero frequency.

[33] Manchester encoding can do equally well when noise levels admitted to the receiver are equal, but it requires approximately twice as much bandwidth. Since the noise admitted to the receiver is approximately proportional to bandwidth, Manchester receivers typically admit twice as much noise power, though.

Figure 2.14   (a) Multiplexer. (b) Concentrator. C_{ik} and C_{ok} refer to the capacities (or rates) of input channels and output subchannels, respectively.

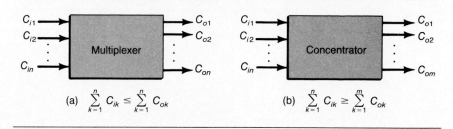

$$(a)\ \sum_{k=1}^{n} C_{ik} \leq \sum_{k=1}^{n} C_{ok} \qquad\qquad (b)\ \sum_{k=1}^{n} C_{ik} \geq \sum_{k=1}^{m} C_{ok}$$

NRZ nor unipolar RZ guarantees pulse transitions within a predetermined interval to aid in maintaining synchronization; a string of either ones or zeros gives no transitions with polar NRZ and a string of zeros gives no transitions with unipolar RZ.

The other three encoding techniques, Miller, Manchester and bipolar RZ have spectrum nulls at *dc* so are better adapted to transmission over channels with limited *dc* response. Miller encoding produces the narrowest spectrum, making it desirable for some applications. It guarantees at least one transition every two bit times, simplifying clocking, though not as well as Manchester encoding, which guarantees a transition in the middle of every bit time.

The two most common encoding techniques for the networks we will be concerned with in succeeding chapters are bipolar RZ and Manchester. Both have the desirable null near *dc.* Manchester encoding makes clock extraction simple but requires a wider bandwidth. It is often used for LANs where bandwidth is not a problem. Bipolar RZ is used for T-1 and many other systems. Clock extraction can be simplified by techniques for ensuring signal transitions within a reasonable period of time. Such techniques are normally used with bipolar RZ.

bandwidth/data rate tradeoff

Note that the horizontal scale for frequency spectra plots represents frequency divided by baud rate. For any technique illustrated, there is a tradeoff between bandwidth and data rate. Some encoding techniques are more efficient in use of the bandwidth than others, but for a given technique data rate can be increased only by increasing bandwidth.

2.4.5   Multiplexing and Concentration

Many communication media have more capacity available than needed for single conversations, so it is possible to share media among different users, reducing cost per user, eliminating extra equipment, and so forth. The two primary approaches to sharing are multiplexing and concentration. In each case, several input channels share a single (higher speed) output channel.

Multiplexing involves fixed assignment of part of the transmission capacity to each user and is performed by devices called multiplexers. The total capacity of channels after a multiplexer cannot be less than the total capacity of its input channels, as illustrated in Fig. 2.14(a). (We consider the portion

Figure 2.15   Example of Frequency Division Multiplexing (FDM) as Used in Assembly of a Channel Group for Telephone Transmission. (a) Idealized version of voice channel spectrum. (b) Corresponding channel group spectrum.

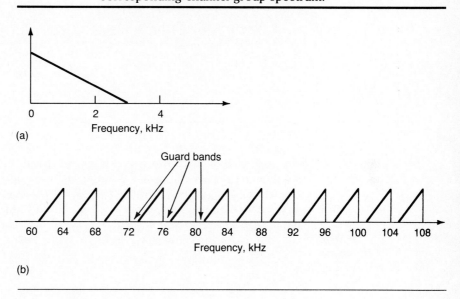

(a)

(b)

of the output channel used by a particular input channel to be a subchannel.) On the other hand, *concentration* involves dynamic assignment of transmission capacity to each user whenever it is needed. Since channel capacity is not assigned to users when they do not transmit, the total output capacity of channels after a concentrator is normally smaller than the total capacity of channels at its input, as illustrated in Fig. 2.14(b). There are two main multiplexing variants, *frequency division multiplexing* and *time division multiplexing*.[34]

frequency division multiplexing

Figure 2.15 illustrates *frequency division multiplexing (FDM)* as it is used in the first level of the telephone FDM multiplexing hierarchy—assembly of 12 voice-grade channels into a channel group.[35] Part (a) shows the spectrum of a voice-grade channel in Fig. 2.2 (idealized to simplify the figure). Part (b) indicates the corresponding spectrum of a channel group, which multiplexes 12 channels into the frequency band 60 to 108 kHz (internationally standard-

---

[34] The term space division multiplexing is also used, especially by telephone companies, to mean provision of separate communication facilities, which are bundled together for installation, to separate users. An example is use of a multipair cable, containing hundreds or even thousands of twisted pairs, to provide facilities for many users.

[35] Standards for channel groups are reasonably well standardized around the world, but higher level groupings are less standardized.

Figure 2.16   TDM Format for T-1 Carrier System.

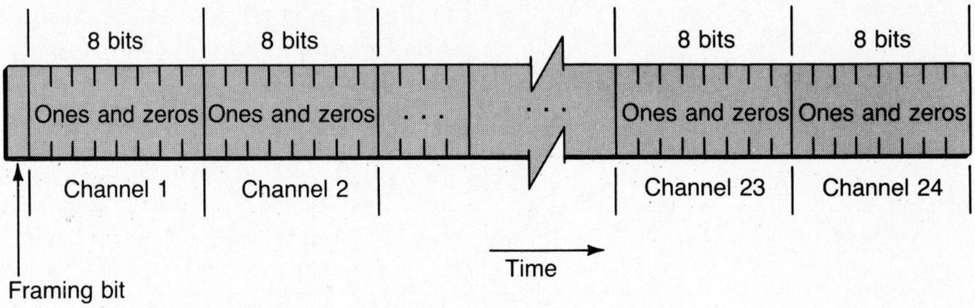

ized frequencies). Each triangle in (b) is flipped horizontally from its representation in (a) since SSB transmission with lower sidebands selected is used, but other variants of AM (or FM or PM) can be frequency division multiplexed as well. Common examples are commercial AM radio, which (in the United States) multiplexes double sideband channels, spaced 10 kHz apart, into the frequency band 540 to 1600 kHz, and commercial FM radio, which (also in the United States) multiplexes channels, spaced 200 kHz apart, into the band 88 to 108 MHz.

FDM is sometimes used to subdivide a voice-grade line among multiple low-speed data communication users. Its efficiency is limited, however, by the necessity to provide guard bands between different users to eliminate overlap of signals. This is a result of the fact that analog filters to eliminate unwanted frequencies cannot be made infinitely sharp.

time division multiplexing

Figure 2.16 illustrates the frame format for T-1, the first stage of the telephone *time division multiplexing* (*TDM*) hierarchy in North America, Japan, and a few other areas. It multiplexes 24 voice-grade channels, each encoded into a digital pulse stream, into a 1.544 Mbps data stream. An alternative TDM format, with the first stage multiplexing 32 voice-grade channels into a 2.048 Mbps data stream, is used in much of the rest of the world.

As the figure indicates, a T-1 frame contains eight bits for each of the 24 channels,[36] plus a framing bit. Thus a total of $24 \times 8 + 1 = 193$ bits constitute a frame. A frame is transmitted every 125 μsec (1/8000 sec), for a data rate of $8000 \times 193 = 1,544,000$ bps.

TDM is also used to multiplex multiple slow-speed data channels onto a single voice-grade channel. Either byte interleaving of individual subchannels, as in T-1, or bit interleaving (transmitting one bit from subchannel one, then one from subchannel two, and so forth), may be used, with a subchannel located by position within a frame. Framing overhead is usually higher

---

[36] One bit from a time slot may sometimes be used to convey telephone signalling information, leaving only seven bits for data.

than the approximately 0.5 percent (1/193) for T-1, but it tends to be less than overhead for guard bands with FDM, so TDM is usually more efficient than FDM.

In contrast to ordinary time division multiplexing, a technique often called asynchronous time division multiplexing (ATDM) involves demand assignment of a high-speed line's capacity, rather than fixed assignment as in ordinary TDM. The term ATDM is actually a misnomer, since devices using demand assignment are concentrators rather than multiplexers, but the term is thoroughly entrenched in the literature. We use the term time division concentrator. Figure 2.17 compares time division concentration, defined in this way, with time division multiplexing.

The figure illustrates synchronized operation of all devices and fixed packet sizes to simplify the drawing. It shows a total of four time intervals for generation of packets, three terminals generating packets during the first time interval, two in the second, none in the third, and two in the fourth. The packets are accepted by the concentrator in the order in which they are generated (with arbitrary ordering, here alphabetical by terminal ID, for packets generated simultaneously). To identify each packet at the receiver, an extra field containing the transmitting terminal's identification or address must be appended to each packet. Queuing of packets for transmission occurs during periods when they are generated more rapidly than they can be transmitted.

For multiplexing, the high-speed line operates at essentially the sum of the rates of the lower speed lines or terminals (ignoring framing overhead, which requires a still higher rate). A slot during each frame is made available for each terminal, regardless of whether it has information to send. For the situation in the figure, nine out of 16 TDM frame slots are wasted. In contrast, the concentrator (using a lower speed line) wastes only one slot, though it requires overhead for addresses.

A variety of other types of demand assignment could also be discussed, but we limit our discussion to time division concentration. A great deal of attention is currently being given to it, since the approach to broadband ISDN receiving the most current emphasis (called asynchronous transfer mode, ATM) is essentially identical to time division concentration.

2.4.6   Relative Capacities of Multiplexers and Concentrators

Some criteria for choosing between time division concentrators and multiplexers should be reasonably obvious. A time division concentrator must add identification of the active device to each block of information carried, but it incurs no overhead from inactive devices since only active devices are granted access to the channel. On the other hand, a time division multiplexer incurs no overhead to identify the device transmitting, since frame position identifies devices but wastes channel capacity every time a device has no data to transmit when its turn comes. Thus we expect a concentrator to be better if data sources are bursty, or active a small percentage of the time,

Figure 2.17   Comparison of Time Division Concentration and Time Division Multiplexing. Framing information for TDM is ignored.

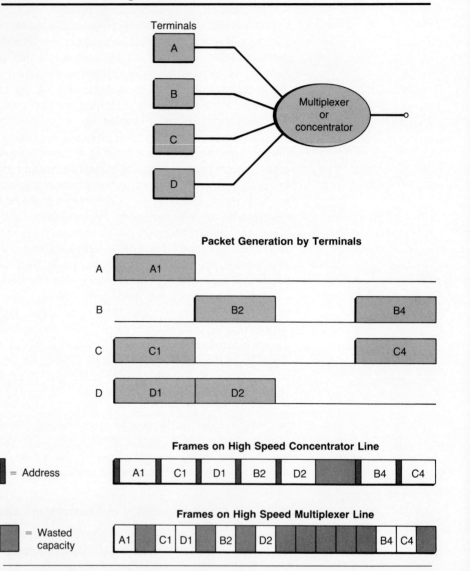

while a multiplexer would be better for data sources with relatively constant utilization. We give formulas to quantify these observations. Specifically, we give formulas for computing the number of devices that can be handled by a time division multiplexer and by a time division concentrator.

Computation of the number of identical devices that can be handled by a multiplexer is simple. Divide the bit rate of the multiplexer (corrected for framing overhead) by the bit rate per device and take the integer part of the result. For a bit rate of 2400 bps at the output of the multiplexer, and low speed devices each operating at 150 bps, a total of 16 devices could theoretically be multiplexed if framing overhead were zero, or 15 devices with framing overhead no greater than the bit rate of one device. A typical implementation would multiplex 15 devices with character (8 bits) interleaving and one framing (8 bit) character per frame, providing an output rate of 2400 bps.

These results can be expressed formally in the form of equations. Assume a channel rate of C bps with fractional framing overhead of o_f, so the bit rate for data is $C(1 - o_f)$ bps. Also assume each slow speed device has a rate of r bps. (This is the rate at which it generates bits when active, but we do not distinguish periods when it is inactive since the multiplexer provides positions for bits regardless of whether they are used or not.) The number, n_{mul}, of devices that can be handled by a multiplexer is then

devices handled by multiplexer

$$n_{\text{mul}} = \left\lfloor \frac{C(1 - o_f)}{r} \right\rfloor, \tag{2.10}$$

with $\lfloor x \rfloor$ denoting the integer part of x. In the example above, $C = 2400$ bps, $o_f = 1/16$, $r = 150$ bps, so $n_{mul} = 15$.

A precise analysis of time division concentration would be based on queuing theory because of the random phenomena involved. Arrivals of messages at data sources are random, and this makes it possible for queues of devices waiting for service to build up during intervals when random fluctuations result in relatively large numbers of arrivals. However, we are interested only in estimating how many devices can be handled by a concentrator. The impact of queuing can be adequately handled by using a rule of thumb—that queues do not tend to become excessive until utilization of a server becomes high. If server utilization (fraction of the time the server is busy) is no higher than approximately 0.8, queues are usually not excessive.[37] If utilization is higher, however, queues may be excessive, with queue lengths approaching infinity as utilization approaches one.

Since a primary factor determining whether concentration is better is per device utilization, we denote utilization (the fraction of time that a device is busy) for slow-speed devices by σ. We again let r denote the bit rate of each

---

[37] Definition of "excessive" is intentionally imprecise. Its definition will depend on the application. If queues for a system designed according to our calculations are felt to be excessive, the factor of 0.8 in the formula can be replaced by a smaller number; if larger queues can be tolerated a larger number might be used.

device during periods when it is active, so the average rate per device is σr. The transmitted rate per device, however, includes overhead for attaching identification to each block; if o_{id} denotes this overhead as a fraction of data in a subframe, the transmitted bit rate per device is $\sigma r(1 + o_{id})$. Our restriction that no more than 0.8 of the capacity of the high-speed line is used implies that the available bit rate is $0.8C$. The maximum number, n_{con}, of devices handled is thus

$$n_{con} = \left\lfloor \frac{0.8C}{\sigma r(1 + o_{id})} \right\rfloor. \tag{2.11}$$

devices handled by concentrator

As an example, using data similar to that for the ordinary TDM example, assume $C = 2400$ bps and $r = 150$ bps, with per device utilization of 15 percent, and o_{id} equal to 0.75. (This allows six-bit addresses in subframes, with each subframe containing one eight-bit character of data.) Our formula yields $n_{con} = \lfloor 48.76 \rfloor = 48$, so six-bit addresses are appropriate. Note the increase in number of devices handled when per-device utilization is low, despite the high *id* overhead.

devices handled by multiplexer

A second example of computations illustrates a situation where device utilization is high enough that a TDM multiplexer would have higher capacity. Assume the same 2400 bps output line, again with r equal to 150 bps. Assume, however, that per-device utilization is $\sigma = 0.80$ (instead of 0.15) and *id* overhead is $o_{id} = 0.50$. Our formula gives $n_{con} = \lfloor 10.67 \rfloor = 10$ (requiring four-bit addresses, so the $o_{id} = 0.50$ assumption is appropriate). This is less than the 15 devices handled with a multiplexer, so a multiplexer would probably be preferred for this situation.

2.5   Fundamental Limits on Transmission Rates

Fundamental theorems discussed in this section give limits on achievable transmission rates in terms of parameters such as bandwidth and noise statistics of communication channels. We discuss two fundamental theorems here: the Nyquist intersymbol interference theorem and the Shannon channel capacity theorem. The first theorem specifies the maximum rate at which pulses can be sent over a channel without different pulses interfering with each other, and hence limits the achievable baud rate in common communication systems. The second specifies the maximum bit rate that can be reliably transmitted. Both are often misinterpreted, however.

2.5.1   Intersymbol Interference and Nyquist Transmission Rate

A typical digital information stream can be represented in the form

$$s(t) = \sum_k a_k p(t - k\tau) \tag{2.12}$$

Figure 2.18   Pulse Shape Used to Illustrate Intersymbol Interference.

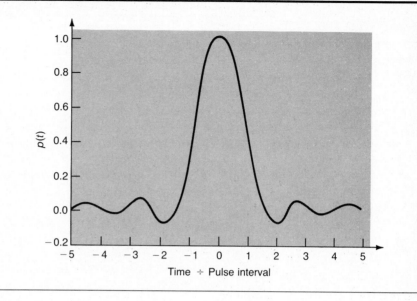

with symbols a_k used to encode digital information and $p(t)$ an appropriately chosen pulse shape,[38] similar to those used in the signal encoding techniques described in Subsection 2.4.4. (As we stated there, rectangular pulses are not actually transmitted, since they cannot be sent across communication links without serious distortion. Rounded pulse shapes are used in practice.)

intersymbol interference

A serious problem with systems using this approach is *intersymbol interference,* which results when preceding or following pulses interfere with reception of a desired pulse. For example, consider using a stream of pulses of the form shown in Fig. 2.18 to convey information.[39]

One way to use this type of pulse to convey binary information is to let $a_k = 1$ if the kth input bit is 1 and $a_k = -1$ if the kth input put is 0. This value is used to multiply $p(t - k\tau)$ and the values summed as in Eq. (2.12). Encoding the binary digits 10011 for transmission at times $k\tau$, $k = 0, 1, 2, 3, 4$, in this manner yields the individual pulses and composite stream in Fig. 2.19.

---

[38] $p(t - k\tau)$ is a pulse identical to $p(t)$ except for the fact that it is centered at the kth pulse time, $k\tau$, instead of 0.

[39] The pulse shape sketched is the result of passing a rectangular pulse of width τ (the interval between pulses), centered around $t = 0$, through a filter that passes all frequencies up to $0.8/\tau$ with no attenuation, but no other frequencies. It is a reasonable approximation of a pulse shape that might be observed at the receiver of a bandwidth limited communication channel.

Figure 2.19   Illustration of Intersymbol Interference. Dashed lines denote individual pulses; solid lines indicate their sum.

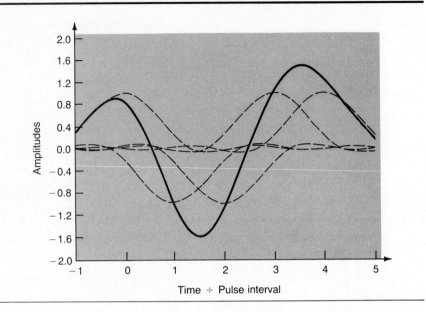

Time ÷ Pulse interval

As the figure indicates, received pulses overlap so much it is difficult to locate individual pulses. In fact, no peak in the received waveform coincides in location with the peak of an individual pulse: the first received peak occurs before the peak of the first pulse, the second (negative) peak between two negative pulses and the final peak between two positive pulses. The original five pulses resemble three pulses at the receiver.

If adequate clocking is available at the receiver, it can try to decipher the pulse stream by sampling at times when peaks of individual pulses should occur and decode the samples as 1 if positive and 0 if negative. This is, in fact, the normal procedure and will correctly interpret each bit in the figure. The errors in received values at sampling instants, called intersymbol interference and plotted in Fig. 2.20 (including values at sampling instants just before and after the pulse stream), are not large enough to cause errors in the situation illustrated. However, they are large enough to cause considerable degradation in tolerance to noise. Other sequences of transmitted symbols might cause errors even in a noiseless situation. Minor perturbations of the pulse shapes would make this even more likely.

Nyquist intersymbol interference theorem

A fundamental theorem, the *Nyquist intersymbol interference theorem*, says it is in theory possible to transmit 2B pulses per second over a channel of bandwidth B with zero intersymbol interference. This is not achievable, since only one pulse shape is suitable and it is not physically realizable. How-

Figure 2.20   Intersymbol Interference Values for Fig. 2.19.

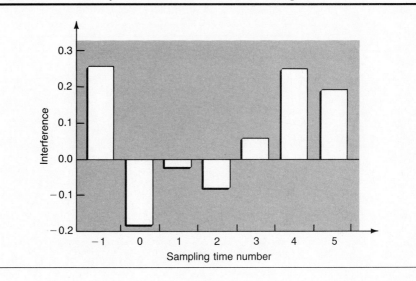

ever, the theorem gives guidelines for modifying the pulse (or equivalently its frequency spectrum) to achieve zero intersymbol interference with realizable pulses and minor increases in bandwidth.

A *Nyquist pulse* is

$$p_{\text{Nyquist}}(t) = \frac{\sin(\pi t/\tau)}{\pi t/\tau} \quad -\infty < t < \infty. \tag{2.13}$$

This is plotted in Fig. 2.21, with its frequency spectrum shown in Fig. 2.22. Note that $p_{\text{Nyquist}}(t)$ is zero if t is any nonzero multiple of the pulse interval, τ. Hence, if pulses are spaced τ seconds apart and sampled at their peaks, there will be zero interference from other pulses at sampling instants. The band-width of the spectrum times the pulse interval equals $B\tau = 0.5$, so the number of pulses per second is $1/\tau = 2B$, as was stated. This pulse rate is known as the Nyquist rate.

Nyquist transmission rate

The spectrum in Fig. 2.22 is not realizable, since it is impossible to realize a spectrum with a perfectly sharp cutoff. A realizable frequency spectrum must have a gradual cutoff, such as that in the *raised cosine spectrum* spectra in Fig. 2.23 with $\alpha > 0$; the raised cosine spectrum is given by

raised cosine spectrum pulses

$$P_{\text{raised cosine}}(f) = \begin{cases} C & f < \dfrac{1-\alpha}{2\tau} \\[2mm] \dfrac{C}{2}\left\{ 1 + \cos\left(\dfrac{\pi(f + (\alpha - 1)/2\tau)}{2\alpha}\right)\right\} & \dfrac{1-\alpha}{2\tau} \le f \le \dfrac{1+\alpha}{2\tau}, \\[2mm] 0 & f > \dfrac{1+\alpha}{2\tau} \end{cases} \tag{2.14}$$

Figure 2.21   Time Domain Plot of Nyquist Pulse over Interval $-3 \leq t/\tau \leq 3$. Complete plot extends for $-\infty < t/\tau < \infty$, with gradually decreasing amplitude of oscillations.

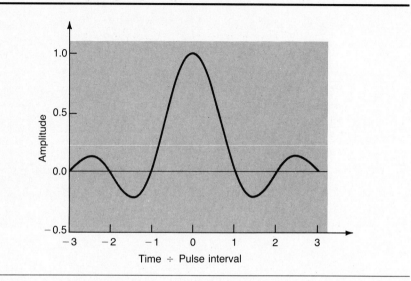

Figure 2.22   Frequency Spectrum of Nyquist Pulse over Interval $0 \leq f\tau \leq 0.5$. Spectrum is constant in this interval and identically zero outside it.[40]

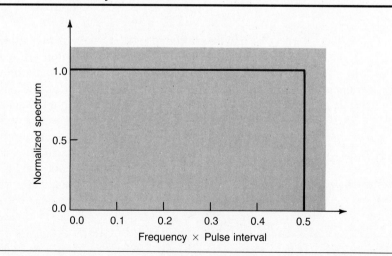

---

[40] As we mentioned in Footnote 28 in Subsection 2.4.2, we plot frequency spectra only for positive frequencies. If negative frequencies were included (to simplify the mathematics), the frequency spectrum of the Nyquist pulse (and those for the raised cosine spectrum pulses discussed below) would be symmetric about zero frequency.

Figure 2.23 Spectra of Raised Cosine Spectrum Spectrum Pulses.

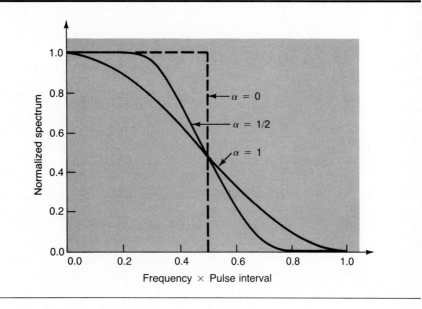

Note the symmetry of the deviations of the raised cosine spectrum spectra from the Nyquist spectrum; in each case if the portion of the spectrum beyond $f = 0.5/\tau$ were flipped both horizontally and vertically about the point $(0.5, 0.5)$ in the figure, it would precisely fill in the area subtracted from the Nyquist spectrum. Any change from the Nyquist spectrum with this type of symmetry preserves zero intersymbol interference. Thus with minor increases in bandwidth it is possible to generate physically realizable pulses yielding zero intersymbol interference. Figure 2.24 shows the time functions.

The time domain description of the pulses in Fig. 2.24 is

$$p_{\text{raised cosine}}(t) = \frac{\sin(\pi t/\tau)}{\pi t/\tau} \frac{\cos(\pi\alpha t/\tau)}{1 - (2\alpha t/\tau)^2} \qquad -\infty < t < \infty. \qquad \textbf{(2.15)}$$

Each pulse is zero for t equal to any nonzero multiple of τ. A variety of other pulses are suitable, but raised cosine spectrum pulses are widely implemented.

It is easy to misinterpret the Nyquist intersymbol interference theorem. The theorem does not say that the maximum number of pulses per second that can be transmitted reliably is $2B$; it says that this is the maximum number that can be transmitted with zero intersymbol interference. Intersymbol interference need not be completely eliminated, especially if it is kept below the level of noise or can be compensated for at the receiver. A technique called duobinary or partial response signaling uses this approach. It uses a pulse

Figure 2.24   Raised Cosine Spectrum Pulses Plotted over Interval $-3 \le t/\tau$ ≤ 3. Complete plots extend for $-\infty < t/\tau < \infty$, with decreasing amplitude of oscillations.

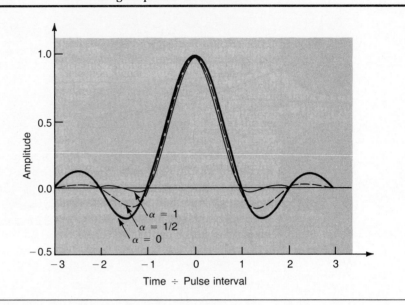

shape that confines intersymbol interference to precisely one previous pulse, making it easy to subtract out its effect.[41]

Rather than trying to completely eliminate intersymbol interference, much current research in digital communication is based on developing algorithms for coping with combined effects of intersymbol interference and noise.

2.5.2   Equalization Techniques

Three communication system components (at least) can be distinguished, as Fig. 2.25 indicates. These are the transmitter, the communication channel, and the receiver.

The shapes and spectra of pulses at the receiver, rather than those at the transmitter, determine feasibility of reliably deciphering transmitted information. By simple extension of Eq. (2.4), the frequency spectrum of pulses at the receiver is

$$P_r(f) = P_t(f)H_t(f)H_c(f)H_r(f), \tag{2.16}$$

---

[41] Duobinary signaling is not actually used to exceed the Nyquist rate, since the pulse it uses occupies twice the Nyquist bandwidth. Instead, it is used to transmit at the Nyquist rate with realizable implementation.

Figure 2.25   System Components Affecting Reception
of Digital Information.

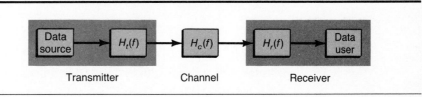

with $P_t(f)$ describing the spectrum of pulses generated at the transmitter, $H_t(f)$ the effect of any spectrum shaping at the transmitter, $H_c(f)$ the channel frequency response, and $H_r(f)$ the effect of spectrum shaping at the receiver. All except for $H_c(f)$ are under the control of the system designer. As long as their product is unchanged, the impact of $H_t(f)$ and $H_r(f)$ on intersymbol interference is independent of the amount of spectrum shaping at each location, but the shapes of the individual spectra can affect sensitivity of the system to noise.

equalization

    The primary difficulty in designing spectrum shaping filters for transmitter and receiver is the variability of the communication channels encountered. Optimal division of responsibility for spectrum shaping, or *equalization,* between transmitter and receiver is complex (see [LUCK68], for example). Compromise equalizers based on average channel characteristics are only satisfactory for slow transmission rates; for rates above approximately 2400 bps adaptive equalizers, which adjust their parameters to match those of communication links, are needed. Within the lower portion of this frequency range, manual adjustment of parameters may be satisfactory, but for higher rates, or when the channels used vary, automatic equalizers are needed. The more sophisticated of these automatically adjust their characteristics as communication lines vary with time. Excellent surveys of these techniques are given in [FRAN86] and [QURE87].

2.5.3 Shannon Channel Capacity Theorem

Probably the most famous, and most widely misinterpreted, theorem in this field is the *Shannon channel capacity theorem.* Since the theorem is widely misstated, we devote some effort to clarifying its meaning.

    Shannon's channel capacity theorem states that for a given communication channel, a rate C, the channel capacity, can be defined such that it is possible to transmit at any rate $R \leq C$ with arbitrarily low probability of error, but this is not possible at higher rates. It does *not* say it is impossible to transmit at rates exceeding C (although numerous books and papers state this is impossible); if high error rates can be tolerated, transmitting at higher rates is possible. In fact, we indicate below how it is possible to "transmit" at higher rates with error rates that might be reasonable for some applications.

The appropriate formula for channel capacity depends on the signal and noise characteristics of a channel, plus its bandwith, but the most commonly quoted formula is

$$C = B \log_2 \left\{ 1 + \frac{S}{N} \right\}, \tag{2.17}$$

with B channel bandwidth in hertz (Hz), S average signal power, and N average noise power at the input to the receiver. As an example of this formula, the useful bandwidth of a typical voice-grade channel may be taken as 3000 Hz (see Fig. 2.2), and a typical signal-to-noise ratio, S/N, may be taken as 1000.[42] This gives $C = 3000 \log_2 (1001) \approx 30{,}000$ bps. Careful computations [LUCK68], based on better models of telephone channels give a capacity closer to 23,500 bps.

Although the channel capacity theorem indicates that it is possible to transmit at rates less than or equal to channel capacity with arbitrarily low error probability, its proof does not indicate realistic ways of achieving this. In fact, the proof is based on limiting arguments, as information symbols are collected together and encoded into long code words with redundancy added. As the length of code words approaches infinity, the probability of error can be shown to approach zero. Unfortunately, as the length of the words approaches infinity, the cost and complexity of encoder and decoder and the delay before received code words can be decoded also approach infinity.

As an indication of the fact that transmission at rates above channel capacity does not necessarily involve tremendously high error rates, we consider a way (obviously suboptimum) of "transmitting" at rates above channel capacity that was suggested by Abramson in 1963 [ABRA63, pp. 174–75].[43] To "transmit" at rates above channel capacity, he suggested encoding and transmitting only part of the bits generated by the source. The number of bits per second encoded and transmitted should be just enough to reach capacity as suggested by Shannon's theorem, with encoding described by the theorem. The receiver simply guesses bits not transmitted since it has no transmitted information to use. The average error rate for transmitted bits will be zero, with that for bits guessed by the receiver 1/2, for an overall error rate of

$$P(\text{error}) = \frac{0 \cdot C + \frac{1}{2}(R - C)}{R} = \frac{R - C}{2R} \tag{2.18}$$

---

[42] The signal-to-noise ratio is normally given in *dB*, with 30 *dB* typical. Since the *dB* ratio is defined to be $10 \log_1{}^0 (S/N)$, 30 *dB* implies a ratio of $10^3 = 1000$.

[43] A converse to the channel capacity theorem, which has been proved, indicates the probability of error in transmitting a long code word approaches one, under the same conditions as in the original theorem, if transmitting at rates above capacity. Since the length of code words approaches infinity, this simply means probability of correct reception for an individual bit is less than one, which is not necessarily unreasonably low.

Figure 2.26   Upper Bound on Bit Error Rates When Using
Abramson's Algorithm.

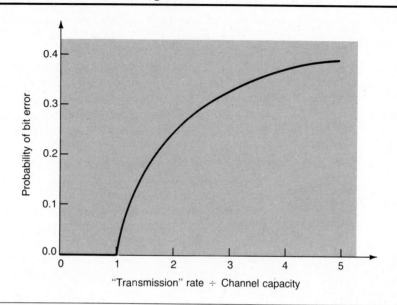

for $R > C$, with zero error probability for $R \leq C$. This is plotted versus R in
Fig. 2.26. Gallager [GALL68] gives more precise, but more complex, bounds
on achievable bit error probabilities when actually transmitting at rates
above channel capacity.

The error probability values in Fig. 2.26 are upper bounds on achievable
error probabilities, since they give values obtained with the suboptimum
algorithm. An optimum algorithm cannot do worse, but might do much better.

2.5.4 Attempting to Reach Shannon Bounds

Shannon's proof that it is theoretically possible to transmit at a finite rate
with arbitrarily low error probability has motivated a tremendous amount of
work trying to approximate the theoretical performance. Most of the work on
error correcting codes during the past 30 or 40 years [BERL68], [CLAR81],
[GALL68], [LIN83], [PETE72] was motivated by this theorem.

coding theory

Early work on *coding theory* was devoted to trying to find the best possible error correcting codes for precisely known channel characteristics. A
wide variety of codes were developed, with a few capable of achieving zero
error rates (in limiting cases) at nonzero transmission rates for some channels; this has been possible when transmitting at only a small fraction of
channel capacity and only for simple channels, however. This initial work was
not overly practical, and the amount of effort put into coding theory dropped

off drastically after essentially all codes that had been developed were shown to be poorly matched to the error statistics on common channels such as telephone channels [BURT72].[44] Error detection and retransmission (ARQ) techniques became (and still are) the standard approach to error control in most data communication systems.

Although codes for error detection have been widely used, until recently the primary uses of error correcting codes have been in situations where retransmissions are impractical (for example, tape recorded data), where using complex equipment for error correction at one location is preferable to requiring retransmissions (for example, deep space communications[45]), or situations where variable response times caused by retransmissions are not acceptable (for example, applications of process control where quick responses are critical). Recently, there has been renewed interest in error correction coding, due to major advances in digital circuitry for implementing such codes as well as advances in coding theory. The most prominent application of error correcting codes for telecommunications has been in modems using trellis codes. Modems using trellis codes and transmitting at 19,200 bps are approaching Shannon's channel capacity bound more closely[46] than any other types of equipment of which the authors are aware, though achieving error rates approaching zero when transmitting at these speeds is well beyond the state of the art.

Rather than attempting to use coding to transmit at rates approaching the Shannon channel capacity limit, much recent work has been devoted to using codes to reduce required signal power, to increasing achievable transmission rates, or to making similar improvements in communication system design. Error correcting codes used for this purpose are often supplemented by error detection and retransmission at higher layers of a networking architecture (most networking architecture standards prescribe protocols that operate in this manner). Hence, achieving error rates approaching zero is not really necessary. We can expect to see more and more use of coding theory in this manner.

---

[44] One of the authors recalls going to a conference during the period of maximum discouragement about coding theory (during the late 1960s and early 1970s) and hearing a speaker comment on managing the "world's largest group of unemployed coding theorists."

[45] Communication achievements in space programs are truly remarkable, especially when limitations on power transmitted by space probes and their distances from the earth are considered. Received signal power at earth stations is greatly exceeded by noise power in situations such as Jupiter or Saturn flybys, but we still see beautiful pictures of the features of these planets published. The most complex equipment for such situations is at earth stations, however.

[46] Another bound on maximum rate, which can be roughly interpreted as the rate beyond which complexity of equipment to reduce probability of error toward zero begins to grow extremely rapidly, has been suggested as a realistic upper bound on achievable rate [MASS74], [UNGE87b]. This bound has been estimated as 19,200 bps for voice-grade telephone channels [LUCK68], so trellis coded modems have reached this barrier (though not with zero error rate)!

2.6 Digital Communications

A major current trend is to transmit information digitally. Data from computer equipment is normally generated in digital form, and other sources of digital data such as alphanumeric codes, graphical display data (in formats recently standardized), and so forth are also common. An important additional source of digital data is digital encoding of analog data. In this section we concentrate primarily on digital transmission of analog information.

Voice and video sources produce analog information, in contrast to the digital information produced by data processing equipment or equipment designed to interface with data processing equipment. This means that waveforms vary continuously with time instead of taking on discrete values. Such waveforms cannot be transmitted digitally without first converting their formats; fortunately, it is feasible to do so. Digital transmission of analog signals is, in fact, often preferred; transmission of all signals in approximately the same format greatly simplifies integration of applications by allowing the same transmission media, protocols, and so forth to be used.

2.6.1 Nyquist Sampling Theorem

The first step in converting analog signals into a form suitable for digital transmission is to sample them at discrete times. This would appear to destroy significant amounts of information, since a continuous time function is represented by sample values at discrete instants of time. Nyquist showed, however, that if a signal is bandlimited to bandwidth B Hz—that is, if it contains no frequency components of a frequency higher than B, it can be precisely reconstructed from sample values if $2B$ samples are taken per second [NYQU28]. This rate of $2B$ samples per second is called the *Nyquist sampling rate*.

Nyquist sampling rate

The *Nyquist sampling theorem* indicates that for a signal, $s(t)$, containing *no* frequency components outside the interval $|f| < B$,

$$s(t) = \sum_{k=-\infty}^{\infty} s(kT) \frac{\sin 2\pi B(t - kT)}{2\pi B(t - kT)}, \qquad (2.19)$$

with T equal to $1/2B$. Thus, a precise representation of $s(t)$ is contained in these samples.

The Nyquist sampling theorem is the primary theorem justifying digital transmission of analog signals. Additional techniques are described below.

2.6.2 Digital Transmission of Analog Signals

Simply sampling waveforms is not enough to convert them to digital form, since samples can take on a continuous range of values and precise representation of each sample would require an infinite number of bits. Before digital transmission is possible, samples must be quantized so that each sam-

Figure 2.27   Sampling and Quantization of Signal for a Typical Pulse
Code Modulation (PCM) System.

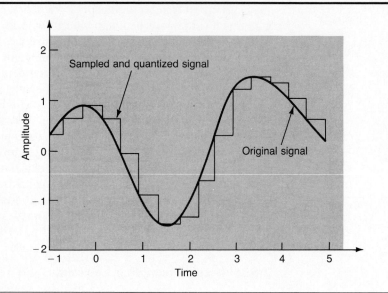

ple can be represented by a finite number of bits. This introduces deterioration so signals cannot be precisely reconstructed afterwards; if the number of quantization levels is great enough, however, no deterioration in quality after reconstruction is noticeable.

*sampling and
quantization*

Figure 2.27 illustrates the processes of *sampling* and *quantization*. The figure illustrates sampling a waveform every 0.4 second and quantizing samples to the nearest multiple of 0.125. For the amplitude range illustrated (peak to peak amplitude variations of less than four), five bits is adequate to encode quantized values. A simple sign/magnitude encoding of samples (magnitude indicating multiple of 0.125 and sign as first bit, with 1 for minus) gives the following values:

00010	00101	01000	00101	10001
11000	11100	11011	10101	00010
01000	01100	01011	01001	00101

pulse code modulation

This approach of sampling, quantizing samples and transmitting binary codes for individual samples is known as *pulse code modulation* (PCM) and is widely used. T-1 is a prominent example of a system using PCM.

As we have noted, it is not possible to precisely reconstruct a waveform from samples after the samples have been quantized. The discrepancy between original sample values and their reconstructed values is called *quantizing noise*

quantizing noise

tizing noise. If a sufficient number of levels are used, in the case of voice signals, the human ear cannot detect any degradation in quality due to quantization. Digital transmission portions of telephone networks use seven or eight bits per sample, which is adequate to give excellent quality. Even fewer bits can be used when available transmission rates are limited and only intelligibility of speech, not fidelity of transmission, is needed.

A number of variants on PCM are common. Most PCM systems for speech transmission do not use uniform spacing between quantized levels, using closer spacing between low amplitude levels at the expense of coarser quantization for high levels. The nonlinear analog to digital and digital to analog characteristics from this have been found to give better subjective speech quality, but they complicate life for modem designers. A technique, called delta modulation, encodes and transmits differences between successive samples rather than actual values of samples. More elaborate algorithms predict sample values based on the past few sample values, using an algorithm common to both transmitter and receiver, and transmit differences between actual samples and predicted values. This can considerably reduce the required transmission rate, with no deterioration in quality.

2.6.3   Characteristics of Major Types of Information Sources

Since several major efforts are today devoted to developing networks satisfying a variety of communication needs, ranging from voice communication through a wide variety of types of data communication to video communication, differences in these sources are of considerable interest. Some relevant characteristics are discussed here. Sources considered include voice, various types of digital data, and video.

Since telephone voice channels are bandlimited to less than 4 kHz (4000 Hz) bandwidth, 8000 samples per second are adequate for precise reconstruction of these signals. This is the standard sampling rate for digital transmission of voice signals in telephone networks throughout the world. On the other hand, in the United States television signals require approximately 5.5 MHz (5.5 million Hz) for full-motion broadcast-quality TV, including audio, luminance (gray scale) and chrominance (color hue and saturation) signals. (Dominant TV standards used in Europe and in some other areas use a slightly greater bandwidth.) Hence, the required sampling rate for TV in the United States is approximately 11 million samples per second.

For voice signals, 128 to 256 levels (encoded in 7 or 8 bits per sample) give excellent quality; for TV signals, 256 to 1024 levels (8 to ten bits per sample) give excellent results. Hence, voice can be transmitted at approximately a 64,000 bps rate,[47] while TV may require close to 100 million bps. High-definition TV systems with much better resolution than commercial TV

---

[47]Much higher rates are required for high quality music or similar audio transmission. For example, compact disc encoding is based on a sampling rate of 44,100 samples per second with 16 bits per sample, for a total data rate of slightly over 700,000 bps.

have been developed, some requiring as much as 35 MHz bandwidth. Digital transmission of high definition TV may require several hundred million bps.

Significant reductions in rates are possible with modern speech and video encoding. For voice, adaptive differential PCM (ADPCM) encoding at 32,000 bps has been adopted as an international standard. More elaborate techniques yield equivalent quality at rates approximating 10 kbps, with "intelligible and acceptable communication quality" possible down to approximately 500 bps [FEHE87b]. Still more gains, for multiple conversations, are possible by taking advantage of the fact that during a conversation speech is present on lines only approximately 40 percent of the time [CAMP87]. Broadcast quality TV is possible at approximately 20 Mbps, with video teleconferencing video encoders at 2 Mbps (or even approximately 500 kbps) commercially available [FEHE87b].

delay and error rate requirements

Different information sources have different *delay and error rate requirements.* A detailed description of performance requirements for satisfactory transmission of speech and data is given in [GRUB83]. Digitized speech, for example, can tolerate relatively high error rates (on the order of 10^{-3}, that is, an average of one out of every thousand bits in error) without significant signal quality degradation. Delays between sender and receiver on the order of 300 ms do not cause significant problems as long as delays are nearly constant, but very little variability in delays can be tolerated. If unusually long delays occur for transmitted speech samples, it may be better to throw these samples away rather than send them to listeners. Standard types of error control, such as detection of errors with automatic request for retransmissions, may introduce variable delays that speech systems cannot tolerate.

Many types of digital data cannot tolerate high error rates, approaching those tolerated by speech, without noticeable degradation. Tolerable delays may depend on whether the system is interactive or not. Some interactive systems may not tolerate long delays but may tolerate variability in delays. On the other hand, other systems, such as those for file transfers, may be able to tolerate long delays, along with considerable variability in delays. There is far too much variability in applications for digital data systems for sweeping generalizations on requirements to be made.

Video (TV) information contains some elements, such as luminance and chrominance values (light intensity and color, respectively) for individual areas on the screen, which can tolerate error rates comparable to those tolerated by voice signals since the human eye will not notice minor perturbations in small picture areas. However, other information, such as that associated with synchronization, can affect the entire picture for a scanning interval if they are not correct.[48] To measure both types of degradation, some

---

[48] Current techniques for digitally encoding TV signals involve separately sampling luminance signals and two different chrominance, or color, signals [GAGG87].

groups have suggested measuring performance of broadcast-quality video service by two parameters [YAMA89]:

- Percentage of 5 sec periods, with bit error rate worse than 10^{-6}
- Percentage of 500 μsec periods, with bit error rate worse than 10^{-2}

Thus either persistent error rates on the order of one bit per million or very brief periods of high error rates can be harmful. Although significant constant delays can be tolerated (delayed broadcasts, after all, are common), circuitry at the receiver imposes stringent restrictions on delay variability.

An important characteristic of information sources is burstiness. A source is considered to be bursty if information tends to be transmitted in bursts with silent periods between bursts. Most information sources for telecommunication networks are bursty. A simple example is a user at a terminal interactively accessing an application program at another location. A large part of the user's time is "think" time between receiving a response and making the next entry. Even data entry while typing is bursty, with widely varying intervals between characters. Communication line utilization averaged over an appreciable period of time may be a fraction of a percent. Utilization of communication facilities is often only a few percent. Even voice communication is bursty, with a typical speaker presenting voice energy to communication facilities 35 to 40 percent of the time.

The bursty nature of communication sources is a major factor making it possible to economize on communication facilities by sharing them among multiple users. We treat a variety of sharing schemes. The multiplexing and concentration techniques in Subsections 2.4.5 and 2.4.6 are examples.

2.7 Computing Power Limitations

In discussing theoretical limitations on performance of telecommunication networks, one should mention yet another kind of restriction that is especially important for large networks. These limitations sometimes make it impossible to follow an optimal strategy for performing network operations. Such situations occur when the time required to find the optimal strategy is too long in comparison with time constraints. There are situations when even the most powerful computers in existence are unable to find optimal solutions in a reasonable time. The study of time constraints characterizing various algorithms is part of the theory of *computational complexity*.

computational complexity

Any algorithm can be considered as a finite sequence of consecutive steps that have to be performed to find the solution of a given problem. The theory of computational complexity provides methods for comparing complexities of various problems and various algorithms invented to solve the same problem. Of course, all problems under consideration have to be solvable, with known algorithms; for a discussion of problems that remain unsolved or even inherently unsolvable, see [MINS67, Chs. 8, 9].

It is obvious that different computers need different times for finding the solution of a problem. To determine *time complexity* of an algorithm independently of the computer executing the algorithm, the number of basic operations performed during the execution of the algorithm is usually analyzed. For example, assume that an ordered list contains N numbers (or addresses of network users) and that this list is searched for a given number (address), X_o. We can start to search the list from the beginning; in the worst case, if X_o is not found, we have to make N comparisons. If X_o is found, on the average we need to make $0.5N$ comparisons. In both cases the number of comparisons is proportional to N. Taking into account the other operations that also have to be performed, we indicate that the time complexity (either in the worst case or in the average case) is of order N. This is normally written as $O(N)$ and read as "big Oh of N" or "of order N."[49]

By applying an alternative algorithm, called binary search, we can conclude (in the worst case) that X_o is not in the list after a number of comparisons proportional to log N. Assume that an ordered list consists of $N = 2^k$ elements. Following the binary search algorithm, we first compare X_o with the element X_m in the middle of the list. If $X_o \neq X_m$, the search is narrowed to the left half of the list (if $X_o < X_m$), or to the right half (if $X_o > X_m$). In the next step, X_o is compared with the middle element in the appropriate half of the remaining list; next X_o is compared with the middle element in the appropriate half of the remaining unchecked sublist, and so forth. In the worst case, $k = \log_2 N$, such comparisons have to be made to find that X_o is not in the list, so the worst case time complexity of the binary search is $O(\log_2 N)$.

In different problems the basic operations and the parameter N may have different meanings. N is usually called the size, or length, of the input to the problem, or simply the size of the problem, and complexity theory is concerned with complexity of algorithms for large N. In the example above, since $\log_2 N << N$ for large N, we say the binary search is more efficient than the linear search as its growth rate in computation time is slower (bounded by a logarithmic function of problem size).

For both algorithms above the next step of the computation is uniquely determined given the results of the previous step. They belong to a class of *deterministic algorithms,* which may be contrasted with nondeterministic algorithms in which the next step can be selected arbitrarily from a set of possible actions. All deterministic algorithms providing solutions at least in time proportional to N^k, $k > 0$, or faster (that is, algorithms with complexity

---

[49] We say that $f(N) = O(g(N))$ if $\lim\limits_{N \to \infty} \dfrac{f(N)}{g(N)} = \text{const} < \infty$. That is, the function $f(N)$ tends to infinity no faster than $g(N)$.

$O(N^k)$ or less) are called *polynomial algorithms,* and problems for which such polynomial algorithms are known constitute the *class P* of problems.

Unfortunately, for many problems of practical importance, deterministic polynomial algorithms are unknown. Such problems belong to the *class NP.* The perennial question is whether $P = NP$—that is, whether deterministic polynomial algorithms for *NP* problems do exist even though they have not been found yet. This problem remains open. In the meantime, many sub-classes of the *NP* problems have been introduced (for example, see [GARE79], [KARP86], [SEDG88]). The class of *NP-complete problems* contains problems that have been shown to be equivalent in the sense that each would have deterministic polynomial algorithms if even one of them had a deterministic polynomial algorithm.

All algorithms known for typical *NP* problems are at least $O(2^N)$—that is, they can provide solutions (in the worst case) in a time growing exponentially (or faster) with the size of the problem. Hence, classification of problems into the classes *P* and *NP* is equivalent to their division into tractable and intrac-table problems. Solutions for *NP* problems can be found only for small values of *N*; see Table 2.2, which lists processing times required for given problem size (assuming one basic operation per μsec). Unrealistically long processing times required for finding optimum solutions of many optimization problems create a need for heuristic algorithms, that is, algorithms with tolerable time requirements, but which do not always provide an optimum solution.

Unfortunately, many optimization problems for networks are computa-tionally difficult. For example, any problem requiring application of integer programming is *NP*-complete. Time complexities of typical network optimi-zation problems are listed in Table 2.3.

2.8 Summary

This chapter is concerned with fundamental limits on the communication process and standard techniques for dealing with these limits. Physical prop-erties of equipment, including communication channels and electronic cir-

Table 2.2 Processing Times for Algorithms of Typical Time Complexity, Assuming 1 μsec per Operation.

Size of Problem	Number of Basic Operations				
	$\log_2 N$	N	N^2	2^N	$N!$
$N = 10$	2.3×10^{-6} sec	10^{-5} sec	10^{-4} sec	10^{-3} sec	3 sec
$N = 100$	6.6×10^{-6} sec	10^{-4} sec	10^{-2} sec	4.1×10^{16} years	$>10^{86}$ years
$N = 1000$	10^{-5} sec	10^{-3} sec	1 sec	$>10^{86}$ years	

Table 2.3   Time Complexities of Network Optimization Problems. N is the number of nodes and L the number of links in the network.

Problem	Complexity
Determining a minimum spanning tree[a]	$O(L \log_2 L)$ or less
Finding the maximum flow between a pair of nodes[b]	$O(N^3)$ or less
Determining shortest paths from one node to all other nodes[c]	$O(N^2)$
Determining all minimal cutsets[d]	$O(2^N)$
Determining minimal Steiner tree[e]	$O(2^N)$
Determining the shortest path linking together all nodes in a single loop[f]	$O(2^N)$

[a]See Chapters 8 through 10.
[b]See Chapter 8.
[c]See Chapters 8 and 9.
[d]A minimal cutset is a minimal set of links that, when cut, disconnects two parts of the network from each other. This computation is typical of most network reliability problems of interest; almost all are *NP*-complete.
[e]A generalization of the minimum spanning tree problem allowing the addition of extra nodes at arbitrary locations as "shortcuts" to minimize total tree length or cost.
[f]This problem is generally known as the traveling salesperson problem.

cuitry—specifically properties such as available bandwidths and noise or other system impairments, limit what is achievable.

Amplitude, frequency, and phase modulation are the most important general techniques for putting information-bearing signals in a form suitable for transmission over communication channels, with each allowing demodulation at the receiver. Transmission of digital information also requires selection of a suitable constellation of signals to encode information plus binary signal encoding or pulse shaping techniques. Sharing of communication facilities via multiplexing or concentration techniques may be included to reduce costs.

Three fundamental theorems are discussesd. The first states that in theory it is possible to transmit up to $2B$ pulses per second, over a communication channel of bandwidth B Hz, with zero intersymbol interference. The second states that it is theoretically possible to transmit at a rate up to a limit known as channel capacity with arbitrarily low probability of error. The third states that a signal of bandwidth B Hz can be perfectly reconstructed from $2B$ samples per second. These three theorems have motivated development of communication techniques, with theoretical limits being approached in some areas.

Other limits on telecommunications are imposed by computing power. Although efficient algorithms may reduce the computing power needed, the theory of computational complexity indicates that power requirements of many important algorithms increase extremely rapidly with network size.

Problems

2.1 Compute estimates of the range of "natural operating speeds" of VLSI circuits, in operations per second, assuming: resistance values range from 100 Ω to 100 kΩ, capacitance values range from 10 pF to 100 pF, and a settling time of five time constants is adequate.

2.2 Assuming a half wavelength antenna is used to give efficient electromagnetic radiation, compute antenna lengths for signals at upper and lower limits of each of the following frequency bands (assigned in the United States):

Commercial AM radio	530–1600 kHz,
Citizens band radio (40 channels)	26.9–27.5 MHz,
VHF commercial television (lower band)	54–88 MHz,
Commercial FM radio	88–108 MHz,
VHF commercial television (upper band)	174–216 MHz,
UHF commercial television	470–890 MHz,
Commercial satellite (downlink)	3.7–4.2 GHz,
Commercial satellite (uplink)	5.9–6.4 GHz.

2.3 How many decimal digits would be necessary to address any one of the telephones in the world if efficient address encoding were used? Compare your answer with the maximum number of decimal address digits (14) in current standards for international addressing.[50] Do you see valid reasons for the excess digits in the international standard?

2.4 Two possible configurations of cells for cellular mobile radio, based on square grids, are illustrated in Fig. P2.4.

Figure P2.4 Figure for Problem 2.4.

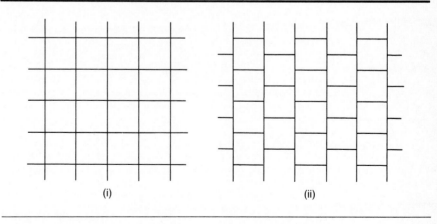

(i)   (ii)

---

[50] CCITT recommendation X.121 defines an international addressing standard with this number of digits allowed.

    a.   For configuration (i) show that two frequencies for carriers are adequate if cells sharing the same frequency may touch at corners, but four frequencies are needed if touching at corners is prohibited. Sketch suitable frequency assignments for each case, using letters, A, B, C, and D to represent frequencies as in Fig. 2.3.

    b.   Show that three frequencies are adequate for the configuration in (ii) and sketch a suitable frequency assignment.

2.5   A simplified version of the "footprint" of a typical geosynchronous communication satellite is a region on the earth's surface extending for $\pm\,60°$ in longitude and $\pm\,60°$ in latitude from the location of the satellite.[51] Consider use of such a satellite to handle communications among the cities listed below (each given with approximate latitude and longitude).

Beijing	116°E, 40°N
Berlin	13°E, 52°N
Buenos Aires	56°W, 35°S
Cairo	31°E, 30°N
Chicago	88°W, 44°N
Delhi	77°E, 28°N
Hong Kong	114°E, 22°N
Johannesburg	28°E, 27°S
London	0°W, 52°N
Mexico City	99°W, 19°N
Moscow	38°E, 56°N
New York	73°W, 41°N
Paris	2°E, 48°N
Rio de Janeiro	43°W, 23°S
Rome	13°E, 42°N
San Francisco	123°W, 37°N
Seoul	127°E, 37°N
Sydney	151°E, 34°S
Tokyo	140°E, 36°N
Zurich	8°E, 47°N

    a.   What is the maximum number of cities in this list that one communication satellite can serve?

    b.   How many satellites are required to establish communications among the 20 cities?

2.6 **a.**   Show that if

$$s(t) = \begin{cases} 1 & -T \le t \le T \\ 0 & \text{otherwise} \end{cases}$$

its Fourier transform is given by

$$S(f) = 2T\,\frac{\sin 2\pi fT}{2\pi fT} \qquad -\infty < f < \infty.$$

---

[51] The actual footprint [SPIL77] is oval-shaped, with the figure described inscribed in the oval.

b.   Show that if

$$S(f) = \begin{cases} 1 & -W \le f \le W \\ 0 & \text{otherwise} \end{cases}$$

its inverse Fourier transform is given by

$$s(t) = 2W \frac{\sin 2\pi Wt}{2\pi Wt} \quad -\infty < t < \infty.$$

2.7   Amplitude modulation with carrier frequency 64 kHz is used to shift the frequency of signals from a voice-grade telephone channel with frequency response sketched in Fig. 2.2. Sketch the frequency response of the equivalent channel after modulation.

2.8   Differential phase shift keying is used with the following encoding for input dibits (pairs of bits): $00 \leftrightarrow +45°$, $01 \leftrightarrow +135°$, $11 \leftrightarrow -135°$, $10 \leftrightarrow -45°$. Assuming initial phase of $0°$, give the transmitted phase sequence for the input sequence 01111110 10001101 01011010.

2.9   Assuming for each case an initial state of **0**, give the state sequence and output sequence for the trellis code in Fig. 2.11 and the following input sequences. In each case also compute the distance between this output sequence and the all 0 output sequence.
a.   00 10 00 00 00
b.   00 11 00 00 00
c.   00 01 01 00 00

2.10   One way to implement a trellis code uses the circuits sketched in Fig. P2.10(i) and called a convolutional encoder. Input digits, m_1 and m_2, are encoded into eight possible outputs, identified by u_1, u_2 and u_3; u_1 is simply m_1, while u_2 and u_3 are determined by a shift register operating as follows. First m_2 is shifted into position s_2, with the previous bit in s_2 moved into s_1, the previous bit in s_1 moved into s_0, and the previous value of s_0 shifted out (and discarded). Next, u_2 and u_3 are computed by taking modulo two sums of the shift register bits indicated. (A modulo two sum is equal to one if an odd number of inputs are equal to one and zero otherwise.) Finally, in a step not illustrated, outputs are mapped into signals, such as the phases in Fig. 2.11(a).
a.   Show that this code is described by the coding trellis in Fig. P2.10(ii).
b.   Find mappings from states **a**, **b**, **c**, and **d** in Fig. P2.10 to states **0**, **1**, **2**, and **3** in Fig. 2.11 and from outputs *000, 001, . . . , 111* in Fig. P2.10 to outputs *0, 1, . . . , 7* in Fig. 2.11 such that the trellis codes are identical. (These mappings are not obtained by simply taking the binary expansions of outputs *0* through *7* or of state numbers in the diagram.)

2.11   Sketch signals at the output of a binary encoder for the input sequence 11001010 01110110, assuming initial conditions in Fig. 2.12 and the following:
a.   Polar NRZ encoding
b.   Unipolar RZ encoding
c.   Bipolar RZ (AMI) encoding
d.   Manchester (biphase) encoding
e.   Miller (delay) encoding

2.12   A time division multiplexer is shown in Fig. P2.12. A frame for the output channel is constructed from a complete scan of the input channels, taking one bit from each, so it contains one bit from each input channel (plus overhead).

Figure P2.10   Figure for Problem 2.10.

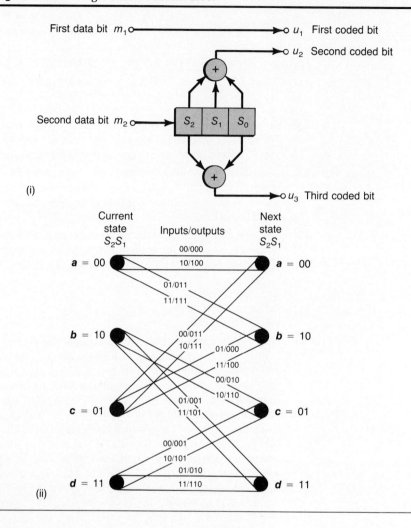

First data bit m_1 u_1 First coded bit

u_2 Second coded bit

Second data bit m_2 S_2 S_1 S_0

u_3 Third coded bit

(i)

Current state S_2S_1 Inputs/outputs   Next state S_2S_1

$a = 00$ 00/000 $a = 00$

10/100

01/011

11/111

$b = 10$ 00/011 $b = 10$

10/111   01/000

11/100

00/010

10/110

$c = 01$ 01/001 $c = 01$

11/101

00/001

10/101

01/010

$d = 11$ 11/110 $d = 11$

(ii)

Figure P2.12   Figure for Problem 2.12.

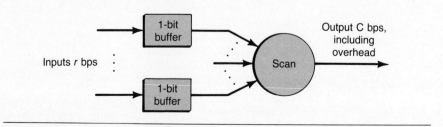

Inputs r bps

1-bit buffer

1-bit buffer

Scan

Output C bps, including overhead

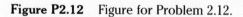

    a. Give an inequality relating the bit period of an input channel to the bit period of the output channel. Use the following notation: n = number of input channels, r = bit rate of input channels (bps), C = bit rate of output channels (bps), overhead = $o_f C$ bps (that is, o_f = fraction of each frame devoted to framing overhead).

    b. Use the inequality of (a) to verify Eq. (2.10) and determine the maximum number of inputs that can be supported by the multiplexer with an overhead of $o_f C$ bps.

    c. How many synchronous 220 bps terminals can be supported by a multiplexer with a 9600 bps output channel and an overhead of 48 bps?

2.13 Using the parameters in the examples in the text comparing the capacities of concentrators and multiplexers (that is, C = 2400 bps, o_f = 1/16, r = 150 bps), find the values of σ for which equal numbers of devices can be handled by a TDM multiplexer and by the corresponding statistical TDM concentrator. Use an appropriate value of o_{id} in your calculations.

2.14 Under what conditions is a deterministic multiplexer using FDM or TDM better than a statistical multiplexer or concentrator. Under what conditions is the concentrator better?

2.15 Why are modems needed to transmit data over long telephone circuits?

2.16 As an illustration of intersymbol interference, assume the pulse shape, $p(t)$, as sketched below.

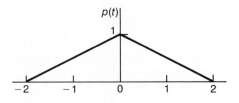

    a. For the data encoding in Subsection 2.5.1, sketch

$$\sum_{k=1}^{5} a_k \, p(t - k)$$

over the interval $-2 \le t \le 7$, for input bits 10110. Assume that no pulses have been transmitted previously (that is, zero initial conditions).

    b. Assuming that the decoding rule decodes to 1 for the kth bit if $s(k) \ge 0$ and to 0 otherwise, compare the decoded bit stream with the transmitted one.

2.17 Data transmission uses the frame format below, with the frames following directly after each other. Estimate the bandwidth required to transmit *data* at a rate of 9600 bps using Nyquist pulses if the following are used:

    a. 2 pulse levels

    b. 16 pulse levels

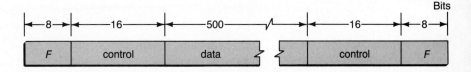

Figure P2.18   Figure for Problem 2.18.

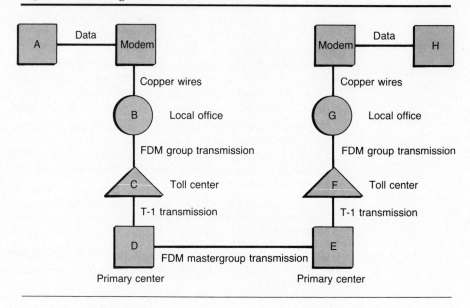

2.18   A data transmission path is set up between locations A and H in Fig. P2.18 via telephone links. The path is via telephone switching offices B, C, D, E, F, and G. Analog carrier systems are used on the B-C, D-E, and F-G links and digital carrier systems are used on the C-D and E-F links. Describe each transformation between analog and digital form (or vice versa) from the time information leaves data processing equipment at A until it reaches data processing equipment at H.

2.19   Information is transmitted in characters, each containing seven data bits and one parity bit. Parity bits are error free and the bit error probability for data bits is 0.0001, with errors in different bits independent. The parity bit is chosen so the total number of ones in each character (with parity) is odd.
   a.   Determine the probability of at least one error in a character.
   b.   Determine the probability of undetected errors in a character.
       Hint: Which error patterns can be detected by a single parity bit?

2.20   A data encoding known as biphase-mark always has a transition at the beginning of a signal interval. Logic "1" also has a transition in the middle of the interval while logic "0" does not. A sequence for 1 0 1 is illustrated.

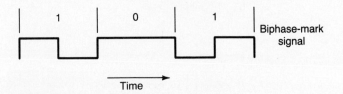

The noisy biphase-mark signal below is received. Comment on the waveform and your estimate of the sequence of logic symbols it represents.

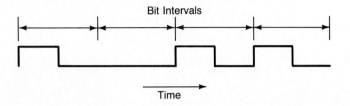

Bit Intervals

Time

2.21 Data to be transmitted are represented in Fig. P2.21(i). (Each rectangle represents the time needed to transmit a character and the number indicates the relative position of the character in sequence, the first number in a double subscript indicating the channel.) Data patterns on two channels, to be combined via a time division concentrator or multiplexer, are given in (ii) and (iii).

a. Select a concentrator or multiplexer for each case, minimizing speed of the "high-speed" line. Assume data arrivals in the figures are typical.

b. Sketch the output of each device (concentrator or multiplexer), including overhead bits, for transmission of the complete input sequences.

Figure P2.21 Figure for Problem 2.21.

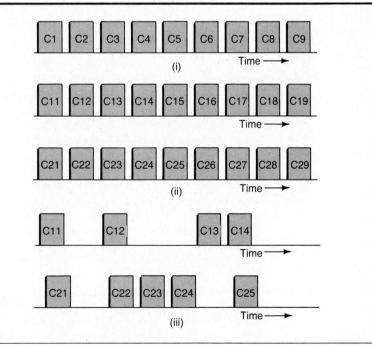

Figure P2.22   Figure for Problem 2.22.

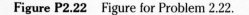

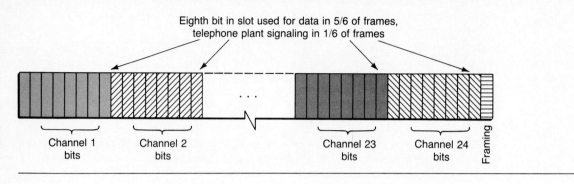

2.22   The format of a T1 frame generated by a D4 channel bank (analog to digital converter and time division multiplexer) is given in Fig. P2.22. Compute each of the following:

a.   Effective data transmission rate, in bps, for each of the 24 channels. (Signaling and framing do not count as data.)

b.   Signaling rate available for telephone company purposes, in bps.

c.   Number of framing bits per second.

d.   Total rate, including data, signaling and framing, in bps.

2.23   Assume a basic pulse shape designed for transmission at interpulse intervals of T seconds (with zero intersymbol interference) with corresponding bandwidth $1/T$ Hz. (Note that this is twice the Nyquist bandwidth.) Such pulses are used to transmit digital data.

a.   What data rate, in bps, is achieved by using these pulses to transmit binary encoded signals over a baseband channel with bandwidth of 3000 Hz?

b.   What is the achievable data rate with the same binary encoded pulses and a modem using amplitude modulation to transmit over a (nonbaseband) channel of bandwidth 3000 Hz?

c.   Explicitly describe a method, using the pulse shapes described, that makes it theoretically possible to transmit at a rate of 21,000 bps over the baseband channel in (a).

d.   Comment on the method of (c) from a practical point of view.

2.24   One technique for protecting transmitted data against unauthorized disclosure is to encode the data with an enciphering algorithm using a key known only to authorized users. Two encryption algorithms are considered to be very secure: the DES (*D*ata *E*ncryption *S*tandard) cipher and the RSA (*R*ivest-*S*hamir-*A*dleman) cipher. The former uses 56-bit-long keys to encipher 64-bit-long messages. Its authors have claimed that an unauthorized person knowing the algorithm can find a key only after exhaustive search over all possible keys, that is, by checking which key has been used to transform a given message into its given secret form (cryptogram). Security of the RSA algorithm is based on the assumption that for finding the key one has to factor a long number into a product of prime numbers. At the present level of knowledge, this can be done in $O(e^{\sqrt{ln(n) \cdot ln(ln(n))}})$ time, where n is the magnitude of the product number. (The second term under the radical is the logarithm of the logarithm of n.) In practical implementations of the RSA cipher, numbers of at least 200 digits are used. Compare security of the two ciphers assuming that a computer used for finding the key can perform one basic operation per µsec and that the product number used for the RSA cipher has 200 digits.

2.25 **Design Problem.** Develop the basic design for a modem operating at 9600 bps over a voice-grade line. Include specification of the following: baud rate, signal constellation, pulse shape before modulation, type of modulation, and type of equalization. Specify your reasons for each choice. Draw a block diagram of the modem, including a block to perform each basic operation and sketch waveforms at each significant location.

3   Introduction to Networks

3.1   Introduction

In this chapter we present basic principles of telecommunicataion networks. Switching techniques for efficient utilization of communications facilities are discussed first, followed by an introduction to major classes of networks and geometries. Naming and addressing conventions and basic structures of networks and protocols are then discussed. An introduction to the protocol architectures treated in this book concludes the chapter.

3.2   Switching Techniques

Three primary switching techniques have been proposed for telecommunication networks: *circuit switching, message switching,* and *packet switching.* Each allows sharing communication facilities among multiple users. We summarize each technique, concentrating on packet switching since it is the one most often used.[1] In each case we illustrate communication across the links in Fig. 3.1. This does not include alternate paths, used with some switching techniques, but it serves to illustrate basic properties of the techniques.

3.2.1   Circuit Switching

Circuit switching is the most familiar technique since it is used for ordinary telephone calls. It allows communications facilities, or circuits, to be shared among users, but each user has sole access to a circuit (functionally equivalent to a pair of copper wires)[2] during network use. Network use is initiated

---

[1] Newer techniques for fast circuit switching are of considerable interest today. This approach could become predominant in the future, though packet switching is currently predominant.

[2] A variety of operations may be performed on information as it flows across a network, and it may be transmitted over a variety of media including copper wires, coaxial cables, microwave links, and optical fibers. This is invisible to users, however; from their viewpoint the net result is equivalent to passing information through copper wires.

Figure 3.1   Network Used for Comparison of Switching Techniques.

Figure 3.2   Circuit Switching.

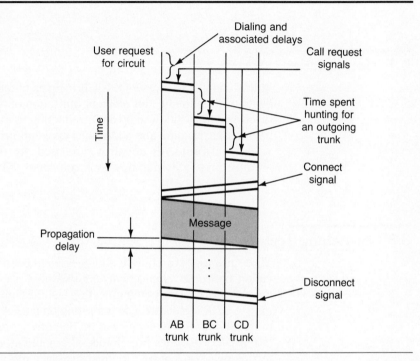

by a connection phase, during which a circuit is set up between source and destination, and terminated by a disconnect phase. These phases, with associated timings, are illustrated in Fig. 3.2.

After a user requests a circuit, the desired destination address must be conveyed to switching equipment in some manner, such as dialing the number. Delays associated with this phase are indicated by the label "dialing and associated delays" in the figure.[3] This is followed by a circuit connection

---

[3] For telephone switching these delays include the time for switching equipment to recognize an "off hook" condition for the calling party's telephone and send a dial tone, time to recognize the dial tone and dial the number, signal processing times, and so forth. The time required to search for the first trunk should also be included in this delay.

phase handled by switching equipment and initiated by transmission of a call request signal from node **A** to node **B**.[4] After a delay to search for a suitable trunk, such as that from **B** to **C**, the call request is passed to node **C**, then on to node **D** after a similar delay. After completion of the connection, a signal confirming circuit establishment (a connect signal in the diagram) is returned; this flows directly back to node **A** with no search delays since the circuit has been established. Data transfer then begins. After data transfer, the circuit is disconnected; a simple disconnect phase is included in the figure.[5]

Delays for setting up a circuit connection can be high, especially if ordinary telephone equipment is used. Call setup time with conventional equipment is typically on the order of 5 to 25 sec after completion of dialing. New fast circuit switching techniques can in theory reduce delays to approximately 140 msec [PICK85b], however. Trade-offs between circuit switching and other types of switching depend strongly on switching times.

3.2.2 Message Switching

Message switching shares a previously established circuit among multiple users. It is also known as store-and-forward switching since messages are stored at intermediate nodes en route to their destinations. Leased communications facilities are used, so no circuit switching delays are involved. Queuing delays occur, however.

Figure 3.3 illustrates message switching; transmission of only one message is illustrated for simplicity. As the figure indicates, a complete message is sent from node **A** to node **B** when the link interconnecting them becomes available. Since the message may be competing with other messages for access to facilities, a queuing delay may be incurred while waiting for the link to become available. The message is stored at **B** until the next link becomes available, with another queuing delay before it can be forwarded. It repeats this process until it reaches its destination.

Circuit setup delays are replaced by queuing delays. Considerable extra delay may result from storage at individual nodes. A delay for putting the message on the communications link (message length in bits divided by link speed in bps) is also incurred at each node en route. Message lengths are slightly longer than they are in circuit switching, after establishment of the circuit, since header information must be included with each message; the

---

[4]The call request signal may be sent over the same trunk later used for transmitting messages or it may be sent over a separate signaling channel. The latter approach, with signaling messages carried over a separate channel (or even a separate network), is known as common channel signaling and is preferred for new installations in most telephone networks.

[5]A precise description of a disconnect phase for telephone connections would include times for recognition of an on hook signal (from either party hanging up) plus explicit modeling of how disconnection frees individual links.

Figure 3.3   Message Switching.

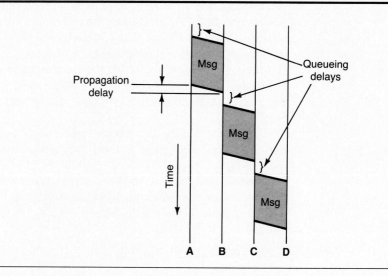

header includes information identifying the destination as well as other types of information discussed later.

3.2.3   Packet Switching

Packet switching is equivalent to message switching for short messages. A maximum message length for transmission as a single entity is imposed, however. Any messages exceeding this maximum length are broken up into shorter units, known as packets, for transmission; the packets, each with associated header, are then transmitted individually through the network. Figure 3.4 illustrates packet switching.

pipelining
   One of the main benefits of packet switching is *pipelining,* an effect visible in the figure. Note that packet 1 is transmitted from **B** to **C** at the same time that packet 2 is transmitted from **A** to **B**; packet 1 is transmitted from **C** to **D** while packet 2 is transmitted from **B** to **C**, and packet 3 is transmitted from **A** to **B**, and so forth. This simultaneous use of communications circuits yields considerable gains in efficiency, so total delays for transmission across a network via packet switching may be considerably less than for message switching, despite the inclusion of a header in each packet rather than in each message.

   Another advantage of packet switching over message switching is a lower probability of retransmission with ARQ error control since individual packets are shorter and less likely to have errors than complete messages. Furthermore, in networks with multiple routes from source to destination, packets

Figure 3.4   Packet Switching.

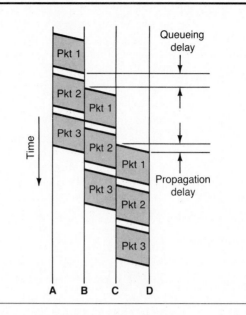

can be routed through a network independently, possibly minimizing congestion by taking different routes to their destinations.

3.2.4   Datagrams Versus Virtual Circuits

Two basic approaches to packet switching are common. In virtual circuit
packet switching, an initial setup phase is used to set up a fixed route for all
packets exchanged during a session between users. Packets can then use relatively short headers, since only identification of the virtual circuit rather
than complete destination addressing is needed. Enhancements such as error
control, guaranteed delivery, and sequencing of packets may also be provided. For datagram transmission, each packet is treated as a separate entity,
with no prior route determination. Packets may follow different routes to their
destination and delivery is not guaranteed. Enhancements of basic service
must be provided by the user.

virtual circuit   A *virtual circuit* appears to the user to be a dedicated physical circuit
between source and destination, although in reality the circuit is shared
among multiple users. The quality of service provided may in some ways be
better than that provided by a dedicated physical circuit, since virtual circuits
may provide powerful error control, enhanced security, or similar features.
Delivery of packets in proper sequence and with essentially no errors is guaranteed, and congestion control to minimize queuing is common. Delays are

more variable than they are with a dedicated circuit, however, since several virtual circuits may compete for the same facilities. An initial connection setup phase and a disconnect phase at the end of data transfer are required.

datagrams

For service using *datagrams,* packets are handled individually, the network making a "best efforts" attempt to deliver them. No guarantee of delivery is given. Packets may arrive out of sequence or be lost or duplicated. Error handling is the responsibility of the user of the service, not the network.

Analogues of these types of service are telephone service for virtual circuit service and postal service for datagram service. As in virtual circuit service, a telephone user must set up a circuit (dial a number), transmit (talk), and finally disconnect the circuit (hang up). Although data transformations inside the telephone network may be complex, two users have the illusion of a direct connection between sender and receiver. In particular, information is delivered to the receiver in the order in which it was transmitted by the sender. On the other hand, the postal service handles letters independently. Each letter must contain the address of the recipient. Letters are not necessarily delivered in the same sequence as mailed; in fact delivery is not guaranteed. If letters are lost, recovery is the sender's responsibility.

Table 3.1 summarizes differences between virtual circuits and datagrams. The choice strongly impacts protocols operating both between intermediate nodes and end-to-end. Some networks use datagrams between intermediate nodes but provide additional functions such as end-to-end error and sequence control to give the user the equivalent of virtual circuits. Use of datagrams between intermediate nodes allows relatively simple protocols at this level, but at the expense of making end-to-end protocols more complex if end-to-end virtual circuit service is desired.

Since the communications subnetwork handles most communications functions when virtual circuits are used, virtual circuits tend to be preferred by unsophisticated users. On the other hand, sophisticated users often prefer datagram service since it gives them more flexibility. They can implement features they need with their own software.

There are important situations where a compromise between virtual circuit and datagram service is best. An example is digitized speech transmission. Although prior establishment of a connection is desirable, features such as ARQ error control and resequencing can be detrimental. A few bit errors in transmitting a speech packet (normally part of one spoken word) are far less disturbing than lengthy delays waiting for retransmission of packets that might be in error. In such situations, a connection may be set up initially as for a virtual circuit, but no error control, resequencing, and so forth may be used.

connection-oriented and connectionless transmission

Recent literature often uses the terms *connection oriented* and *connectionless*. These terms only distinguish whether connections are set up before data transfer and disconnected afterward. Virtual circuit transmission is a special case of connection-oriented transmission, and datagram transmission is a special case of connectionless transmission.

Table 3.1   Comparison of Virtual Circuit and Datagram Service

Issue	Virtual Circuit	Datagram
Initial setup	Required	Not applicable
Destination address	Only needed during setup; after setup, only short virtual circuit number needed	Needed in every packet
Error handling	Transparent to data processing equipment (done by communications subnet)	Explicitly done by data processing equipment
End-to-end flow control	Provided by communications subnet	Not provided by communications subnet
Packet sequencing	Messages always passed to data processing equipment in order sent	Messages passed to data processing equipment in order of arrival

(Adapted from [TANE81], Andrew Tanenbaum, *Computer Networks, 2/e,* © 1988, p. 260. Reprinted by permission of Prentice-Hall, Inc., Englewood Cliffs, NJ).

3.2.5   Performance Trade-offs

Any of the three approaches—circuit switching, message switching, and packet switching—could yield minimum delay in a particular situation, though situations where message switching yields minimum delay are rare.[6] The relative performance of circuit switching and packet switching depends strongly on the type of circuit switching used. With slow circuit switching, typical of today's telephone networks, packet switching usually yields minimum delay for transmission of one or a few messages. This is illustrated by Fig. 3.5, which compares the three techniques for transmitting a single message under approximately the conditions assumed for Figs. 3.2, 3.3, and 3.4. Figure 3.6 indicates the potential advantages of fast circuit switching by showing how increasing switching speed gives the advantage to circuit switching. Circuit switching also yields lower delays if large amounts of data are exchanged, even if slow circuit switches are used, since elimination of delays at intermediate nodes and the elimination of headers on messages more than compensate for circuit switching delays if enough data are transmitted.

An advantage of packet switching cited by some authors [TURN86] is its flexibility in giving the user the data rate needed. For example, if a user needs a 75 kbits per sec rate and only 64 kbits per sec channels are available, use of packet switching and shared channels could meet these needs more easily than could circuit switching. On the other hand, current developments in

---

[6] The primary situation of this type is when only one communication link is involved, which means that there is no storage at intermediate nodes.

Figure 3.5   Comparison of Circuit Switching, Message Switching, and Packet Switching with Slow Circuit Switches.

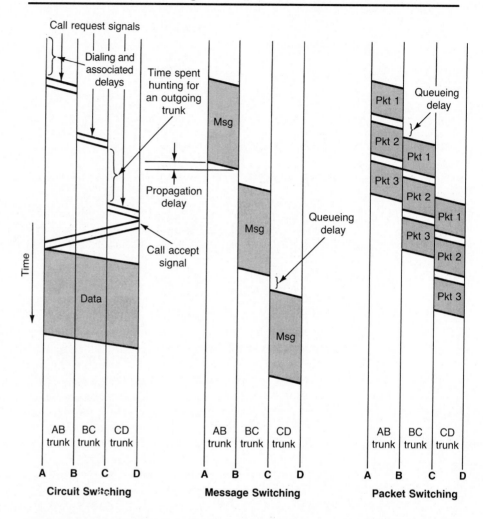

optical fibers and high-speed switching could eventually provide very high bandwidth to all users, say gigabits per sec, at low cost [OREI88], making concerns about efficiency of communication irrelevant.[7] If it does not cost extra money to provide more bandwidth than a user needs, the bandwidth may well be provided.

[7]Readers who feel efficient utilization of communications facilities is as important as doing the desired job at minimum cost may want to consider the best ways to increase utilization of personal automobiles. Possibilities include buying a one passenger vehicle and driving it at minimum speed so its utilization is as high as possible. This will not appeal to many. For personal automobiles, and many other things, efficiency of utilization ranks far below other criteria for choices.

Figure 3.6 Comparison of Circuit Switching, Message Switching, and Packet Switching with Fast Circuit Switches.

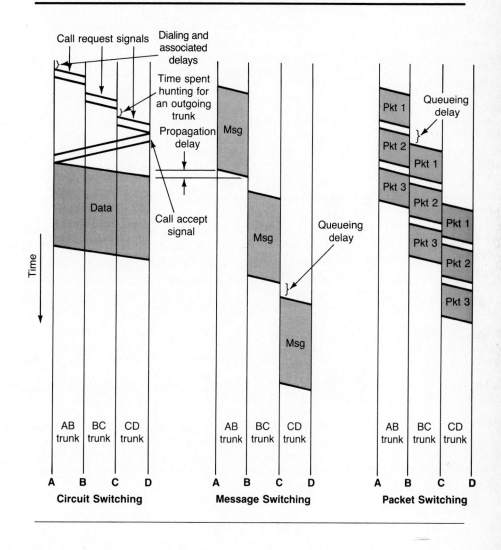

3.2.6 Optimal Packet Lengths to Maximize Pipelining

The relative performance of packet switching can be improved by using an optimal packet length. We indicate how to choose packet length to minimize delay under light load [BERT87], including effects of pipelining but ignoring queuing delays or retransmissions due to errors. An alternative approach of choosing an optimum length to maximize effective transmission rate in the presence of errors, but ignoring pipelining, is discussed in Section 6.5 in

Chapter 6. Since a major advantage of packet switching is its potential for pipelining, the analysis here is based on what is often the dominant factor.

Let M represent total length of a message in bits, including fixed overhead[8] but excluding overhead bits added to each packet, and let n_h be the number of overhead bits added to each packet. Also, let K_{max} be the maximum packet length including packet header. The total number of bits that must be transmitted to send a message is then

$$N_{bits} = M + [M/(K_{max} - n_h)]n_h, \tag{3.1}$$

with $[x]$ the smallest integer greater than or equal to x. The first term in this equation represents original message bits and the second represents the total number of header bits added for transmission.

We assume that the algorithm to divide messages into packets generates packets containing $K_{max} - n_h$ data bits for all except the last packet, with the last packet of the length needed to complete message transmission. Ignoring queuing and processing delays at nodes, the total time (T) required to transmit a message over j equal capacity links is then the time it takes for the first packet to travel over the first $j - 1$ links plus the time it takes the entire message to travel over the final link (see Fig. 3.4). If R represents the transmission rate of each link in bits per second, T is thus given by

$$T = \frac{(j-1)K_{max} + M + [M/(K_{max} - n_h)]n_h}{R}. \tag{3.2}$$

The only random quantity on the right side of this equation is M, the message length. Hence the expected value of T is given by

$$E[T] = \frac{(j-1)K_{max} + E[M] + E[[M/(K_{max} - n_h)]]n_h}{R}, \tag{3.3}$$

with $E[X]$ notation for the expected value of random variable X. Unfortunately, the second expected value on the right cannot be evaluated without prior knowledge of the probability distribution of M. A reasonable approximation, however, is that

$$E[[M/(K_{max} - n_h)]] \approx E[M/(K_{max} - n_h)] + \frac{1}{2} \tag{3.4}$$

$$= \frac{E[M]}{K_{max} - n_h} + \frac{1}{2},$$

which would be exact for a probability distribution of M such that on average $M/(K_{max} - n_h)$ falls halfway between integer values. This yields

$$E[T] \approx \frac{(j-1)K_{max} + E[M] + E[M]n_h/(K_{max} - n_h) + n_h/2}{R}. \tag{3.5}$$

---

[8] This would normally represent overhead added by higher layers in a protocol architecture (see Fig. 1.3) and not under control of the layer doing the packetizing.

optimum packet size   An optimum packet size can be approximated by differentiating Eq. (3.5) with respect to K_{max} (ignoring the fact that K_{max} must be an integer) and setting the derivative equal to zero. This yields an optimum packet size of

$$K_{max}^{opt} \approx n_h + \sqrt{\frac{E[M]n_h}{j-1}}, \tag{3.6}$$

which increases as message length increases or as the number of overhead bits per message increases. On the other hand, as the number of communications links increases the optimum size decreases to allow more parallel data transfers. Link transmission rate (R) does not affect the optimum. Values of K_{max}^{opt} from this formula are plotted versus $E[M]$, for $j = 3$ and 5, in Fig. 3.7.

3.3   Classes of Networks

Telecommunication networks are commonly classified into categories according to geographic extent, purpose, or implementation. We briefly consider major categories of networks that have been widely implemented.

Figure 3.7   Optimum Packet Lengths to Maximize Gains from Pipelining in Multihop Networks.

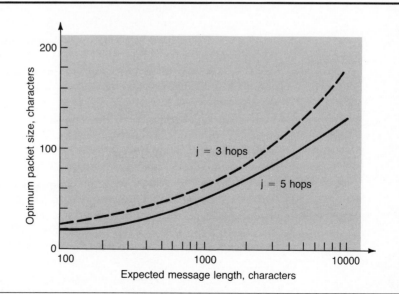

3.3.1   Local Area Networks

LANs

The *local area network* (*LAN*) is by far the most common type of telecommunications network. As was mentioned in Chapter 1, the number of installed LANs was estimated as 423,000 in 1987, with the number predicted to double again by the end of 1988. As the name suggests, a LAN serves a local area. The maximum distance between nodes is normally not more than a few kilometers. They also tend to be owned by a single organization. Typical installations are in industrial plants, office buildings, college or university campuses, or similar locations. Under these circumstances, it is feasible for the owning organization to install high quality, high-speed communication links interconnecting nodes. Typical data transmission speeds are one to 10 megabits per second, with even higher speeds (hundreds of megabits per second or even more) coming in as fiber optic links become more common.

A wide variety of LANs have been built and installed, but a few types seem likely to be dominant in the near future. These include some for which the IEEE[9] has been developing standards. Three architectures have been standardized: a *carrier-sense multiple access/collision detection* (*CSMA/CD*) architecture (IEEE 802.3 standard); a *token bus* architecture (IEEE 802.4 standard); and a *token ring* architecture (IEEE 802.5 standard). We concentrate on these in our treatment of local area networks.

IEEE 802 LANs

3.3.2   Wide Area Networks

WANs

The *wide area network* (*WAN*) normally covers larger geographical areas than a LAN does, though the precise dividing line between the two types is vague. A major factor impacting WAN design and performance is a requirement that they obtain communications links from telephone companies or other communications common carriers. This restricts the communications facilities, and transmission speeds, to those normally provided by such companies. Transmission rates are typically 56 kbps or less (often 9.6 kbps, 4.8 kbps, or slower). Transmission quality, measured by such parameters as line error rate, also tends to be poorer than it is with LANs, and transmission delays are greater.

The characteristics of the transmission facilities lead to emphasis on efficiency of communications techniques in the design of WANs. Flow control to limit traffic and avoid excessive delays is important, as is recovery from transmission errors. Since the geometries of WANs are likely to be more complex than those of LANs, routing algorithms also receive more emphasis. Furthermore, equipment located at nodes in the network is more likely to be of diverse types, so the overall networking architecture tends to be complex.

A number of WANs are very extensive, a few spanning the globe. One of the most influential such networks is the ARPANET, mentioned originally in

---

[9]Institute of Electrical and Electronics Engineers.

Chapter 1 and discussed throughout this text. Many installations of vendor architectures, such as SNA and DNA, are also WANs. Numerous other networks, including public packet networks, corporate networks, networks used by government agencies, banking networks, stock brokerage networks, and airline and other reservation networks are also included.

3.3.3   Metropolitan Area Networks

MANs

The *metropolitan area network* (*MAN*) is a relatively new class of networks falling somewhere intermediate between LANs and WANs. Although the term has appeared recently in the communications literature and standards activities, few current networks fit the definitions. Networks are being developed to support data, voice and video at speeds above 1 Mbps, and to span distances between 5 and 50 km.

A variety of techniques for implementing MANs have been proposed, with coaxial cables or optical fibers the most common communications media. The most prominent current standardization activities are being conducted by the IEEE under the same project leading to the IEEE 802 LAN standards mentioned above. An IEEE 802.6 standard for MANs is under development.

3.3.4   Radio and Satellite Networks

Another important class of networks is one using radio propagation. There are two basic subclasses of this type of network: terrestrial radio networks and satellite networks. The most important difference between the subclasses is relative propagation times. Communications satellite propagation times (for earth-satellite-earth propagation at the speed of light, 300,000 km/per sec) are typically around a quarter of a second, while propagation times for terrestrial radio networks seldom exceed a few milliseconds.

broadcast transmission

The most important feature distinguishing radio and satellite networks from the other types is their use of *broadcast transmission*.[10] Radio transmission tends to be omnidirectional. This means a transmission from one node may be heard by several, if not all, other nodes. In networks involving communications satellites, with all communications relayed via the satellite, all nodes are especially likely to hear each transmission since the "footprint" of the satellite (area covered by its antenna beam) may cover as much as one third of the earth's surface.

A hypothetical radio network is sketched in Fig. 3.8, with lines indicating the interconnection pattern, that is, which nodes hear each other. Although this looks similar to a corresponding network with "wired connections" between nodes, it is distinguished by the fact that *all* nodes connected to a transmitting node hear its transmissions essentially simultaneously. Thus

---

[10]Broadcast transmission is also characteristic of many LANs.

Figure 3.8   Hypothetical Radio Network.

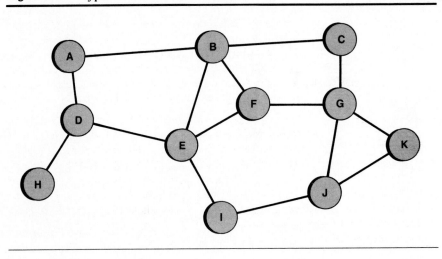

transmissions from node **F** in the figure are heard simmultaneously by nodes **B**, **E**, and **G**.

Broadcast transmission requires modifications to standard algorithms. Routing algorithms must be modified since receipt of a message or packet by a node other than the destination no longer means that node is supposed to relay it. The message could have also reached the destination on the same transmission, or another node that also received it could be a more appropriate relayer. Furthermore, deciding which node should acknowledge a transmission for error recovery purposes is complex when several nodes receive it simultaneously.

3.3.5   Internetworks

Interconnection of computer networks to form larger *internetworks* is evolving rapidly. As was mentioned in the introduction to Chapter 1, the DARPA Internet, the world's largest internetwork, links several thousand networks and in late 1987 had a growth rate estimated to be 15 percent per month. Other prominent internetworks are also being developed, with interconnections between internetworks. Together, these networks form a metanetwork (sometimes called Worldnet) used by communities throughout the world.

Internetworking is complicated by the fact that different networks are often far from compatible. A few examples of variations include the following: addressing and naming schemes, routing techniques, congestion control techniques, information quanta, basic units used for numbering schemes, and packet sizes. More details and techniques for internetworking are covered in Chapter 10.

3.3.6   Special Purpose Networks

A large percentage of the telecommunication networks in existence have been designed and implemented for special purposes. The first networks, such as the SAGE network implemented in the early 1950s for air defense [EVER83] and the SABRE airline reservation system for American Airlines, implemented in 1964 [JARE81], were specialized networks. Other important networks implemented in the early 1970s included the NASDAQ (National Association of Securities Dealers Automated Quotation) System [MILL72b], [SCHW72] to automate trading of over-the-counter stocks (stocks not listed on regular exchanges) in the United States and a variety of networks that originated as time-sharing computer networks and evolved into more general purpose networks. An excellent example of this type of network is the GE Information Services Network [SCHW72].

A tremendous variety of specialized networks have evolved. They include remote teller banking systems, electronic funds transfer networks, stock brokerage networks, grocery and retail store checkout systems, reservation systems, home shopping networks, office and factory automation networks, credit verification systems, corporate networks, military networks, and electronic mail systems. As one example of the proliferation of special purpose networks, Table 3.2 lists major current networks for research [LAND86], [QUAR86].

Similar lists could be given for networks for a wide variety of other applications. Architectures for these networks vary widely, but a large percentage use protocols we discuss, especially protocols growing out of the ARPANET project and protocols being standardized by ISO.

3.3.7   Major Types of Geometries

A variety of ways of interconnecting the nodes in a network are possible. We call the particular interconnection pattern used for a network its geometry.[11] Important classes of geometries are illustrated in Fig. 3.9.

The simplest geometry, which hardly deserves to be called a geometry since it only applies for "networks" with two nodes, is a point-to-point geometry in which a single link connects two nodes. A second, more interesting geometry is a bus geometry in which all nodes are connected via short drops to a common bus. A third common geometry is a ring or loop geometry with all nodes connected in series along a closed path, usually implemented with all data transmission in one direction (clockwise or counterclockwise). A star geometry uses one node as a central switching point with all other nodes connected to this central node via point-to-point links.

The last two geometries illustrated are general configurations allowing alternate routing. Any geometry of the general form in (e) is known as a mesh

---

[11] The term topology is often used, but we use geometry as being more descriptive.

Table 3.2   Today's Major Research Networks.

ARPANET (*A*dvanced *R*esearch *P*rojects *A*gency *NET*work): The original such
  network; see Chapter 1 for origins.

CSNET (*C*omputer *S*cience *NET*work): Developed to facilitate research and advanced
  development in computer science; mostly confined to United States and Canada
  but with links to international affiliates in Australia, France, Germany, Israel,
  Japan, Korea, Sweden, and the United Kingdom.

NSFNET (*N*ational *S*cience *F*oundation *NET*work): Developed (under sponsorship of
  American National Science Foundation) to provide the general academic
  community with the kind of resources CSNET provides to computer science
  researchers; emphasis on access to supercomputer centers.

MFENET (*M*agnetic *F*usion *E*nergy *NET*work): Developed to connect physics
  departments doing research in nuclear fusion, specifically in magnetic fusion
  energy; connects most nodes in United States with one in Japan.

SPAN (*S*pace *P*hysics *A*nalysis *N*etwork): Network serving projects and facilities of
  American National Aeronautics and Space Administration (NASA) and European
  Space Agency (ESA).

JANET (*J*oint *A*cademic *NET*work): Interconnects centrally funded university
  computer centers and research establishments in United Kingdom.

COSAC (*CO*mmunications *SA*ns *C*onnections): Network developed by Centre
  National d'Études des Télécommunications (CNET) as French research network.

DFN (*D*eutsche *F*orshung/*N*etz): National research network in Germany.

ROSE (*R*esearch *O*pen *S*ystems for *E*urope): Network developed by ESPRIT
  (European Strategic Program in Information Technology) to provide infrastructure
  for collaborative research and development projects within ESPRIT and eventually
  for other projects in Europe.

ACSNET (*A*ustralian *C*omputer *S*cience *NET*work): Supports mail traffic and file
  transfers among research, industry, and academic users in Australia.

JUNET (*J*apanese *U*NIX *NET*work): UNIX-based network supporting information
  exchange among Japanese researchers and with researchers outside Japan via
  links to Europe, North America, Australia, and Korea.

SDN (*S*ystem *D*evelopment *N*etwork): Network interconnecting research sites in
  South Korea, with links to other international networks.

AUSEANET: Network developed to join microelectronics (VLSI) projects in ASEAN
  (Association of South East Asia) and Australia; ASEAN countries include Thailand,
  Indonesia, Malaya, Singapore, Brunei, and the Philippines.

CDNNET, FUNET, HEATNET, SUNET, UNINETT and NORDUNET: Networks in Canada,
  Finland, Ireland, Sweden, Norway, and Nordic countries, respectively.

BITNET (*B*ecause *I*t's *T*ime *NET*work): Network serving hosts at sites around globe
  (over 1300 hosts at several hundred sites in 21 countries by 1986); few
  requirements for connection.

EARN (*E*uropean *A*cademic *R*esearch *N*etwork): European-based network analogous
  to BITNET.

UUCPNET (*U*NIX to *U*NIX Co*P*y *NET*work): Informal network of UNIX systems in
  North America, Europe, and Asia; probably the largest network in the world, with
  about 10,000 machines and a million users, but administration can be described
  as near total anarchy [TANE88].

EUNET (*E*uropean *U*NIX *NET*work): Network interconnecting UNIX-based hosts
  throughout Europe

Figure 3.9   Major Classes of Geometries for Networks. (a) Point-to-point.
(b) Bus or multipoint.* (c) Ring or loop. (d) Star.
(e) General or mesh. (f) Fully connected.

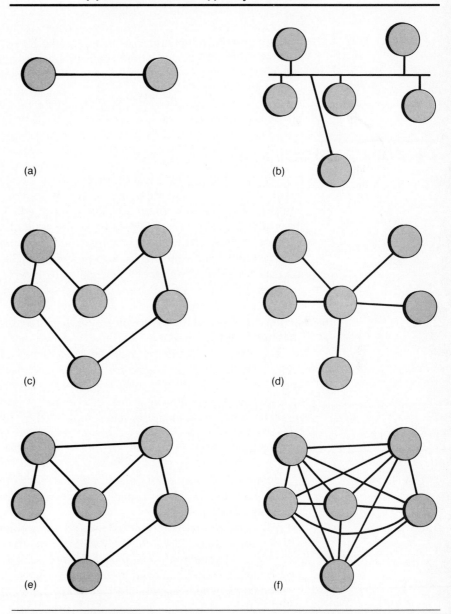

(a)

(b)

(c)

(d)

(e)

(f)

*Also called multidrop or tree. In general a tree is any configuration with only one path between any pair of nodes.

gemoetry, with the special case in which each node is connected to all other nodes, (f), called fully (or completely) connected. Although this is the most failure tolerant of the geometries indicated, it is impractical for large networks; with N nodes, there are $N(N-1)/2$ links for a fully connected network with bidirectional links.[12] Each node would also require $N-1$ ports for interconnection with other nodes, which can also be excessive. For example, a network with 1000 nodes would require approximately half a million links, with 999 ports at each node, if a fully connected geometry were used.

3.4   Naming and Addressing Conventions

A basic function is identification of communicating entities. Three related terms are names, addresses, and routes. In general, a name identifies an entity, an address tells where it is, and a route tells how to get there. These descriptions are oversimplified, but give the general nature of the terms.

names, addresses, and routes

More precisely: A *name* is a symbol, such as human-readable text string, which identifies a resource such as a process, device, or service. An *address* is a data structure, understood by the network, used to specify the destination of a message. A *route* is the information needed to send a message to a specified address. We concentrate on names and addresses. Approaches to determining routes and using them to forward messages are discusssed in Chapters 8 and 9.

The major distinction we make between names and addresses is whether they are intended to be human-readable or machine-readable. Names range from simple names of only local applicability, such as *mail* used to access mail service after gaining access to a computer providing this facility, to universal names. A truly universal name suggested by Tanenbaum [TANE81] is

$$<galaxy><star><planet><country><network><host><port>.$$

Host used in this way refers to a particular computer attached to the network specified, while *port* specifies a process or user program running on the accessed computer.

A typical address could be 8036565901, or a binary expansion of such a number, for example, 1000 0000 0011 0110 0101 0110 0101 1001 0000 0001. Techniques for efficiently mapping human-readable names onto machine-readable addresses are needed in implementations.

Two types of name spaces are *hierarchical* and *flat*. A typical hierarchical

---

[12] Each of the N nodes is connected to $N-1$ other nodes; however, with bidirectional links each link should be counted only once, not twice, when enumerating them in this manner.

name is Tanenbaum's universal name. Such names consist of a sequence of fields that jointly identify the entity. Fields often help to determine where an entity is located. This is not necessarily so; for example, the name <*manufacturer*><*model*><*serial number*> may have little relation to a computer location. Flat name spaces have no particular relation to geography or any hierarchy. Such names are normally assigned by a system-wide authority that assigns a unique identifier to each entity named.

The difference between hierarchical and flat name spaces may be clarified by considering international telephone numbers, which form a hierarchical space, and U.S. social security numbers, which form a flat space. For example, the international phone number 643482009 can be parsed as 64-3-48-2009 where 64 designates a country code (New Zealand), 3 an area within the country (city of Christchurch), 48 a central office, and 2009 a subscriber line within that central office. A social security number, such as 067342369, cannot be broken down into a sequence of fields to tell where the person identified is located, as social security numbers form a flat space.

Both hierarchical and flat name spaces have advantages. Hierarchical names can make routing easy since successive steps may depend on individual fields, for example, successively locating a country, an area, a central office, and a subscriber. It is also simple to assign hierarchical names, since they may be assigned without a central authority. Abbreviation of names for local use is easy; for example, country and area codes do not need to be dialed for local phone calls. On the other hand, hierarchical names must be changed if subscribers move and the name space may be inefficiently used since the number of names available at a hierarchy level is largely independent of how many are needed. Both types of names are common.

3.5   The Structure of Telecommunication Networks

Despite superficial differences, all computer networks perform a well-defined set of tasks. Standards capitalize on this to establish uniform techniques for accomplishing these tasks. The primary tasks are summarized in [GREE80a], [GREE82b].

node

The basic function performed by a network is provision of an effective access path between users at different locations. We use the term *node* to indicate the physical location of a user. Users can be either human users or computer programs; a typical example might be a terminal user interfacing with an application program (the second user). A telecommunication network must provide communication between nodes and between specific users at nodes by using communication *links* between the nodes. Table 3.3 lists major functions that must be performed to provide the access path and make it effective for the intended purposes.

link

Table 3.3   Functions Required to Set up and Use an Access Path

To give one end user access to another end user, the following functions must be performed:		
Ensure that transmission facilities between end points exist.	USING	Common carrier provided lines or other communications links
Ensure that bits can be transmitted over the facilities.	USING	Modems or digital adapters
Provide electrical connections to modems or digital adapters and control their operation.	USING	Physical interfaces or adapters
Share equipment to reduce costs during intermittent use.	USING	Switching techniques; line sharing; other multiaccess schemes
Group bits into characters and messages and move them across communications links without error.	USING	Data link control; transmission error protection techniques
Ensure that messages get to correct nodes and to correct subsidiary addresses within nodes; bypass failed or congested nodes.	USING	Addressing and routing techniques
Accommodate available buffer sizes: avoid need to resend long messages.	USING	Packetizing and depacketizing
Resolve mismatches between feasible rates at which messages can flow across network and rates desired by users.	USING	Buffering and flow control techniques
Accommodate dialogue patterns peculiar to the user pair.	USING	Techniques for setting up, taking down, and managing session dialogue
Make it possible for each user to interpret and use code, format and command conventions, and so forth used by the other users.	USING	Protocol conversions

(Adapted from [GREE82b] P.E. Green, Jr., "An Introduction to Network Architectures and Protocols," *IEEE Transactions on Communications,* © 1980 IEEE. Reprinted by permission.)

3.5.1   Access Path Requirements[13]

The first step in providing an effective access path is to ensure that physical transmission resources exist between origin and destination, often via intermediate nodes. A protocol is executed across links in the path in order for two pieces of hardware to set up the connection.

Transmission is often provided by telephone lines. If telephone lines or similar lines are used, equipment to *convert digital data* to a form suitable for analog transmission and back to digital at the receiver are needed.[14] *Modems* (modulator-demodulator units) normally provide this function as well as others; in purely digital systems their place is taken by adapters to convert digital formats between those expected by data processing equip-

---

[13] Adapted from P.E. Green, Jr., "An Introduction to Network Architectures and Protocols," *IEEE Transactions on Communications,* © 1980 IEEE.

[14] Purely digital facilities are becoming more common and should largely replace analog facilities by the end of the century. No modems are needed with such facilities.

ment and those expected by communications equipment. Modems or digital adapters at the two ends of a connection execute their own protocol to synchronize themselves and exchange data.[15]

With either analog or digital communication facilities, the state of a modem or adapter must be controlled by equipment at the node to which it is attached. An electrical or *physical interface* performs this function, with appropriate protocols executed to accomplish it.

Since many information sources are bursty, a problem is to exploit the bursty nature of traffic to economize on line costs. A variety of techniques for doing this by sharing communications facilities have been developed. Such sharing involves giving multiple users access to the same facilities, so techniques for controlling access are needed. *Multiple access* techniques must be adapted to the facilities used. These functions are also provided by protocols executing between equipment at either end of communications paths.

Another requirement is to *assemble bits* into characters, frames, messages, and so forth, and (in most networks) to provide automated *error handling* to ensure that the received bit stream is an accurate replica of the transmitted one. Although some networks rely on error correction at the receiver, using redundant bits in transmitted data to make this possible, the great majority use ARQ protocols, with automatic detection of errors followed by requests for retransmission if errors are detected.

Control over access to communications facilities, assembly of bits into characters, messages, and so forth and automated error handling techniques are handled by protocols operating between equipment at either end of a communications link. Such protocols handle communications over a particular link or channel and specify formats for messages or frames, including bits or fields used for synchronization, link control, addressing information, error checking overhead, and data, in addition to procedures for setting up, managing, and taking down link connections.

It is necessary to ensure that messages reach correct destinations. This may include more than reaching the correct node since some nodes (such as host computers) may simultaneously serve multiple applications or users. *Addressing* and *routing* protocols handle these functions. A wide variety of techniques have been developed, ranging from simple fixed techniques to highly adaptive distributed techniques to provide nearly optimal routes and avoid failed or congested nodes. Intranode routing directs information to the correct location within a node.

Another function is *buffering* received messages until they can be served and outgoing messages until they can be transmitted. Limitations on buffer

---

[15] The portions of a network discussed so far are often considered to be part of the purview of the communications system designer, with computer communications network architectures built on top of their communications systems. We concentrate in this book on the levels considered to be the purview of the computer communications network designer.

size, desire for fast response time, and the need to do error checking on incoming frames without excessive time for retransmissions, have led to techniques to segment or packetize long messages into packets of reasonable maximum size and to reassemble packets in correct order at destinations.

The *flow* of packets must be regulated to avoid either overflowing buffers at the receiving station or leaving the receiving end user waiting for traffic. Regulating flow can also avoid excessive congestion of virtually any facilities used. It may be necessary to control the rate of traffic on an individual link to protect a buffer dedicated to operation of that link at the same time that a different mechanism is used to control end-to-end flow to protect a buffer dedicated to a particular user. These functions are accomplished by flow control protocols. Numerous options are available, with flow control often used at several levels in a protocol architecture.

A way for a user to conduct a *dialogue* with a user at the other end of a path is necessary. It must be possible for the dialogue to have the request-response pattern users wish. This may consist of individual packets flowing in one direction without coordination with other packets, or it could be highly structured with fixed request-response patterns of flow. Another possibility is a session between users in which flow of packets is part of a related series of transactions; techniques for setting up, taking down, and managing the session are required. Significant amounts of information may need to be exchanged, including negotiating parameter choices. Management may also provide techniques to associate related packets with each other and to determine when users should listen and talk.

A final function is to make sure the access path *accommodates peculiarities of users* with respect to such things as format of messages, character codes, device control, and data base access. Protocol conversion functions are commonly provided to handle these peculiarities and provide the correct form of presentation of data to the user.

Once these functions have been provided, the access path is complete. Additional functions at the user or application level are being considered for standardization, however. These include *management functions* and mechanisms considered useful to support *distributed applications*. An example of a useful mechanism is standard interprocess communications techniques.

Each function listed is normally provided at each end of an access path— that is, functions occur in pairs, with one member of the pair at each location or at each end of a communications link for functions provided on a link-by-link, or hop-by-hop, basis. Furthermore, the two members of a pair talk primarily to each other. Thus one modem or communications adapter talks to the other, ignoring details of the transmission link and the meaning of the bits it forwards. Extra fields added to a message by a protocol at one location are for interpretation by the corresponding protocol at the other end. For example, fields to aid in addressing and routing are used only for addressing and *peer-to-peer* routing and ignored by other functions. This type of *peer-to-peer communication* is almost always used for each function we have discussed.

peer-to-peer communication

3.5.2 Characteristics of Protocols

protocols

Conventions for communications are commonly called *protocols.* A protocol is a set of mutually agreed upon rules of procedure stating how two or more parties are to interact to exchange information. Relatively informal protocols may be satisfactory for some situations, such as exchange of information among people, but protocols need to be carefully defined for communication among machines. What is communicated, how it is communicated, and when it is communicated must all be specified.

The key elements of any protocol are as follows:

- *Syntax:* The structure of information communicated, including such things as data format, coding, and representation in terms of signal levels.
- *Semantics:* The meanings of signals exchanged, including control information for coordination and error handling.
- *Timings:* Times at which data should be transmitted or looked for by a receiver, sequencing of information, speed matching, and so forth.

We will examine a variety of protocols, including those most commonly used in telecommunication networks. Major efforts to standardize telecommunication protocols are underway, and most protocols we treat are results of these efforts. A few others are included because of wide use, their being pushed by major groups, or for information they reveal about how techniques evolve as designers learn from experience with earlier techniques.

3.5.3 Layered Protocol Architectures

All major telecommunication network architectures currently used or being developed use *layered* protocol architectures to accomplish their functions. Precise functions in each layer vary. In each case, however, there is a distinction between functions of lower layers, which are primarily designed to provide a connection or path between users that hides details of underlying communications facilities, and the upper layers, which ensure data exchanged are in correct and understandable form.

Important network architectures are discussed in detail in the rest of this book. We use the Open Systems Interconnection (OSI) architecture developed by the International Organization for Standardization (ISO), and introduced in Subsection 1.3.2, to describe layered architectures. Many aspects of the OSI architecture are similar to those of other layered architectures, which means that much of the discussion is applicable to other architectures. The architectures introduced here are discussed in detail in later chapters.

The basic idea of a layered architecture is to divide the architecture into small pieces. Each layer adds to services provided by lower layers in such a manner that the highest layer is provided a full set of services to manage communications and run distributed applications. A basic principle is to ensure independence of layers by defining services provided by each layer to the next higher layer without defining how services are performed. This permits changes in a layer without affecting other layers. Prior to the use of layered protocol architectures, simple changes such as adding one terminal type to the list of those supported by an architecture often required changes to essentially all communications software at a site.

3.6   The OSI Reference Model

The *OSI Reference Model* is the most prominent current networking architecture. The basic goal has been to develop standards for "Open Systems Interconnection," with "the term 'open' chosen to emphasize the fact that by conforming to these international standards, a system will be open to all other systems obeying the same standards throughout the world" [ZIMM80]. We limit use of ISO terminology, which Tanenbaum calls "internationalbureaucratspeak" [TANE88, p. 22], however, since this phrase is a good description of the terminology.

Principles for defining layers, agreed upon by ISO and applied in defining the architecture, are given in Table 3.4 (see [ZIMM80] for the reasoning followed). No optimal way of defining layers is known, and other architectcures use other definitions.

3.6.1   OSI Architecture Layers

The structure of the OSI architecture is given in Fig. 3.10, which indicates data exchange between applications X and Y, with one intermediate node.[16] The figure shows bidirectional information flow; information in either direction passes through all seven layers at end points, but only through the lower three layers at intermediate nodes.[17] The figure also indicates peer-to-peer communication between corresponding layers.

Brief descriptions of the layers follow. Details are given for each when the relevant part of the architecture is discussed in later chapters.

- *Physical layer:* Provides electrical,[18] functional, and procedural characteristics to activate, maintain, and deactivate physical links that transparently pass the bit stream for communication; only recognizes individual bits, not characters or multicharacter frames.

---

[16] Although a user may be a person accessing the system via a terminal, the network sees an application, such as a terminal handling a program allowing the user to access the network.

[17] If there were more than one intermediate node, information would flow up through the physical and data link layers to the network layer and back down at each intermediate node.

[18] Electrical characteristics are replaced by optical characteristics if optical media are used.

Table 3.4 ISO Criteria for Choosing Layers

1. Do not create so many layers that the system engineering task of describing and integrating layers is made difficult.
2. Create a boundary at a point where the description of services can be small and the number of interactions across the boundary minimal.
3. Create separate layers to handle functions that are manifestly different in process performed or technology involved.
4. Collect similar functions into the same layer.
5. Select boundaries at a point that past experience has demonstrated to be successful.
6. Create a layer of easily localized functions so the layer could be totally redesigned and its protocols changed in a major way to take advantage of advances in architectural, hardware, or software technology without changing services and interfaces with adjacent layers.
7. Create a boundary where it may be useful to at some point in time have the corresponding interface standardized.
8. Create a layer when there is a need for a different level of abstraction in handling of data.
9. For each layer, create interfaces only with its upper and lower layers.
10. Create further subgrouping and organization of functions to form sublayers within a layer in cases where distinct communications services need it.
11. Create, where needed, two or more sublayers with a common, and therefore minimum, functionality to allow interface operation with adjacent layers.
12. Allow bypassing of sublayers.

From [ZIMM80], H. Zimmerman, "OSI Reference Model—The ISO Model of Architecture for Open Systems Interconnection," *IEEE Transactions on Communications,* © 1980 IEEE. Reprinted by permission.

- *Data link layer:* Provides functional and procedural means to transfer data between network entities and (possibly) correct transmission errors; provides for activation, maintenance, and deactivation of data link connections, grouping of bits into characters and message frames, character and frame synchronization, error control, media access control, and flow control.

- *Network layer:* Provides independence from data transfer technology and relaying and routing considerations; masks peculiarities of data transfer medium from higher layers and provides switching and routing functions to establish, maintain, and terminate network layer connections and transfer data between users.

- *Transport layer:* Provides transparent transfer of data between systems, relieving upper layers from concern with providing reliable and cost effective data transfer; provides end-to-end control and information interchange with quality of service needed by the application program; first true end-to-end layer.

Figure 3.10   Layered Structure of OSI Reference Model Architecture.

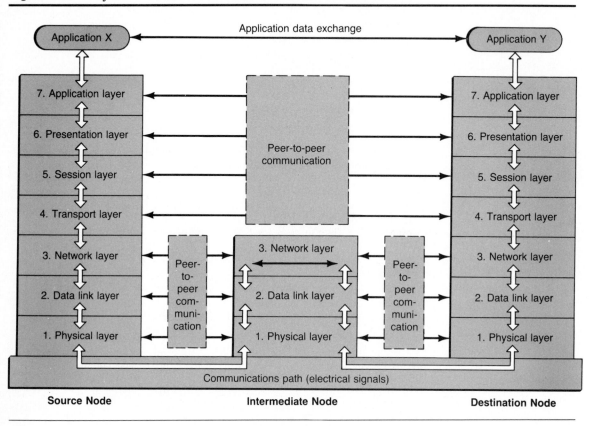

- *Session layer:* Provides mechanisms for organizing and structuring dialogues between application processes; mechanisms allow for two-way simultaneous or two-way alternate operation, establishment of major and minor synchronization points, and techniques for structuring data exchanges.
- *Presentation layer:* Provides independence to application processes from differences in data representation, that is, in syntax; syntax selection and conversion provided by allowing the user to select a "presentation context," with conversion between alternative contexts.
- *Application layer:* Concerned with semantics of application. All application processes reside in the application layer but only part are in the real OSI system. Those aspects of an application process concerned with interprocess communication, common procedures for constructing application protocols and for accessing the services of OSI, plus a few application protocols of general interest, are being standardized.

The lower three layers—physical, data link, and network—are concerned with providing transparent connections between users and operate primarily on a hop-by-hop basis, handling communications over individual data links between nodes. The network layer uses algorithms involving cooperation among nodes to handle routing, control of congestion, and so forth, but only controls communication over individual hops, relying on consistent operation of other nodes to achieve its goals. The upper three layers—application, presentation, and session—are primarily concerned with ensuring that information is delivered in correct and understandable form. The middle (transport) layer is an interface between these groupings. It is the first end-to-end layer and ensures a transparent end-to-end connection meeting user requirements for quality of service. It also provides information to the upper three layers in suitable form.

In this book we concentrate on protocols operating at the equivalent of the bottom four layers (physical, data link, network, and transport) in the OSI model and other communications architectures. These layers provide the basic communications service; techniques for ensuring that information exchanged is in correct and understandable form deserve separate treatment.

3.6.2 Peer-to-Peer Communication

The manner in which peer-to-peer communication in the OSI architecture takes place is indicated by Fig. 3.11, drawn for the same situation as Fig. 3.10 but with the intermediate node omitted for simplicity. Communication from application X to application Y is illustrated. As the figure indicates, application data (AP data) are exchanged between applications. Application X presents this data to the system for transmission to application Y.

frame construction   In constructing an outgoing frame, each layer (except for the physical layer) adds one or more fields to information from higher layers, with corresponding field(s) stripped off by the corresponding peer layer during incoming frame reduction. The added fields are used for peer-to-peer communication. The application layer at the source adds an *application header* (*AH*), containing information it wants to send to its peer application layer, to application data. The AH is passed unchanged (ignoring transmission errors) to the receiver where the application layer strips it off and takes actions indicated by its contents. The remaining portion of the received packet is AP data and is passed up to application Y. Similarly, a *presentation header* (*PH*), *session header* (*SH*), *transport header* (*TH*), *network header* (*NH*), and *data link header* (*DH*) are added by the corresponding layer at the transmitting end and stripped off by the peer layer at the receiving end, with each used for peer-to-peer communication. Each layer treats the assemblage of information from higher layers as data, and does not worry about its contents.

Minor modifications of this scheme are as follows. The data link layer also adds a *data link trailer* (*DT*) to the end of the message. This is also for

Figure 3.11   Frame Construction in OSI Architecture.

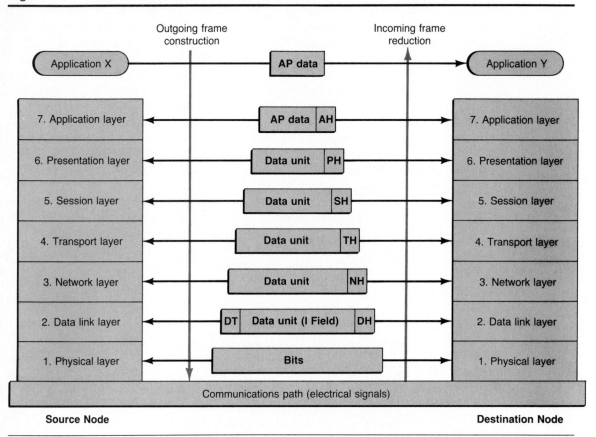

peer-to-peer communication and is primarily error control information most readily put at the end of a frame. If we had shown intermediate nodes, we would have shown the data link and network layers stripping off their headers as the packet flows up at such nodes, adding new headers as it flows back down. No header or trailer is added at the physical layer, which does not recognize data units larger than a bit and views data as a string of bits. Similarly, the communications path views information as a sequence of electrical (or optical) signals used to transmit bits.

In addition to data messages that flow across the network as indicated, control messages may be exchanged between peer processes at any layer. Each peer-to-peer protocol defines control messages for purposes such as setting up connections at that layer, flow control, and error control. Control messages generated by layers originate at those layers and have headers (occasionally trailers) added by lower layers.

Figure 3.12   Peer and Interface Protocols in OSI Architecture.

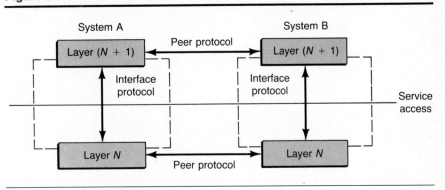

3.6.3   Interfaces between Layers

The protocol for each layer is concerned with peer-to-peer communication with the corresponding layer at the other end of the access path (hop for the lower three layers, end-to-end for the upper four), but each layer uses services of layers below it. During peer-to-peer communication, information flows down through lower layers in the same node, across the communications path and up through layers in the other node until it reaches the peer layer. Boundaries between adjacent layers in the same system are called *interfaces* in OSI terminology, with an *interface protocol* operating across the boundary.[19]

interface

Figure 3.12 illustrates the difference between peer and interface protocols. The interface is used to access services provided by a lower layer to a higher layer. The point at which a service is provided is called *service access point* (*SAP*). The abbreviation *N-SAP* is used for a service provided by the *N*th layer. The (*N* + 1)st layer can access this service at an *N-SAP address*. Services provided by peer protocols at the *N*th layer can be accessed only across the network, using the information flows described above.

service access point

Interactions between adjacent layers are managed by passing messages called *primitives* between layers. A primitive initiates an action or advises of the result of an action. Each primitive may contain parameters to convey control information needed to perform its functions. Four basic types are defined:

primitives

- *Request:* A primitive sent by layer (*N* + 1) to layer *N* to request a service. It invokes the service and passes those parameters needed to fully specify the request.

---

[19]This is a specialized use of the term interface, which does not indicate a connection between different devices or equipment.

Figure 3.13   OSI Use of Four Types of Primitives.

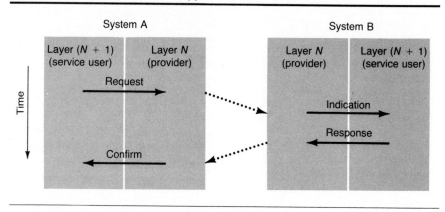

- *Indication:* A primitive returned to layer $(N + 1)$ from layer N to advise of activation of a requested service or of an action initiated by the layer N service provider.

- *Response:* A primitive provided by layer $(N + 1)$ in reply to an indication primitive. It may acknowledge or complete a procedure previously invoked by the indication primitive.

- *Confirm:* A primitive returned to the requesting $(N + 1)$st layer by the Nth layer to acknowledge or complete a procedure previously invoked by the request primitive.

Each type of primitive has a variety of forms for different purposes. For example, there are Connect Request, Data Request, Flow Control Request, Disconnect Request, and a variety of other forms of Request primitives plus similar forms for Indication, Response, and Confirm primitives. Primitives are also identified by the protocol layers using them.

Since primitives flow between layers implemented on one system, for example, a communications processor, *primitive transmission is not externally visible.* Hence, the primitive passing sequences that we discuss do not have to be implemented in the manner implied. The real requirement is that externally visible results, such as frames of data transmitted, must be in the form indicated. There are, in fact, often strong arguments for not implementing sequences precisely as sketched since excessive data processing, memory accesses, and so forth may result if each end-to-end transmission requires separate processing for each of the seven protocol layers at each end of the path.

Figure 3.13 illustrates the use of the four types of primitives for a situation in which communication between system A and system B is necessary to satisfy a Request. Dotted arrows indicate transmissions from system to system or externally visible data flows.

3.6.4 Typical Communications Scenario

A typical OSI communications scenario can now be sketched. Assume that application X in system A wants to send AP Data to application Y in system B, across a direct path with no intermediate nodes. In the scenario connection-oriented service is required at each layer, but initially no connection exists at any layer.

connection phase

Initially application X in system A sends a Connect Request primitive across its interface with the application layer at its location (system A) requesting a connection to application Y in system B. The application layer, after determining no Presentation Layer Connection exists, sends a Connect Request to the presentation layer in system A. Similarly, the presentation layer, session layer, transport layer, network layer and data link layer in system A each send Connect Requests to adjacent lower layers. After the Connect Request is passed from the data link layer to the physical layer at A,[20] the physical layer sets up a Physical Layer Connection with its peer physical layer at B, which informs the data link layer at B of this via a Connect Indication. The data link layer at B returns a Connect Response, causing an appropriate signal to be passed across to the physical layer at system A and a Connect Confirm to be sent up to the data link layer at A.

At this point a Physical Layer Connection has been established and the data link layers at each end are aware of its existence. This allows the data link layer at A to respond to the Connect Request it previously received by passing a control frame to establish a Data Link Connection across the link between A and B. When the data link layer at B receives this frame, it sends an Indication primitive to its network layer, which returns a Connect Response primitive. This causes the data link layer at B to generate a control frame and pass it across the link to the data link layer at A, causing this layer to send a Confirm primitive to its network layer. A Data Link Connection is now established, with both network layers aware of it.

In a similar manner, Network Layer, Transport Layer, Session Layer, Presentation Layer, and Application Layer Connections are established and verified; the access path is then ready to transmit data.

data transfer phase

Transmission of data is accomplished similarly. Application X sends a Data Request primitive, along with AP Data, to the application layer, which adds its header and passes a Data Request primitive and Data Unit to the presentation layer.[21] This process continues, with addition of appropriate headers, until it reaches the data link layer, which adds its header and trailer and passes individual bits, each with a Data Request, to the physical layer for transmission. The bits are received at the system B physical layer, each gen-

---

[20] The primitives between the data link layer and physical layer may be referred to as Activate Request, Indication, Response, and Confirm primitives, but we use the term Connect for consistency.

[21] The data unit is the AP data plus application header.

erating a Data Indication primitive (including the received bit) for the system B data link layer. When the data link layer recognizes the end of the data trailer terminating its data unit, it strips off the header and trailer and sends the remaining data unit to the network layer (by including it in a Data Indication primitive).

Data Indication primitives, with reduced data units, cascade up the layers until the AP data reaches application Y. Application X only learns of successful receipt if application Y returns an Acknowledgment. The Acknowledgment makes its way across the access path in the same way as the data, but in the reverse direction and with Response and Confirm primitives replacing Request and Indication primitives.

disconnect phase

After data transfer is complete, a disconnect phase similar to the connect phase will occur.

3.6.5   Limitations of Reference Model

Limitations of the OSI Reference Model are becoming obvious, with a number of them discussed in later chapters. Despite the tremendous current momentum of the model, we expect these limitations to lead to modifications of it in the future.

The OSI Model is complex, with a number of thick documents used to describe it. Furthermore, this documentation is extraordinarily complex and very little of it is written in a readable form; it focuses almost entirely on listing of details, giving virtually no motivation for techniques adopted. Since the normal procedure for dealing with alternative techniques to solve the same problem followed by committees developing the model has been to adopt all reasonable suggestions as options, there are an excessive number of options for a protocol architecture supposed to be an international standard. Hence different implementations of the OSI Model are not necessarily compatible, violating the fundamental principle of "open systems interconnection" motivating the whole architecture.

The layering structure of the OSI Model is arbitrary, despite the attempts of its developers to create a logical layering structure. Appropriate placement of features in layers is not always obvious, but is tightly prescribed in the model. Furthermore, layers differ significantly in their complexity. There are many occasions when some layers are superfluous, but the architecture does not at present provide for bypassing of layers.

The most serious weaknesses of the model result from the fact that its development has been dominated by persons in the communications field rather than the computer field. Techniques used in communications tend to be used even in places where other techniques developed for computer applications would be more satisfactory. An example is the fact many OSI Model techniques are analogous to use of interrupts when implementation efficiency could be improved through use of procedure calls or similar techniques.

The "communications mentality" of most OSI Model developers seems to be one main reason it has tended to focus almost entirely on connection-oriented rather than connectionless service despite the facts that many applications are better handled by connectionless service and some types of networks (especially LANs) normally work in a connectionless mode. (Telephone networks do operate in a connection-oriented mode.) This is being remedied by modifications of the model, but connectionless operation still receives far less emphasis than connection-oriented operation.

The OSI Model can also lead to inefficiencies because it does not explicitly allow for bypassing of layers, even in situations where some layers are not needed. Furthermore, it has considerably less flexibility than some alternative architectures. This should become apparent as we discuss alternative architectures.

3.7 The DoD Reference Model

An alternative architecture, which deserves attention, is the architecture developed for the U.S. Department of Defense as an outgrowth of the ARPA-NET project. Far more implementation experience has been gained with this architecture than with the ISO architecture. The DoD has issued internal standards within this architecture and has chosen to develop its own protocols and architecture rather than adopting OSI standards.

Although development of what we call the *DoD Reference Model* was less systematic than development of the OSI Model, there are excellent ideas embodied in the DoD architecture. Criticisms of the OSI Model by developers of the DoD Model also merit careful evaluation.

There are three primary reasons for the DoD decision to develop its own protocols and architecture [STAL88b]. DoD protocols have been specified and used extensively while large parts of the OSI Model have not yet been implemented; communications requirements specific to DoD have not been reflected in the OSI Model; and there are philosophic differences concerning the appropriate nature of a communications architecture and protocols.

3.7.1 Differences in Emphases

There are four fundamental differences between the DoD Reference Model and the OSI Reference Model [ENNI83], [STAL88b]:

- Hierarchy versus layering
- The importance of internetworking
- The utility of connectionless services
- The approach to management functions

Good software design practice dictates that complex tasks be broken up into modules or entities that may communicate with peer entities in another system. Furthermore, entities should be arranged hierarchically so that no entity uses its own services either directly or indirectly. Grouping entities into layers, as is done in the OSI Model, is not the most flexible way to implement a hierarchical structure. The developers of the DoD Model feel the strict layering of the OSI Model is too restrictive. Furthermore, they feel the OSI Model has been interpreted in an overly prescriptive, rather than descriptive, manner—that is, as prescribing how to design protocols more precisely than should be done in an architecture. For example, the OSI Model requires that information pass through all lower layers en route to the destination node, even if some do not perform useful tasks for a particular type of peer-to-peer communication, and it is very restrictive in prescribing the ways in which information may be passed.

The DoD Model is less restrictive and allows such things as an entity directly using services of a hierarchically lower entity that is not in an adjacent layer, use of separate control and data connections and use of lower level control information to accomplish higher level control (for example, closure of a lower-level connection implicitly closing a corresponding higher-level connection without requiring the higher-level entity to pass control information).

The primary difference between the DoD Model and the OSI Model is that the former is much more flexible. Simply requiring that protocols be modular and hierarchical puts fewer restrictions on designers than specifying that protocols have a layered structure with predefined assignments of functions to different layers.

internetworking

The DoD Model also puts more emphasis on *internetworking* than does the ISO Model. Internetworking occurs when two communicating systems are not attached to the same network. Strong emphasis on internetworking in the DoD Model can be traced to the variety of communications networks used by the armed services; cooperative use of networks should greatly enhance their effectiveness. Because of the emphasis on internetworking in the DoD Model, a separate layer is devoted to it in the architecture. Within the ISO Model, internetworking is considered to be handled by a sublayer of the network layer; it is mentioned regularly in ISO documentation, but is usually treated as a complication of the network layer.

utility of connectionless services

Most of the emphasis in development of the OSI architecture has been placed on connection-oriented services, which require an initial connection setup phase and a final disconnect phase in addition to a data transfer phase. Connectionless services, for which data are presented to the network without prior connection establishment, have for historical reasons been considered equally important by the DoD, since important applications of DoD networks involve short interactive data transfers of types best handled without the overhead of connection establishment.[22]

---

[22] Connectionless services are also being developed for the OSI Reference Model, but to date essentially all work on connectionless services has been as an afterthought resulting in modification of standards initially defined only for connection-oriented services.

Figure 3.14 Network Configuration Leading to Development of DoD Reference Model.

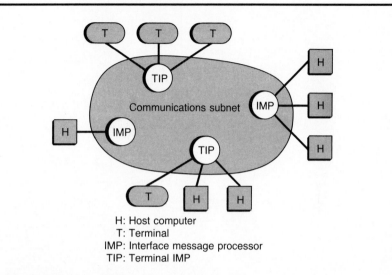

H: Host computer
T: Terminal
IMP: Interface message processor
TIP: Terminal IMP

Management functions include naming or identification of resources, control of access to resources and accounting for resource and network usage. This has also received more emphasis in the DoD Model. The original OSI documentation indicates most of these functions will be in the application layer, but the DoD architecture puts most of them in what corresponds to the OSI session layer since their use is normally associated with sessions.

3.7.2 DoD Reference Model Architecture

The DoD Reference Model was developed for use in the ARPANET with the basic configuration depicted in Fig. 3.14. DoD Model networks use a wide variety of heterogeneous host computers. These are interfaced to a communications subnetwork via communication processors, called IMPs (interface message processors). Direct connection of terminals is possible through TIPs, or Terminal IMPs, which contain additional functions for terminal handling. In later discussions we use the term *CSNP* (for communications subnet processor) for either IMPs or TIPs. Internetworking is accomplished via gateways, as Fig. 3.15 illustrates.

CSNPs

The fundamental concept leading to the DoD Reference Model is *interprocess communication*. Active entities in hosts (programs in execution) are referred to as processes, with the primary goal of the architecture being to enable processes to communicate with each other. Processes execute on hosts that often support multiple processes simultaneously.

Figure 3.15   Internetworking Form of DoD Architecture.

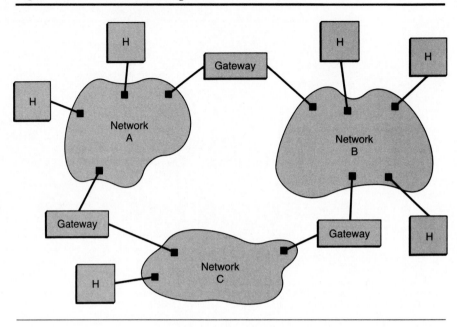

The hierarchical structure of the architecture is illustrated in Fig. 3.16. Layers are defined from the viewpoint of individual hosts—that is, they are based on functions visible to hosts in at least some implementations.[23]

The architecture contains four layers: network access (I), internet (II), host-host (III), and process/application (IV).[24] Another protocol operates between CSNPs. Some of the most interesting aspects of the architecture are handled by this CSNP-CSNP protocol, which we include in our discussion. A layer corresponding to the OSI physical layer is also needed. The layers can be defined as follows:

- *Network access layer* (Layer I): Contains protocols, between hosts or terminals and CSNPs, to provide network access. It allows a host to pass data to its attached CSNP (plus an indication of where it is to be sent), receive data reaching the CSNP from other nodes, and regulate flow of data over the host-CSNP link.

---

[23] Some may be offloaded to front-end processors in other implementations.

[24] The precise layers differ slightly in different presentations of the DoD Reference Model. (For example, the internet layer is considered to be part of the host-host layer in [PADL83].) The term "level" is also used in these presentations instead of "layer," which we use for consistency with our discussions of other architectures.

Figure 3.16 Hierarchical Structure of DoD Reference Model (from [PADL83]). © 1983 IEEE. Reprinted by permission.

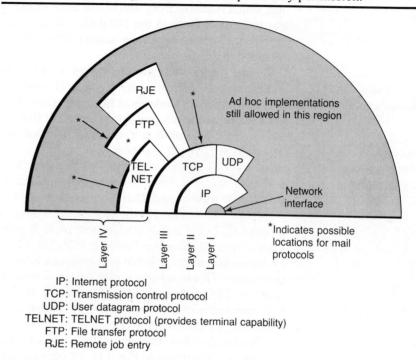

IP: Internet protocol
TCP: Transmission control protocol
UDP: User datagram protocol
TELNET: TELNET protocol (provides terminal capability)
FTP: File transfer protocol
RJE: Remote job entry

- *Internet layer* (Layer II): Contains functions necessary to allow data to traverse multiple networks between hosts. Protocols are defined between hosts and gateways. A major aspect is use of appropriate addressing to allow communication with processes on hosts attached to other networks. Functions such as message frame identification, message segmentation and reassembly, and error handling may also be included.

- *Host-host layer* (Layer III): Contains functions needed for interprocess communications between processes on different hosts. This may or may not involve setting up logical connections between higher-level entities. Other possible services include error control, flow control, and so forth.

- *Process/application layer* (Layer IV): Contains protocols that perform specific resource-sharing and remote-access functions such as allowing users to log on to remote hosts, transferring files, and exchanging messages.

Figure 3.18   Typical Phase IV DNA Network.

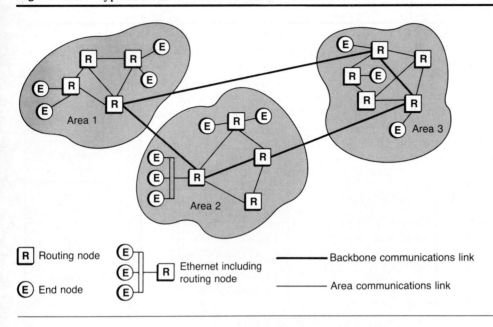

The DNA architecture is *symmetric* and *peer-to-peer,* with any two processes free to communicate as long as naming and security constraints are satisfied. The primary motivation for the architecture was to support networks of resource-sharing computers, primarily minicomputers, since Digital has been, historically, a minicomputer vendor.[27] Minicomputer users have often had greater needs for resource sharing than users of larger computers.

end nodes and routing nodes

Two types of nodes are distinguished: *end nodes* and *routing nodes.* End nodes have one attachment to the network and do not calculate routes or forward packets on behalf of other nodes. Routing nodes support multiple links and forward traffic; so they compute routes. Phase IV networks have an architectural limitation of 64,000 nodes, with a practical limitation approximately one half of that.[28] A typical network is illustrated in Fig. 3.18.

Large networks may be broken up into areas connected by backbone links. Routes can pass through two levels: Level I carrying traffic within an area and Level II carrying traffic between areas. Nodes attached to both area

---

[27] The minicomputer market is now becoming almost indistinguishable from the general purpose computer market, but this is a recent phenomenon.

[28] DEC's internal network is already approaching this limit [JOHN86].

Figure 3.19   DNA Layered Architecture.

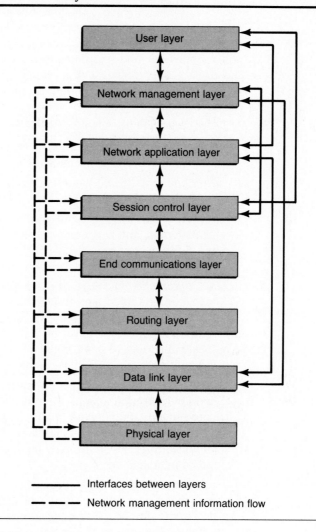

Interfaces between layers

— — — Network management information flow

and backbone links serve both types of routes. Phase IV DECNETs also allow nodes to be connected to Ethernets.[29]

DNA uses the "hierarchically layered" architecture in Fig. 3.19. Although eight layers are shown, the architecture is basically a seven-layer architecture with addition of a network management layer that communicates with every

---

[29]Ethernet is DEC's local area network architecture; it is discussed in a variety of locations in this text. We also note that in Phase IV some communications links can be X.25 links; X.25 is discussed in Chapter 9.

other layer. This gives even more emphasis to network management than in the DoD Reference Model (which in turn emphasizes management more than does the OSI model). Interfaces between layers are more flexible than in the OSI model; the user layer has direct access to the network management, network application, and session control layers; the network management layer directly accesses the network application, session control, and data link layers; and the network application layer accesses the session control and data link layers.[30]

The layers can be defined as follows:

- *Physical layer:* Same definition as for OSI Reference Model. Any of a wide variety of standard physical layers may be used.

- *Data link layer:* Same definition as for OSI Reference Model. Three protocols are used: DDCMP, Ethernet, and X.25.[31]

- *Routing layer:* Provides network-wide message delivery service. It implements a datagram service that delivers packets on a "best-efforts" basis, making no guarantees against packets being lost, duplicated, or delivered out of order. This layer is discussed in Chapter 9.

- *End communications layer:* Provides a standardized reliable, sequential, connection-oriented, end-to-end communications service to higher layers. The layer isolates higher layers from transient errors or reordering of data introduced by lower layers. It also multiplexes multiple connections, called logical links, between pairs of nodes or between one node and multiple nodes. This layer is discussed in Chapter 11.

- *Session control layer:* Provides system-dependent, process-to-process communications functions for processes residing in the user, network management, and network application layers. Its service is connection-oriented. Once a connection is established, data flows between processes without further intervention by the layer, using facilities provided by the end communications layer.

- *Network application layer:* Provides generic services to the user and network management layers. Services include remote file access and transfer, remote interactive terminal access, and gateway access to non-DNA systems. Modules in this layer operate independently and asynchronously. The layer supports modules supplied by both DEC and users and can be bypassed if not needed.

- *Network management layer:* Provides decentralized management for a DNA network. Modules within a node are responsible for two functions:

---

[30] Earlier versions of the architecture used different termonology: what is now the routing layer was called the transport layer and what is now the end communication layer was called the network services layer. Since these correspond approximately to the network layer and the transport layer, respectively, in the OSI architecture, the older terminology caused a great deal of confusion.

[31] DDCMP is discussed in Chapter 7; Ethernet in Chapter 5; and X.25 in Chapter 9.

Figure 3.20 Frame Construction and Reduction in DNA Architecture.

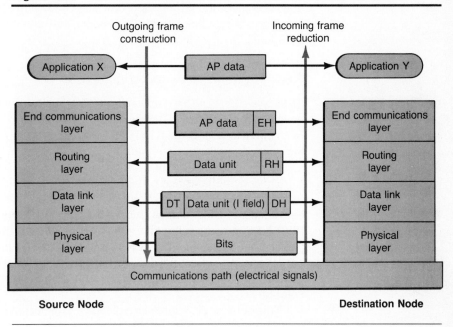

They coordinate management of that node and communicate with peer modules in remote nodes to accomplish decentralized management. This layer uses services of three lower layers: the network application layer to provide network services such as remote file access, the session control layer to communicate with peer entities for decentralized management functions, and the data link layer to handle simple management functions between adjacent nodes.

- *User layer:* Contains user-defined functions such as application programs. Only one such function is currently specified—the network control program (NCP), which implements a network command language providing a user interface to network management functions. Interfaces to three lower layers are provided. Application programs can directly access the session control layer for program-to-program communication, the network application layer to gain access to services such as file access, or the network management layer to enhance the capabilities of NCP.

Typical information flow through a DNA network after session initiation is shown in Fig. 3.20, which is similar to Fig. 3.11 (the corresponding figure for the OSI architecture) but simpler. The figure has been drawn under an assumption that facilities of the network application layer (remote file access, and so forth) are not needed, so this layer is bypassed. Since the session control layer is essentially transparent after a session has been initiated, it does not affect the information flow sketched. Only the end communication, routing, and data link layers add headers, with the data link layer also adding a trailer.

3.8.2   Systems Network Architecture

IBM's *Systems Network Architecture* (*SNA*) is the dominant vendor-supplied networking architecture. It was also the first vendor architecture at the time of its initial announcement in 1974. New releases and modifications have been announced regularly since then.

A major motivation for SNA was proliferation of telecommunication software access methods and protocols. At the time of SNA's announcement, IBM had more than 200 communications products in the field, requiring 35 different teleprocessing access methods and 15 different data link controls [PIAT77]. Networks using different communications products were so incompatible that a user wanting access to two different applications in two different computers often needed two terminals (possibly identical) and two communications lines, each connected from one terminal to the appropriate computer.

This situation led some large leading-edge users to make their own modifications or extensions of IBM products. Modifications were also made to improve efficiency of network protocols, and some gained semistandard or even standard status. IBM was also vulnerable to vendors of intelligent front-end processors and minicomputers, which could be used to solve users' teleprocessing problems. Furthermore, users often found themselves locked into equipment when running complex software, sometimes with heavily modified software; this caused problems for IBM in managing the growth and evolution of its customer base.[32]

These factors greatly influenced the development and evolution of SNA, whose goals were clearly established:

- To provide consistent network access methods, making network and terminal characteristics transparent to application programs.
- To allow a terminal to access more than one application in a computer.
- To improve efficiency of communications protocols.
- To allow different types of terminals to share a multipoint link.

SNA was also developed to handle a wide variety of previously existing applications in an evolutionary fashion. It is the most comprehensive, and most complex, networking architecture in current use. It is also the most thoroughly documented (though with much documentation incomprehensible to anyone who has not thoroughly mastered SNA) and the most carefully veri-

---

[32] A sumary of major motivations for development of SNA is given in [BOOT81, pp. 81–83].

fied architecture. Our treatment of SNA does not do full justice to the rigor with which it has been defined, both because of space limitations and because we attempt to describe various architectures in a reasonably consistent manner.

SNA was initially structured using a *hierarchical control* philosophy. Overall control was centered in a single host system, since this was the approach in systems motivating its development. It has evolved toward a more flexible approach tailored to true distributed systems, but it is still more hierarchical and more tightly specified than other architectures. It is a comprehensive architecture implemented in products ranging from mainframes to personal computers, and it makes explicit provisions for attachment of equipment such as simple terminals. In addition to protocols, SNA includes a collection of IBM software and hardware products used in networking.

Rather than treating all nodes uniformly, SNA distinguishes among node types according to their abilities. It uses the idea of subsetting functions to do this, since not all SNA functions can be implemented in all products.[33] It also makes a variety of similar distinctions, such as distinguishing among different types of communications sessions.

Most SNA concepts are straightforward, but as [BARK86] states, "The SNA terminology and acronyms are horrendous!"[34] In keeping with Tanenbaum's descriptive terminology (see Section 3.6), we dub this *"systemsnetworkarchitecturecommitteespeak."* It is far worse than the "internationalbureaucratspeak" Tanenbaum criticizes—hence the even longer "word."[35] A large part of the complexity of the terminology and acronyms arises from the fact that the architecture is comprehensive, but the members of SNA design committees also have had an unfortunate proclivity to create a new acronym or define a new term whenever possible.

Some terminology can be presented by discussing a typical SNA configuration, such as that in Fig. 3.21. Part (a) illustrates four distinct types of nodes and terminology involved with these node types, while part (b) shows a typical way in which nodes could be interconnected, along with more of the almost limitless SNA terminology and acronyms.

peripheral nodes
subarea nodes

Node types include two types of *peripheral nodes* and two types of *subarea nodes.* Simple peripheral nodes are terminal nodes, (simple terminals

---

[33] Low-function nodes may also have some functions performed for them by nearby nodes with more capability.

[34] The horrendous complexity of the terminology and the architecture has led to such monstrosities as a single manual running more than 1400 pages long and being almost unreadable [IBM80]. The simplicity and elegance of a number of SNA concepts is thoroughly disguised by the documentation, but the documentation is the most complete available—if anyone can understand it. IBM's complete documentation of SNA in terms of finite state machines is especially noteworthy.

[35] It is only fair to note that SNA terminology is the most precise terminology used in the field, but we feel that it could be just as precise, and documentation more understandable, with far fewer acronyms, abbreviations, and so forth.

Figure 3.21   SNA network. (a) Node types. (b) Typical configuration.
See text for definitions of terms.

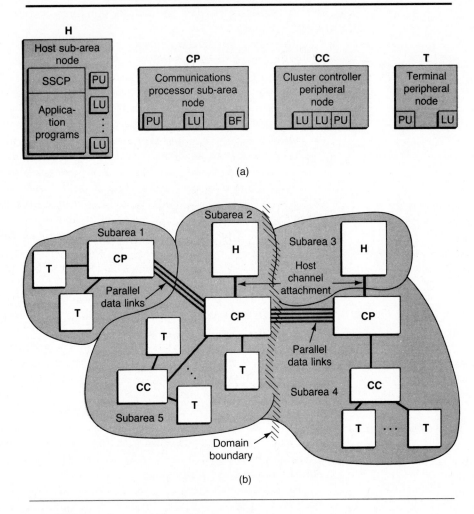

(a)

(b)

or cluster controllers). Intelligent cluster controllers are the second type.[36]
Subarea nodes have extensive processing capabilities and include communications processors, which handle communications functions, and hosts,
which control the network. Subarea nodes handle complete network

---

[36] In revisions of SNA announced in 1982, a subcategory of terminal controller nodes with
more capability, including a limited number of network control functions, was added. Support for the simplest type of terminal nodes is being phased out, with no new nodes of this
class to be added.

addresses but peripheral nodes recognize only local addresses defined within subareas. Each subarea contains one subarea node plus any peripheral nodes it supports. Any functions that peripheral nodes cannot handle, such as conversion between subarea and network addresses, are handled by a boundary function (BF), in the closest subarea node.

system services control point

logical units

Three additional acronyms are important. A *system services control point* (SSCP), in each host subarea node exercises control over activities in its domain, with cross-domain links to other domains. *Logical units* (LUs), are user ports into the network and perform functions to help users[37] communicate; it comprises the upper layers of the architecture.[38] Several LUs may exist in one node, one for each session active, but a node with no user may have no LUs. *Physical units* (PUs) interface control functions in physical equipment to the network; each node contains one PU. PUs are not normally noticed by users, but they control device functions necessary for proper operation. All information flow is to or from SSCPs, LUs, and PUs, which constitute *network addressable units* (NAUs).

physical units

network addressable units

A major characteristic of SNA is its hierarchical structure with multiple classes of nodes (multiple classes of sessions, and so forth) and recognition of varying equipment capabilities. Network operation is tightly controlled by SSCPs. This is a major difference compared with the peer coupled approach of DNA and other architectures. SNA is also almost paranoid in attempting to correct problems as soon as they occur rather than passing the job on to higher layers, as is done in some other architectures.[39]

Communications links may be any of several types. Some devices, especially communications processors at host locations, attach to hosts via channels and use channel access protocols. Other links are serial communications links (usually telephone channels). Parallel combinations of links between locations are also allowed, with special functions in SNA making them appear to users to be single links with higher capacity than the individual links.

As with the other architectures we have discussed, SNA is a layered architecture. SNA layers are illustrated in Fig. 3.22. Definitions of the layers are as follows:

- *Physical control layer:* Same definition as that in OSI Reference Model physical layer. A variety of standard physical layer interfaces may be used.

---

[37] Users can be either human users or application programs.

[38] The precise definitions of SSCPs and PUs are similar. All three NAUs are defined in SNA by the logical functions in the upper layers they provide for their types of communication.

[39] There are valid arguments for both approaches. Passing problems on to higher layers may reduce processing by allowing functions to be performed only once. On the other hand, problems tend to be easier to diagnose before their effects have propagated; this propagation can occasionally cause problems to manifest themselves in ways that are extremely hard to diagnose.

Figure 3.22   SNA Architecture Layers.

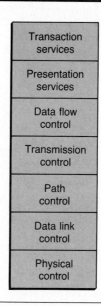

- *Data link control layer:* Same definition as that in OSI Reference Model. SDLC is used over serial data links with S/370 channel protocol over channel attachments.[40]

- *Path control layer:* Creates logical channels between endpoints, or NAUs. Its primary functions are routing and flow control. All communication in SNA is via virtual circuits, which path control helps establish, operate, and clear. It resolves addressing for subareas and elements, provides for transmission priorities and various classes of service such as fast response, secure routes, or reliable connections, and segments and blocks messages. It may make parallel transmission links appear to higher layers to be one higher speed link.

- *Transmission control layer:* Responsible for establishing, maintaining, and terminating SNA sessions, sequencing of data messages, and session level flow control. It encapsulates messages with appropriate headers and routes data to appropriate points within NAUs.

- *Data flow control layer:* Provides session-related services visible and of interest to user processes and terminals. This includes determining send/

---

[40]SDLC is discussed in Chapter 7. S/370 channel protocol is discussed in a wide variety of computer science texts, as for example in [SIEW82].

receive mode, chaining transmissions to facilitate error recovery, grouping related messages together, and specifying response options.

- *Presentation services layer:* Handles presentation or formatting of data. This includes translations between formats, data compression, formatting, and control character translation for different terminal types. Transaction support functions are included.

- *Transaction services layer:* Provides network management services used directly by users, including configuration services allowing operators to start up or reconfigure networks, operator services allowing collection and display of network statistics, or communication from users and processes to network operators; also provides a user interface to transmission control and maintenance and management services.

As was mentioned earlier, the upper layers define logical units (LUs). The layers defining an LU are transmission control, data flow control, presentation services, and transaction services. They provide the user's port into the network.

Although Fig. 3.22 represents SNA layers in a manner similar to the OSI layers in Fig. 3.10, the layering in SNA is not as strict. The data flow control layer works more nearly in parallel with the transmission control layer than above it, the former providing end user–oriented aspects of session control and the latter transmission-oriented aspects. The top two layers (transaction services and presentation services) were until recently considered to be a single layer known as the function management (FM) layer.

SNA does not require a different header at each layer of the hierarchy. Figure 3.23 shows SNA frame construction and reduction. User data, or control information originating in an application process, is converted into a request response unit (RU) by the presentation services and transaction services layer.[41] "Request" refers to a message containing user data or SNA commands, regardless of whether it asks the message recipient to do anything or not. By convention, a "response" contains an acknowledgment of a request.

Rather than creating a separate header, data flow control passes parameters to transmission control[42] to be included in a transmission control header (TCH) the latter constructs.[43] Path control adds a path control header (PCH) and passes the resulting data unit to data link control, which adds a

---

[41] A function management data (FMD) header may be included; however, if it is used, it will appear only occasionally and is considered part of the RU. Hence it is not shown.

[42] On occasion, some still higher layers may pass parameters down in this way.

[43] In an attempt to simplify SNA's horrendous terminology, we have identified headers according to the layer adding them. SNA terminology uses terms bearing less relationship to these layers. Request/response header (RH) is used for what we call the TCH; transmission header (TH) for the PCH; and link header (LH) and link trailer (LT), respectively, for the DLH and DLT.

Figure 3.23   SNA Frame Construction and Reduction.

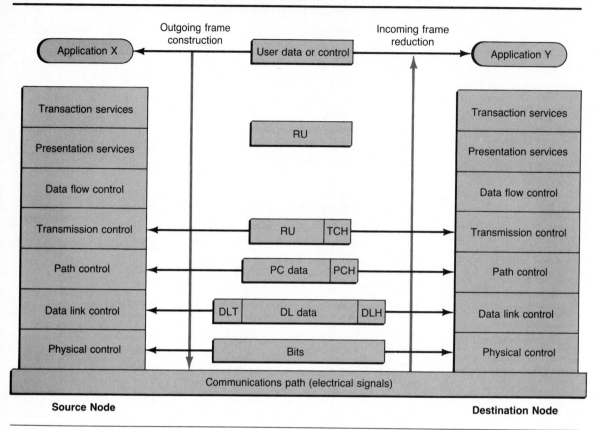

data link header (DLH) and data link trailer (DLT) and passes the resulting frame to the physical control layer for transmission as a string of bits.

3.9   Other Important Standards

Two other networking architecture standards are prominent in our discussions: standards for LANs and standards for ISDNs.

3.9.1   Local Area Network Protocol Architectures

IEEE 802 standards

An architecture is being developed by IEEE (Institute of Electrical and Electronics Engineers) for local area networks (LANs). LAN standards are being developed by IEEE 802 LAN Standards Committees and are called IEEE 802

Figure 3.24   Comparison of IEEE 802 Architecture Layers with
Those for OSI Reference Model.

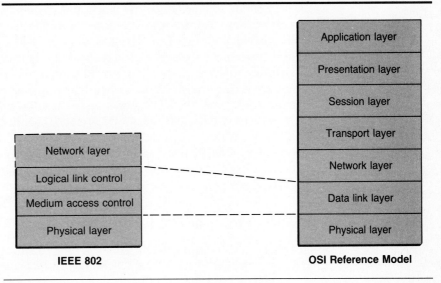

IEEE 802

Network layer
Logical link control
Medium access control
Physical layer

OSI Reference Model

Application layer
Presentation layer
Session layer
Transport layer
Network layer
Data link layer
Physical layer

standards; they are similar to OSI Reference Model standards,[44] but less comprehensive. Before 1988, IEEE 802 standards included only the equivalent of the first two layers of the OSI Reference Model, but the 802 committee is working on a partial third layer. The second layer (Data Link) in the OSI architecture is split into two sublayers in the IEEE 802 architecture. Figure 3.24 compares IEEE 802 layers with OSI layers. The dotted boundary around the network layer indicates that it is not yet available.

medium access control
logical link control

The physical layer and the *medium access control* (*MAC*) sublayer depend on the LAN medium access technique implemented, but the *logical link control* (*LLC*) sublayer is the same for all access techniques for which IEEE 802 standards have been approved. Three architectures have been standardized so far: *carrier sense multiple access/collision detection* (*CSMA/CD*) (802.3), *token bus* (802.4,) and *token ring* (802.5).[45]

Frame construction and reduction are analogous to those for the other architectures. Both the MAC and LLC sublayers add headers and the LLC sublayer adds a trailer[46] at the source, with these removed by corresponding

---

[44] The ISO is accepting the IEEE 802 standards under the number ISO 8802.

[45] These medium access techniques are discussed in Chapter 5 and the physical layer and data link layer protocols are discussed in Chapters 4 and 7, respectively.

[46] The LLC sublayer implements a version of HDLC, one of the major data link layer protocols discussed in Chapter 7.

sublayers at the destination. The physical layer is, as usual, responsible for transmission of frames, as strings of bits, from source to destination.

Routing is unnecessary in single LANs, since frames reach all possible destinations. High transmission rates, typically 1 to 20 Mbps, also minimize the necessity for network layer flow control. As long as no data are transmitted between a LAN and other networks, the normal functions of a network layer are not necessary. Thus the IEEE 802 protocols provide a reasonably complete data transmission service that can be directly accessed by application programs. Many LANs operate in essentially this manner.

Ethernets (IEEE 802 CSMA/CD LANs) are included in the DNA architecture since this is DEC's preferred LAN architecture. Similarly, token ring IEEE 802 LANs are included in SNA, as this is IBM's preferred architecture. Any of these LANs, as well as many others, can be incorporated into the DoD architecture via internetworking techniques discussed in Chapter 10.

3.9.2   MAP and TOP

Manufacturing Automation Protocol

Technical and Office Products System

Two architectures developed by users, rather than vendors, use IEEE 802 protocols for their lower layers. The *Manufacturing Automation Protocol* (*MAP*) architecture developed by General Motors, uses IEEE 802.4 token bus protocols for lower layers and OSI Reference Model protocols for upper layers; the *Technical and Office Products System* (*TOP*) architecture developed by Boeing Computer Services uses IEEE 802.3 CSMA/CD or 802.5 token ring[47] protocols at lower layers and the same OSI protocols at upper layers. The companies involved have worked together to ensure compatibility at upper layers by using consistent standards. Both architectures were developed to guarantee that products from vendors meet uniform interface requirements, and each has been adopted by a significant number of companies aside from the one originally developing it. Current development efforts are largely handled by user groups.

3.9.3   Integrated Services Digital Network Architecture

integrated services digital networks

The architecture being developed for integrated services digital networks (ISDNs) is currently the least complete of the architectures we discuss. The ISDN approach is to provide user support through the seven layers of the OSI Reference Model. In other words, ISDN networks are to basically adopt the OSI model, but with new layer definitions. Major functions to be performed by the layers are indicated in Fig. 3.25.

ISDN services are classified in two categories: *bearer services,* which provide support for the lower three layers of the standard, and *teleservices* (such as telephone, Teletex, Videotex, and message handling), which are based on the upper four layers while making use of the underlying bearer services.

---

[47]Added in 1987.

Figure 3.25   Assignment of Major Functions to Layers in ISDN Architecture (adapted from [BLAC87a]). © 1987, p. 274. Reprinted by permission of Prentice-Hall, Inc., Englewood Cliffs, NJ.

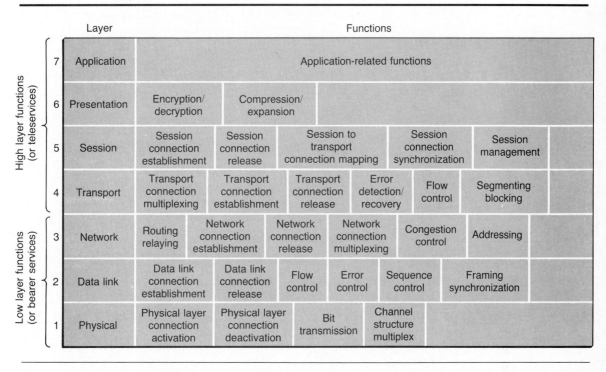

These services are also referred to as low-layer and high-layer services (or functions), respectively.

ISDN differs significantly from other architectures we study in one important way. Circuit switching and packet switching technologies are receiving roughly equal emphasis in ISDN work in contrast with heavier emphasis on packet switching in the other architectures.

3.10   Comparison of Architectures

Figure 3.26 compares layers of the primary architectures discussed in this text. The figure should be interpreted with caution, since the architectures differ significantly enough that it is impossible to give a completely valid comparison in this manner. Two layers occupying the same relative positions in different columns do not imply that functions performed by these layers are close to the same even if the layers have the same, or nearly the same, name.

Figure 3.26   Comparison of Architectures.

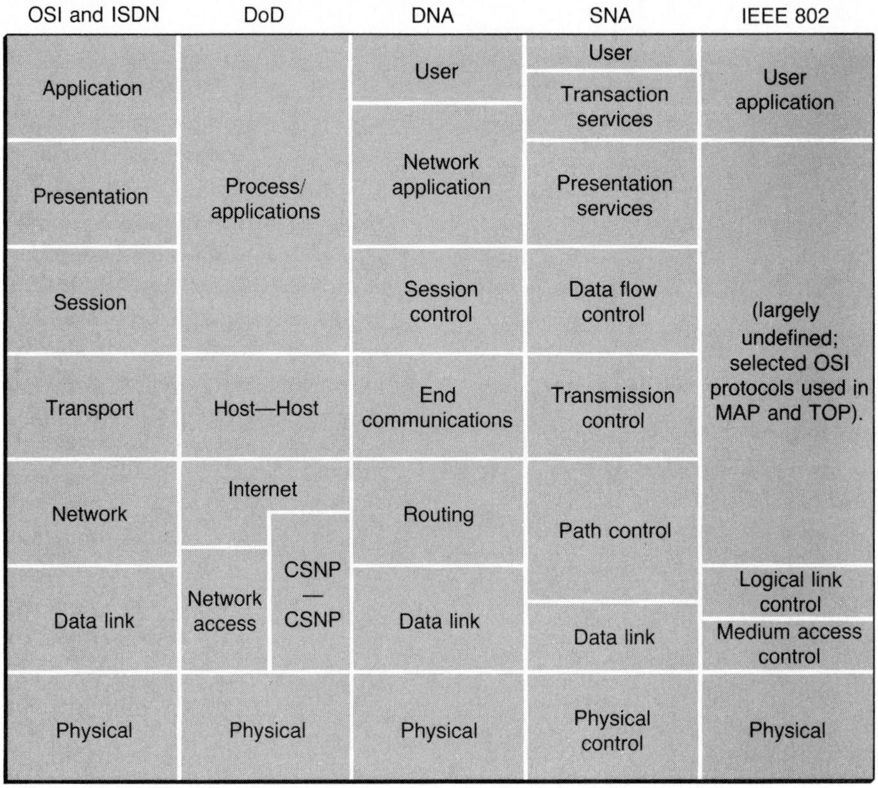

OSI and ISDN	DoD	DNA	SNA	IEEE 802
Application	Process/ applications	User	User	User application
Presentation		Network application	Transaction services	
			Presentation services	(largely undefined; selected OSI protocols used in MAP and TOP).
Session		Session control	Data flow control	
Transport	Host—Host	End communications	Transmission control	
Network	Internet	Routing	Path control	
	Network access / CSNP — CSNP			Logical link control
Data link	Network access	Data link	Data link	Medium access control
Physical	Physical	Physical	Physical control	Physical

The network access and CSNP-CSNP layers of the DoD model are shown essentially in parallel with the data link and part of the network layer levels of the OSI model; the DoD network access layer operates between host computers and attached CSNPs, while the CSNP-CSNP protocols operate between CSNPs. The CSNP-CSNP protocols include such aspects of the network layer as routing. A physical layer has been included for all architectures, even if it is not mentioned in architectural documents.

A sublayer labeled "user" has been added at the top of the SNA hierarchy. Otherwise SNA would be less complete than the first three architectures, which include user applications. The path control layer of SNA overlaps part of the OSI data link layer, as well as all of the OSI network layer. In general, agreement between SNA layers and OSI layers is less precise than for other architectures, but it is difficult to show this.

The portion of the IEEE 802 architecture labeled "largely undefined" is, as we indicated above, often unnecessary when LANs are used without internetworking with other networks. It is filled in if LANs are used in DNA or SNA networks, and either the MAP or the TOP architecture fills in these portions with OSI protocols. Structures of MAP and TOP are indicated in the right column. At the OSI network and transport layer levels specific subsets of the OSI protocols, described in Chapters 9 and 11 respectively, are indicated.

There are significant differences in the general orientation of the architectures, in flexibility for interfacing with different layers, in bypassing layers, in approaches to network control, and so forth. These differences have been partially addressed in the preceding subsections.

3.11 Summary

This chapter has presented an introduction to basic principles of telecommunications networks. Major types of switching techniques and important classes of networks were discussed. Standard problems that must be addressed by a network designer were used to motivate protocol architectures studied in the rest of the text. These architectures were then introduced and the basic nature of each pointed out. Much of the remainder of the book is devoted to further development of these architectures.

Problems

3.1 Briefly describe at least six different telecommunication networks and give their structures and applications. (See hint for Problems 1.1 and 1.2.)

3.2 Discuss the relative advantages and disadvantages of common channel signaling for circuit-switched networks. Why do you think telephone networks are currently converting most of their equipment to use common channel signalling?

3.3 Figure P3.3 illustrates the first two messages exchanged in the course of a telephone call from user A to user B. Extend this diagram to include the following messages (with the first two already illustrated: A Off Hook; Dial Tone to A; Dialed Digits from A; Call Request signals; Call Accept signal; Conversation with B answering, A responding, and one more response from B; B On Hook, Disconnect signals (originating from location of first user to hang up); and finally A on hook. (This is an arbitrary choice of the order of actions during disconnect; for example, either user could equally well hang up first.)

3.4 Compare virtual circuit and datagram service with respect to the following:
a. Ability to survive link or node failures
b. Ability to deal with congestion in the network
c. Applicability for use with the following types of traffic: (i) voice, (ii) interactive data traffic, and (iii) file transfers.

Figure PS3.3   Figure for Problem 3.3.

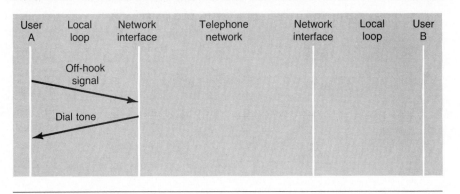

3.5   Refer to the timing diagram in Fig. 3.5 illustrating a situation where packet switching gives less delay than either circuit or message switching.

    a.   What are the major features of packet switching that allow it to achieve this advantage in response time?

    b.   State reasonable conditions under which circuit switching might give minimum delay.

    c.   State reasonable conditions under which message switching might give minimum delay.

3.6   **a.**   Compute the total delays to transfer a message 10,000 (eight bit) characters long across the three-hop communications path in Fig. 3.1 under the following timing assumptions. (See Fig. 3.5 for an illustration of delays.) Assume that all links operate at 4800 bps, with a 20 msec delay per link for propagation delays plus reaction time at the receiver. (Timings given are typical, though they are so highly variable that results of this problem should be treated with caution.) (i) For circuit switching assume 3 sec per link connection delay (time hunting for an outgoing trunk or dialing and associated delays) and 100 msec duration of call request or accept signal (times to put signals on the communications lines, for example, number of bits divided by line rate in bps). (ii) For message switching, assume 50 bytes per message of header (routing information, identification, sequencing information, and so forth) and (arbitrarily) 300 msec processing plus queueing time at each node. (iii) For packet switching, assume the message is divided into 20 equal length packets with 50 bytes of header information per packet, 300 msec processing plus queueing time at each node and 20 msec between packets.

    b.   Discuss your results and indicate reasonable conditions under which the relative rankings of the delays might change.

3.7   A "reliable" datagram service, which adds such functions as error control and resequencing to standard datagram service, has been proposed for some applications. Discuss how standard datagram service might be altered to provide these extra features. What applications for such services can you visualize?

3.8   Assume a 1000-character message is to be divided into packets and transmitted over a five-hop communications path, with a bit rate of 4800 bps on each hop. Each packet contains six characters of overhead, with eight bits per character. Compute and plot total time (T) for transmitting the message across the network for maximum packet sizes, including overhead, ranging from 10 to 1006 characters. Compare the packet size yielding minimum total delay with values from Eq. 3.6 and explain any discrepancies.

3.9   Buffer allocation in the main memory of a computer system can be analyzed by a model similar to our model of pipelining in packet networks. Assume a fixed buffer size, B bytes, is used as a basic memory allocation, including n_o bytes of overhead (pointers to next buffer and so forth). The total number of memory bytes to store a packet of length P bytes is thus $M = B[P/(B - n_o)]$.

a.   Assuming P is a random variable, find an approximate expression for $E[M]$ in terms of B, n_o and $E[P]$.

b.   Find the optimal value of B in order to minimize the average amount of main memory required for buffers.

3.10   Prove that for any fixed set of node locations the minimum possible total length of a tree interconnecting all nodes is less than the minimum possible length of a ring interconnecting all nodes.

3.11   Assuming the 15 percent per month DARPA Internet growth rate mentioned in Subsection 3.3.5 can be kept up long enough, compute the length of time it would take before the Internet served one network per man, woman, or child on earth. For the purposes of computation, you may assume that there were 2000 networks connected to the Internet by year end 1987 and the world population is constant at 5 billion.

3.12   Assume that the $(N + 1)$st layer in a network based on the OSI architecture desires a particular type of service from the Nth layer, with communication across the network necessary to obtain this service.

a.   What two types of protocols will local layer N use to provide the requested service?

b.   How does local layer N convey appropriate information to the entity it communicates with for each of the two types of protocols?

3.13   Sketch the sequence of primitives and frames exchanged to set up each of the OSI connections described in Subsection 3.6.4. Your sketch should be in the form of Fig. 3.13, but with labeling of primitives and frames. Label primitives in the general form d-Connect-Request, for a connection request from the data link layer to the physical layer. (Use labels a, pr, s, t, n, d, and ph for primitives sent by the application, presentation, session, transport, network, data link, and physical layers, respectively.) Only general labeling of frames is required, since we have not studied standard frames yet.

3.14   Sketch the sequence of primitives and frames exchanged, using the conventions described in Problem 3.13, for the data transfer phase of an OSI connection in which application X at system A sends one frame to application Y at system B and then receives a response frame from application Y.

3.15   **a.**   For a typical OSI communication scenario, discuss the disconnect phase after data transfer from application X at location A to application Y at location B has been completed. Include three possibilities for initiation of the disconnect: (i) By application X, (ii) By application Y, and (iii) By the network.

b.   Sketch typical sequences of primitives and frames exchanged, using the conventions described in Problem 3.13 for each of the three cases considered in (a).

3.16   Compare the OSI Reference Model layered structure with the criteria in Table 3.4. Indicate which functions of those we have listed for various layers could go equally well into other layers.

3.17   Compare the DoD Reference Model layered structure with the criteria in Table 3.4. Indicate which functions of those we have listed for various layers could go equally well into other layers.

3.18   Compare the DNA layered structure with the criteria in Table 3.4. Indicate which functions of those we have listed for various layers could go equally well into other layers.

3.19   Compare the SNA layered structure with the criteria in Table 3.4. Indicate which functions of those we have listed for various layers could go equally well into other layers.

3.20   We have indicated that protection against errors may be needed at a variety of levels throughout a network architecture. For each of the layers of the OSI architecture, from lowest to highest, list some types of errors for which this layer needs to provide protection since they are not adequately handled by lower layers.

3.21   In cases where data encryption is used for protection against unauthorized access, the best layer in a networking architecture at which to do encryption and decryption is not always clear. Discuss the relative advantages and disadvantages of doing encryption and decryption at the physical, data link, transport, presentation, and application layers of the OSI Reference Model architecture.

3.22   **Design Problem.** Design your own set of layers for a networking architecture, taking advantage of the principles we have discussed and including as many as possible of the best features of each networking architecture discussed. Present arguments for the merits of your networking architecture.

4 Computer Equipment–Communications Equipment Interface Design

4.1 Introduction

The boundary between communications equipment and data processing equipment is an important interface in a telecommunication network. Figure 4.1 illustrates the positions of this and other major interfaces in a typical network.

The term "interface" often describes a boundary between two pieces of equipment, across which messages flow. Many interfaces are used in complex networks. In Subsection 3.5.1, we stated that in analog or digital communications, the state of a modem or adapter must be controlled by equipment at the node to which it is attached; an electrical or *physical interface* performs this function, with *protocols* to accomplish it. The boundary between protocol layers in an architecture such as the OSI model is another type of interface (see Subsection 3.6.3). Interfaces discussed here include the physical layer of the OSI architecture and other architectures plus interfaces interconnecting the physical layer with computer equipment and communications equipment with communications channels.

design problems addressed

Major *design problems addressed* in design of interfaces include resolving incompatibilities of equipment, and coordination of sender and receiver, where sender and receiver refer to equipment on the two sides of the interface. Other design problems such as maximizing reliability and freedom from errors, optimizing performance, minimizing costs, and network management, are also considered (for example, an interface that is overly expensive or causes performance problems is not satisfactory), but are less directly influential in determining the design approach.

physical layer interface

As Fig. 4.1 indicates, when data are received from a communications link, serial data mixed with control information come into a *physical layer inter-*

153

Figure 4.1 Interface Between Computer Equipment and Communications Equipment.

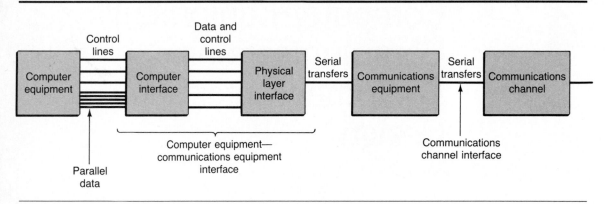

face from communications equipment, and the same type of data are sent to communications equipment when data are transmitted. Received data correspond to strings of bits presented to the interface by a modem if a modem is used, or by a line adapter in purely digital transmission. During transmission, strings of bits are sent to communications equipment. Information sent, in either direction, between the physical layer and computer interface flows over several wires, some for control and others for data; data (transmitted or received) are usually in serial form, however. Data transfers must be coordinated and use signals decodable by the receiving end.

computer equipment interface

The *computer equipment interface* translates information from the physical layer interface to or from the format used by computer equipment, where data transfers are in parallel (character or word) form; it also passes control information to the computer. In addition, there is a change in signal formats at the interface, since voltages, currents, and signal representations used in communications equipment are not compatible with those used in computer equipment.

communications channel interface

The *communications channel interface,* occurring between the communications equipment (or modem) and the communications channel, is shown by an arrow at the right side of Fig. 4.1. We treat this interface briefly; it determines sequences of signals flowing back and forth between pairs of modems or other communications equipment to allow communication to take place.

handshaking sequence

Data and control information flow over each interface. Careful design is necessary if all signals are to be understood by both sender and receiver. *Handshaking sequences*—sequences of messages sent to both convey information and verify correct receipt—are common.

After giving an overview of physical layer interfaces in the next section, we discuss important examples: EIA-232D/RS-232C/V.24, X.21, and I.430/I.431.

This is followed by a discussion of computer interfaces (the second block from the left). Synchronous/asynchronous adapters, which have been standardized by integrated circuit manufacturers, are discussed and typical interfaces presented. We then briefly discuss the final interface in Fig. 4.1, the communications channel interface. The effects of parameters on network performance are illustrated. Details of the EIA-232D/RS-232C/V.24 interface are given in the appendix to this chapter.

4.2   Overview of Physical Layer Protocols

DCEs and DTEs

physical layer

Physical layer protocols resolve equipment incompatibilities. They are concerned with physically interconnecting communicating devices and transmission of bits between devices. Normally the devices interconnected are modems or communications adapters (called *DCEs,* for *D*ata *C*ircuit-terminating *E*quipment, in much of the literature) and computer equipment or terminals (called *DTEs,* for *D*ata *T*erminal *E*quipment).[1]

The *physical layer* is the lowest layer considered to be part of a communications network architecture. The bit stream generated by the physical layer at the transmitting end of a communications channel is designed to be interpreted by the physical layer at the receiving end, which regenerates a corresponding (hopefully identical) bit stream and presents it to equipment at that end via the computer equipment interface. The meaning of the bit stream is not of concern to the physical layer. The meaning, and other factors, are the responsibility of higher layers of the architecture.

Standards manuals present standards in complete—and boring—detail. In this book we give far fewer details than the manuals do but provide all the essentials needed to obtain their general flavor.

Physical layer protocols describe four types of characteristics [BERT82a]: mechanical, electrical, functional, and procedural. Some persons advocate considering mechanical and electrical characteristics to be outside the scope of these standards, so they apply equally well to media such as fiber optics, but we follow this subdivision, with emphasis on procedural characteristics.

Mechanical characteristics include physical dimensions of plugs or connectors, assignment of circuits to pins, connector latching and mounting arrangements, and so forth. Such characteristics make it physically possible to interconnect equipment.

Electrical characteristics include voltages or current levels, timings of signals, and their interpretations as zeros or ones. They ensure that electrical

---

[1] These are typical of the terms coming out of standards committees. We avoid using such terms except when (as in the case of DCE and DTE) they are so widely used in the literature that not knowing them would cause problems for someone working in the field.

levels are compatible. Timings include pulse rise times and durations and largely determine transmission speeds.

Functional specifications assign meanings to circuits (or pins). It is common to classify circuits into categories of data, control, timing, and grounds. Data, control, and timing signals are represented by voltage pulses corresponding to ones and zeros.[2]

Finally, *procedural specifications* indicate sequences of control and data messages to set up, use, and deactivate physical level connections. Such sequences must be agreed upon by both ends if data transfers are to be intelligible. A major aspect of interface design is developing sequences capable of handling all situations that may occur, including error recovery sequences. Procedural specifications are the specifications of most interest to us, as they involve protocols similar to those at other architecture layers.

4.3   EIA-232D/RS-232C/V.24

By far the most commonly used physical layer interface in the United States is the RS-232C interface standardized by the EIA.[3] The CCITT[4] V.24 interface (with V.28 electrical interface) is almost identical and is a corresponding standard in much of the rest of the world. A new version of RS-232C, known as EIA-232D, was adopted in January 1987, but it is far less widely implemented than RS-232C. Although EIA-232D, RS-232C, and V.24/V.28 are based on obsolescent technology (the first version was adopted in 1960), the installed base of equipment using them guarantees that they will continue to be used and that equipment will be manufactured to them, for the foreseeable future.

In the following section we emphasize RS-232C rather than V.24 or EIA-232D, but the interfaces are virtually the same. The aspects we discuss here are identical for all three interfaces. Minor differences, such as nomenclature for pins, are covered in the appendix to this chapter.

4.3.1   Overview of RS-232C Interface

The RS-232C specifications list sequences of control and data line activations to *set up, maintain,* and *deactivate* physical level interconnections and accomplish *data transfer.* These involve handshaking sequences to initiate control actions and to verify that they have been accomplished. Allowable sequences and timings have significant impact on performance, with timings often determined more by equipment than by the standard.

---

[2] Alternative representations are, of course, needed with media such as fiber optics.

[3] The EIA is the Electronic Industries Association.

[4] The CCITT is the International Telegraph and Telephone Consultative Committee.

A total of 25 pins are available. Multiconductor cables with appropriate numbers of conductors interconnect devices. In almost all situations fewer than 25 pins are used, so fewer than 25 conductors are needed.

One point about the electrical specifications deserves mention. The signaling mode is known as unbalanced mode with each signal transmitted on a single conductor or circuit and all circuits sharing a common ground. Balanced mode transmission, with a pair of wires used for each circuit and voltage differences between wires conveying information, is common in newer standards. This largely eliminates problems due to coupling of voltages from other sources into wires, since essentially the same voltage is induced into each wire in a pair and voltage differences preserved; hence possible transmission speeds are increased.

4.3.2   Major Signals

All 25 RS-232C pins, or signals, are defined in the appendix to this chapter. At this point we define only signals used in examples of timing and state diagrams. Table 4.1 defines the signals treated; they are the most important ones during normal operation.

4.3.3   Typical States and Timing Diagrams

Defining allowable states and state transitions in a manner avoiding ambiguities is the most complex part of designing an interface. Figure 4.2 illustrates state definitions and allowable state transitions for RS-232C

Table 4.1   Major Signals for RS-232C Interface.

Signal Ground. All signal levels on other circuits are measured relative to this point.

Transmitted Data (TD). The sequence of pulses representing data sent out from the DTE is placed on this circuit at appropriate times dictated by procedural specifications in the protocol.

Received Data (RD). The sequence of pulses representing data received by the DCE is placed on this circuit.

Request to Send (RTS). The data processing equipment (DTE) activates this circuit when it is ready to send data.

Clear to Send (CTS). The communications equipment (DCE) activates this circuit when it is ready to accept data from the DTE.

Data Set Ready (DSR). The modem (data set) activates this circuit when it is powered up and ready for use.

Data Terminal Ready (DTR). The DTE (terminal or computer) activates this circuit to let the modem know it is ready to send and receive data.

Ring Indicator (R). The modem (DCE) activates this circuit to tell the DTE that a ringing signal due to an incoming call is being received.

Received Line Signal Detector (RLSD). The modem (data set) activates this circuit when it is detecting a carrier signal from another modem.

Figure 4.2(a)   Allowable state Transitions for RS-232C Interchange Circuit (Control Pin) Sequences for Half Duplex Operation with Switched Telephone Service. X represents a "don't care" bit combination, e.g. (7,X) means state (7,0) or (7,1) or ... (7,7). Disconnect sequences may originate from any states in dashed rectangles.

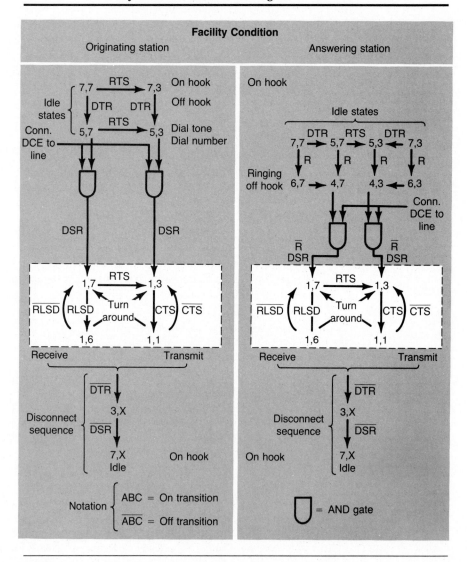

Figure 4.2(b)   State Definitions for RS-232C Control Interchange Circuit
                     (Control Pin) Sequences for Half Duplex Operation on
                     Switched Telephone Service. On a circuit 0 denotes ON, 1
                     denotes OFF.

First Digit of State Encoding				Second Digit of State Encoding			
Octal Code	DSR, Data Set Ready	DTR, Data Terminal Ready	R, Ring Indicator	Octal Code	RTS, Request to Send	CTS, Clear to Send	RLSD, Rec. Line Signal Detector
0	0	0	0	0	0	0	0
1	0	0	1	1	0	0	1
2	0	1	0	2	0	1	0
3	0	1	1	3	0	1	1
4	1	0	0	4	1	0	0
5	1	0	1	5	1	0	1
6	1	1	0	6	1	1	0
7	1	1	1	7	1	1	1

implementation for half duplex operation with switched telephone service;
the state in Fig. 4.2(a) is defined by the status of control lines, as is indicated
in Fig. 4.2(b). The circuits used are indicated in Fig. 4.3. A variety of similar
state transition diagrams are defined, each applicable under a different set of
operating conditions. Standards documents give details.

Figure 4.4 is a *timing diagram* showing typical control circuit activations
chosen from those in Fig. 4.2(a); time periods when data is transmitted over
TD and received over RD are also shown. Relative lengths of time intervals
are not necessarily accurate; they are highly application and equipment
dependent. Figure 4.2(a) shows optional sequences, but reasonable choices
for options have been made in Fig. 4.4. State transitions are labeled in Fig. 4.4
and tabulated, with control signal changes causing transitions as shown in
Table 4.2. These state transitions also occur in Fig. 4.2(a).

At the initial instant both originating station and answering station are
idle, with all control circuits off. The sequence begins with the computer
equipment, DTE, at the originating station signaling the interface that it wants
to transmit by activating DTR (originating station transition 1). The following
time interval, during which the receiver is dialed, is not covered in the diagram (or the standard); either manual or automatic dialing would be possible.
The next step at the originating station is for its DCE to activate DSR (originating station transition 2).

The first event to occur at the answering station is reception of a ringing
signal (answering station transition 1), after a delay from circuit switching.
This causes the DTE at the answering station to activate DTR (answering

Figure 4.3   Circuits Used for RS-232C Example in Figs. 4.2 and 4.4.

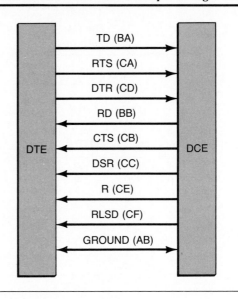

station transition 2) and connect its DCE to the line. (This corresponds to answering the phone and causes ringing to stop.) The DCE simultaneously activates its DSR circuit (answering station transition 3). At this point both stations are active and ready to send data to and receive data from each other.

The diagram in Fig. 4.2 allows either station to be first to transmit, but Fig. 4.4 is drawn for a case where the answering station transmits first. This corresponds to a called party saying "hello" in a telephone conversation and avoids having the originating station guess when it is safe to transmit.

Transmission of data is initiated by the computer equipment, DTE, at a transmitting station activating its RTS circuit (answering station transition 4, originating station transition 5). After a delay characteristic of the modem, the corresponding DCE activates CTS (answering station transition 5, originating station transition 6).[5] When it senses CTS, the DTE places a string of pulses representing data to be transmitted (indicated by a crosshatched rectangle) on TD.[6] At the end of transmission, RTS and CTS are turned off as the state in Fig. 4.2(a) moves diagonally across the transmit-receive rectangle in the figure (answering station transition 6, originating station transition 7). A

---

[5] The RTS-CTS delay is one of the more important parameters affecting data communications efficiency; we discuss this later.

[6] For synchronous transmission, clock pulses indicating times of occurrence of data pulses would be simultaneously placed on a transmitter signal element timing circuit.

Table 4.2 State Transitions for Figs. 4.2 and 4.4.

	Originating Station		
Transition Number	**Original State**	**Signal Changes**	**New State**
1	(7,7)	DTR	(5,7)
2	(5,7)	DSR	(1,7)
3	(1,7)	RLSD	(1,6)
		Data received	
4	(1,6)	$\overline{\text{RLSD}}$	(1,7)
5	(1,7)	RTS	(1,3)
6	(1,3)	CTS	(1,1)
		Data transmitted	
7	(1,1)	$\overline{\text{RTS}}$, $\overline{\text{CTS}}$	(1,7)
8	(1,7)	$\overline{\text{DTR}}$	(3,7)
9	(3,7)	$\overline{\text{DSR}}$	(7,7)

	Answering Station		
Transition Number	**Original State**	**Signal Changes**	**New State**
1	(7,7)	R	(6,7)
2	(6,7)	DTR	(4,7)
3	(4,7)	DSR, $\overline{\text{R}}$	(1,7)
4	(1,7)	RTS	(1,3)
5	(1,3)	CTS	(1,1)
		Data transmitted	
6	(1,1)	$\overline{\text{RTS}}$, $\overline{\text{CTS}}$	(1,7)
7	(1,7)	RLSD	(1,6)
		Data received	
8	(1,6)	$\overline{\text{RLSD}}$	(1,7)
9	(1,7)	$\overline{\text{DTR}}$	(3,7)
10	(3,7)	$\overline{\text{DSR}}$	(7,7)

station can go either into transmit mode again or into receive mode. Figure 4.4 is drawn for the latter case.

After a delay due to propagation of signals between stations, transmitted data are received at the other station. This activates the RLSD circuit at this station (originating station transition 3, answering station transition 7) and alerts it to look for data on RD. After reception is complete, RLSD becomes inactive (originating station transition 4, answering station transition 8) and no more data come in on RD.

Data transmissions between the two stations could continue indefinitely from this point, as long as the sequence of circuit activations is allowed by

Figure 4.4 Typical Timing Diagram Corresponding to Control Sequences in Fig. 4.2. High levels correspond to 1 or OFF, low levels to 0 or ON.

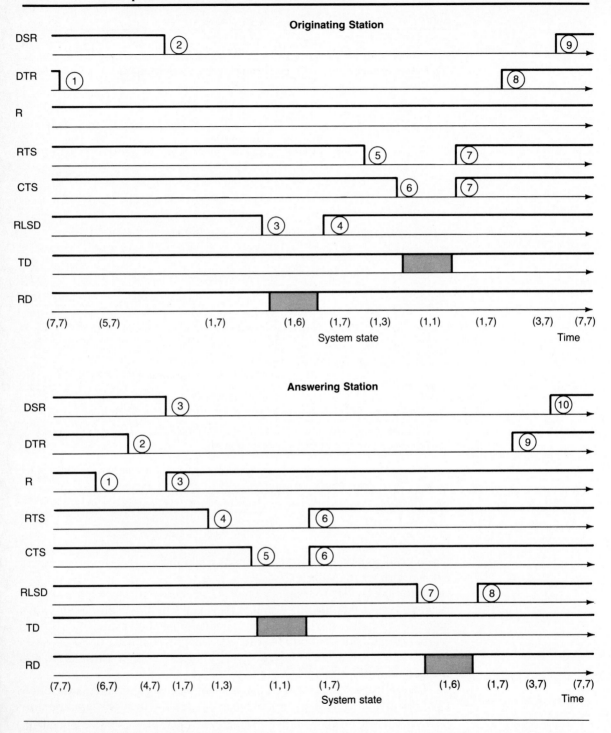

Fig. 4.2(a). The sequence in Fig. 4.4, however, terminates after each station transmits once, with both stations going through disconnect sequences; each deactivates its DTR (originating station transition 8, answering station transition 9) and DSR circuits (originating station transition 9, answering station transition 10) to implement disconnection.

4.3.4 Null Modems

Direct interconnection of two DTEs, each with a RS-232C port, is common for local connections that do not require telecommmunication links and modems. This is the normal way to connect a local terminal to a computer. Since each DTE expects to talk to a DCE (that is, a modem), each DTE must "fool" the other by acting like a DCE. An appropriately wired RS-232C cable can be used to make this possible; such a cable is called a null modem.

For a null modem, the Transmitted Data (TD) circuit on one DTE should be routed to Received Data (RD) on the other, and vice versa, and essential modem control signals provided. A simple way of providing control signals is to connect the Request to Send (RTS) circuit on each DTE to Clear to Send (CTS) on the same DTE (providing essentially zero RTS-CTS delay) and to interconnect the Data Terminal Ready (DTR), Receive Line Signal Detector (RLSD), and Data Set Ready (DSR) circuits on the same DTE. This causes the DTE to provide its own full duplex modem controls with zero delays.

A variety of alternative configurations for null modems are available (see Problem 4.4 for one example). These are covered in numerous manuals. More sophisticated modem eliminators are commercially available.

4.3.5 Limitations of RS-232C

RS-232C originated during early development of data networks and was designed to use technologies available at that time, discrete electrical components rather than integrated circuits, and for low speeds. Its primary limitations stem from this history; although it met requirements at the time of its development, technology has evolved so much since 1960 that it is now obsolescent. Even the voltage levels it uses cause difficulties since they are better suited for discrete components than for integrated circuits.

The interface is normally limited to speeds of less than 20 kbps and distances between DTE and DCE of less than 15 meters. It cannot be modified easily to operate at higher speeds and distances because of limitations on pulse rise times with interconnecting cables often used. Newer standards achieve significant performance improvements by including tight restrictions on cabling.

The unbalanced signaling technique, with all signal voltage levels measured with respect to the same circuit (AB-signal ground), can cause problems if the voltage level of AB at the DTE differs from its level at the DCE. A difference in ground voltages at two locations is common and is independent of signals applied. It is easy to have voltage differences sufficient to cause erroneous decoding of pulses. Better performance can be obtained with balanced signaling, so higher speed interface standards use balanced signaling even though it doubles the number of interconnecting wires.

Using a separate circuit for each control function in RS-232C also leads to a more expensive design than is necessary. More recent standards, such as X.21 (see the next section) achieve reductions in number of pins, despite balanced signaling, by using pulse streams on an interface circuit or the status of more than one interface circuit to represent control functions. This requires more logic at the interface, but rapid decreases in cost of logic make the approach attractive.

Finally, unused pins in many applications of the RS-232C interface have proved to be a great temptation to some designers, who use these pins in ways that are not part of the standard. Some nonstandard ways of using pins have become so common they are pseudo standards; a few are listed in references such as [ALIS85, pp. 64–67] for this reason. The impact of nonstandard pin usage on equipment interoperability should be obvious.

Minor differences between the EIA-232D, RS-232C, and V.24 standards, aside from differences in pin or circuit identifications indicated in Tables A4.1 and A4.2, will not be discussed. The differences can cause problems when attempting to interconnect equipment, however. Even interconnecting devices using RS-232C interfaces may not be successful since the standard allows a variety of options and a consistent choice of options is necessary for successful interconnection. A total of 15 different options available with RS-232C, each usable for either asynchronous or synchronous service and either switched or leased line configurations, are listed in [DOLL78, pp. 250–251]. This gives at least 60 different options for systems using RS-232C, with more possibilities available by choosing from a wide range of data transmission rates plus options of signals to be interchanged in the procedural specifications, options for transmitter timing in synchronous transmission, and so forth!

An interface standard closely related to RS-232C is RS-449, adopted by EIA as a higher speed alternative to RS-232C with added features. The major techniques for improving performance are the use of balanced signaling, careful specification of cabling, and tight electrical specifications. These changes allow RS-449 interfaces to be used at speeds as high as 2 Mbps, but the number of pins is increased. The interface uses a 37-pin connector for standard configurations with a second nine-pin connector for situations where a secondary channel is available.[7] Aside from improvements in signaling and more features, the approach does not differ greatly from that used for RS-232C.

---

[7] RS-232C and RS-449 explicitly provide options for handling situations in which a primary channel is used in conjunction with a second slower speed channel called a secondary channel.

4.4 X.21

A prominent "new" physical layer interface (actually dating back to 1972) is the CCITT X.21 interface forming the physical layer of the OSI model.[8]

4.4.1 Overview of X.21

The *X.21 interface* uses techniques to eliminate the limitations of RS-232C mentioned in the previous section. Balanced signaling, with a pair of wires for each circuit, is used, with careful specification of allowable cabling. Furthermore, the number of circuits is reduced by encoding state information by signals on multiple circuits and using sequences of pulses to convey control information.

The standard is broader than the physical layer standard in the OSI architecture; it is a full protocol for circuit-switched networks, including elements of OSI data link and network layers when applied in circuit switching. As we find throughout this book, such blurring of layers is common. We are primarily concerned with X.21's use as a physical layer standard, but some aspects of our treatment overlap into its data link and network layers.

Although X.21 is less common than RS-232C, it is used in a number of networks, including national digital networks in Japan, Scandinavia, and Germany, and it is becoming more common. The interface uses a smaller connector in the same connector family as that in Fig. A4.1 (see the appendix to this chapter) for RS-232C; 15 pins are provided.

4.4.2 Major Signals

Table 4.3 lists pin assignments. Six balanced circuits plus signal ground are defined, with pairs corresponding to one circuit, designated by A and B, for example T(A) and T(B). Two other pins are used for special applications or reserved. Figure 4.5 shows circuits and directions (DTE to DCE or vice versa).

The primary electrical specification[9] uses balanced mode transmission. It contains detailed specifications for transmitter and receiver and guidelines for interconnecting cables. Differential voltages at the receiver, $X(A) - X(B)$ ≤ -0.3 volt, are interpreted as logical ones while voltages $X(A) - X(B) \geq +0.3$ volt are interpreted as zeros. Also rise times of signals must not require more than 10 percent of a bit time. The interface has, so far, been defined to operate at bit rates up to 48 kbps. Cable lengths up to 1 km are allowed.

We are primarily concerned with the transmit (T), control (C), receive (R), and indication (I) circuits. The signal element timing circuit is used for

---

[8] To be precise, the X.21 standard describes the procedural portion of the interface, with other standards describing the mechanical, electrical, and functional portions. We use X.21 as a generic term to simplify discussion.

[9] Recommendation X.27 (also called V.11) is the normal electrical specification. An alternative, unbalanced mode specification, is X.26 (V.10).

Table 4.3   Pin Assignments for X.21 Interface.

Pin Number	Circuit	Description
1		Connects shields between shielded cable sections
2	T(A)	Transmit (A)
3	C(A)	Control (A)
4	R(A)	Receive (A)
5	I(A)	Indication (A)
6	S(A)	Signal element timing (A)
7	B(A)	Byte timing (A)
8	G	Signal ground
9	T(B)	Transmit (B)
10	C(B)	Control (B)
11	R(B)	Receive (B)
12	I(B)	Indication (B)
13	S(B)	Signal element timing (B)
14	B(B)	Byte timing (B)
15	—	Reserved for future international use

Figure 4.5   X.21 Interface.

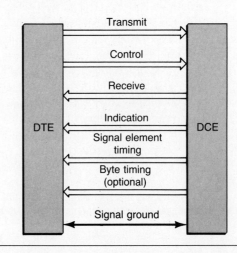

bit timing while the optional byte timing circuit provides byte timing for a second level of synchronization.

4.4.3   Typical States and Timing Diagrams

X.21 procedural specifications are complex and detailed. *Allowable states* are defined by signals on all of the T, C, R, and I circuits, with constant signals or sequences of binary signals on a circuit allowed, so an arbitrary number of states can be defined. *Permissible transitions* between states are also defined. The primary states are listed in Table 4.4, which illustrates the type of state definitions in carefully defined interface standards.[10]

Control sequences are either simple sequences, alternating zeros and ones or slightly more complex patterns, or they are expressed in terms of characters in the CCITT International Alphabet No. 5.[11] An entry "IA5" in the table denotes characters from International Alphabet No. 5. "D" corresponds to data, and "X" to "don't care," "+," "*," "BEL," and "SYN" are characters from IA5.[12] When these define a state, they are transmitted continuously.[13] At least two SYNs also precede any sequence of characters from IA5 to allow byte synchronization.[14] The values, "0" and "1" in the table correspond to constant signals, while "01" refers to an alternating sequence of binary zeros and ones.[15]

The meanings of most states in Table 4.4 are obvious from their names. Typical details include the following. Full duplex data transfers, with information flowing in both directions simultaneously, are allowed in state 13 (Data Transfer). Information provided by the DCE in state 7 (Call Progress Signal) concerns the progress of the call, including whether it is successful or not and the reason for failure if unsuccessful; in state 25 the DTE provides this type of information. In state 4 (Selection Signal) the DTE provides identification of the called DTE plus other information related to setting up the call. Thus DCE and DTE provide each other with detailed descriptions of what is happening.

A timing diagram, illustrating a typical sequence of transmissions for setting up and clearing a circuit-switched call, is given in Fig. 4.6. At the start both calling and called stations are in state 1, Ready (1, OFF, 1, OFF). (Recall that the state is defined by the status of T, C, R, and I.) The calling DTE initiates the call by changing T and C to 0 and ON, respectively (state 2, Call

---

[10] This is not a complete listing of states. Additional states defined for specialized situations have not been included; the state descriptions included are complex enough!

[11] ASCII is the U.S. version of International Alphabet No. 5.

[12] "BEL" and "SYN" are control characters in the IA5 alphabet, not three-character sequences.

[13] Except for "SYN," which must be transmitted at least twice.

[14] In Chapter 7 we show how SYN is used in exactly this way in some data link protocols.

[15] This must persist for a minimum of 24-bit intervals.

Table 4.4   Primary States of X.21 Interface.

State Number	Name	T	C	R	I
1	Ready	1	OFF	1	OFF
2	Call Request	0	ON	1	OFF
3	Proceed-to-Select	0	ON	+	OFF
4	Selection Signal	IA5	ON	+	OFF
5	DTE Waiting	1	ON	+	OFF
6A	DCE Waiting, Calling Procedures	1	ON	SYN	OFF
6B	DCE Waiting, Called Procedures	1	ON	SYN	OFF
6C	DCE Waiting, for DTE Information	*	OFF	SYN	OFF
6D	DCE Waiting, for Call Acceptance	1	OFF	SYN	OFF
7	Call Progress Signal	1	ON	IA5	OFF
8	Incoming Call	1	OFF	BEL	OFF
9	Call Accepted	1	ON	BEL	OFF
9B	Proceed with Call Information	*	OFF	BEL	OFF
9C	Call Accepted, Using Subaddressing	1	ON	SYN	OFF
10A	DCE-Provided Information, Calling DTE	1	ON	IA5	OFF
10B	DCE-Provided Information, Called DTE	1	ON	IA5	OFF
10C	Call information	*	OFF	IA5	OFF
11	Connection in progress	1	ON	1	OFF
12	Ready for Data	1	ON	1	ON
13	Data Transfer	D	ON	D	ON
13R	Receive Data[a]	1	OFF	D	ON
13S	Send data[a]	D	ON	1	OFF
14	DTE Controlled Not Ready, DCE Ready	01	OFF	1	OFF
15	Call Collison	0	ON	BEL	OFF
16	DTE Clear Request	0	OFF	X	X
17	DCE Clear Confirmation	0	OFF	0	OFF
18	DTE Ready, DCE Not Ready	1	OFF	0	OFF
—	DCE Not Ready	D	ON	0	OFF
19	DCE Clear Indication	X	X	0	OFF
20	DTE Clear Confirmation	0	OFF	0	OFF

Table 4.4   *continued*

State Number	Name	T	C	R	I
21	DCE Ready	0	OFF	1	OFF
22	DTE Uncontrolled Not Ready, DCE Not Ready	0	OFF	0	OFF
23	DTE Controlled Not Ready, DCE Not Ready	01	OFF	0	OFF
24	DTE Uncontrolled Not Ready, DCE Ready	0	OFF	1	OFF
—	DCE Controlled Not Ready	X	X	01	OFF
25	DTE Provided Information	IA5	OFF	SYN	OFF

[a]13R and 13S are used only in leased point–to–point and packet–switched service.

Request). The local DCE responds by putting a continuous " + " signal[16] on R (state 3, Proceed-to-Select). The calling DTE next puts information identifying the call on T (state 4, Selection Signal), with the call identification terminated by a " + " character. The DCE then proceeds to establish the call.

While the DCE is establishing the call, the DTE puts a 1 on T, maintaining C ON, to indicate it is waiting (state 5 DTE Waiting). The DCE responds by sending SYNs (state 6A, DCE Waiting, Calling Procedures) followed by IA5 information (state 7, Call Progress Signal), indicating the status of the call attempt. It fills in time until it has more information to return with more SYNs (state 6A) before returning more information (state 10A, DCE-provided Information, Calling DTE), indicating charging or similar information. After one more visit to state 6A, the DCE places a 1 on R (state 11, Connection in Progress), then changes I to ON to indicate the connection has been completed (state 12, Ready for Data).

As soon as the DCE at the calling end succeeds in placing the call, the DCE at the called end places a BEL (bell or ringing) signal on R (state 8, Incoming Call). Its DTE responds by turning C ON (state 9, Call Accepted). The DCE returns information about the call to its DTE (state 10B, DCE-provided Information, Called DTE) with brief delays (state 6B, DCE Waiting). It then places 1 on R (state 11, Connection in Progress), and turns I ON (state 12, Ready for Data). At this point, both ends may exchange data (state 13,

---

[16] Preceded by two SYNs for byte synchronization.

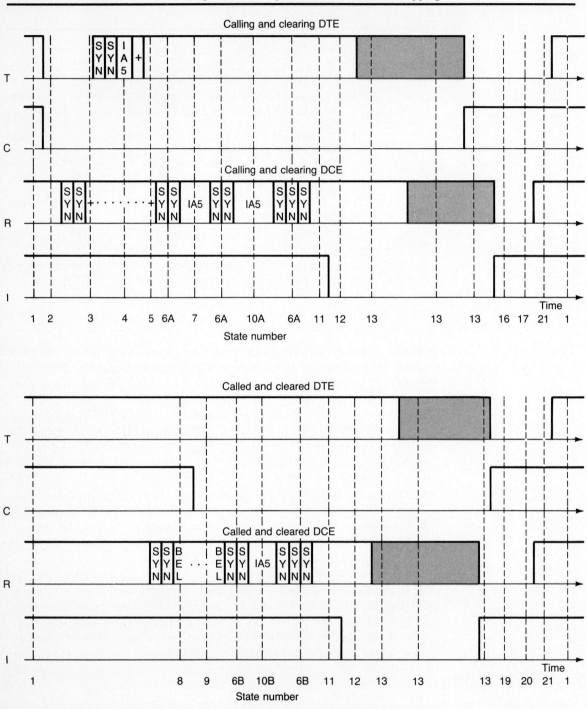

Figure 4.6 Typical Timing Diagram for X.21 Interface. High levels correspond to 1 or OFF, low levels to 0 or ON.[32] Reprinted with permission from ITU, copyright © 1984.

Data Transfer),[17] with each side transmitting data on its T circuit; the same data is received at the other end, on R, after a propagation delay.

After transmitting data, the calling DTE decides to clear the call by turning C off (state 16, DTE Clear Request). Its DCE then begins to clear the call, and verifies receiving the request by turning I OFF immediately after completing reception of incoming data (state 17, DCE Clear Confirmation). Upon receiving the clear request and completing reception of incoming data, the DCE at the cleared end turns I OFF (state 19, DCE Clear Indication), with its DTE acknowledging this by turning C OFF (state 20, DTE Clear Confirmation). Soon afterward, each DCE puts 1 on R (state 21, DCE Ready), and the system returns to the original state (state 1, Ready) after its DTE places 1 on T.

4.4.4   Limitations of X.21 Interfaces

X.21 represents a significant advance over RS-232C, but it has limitations. X.21 has not been as widely adopted as some other standards developed by the CCITT. It is applicable only for networks using digital transmission, and such networks are still far less common than networks using analog transmission. Its lack of compatibility with standards such as RS-232C has made it difficult to evolve toward use of X.21. An interim standard, X.21bis, to allow evolution from EIA-232D, RS-232C, or V.24, has been defined.

There are situations where RS-232C has advantages. Recognition of sequences of pulses to convey control information in X.21 is inevitably slower than recognition of activation of a single control line in RS-232C. This can lead to inefficient use of transmission facilities with X.21 if half duplex transmission is used with frequent changes in direction of transmission.

A significant limitation of X.21 is its inability to pass control information during data transfer. We give two examples [BERT82a], both applicable when data are being received. During data reception, both circuits under control of the DCE are in use, circuit I ON to indicate data transfer and circuit R carrying data. There is then no way for the DCE to send the DTE information about things such as signal quality. (A CG circuit in the RS-232C interface, described in the appendix to this chapter, is used for this.) Furthermore, if a DTE wishes to transmit while in receive mode, it can turn C ON to indicate it wishes to transmit (equivalent to a Request to Send with RS-232C), but there is no way for the DCE to return an equivalent of Clear to Send since both circuits it controls are in use. This requires the DTE to guess when it can start transmitting, and such a guess is risky since RTS-CTS (or equivalent) delays differ widely for different modems.

The X.21 interface also lacks other flexibilities of the RS-232C interface, such as separate transmit and receive signal element timing circuits and sig-

---

[17] States 13R and 13S, receive and send data respectively, are not allowed for switched connections, so only state 13 is used when either is transmitting or receiving data. For point-to-point leased circuits or packet-switched networks, 13R and 13S could be used, with appropriate modifications in control signals.

nal rate selectors. Furthermore, it is not well adapted for use with systems providing a secondary channel.

As was mentioned earlier, the X.21 standard covers considerably more than functions considered to fall in the physical layer in the OSI architecture. Its procedures for placing, setting up, managing, and clearing circuit-switched calls, and for character synchronization of transmitted data are not part of what is commonly considered to be the physical layer.

New interfaces are being developed, with some of the most prominent part of the ISDN standardization effort. Two are described in the next section.

4.5   I.430 and I.431 ISDN Interfaces

As we have already indicated, a major current effort is development of standards for Integrated Services Digital Networks (ISDNs). Such networks will serve a wide variety of user needs through integration of digital transmission and switching. ISDN networks will simultaneously carry digitized voice, a variety of data traffic, and possibly video on the same digital transmission links and will use the same digital switching exchanges. Details are presented in Chapter 12.

4.5.1   Overview of I.430

In this section we discuss the *ISDN I.430 interface* being developed for *basic access* and introduce the *I.431 primary rate interface.* Basic access provides access to two 64 kbps full duplex basic access (B) channels for user data and one 16 kbps full duplex (D) channel for transmission of control data and low-speed user data, for a total data rate of 144 kbps; 48 kbps of framing and housekeeping information are also transmitted for a total bit rate of 192 kbps. The primary rate interfaces provide access at either 1544 or 2048 kbps; they are discussed in Subsection 4.5.4.

The I.430 interface uses a small eight-pin connector. A unique feature is its capability to transfer power across the interface. This is similar to remote powering of traditional telephone systems.

Electrical specifications give masks within which pulse shapes must fall. This is approximately equivalent to rise times no greater than 10 percent of a bit time. Nominal pulse amplitudes at the transmitter are 750 mV.

Wiring is either point-to-point (one DCE connected to one DTE) or it is in a star or multipoint configuration (one DCE may be used with several DTEs). Possibilities are indicated in Fig. 4.7.[18] Both DTEs and DCEs must be

---

[18] The standards use the terminology NT, for network termination, for equipment equivalent to what we call a DCE, and TE, for terminal equipment, for what we call a DTE. We maintain our DCE and DTE terminology for clarity even though the NT and TE terms are more descriptive.

Figure 4.7 DCE-DTE Configurations for I.430. (a) Point-to-point configuration. (b) Star configuration. (c) Multipoint configuration.[32] Reprinted with permission from ITU, copyright © 1984.

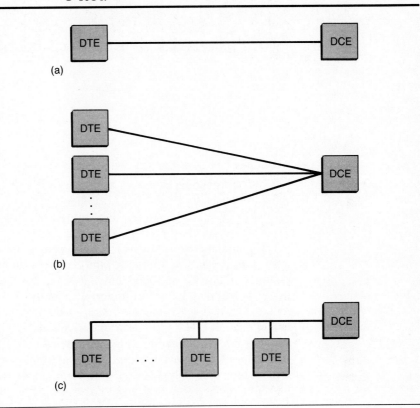

designed for use with ISDN networks. Non-ISDN-compatible DTEs may be used, however, if appropriate adapters are provided.

Two types of DCEs are defined, one providing only physical layer functions and the other providing some data link and network layer functions as well. Since we are concerned with physical layer functions, we discuss the first type (equivalent to physical layer portions of the second).

4.5.2 Signal Formats

Balanced transmission is used. Performance is further enhanced by the signal encoding in Fig. 4.8. A binary "1" is represented by zero voltage and a "0" by either a positive or a negative voltage, with polarity alternating for successive

Figure 4.8   AMI Signal Format Used in I.430 Interface.

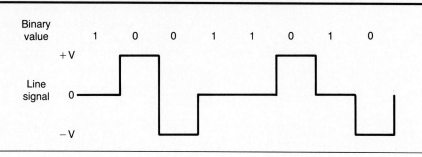

"0s."[19] This encoding ensures that the pulse stream has no *dc* (average) value; most transmission circuits do not reliably transmit *dc*. Violations of the AMI format are used to locate framing and housekeeping bits, but the overall *dc* level is maintained at zero by balancing out the effects of violations.

The B channels and the D channel are time division multiplexed together with housekeeping and framing information in the format in Fig. 4.9. A frame contains 16 bits for each B channel, four bits for the D channel, plus 12 framing and overhead bits.[20] An important feature is *separation of control* (signalling and management) information from application information through use of distinct bits in the TDM format. Situations similar to those when X.21 cannnot convey signaling information to the DTE (if the DCE is receiving information) never occur since control information can always be transmitted between DCE and DTE (and vice versa).

separation of control from application information

Framing bits (the first and fourteenth bits in each frame) occur in patterns that allow synchronization to be achieved by hunting for and locking onto these patterns. E bits (used only for DCE to DTE communication) repeat the most recent D bits received from the DTE. Arrows in the figure indicate repeated bits; the bit at the head of each arrow is identical to that at the tail. These bits are used in the D channel access protocol discussed in the next subsection.

---

[19] A variety of terms for this encoding are used in the literature; these include bipolar, AMI (alternate mark inverted) and pseudoternary. Often, however, the encoding meant when these terms are used is the inverse of that in Fig. 5.8, with "zero" encoded into no voltage and "one" encoded into alternating positive and negative voltages.

[20] Figure 4.9 does not indicate the pulse levels for each bit. All data bits can be either "ones" or "zeros," but the polarity of some "zero" data bits may be restricted to maintain zero *dc* components. Overhead bits are also restricted, with *dc* balance in specified portions of the frame. This is achieved by *dc* balance bits in the frame. As the figure indicates, there is a two-bit time offset between the start of DCE to DTE frames and the start of corresponding DTE to DCE frames.

Figure 4.9 TDM Format for I.430, ISDN Basic Interface, Frames.

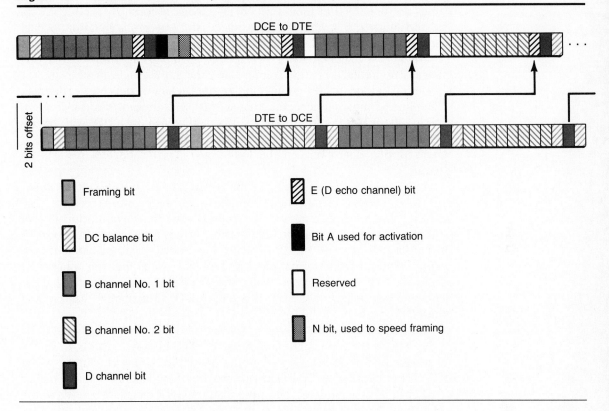

4.5.3 I.430 Bidding for Access to D Channel

The I.430 procedural specifications state procedures for activating, transferring data over, and deactivating the D channel. Once the D channel has been activated, full duplex communication over it is possible. Since the D channel can be used for control, the DCE and DTE(s) can use it to control either or both of the B channels, as well as for data transfers over the D channel itself, in any manner they wish. Specification of how the D channel is to be used for such functions is, so far, outside the scope of the standard.

Specifications of procedures to activate and deactivate D channel communications circuits are in the standard.[21] The most interesting and unique aspect is a procedure allowing active DTEs to bid for control of the D channel.

The bidding procedure relies on a requirement, in ISDN data link layer specifications, that D channel data transmissions use the LAPD (HDLC) for-

---

[21] The A bit in the DCE to DTE frame format is set to "one" during normal frame transmission whenever the frames transmitted—on B, D, or D-echo channels—contain operational data.

mat discussed in Chapter 7. For current purposes, we need to know LAPD frames are delimited at their start by Flag bytes of form 01111110 ("0," six "1s" and another "0"), with this many successive "1s" prohibited from occurring elsewhere in LAPD frames.[22] However, an inactive DTE or DCE transmits a continuous string of binary "1s" (zero voltage in Fig. 4.8), rather than LAPD frames. The first field in a LAPD frame after the initial flag is an address.

During bidding, a DTE contending for the D channel puts successive bits on the line, comparing echoed E channel bits with those it has transmitted, until it either gains control or senses it is bidding against a DTE with higher priority. (Two priority classes are allowed, each subdivided into two subclasses during operation.) In the latter case it gives up and restarts its attempt.

A DTE, say DTE_i, contending for the D channel first monitors the E channel, which echoes D channel bits (from the DCE), and counts successive "1s." When the count equals a number (X_i) representing the current priority level of its data request, it begins to bid. Possible values for X_i range from 8 to 11 (8 or 9 for high priority and 10 or 11 for low priority, with adjustment procedure below), but the format of LAPD frames ensures that this number of successive "1s" never occurs when the D channel is transmitting data. Hence bidding begins only when it is idle. A DTE with small X_i has bidding priority.

After counting a number of successive "1s" on the E channel equal to its X_i, the DTE starts to transmit its first frame, $01111110A_1A_2 \ldots$ with $A_1, A_2, \ldots$ successive bits of its address. When the first "0" of a Flag is echoed by the DCE, DTEs counting "1s" start over (a process they cannot complete before the D channel is again free), but more than one DTE can start bidding at the same time. As long as all bits such DTEs transmit are identical, the DCE receives them properly, and echoes them; "0s" reinforce and "1s" have zero amplitude.

Bidding DTEs monitor the E channel and transmit as long as each echoed bit is identical to that sent, but give up when it differs. This implies one DTE sent "0" and another sent "1" with the echoed bit "0" (since total amplitude at the DCE is nonzero), so the DTE that transmitted "1" gives up. Addresses are distinct, so a DTE always wins. Upon conclusion of transmission, the winning DTE's X_i is set to the maximum for its data type (9 for high and 11 for low priority), so other DTEs are likely to win next; the values are decremented to 8 or 10 after all DTEs with higher priority have had a chance to transmit.

Figure 4.10 illustrates bidding. DTE1, DTE2, and DTE3 are monitoring E bits to gain access for high-priority traffic. DTE2 and DTE3 both have X_i values of 8, but DTE1 has an X_i of 9. The figure starts at a moment when the first "1" in an eight-bit string of "1s" occurs on the E channel. After observing eight

---

[22]A process known as zero insertion guarantees this. This is described in Chapter 7.

Figure 4.10 Illustration of D Channel Access Algorithm.

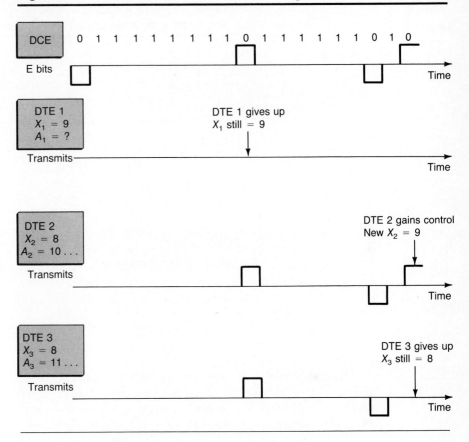

"1s," DTE2 and DTE3 put Flags on the DTE to DCE D channel, and the DCE echoes the bits on the E channel, the initial "0" causing DTE1 to give up.[23]

Both DTE2 and DTE3 are still competing at this stage, but the conflict is resolved in favor of DTE2 when the second bit in its address field turns out to be "0" while the second bit in the address field for DTE3 is "1." The returned E bit is "0," so DTE3 gives up and DTE2 gains control. The X_2 value is then increased to 9, so DTE3 will have higher priority during the next cycle.

After completion of the bidding cycle, the winning DTE and the DCE have a full duplex 16 kbps connection over the D channel. They can use this for data or they can use it to control communications over either or both of the

---

[23] X_1 = 9 since DTE1 has transmitted recently. It will be set back to 8 after DTE3 also has a chance to transmit (assuming no other DTEs are involved).

Figure 4.11   Frame Formats for Digital Carrier Systems. (a) T-1 frame
format. (b) Corresponding format recommended by CCITT.[33]
Reprinted with permission from ITU, copyright © 1984.

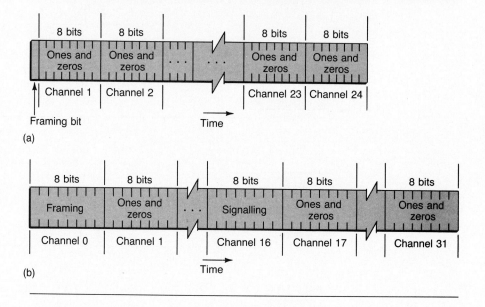

(a)

(b)

full duplex 64 kbps B channels. The connection is maintained only as long as
the DTE uses it continuously, however. If it quits, it transmits continuous "1s"
and bidding restarts. By using the D channel in this way, it is possible to
maintain extremely flexible control over data transmission, even when both
the DTE and the DCE are transmitting simultaneously in a full duplex manner.

4.5.4   I.431 Interface

The I.431 interface operates at either 1.544 Mbps or 2.048 Mbps. It is an adaptation of the 1.544 Mbps, 24-channel, T-1 carrier system used in North America and Japan or the 2.048 Mbps, 30-channel carrier system standardized by
the CCITT and used in much of the rest of the world. Frame formats of these
two digital carrier systems are illustrated in Fig. 4.11.

Primary rate interfaces can carry any of the following channel types:

ISDN
channel types

B channel	64 kbps; equivalent to B channel for basic rate interface.
D channel	64 kbps; higher speed version of basic rate access D channel; used primarily for signaling, but also for telemetry or packet-switched data.
E channel	64 kbps; signaling for circuit switching; used only with multiple access configurations; not the same as the E-echo

channel of basic rate interface, though both are called E channels.

H channels H0: 384 kbps

H1: H11-1536 kbps for 1544 kbps primary rate

H12-1920 kbps for 2048 kbps primary rate; used for a variety of user information streams, but not for signaling.

In the T-1 carrier system, signaling (such as on-hook/off-hook or ringing signals, switching information, and billing information) is conveyed by stealing the least significant bit from each channel in every sixth frame; this provides an average of four signaling bits per frame, or a signaling rate of 32 kbps. One bit per frame is used for framing. For ISDN no bits are stolen for signaling, so eight bits are always available for data in each channel.[24] For the CCITT format, one channel per frame is used for signaling plus one for framing. If a D or E channel is included, Channel 24 is used in T-1 or Channel 16 in the CCITT format.

primary rate configurations

Any of the following configurations are possible for primary rate access:

> At 1.544 Mbps:
> 23 B + D
> 23 B + E
> 4 H0
> 3 H0 + D
> 1 H11
> At 2.048 Mbps:
> 30 B + D
> 30 B + E
> 5 H0 + D
> 1 H12 + D

Signaling for 1.544 Mbps configurations without a D or E channel is obtained from another primary rate interface that has such a channel.

Only point-to-point configurations are defined for the primary rate interface, so no channel access algorithms comparable to the contention algorithm for the basic rate interface are needed.

4.5.5   Limitations of I.430 and I.431

It is difficult to be precise about limitations of interfaces just being implemented. The major limitation is the fact that essentially no equipment now available supports the interfaces. Current equipment can be used only in conjunction with adapters—but adapters are not available either! ISDN

---

[24]Always using eight bits for data in each channel eliminates restrictions on sequences of ones and zeros that T-1 relies on; techniques for getting around problems have been devised, however.

equipment will be relatively expensive until market size justifies volume production.

The I.430 interface can be used only on channels with end-to-end digital transmission. Although end-to-end digital channels are becoming available, they are less common than channels with analog transmission over at least part of their routes. The interface is sophisticated and has many good features, but time will tell how successful it will be.

The I.431 interface was developed under restrictions imposed by large amounts of installed equipment. It involves minimal modifications to existing standards but is hampered by incompatibility of standards in various countries. As we discuss in Chapter 12, incompatible standards in different parts of the world have caused major problems in the development of ISDN.

4.6   Physical Layer in DNA and SNA

Neither DNA nor SNA specifies a physical layer. In each case a wide variety of standard physical layer protocols may be used. These include most of the protocols in common use.

4.7   Physical Layer in IEEE 802 and MAP/TOP

The physical layers for IEEE 802 and MAP/TOP (identical at this level) implement medium access procedures for these architectures (see Chapter 5) as well as handling functions handled by other physical layer protocols. Since their functions are specialized, we only briefly outline such interfaces.

The physical layer for IEEE 802.3 contains several components. A component attached to computer equipment generates electrical signals representing a bit stream transmitted over the LAN and interprets incoming signals from the LAN. It is connected to the main LAN cable by a multiwire cable, similar to those interconnecting devices with RS-232C or similar interfaces. Fifteen pin connectors are used, with a pair of wires each for balanced Data Out, Data In, Control Out, and Control In circuits, plus shields for each, voltage references, and ground. Circuitry converts signals between formats on these multiwire cables and those on the main LAN cable. Additional components interconnect equipment with the LAN cable; for the coaxial cables commonly used this usually includes a tapping screw to make contact with the center coaxial cable conductor plus a tap block to make contact with the outer shield.

IEEE 802.4 and IEEE 802.5 physical layer interfaces are similar, but with short coaxial cables instead of multiwire cables between the LAN cable and computer equipment. Three distinct interfaces, for different signaling techniques and media, are specified in the IEEE 802.4 standard.

Special functions provided by the IEEE 802.3 physical layer include recognition of the presence or absence of carrier for CSMA/CD,[25] recognition of collisions, and provision of signals conveyed to the MAC sublayer of the data link layer (see Subsection 3.9.1). These are in addition to bit streams handled by all physical layers. IEEE 802.4 and 802.5 use "non-data" symbols (bits with encoding violating normal formats) to locate frames. Hence the physical layers recognize such symbols as well as normal data symbols. Special functions, analogous to those for IEEE 802.3, are also provided.

4.8   Invoking Physical Layer Services

Procedures for invoking physical layer services vary widely. We sketch them using the OSI Reference Model framework, with services invoked by *passing primitives* across an interface defining a boundary between the layer requesting services and the layer providing them. This is not the type of interface that is the primary subject of this chapter; the OSI interface is between architecture layers, but the chapter focuses on device interfaces.

Figure 4.12 illustrates how primitives could be used to activate a physical layer connection, transfer data across it (in one direction of transfer for simplicity), and deactivate the connection. The sequence could be used with a protocol such as X.21. Primitives (from ISO working papers [MCCL83]), are as follows:

- Ph-Activate Request
- Ph-Activate Indication
- Ph-Activate Response
- Ph-Activate Confirm
- Ph-Data Request
- Ph-Data Indication
- Ph-Deactivate Request
- Ph-Deactivate Indication
- Ph-Deactivate Response
- Ph-Deactivate Confirm

The figure shows a sequence in which DTE A requests a connection. This causes a Ph-Activate Request primitive to be passed across the data link layer–physical layer interface and DTE A, DCE A, DTE B, and DCE B to go through a sequence, such as that described in Subsection 4.4.3, to set up a

---

[25] For details of carrier sense functions and so forth, see Chapter 5.

Figure 4.12   Use of Primitives to Set Up Physical Layer Services.
Ph- prefixes designate physical layer services invoked across
data link–physical layer interface.

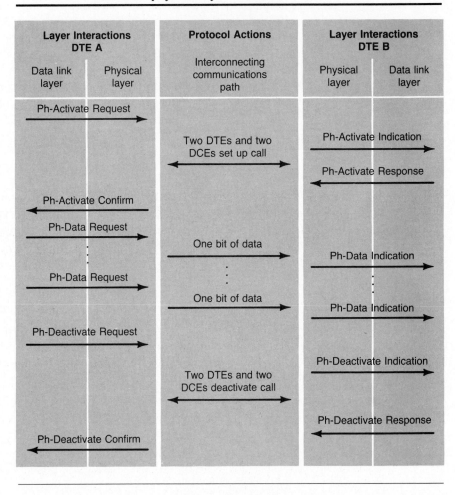

physical layer connection. A Ph-Activate Indication primitive is passed across
the physical layer–data link layer interface at B when DTE B is informed of
the request for connection. Ph-Activate Response and Ph-Activate Confirm
primitives inform DTE A that the connection has been established.[26]

During data transfer, only one directional data transfer is shown for simplicity. Two primitives are used, Ph-Data Request (asking the physical layer
at DTE A to transmit a bit) and Ph-Data Indication (telling the data link layer

---

[26]Primitives passing results of X.21 call progress signals would suffice.

at DTE B that a bit has been received). Both are used each time a bit is transmitted, since the physical layer transfers bits and does not recognize groupings of bits into characters or frames.

After DTE A has finished data transfer, the physical layer connection is deactivated. Deactivation is similar to activation. Ph-Deactivate Request, Ph-Deactivate Indication, Ph-Deactivate Response, and Ph-Deactivate Confirm primitives are illustrated.

This exact sequence of primitives, with the names indicated, may not be found in any network implementation. However, procedures to pass requests for physical layer activation and deactivation and to send bits through the network and indicate their receipt are needed in any network.

4.9 Computer Equipment Interface

Information on the computer equipment (DTE) side of a physical layer interface is in the form of serial strings of pulses on transmitter or receiver data lines, voltages or sequences of voltages on separate control lines, and possibly clocking pulses (in interfaces for synchronous communication). This is not the format for data in computer equipment, where data transfers are parallel and the voltage formats for representing information at the physical layer interface are not appropriate. An interface with computer equipment is needed to resolve discrepancies. This interface is often called a computer port and is implemented with a microcomputer (or a minicomputer). We discuss simple versions of interfaces that illustrate basic principles of design.

The interface must perform at least the following functions:

1. Conversions between voltage levels appropriate for representing data at the physical layer interface and those appropriate in computer equipment

2. Parallel-to-serial conversion of data from computer equipment in preparation for transmission on a communications channel

3. Serial-to-parallel conversion of received data in preparation for storage and processing by computer equipment

4. Provision of a means for the receiver to achieve bit, character and (for synchronous transmission) frame synchronization

5. Generation of suitable error check digits for error detection

6. Passing appropriate control information across the interface

Because of the tremendous variety of types of computer equipment available, with little compatibility among different equipment, it is impossible to give a general treatment of computer interfacing in a brief overview. Limited parts of the interface are standardized, with standards due as much to efforts of integrated circuit manufacturers as to standards bodies. We present typical computer interfacing techniques, however.

4.9.1   A Typical Computer Equipment Interface

Figure 4.13 is a diagram of a typical computer equipment interface. The interface is a slightly idealized version of one designed for synchronous communications using RS-232C, possibly on a communications link employing a data link control similar to Binary Synchronous Communications (Bisync) at the data link layer,[27] but it could, with minor modifications, be used for other types of communications. Asynchronous communications using RS-232C would involve particularly simple modifications.[28]

USART

The standardized portion of the interface is the box labeled *USART* (*U*niversal *S*ynchronous/*A*synchronous *R*eceiver *T*ransmitter), often denoted "UART." Integrated circuit manufacturers offer USARTs, which allow users to specify operating characteristics of equipment supported by loading a predefined code word or bit pattern. The USART handles most of the format conversions needed; they are discussed in the next subsection.

The "level converters" convert voltages from those used by the physical layer to those for computer equipment. Signals on the left side of this box are identical to corresponding signals on the right, except for level. Control signals, such as Data Terminal Ready and Request to Send (computer to physical layer), Ring, Clear to Send, Data Set Ready, and Carrier Detect (physical layer to computer) pass across the interface.

The USART contains a status register (discussed below), but the implementation uses program addressable Receiver Status and Transmitter Status Registers to store precise status. This includes the status of Ring, Request to Send, Clear to Send, Data Set Ready, Data Terminal Ready, and Carrier Detect lines, bits to enable interrupts and cause them if they are enabled, and so forth. Important bits include one for Received Data Available (RDA) in the Receiver Status Register and one for Transmitter Buffer Empty (TBMT) in the Transmitter Status Register. Both are set, when appropriate, by the USART.

operation of interface

When a register is read, the Address Selection Logic gates contents of the appropriate register (Receiver Buffer in the USART, Receiver Status Register, or Transmitter Status Register) onto the Parallel Data leads. This causes the Bus Drivers to place data onto the CPU bus data lines. Conversely, when a register is written, data on the CPU bus come into the Bus Receivers and are presented to the appropriate register via the Parallel Data lines. The Address Selection Logic then strobes the data into the register.

During receiving, all bits comprising a character are assembled in the USART. It then sets the RDA bit in the Receiver Status Register, causing an interrupt to be generated as soon as interrupts are enabled. The computer

---

[27] See Chapter 7 for a description of Bisync.

[28] See Chapter 7 for a description of asynchronous data link controls.

Figure 4.13 Typical Computer Equipment Interface (adapted from [MCNA82]). Reprinted with permission from *Technical Aspects of Data Communication* by John E. McNamara, copyright © Digital Press/Digital Equipment Corporation, 12 Crosby Drive, Bedford, MA 10730.

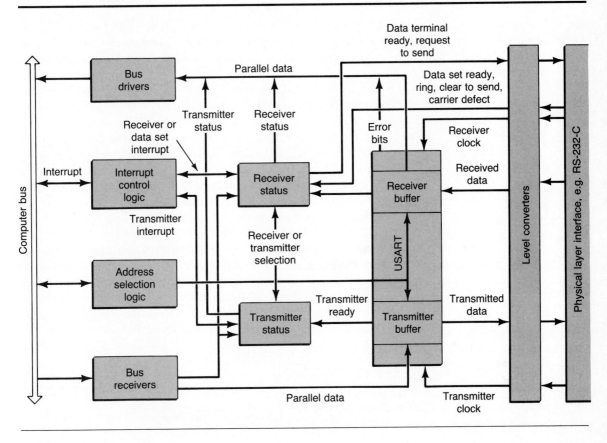

program responds to the interrupt and reads the Receiver Buffer. It next uses a circuit, Reset Data Available, to assure the USART that it is safe to replace data in its Receive Buffer with new data. If this is not finished before the USART needs to shift a character into the buffer (in order to have space to assemble a new character), a bit indicating Overrun Error is set and passed to the CPU. The CPU can deal with this as it wishes.

During transmission, the USART sets the TBMT bit when the Transmit Buffer is empty. This is passed to a Transmitter Status Register in the CPU and causes an interrupt. The CPU checks to see if it has anything to send on this line. If it does, it loads it into the USART's Transmit Buffer and clears the TBMT bit. The USART shifts the character from the Transmit Buffer into another register, resetting the TBMT bit, and shifts the character out a bit at a time.

4.9.2   USARTs

A diagram of a typical USART is given in Fig. 4.14. The USART can operate as either a synchronous or an asynchronous receiver/transmitter, depending on bits in the Mode Register. Bits in this register also determine the number of bits per character, whether even parity, odd parity, or no parity is used, and details of treatment of SYN characters. In addition, for asynchronous mode, the register determines speed of the transmit and receive clocks with respect to the bit rate.

The Status Register contains the Received Data Available (RDA) and Transmitter Buffer Empty (TBMT) bits discussed above, plus bits indicating Parity Error, Framing Error, and Overrun Error and some for modem control. Since the USART is programmable, simple commands cause it to examine status bits and execute transmit and receive functions.

During synchronous data reception, the USART uses incoming clock signals (see Fig. 4.13) to determine when to sample voltage on the Received Data line; each sample is classified as a "0" or "1" and shifted into the Receive Register. When this register is full, the bits are shifted in parallel into the Receive Buffer and the RDA bit set, causing an interrupt to the CPU and processing discussed above. The USART also searches for characters that indicate frames are starting, using information in its Mode Register. Error bits in the Status Register are set when errors are detected.

Figure 4.14   USART Diagram.

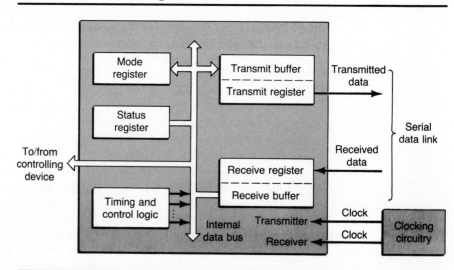

During synchronous data transmission, characters obtained from the CPU, after interrupts initiated by the TBMT bit, are shifted in parallel from the Transmit Buffer to the Transmit Register and out onto the line a bit at a time. The USART automatically inserts control characters to start frames. As soon as a character has been shifted into the Transmit Register, the TBMT bit is set and the process restarts. Data reception and transmission can go on simultaneously in a full duplex manner.

The primary difference between synchronous and asynchronous operation is in clocking; no external clocking is available for asynchronous operation. A clock running at approximately a multiple of the bit rate (16 to 64 times) is used. Voltage on the Received Data line is examined each clock time until a transition from "1" to "0" is seen. This is a transition from an idle line condition to the voltage for a START bit. If the clock is running at 16 times the bit rate, further samples are taken 8, 24, 40, and so forth clock times later. The first should be approximately at the middle of the START bit, the second near the middle of the first data bit, and so forth. With any reasonably accurate clock, sampling times are accurate enough to sample all bits in a character. The sampled bits are shifted into the Receiver Register, with the rest of the operation essentially the same as for the synchronous case.

Transmission of asynchronous data using the USART is done in the same manner as transmission of synchronous data, except for using the internal clock instead of an externally derived clock and shifting a bit out once every 16 (to 64) clock cycles instead of every clock cycle.

4.9.3 More Powerful Computer Equipment Interfaces

Although the interface described is an approximation of some commonly used, it has limitations. It can use excessive CPU cycles responding to interrupts, since it interrupts the CPU for every character received or transmitted. Interrupt service routines for typical communications interfaces take 25 to 200 μsec, with getting into the service routine and out of it again sometimes taking nearly as much time as the service routine itself [MCNA82]. Sophisticated interfaces use *direct memory access (DMA)* to read or write blocks of data without interrupting the CPU, except to initiate or conclude block transfers or deal with exceptional conditions.

direct memory access (DMA)

The simple interface has limited capabilities for dealing with fill characters often transmitted at the beginning or end of frames or when there is nothing else to transmit. The USART is, to a limited extent, capable of stripping off fill characters, but leaves much of this job for the CPU. More complex interfaces do a thorough job. They may also be able to pull out data portions of received frames, discarding overhead information, or insert overhead information in transmitted frames. Error detection with polynomial codes commonly used with data link controls can also be implemented in the interface.

A major limitation of the simple interface is the fact it is a single line interface. Multiple line interfaces sharing significant amounts of equipment can be more cost effective. Multiline controllers are common where a number of communications lines come into one CPU.

4.10   Interface to Communications Channel

The final interface in Fig. 4.1 is between the DCE and the communications channel. The DCE (modem for analog channels, adapter for digital channels) puts signals going out over this interface in form for transmission over the communications channel and incoming signals in form for transmission from DCE to DTE. A DCE-DCE protocol may also be implemented.

For analog transmission, signals between DCEs are modulated waveforms using the modulation type employed by the modems; this can be any type discussed in Chapter 2. For digital transmission, signals between DCEs are pulses of appropriate shape, employing the coding used by the channels.

A protocol between modems handles functions such as synchronizing carrier oscillators, initializing equalizers, establishing gains for amplifiers, and similar functions. For channels with long propagation delays—that is, terrestrial circuits of more than 2500 km in length—it may disable echo suppressors.[29] A training sequence, with predetermined signal patterns transmitted between modems, is often used in accomplishing these functions. A scrambler at the transmitter and descrambler at the receiver are often used. A scrambler alters the transmitted bit stream in a manner that makes it appear to be random, but the transformation is reversible by the descrambler. This eliminates periodic bit streams, such as long sequences of ones or zeros; these could cause problems with receiving circuitry, especially when automatic equalization is used.

For digital channels, functions performed by the DCE at this interface are simpler. They consist primarily of converting from the digital pulse streams used between DCE and DTE to those used between DCEs over the digital channel, and maintaining appropriate timings for pulse streams. Protocols to lock onto appropriate synchronization, resynchronize if synchronization is lost, and so forth are also needed.

---

[29] Echo suppressors are one-way amplifiers allowing signal energy to propagate only from talker to receiver, with a reversal in direction when they sense a reversal in direction of energy flowing across the channel. They are used on telephone circuits with propagation delay long enough for echoes to be psychologically disturbing, but can cause problems with digital data transmission. They can be disabled by transmitting appropriate waveforms for 400 msec, then ensuring there are no periods longer than 100 msec without energy present on the circuit. If they are not disabled, reversing this direction takes approximately 100 msec.

4.11   Impact of Interfaces on Performance

Parameters related to the interfaces described in this chapter can have significant impact on performance. These parameters are often determined by equipment (modems, and so forth) used, rather than by interface standards, but their effects are conveyed via their influence on timings of events such as those portrayed in Figs. 4.4 and 4.6.

An important parameter is the time interval between a DTE's requesting permission to send information over a communications line and permission being granted. This may depend on several factors. For networks using EIA-232D/RS-232C/V.24 interfaces, however, significant delays may be lumped together into a delay between activating a Request-to-Send circuit and sensing activation of the Clear-to-Send circuit (RTS-CTS delay). This is often called *line turnaround time,* since the delay is incurred every time the direction of transmission is reversed. It can vary from around 10 to 250 ms, or even longer. Line turnaround times can be large for systems employing two-wire half duplex communication, especially those spanning long distances; modems for lines with echo suppressors often have a RTS-CTS delay of 200 ms or more.

line turnaround time

A second component of line turnaround time that can be significant is equalization time for modems. Equalization times can be especially significant for high-speed, high-performance modems, which may use long training sequences and adaptive techniques to achieve the best equalization possible. However, new equalization techniques are making it possible to reduce equalization time.

Other delays that may be encountered include a modem delay (defined to be the time from when a digital signal is presented to the physical layer interface until the modulated carrier appears on the line), plus a similar delay while the receiving modem demodulates the incoming signal, and reaction time of the terminal equipment or computer at each end of the line (defined to be the time it takes for equipment to realize it has received some data and some type of action has to be taken). These delays are also highly variable, but may total around 20 to 30 msec at each end of the line. For simplicity, however, we lump them in with line turnaround times.

A consequence of long line turnarounds is that it may be cost effective to pay more money for four-wire, potentially full duplex, circuits even if operation is to be half duplex. True line turnarounds are then avoided and only the data processing equipment and modem turned around; this may take no more than 5 to 10 msec.

Simple calculations illustrate the impact of line turnaround times. Consider a system transmitting at R bps. All information flows in one direction, with brief acknowledgment messages after each data block. Transmission errors are ignored, so all data are assumed to be received correctly. Assume that all messages are K bits long, with n_h of these overhead bits, acknowedg-

Figure 4.15   Impact of line turnaround times on effective transmission
rates for 4800 bps line with short propagation delays
of 10 msec.

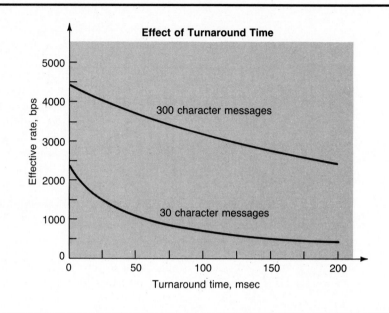

ment messages require n_a bits, line turnaround time is t_{ta} seconds, and propagation delay between transmitter and receiver is t_p seconds. Then a total of
$K - n_h$ bits of data are transferred every $[(K + n_a)/R] + 2(t_p + t_{ta})$ seconds.
(Two line turnaround times and propagation times in both directions are
required for each transmitted block of data.) This gives an effective data rate
in bps of

$$R_e = \frac{K - n_h}{\dfrac{K + n_a}{R} + 2(t_p + t_{ta})} = \frac{(K - n_h)R}{K + n_a + 2(t_p + t_{ta})R}. \tag{4.1}$$

Figure 4.15 contains plots of values from this formula versus line turnaround times from 0 to 200 msec for two different cases. In both cases line
speed is 4800 bps, propagation delays are 10 msec,[30] acknowledgments consist of six eight-bit characters and overhead in message blocks is six eight-

---

[30] A rule of thumb for computing propagation delays in terrestrial circuits is to assume 6 to
10 μsec per kilometer, or propagation at roughly one half to one third of the speed of light
in free space. Hence, the propagation delay assumed corresponds to a terrestrial circuit
1000 to 1500 km long.

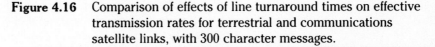

Figure 4.16 Comparison of effects of line turnaround times on effective transmission rates for terrestrial and communications satellite links, with 300 character messages.

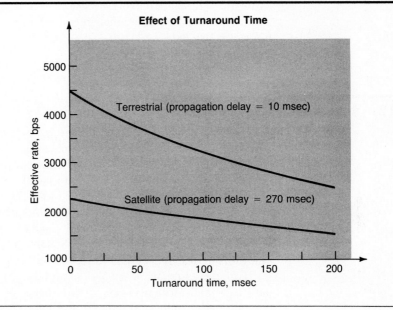

bit characters.[31] In the first case messages consist of 300 eight-bit characters including overhead, and in the second case messages consist of 30 eight-bit characters including overhead. The impact of long turnaround times is much more severe for short messages.

Figure 4.16 indicates the impact of long propagation delays such as those for communications satellites. The upper curve in this figure is the same as the 300-character message curve in Fig. 4.15, with the lower curve differing only in propagation delay, assumed to be 270 msec. This is a reasonable estimate for propagation delay for communications satellite links. The long propagation delay masks out much of the effect of different turnaround times. Eliminating this delay is physically impossible, since it results from the speed of light for propagation of radio waves, but techniques in later chapters indicate ways to mask some of its effects.

A second important performance factor normally manifested via the interface is time to set up a call or establish a communications circuit from source to destination. With conventional telephone equipment, call setup

---

[31] We show, in Chapter 8, that these are typical lengths of acknowledgment messages and typical numbers of overhead characters.

time is typically 5 to 25 sec after completion of dialing and clear down time is on the order of 1 sec [PICK85b]. New networks employing fast circuit switching (and possibly X.21 interfaces) could in theory reduce this to as low as 140 ms [PICK85b], making it feasible to hold a line only for the duration of a short message transfer, then reconnect the circuit the next time a data transfer is needed.

4.12   Summary

A number of equipment interfaces are important in telecommunication network architectures, though only the physical layer (among the interfaces we have studied) is normally considered to be part of a networking architecture. X.21 is the standard physical layer interface for the OSI Reference Model, but other interfaces such as EIA-232D/RS-232C/V.24 are more common. The I.430 and I.431 ISDN interfaces are likely to be among the more dominant interfaces when ISDN becomes more prominent. Each of these physical layer interfaces has advantages and disadvantages, though the newer ones show some definite gains in design effectiveness as a result of past experience.

Two other interfaces are also important. Computer interfaces transform data back and forth between formats used by DCEs and formats used by computer equipment and pass control information between DCEs and computers. A communications channel interface transforms back and forth between formats on communications lines and formats used for DCE-DTE communications.

Timings of protocol sequences implemented by interfaces may have significant impact on performance. Delays between requesting permission to send and obtaining it and delays to establish circuits are especially important.

Appendix 4.A   Details of EIA-232D/RS-232C/V.24 Interface

Figure A4.1 shows RS-232C connectors. Such details constitute the mechanical specification of the interface. Other details such as dimensions of pins and their spacing must also be specified. Pin numbers and functions, plus EIA RS-232C and CCITT V.24 nomenclature for them are in Table A4.1. Functions of signals on common circuits are in Table A4.2.

The standard also specifies signal voltages. Voltage drivers are limited to a range of -5 to -15 volts for a logic 1 on a data line, or OFF for a control line, and $+5$ to $+15$ volts for a logic 0 or ON signal, with received voltages less than -3 volts interpreted as logic 1 or OFF and voltages greater than $+3$ volts as logic 0 or ON. Rise times are also specified. For example, the time for the signal on a data and timing interchange circuit to pass through the transition region between -3 and $+3$ volts is

Figure A4.1   Pin Layout and Connector Dimensions for RS-232C Interface.

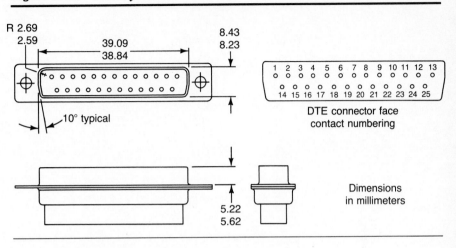

DTE connector face
contact numbering

Dimensions
in millimeters

Table A4.1   Circuit Assignments for RS-232C and
V.24 Physical Layer Interfaces.

Pin Number	EIA Circuit	CCITT Circuit	Description
1	AA	101	Protective ground
2	BA	103	Transmitted data (TD)
3	BB	104	Received data (RD)
4	CA	105	Request to send (RTS)
5	CB	106	Clear to send (CTS)
6	CC	107	Data set ready (DSR)
7	AB	102	Signal ground
8	CF	109	Received line signal detector (RLSD)
9	—	—	Reserved for test
10	—	—	Reserved for test
11	—	—	Not assigned
12	SCF	122	Secondary received line signal detector
13	SCB	121	Secondary clear to send
14	SCA	118	Secondary transmitted data
15	DB	114	Transmitter signal element timing (from DCE)
16	SBB	119	Secondary received data
17	DD	115	Receiver signal element timing (from DCE)
18	—	—	Not assigned

continued

Table A4.1 *continued*

Pin Number	EIA Circuit	CCITT Circuit	Description
19	SCA	120	Secondary request to send
20	CD	108	Data terminal ready (DTR)
21	CG	110	Signal quality detector
22	CE	125	Ring indicator (R)
23	CH	111	Data signal rate selector (DTE driver)[a]
23	CI	112	Data signal rate selector (DCE driver)
24	DA	113	Transmitter signal element timing (from DTE)
25	—	—	Not assigned

[a]Either the DTE or the DCE can be the driver for the data signal rate selector.

Table A4.2 Functions of Signals on RS-232C Circuits, with EIA Terminology.

AB—Signal Ground. All signal levels on other circuits are measured relative to this point.

BA—Transmitted Data (TD). Pulses representing data sent out over the communications channel are placed on this circuit at times dictated by procedural specifications.

BB—Received Data (RD). Pulses representing data received by the DCE from the communications channel are placed on this circuit.

CA—Request to Send (RTS). The DTE activates this circuit when it is ready to send data.

CB—Clear to Send (CTS). The DCE activates this circuit when it is ready to accept data.

CC—Data Set Ready (DSR). The DCE activates this circuit when it is ready for use.

CD—Data Terminal Ready (DTR). The DTE activates this circuit to let the DCE know it is ready to send and receive data.

CE—Ring Indicator (R). The DCE activates this circuit to tell the DTE a ringing signal due to an incoming call is being received.

CF—Received Line Signal Detector (RLSD). The DCE activates this circuit when it detects a carrier signal from another modem on the communications channel.

DA or DB—Transmitter Signal Element Timing. Either DCE or DTE may provide a signal to determine timing of pulses sent to the modem for transmission over the channel.

DD—Receiver Signal Element Timing. The DCE may provide this signal to the DTE to tell it where to sample pulses.

Table A4.3   Circuit Assignments for EIA-232D.

Pin Number	EIA Circuit	CCITT Circuit	Description
1			Shield
2	BA	103	Transmitted Data (TD)
3	BB	104	Received Data (RD)
4	CA	105	Request to Send (RTS)
5	CB	106	Clear to Send (CTS)
6	CC	107	Data Set Ready (DSR)
7	AB	102	Signal Ground
8	CF	109	Received Line Signal Detector (RLSD)
9	—	—	Reserved for Test
10	—	—	Reserved for Test
11	—	—	Not Assigned
12	SCF	122	Secondary Received Line Signal Detector
12	CI	112	Data Signal Rate Selector[a]
13	SCB	121	Secondary Clear to Send
14	SCA	118	Secondary Transmitted Data
15	DB	114	Transmitter Signal Element Timing (from DCE)
16	SBB	119	Secondary Received Data
17	DD	115	Receiver Signal Element Timing (from DCE)
18	LL	141	Local Loopback
19	SCA	120	Secondary Request to Send
20	CD	108	Data Terminal Ready (DTR)
21	RL	140	Remote Loopback
21	CG	110	Signal Quality Detector[b]
22	CE	125	Ring Indicator (R)
23	CH	111	Data Signal Rate Selector (DTE driver)
23	CI	112	Data Signal Rate Selector (DCE driver)
24	DA	113	Transmitter Signal Element Timing (from DTE)
25	TM	142	Test Mode

[a]If Secondary received line signal detector is not used, then pin 12 is used for data signal rate selector.
[b]Signal quality detector if received line signal detector not used.

specified as less than or equal to one msec or 4 percent of the nominal duration of a signal element on that interchange circuit, whichever is less.

Pin definitions for EIA-232D are given in Table A4.3. The EIA-232D version was adopted in January 1987 and brings the specification in line with CCITT V.24/V.28. It includes addition of local loopback, remote loopback, and test mode circuits. Protective ground has been redefined and a shield added. Furthermore, the term DCE is substituted for data set, with the definition changed from data communications equipment to data circuit-terminating equipment. Some other terminology has also been changed.

Problems

4.1 This problem involves construction of approximate state and timing diagrams for an ordinary telephone conversation using major states that occur in such a conversation. The following "signals" will be used to define states:

$$OH = \begin{cases} \text{OFF if phone off hook} \\ \text{ON if phone on hook} \end{cases} \quad B = \begin{cases} \text{OFF if no busy signal} \\ \text{ON if busy signal} \end{cases}$$

$$DT = \begin{cases} \text{OFF if no dial tone} \\ \text{ON if dial tone} \end{cases} \quad T = \begin{cases} 1 \text{ if user talking} \\ 0 \text{ if user not talking} \end{cases}$$

$$R = \begin{cases} \text{OFF if no ringing signals} \\ \text{ON if ringing signal} \end{cases} \quad L = \begin{cases} 1 \text{ if user listening} \\ 0 \text{ if user not listening} \end{cases}$$

States will be defined by the two octal numbers (OH, DT, R) and (B, T, L), where OFF = 1 and ON = 0 as in the text.

a. Construct a state diagram based on these state definitions and showing legitimate transitions between states. Include all state transitions normally involved in placing a call or receiving one. (As in the diagram in Fig. 4.2(a), indicate when a number is dialed without specific states for dialing.)

b. Construct a timing diagram similar to Fig. 4.4 and showing Telephone A placing a call to Telephone B (see Fig. P4.1). The user of Telephone B says, "Hello," after which the user of A starts a sales pitch. The user of B, who is not interested in buying, hangs up during the sales pitch.

4.2 Two DTEs are communicating via modems and RS-232C interfaces over a half duplex link. Data transmission is one directional with each message of length M. After a message is transmitted, the sender waits for an acknowledgment message of length n_a

Figure P4.1   Telephone Interconnection.

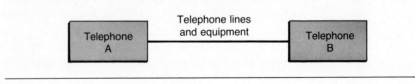

Figure P4.4   Figure for Problem 4.4.

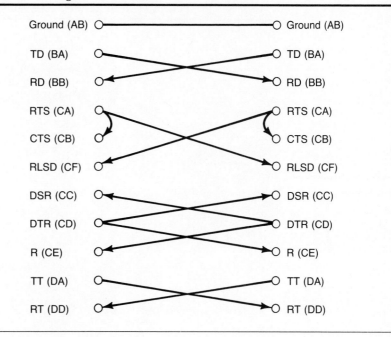

before sending another message. Assume error-free operation and the following parameter values:

R (line bit rate) = 9600 bps
M (message length) = 1000 bits (all data bits)
n_h (overhead in each message packet) = 24 bits
n_a (acknowledgment length) = 24 bits
t_p (propagation time in one direction) = 1 msec
Minimum transition time between states (1,3) and (1,1) (RTS-CTS delay) 10 msec at message transmitting end, 20 msec at message receiving end (acknowledgment transmitting end)
All other delays during message transmission and reception negligible

For the stated conditions, determine the effective data rate in bps.

4.3   Sketch the interconnection of two DTEs via the type of null modem discussed in Subsection 4.3.4 and discuss the purpose of each interconnection.

4.4   A slightly more elaborate type of null modem is sketched in Fig. P4.4. Discuss the purpose of each interconnection. (Circuits in Fig. P4.4 but not in Fig. 4.3 are TT–Transmitter Timing, and RT–Receiver Timing.)

4.5   Compare X.21 control lines with major RS-232C control lines on a circuit-by-circuit basis. Comment on the differences.

Figure P4.6   State Diagram for X.21 Leased Point-to-Point Service.[32]
Reprinted with permission from ITU, copyright © 1984.

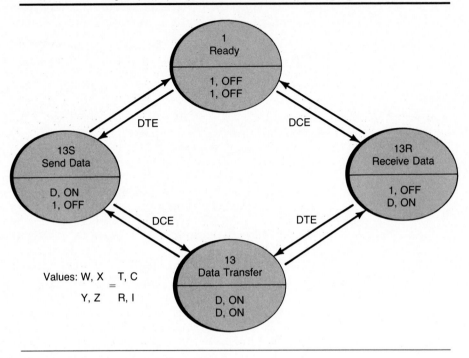

4.6   The diagram in Fig. P4.6 is the simplest state diagram in the X.21 standard. Its simplicity is due to the fact that it is applicable for leased point-to-point service, so no connection and disconnection phases are necessary.

On axes equivalent to those in Fig. 4.6 in the text, sketch a timing diagram based on this state diagram. Show stations A and B both starting in the Ready state, with station A first transmitting to B and station B replying. Your sketch should be for half duplex (one direction at a time) transmission. Indicate state numbers for each station initially and after each state transition.

4.7   Two state diagrams from the X.21 standard are in Fig. P4.7, one valid during call control for circuit switched service and the other during clearing.

   a.   Verify that the sequence in Fig. 4.6 in the text is valid by tracing the sequence of states entered on a copy of these figures.

[32] (Reprinted with permission from ITU [International Telecommunications Union], CCITT [International Telegraph and Telephone Consultative Committee], "Interface between Data Terminal Equipment (DTE) and Data Circuit-Terminating Equipment (DCE) for Synchronous Operation on Public Data Networks." *CCITT Red Book,* vol VIII.3 1984. Updated material can be found in the *CCITT Blue Book* and may be obtained from the ITU General Secretariat, Place des Nations, CH-1211 Geneva 20, Switzerland.)

b. If an incoming call comes in at the same time as a request for an outgoing call, which call wins out?

c. What sequence of states is used when the DCE clears the call after informing the DTE of its inability to complete a call via an appropriate call progress signal?

4.8 Although they were not discussed in this chapter, timeouts followed by branches to appropriate clearing or recovery routines are an essential part of physical layer protocols. Three (out of more than 20) X.21 timeout situations are listed below. In each case propose an appropriate clearing routine. You may find it helpful to refer to Fig. P4.7.

a. The DTE sends appropriate signals to enter State 5, DTE Waiting, which is normally followed soon by reception of Call Progress Signals, DCE Provided Information, Ready for Data or DCE Clear Indication (states 7, 10A, 12, or 19), but none of these is received within the timeout period.

b. The DTE sends signals to enter states 7 or 10A, Call Progress signals or DCE-Provided Information. This is normally followed soon by other Call Progress Signals or DCE Provided Information or by Ready for Data or DCE Clear Indication (states 12 or 19) but none is received within the timeout period.

c. The DCE sends appropriate signals to enter state 3, Proceed to Select, which is normally followed soon by reception of selection characters, but the selection characters do not arrive within a timeout period.

4.9 (**Design Problem**) Design an automatic answering device for a DTE using X.21 interface to its DCE. It should be possible for the user of the DTE to set up a list of calling party addresses from which calls will not be accepted, or to set up a list of all those from which calls will be accepted. It should also be possible to set up a list of addresses from which reverse charged (collect) calls will be accepted, or to block all reverse charged calls. Calling party addresses and charging information can be assumed to be provided to the DTE while the interface is in state 10B. If a call is rejected, the DTE immediately initiates circuit disconnection; otherwise it completes the normal call acceptance procedures with circuit disconnection at the end of the call. Draw a diagram of your device with descriptions of the functions performed by each component.

4.10 Assume four DTEs are bidding for access to the D channel using an I.430 multipoint interface. They are characterized by the following parameters: $X_1 = X_3 = 9, X_2 = 10, X_4 = 11, A_1 = (10110010), A_2 = (00001011), A_3 = (10001110), A_4 = (00000001)$. Sketch the bidding sequence in a manner similar to Fig. 4.10. Indicate which DTE wins the bidding and what X_i values are applicable for each after completion of bidding.

4.11 The flow chart in Fig. P4.11, from the I.430 standard, describes the D channel access algorithm.

a. Compare this flow chart with the behavior of each DTE in the bidding sequence illustrated in Fig. 4.10.

b. Explain how X_1 for DTE1 is set back to 8 after all priority 8 DTEs have had a chance to transmit.

4.12 Assume a DTE (computer) uses the simple type of computer interface sketched in Fig. 4.13. The cycle time of the computer is 1 μsec and each interrupt service routine requires 100 cycles of processing time, including time spent getting into the service routine and out of it. Plot percentage of CPU cycles used for communications I/O

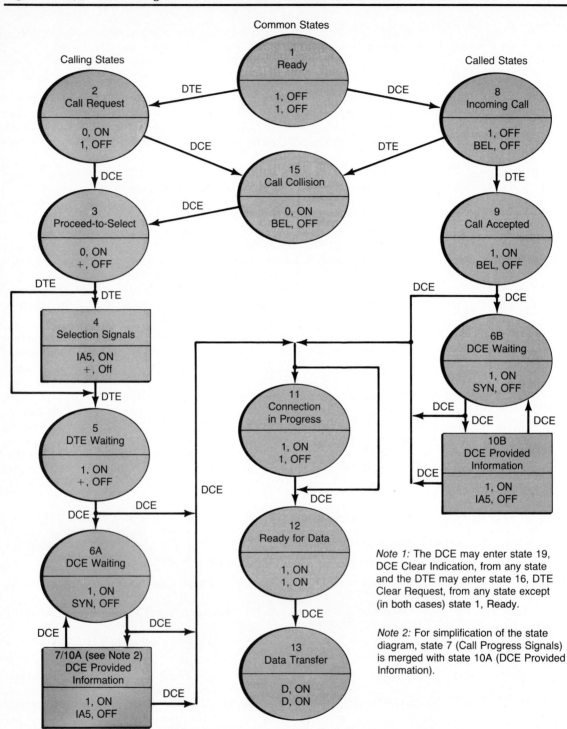

Note 1: The DCE may enter state 19, DCE Clear Indication, from any state and the DTE may enter state 16, DTE Clear Request, from any state except (in both cases) state 1, Ready.

Note 2: For simplification of the state diagram, state 7 (Call Progress Signals) is merged with state 10A (DCE Provided Information).

Figure P4.7(b)   State Diagram for X.21 Clearing Phase.[33]

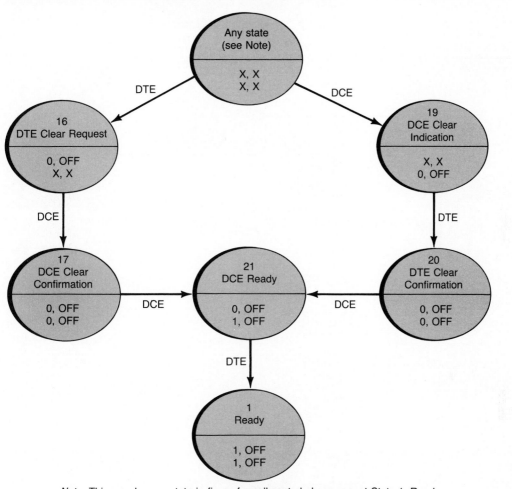

Note: This may be any state in figure for call control phase except State 1, Ready

[33](Reprinted with permission from ITU [International Telecommunications Union], CCITT [International Telegraph and Telephone Consultative Committee], "Interface between Data Terminal Equipment (DTE) and Data Circuit-Terminating Equipment (DCE) for Synchronous Operation on Public Data Networks." *CCITT Red Book,* vol VIII.3 1984. Updated material can be found in the *CCITT Blue Book* and may be obtained from the ITU General Secretariat, Place des Nations, CH-1211 Geneva 20, Switzerland.)

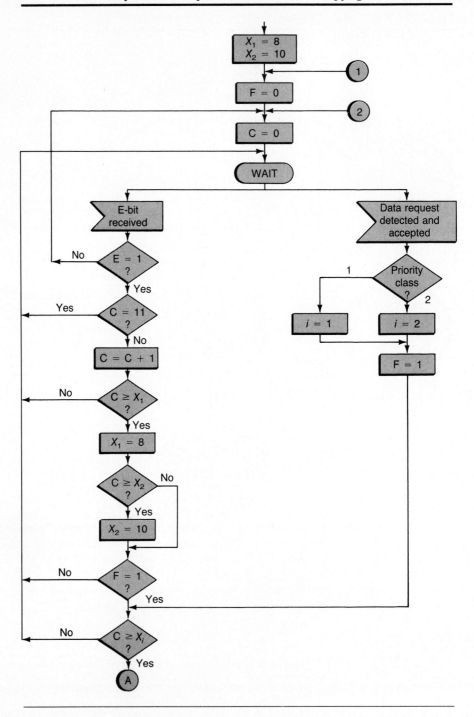

Figure P4.11 *continued*

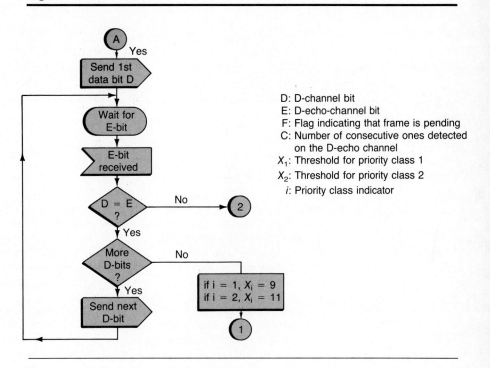

D: D-channel bit
E: D-echo-channel bit
F: Flag indicating that frame is pending
C: Number of consecutive ones detected
on the D-echo channel
X_1: Threshold for priority class 1
X_2: Threshold for priority class 2
i: Priority class indicator

versus input channel rate for rates varying from 0 to 8000 characters per second (0 to 64 kbps with 8 bit characters).

4.13 Assume that the computer interface in Problem 4.12 is replaced by one using DMA. Cycle time of the computer is again 1 μsec. On the average a block of 32 characters is transferred directly into the CPU's memory each time a transfer occurs, with one memory cycle required for each word transferred (assume four character words). Ten per cent of these DMA memory cycles delay CPU cycles (which need to access the memory bus). In addition, 100 cycles of CPU time are required to service an interrupt after each DMA transfer. Again plot percentage of CPU cycles used for communications I/O versus input channel rate for rates varying from 0 to 8000 characters per second.

4.14 (**Design Problem**) Design a multiline computer interface similar to that in Fig. 4.13 but using sharing of any components for which sharing is reasonable. Give a block diagram of your interface and describe its operation.

[34](Reprinted with permission from ITU [International Telecommunications Union], CCITT [International Telegraph and Telephone Consultative Committee], "Basic User-Network Interface-Layer 1 Specification." *CCITT Red Book,* vol III.5 1984. Updated material can be found in the *CCITT Blue Book* and may be obtained from the ITU General Secretariat, Place des Nations, CH-1211 Geneva 20, Switzerland.)

Figure P4.15   Figure for Problem 4.15.

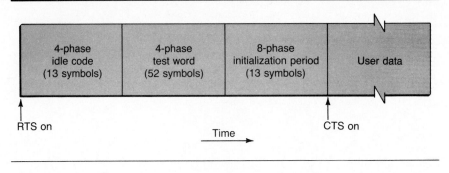

4.15   The training sequence for a Bell System 208 modem, a synchronous 4800 bps modem using eight-phase differential PSK, is shown in Fig. P4.15. Each symbol is three bits. The four-phase idle code allows the receiver to recover carrier and clocking information; the four-phase test word allows an adaptive equalizer in the receiver modem to adjust; the eight-phase initialization period prepares descrambler circuits for eight-phase operation.

a.   Compute the duration of the training sequence.

b.   Compute the effective rate with this modem, assuming that the training period is equivalent to modem turnaround time and that other parameters are the same as for Fig. 4.15 (that is, propagation delays are 10 msec, acknowledgments consist of six eight-bit characters and overhead in message blocks is six eight-bit characters). Consider both messages consisting of 300 eight-bit characters including overhead, and messages consisting of 30 eight-bit characters including overhead.

4.16   Plot the effective data rate (R_e) versus line rate (R) in bps over the range $0 \le R \le 64$ kbps, assuming: propagation delays = 10 msec, line turnaround time = 100 msec, acknowledgments consist of six eight-bit characters, overhead in message blocks is six eight-bit characters, and messages consist of 300 eight-bit characters including overhead.

4.17   Repeat Problem 4.16 but use a propagation delay of 270 msec rather than 10 msec. All other parameters remain the same.

4.18   Plot the effective transmission rate (R_e) versus the data portion of message length ($K - n_h$) over the range $0 \le K - n_h \le 10{,}000$ eight-bit characters. Assume the following conditions: line speed = 4800 bps, propagation delays are 10 msec, line turnaround time = 100 msec, acknowledgments consist of six eight-bit characters, and overhead in message blocks is six eight-bit characters. Comment on any problems you see in using the message length giving maximum value of R_e.

5 Medium Access Control

5.1 Introduction

To communicate between two points, a communications medium must be used. If the medium is controlled by one user, there is no need for medium access control. This is the case for store-and-forward computer networks with full duplex links; for a store-and-forward network, all traffic on a link from node **A** to node **B** is intended to go between these locations, with **A** controlling link access.

In contrast to store-and-forward networks, there are networks for which many users share a common channel or *multiaccess medium.* Examples are local area networks, metropolitan area networks, satellite networks, and packet radio networks. For a multiaccess medium, the signal received at any one station can come from a variety of other stations. Medium access control is needed to keep signals from multiple stations from interfering with one another.

In a typical network there may be a wide range of requirements for users connected to a multiaccess medium. Some users may require frequent transmissions of short messages while others may require lengthy file transfers. During some time intervals, a small percentage of users will desire to transmit, but during others most will be active.

In this chapter we discuss medium access methods that can be used under conditions for which service required by users at different locations is highly variable. For completeness, and to give an overall perspective, we also include protocols used primarily when required service is less variable. The performance modeling techniques in Section 5.3 are more mathematical than most of the book, though elementary versions of models are used; they are not necessary background for later material. A separate coverage of these aspects can be found in [HAMM86] or, at a higher level, in [BERT87] or [SCHW87].

multiaccess geometries

Geometries assumed by multiaccess media are restricted by a requirement that each station be able to hear the signal from every other station.[1]

---

[1] Although there are important situations where only a subset of stations can be heard by a given station, we assume all stations can hear each other unless we state otherwise.

Figure 5.1   Common Geometries for Multiaccess Media.

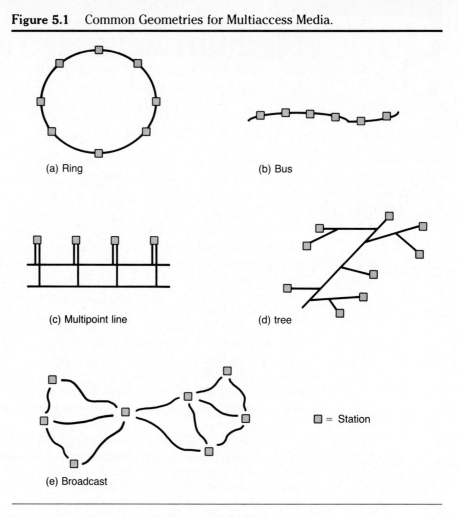

(a) Ring

(b) Bus

(c) Multipoint line

(d) tree

(e) Broadcast

▨ = Station

Five common geometries for which this is possible are shown in Fig. 5.1. For broadcast media, a situation in which some groups of stations cannot hear the signals of others directly, without relaying, could occur.[2] As we discuss below, the network geometry often has an influence on the medium access method chosen.

As was noted above, there are a number of applications that use multi-access media and hence require medium access control. Introductory discussions of several such applications are given in Chapter 3: local area networks in Subsection 3.3.1, metropolitan area networks in 3.3.3, and radio

---

[2]An important example of this is packet radio.

and satellite networks in 3.3.4. Both local area networks and metropolitan area networks are highly structured, and standards have been developed for their protocol architectures; IEEE 802 standards are currently the best known. Figure 3.24, which shows the IEEE 802 architecture layers for local area networks, indicates that a medium access control layer has been identified; this chapter is concerned with protocols for medium access.

5.2 Modes of Accessing Communications Media

In this section we discuss common medium access methods from the perspective of their operating strategies. We first list general approaches and then consider specific methods. In subsequent sections we develop performance models for the more important cases.

There are two general approaches to sharing a common channel:

scheduling
random access

- Scheduling, in which ready stations transmit in an orderly scheduled fashion

- Random access, in which stations transmit when ready in a random free-for-all approach.

In a network environment, both approaches can have undesirable overhead. In scheduling, time spent in finding that some stations in the scheduled sequence do not desire to transmit is wasted. At the other extreme, using random access, no time is wasted with stations not desiring to transmit, but difficulty arises due to *collisions* that occur when more than one station attempts to transmit at the same time. Such collisions must be resolved using channel resources. A variety of media access techniques using each general approach have been developed; all attempt to minimize overhead.

collisions

fixed assignment and demand assignment

Scheduling approaches can be further subdivided into *fixed assignment* and *demand assignment*. Demand assignment can be further classified as central control and distributed control. The random access methods have a number of variations. One useful classification subdivides these methods according to the type of information sensed from the transmission medium. Resulting categories are *no sensing, sensing before transmission,* and *sensing both before and during transmission.* Algorithms identified with these categories are, respectively, ALOHA, carrier-sense-multiple-access (CSMA), and carrier-sense-multiple-access with collision detection (CSMA/CD).

In passing we note that circuit switching of the type used in a computerized branch exchange, or PBX, can be regarded as a method for handling medium access. In contrast to the methods discussed above, circuit switching provides a complete physical connection between a pair of users for the duration of a call. This approach, however, has decidedly different characteristics from those listed above and will not be considered here.

Having identified the major categories of medium access methods, the next step is to consider specific implementations. We first list methods to be considered and then discuss each in detail. Three representatives are chosen from each of the two general methods to illustrate the range of possibilities. These six specific methods, classified by their general approach are as follows:

- General Approach: Scheduling
  - TDMA/reservations
  - Polling
  - Token passing
- General Approach: Random Access
  - No sensing (ALOHA)
  - Sensing before transmission (CSMA)
  - Sensing both before and during transmission (CSMA/CD)

As Chapter 3 points out, IEEE 802 protocols are defined for CSMA/CD and two types of token passing networks, token rings and token buses.

5.2.1   TDMA/Reservations

The first access protocol we discuss, from the general class of scheduling methods, falls into the subdivision of fixed assignment.

TDMA

As the section heading implies, two variations of the approach are considered. One of the most basic of all possible medium access protocols is *time division multiple access* (*TDMA*). This method, which is closely related to time division multiplexing, assigns a fixed time slot to each station. Time is divided into frames and one unit of data is transmitted by each station as its turn comes up in each frame. The unit of data is arbitrary: one character, one packet, and one bit are frequent choices.

frame structure

Figure 5.2 shows a common *frame structure* for use with a satellite channel. Each frame has a control segment, which contains a synchronization pattern to keep stations in synchronization, as well as other control data. The remainder of the frame contains a time slot allocated to each station and guard bands to ensure that these are separated. The typical station time slot allows for transmission of both overhead and data signals.

A frame structure for use with nonsatellite media, such as a coaxial cable, could be somewhat simpler than that in Fig. 5.2. The guard bands might not be necessary and control overhead might not be required for every slot. A flow diagram showing logical steps in the algorithm followed by each station for TDMA operation is shown in Fig. 5.3.

The major disadvantage of TDMA is the requirement that each station must have a fixed allocation of channel time whether or not it has data to transmit. In most networking environments transmission requirements are bursty, in the sense that stations have busy periods during which they trans-

Figure 5.2 Basic TDMA Frame Structure.

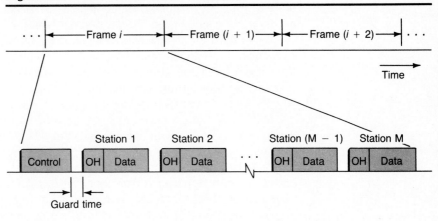

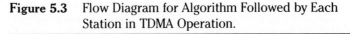

Figure 5.3 Flow Diagram for Algorithm Followed by Each
Station in TDMA Operation.

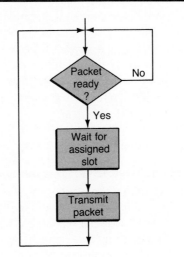

mit significant amounts of traffic and idle periods when they transmit little or
no traffic, interspersed at random. For this common type of environment, a
fixed allocation of channel time to each station is wasteful of resources.

There are environments, however, in which all stations have consistent
traffic patterns over long periods and for this type of "stream" traffic TDMA
is an excellent access protocol. If all stations require regular service, the fixed
allocation of resources is not a drawback and other characteristics of TDMA

are desirable. For example, it does not consume overhead in selecting stations to transmit or in resolving collisions between stations. It is, however, necessary to maintain synchronization between the stations.

reservations

A modification to the basic TDMA strategy, termed *reservations,* attempts to reduce the inefficiencies associated with completely fixed channel allocation. The reservations strategy works well when not all stations are transmitting and the transmitting stations have predominately stream traffic (that is, when stations transmit, they tend to transmit long messages).

In its basic form, a reservation type protocol uses a frame structure similar to that of Fig. 5.2, with two modifications. There are fewer available time slots than there are stations and some channel resources are used for making reservations, possibly through allocation of a small number of time slots in each frame to this purpose. After reservations are made by stations wishing to transmit, TDMA operation is used with only those stations needing to transmit assigned time slots in each frame. Making reservations is a dynamic process, and slot allocations can be changed as stations experience changes in their traffic demands.

Since the multiaccess channel is normally the only connection between stations, it must also be used (with some medium access protocol) during the reservation period. The choice of protocol during the reservation period is independent of the choice of protocol for transmitting data and thus any medium access protocol could be used for reservations. The length and frequency of reservation periods depends on the number of stations involved and the type of traffic. Bursty traffic, for which stations have short messages transmitted at random times, requires frequent reservations whereas stream traffic, such as file transfers with long messages, requires reservations much less often.

A flow diagram for the basic steps in the algorithm followed by each station for a reservation protocol is given in Fig. 5.4. Details of making reservations and waiting for assigned slots depend on implementations and are discussed below. Differences in types of reservation protocols depend primarily on the way in which reservations are made. Two major categories are identified as *implicit* and *explicit* reservations. For explicit reservations, some type of channel resource is set aside for stations to use in reserving slots for transmitting data. Implicit reservations are made without the use of dedicated channel resources.

R-ALOHA

A scheme referred to as *R-ALOHA* [CROW73] is an example of an implicit reservation method. For this method there are typically fewer slots in a TDMA frame than stations, although this is not a requirement and the number of stations can vary dynamically. Each slot provides time to transmit one packet. Any slot position in a current TDMA frame that is either empty or contains a collision (that is, transmissions from more than one station) can be captured for use in the next frame. Stations monitor the channel and contend for the "open" slots using slotted-ALOHA, a random access method discussed below. By repeatedly using a given slot time, a station retains the

Figure 5.4   Flow Diagram for the Algorithm Followed by Each Station
for a Reservation Type Protocol.

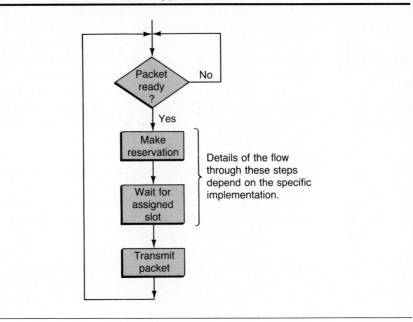

ability to transmit and can transmit multipacket messages. Under conditions
such that most stations are transmitting stream traffic, the method has characteristics similar to those of fixed-assignment TDMA. For bursty traffic,
behavior is closer to that of slotted-ALOHA.

A variation of R-ALOHA [BIND75] requires a fixed number of stations with
the number less than or equal to the number of slots in a frame. If the frame
contains more slots than there are stations, the stations contend for the extra
slots using slotted-ALOHA. Each station has first call on use of one slot, and
uses it as long as it has data to transmit. If the slot is left idle, another station
can capture it; however, that station must release the slot if a collision occurs.
Thus a station can regain its own slot just by transmitting in it. Of course, in
both versions of R-ALOHA, collided packets must be retransmitted.

The implicit reservation schemes just discussed have the property that
they are "distributed" in the sense that there is no central control over all
stations. The explicit reservation schemes, to be discussed next, can have
either distributed or centralized control. Before discussing explicit reservation schemes in detail, we note that channel resources for reservations can
be set aside either as reserved time slots or by dedicating one frequency band
to reservations and another to TDMA data frames. There seems to be no
major difference between the two approaches.

As was just noted, explicit reservation schemes can be either distributed or centralized. In one *distributed reservation scheme* [ROBE73] a frame is subdivided into equal length slots and the first slot is further subdivided into minislots. In order to transmit, a station must make a reservation for some number of slots, up to a specified maximum. Reservations are made using slotted-ALOHA to compete for use of reservation minislots. A reservation is successfully made if the reservation request is collision-free.

Having successfully made a reservation, a station has the additional responsibility of determining the future frame and slot in which it can transmit. The method allows transmissions on a first-come-first-served basis in order of successful reservations. Thus a successful reservation entitles a station to begin transmitting in the first data slot occurring after all previously reserved slots have been used. Reservation minislots are monitored by all stations. Thus each station can keep a running sum of the number of slots successfully reserved and which will be used in sequence for transmissions before it is entitled to transmit (following its own successful reservation). The number of slots that must be counted off and waited before a station can begin transmitting, following its successful reservation, is equal to the sum of the number of slots successfully requested by all stations, less the number of slots containing data that have already been transmitted.

Distributed reservation schemes, such as that just discussed, require that all stations develop and maintain the correct state of the complete network from normal transmissions. This imposes a computational burden on stations and makes the network vulnerable to loss of synchronization. These drawbacks of distributed reservations can be removed by using *centralized reservation schemes* for which one station maintains a network clock and assigns channels. Of course, centralized methods have their own shortcomings, which include vulnerability to loss of the central station and additional time delays in processing requests and assigning channels.

fixed-priority-oriented-demand assignment

For a small and fixed number of stations, a centralized reservation method can be unsophisticated; an example is *fixed-priority-oriented-demand assignment* (FPODA) [JACO79]. One station acts as the master and controls the channel based on requests from stations. A frame is subdivided into a reservation slot and one slot for each station; a well-known application has only six stations [WATE84]. The reservation slot has a minislot for each station, and (using its minislot) each station requests a particular allocation of data slots for three categories of traffic: priority, normal, and bulk. The station also transmits an estimate of future requirements.

The master station maintains a directory of requested service and allocates available slots. Priority traffic is served first on a first-come-first-served basis. Normal traffic is serviced next, and bulk traffic last if channel time is available. If throughput requirements of a station remain constant, it does not need to continually make reservations. This means that in steady state operation some minislots can be used for short data blocks or control messages.

If it is necessary to accommodate a large and variable number of stations, the most significant change required in the centralized reservation method described is to allow contention for reservation minislots. A master station still manages information slots based on reservations made by other stations. Several variations in the structure of the user data part of the frame are possible. These include having a variable number of slots for data and having two types of data slots, one type reserved and one type open to be contended for by the stations.

Recall that TDMA and TDMA/reservations protocols are scheduling methods. TDMA is a (strictly) fixed assignment strategy whereas reservation protocols could be termed quasi-fixed assignments. Under heavy traffic conditions, reservation protocols have characteristics similar to those of TDMA.

The next subsection treats polling, a scheduling method that uses demand assignment and central control to assign channel time to the stations.

5.2.2 Polling

As was just noted, *polling* is an illustration of a central control type of protocol. Such geometries as the tree and multipoint line in Fig. 5.1 are particularly suited to polling, although in principle it could be used with any geometry. Regardless of geometry, polling is carried out by a control station connected to all other stations on the network by a common link. Other texts that discuss polling include [BERT87], [HAMM86], [HAYE84], and [SCHW87].

polling and go-ahead messages

The essence of the polling procedure can be described as follows. The central computer sends a *polling message* to the first station in the polling sequence. After receiving the polling message, this station transmits its data to the central computer and indicates that it has completed transmitting by adding a *go-ahead message*. The central computer then polls the next station in sequence and the process continues until all stations have had an opportunity to transmit. The polling sequence is used over and over for as long as desired. The stations can receive messages from terminals or other sources and store these messages locally when not transmitting.

The algorithms followed by the central controller and each station can be represented by the flow diagrams in Figs. 5.5 (a) and (b). In part (b) note that if a station is polled when it has no data to transmit, it immediately sends a go-ahead back to the control station. Other steps in the flow diagram are apparent or have been discussed above. Figure 5.6 (a) gives the message flow for the basic polling network and is another useful way to represent operation.

This description of polling is intended to present basic ideas behind the method and it omits details that might obscure basic ideas. One such detail concerns the nature of the link between the stations and the central computer. If this link is full duplex, the central computer can transmit to a station

Figure 5.5   Flow Diagrams for Algorithms for a Roll Call Polling Network. (a) Central controller. (b) Each station.

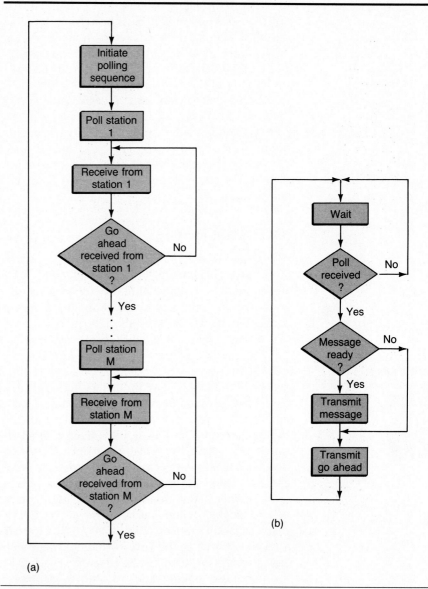

(a)

(b)

Figure 5.6 Message Flow Diagrams for Polling Networks.
(a) Roll call. (b) Hub polling.

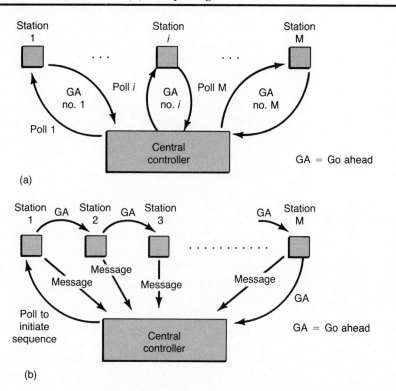

(a)

(b)

on its outgoing link while it is receiving from a different station on the incoming link.[3] On the other hand, if the link is half duplex, transmission can be in only one direction at a time, and the link must be "turned around" when the direction of transmission is changed.

Another detail concerns synchronization. When half duplex links are used, it is necessary to carry out synchronization on both the polling message and the data transfer from the station to the central computer. If full duplex links are used, it is possible for the central computer to maintain continuous synchronization on its outgoing link to all stations. This means that synchronization on a message-by-message basis is not required for transmission in this direction.

roll call polling
hub polling

As a final detail, the type of polling described above is referred to as *roll call polling.* A modification is termed *hub polling.* Hub polling requires a sen-

---

[3] In theory it could simultaneously transmit to and receive from one station, but instances where this is worthwhile are rare.

sor at each station and a link configuration and polling sequence such that after the polling sequence is initiated the next station in the polling sequence that has not yet transmitted can listen to the transmissions of the station transmitting immediately before it.[4] Transmitting stations affix a next station address to their go-ahead messages. Using the added sensors (and links), the next station reads its address and the go-ahead message and can thus begin transmitting without waiting to be polled by the central computer. Message flow for this modified polling operation is shown in Fig. 5.6 (b).

When the polling strategy is compared with the other methods discussed below, it is found to be the least structured in terms of interactions between stations, format of packets, and so forth. Additional comments on polling, including typical packet formats, are given in Chapter 7 on data link protocols.

5.2.3   Token Passing

Another member of the class of scheduling medium access methods is termed *token passing.* Token passing provides distributed control of the sequence in which stations transmit, unlike polling which has a centralized controller. Although the type of control is different, operation of a token passing network is much like that of hub polling.

Use of distributed control necessitates a much more restricted environment than polling, with more structured packets and more structured interactions between stations and between stations and the transmitting medium. Typically, all packets in a token passing environment make a complete circuit of the network and return to the originating station to be removed.

IEEE 802.4 and 802.5

As was pointed out above, token passing is a type of medium access that is often used in local area networks, and two versions are included in IEEE 802 standards: *IEEE 802.4* for the token bus and *IEEE 802.5* for the token ring. A newer standard, FDDI [ROSS86], being developed by the American National Standards Institute (ANSI), is finding applications in high-speed local or metropolitan area networks including those using token passing. The publications of standards committees describe the various methods in precise detail. Typical texts that contain material on token passing medium access methods include [HAMM86], [STAL87a], [SCHW87], and [BLAC89].

Our purpose in this section is to present the major ideas and characteristics of classes of medium access protocols. Reference to the standards literature and to papers such as [BUX89] can be made to obtain details for specific implementations.

---

[4]Although most polling networks are sketched in a manner implying that this type of communication between stations is possible, the actual link connections are often such that direct communication is possible only between the central station and the subsidiary stations. Special wiring (at extra cost for telephone networks) is necessary for hub polling to be feasible.

Figure 5.7   Ring Network Used for Token Passing.

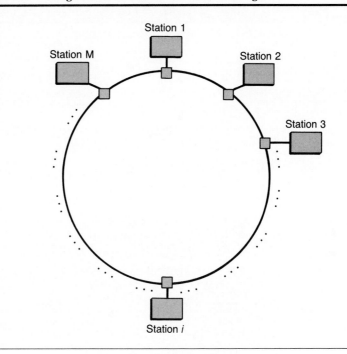

The bus or ring geometries of Fig. 5.1 are normally used for token passing. For a bus geometry, the stations are ordered in a logical sequence and packets are directed through this sequence back to the initiating station. A ring, such as that shown in Fig. 5.7, is the easiest geometry to visualize since there is a natural sequence of stations around the ring; such a ring will be discussed first, with only the essentials included in order to explain basic ideas. Refinements will then be added.

token ring

Each station on a *token ring* network is coupled to the ring through an active interface, which allows the station to be in either a "listen" or a "transmit" condition, as shown in Fig. 5.8. The fact that the interface is active means that the analog signals representing bits arriving from the ring are actively reconstructed as they are retransmitted. This prevents bit shape deterioration over large network spans but requires that all interfaces on the network operate properly for the network to function. Some network interfaces have a "bypass" connection to make it possible to remove a station from the ring.

A delay of at least one bit time in each interface is necessary to allow time for a station to obtain information from an incoming packet and act on any necessary decision before producing its output packet. This delay makes it possible to manipulate tokens, as dicussed below, and remove or pass on packets based on header information. In the listen mode, shown in Fig. 5.8

Figure 5.8   Station/Ring Interface for a Token Passing Ring Network. (a) Logical operation of interface. (b) Attached device and components of interface.

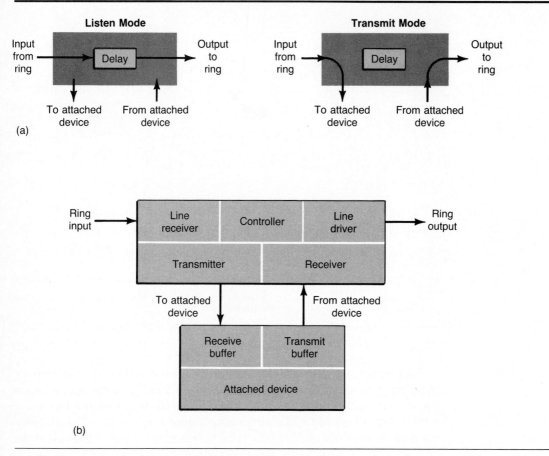

(a)

(b)

(a), the ring is closed through the station. The station reads headers on incoming packets and duplicates, at its interface output after the station delay, all bits arriving from the ring. In the transmit mode, the station opens the ring, removing the bits coming in from the ring, while transmitting bits from the attached station over the output port to the ring. A typical interface structure is shown in Fig. 5.8 (b).

token   A *token* is a distinctive bit or bit pattern different from all other bits or patterns, or if not different, protected by expedients such as "bit stuffing." For example, a token could be a sequence of a zero followed by seven ones.[5] To

[5] A token defined in IBM's SDLC Protocol (see Chapter 7) is of precisely this form.

protect this pattern, if a sequence of this sort occurs in a data pattern, it is broken up by adding, or stuffing, an extra bit (in this case a zero) to break up the pattern. When stuffing is used, the stuffed bits are removed at the receiver before passing data to the device attached to the ring.

In operation, the token has two states: busy and free. If no station is transmitting, the token is set to "free" and allowed to circulate around the ring from one station to another. A station with data to transmit "captures" the ring by changing the token from free to busy as it passes through the station interface. In the example above, the pattern representing a free token is modified by changing the last bit of the token pattern to a zero to indicate a busy condition.[6] The transmitting station transmits the bits of its data packets immediately after the bits of the busy token. When its data packets have been transmitted, the station that claimed the token is responsible for replacing the free token, which then circulates to allow another station to transmit.

multiple token, single token, and single packet operation

The precise time at which the free token is put back on the ring by the transmitting station is commonly determined by one of three token management schemes: *multiple token operation,* for which the free token is transmitted immediately after the last bit of the data packet; *single token operation,* for which the free token is transmitted after the transmitting station receives the last bit of its busy token (and has also transmitted the last bit of its data packet); and *single packet operation,* for which the free token is transmitted after the transmitting station has received the last bit of its transmitted packet. The single token mode has been chosen for the IEEE 802 standards. A flow diagram for the algorithm followed by each station, indicating the three different modes of token management, is given in Fig. 5.9.

The three operational modes are illustrated with a simple example in Fig. 5.10. In this diagram discrete time is displayed in the vertical direction and progress through the four stations on the ring is given along the horizontal axis. Propagation delay is ignored, so the output of one station is the input of the adjacent station next in order around the ring. This is emphasized in the diagram by showing both the outputs of station 4 and the inputs of station 1, which are the same. The reader is encouraged to trace the flow of bits around the simple ring for all three types of token management.

The description above covers basic operation of the token ring access protocol. Practical implementations require attention to additional details. Basic approaches to some of the more important details are discussed below, although the standards should be consulted for exact implementations. More details on most features mentioned can be found in the excellent survey papers on token rings [BUX89] and [STRO87].

The length of time a station is allowed to transmit when it has captured the free token is frequently determined by a token-holding time specified as a part of the protocol. If the token-holding time is long, a station can transmit

---

[6] For SDLC, this produces a pattern of a zero followed by six ones and another zero. This is an SDLC flag, which is the initial character in all SDLC transmissions.

Figure 5.9   Flow Diagram for the Algorithm Followed by Each Station on a Token Ring. (Note three different operating modes.)

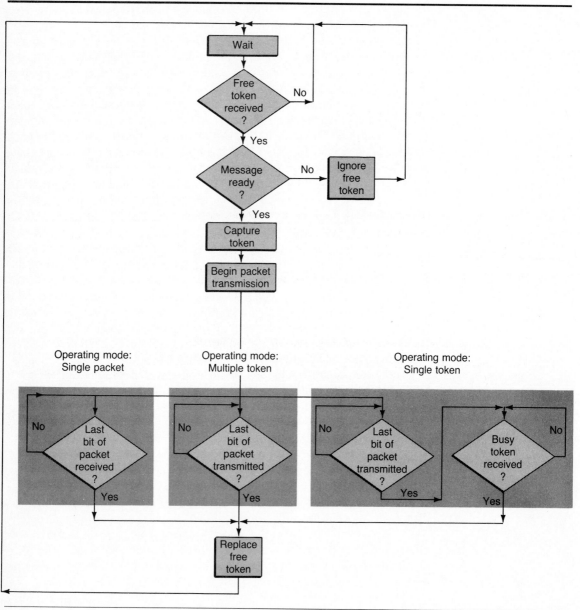

Figure 5.10   An Illustration of Packet Flow Around a Token Ring for Three Operating Modes (adapted from [HAMM86]).

Discrete time	Multiple token					Single token					Single packet					Discrete time
	In 1	Out 1 / In 2	Out 2 / In 3	Out 3 / In 4	Out 4	In 1	Out 1 / In 2	Out 2 / In 3	Out 3 / In 4	Out 4	In 1	Out 1 / In 2	Out 2 / In 3	Out 3 / In 4	Out 4	
1	□				□	□				□	□				□	1
2		■					■					■				2
3		d	■				d	■				d	■			3
4		d	d	■			d	d	■			d	d	■		4
5	■	d	d	d	■	■	d	d	d	■	■	d	d	d	■	5
6	d	d	d	d	d	d	d	d	d	d	d	d	d	d	d	6
7	d	d	d	d	d	d	d	d	d	d	d	d	d	d	d	7
8	d	□	d	d	d	d	□	d	d	d			d	d	d	8
9	d		□	d	d	d		□	d	d	d			d	d	9
10	d			□	d	d			□	d	d				d	10
11	■			■		■				■	□					11
12	d	■			d	d	■			d		□				12
13	d	d	■		d	d	d	■		d			□			13
14	□	d	d	■	□			d	d	■	■				■	14
15		□	d	d		□			d	d	□	■			d	15
16			□	d			□			d	d	d	■		d	16
17				□				□				d	d	■		17
18	□				□				□				d	d		18
19						□				□					d	19
20											□				□	20
21												□				21
22													□			22
23														□		23
24											□				□	24

Out 4 / In 1   Out 1 / In 2
Out 3 / In 4   Out 2 / In 3

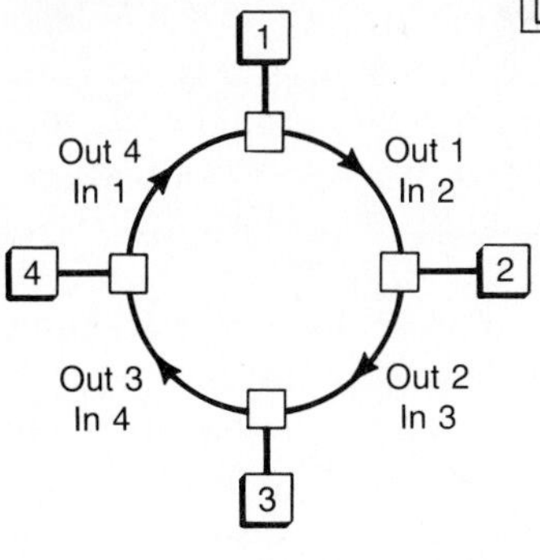

□ Free token
■ Busy token
d Data bit

all of its backlog in each use of the channel, effectively providing "exhaustive service." On the other hand, for short token-holding times only one packet may be transmitted per use of the channel.

The operation of the token ring as described above is synchronous, and maintaining synchronization is a design problem. The IEEE 802.5 standard specifies a central clock maintained by one station on the ring, designated as a monitor station. All stations on the ring are frequency- and phase-locked to the monitor. The FDDI approach is more flexible with regard to synchronization. It is easier to implement at high transmission rates but requires a larger ring latency.

The frame format specified by the IEEE 802.5 standard contains a number of fields including starting and ending delimiters, a frame check sequence, source and destination addresses, an information field, and an access control field.[7] The access control field, which contains a one-bit token, a single monitor bit, and three bits each for priority and for priority reservation, is of particular interest because these bits must be read by a station before a packet can be passed on to the next station. The station latency, or the number of bit times required for a packet to pass through a station interface, is thus determined by the access control field. For the IEEE 802.5 standard just described, a station latency of eight bits is required.

Clearly, a token ring requires integrity of the token for proper operation. Loss of token, unwanted multiple tokens, and loss of a station with the token cause abnormal operation, and such faults must be corrected. Detection and recovery from error is the function of the monitor station according to IEEE 802.5 standards. The standards prescribe an arbitration process for selecting one station as an active monitor (which also maintains the clock).

As a final comment on details of ring networks, the basic token ring operation can be extended to provide levels of priority. This is accomplished through use of the priority and priority reservation bits of the access control field. The procedure is briefly described as follows.

priority operation

For *priority operation,* the token has a priority, specified by its priority bits, for each trip around the ring. Assume that a station (station I) not currently transmitting wishes to transmit at a priority higher than the current transmission from station J. As the next packet passes through the interface of station I, it sets its desired priority in the reservation bits of the access control field of the packet unless another station has already set these bits to a priority at least that of station I. When station J, which initiated the packet, receives it again (with the priority reservation indicated), it transmits a free token with the indicated priority, while remembering the earlier priority. Station I receives the free token with the higher priority and transmits its high priority packet. When the high priority packet has traversed the ring, station I receives it and transmits a free token with the same high priority. Station J

---

[7] This format is given in Chapter 7, Fig. 7.18 (c.ii).

ultimately receives the free token (with high priority) and changes the priority back to the earlier level before passing it on. Further details on the implementation of priorities can be found in [BUX89] or in the standards [INST85d].

token bus

Modifications of the token passing methods used for the ring geometry can be used for *token buses*, although a bus does not lend itself to quite as structured an environment as a ring. Since a bus is not closed on itself, stations on a bus are structured into a logical ring by the manner in which the token is transferred from one station to another; this is accomplished by using the destination address in the frame structure. Stations are assigned an ordered sequence of addresses with each station knowing the addresses of the stations preceding and following it. The station with the token can use the medium for a specified maximum amount of time to transmit data, poll other stations and receive responses from stations. When the station with the token has completed its tasks or used up its time, it sends the token to the next station in the logical sequence.

A major type of flexibility enjoyed by the token bus is the ability to add and delete stations while the network is in normal operation. For the token ring, adding and deleting stations is only feasible when switches to bypass potential station locations are available; otherwise it can be accomplished only through significant changes, involving hardware and requiring that the ring be nonoperational in the process.[8] For the token bus, the price paid for the ability to add and delete stations in normal operation and for using distributed control is complexity in the operational algorithms used by each station.

Token bus algorithms typically provide for normal steady state operation and four types of overhead: initialization, adding stations, deleting stations, and error recovery. In normal steady state operation, after the logical ring is set up and operating correctly, the token bus operates in either a data transfer or token passing mode and can provide multiple priorities in essentially the same manner as token rings.

Initialization is required when operation of the network is started up and when, for some reason, logical operation has broken down. During initialization, stations learn which stations are active and set up a logical order for these stations. Doing this in a noncentralized fashion requires a distributed algorithm; for example, a sequence of information exchanges can take place between active stations, with addresses for a logical ring constructed in an iterative fashion.

In order to add stations, each station must periodically allow time, when it has the token, for new stations to request that they be added. The algorithm must allow for zero, one, or multiple requests and provide a means for expanding the sequence of addresses to include new stations.

Deleting an operational station from the logical ring is simpler since the

---

[8] Such bypass switches are standard for many token rings, including IBM token rings.

station wishing to remove itself can revise the address sequence by removing its own address when it has the token.

Error recovery requires dealing with the same faults as for token rings, and essentially the same methods are employed. Faults unique to token buses (that is, initialization plus adding and deleting stations) are also dealt with in these procedures. Details of methods for error recovery are discussed in [INST85c] and [PHIN83].

For basic steady state operation, the stations on the token bus can follow an algorithm very similar to that of Fig. 5.9 for a token ring. In fact if single priority operation is desired, no major changes are required. For multiple priority class operation, the actions of a station after it captures the token must be adjusted, as was discussed for ring networks, to allow for the multiple priorities. The priority algorithms control the amount of time spent in transmitting each class of data—guaranteeing that the highest priority receives the greatest amount of time, assigning a lesser amount of time to the next lower priority, and so forth.

The medium access methods just discussed—TDMA, polling, and token passing—are all scheduling type algorithms that illustrate, respectively, fixed assignment, central control, and distributed control of the scheduling process. The next section begins the discussion of random access type algorithms.

5.2.4 Random Access with No Sensing

The scheduling type medium access algorithms all provide some method of allowing ready stations to use the transmission medium one at a time in an orderly fashion. In contrast, the random access methods make no attempt at scheduling but allow stations to transmit when ready. The penalty for allowing stations to transmit in this haphazard fashion is the necessity to detect collisions and retransmit collided packets.

Random access algorithms are typically used with the bus, tree, and broadcast geometries in Fig. 5.1. Packets transmitted from any station can be received by any other station and, as in other multiaccess networks, must have a destination address in their header to ensure proper delivery. Over coaxial cable, signals propagate at approximately 65 percent of the speed of light, so the propagation delay is approximately 5 μsec per km. Propagation delays can have a pronounced effect in random access networks.

ALOHA

The most basic random access algorithm is termed *ALOHA* [ABRA70]. It was the first random access algorithm and is discussed in a number of texts such as [BERT87], [BLAC89], [HAMM86], [SCHW87], and [STAL88b]. Stations using this algorithm transmit immediately when they have a ready packet, with no sensing to determine whether or not the channel is busy. Having transmitted a packet, a station must then determine whether the packet has "collided" with packets from other stations or has reached its destination successfully. Lost packets, of course, cause no response from the receiving station. Packets with errors due to collisions or other sources are detected with error detection code bits, which are added as overhead bits, usually in the packet trailer. Use of codes for error detection is discussed in Section 6.3.

A successful transmission is normally indicated by a positive acknowledgment packet sent from the destination back to the transmitting station. The acknowledgment packet is much shorter than the information packet and can be sent over the channel used by the information packet or, in some cases, over a separate channel of reduced bandwidth. The information packet must propagate from the transmitting to the receiving station over the channel and the acknowledgment packet must travel over the reverse route. Thus two source-to-destination propagation times are required before a positive acknowledgment can be received. When such an acknowledgment is received, however, successful transmission of the packet is completed.

If a positive acknowledgment is not received by the transmitting station, it infers that a collision has taken place with packets from another station or stations and the packet must be retransmitted. If the collided packet is immediately retransmitted, and the same strategy is used by all competing stations, another collision is inevitable. To avoid this eventuality, a station determines

backoff algorithm

when it will retransmit a packet through use of a *backoff algorithm*. In its simplest form, a backoff algorithm involves choosing a random number and delaying retransmission by this random number times a packet transmission time. Zero is typically included in the range of the random numbers chosen so there is a nonzero probability that retransmission will be immediate. If all colliding stations make independent random choices of backoff times, there is a high probability they will not collide again when they retransmit.

The number of collisions that occur in transmitting a packet depends on the traffic load on the network, which in turn determines how frequently stations transmit. Regardless of the number of tries necessary, in most implementations of the ALOHA algorithm, a station continues to backoff and then retransmit until its packet is successfully received. A flow diagram for the algorithm followed by each station for the ALOHA medium access method is given in Fig. 5.11.

The basic ALOHA method is efficient for very light network loads but has several drawbacks, discussed in Subsection 5.3.4 and Section 5.4, for medium and heavy loads. As the number of attempted transmissions by all stations increases, more and more channel time is spent in resolving collisions and a point is reached past which the number of successful packets actually decreases as the number of attempted transmissions increases. This occurs at approximately 18 percent of the capacity of the channel. There is also a stability problem pointed out in the performance analysis in Subsection 5.3.4.

A minor modification of the ALOHA method can effectively double, to approximately 36 percent, the fraction of the capacity of the transmission channel that can be used. The modified method, called *slotted-ALOHA*

slotted-ALOHA

[ROBE72], requires that all stations be synchronized and that packet transmissions all start at the beginning of agreed-upon slots. Time slots are typi-

Figure 5.11   Flow Diagram for Algorithms Followed at Each Station for ALOHA and Slotted-ALOHA Medium Access. (Portions in dotted box and below dotted line in propagation delay wait box are executed only for slotted version.)

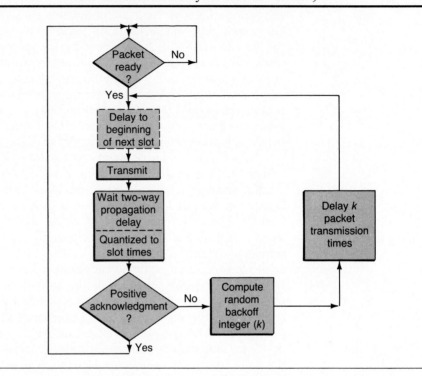

cally one packet transmission time in length. As is discussed in Subsection 5.3.4, slotting reduces the time during which collisions can occur, and thus increases the achievable throughput. The dotted lines in Fig. 5.11 indicate the modifications to the basic ALOHA algorithm needed for slotted-ALOHA.

Improvement with respect to some of the drawbacks of the ALOHA algorithms, especially that of underutilizing the transmission channel capacity, can be made by causing each station to monitor the channel before transmitting. Random access algorithms that monitor before transmission are discussed in the next subsection.

5.2.5   Random Access with Sensing Before Transmission

CSMA

A class of random access algorithms, called *carrier sense multiple access* (*CSMA*) algorithms, uses hardware at each station to sense the state of the channel before transmitting. Basic operation of these algorithms is similar to

Figure 5.12   Flow Diagram for the Algorithm Executed at Each
                  Station in CSMA Medium Access.

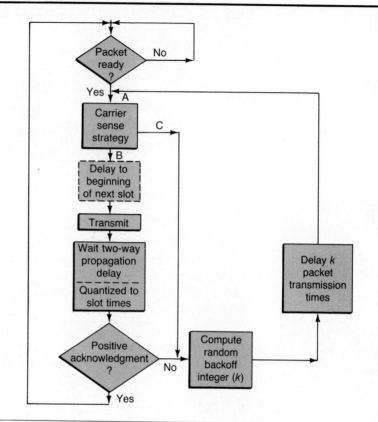

that of ALOHA, but by making use of the sense information a station is able
to defer from transmitting if the channel is busy. This ability to defer reduces
the number of collisions and typically improves performance. CSMA algorithms can operate in a slotted or unslotted fashion similar to that of pure
and slotted-ALOHA

A flow diagram showing the basic CSMA medium access strategy is given
in Fig. 5.12. The only difference between CSMA and ALOHA is addition of
the "carrier sense strategy." There are two versions of CSMA, nonpersistent
and *p*-persistent, which make slightly different use of the carrier sense
information.

nonpersistent CSMA   For *nonpersistent CSMA,* using the carrier sense operation shown in Fig.
5.13, the channel is sensed and the following algorithm is used when a station
has a ready packet:

Figure 5.13   Flow Diagram of the Carrier Sense Strategy for the
Nonpersistent CSMA Algorithm. A, B, and C
correspond to points in Fig. 5.12.

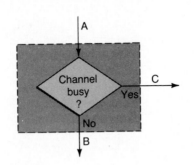

- If the channel is idle, the packet is transmitted.
- If the channel is busy, the station uses the backoff algorithm, following the path through C in the figure, to reschedule the transmission. At the time of retransmission, the channel is sensed again and the algorithm is repeated.

p-persistent CSMA

An alternate algorithm is *p-persistent CSMA*. Two constants are used with this algorithm: τ, the end-to-end propagation delay of the bus, and *p,* a specified probability. A station using the *p*-persistent algorithm senses the channel and then the following occurs:

- If the channel is sensed idle, a random number uniformly distributed on [0,1] is chosen. If the selected number is less than *p,* the packet is transmitted; if not, the station waits τ seconds and repeats the complete algorithm (which includes the contingency that the channel may be busy);
- If the channel is busy, the station persists in sensing the channel until it is found to be idle and then proceeds as described above.

A flow diagram for the carrier sense strategy of *p*-persistent CSMA is given in Fig. 5.14.

A quantitative discussion of the performance of the nonpersistent CSMA algorithm is given in Subsection 5.3.4. Intuitively, the reasoning behind this algorithm is as follows. The nonpersistent algorithm attempts to reduce the number of collisions by sensing the channel before transmitting. If the channel is in use, a ready station immediately backs off, delaying its next transmission a random amount of time. Other stations sensing that the channel is busy back off and also delay their retransmissions a random amount of time. Because of the random delays before retransmissions, repeated collisions are unlikely.

Figure 5.14   Flow Diagram for the Carrier Sense Strategy of the
p-Persistent CSMA Algorithm. A, B, and C correspond
to points in Fig. 5.12.

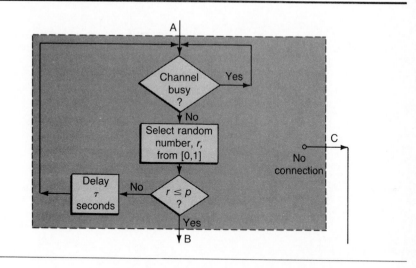

The *p*-persistent algorithm also uses the sense information to avoid
transmitting when the channel is busy. It differs from the nonpersistent algorithm (which never persists in transmitting since it always backs off), by per-
sisting in attempting to transmit. When the channel becomes free, this fact is
immediately sensed by a station using the *p*-persistent algorithm. To avoid
collisions with other ready stations that are employing the same strategy, or
at least to inject flexibility into the algorithm, a station transmits only with
probability *p* when the channel becomes free. With probability $1 - p$ it delays
by the time, τ, required for the first bit of a packet to traverse the complete
length of the bus, and then samples the channel again.

The delay τ is chosen so that two stations sensing the channel to be free
at the same time will not collide if one transmits and the other delays. In time
τ the first bit of the packet from the transmitting station will reach the sensor
of the second station and the sensed signal will prevent it from transmitting.
The parameter *p* can be chosen to optimize the algorithm.

5.2.6 Random Access with Sensing Before and During Transmission

As noted in the previous subsection, the CSMA algorithm differs from ALOHA
only in the carrier sense strategy shown in the flow diagram of Fig. 5.12. Both
CSMA and ALOHA must wait a two-way propagation delay before knowing
whether a transmitted packet has successfully reached its destination or not.

Figure 5.15   Flow Diagram for the Algorithm Executed at Each Station in CSMA/CD Medium Access.

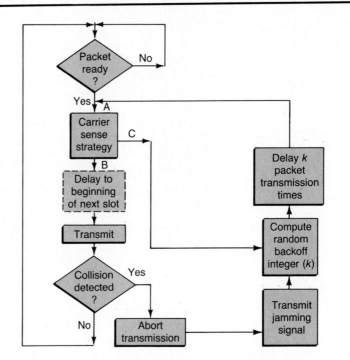

A further refinement of the random access algorithms results from the use of hardware to sense the channel during transmission of a packet as well as before transmission. The method is termed carrier sense multiple access with collision detection, or CSMA/CD. By observing the analog waveform of the signal on the channel during transmission, the presence of a collision can be detected as soon as a distorted waveform reaches the station monitor.[9] This additional monitoring action normally makes it possible to detect a collision much sooner than waiting for a positive acknowledgment, as is necessary for ALOHA or CSMA.

As was noted above, CSMA/CD is frequently used with local area networks and, as such, is one of the standards chosen by the IEEE 802 committee. The *IEEE 802.3* standards [INST85b] give precise details of the algorithm. Texts

IEEE 802.3

---

[9] This is possible only for appropriately chosen signals and transmission media. In other cases, including virtually all radio links, amplitudes of transmitted signals are so much higher than those of received signals that they swamp the latter signals and prohibit collision detection.

providing additional information include [BERT87], [BLAC89], [HAMM86], [SCHW87], and [STAL88b].

A flow diagram of the algorithm followed at each station for CSMA/CD operation is given in Fig. 5.15. Note that the carrier sense operation is the same as for CSMA and uses the same variations—nonpersistent and *p*-persistent. In typical implementations, CSMA/CD algorithms require each *jamming signal*   station to transmit a short *jamming signal* immediately after detecting a collision to ensure all stations know that a collision has occurred.

Both CSMA and CSMA/CD are very sensitive to propagation delays since monitoring, either before or during transmission, cannot take place until signals reach the monitor. Thus if propagation delays are long, the presence of a packet from another station or of a distorted signal due to a collision may not be detected until it is too late to take effective corrective action. Propagation delay is a parameter in the quantitative performance equations in Subsection 5.3.4 and its effect is discussed in Section 5.4.

5.3 Performance Modeling

In Section 5.2 six methods for medium access, in two general categories, were discussed. The purpose of the present section is to examine the performance of these medium access methods in a manner that will make comparisons possible. In order to accomplish this, performance parameters and a common environment must be established.

5.3.1 Performance Parameters and Environment

The medium access protocols being discussed are used when a number of stations share a common medium. The protocols are designed to accommodate communication between any pair of stations and to function when any or all of the connected stations are active. Our performance modeling objective is to obtain a general comparison of the methods under average or typical conditions. Although many special routing or loading conditions can be identified, it is beyond the scope of this book to deal with a large number of special cases. Instead a list of conditions considered to be representative, and to not favor any one method, are defined below and used in each of the cases to be studied.

The following notation and conditions are assumed for each type of medium access method:

- Number of stations, M

- Average arrival rate to each station, λ packets per second (the same for each station)

Figure 5.16   Generic Multiaccess Network for Performance Studies.

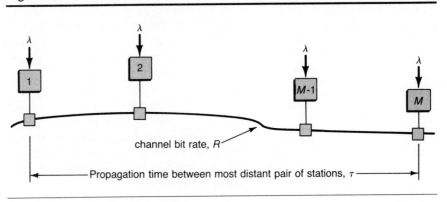

channel bit rate, R

Propagation time between most distant pair of stations, τ

- Arrival process, Poisson
- Medium bit rate, or clock rate, R bits per second
- Average packet length, $\bar{X}$ bits per packet
- Service at each station is exhaustive (that is, all stored packets are transmitted each time the station has access to the channel)
- Propagation and other delays between stations equal
- Maximum propagation delay between stations, τ seconds

Figure 5.16 is a diagram showing a multiaccess network incorporating the above assumptions. Several parameters common to all of the access protocols, in addition to those listed above, are used in the study. The time to transmit a packet of length X is X/R, which depends on packet length and can be random. Of course, the *average* time to transmit a packet is $\bar{X}/R$. For several medium access strategies, the channel cannot be used at its maximum rate, but it is used at a lower effective rate (R'). It is convenient to introduce a symbol, E, for *effective packet transmission time*, defined as

effective packet transmission time (E)

$$E = \frac{X}{R'} . \tag{5.1}$$

Note that E is a random variable.

average transfer delay (T)

Performance is measured in terms of two parameters, average transfer delay and throughput, which are now defined. *Average transfer delay* (T) is the average time between the arrival of a packet to a station interface and its complete delivery to the destination station. Throughput, in general, is the correctly received traffic, in packets per second, passing any network bound-

normalized network throughput (S)

ary. The network throughput is the total traffic entering (or exiting) the complete network per unit time. *Normalized network throughput (S)*, often simply called "throughput," is network throughput divided by the clock rate of the medium. S is used as a reference quantity for all of the medium access methods considered. In terms of the symbols specified above, S can be expressed as

$$S = \frac{M\lambda\bar{X}}{R}. \tag{5.2}$$

effective network throughput (S′)

If the medium is not used at its maximum rate, normalized network throughput is reduced to an *effective network throughput (S′)*, expressed by using Eq. (5.2) with R replaced by R', as

$$S' = \frac{M\lambda\bar{X}}{R'}. \tag{5.3}$$

Since S' contains $\bar{X}/R'$, the average value of E, or $\bar{E}$, can be used in Eq. (5.3) to express S' as

$$S' = M\lambda\bar{E}. \tag{5.4}$$

In Eq. (5.4) S' is determined by the effective packet transmission time as seen by the whole network, and it is typically determined by the time the channel is kept busy by all stations. On the other hand, the effective packet transmission time for a single station depends on how that station can access the channel and hence may be different from that seen by the complete network. The two different uses of E can be distinguished by context.

offered traffic (G)

In random access networks that require retransmissions, the concept of offered traffic (G) is required. Offered traffic (also normalized) is defined as the average number of attempted packet transmissions per second divided by the average number of packet transmissions per second possible at the clock rate of the medium.

In a stable network for which packets can be neither created nor destroyed, the total arrival rate to the network in the steady state must not exceed the total effective transmission capacity of the medium. This is true because if the arrival rate does exceed the transmission capacity, packets would build up without limit and no steady state could result. In symbols the above statement is written as $M\lambda < R'/\bar{X} \leq R/\bar{X}$, with the first term arriving packets per second and the second term transmitted packets per second at the effective bit rate $R' \leq R$.

These results can be stated in an equivalent fashion as

$$S = \frac{M\lambda\bar{X}}{R} \leq \frac{M\lambda\bar{X}}{R'} = S' < 1. \tag{5.5}$$

5.3.2   Basic Performance Relations

normalized average transfer delay ($\hat{T}$)

Average transfer delay depends on network loading or throughput, therefore an equation or relation between network throughput and average transfer delay is the desired result. Average transfer delay is proportional to the clock rate of the medium, so it is often desirable to define *normalized average transfer delay ($\hat{T}$)* as average transfer delay divided by packet transmission time at the clock rate of the medium, $\bar{X}/R$, or

$$\hat{T} = \frac{TR}{\bar{X}}.\tag{5.6}$$

Regardless of the medium access method used, the average time to transmit a packet is at least the average time for transmission at the clock rate of the medium, or $\bar{X}/R$. Thus for all medium access methods

$$T \geq \frac{\bar{X}}{R},\tag{5.7}$$

or

$$\hat{T} \geq 1.\tag{5.8}$$

Each medium access method adds its characteristic delays to this minimum. Results for specific methods are given below.

One other reference point for average transfer delay is shared by all medium access methods. Recall that effective throughput S' is limited to less than unity by the constraint that the flow of packets into a network must not exceed the flow of packets out of the network in the steady state. As the capacity of the network to transmit packets is reached, the average transfer delay, T, for a typical packet increases without limit. Thus one symptom of S' approaching its limit is increase of T without bound. The limiting value of S as S' approaches unity or as T approaches infinity is denoted S_{max} and is

channel capacity

termed *channel capacity* by some authors. Channel capacity measures the largest steady-state normalized throughput possible for a complete network.[10]

In summary, we can characterize the general shape of the relation between throughput and average transfer delay for any multiaccess protocol: For small throughput, the average transfer delay is close to, but cannot be less than, $\bar{X}/R$, whereas for throughput near the maximum value, S_{max}, average transfer delay increases without bound.

In addition to channel capacity, a relation between throughput and offered traffic (that is, S versus G) is useful for random access networks; such curves complement average transfer delay-throughput curves in describing medium access protocol behavior.

---

[10] This should not be confused with the Shannon channel capacity theorem, defined in Chapter 2. Shannon channel capacity is the maximum rate of transmission at which it is theoretically possible to reduce the error rate to zero. We are ignoring errors in the current discussion.

Development of accurate mathematical models, which relate average transfer delay to throughput and throughput to offered traffic, for any medium access procedure requires extensive study. Mathematical tools required include renewal theory, Markov chains, and queueing theory. Detailed development of the performance equations for the medium access procedures is thus beyond the scope of the present book. The literature contains detailed analyses of the important medium access methods.[11]

Instead of a rigorous development, the objective of the next two subsections is to make plausible certain performance results that are stated in reasonably accurate form. Performance equations are plotted for a number of special cases to aid in visualizing the results.

5.3.3 Performance of Scheduling Methods

For scheduling methods, to a first approximation, packets are not lost and thus throughput is equal to offered traffic. This neglects packets lost because of bit errors and so forth, but these are second order effects and largely independent of scheduling methods.

For token rings using single token or single packet token management methods, there are intervals during which the medium is forced to be inactive while waiting for the busy token or the packet to complete traversal of the ring. In such cases the effective bit rate is less than the clock rate and throughput has a limit, S_{max}, less than one.

An idealized plot of S versus G for scheduling methods is given in Fig. 5.17. A specific relation for S_{max} for a single token ring is given later in this section along with the equation for T versus S. Given an equation relating T and S, S_{max} can be obtained from the S-axis intercept of the asymptote of the curve of T as it approaches infinity.

The relations between average transfer delay and throughput vary for different scheduling methods. Each relation adds terms characteristic of the particular method to the minimum delay in Inequality (5.7) or (5.8). All such relations include an added term equivalent to the waiting time a packet would experience if it joined a single queue at some station and had to wait for the packets ahead of it to be served. This waiting time depends on network loading and on the effective time required for transmission of packets from the station when they reach the head of the queue and the station has use of the channel. Effective packet transmission time, denoted as E above, depends on the distribution of packet lengths and on detailed operation of the medium access method.

---

[11] Typical references to multiaccess network performance, classified according to the particular method, are TDMA: [LAM77], [ANDE79], [HAYE84]; Polling: [SPRA77b], [LAM83b], [KONH74], [TAKA86a]; Ring: [BUX81], [BUX89], [FERG85], [RUBI83]; ALOHA: [ABRA70], [ABRA73], [KLEI73], [KLEI75a], [ROBE72]; CSMA: [KLEI75b], [KLEI75c], [LAM75], [TOBA77]; CSMA/CD: [TOBA80a].

Figure 5.17   Idealized S Versus G Plot.

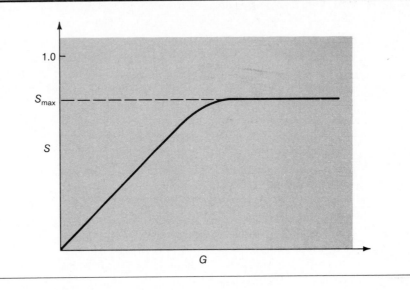

The queueing theory formula for the waiting time, W, for packets at a single queue is[12]

$$W = \frac{\rho}{1 - \rho} \frac{\bar{Y^2}}{2\bar{Y}},$$ **(5.9)**

where ρ is the ratio of packet arrival rate to service rate at the queue, and Y is a random variable denoting service time. (Of course, $\bar{Y^2}$ and $\bar{Y}$ are, respectively, the mean square and mean values of Y.)

A heuristic formula for a packet waiting time component of average transfer delay for medium access strategies can be determined by modeling each station as an independent queue, with W in Eq. (5.9) the average wait for a typical packet. The service time, Y, of a packet is the effective packet transmission time seen by a single station. The quantity ρ is the ratio of packet arrival rate to effective service rate at a single station. Note that ρ can also be equated to S' in Eq. (5.4) if $\bar{E}$ is interpreted as the effective packet transmission time seen by the whole network. Thus, W can be expressed as

$$W = \frac{S'}{1 - S'} \frac{\bar{E^2}}{2\bar{E}}.$$ **(5.10)**

As Eq. (5.10) indicates, waiting time also depends on the mean square value of packet transmission time, which depends on the packet transmission

---

[12] This is the expression for the waiting time of an $M/G/1$ queue; see for example [KLEI75b].

time distribution function. For an exponential distribution of packet transmission times $\bar{E}^2 = 2(\bar{E})^2$ and for constant packet transmission times $\bar{E}^2 = (\bar{E})^2$.

This determines two components of average transfer delay common to all scheduling medium access methods, namely waiting time and packet transmission time. We now examine the access methods separately.

For the time division multiple access (TDMA) type of fixed assignment method, each station is allowed use of the channel $(1/M)$th of the time, and thus, on the average, the capacity available to each station is R/M.[13] For this method, constant length packets are normally used so that $\bar{E}^2 = (\bar{E})^2$ and since channel capacity seen by each station is R/M, $\bar{E} = M\bar{X}/R$. Even though each station receives use of the channel only $(1/M)$th of the time, the full channel is always busy at its maximum rate, R. Therefore the effective packet transmission time seen by the whole network is $\bar{X}/R$, and S' in Eq. (5.10) is equal to S. Thus W for the TDMA method is given by

$$W = \frac{S}{2(1 - S)}\left(\frac{M\bar{X}}{R}\right). \tag{5.11}$$

In TDMA, packets experience an additional average delay of $M\bar{X}/2R$ to account for the fact that a representative packet arrives at random relative to a frame of length $M\bar{X}/R$ seconds and must wait for the appropriate time slot; if the packet arrival time is uniformly distributed over the length of the frame,[14] the average wait is one-half frame time or $M\bar{X}/2R$.

average transfer delay
for TDMA

The final result for the *average transfer delay for TDMA* is then given by the sum of these items as

$$T = \frac{\bar{X}}{R} + \frac{M}{2}\frac{\bar{X}}{R} + \frac{S}{2(1 - S)}\left(\frac{M\bar{X}}{R}\right), \tag{5.12}$$

and normalized average transfer delay by

$$\hat{T} = 1 + \frac{M}{2} + \frac{MS}{2(1 - S)}. \tag{5.13}$$

Curves of $\hat{T}$ versus S, with M as a parameter, are given in Fig. 5.18.

Polling and the *token ring*, which are demand assignment methods, differ in several respects from the fixed assignment methods typified by TDMA. In terms of average transfer delay, they share the packet transmission time and the term caused by a queuing wait as given by Eq. (5.10), although for the demand assignment methods the channel capacity available to each station, on the average, is R'. The term in Eqs. (5.12) and (5.13) caused by waiting on

---

[13] This ignores any overhead for maintaining synchronization of TDMA frames. Normally such overhead is minor. If it is significant, it can be included in the models (to a reasonable degree of approximation) by appropriately decreasing the transmission rate, R.

[14] Under the assumption of Poisson arrivals for packets almost always made in queuing theory analyses, this is valid.

Figure 5.18   Average Normalized Transfer Delay Versus
Throughput for TDMA.

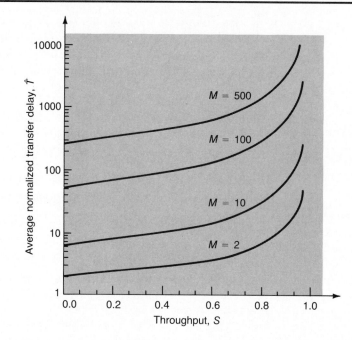

the average one-half frame time or until one half of the channels are deter-
ministically served is replaced by a term more accurately accounting for the
demand assignment.

To derive an approximate expression for the added wait for polling and
token passing, consider the fact that for these methods an average cycle time,
T_c, can be determined as the average time to serve all M stations. Once T_c is
known, it plays the same role as the frame time for TDMA. A representative
packet arrives at random relative to the cycle and on the average must wait
one-half cycle, $T_c/2$, until its station is served.[15]

Let t' be the average time required to transfer use of the channel from
one station to the next. The cycle time is then computed as follows. During
a cycle, an average of λT_c packets arrive at a station and (assuming exhaustive service) all must be transmitted while the station has use of the channel.
Thus T_c can be expressed as

$$T_c = M[\lambda T_c \bar{X}/R' + t'], \tag{5.14}$$

---

[15]This is an approximation, which is only exact for constant cycle times, but it has been
shown to give excellent results in this situation.

where the term in brackets is the time used on the average by each station. The bracketed quantity consists of $\lambda T_c \bar{X}/R'$ (the average time to transmit the packets that have arrived since the last service of a station, at the effective bit rate R'), plus t' (the time to transfer use of the channel from one station to another). The average time used by each station is multiplied by M, the number of stations, to compute cycle time. Solving Eq. (5.14) for T_c gives

$$T_c = \frac{Mt'}{1 - \dfrac{M\lambda\bar{X}}{R'}} = \frac{\tau'}{1 - S'}, \tag{5.15}$$

where Mt' is replaced by τ', the total average time required to transfer the use of the channel from one station to another summed over the whole sequence of stations. The average wait until a typical station is served is thus given by

$$\frac{T_c}{2} = \frac{\tau'}{2(1 - S')}. \tag{5.16}$$

ring latency, τ'

For ring networks, τ' is termed the *ring latency* and is given by

$$\tau' = \tau + \frac{MB}{R}, \tag{5.17}$$

where τ is propagation time around the ring and B is the latency, in bits, of each station. A normalized ring latency a' is defined as

$$a' = \frac{\tau'}{\bar{X}/R} = a + \frac{MB}{\bar{X}}, \tag{5.18}$$

where

$$a = \frac{\tau}{\bar{X}/R} \tag{5.19}$$

is a normalized time delay.

walk time

For polling networks, τ' is referred to as the total *walk time* for the polling network. Walk time per station consists of a propagation time, a time to transmit control packets (polling and go-ahead), and possibly synchronization and turnaround times.

average transfer delay for polling and token rings

A final result for *average transfer delay for polling and token rings* has the three terms just discussed and another, denoted τ_{average}, which is the average propagation time of packets from a source station to the destination station. For token rings, τ_{average} is often taken as $\tau'/2$, while for polling the maximum delay between stations is sometimes used. The resulting expression for average transfer delay for either polling or token ring networks is

$$T = \frac{\bar{X}}{R} + \tau_{\text{average}} + \frac{\tau'(1 - S'/M)}{2(1 - S')} + \frac{S'}{2(1 - S')} \frac{\bar{E}^2}{\bar{E}}. \tag{5.20}$$

Figure 5.19   Normalized average transfer delay for polling with $M = 10$ stations and different values of normalized propagation time, a', and a_{average} assumed to be zero.

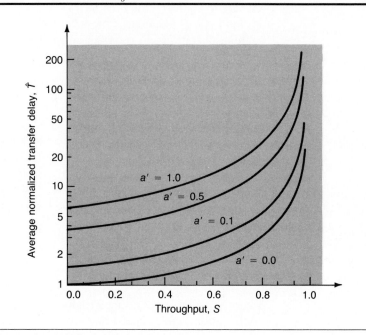

The quantity S'/M in the third term, which is often negligible, appears in more detailed quantitative analysis (see [HAMM86]) and does not have a simple heuristic explanation.

For polling $\bar{E} = \bar{X}/R,$ and for constant packet lengths, a specific result for normalized average transfer delay is

$$\hat{T} = 1 + a_{\text{average}} + \frac{a'(1 - S/M)}{2(1 - S)} + \frac{S}{2(1 - S)}, \tag{5.21}$$

where $a_{\text{average}} = \tau_{\text{average}}/(\bar{X}/R)$. A plot of $\hat{T}$ for several values of a', $M = 10$ stations, and a_{average} assumed to be zero is given in Fig. 5.19. If a_{average} is nonzero, it can be added to values read from the plot.

For token rings with single token operation and exponential packet length distributions, the first and second moments of effective service time can be shown [HAMM86] to be given, respectively, by

$$\bar{E} = \frac{\bar{X}}{R}e^{-a'} + \tau', \tag{5.22}$$

and

$$\bar{E}^2 = (\tau')^2 + 2(\bar{X}/R)^2 e^{-a'}(1 + a'). \tag{5.23}$$

The corresponding normalized average transfer delay is then given by

$$\hat{T} = 1 + \frac{a'}{2} + \frac{a' \, [1 \, - \, S(e^{-a'} \, + \, a')/M]}{2[1 \, - \, S(e^{-a'} \, + \, a')]} + \frac{S[(a')^2 \, + \, 2(1 \, + \, a')e^{-a'}]}{2[1 \, - \, S(e^{-a'} \, + \, a')]}. \qquad (5.24)$$

Under these conditions

$$S_{\max} = \frac{1}{e^{-a'} \, + \, a'}, \qquad (5.25)$$

can be determined from the value of S which causes the denominator of two terms in Eq. (5.24) to become zero or $\bar{T}$ to approach infinity.

A plot of the results of Eq. (5.24) for several values of a' and $M = 50$ stations is given in Fig. 5.20. The curves of delay approach infinity along an asymptote that is a vertical line at $S_{\max}$. Figure 5.21 shows the effect of normalized ring latency, a', on normalized average transfer delay, for different values of network throughput, S. Delay is not a strong function of a' until a' reaches a value causing S to approach $S_{\max}$, in which region delay increases sharply.

Figure 5.20   Normalized Transfer Delay Versus S for Single Token Ring. $M = 50$ stations, exponential distribution of packet length, and various values of normalized ring latency, a' have been assumed.

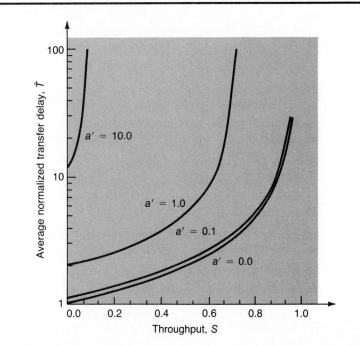

Figure 5.21   Average Normalized Transfer Delay as a Function of
Normalized Ring Latency. A ring with 50 stations, single
token operation, and exponential packet length
distribution have been assumed.

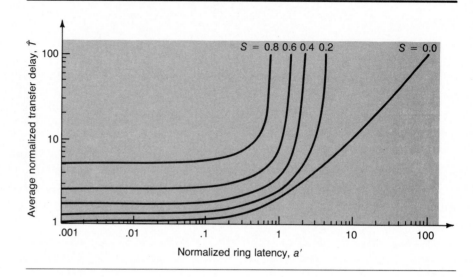

Expressions comparable to Eq. (5.24) can be derived for ring networks
with multiple token and single packet token management strategies (see
Problems 5.2–5.5). Comparable expressions can be derived for token bus
access methods.

5.3.4   Performance of Random Access Methods

Performance calculations for random access methods differ from those for
scheduling methods in several significant respects. Average transfer delay
versus mean throughput (T versus S) and throughput versus offered traffic (S
versus G) are still the primary curves to depict performance. For random
access methods, however, S versus G takes on greater significance, because
this tends to fall far short of the ideal curve (that is, S equal to G from $S = 0$
to $S = 1$, then remaining at 1 for larger values of G). The same value of S is
obtained for more than one value of G, which suggests the accurate conclu-

stability problem   sion that the methods also have a *stability problem* not experienced with
scheduling methods.

   As with scheduling methods, average transfer delay is the sum of several
components with packet transmission time, $\bar{X}/R$, representing the absolute
lower limit on T as it did for scheduling methods. Propagation time of packets
from source station to destination station is also still a factor.

A new type of delay caused by the necessity for retransmissions after collisions is characteristic of random access methods and its behavior is quite different from that of any of the delay components for the scheduling methods. Delay due to retransmissions can be the dominant factor for random access methods.

Unfortunately operation of random access protocols is much more difficult to model mathematically than that of the scheduling methods, and tractable approximations are available only for ALOHA and slotted ALOHA. Our plan for this section is to explain random access performance as completely as possible, within the mathematical background assumed for the reader, by identifying the components of average transfer delay, but not in all cases expressing these as equations in terms of basic parameters. Instead, final T versus S curves from simulation are given as the concluding result. An exception is average transfer delay for ALOHA, which is derived to illustrate procedures and difficulties. In the same vein, approximate S versus G equations are derived for ALOHA, slotted-ALOHA, and one version of CSMA, but such results are not derived for all versions of CSMA or for CSMA/CD in any of its forms. Final curves of S versus G for the latter cases are given for reference, but without derivation, however.

The basic assumptions listed in Subsection 5.3.1 still apply. Some additional assumptions are needed specifically for random access methods:

- The number of stations is large and the arrival process from this large (actually assumed to be infinite) population is Poisson;[16] the process includes both new and previously collided packets.

- The maximum propagation delay is τ seconds, and this value is used for the delay between any two stations.

- All packets have the same length and the same transmission time, P seconds.

- At any moment each station has at most one packet ready to transmit, including packets for retransmission.

- Carrier sensing and collision detection take place instantaneously, and there is no turnaround time between transmitting and receiving.

- The channel is noiseless so that transmission failures are due exclusively to collisions.

- The overlap of any fraction of two packets causes a collision so that all colliding packets must be retransmitted.

The three classes of methods—ALOHA, CSMA, and CSMA/CD—make use of three levels of sensed data. ALOHA uses no sensed data, CSMA adds to the ALOHA procedure sensing the channel before transmitting, and CSMA/CD

---

[16]An assumption of an infinite population is necessary for the Poisson assumption to be precise.

Figure 5.22   A Bus-Type Random Access Network Used for Studying
ALOHA, CSMA, and CSMA/CD Access Algorithms.

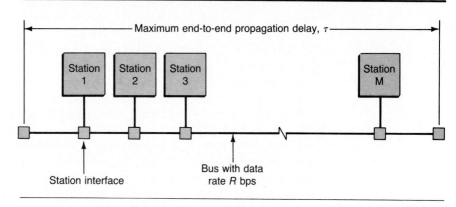

adds to the CSMA procedure sensing the channel during transmission. The
effect of the levels of sensing carries over to performance. There is a basic
behavior for ALOHA, which is refined by the sensing used in CSMA and further
refined by the additional sensing of CSMA/CD.

We begin the analysis by considering the basic ALOHA behavior. Consider
a bus-type network such as that in Fig. 5.22.

An arrival to any station, say station i, causes a packet to be transmitted.
If station i has exclusive use of the channel during the time required to transmit its packet, then the transmission is successful. On the other hand, if the
transmissions from any other stations are on the channel simultaneously with
that of station i, all collide and must be retransmitted.

Packet lengths are assumed to be constant and equal to $\bar{X}$, and packet
transmission time is $\bar{X}/R$; for convenience we denote this time by P. Figure
5.23 shows on a time axis the transmission of a packet from station i. Also
indicated on the figure is the vulnerable period, during which arrivals at other
stations can initiate transmissions that will collide with that of station i. As
was noted in the flow chart for operation of the ALOHA algorithm (Fig. 5.11),
after transmitting a packet station i must wait for a positive acknowledgment
from the destination station or for a time out indicating that the packet has
suffered a collision. Assuming a propagation time, τ, between stations, a
period 2τ is required for station i to know the *fate* of its packet. Thus at time
$t_0 + P + 2\tau$ the station is aware that it has successfully transmitted its packet
or that it must choose a random backoff time, delay the random amount of
time selected, and then attempt to transmit again. This sequence, including
a possible retransmission, is also shown in Fig. 5.23.

Figure 5.23 gives sufficient information to derive a reasonable approximation to the behavior of the ALOHA algorithm. *Throughput* can be
expressed in terms of G and the probability of successful transmission as

Figure 5.23   Transmission of a Packet from Typical Station i Using the
ALOHA Random Access Protocol. Also shown are the
vulnerable period and the backoff period with a
retransmission, assuming that retransmission is necessary.

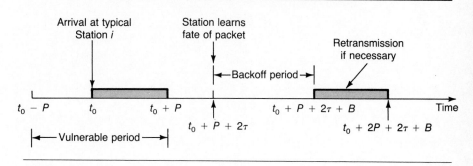

$$S = G \, Pr\{\text{successful transmission}\}. \tag{5.26}$$

As the figure indicates, a successful transmission requires that no station
other than station i initiate a transmission during the vulnerable period of
length $2P$. For ALOHA, subject to our present assumptions, transmissions
coincide with arrivals, either new arrivals or packets for retransmission
whose backoff time has expired. Also, according to our assumptions, the total
arrival distribution is Poisson with average rate $\Lambda = G/P$. For Poisson arrival
processes, the number of arrivals in an interval is independent of the starting
time of the interval and is determined only by the length of the interval.

Poisson distribution   The *Poisson distribution with rate* Λ gives the probability of k arrivals in
t seconds as

$$Pr[k \text{ arrivals in } t \text{ seconds}] = \frac{(\Lambda t)^k}{k!} \, e^{-\Lambda t}. \tag{5.27}$$

For our purposes, the probability of a successful transmission is the probability of zero arrivals in the vulnerable interval of $2P$ seconds, or $e^{-2\Lambda P}$ so

$$S = G \, e^{-2\Lambda P} = G \, e^{-2G}. \tag{5.28}$$

Slotted-ALOHA differs from ALOHA only in the respect that packet transmissions must be initiated at the beginning of a slot. For this case any arrival
in the slot preceding the slot in which station i transmits will result in a
collision. Thus the vulnerable interval is reduced to one slot of length P and
throughput can be expressed as

$$S = G \, e^{-G}. \tag{5.29}$$

Figure 5.24 shows S versus G for pure and slotted-ALOHA. The maximum
possible values of S are $1/2e = 0.184$ and $1/e = 0.368$, respectively.

Figure 5.24   Throughput Versus Offered Traffic for ALOHA
                   and Slotted ALOHA.

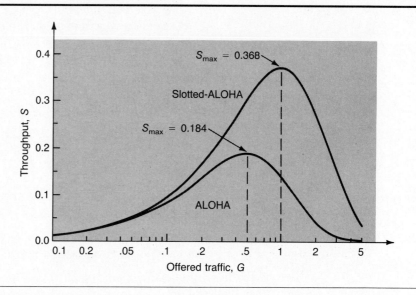

The transmission history shown in Fig. 5.20 can also be used to obtain an approximate expression for average transfer delay, T. Each successful transmission requires at a minimum a packet transmission time, P, and a time depending on propagation from the source to the destination station. An expression for the latter time depends on the point of view taken. Here we assume that packet transmission is not complete until an acknowledgment is received, so two propagation times, or 2τ, is the added factor.

Each retransmission attempt, if retransmissions are needed, consists of a backoff time (B), a packet transmission time (P), and a time (2τ) to determine the fate of the transmission. Denoting the average number of retransmissions by H and the average backoff time by $\bar{B}$, T can be expressed as

$$T = P + 2\tau + H[\bar{B} + P + 2\tau]. \qquad (5.30)$$

The average number of retransmissions can be expressed by noting that G/S is the number of attempted transmissions for each successful one, with $G/S - 1$ of these retransmissions. Thus, using Eq. (5.28) for ALOHA, H can be expressed as

$$H = e^{2G} - 1, \qquad (5.31)$$

with a similar expression for slotted-ALOHA derived from Eq. (5.29).

As in the case of scheduling type algorithms, it is convenient to use normalized delay, which in this case can be expressed as $\hat{T} = T/P$, since $P = \bar{X}/R$. Normalized average transfer delay for ALOHA thus becomes

$$\hat{T} = 1 + 2a + (e^{2G} - 1)\left[\frac{\bar{B}}{P} + 2a + 1\right], \qquad (5.32)$$

where $a = \tau/P$ is normalized propagation delay.

The average backoff delay, $\bar{B}$, is computed as follows. The backoff time is determined by selecting an integer, k, at random from a set $0, 1, 2, \ldots, K - 1$, and then backing off an amount $B = kP$. For equally likely backoff integers,[17] the average backoff delay is given by

$$B = \bar{k}P = \frac{1}{k}\sum_{k=0}^{K-1} kP = \left(\frac{K-1}{2}\right)P. \qquad (5.33)$$

$\hat{T}$ for ALOHA

Using this result in Eq. (5.32) and simplifying the expression gives a final result for $\hat{T}$ for ALOHA of

$$\hat{T} = (1 + 2a)e^{2G} + \left(\frac{K-1}{2}\right)(e^{2G} - 1). \qquad (5.34)$$

Plots of $\hat{T}$ versus S for $a = 0$ and several values of K are given in Fig. 5.25.

Both the S versus G plots of Fig. 5.24 and the $\hat{T}$ versus S curves in Fig. 5.25 show a multivalued behavior (that is, two values of $\hat{T}$ or G for a single value of S). This is suggestive of instability. It is beyond the scope of this book to investigate details of the stability problem, which is characteristic of all of the random access methods discussed here. It is always possible to make the backoff factor K large enough to give stable operation. More useful adaptive control procedures have been introduced by several researchers, notably Lam and Kleinrock [LAM75].

With respect to the $\hat{T}$ versus S curves of Fig. 5.25, it should be noted that delay in an actual network will not follow a curve in the figure after the knee of the curve. Instead, when S reaches $S_{\max} = 1/2e$, the saturation value of S, delay will approach infinity along the dotted line shown in the figure.

The approximate result derived above and given by Eqs. (5.28), (5.29), and (5.34), along with the associated plots, have been compared to more accurate results for slotted-ALOHA [HAMM86]. This comparison indicates that the approximate results are reasonably accurate for large values of K and for a large number of stations, M. In both cases "large" can be taken as 20 or greater.

As more features are added to ALOHA to produce CSMA or CSMA/CD, it becomes more difficult to obtain analytic expressions for $\hat{T}$ versus S or S versus G, even with the approximations used for ALOHA and slotted-ALOHA. The literature does, however, contain results, using the type of approximations above, for S versus G for all cases. These results can be plotted to give

---

[17] It has been shown [LAM75] that delay performance for ALOHA algorithms depends principally on the mean retransmission delay and is not sensitive to the distribution of the backoff integers.

Figure 5.25   Average Normalized Transfer Delay Versus Throughput
for the ALOHA Access Algorithm, with $a = 0$ and
$K = 5$, 20, and 100.

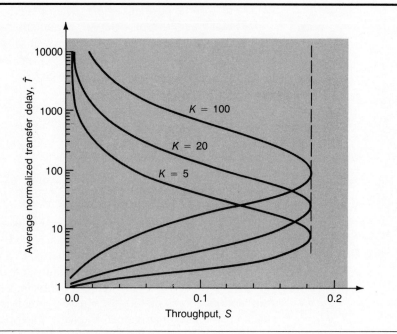

useful curves, but derivations of the equations are tedious. It is even more
difficult to obtain meaningful analytical results for $\hat{T}$ versus S, since stability
must be dealt with. For this reason, curves from simulation often replace
analytical results in studies of CSMA and CSMA/CD.

The approach commonly used to obtain approximate S versus G curves
for the CSMA and CSMA/CD methods differ to some extent from that used for
ALOHA and slotted-ALOHA. The general approach is presented below and
applied to the tractable case of nonpersistent CSMA. We then conclude this
section on performance analysis by giving performance curves for other
important random access methods.

Relations between S and G for the more sophisticated random access
methods can be obtained by considering time as consisting of *cycles of alternating busy and idle periods*. A cycle ends with an idle period during which
no station is attempting to transmit. The cycle is initiated by the arrival of a
packet to some station following an idle period. A cycle containing multiple
transmissions is shown in Fig. 5.26.

In the figure the packet initiating the cycle is denoted packet 0. It takes τ
seconds for the first bit of this packet to propagate the full length of the bus
and for all stations to sense its presence. Before the end of this interval of

Figure 5.26   A Typical Cycle Used to Analyze Random Access Protocols that Use Carrier Sensing.

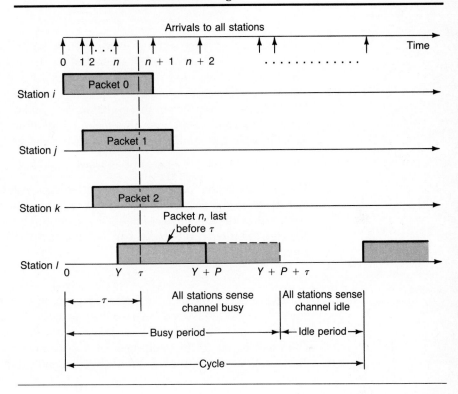

duration τ, when other stations have not yet sensed the presence of packet 0, any station can transmit its own packet if an arrival comes in, as is indicated for stations j, k, and l in the figure. The arrival labeled $n + 1$ in the figure arrives after all stations sense the channel to be busy and thus is not transmitted. The vulnerable interval for methods using carrier sensing is thus τ seconds.

As the figure indicates, a cycle has two components: a busy period (including a time τ after the last packet is transmitted to ensure this packet has propagated the full length of the bus) and an idle period. If the busy period contains only packet 0, transmission is successful; otherwise collision occurs.

In the steady state all cycles are statistically similar; thus throughput can be determined as the ratio of the average amount of time during a cycle devoted to successful transmission to the average total length of the cycle. Denoting the average duration of the period devoted to successful transmissions as $\bar{U}$, the average length of the idle period as $\bar{I}$, and the average length of the busy period as $\bar{B}$, S can hence be expressed as

$$S = \frac{\bar{U}}{\bar{I} + \bar{B}}. \tag{5.35}$$

Because of carrier sensing, every cycle is limited to a busy period of no more than $2\tau + P$ seconds. This contrasts with the busy period of ALOHA, which has no upper limit since the channel is not sensed.[18]

We now illustrate the use of Eq. (5.35) by applying it to nonpersistent CSMA. To determine S from this equation, $\bar{U}$, $\bar{I}$, and $\bar{B}$ must be determined for the nonpersistent CSMA algorithm given by the flow charts of Figs. 5.12 and 5.13. To begin the calculation, $\bar{U}$ can be expressed as

$$\bar{U} = P \cdot Pr\{\text{packet 0 is transmitted successfully}\}. \tag{5.36}$$

The probability that packet 0 is transmitted successfully is the probability of no arrivals in the vulnerable period of length τ. Using the same approach as was used in the calculation for ALOHA (for which the vulnerable period is $2P$), the required probability can be evaluated as $e^{-(G/P)\tau}$, so $\bar{U}$ can be expressed as

$$\bar{U} = P e^{-G\tau/P}. \tag{5.37}$$

To evaluate $\bar{I}$, recall that transmissions are modeled as a Poisson process. Poisson processes have a "memoryless property,"[19] which implies that the average time until the next arrival from any reference point is always the same and equal to the reciprocal of the average rate for the process. In the present case this average time is P/G seconds, and $\bar{I}$ is thus given by

$$\bar{I} = P/G. \tag{5.38}$$

To compute $\bar{B}$, denote the arrival time of the last packet in the vulnerable period shown in Fig. 5.23 by Y. Then, the length of the busy period is

$$B = Y + P + \tau, \tag{5.39}$$

and the next step is to find its average value. The quantity Y is the only random quantity in Eq. (5.39), and its average value can be obtained by making use of the Poisson distribution of arrivals to the network. If Y denotes the time of arrival of the last packet in the interval $[0, \tau]$ then there must be zero arrivals in the interval $[Y, \tau]$, which is $\tau - Y$ seconds long. Thus, for any fixed y

$$P[Y \leq y] = F_Y(y) = P[0 \text{ arrivals in } (\tau - y) \text{ seconds}] = e^{-G(\tau - y)/P}. \tag{5.40}$$

---

[18] The fact that the busy period is not bounded prevents the approach using cycles from being useful with ALOHA.

[19] See, for example, [KLEI75b] for a discussion of this property.

Figure 5.27 Throughput (S) Versus Offered Load (G) for Nonpersistent CSMA for Indicated Values of Normalized Propagation Time, a.

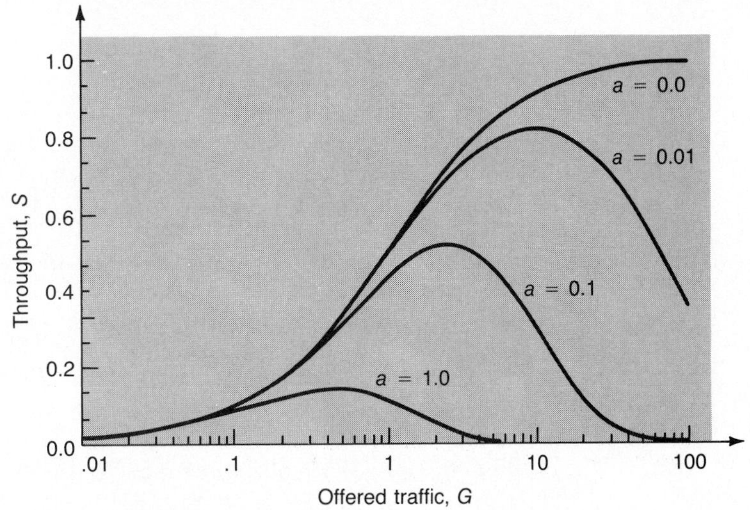

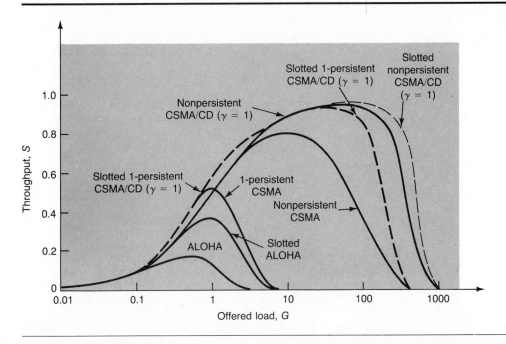

Figure 5.28 Throughput (S) versus offered load (G) for slotted nonpersistent and slotted 1-persistent CSMA/CD, nonpersistent and 1-persistent CSMA, slotted-ALOHA and ALOHA, all with $a = 0.01$ (from [HAMM86]).

Here $F_Y(y)$ is the distribution function of the random variable Y. It is straightforward (but tedious) to obtain $\bar{Y}$ from this as

$$\bar{Y} = \tau - \frac{P}{G}(1 - e^{-G\tau/P}), \tag{5.41}$$

and $\bar{B}$ as

$$\bar{B} = P + 2\tau - \frac{P}{G}(1 - e^{-G\tau/P}). \tag{5.42}$$

S versus G for nonpersistent CSMA

Recalling that a is defined as τ/P, S for nonpersistent CSMA can finally be expressed as

$$S = \frac{\bar{U}}{\bar{I} + \bar{B}} = \frac{Ge^{-aG}}{(1 + 2a)G + e^{-aG}}. \tag{5.43}$$

Figure 5.29 Normalized Average Transfer Delay ($\hat{T}$) versus Throughput (S) for Slotted-ALOHA, slotted nonpersistent CSMA and CSMA/CD for $a = 0.01$ (From simulation results by [KLEI75c] and numerical solutions by [TOBA80a].)

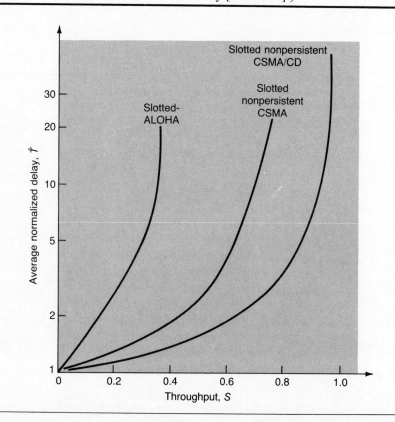

A family of curves for S versus G obtained from Eq. (5.43) with $a = 0.0, 0.01,$ 0.1, and 1.0, is given in Fig. 5.27.

The procedure for analyzing the ordinary carrier sense methods also applies for the carrier sense methods with collision detection. Each added feature, however, adds to the complexity of the calculation. Figure 5.28 gives S versus G curves for all of the common random access methods with $a = 0.01$. The parameter γ for the CSMA/CD curves is equal to J/τ, with J the duration of the jamming signal used. The equations from which the curves are plotted can be found in the literature, for example in Hammond and O'Reilly [HAMM86]. Simulation results for $\hat{T}$ versus S for three of the methods are given in Fig. 5.29, also for $a = 0.01$.

As is discussed briefly in the next section, the behavior of multiaccess protocols with respect to normalized propagation delay, a, is of significant interest. The channel capacity, or S_{max}, for most of the important random access methods, is plotted versus a in Fig. 5.30.

Figure 5.30 Channel capacity S_{max}, versus normalized propagation delay (a) for ALOHA, CSMA, and CSMA/CD protocols (from [HAMM86]).

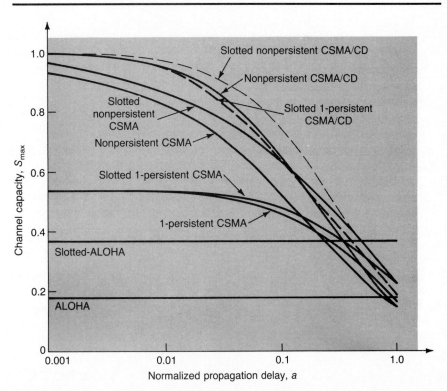

5.4   Assessment of Medium Access Protocols

In the first three sections of this chapter we have discussed a representative selection of multiaccess protocols. The protocols have been selected because they represent several general types and include those specified by IEEE 802 standards. We have also obtained performance results for protocols in terms of throughput, delay and, where appropriate, channel capacity, $S_{\max}$. In this section we assess the protocols in order to provide guidelines for making a choice of medium access protocol for specific applications.

Medium access protocols are selected on the basis of two types of properties, which we call *physical attributes* and *performance attributes*. Performance attributes, quantified in Section 5.3, are important and certainly acceptable values must be provided. However, such attributes do not completely override the physical attributes we now discuss.

Fixed assignment scheduling methods are robust and simple to implement but suffer from the fact that these methods assign channel capacity to stations that are inactive. In many applications this makes fixed assignment algorithms inefficient. Modifications, such as a reservation feature, can improve their efficiencies and will cause the modified algorithms to tend toward the class of demand assignment scheduling methods.

Demand assignment scheduling methods are associated principally with polling, a central control method, and token passing, which has distributed control. The type of control affects the complexity of equipment at each station and the reliability. A polling network requires a minimum of intelligence at each station but is not robust with respect to loss of the central controller. The two characteristics are reversed for token passing, which is not vulnerable to loss of a single station because of distributed control but requires more intelligence at each station.

Token passing rings have other characteristics that can be undesirable. Interfaces of stations to the ring are active and therefore, in a basic network, single station failures can interrupt the complete network. However, methods, such as the "star" wiring and bypass switches used by IBM [BUX89], can prevent this dependence on single stations and significantly improve ring reliability. In any type of active token passing network, integrity of the token is a key requirement. Since synchronous action is required, token passing protocols also require that synchronization be maintained.

Random access is another major type of medium access. This approach is robust and simple to implement. Interfaces are passive and thus are more reliable. Adding and deleting stations is typically much easier for random access than for token passing methods. A drawback to random access, however, is the channel capacity wasted in resolving collisions.

We turn now to performance results to complete the assessment of medium access methods. Performance is typically measured in terms of throughput (S) and normalized average time delay ($\hat{T}$). Performance curves relating S to G and $\hat{T}$ to S for most of the methods are given in Section 5.3.

Before using these results for average quantities, it is useful to note an important difference between the scheduling and random access methods with respect to delay of a specific packet. Although average delay for random access networks may be reasonable, the delay of a single packet cannot be guaranteed to be bounded as there is a nonzero probability of a continuing sequence of collisions.[20] Scheduling algorithms, on the other hand, have bounded maximum transmission delays.

A classification of the types of physical networks into two categories is useful for comparative performance assessment. Examination of Fig. 5.30 shows that random access protocols are quite sensitive to normalized propagation delay, a, defined as the ratio of end-to-end propagation delay, τ, to average packet transmission time, $\bar{X}/R$. Bertsekas and Gallager [BERT87] classify networks with respect to a, using the term *"higher speed"* if $a > 1$. For $a > 1$, the time to transmit a packet, $\bar{X}/R$, is less than the time for the packet to propagate to the most distant stations, so a packet is completely transmitted before its leading bit reaches some station. Thus for higher speed networks, as Fig. 5.30 indicates, carrier sensing and collision detection are ineffective and channel capacity is small, close to that of ALOHA.

Note that a "higher speed" ($a > 1$) network results if propagation delay is large with respect to packet transmission time. This can be the result of one or more of the following: large network expanse (for example, use of communications satellite links), high channel clock rate, or short packet length. The results for token rings show that this scheduling algorithm is also sensitive to a, or more specifically to $a' = a + MB/\bar{X}$, as shown in Eq. (5.18). In fact, as given by Eq. (5.25), $S_{\max} = 1/(e^{-a'} + a')$ for token rings. A plot of $S_{\max}$ for a token ring with exponentially distributed packet lengths is given in Fig. 5.31, with the result for nonpersistent CSMA/CD included for comparison. The curves show that token passing is useful for values of a' up to essentially a factor of 10 times the maximum values of a for which CSMA/CD is useful. Of course, a' is naturally larger than a by the factor $MB/\bar{X}$ given above; if either M or B is large, this term can dominate and eliminate any token ring advantages.

If networks are classified as higher speed and lower speed, it is clear that random access is primarily useful for lower speed networks. In the intermediate range near higher speed, token passing is useful for a' up to on the order of 10.

For lower speed networks, for which both token passing and random access can be used, the performance curves show relatively small differences. Average transfer delay for random access protocols is given for typical parameters in Fig. 5.29, while similar results for a token passing ring are given in Fig. 5.20. The curves are generic, however, so precise comparisons cannot

---

[20] Ethernet limits the number of retransmissions to 16, after which the packet may be discarded.

Figure 5.31   Channel Capacity, S_{max}, Versus Normalized Ring Latency (a')
or Normalized Propagation Delay a, for a Token Ring
and for Nonpersistent CSMA/CD.

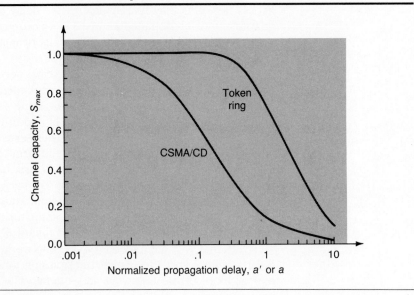

be made. Figure 5.31 shows that either approach can essentially achieve unity
for S_{max} if a or a' is small.

From the point of view of performance, as measured by $\hat{T}$ and S_{max}, a
significant factor in the choice of access algorithm is based on the "speed" of
the network as defined by a. Both scheduling methods and random access
methods are suitable for $a < 1$ ($a' < 10$ for scheduling), and a choice between
these methods depends on physical attributes or second order details of the
methods. For "higher speed" networks such as satellite networks or networks
using high speed channels such as fiber optic links, the scheduling and random access methods above are not suitable. The remaining choices are
TDMA with improvements due to reservations or similar expedients. ALOHA
or slotted-ALOHA can also be used in spite of inefficient channel use.

Increasing use of fiber optic links and interest in expanding local area
networks into larger metropolitan area networks have increased the interest
in medium access algorithms suitable for "higher speed" networks. As a result, a number of demand assignment scheduling methods have recently been
devised. In a survey paper, Fine and Tobagi [FINE84] discuss many of these
methods in a unified fashion. Most of the methods can be classified as token
passing; however, the "token" is implicit rather than explicit and is defined
by some event in channel activity, such as an idle period.

The new algorithms have many of the properties of the explicit token
methods discussed above, except that their performance deteriorates less

with increasing normalized propagation delay. The absence of an explicit to-ken makes them somewhat more robust than methods with explicit tokens. For more details on implicit token methods, the reader is referred to [FINE84].

5.5 Summary

Medium access control protocols are classified into scheduling and random access categories. Representative members of these two general classes are discussed by first defining the algorithm used by each protocol and then mod-eling the protocol and evaluating its performance.

Six general types of protocols were chosen for discussion: TDMA/reser-vations, polling, token passing, ALOHA, CSMA and CSMA/CD. IEEE 802 stan-dards exist for two types of token passing and for CSMA/CD. Each protocol is described and a flow chart given to define its algorithm.

Average transfer delay and throughput are defined and used as perfor-mance measures for the algorithms. Equations relating average time delay and throughput and throughput and offered traffic are derived, in a heuristic fashion, for most of the methods; the results are plotted to give a useful guide to performance. For several methods, an elementary derivation of perfor-mance equations is not possible; for these cases plots of time delay versus throughput and throughput versus offered traffic from the literature are given.

A final section compares the medium access protocols, discussing the characteristics of each class in a manner that should aid in making a choice for a given application.

Problems

5.1 An alternate to the TDMA fixed assignment method is FDMA. For this medium access method, each station is allocated a fixed bandwidth. For modulation at 1 bps/Hz, the channel bandwidth, B Hz, is equal to the clock rate, R bps, and each station is allo-cated a bandwidth $B/M = R/M$, where M is the number of stations.

a. Assume noise-free operation and no loss of bandwidth due to guard bands. Follow the approach and notation used with TDMA and show that average transfer delay is given by

$$T = \frac{M\bar{X}}{R} + \frac{S}{2(1-S)}\left(\frac{M\bar{X}}{R}\right).$$

b. For what values of M does TDMA yield a smaller value of T than does FDMA?

5.2 For medium access methods for which the channel cannot be active at all times, we have introduced effective service time E defined by Eq. (5.1). In many cases the effec-tive average channel rate R' is equal to R times the fraction of time the channel is in use.

       a.   Assuming constant length packets and that network constants are such that $a' > 1$, find $\bar{E}$ for single token operation.

       b.   Repeat (a) if $a' < 1$.

5.3   Repeat Problem 5.2 for multiple token operation.

5.4   Repeat Problem 5.2 for single packet operation.

5.5   Use the results of Problems 5.2, 5.3, and 5.4 to express $\hat{T}$ as a function of S for a token ring network with constant packet lengths and the following:

       a.   Single token operation

       b.   Multiple token operation

       c.   Single packet operation

5.6   Consider the elementary ring network with four stations shown in Fig. 5.10. Modify the example used for the figure by increasing all station latencies to two bits. Reconstruct the figure for the new values of station latency.

5.7   A channel using random access algorithms has stations at two ends of a bus with end-to-end propagation delay τ. Station A is at one end of the bus and stations B and C are essentially together at the other end. Packets arrive at the three stations and are available to be transmitted at stations A, B, and C at the respective times $t_A = 0$, $t_B = \tau/2$, and $t_C = 3\tau/2$. Packets require transmission times of 4τ. In appropriate figures, with time as the horizontal axis, show the result for each packet at the time it is available for transmission using the algorithms below:

       a.   ALOHA

       b.   Nonpersistent CSMA

       c.   Nonpersistent CSMA/CD

5.8   Repeat Problem 5.7 for p-persistent CSMA, with $p = 0.5$. Behavior is nondeterministic in this case; therefore show at least three possible time histories.

5.9   Consider a polling network with constant packet lengths for which $\hat{T}$ is given by Eq. (5.21).

       a.   How large must a' be for the third term to be the dominant factor in $\hat{T}$ (less than 10 percent error if $\hat{T}$ is approximated by just this factor). Assume that M is large so that $S/M \approx 0$ and that $a_{\text{average}} \approx 0$.

       b.   Define $\hat{T}_c = T_c/(\bar{X}/R)$ and express the approximation for $\hat{T}$ from (a) in terms of $\hat{T}_c$

5.10   Consider a polling network with M stations using roll-call polling. The clock rate of the connecting channel is 19.2 kbps. Data messages to the central computer are a constant length of 30 eight-bit characters. Each station connects 10 terminals, which each generate 0.05 packets per sec with Poisson arrival times. Propagation time and modem synchronization time may be neglected. Polling packets are five eight-bit characters and go-ahead messages are one eight-bit character.

       a.   Use the tractable approximate relation from Problem 5.9 to determine the maximum number of stations that the network can support if the average transfer delay is not to exceed 0.2 sec.

       b.   Check the approximation used in (a) by computing the average transfer delay corresponding to the number of stations found in (a) using an accurate formula for T.

       c.   What is the ultimate limit on the number of stations imposed by stability?

5.11   Consider a network using roll call polling. The central computer is located at the center of a configuration such that the distance from this computer to each station is

1 km. The polling packet is 16 bits long and the station must synchronize before reading this packet. The go-ahead packet, affixed to the end of each data packet, is four bits long. The central computer must synchronize before reading the data packet. Propagation delay is 5 μsec per km and each sync time is 10 μsec. The communications lines have a bit rate of 1 Mbps in both directions.

a. Find the walk time per station.

b. For a network with 100 stations, determine a complete expression for normalized average transfer delay in terms of normalized throughput.

c. For $S = 0.5$, determine the unnormalized throughput of 1000 bit packets for each station. (Assume that all stations have identical behavior.)

d. For $S = 0.5$, what is the average transfer delay for the network in seconds?

5.12 The IBM token ring uses "single token" operation. For such operation neglect overhead, assume that the packet length is constant at d bits, ring latency is 90 bits, and the ring bit rate is 1 Mbps.

a. Determine an expression for the maximum achievable throughput as a function of d and sketch S_{max} versus d. (Note that the general formula for token passing and polling must be used here rather than the special formula derived for single token operation, since the latter is for exponential packet lengths.)

b. For $d = 100$ bits, express average normalized transfer delay as a function of S. S/M and $\tau_{average}$ can be assumed to be negligible.

c. For what minimum value of S will T exceed three times the packet transfer time $\bar{X}/R$?

5.13 This problem concerns a token ring network with the following parameters:

τ: negligible
M: 100
B: 2 bits
R: 1 Mbps
$\bar{X}$: 1000 bits per packet
Exponential packet length distribution
Poisson arrivals with the same average arrival rate to each station

a. If a new free token is returned to the ring immediately after the busy token is removed, what is the maximum throughput per station (in packets per second) that the ring can support?

b. What is the largest percentage of increase in maximum throughput per station that can be achieved by using another token management strategy?

5.14 A bus-type network has an average arrival rate $\lambda = 1.5$ packets per second to each station. Packets are 1000 bits long and the channel clock rate is 1 Mbps. Assuming ideal operation, what maximum number of stations can be supported by using the following:

a. ALOHA,

b. Slotted-ALOHA.

5.15 A bus network using nonpersistent CSMA/CD must support (at most) 500 stations. The average arrival rate, λ, to each station is the same and equal to 3.2 packets per second. Packets are 5000 bits long and the channel clock rate is 10 Mbps. What is the maximum end-to-end bus length that can be tolerated if the propagation speed is 5 μsec per kilometer? Assume ideal operation and use Fig. 5.30 to obtain an approximate answer.

5.16 Derive an expression, similar to that of Eq. (5.34), giving $\hat{T}$ versus G for slotted-ALOHA.

5.17 Figure 5.30 indicates that if a is large enough, the maximum possible throughputs for CSMA and for CSMA/CD are less than those for either ALOHA or slotted-ALOHA. It seems intuitively reasonable, however, that using additional information, as is done in CSMA and CSMA/CD, should not make performance worse. Briefly discuss what the curves imply with respect to this proposition.

5.18 (**Design Problem**) A ring network is to have 100 stations connected over a total distance of 1 kilometer. The minimum station latency is at least three bits. Packet lengths have an exponential distribution with mean length $\bar{X} = 1000$ bits and single-token type operation is to be used. The maximum average arrival rate, assumed to be the same for each station, is 0.5 packet per second at each station. Choose the remaining parameters and design a network for which the normalized average transfer delay is less than 2.

5.19 (**Design Problem**) A bus network uses slotted nonpersistent CSMA/CD. The maximum average transfer delay that can be tolerated is 1.2 msec. The number of stations is 1000, $\bar{X} = 1000$ bits and the average arrival rate, assumed the same for each station, is 1.0 packet per second. Choose the remaining parameters and design an appropriate network. Specify the maximum possible length for the bus.

5.20 (**Design Problem**) A network is to be designed in the "higher speed" category and is to handle bursty traffic.

   a. What medium access algorithms are appropriate for use?

   b. Assume the following parameters:

$$\tau = 10^{-1} \text{ second}$$
$$\bar{X} = 1000 \text{ bits}$$
$$R = 100 \text{ Mbps}$$

   Discuss limitations on network throughput in terms of the number of stations as related to the average arrival rate per station.

   c. Are there significant restrictions on average transfer delay above those imposed if a is less than 1?

6 Synchronization and Error Control

6.1 Introduction

The data link layer—the second layer in the OSI architecture and most other network architectures—implements *data link control (DLC) protocols.* Chapter 7 is devoted to discussion of these protocols. In this chapter we discuss elements of data link control, concentrating on two important functions—*synchronization* and *error control.* We review major types of synchronization and techniques for providing synchronization at this protocol level. This is followed by discussion of techniques for protecting the network against effects of transmission errors. A final section briefly reviews techniques for flow control used by DLC protocols; such techniques help to avoid problems that are the result of excessive congestion of network facilities.

The primary design problems addressed in this chapter are coordination of sender and receiver, and maximizing reliability and freedom from errors, though the other design problems listed in Section 1.4 are also addressed.

☐ Subsections 6.3.10, on implementation of polynomial codes, 6.3.11, on error detection capabilities of polynomial codes, and 6.4.4, on communications efficiency calculations, plus Section 6.5, on optimum frame or packet sizes, require more mathematical sophistication than most of the book. For this reason they are marked as advanced sections. They can be omitted with no serious loss of background needed for later chapters. In addition, Subsections 6.3.6 and 6.3.7, on error correcting and detecting capabilities of Hamming (7,4) codes, are somewhat specialized.

6.2 Types of Synchronization

A variety of types of synchronization are important in telecommunication networks. *Synchronization of waveforms* is a function of the communications link. As we saw in Chapter 4, the physical layer is responsible for *bit synchro-*

nization. The next level of synchronization is *byte or character synchronization*; this is normally provided by the data link layer, although it may be provided by the physical layer (for example, when the byte timing circuit in X.21 is used). *Message or frame synchronization* is the responsibility of the data link layer, as is *content synchronization,* or distinguishing which fields in a message or frame contain synchronization, control, addressing, or error protection information and which contain user data. A related function is sequencing of frames or messages so they are either received in appropriate order or can be resequenced at the receiver. We sketch how these functions are commonly handled. Examples of DLC protocols in Chapter 7 should further clarify techniques.

synchronization characters

The normal approach to character synchronization is through use of special bit patterns or *synchronization* (sync) *characters.* SYN, Synchronous Idle, characters mentioned in Chapter 4 in conjunction with the X.21 interface, and Flag characters, used in the I.430 interface, are the characters most commonly used in this manner. Special logic circuitry (or software) at the receiver continually looks for synchronization characters when it is not in character sync. After finding an appropriate number of synchronization characters, the receiver assumes it is in character synchronization and "locks on" to the timing pattern in order to assemble the bits into characters.

control characters and count fields

Although messages of arbitrary length may be transmitted with message switching protocols, the usual procedure is to break long messages up into packets or frames of fixed maximum length. The beginning and end of frames are often delimited by *control characters.* A *count field* giving the length of a frame or of certain portions of a frame is an alternative technique for maintaining frame synchronization. Examples of these techniques are given by the DLC protocols discussed in the next chapter.

content synchronization

Standard approaches to *content synchronization* are based at least partially on the position of fields within a frame. Fields containing known types of information can be located by counting characters from the beginning (or the end) of a frame. A second approach is to delimit fields by special characters. For example, Chapter 7 shows how the control characters STX (Start of Text) and ETX (End of Text) begin and end text fields in some protocols. A third basic approach to content synchronization is to use count fields (often located by position within a frame) to indicate the length of particular fields. The ends of these fields can then be found by counting. Each of these approaches is used in at least one DLC protocol discussed in Chapter 7.

synchronization of access

Still another function of a data link control protocol is *synchronization of access* to the communications medium. This is especially important for data links, such as multipoint or broadcast links or LANs, where more than two end points exist on the communications link. However, even point-to-point links require control of which end is allowed to transmit and which must receive at a given instant. Medium access control was covered in Chapter 5.

6.3 Control of Transmission Errors

Error control functions handled by the data link layer involve ensuring the accuracy of transmitted bit or character streams without regard to frame content; content is ignored by the data link layer.

6.3.1 Sources of Transmission Errors

Noise

Errors in bit streams result from *noise* and other *distortions.* Some types of noise are inevitable results of factors such as random motion of electrons in dissipative elements and quantization of electronic charges due to electrons or holes in electronic circuits. Other types result from natural phenomena such as lightning, human phenomena such as electric ignition systems or motors, and the actions of switching systems or system repair personnel.

intersymbol interference

Another major source of errors is *intersymbol interference*, which results from the fact that voltages observed when sampling a pulse may be influenced by neighboring pulses. In theory, it is possible to eliminate intersymbol interference by proper choice of pulse shapes and matching equipment to channels (see Chapter 2), but channel characteristics are too variable for fixed pulse shapes to accomplish this at high rates. Dynamic and adaptive equalization techniques, which adjust characteristics to match channels, can largely eliminate intersymbol interference [QURE87] but are complex and expensive.

crosstalk

echoes

Still other errors result from signals carried over communications links, or from signals on other links in close proximity to those used. *Crosstalk* from some communications circuits to others, due to electromagnetic coupling of signals, can be a serious problem, especially if signal amplitudes are high. *Echoes*, due to reflection of electromagnetic energy from the far end of improperly terminated circuits, or to coupling energy from transmit to receive circuitry at the transmitting end of a circuit, are fundamental factors currently limiting bit rates over communications channels.

It is impossible to eliminate all transmission errors. Observed error rates are *highly variable,* varying by three to four decimal orders of magnitude for different data links. The fraction of bits received in error is typically in the 10^{-3} to 10^{-7} range, though new data links may achieve better error rates (if not used in conjunction with older equipment, for example, vintage terrestrial links at each end of a communications satellite path). A reasonable value to use for bit error rate on a communications link employing telephone lines is 10^{-5}, but significant numbers of communications links have much worse error performance [BALK71], [BURT72]. Errors also tend to occur in bursts, rather than being independent of each other, and error rate calculations based on independent errors can be seriously incorrect.

6.3.2   Approaches to Error Control

An *error control algorithm* used at the receiving end of a data link takes the bit stream presented to it by the physical layer and computer interface, decides whether the bit stream appears to be legitimate, and tries to correct it or request retransmission if it does not appear legitimate.

If bit errors are to be detected and/or corrected, restrictions limiting allowable bit streams must exist; if all possible bit streams were legitimate there would be no basis for rejecting some. Hence, a *coding* technique that restricts allowable bit streams must be used, or redundancy introduced into the communications process.

6.3.3   Early DLC Error Control Techniques

Early DLC protocols used rudimentary error control. The most common techniques were echoing and a simple parity check on each character. Both are still common, though more powerful techniques are replacing them.

echoing

Echoing has been used in numerous systems with keyboard data entry. In such systems a character printed or displayed at a terminal is not obtained directly from the keyboard but is the result of transmitting the keyed character to the receiver and returning (echoing) it back. A full duplex data link is necessary to allow the echo to come back with minimal delay, but a number of modems provide this capability.

Error correction for such systems is the responsibility of the keyboard operator, who compares printed characters with those entered (or supposedly entered). A discrepancy can result from any of three causes (not mutually exclusive): Either the character was mistyped, or there was an error on the way from transmitter to receiver, or there was an error in the echo path. If only the third cause is applicable, the character reached the receiver correctly, but in the other two cases it was incorrect at the receiver. Which occurred is unknown, so it is always best to reenter the character or message.

Although echoing is reasonable for error control in simple systems, it has disadvantages. One is the need for full duplex transmission facilities. Also, it relies heavily on human detection of errors. There is a natural tendency to see expected characters when proofreading your own entries, so the error correction capability is weak. Finally, echoing does not work well on long distance connections, where the time delay between data entry and printout can be long enough to be disturbing.

parity check

In a simple *parity check*, a bit is added to each character and chosen so that the total number of ones is even (even parity) or odd (odd parity). (Both are used.) The parity bit is

$$\phi_{\text{even}} = \sum_{i=1}^{n} d_i \quad modulo \ two \tag{6.1a}$$

if even parity is used, with d_i the ith data bit (0 or 1) and ϕ the parity bit, or

$$\phi_{\text{odd}} = \phi_{\text{even}} \oplus 1 \tag{6.1b}$$

if odd parity is used. In each case addition is modulo two, that is, using

$$0 \oplus 0 = 0$$
$$0 \oplus 1 = 1$$
$$1 \oplus 0 = 1 \tag{6.2}$$
$$1 \oplus 1 = 0.$$

The symbol $\oplus$ represents modulo two addition, equivalent to the EXCLUSIVE OR operation in logic.

A single transmission error causes recomputed parity at the receiver to be invalid, so the error is detected. A double error, however, results in the same parity and cannot be detected. More generally, any odd number of errors can be detected but any even number cannot be detected. This is true for either odd or even parity.

If errors occurred independently, with probability p of an error on any individual bit, the probability of k errors in n bits transmitted would be

$$P[k \text{ errors}] = \binom{n}{k} p^k (1 - p)^{n-k}, \tag{6.3}$$

with $1 - P[0 \text{ errors}]$ the simplest way to compute the probability of at least one error. Equation (6.3) reflects the facts that a specific pattern with k errors in n bits transmitted has probability $p^k(1 - p)^{n-k}$ of occurring and that there are

$$\binom{n}{k} = \frac{n!}{k!(n - k)!}$$

ways in which k errors in n bits can occur.

The probability an error is detected with a single parity bit is the sum of the probabilities of all odd numbers of errors or

$$P[\text{error detected}] = \sum_{k=1}^{\lceil n/2 \rceil} \binom{n}{2k - 1} p^{2k-1}(1 - p)^{n-2k+1} \tag{6.4}$$

with $[x]$ indicating the smallest integer $\geq x$. The probability that the parity check succeeds in detecting errors when they occur is the conditional probability of an error being detected given that at least one error occurs. This is given by

$$
\begin{aligned}
P[\text{parity check successful}] &= P[\text{error detected} \mid > 0 \text{ errors}] \\
&= \frac{P[\{\text{error detected}\} \cap \{> 0 \text{ errors}\}]}{P[> 0 \text{ errors}]} \\
&= \frac{P[\text{error detected}]}{1 - P[0 \text{ errors}]}
\end{aligned}
\tag{6.5}
$$

Table 6.1   Error Detection Performance of Single Parity Bit on
Eight-bit Characters (including parity), Assuming Independent
Errors on Communication Link.

Bit Error Rate	Probability of Error in Transmission	Probability of Detection Failure given Error	Probability of Undetected Error
10^{-5}	8.0×10^{-5}	3.5×10^{-5}	2.8×10^{-9}
10^{-4}	8.0×10^{-4}	3.5×10^{-4}	2.8×10^{-7}
10^{-3}	8.0×10^{-3}	3.5×10^{-3}	2.8×10^{-5}
10^{-2}	7.7×10^{-2}	3.4×10^{-2}	2.6×10^{-3}
10^{-1}	5.7×10^{-1}	2.7×10^{-1}	1.5×10^{-2}

with denominator and numerator obtainable from Eqs. (6.3) and (6.4), respectively. The final form results from the fact that an error can be detected only if at least one error occurs, so the intersection of two events in the previous numerator is equivalent to the event that an error is detected.

Typical values from these formulas, for the case $n = 8$ (one byte characters, including parity), are given in Table 6.1, which lists bit error rate (p), probability of at least one error in a character, probability of detection failure given at least one error (one minus probability detection is successful), and probability an undetected error occurs when transmitting a character.

Values in Table 6.1 should be treated with caution; burstiness of errors causes the number of multiple errors to be an appreciable fraction of the number of single errors. Curves plotting $P[\geq m, n]$ (probability of at least m errors in an n-bit block) are given for telephone network data transmission in [BALK71]. $P[\geq 2, 10]$ is one half to one third $P[\geq 1, 10]$, so double errors in ten-bit blocks are not much less common than single errors. Thus, *actual performance* of a single parity check *is often far inferior* to that indicated by Table 6.1. The majority of transmission errors are detected by single parity checks, but they may not reduce undetected error rate to an acceptable level.

6.3.4   Two-Dimensional Parity Checks

Stronger error protection is possible with two-dimensional parity checking, including an overall block check character, each bit computed as a parity check on the corresponding bit position in each data character, appended to the end of transmitted data. This is illustrated in Fig. 6.1.

*row and column
parity bits*

Row and column parity bits, with m characters, n bits per character, and even parity, are given by

$$\phi_{ir} = \sum_{k=1}^{n} d_{ik} \quad \text{modulo two} \tag{6.6a}$$

Figure 6.1   Two-dimensional Parity Check.

		Data bits			Character parity bits
Character 1	d_{11}	d_{12}	$\cdots$	d_{1n}	ϕ_{1r}
Character 2	d_{21}	d_{22}	$\cdots$	d_{2n}	ϕ_{2r}
$\cdots$	$\cdots$	$\cdots$	$\cdots$	$\cdots$	$\cdots$
Character m	d_{m1}	d_{m2}	$\cdots$	d_{mn}	ϕ_{mr}
Block check character	ϕ_{c1}	ϕ_{c2}	$\cdots$	ϕ_{cn}	ϕ_{cr}

$$\phi_{cj} = \sum_{k=1}^{m} d_{kj} \quad \text{modulo two} \tag{6.6b}$$

respectively, with addition defined as in Eq. (6.2). For odd parity, 1 is added modulo two to each total. The final parity bit, ϕ_{cr}, is usually computed by summing over ϕ_{ir} values rather than ϕ_{cj} values; for even parity, the same value is obtained either way, but this is not always true for odd parity (see Problem 6.4).

Two-dimensional parity checks can correct single errors, located by the intersection of a row and a column with invalid parity. Double errors in any row or column can be detected, but not corrected; recomputed parity checks on columns or rows, respectively, reveal the presence of errors. Many other patterns of errors can be corrected or detected but any that result both in even numbers of errors in each row and even numbers of errors in each column cannot be detected. Patterns that can or cannot be detected or corrected are complex, so we will not attempt to fully analyze error handling properties. Problem 6.6 involves partial analysis of undetectable error patterns.

6.3.5   Simple Block Codes

Although simple parity checks, possibly supplemented with block check characters, are still used in many systems, much more powerful codes are available. The most common are *block codes*, often defined in an algebra of polynomials (see Subsection 6.3.9). Other codes, such as *convolutional codes*, are used in circumstances requiring even more powerful codes.

Block codes use sequences of n channel symbols (or n-tuples) as code words. A block code is defined by defining a subset of the possible n-tuples to be legitimate; only words in the subset are transmitted. Due to transmission errors, however, any n-tuple may be observed by the receiver. Moreover, the receiver can use only the received sequence for decoding. If the received word is a word in the set used by the transmitter, it will be decoded as this word. This could result from transmission of a different word with error pattern mapping it over into the received word, but the receiver has no way of determining this. If a received word is not legitimate, error recovery is invoked.

Error handling capabilities of block codes are determined largely by "distances" between code words. Codes should be chosen to minimize the probability errors will transform transmitted words into received words that will not be handled properly. A more precise statement is given below.

Hamming distance

Error detecting and correcting capabilities of block codes can be described in terms of *Hamming distance* between code words. If i and j are two code words, the Hamming distance, $d_H(i,j)$ between them is the number of bit positions in which they differ. A code can correct any c or fewer errors if

error correction capability

$$\min_{i,j} d_H(i,j) \geq 2c + 1. \tag{6.7}$$

This is true since with c or fewer errors the received word has Hamming distance from the transmitted word of c or less, while its Hamming distance from any other possible transmitted word will be $c + 1$ or greater. A decoder that selects the legitimate word with minimum Hamming distance will decode into the only such word corresponding to c or fewer errors. The probability of a bit error is normally much less than one half, so this is usually correct decoding.

error detection capability

On the other hand, if the code is used purely for error detection, with no attempt at correction, it can detect any combination of d or fewer errors if

$$\min_{i,j} d_H(i,j) \geq d + 1. \tag{6.8}$$

This is true since it requires at least $d + 1$ errors to transform a transmitted word into another legitimate word, so with more than zero but fewer than d errors the received bit pattern cannot be legitimate. Many error patterns corresponding to more than d errors can also be detected, since they will not transform the transmitted word into another legitimate word, but it is difficult to fully describe such patterns. If the code is used for any error correction, however, some patterns with d or fewer errors will not be detectable (see Problem 6.7).

simultaneous error correction and detection capability

A combination error correcting and error detecting code, which corrects all combinations of c or fewer errors and simultaneously detects all combinations of $c + 1$ to d errors, with $d > c$, requires that

$$\min_{i,j} d_H(i,j) \geq d + c + 1. \tag{6.9}$$

The proof is similar to those just stated (see Problem 6.8).

weight

A term closely related to Hamming distance is *weight*. The weight of a code word is its Hamming distance from the all zero word (a valid word in almost all codes); equivalently, the weight is the number of ones in the word. It can be proved that minimum distance between any two code words in important codes is the minimum weight of a nonzero code word (see Problem 6.9).

block code

A wide variety of error correcting and error detecting codes have been studied. In an (n,k) *block code* each code word has k data bits and $n - k$ parity bits, for a total of n bits. In many implementations the first k bits transmitted are unmodified data bits; $n - k$ parity bits then follow. Since k of n bits correspond to data, the *effective rate* of the code (fraction of bits used for data) is k/n. Such codes can give strong error protection while using small fractions of their bits for parity; typical codes have on the order of 16 parity bits but give strong error detection protection for thousands of data bits.

For common (n,k) block codes, each parity bit is the modulo two sum of a subset of the data bits. That is,

$$\phi_k = d_{i_1} \oplus d_{i_2} \oplus \ldots \oplus d_{i_{rk}}, \tag{6.10a}$$

with r_k denoting the number of data bits involved in the kth parity check. Equivalently, using the definition of modulo 2 addition in Eq. (6.2),

$$\phi_k \oplus d_{i_1} \oplus d_{i_2} \oplus \ldots \oplus d_{i_{rk}} = 0. \tag{6.10b}$$

An example in Problem 6.13 is a Hamming single error correcting code with three parity bits satisfying

$$\phi_1 \oplus d_4 \oplus d_6 \oplus d_7 = 0$$

$$\phi_2 \oplus d_4 \oplus d_5 \oplus d_6 = 0 \tag{6.11}$$

$$\phi_3 \oplus d_5 \oplus d_6 \oplus d_7 = 0 \,.$$

(The subscript on ϕ_i or d_i indicates position in a code word.) Hamming single error correcting codes exist for any n of form $2^r - 1$; each has r parity bits and $k = 2^r - 1 - r$ data bits; the (7,4) case ($r = 3$) is treated in Problems 6.13 and 6.14. Each such code has minimum distance 3, and can be used as a double error detecting code if no attempt to correct errors is made. A wide variety of other codes are known, some capable of correcting substantial numbers of errors.

6.3.6 Error Correcting Capabilities of Hamming (7,4) Code

The error protection capabilities of simple codes used purely for error correction can be analyzed reasonably simply if the assumption that errors on different bits are independent is made.[1] When the code is used purely for error correction, it *always decodes* a received word. If the error pattern is one the code is designed to correct, it is corrected; otherwise the decoder

---

[1] As we have pointed out, this assumption is often unrealistic.

inserts an erroneous word into the message stream. For a single error correcting code, with minimum distance of three, this means decoding is done correctly if and only if fewer than two errors occur during transmission.

probability of decoding error for error correcting code

Hence

$$P[\text{decoding error}] = 1 - P[0 \text{ error}] - P[1 \text{ error}], \qquad (6.12)$$

with probabilities of 0 error and 1 error given by Eq. (6.3). Table 6.2 gives the probability of (at least one) error in transmission of a character, the probability of erroneous decoding given (at least one) error occurs, and the residual error rate for this code. The probability of erroneous decoding given (at least one) error is the ratio of the probability of a decoding error to the probability that at least one error occurs (see the derivation of Eq. 6.5), while residual error rate is the probability of decoding error.

6.3.7   Error Detecting Capabilities of Hamming (7,4) Code

Analysis of error handling performance of the Hamming (7,4) code used purely as an error detecting code is more complex. We sketch the analysis, using the results of Problems 6.9 and 6.13.

Since an error detecting code indicates an error any time a received word is not legitimate, the only error patterns causing it to fail are those that transform one legitimate code word into another legitimate one. Hence we need to know the distance structure of the code. This can be simplified for a Hamming code (and similar codes) since results of Problem 6.9 show that the number of code words at distance d from a designated code word is independent of the word designated. Thus we can analyze distance of other words from any word we wish, such as the all zero word. Furthermore, there are seven code words at distance 3 from the all zeros word, seven at distance 4

probability of decoding error for error detecting code

and one at distance 7 (see Problem 6.13). Hence,

$$P[\text{decoding error}] = 7\,p^3(1 - p)^4 + 7\,p^4(1 - p)^3 + p^7. \qquad (6.13)$$

Table 6.2   Error Handling Performance of Hamming (7,4) Code When Used as Error Correcting Code, Assuming Independent Errors on Communication Link.

Bit Error Rate	Probability of Error in Transmission	Probability of Erroneous Decoding Given Error	Residual Error Rate
10^{-5}	7.0×10^{-5}	3.0×10^{-5}	2.1×10^{-9}
10^{-4}	7.0×10^{-4}	3.0×10^{-4}	2.1×10^{-7}
10^{-3}	7.0×10^{-3}	3.0×10^{-3}	2.1×10^{-5}
10^{-2}	6.8×10^{-2}	3.0×10^{-2}	2.0×10^{-3}
10^{-1}	5.2×10^{-1}	2.9×10^{-1}	1.5×10^{-1}

Table 6.3 Error Detection Performance of Hamming (7,4) Code Assuming Independent Errors on Communications Link.

Bit Error Rate	Probability of Error in Transmission	Probability of Detection Failure Given Error	Probability of Undetected Error
10^{-5}	7.0×10^{-5}	1.0×10^{-10}	7.0×10^{-15}
10^{-4}	7.0×10^{-4}	1.0×10^{-8}	7.0×10^{-12}
10^{-3}	7.0×10^{-3}	1.0×10^{-6}	7.0×10^{-9}
10^{-2}	6.8×10^{-2}	1.0×10^{-4}	6.8×10^{-6}
10^{-1}	5.2×10^{-1}	1.0×10^{-2}	5.1×10^{-3}

Table 6.3 lists the probability of at least one error in transmission, the probability of detection failure given at least one error, and the probability of undetected error.

Note that the code is much more powerful at detecting errors than it is at correcting errors. This is true for most codes. Error performance when the code is used for error detection will, of course, depend on successful retransmission as well as on successful error detection.

This general approach to analysis of error handling performance can be used for studying various types of codes, but it becomes complex for powerful codes. In any situation where the approach is used, the precise results should be treated with caution, especially because the basic assumption of independent errors needed for the analysis is seldom valid.

6.3.8 More Powerful Codes

Convolutional codes [CLAR81] are an important alternative class of codes used in telecommunications. They differ from block codes by encoding data bits in a continuous stream rather than k bits at a time. After each k input bits are fed into the encoder, a total of n output bits are obtained. Each output bit depends on the current group of k input bits as well as $N - 1$ previous groups of k input bits, which determine the current state of the encoder. The effective rate of a convolutional encoder is thus k/n. Such codes can be extremely powerful, but often use substantial amounts of overhead; rate 2/3 and rate 1/2 codes are common, with 50 percent and 100 percent overhead, respectively. We do not treat such codes in detail, although they are often used for error correction in situations where this is necessary.

Bose-Chaudhuri-Hocquenghem, Reed-Solomon codes

Two classes of block codes that contain codes powerful enough to compete with convolutional codes are *Bose-Chaudhuri-Hocquenghem (BCH)* codes and *Reed-Solomon* codes [BERL68], [CLAR81], [LIN83], [PETE72]. Reed-Solomon codes are actually a subclass of BCH codes, but were discovered earlier; the subclass includes the ones most commonly applied. In some ways they are simpler to implement than convolutional codes, especially at high speeds. They have been adopted in such situations as compact disc players and are specified in a few standards.

speeds. They have been adopted in such situations as compact disc players and are specified in a few standards.

trellis codes

Another important class of codes is *trellis codes,* which are commonly used with more complex signal sets than would be feasible without coding [UNGE87a]. A simple example was treated in Subsection 2.4.3. Although different signals are too close together to be reliably distinguished without coding, the codes limit sequences in which signals can be used. The extra redundancy makes use of complex signal sets feasible without high error rates and provides higher transmission rates than would be possible without coding. Development of trellis codes has been largely responsible for recent major advances in modem data rates.

6.3.9   Polynomial Codes

polynomial codes

The most common block codes used for error detection are called *polynomial codes* since their mathematical structure has been developed with the aid of a polynomial representation of code words. In order to present basic ideas behind these codes, we briefly discuss this representation and polynomial manipulations used in the theory.

Consider a binary code word expressed as the vector

$$\mathbf{v} = [v_0, v_1, \ldots, v_7] = [1\,0\,0\,1\,1\,0\,1\,1].$$

This eight-bit vector can be represented by a seventh order polynomial with binary coefficients

$$V(X) = 1{\cdot}X^0 + 0{\cdot}X^1 + 0{\cdot}X^2 + 1{\cdot}X^3 + 1{\cdot}X^4 + 0{\cdot}X^5 + 1{\cdot}X^6 + 1{\cdot}X^7$$
$$= 1 + X^3 + X^4 + X^6 + X^7.$$

The second form is normally used, since it is customary to omit terms with zero coefficients and not write out ones as multipliers.

Addition, subtraction, multiplication, and division of polynomials are defined as in ordinary algebra, with modulo two arithmetic to compute coefficients.[2] For example, if $V(X)$ is as defined above and

$$W(X) = X + X^4 + X^5,$$

then their sum is

$$S(X) = V(X) \oplus W(X) = 1 + X^3 + X^5 + X^6 + X^7,$$

and their product is

$$P(X) = V(X){\cdot}W(X) = X + X^8 + X^9 + X^{10} + X^{12}.$$

(Readers are asked to verify these calculations.) Division of polynomials is best accomplished by long division, writing polynomials with higher order terms on the left. For example $S(X)/W(X)$ is found from

---

[2] Note that subtraction modulo two is equivalent to addition modulo two.

$$
\begin{array}{r}
X^2 \qquad\qquad + 1 \\[2pt]
\hline
X^5 + X^4 + X \,\overline{)\, X^7 + X^6 + X^5 \qquad\quad + X^3 \qquad\quad + X + 1\,} \\
\underline{X^7 + X^6 \qquad\qquad\quad\; + X^3 \qquad\qquad\qquad} \\
X^5 \qquad\qquad\quad + X + 1 \\
\underline{X^5 + X^4 \qquad\qquad + X \qquad} \\
X^4 \qquad\qquad\qquad + 1 .
\end{array}
$$

Thus division gives a quotient and remainder of

$$Q(X) = X^2 + 1$$
$$R(X) = X^4 + 1 .$$

Note that $S(X) = W(X){\cdot}Q(X) \oplus R(X)$ and the degree of $R(X)$ is less than that of the divisor, $W(X)$. Equivalent relationships hold for any polynomials.

generator polynomial   A polynomial code is defined by a *generator polynomial* $G(X)$, with unity X^0 coefficient (that is, the lowest order term of 1) and of degree r, with r representing the number of parity bits added to data messages. A message represented by

$$M(X) = m_0 + m_1 X + m_2 X^2 + \dots + m_{k-1} X^{k-1} \qquad \textbf{(6.14)}$$

is encoded as follows:

- *Step 1:* The data message polynomial, $M(X)$, is multiplied by X^r, giving r 0's in the low-order positions.

message encoding
- *Step 2:* The result is divided by $G(X)$, giving a unique quotient $Q(X)$ and remainder $R(X)$ satisfying

$$X^r{\cdot}M(X) = Q(X){\cdot}G(X) \oplus R(X) \qquad \textbf{(6.15a)}$$

or

$$X^r{\cdot}M(X) \oplus R(X) = Q(X){\cdot}G(X). \qquad \textbf{(6.15b)}$$

- *Step 3:* The remainder is added to $X^r{\cdot}M(X)$, giving

$$T(X) = X^r{\cdot}M(X) \oplus R(X) \qquad \textbf{(6.16)}$$

for the polynomial describing the transmitted code word.

Since the degree of the remainder is less than r, the bits corresponding to the remainder are placed in positions where $X^r{\cdot}M(X)$ has zeros. The added bits correspond to r parity check bits.

Note that Eq. (6.15b) indicates that any legitimate code word is a multiple of $G(X)$; in fact, it can be shown that the set of legitimate code words is identical to the set of multiples of $G(X)$.[3] Hence the receiver determines whether or not an error has occurred by dividing $T(X)$ by $G(X)$ and examining

---

[3] To be precise, when a polynomial code is defined in this manner, polynomial multiplication is defined modulo $X^n + 1$, or equivalently $X^n - 1$, that is, $X^n - 1 = 0$ or $X^n = 1$.

the remainder. If the remainder is zero, the received word is accepted as correct; otherwise an error has been detected and error recovery is initiated.

example of encoding

As an example, consider a (7,4) code with generator polynomial

$$G(X) = 1 + X + X^3. \tag{6.17}$$

Thus $n = 7$, $k = 4$, and $r = n - k = 3$. Encoding of message $(m_0, m_1, m_2, m_3) = (0\ 1\ 0\ 1)$ proceeds as follows:

$$M(X) = X + X^3,$$
$$X^3 \cdot M(X) = X^4 + X^6.$$

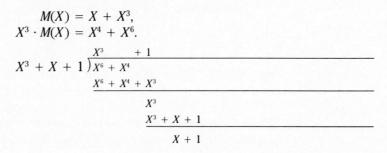

Thus, $R(X) = 1 + X$ and $T(X) = 1 + X + X^4 + X^6$, yielding a code word vector

$$\mathbf{t} = (t_0, t_1, \ldots, t_6) = (1\ 1\ 0\ 0\ 1\ 0\ 1).$$

This code is identical to the Hamming code defined by the parity checks in Eq. (6.11) (see Problem 6.14). Any Hamming code can be represented as a polynomial code with some generator polynomial. However, the class of polynomial codes is broader than that of Hamming codes.

☐ ### 6.3.10 Implementation of Polynomial Codes

shift registers

An advantage of polynomial codes, in addition to excellent error detecting capabilities, is the fact that they can be implemented simply. *Shift registers* with feedback connections determined by the generating polynomial can be used. We illustrate shift registers for $G(X) = 1 + X + X^3$. Figures 6.2 and 6.3 illustrate encoder and decoder circuits for the code defined by $G(X)$.

The circles containing $\oplus$ in the figures represent modulo two adders, or EXCLUSIVE OR gates, and stages 0, 1, and 2 represent memory elements. A feedback connection, with a modulo two adder, is made for each g_i that is nonzero; thus another feedback connection to a modulo two adder between stage 1 and stage 2 would have been made if g_2 were nonzero. Each circuit is clocked, with the output of stage i (possibly added modulo two to a value obtained via feedback) fed into stage $i + 1$ after each clocking pulse.

operation of encoder

Operation of the encoder is as follows. The four message bits for a message to be encoded are fed into the circuitry in Fig. 6.2 with both switches in position 1. Thus each bit is sent both directly to the output line and into the feedback connection. Both switches are next switched to position 2, disabling feedback and connecting the output of stage 2 to the output line. The shift register is shifted three more times, shifting the contents of stages 2, 1, and 0 onto the output line; the circuit effectively divides

Figure 6.2   Shift Register Encoder for Polynomial Code with
$G(X) = 1 + X + X^3$.

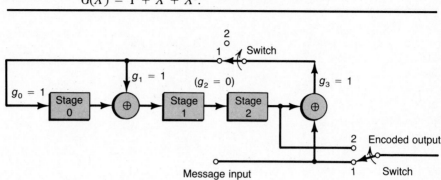

Figure 6.3   Shift Register Error Detection Circuit for Polynomial
Code with Encoder in Fig. 6.2.

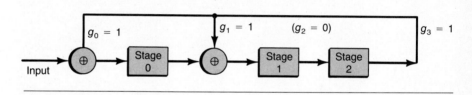

$X^3 \cdot M(X)$ by $G(X)$ and leaves $R(X) = r_2 X^2 + r_1 X + r_0$ in stages 2, 1, and 0, which means that these are check bits.

operation of decoder

　　The shift register in Fig. 6.3 is simpler since it is a division circuit without pre-multiplication by X^3 and with no switching. The received code word, $C(X) = c_0 + c_1 X + c_2 X^2 + \ldots + c_6 X^6$ is shifted into the input, c_6 first, with feedback during each shift. The remainder is then in stages 2, 1, and 0. If all three contain zero, the word is accepted; otherwise an error has been detected.[4] We verify operation of this circuit by an example; comparable verification of operation of the circuit in Fig. 6.2 is the topic of Problem 6.15.

example of decoding

　　Assume that a received code word is $\mathbf{c} = (c_0, c_1, \ldots c_6) = (1\ 0\ 1\ 0\ 0\ 1\ 1)$. Thus the decoder should divide the polynomial $X^6 + X^5 + X^2 + 1$ by $G(X) = X^3 + X + 1$. This is illustrated by long division below and by tabulation of shift register parameters (for the shift register in Fig. 6.3) in Table 6.4. All coefficients, including zeros, have been included in the long division and corresponding values in the two representations labeled.

---

[4] If the code were being used as a single error correcting code, nonzero values in the stages would be used to try to identify the error as well as to detect an error.

Table 6.4   Tabulation of Steps in Operation of Division Circuit in Fig. 6.3 when Dividing $X^6 + X^5 + X^2 + 1$ by $X^3 + X + 1$.

j	Shift Register Contents after j Shifts, Stages 210	Output Symbol after jth Shift	Feedback on jth Shift into Stages 210	Input Symbol on jth Shift
0	000	0	—	—
1	001	0	000	1
2	011	0	000	1
3	110 (A)	0	000	0
4	111 (B)	1	011 (F)	0 (J)
5	100 (C)	1	011 (G)	1 (K)
6	011 (D)	1	011 (H)	0 (L)
7	111 (E)	0	000 (I)	1 (M)

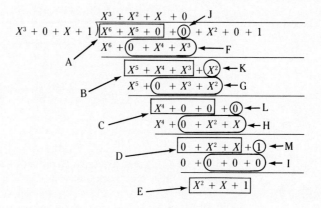

The first three shifts in the shift register implementation are not explicitly depicted in long division; they involve shifting the first three input digits into the shift register; afterward the contents, 1, 1, and 0, of stages 2, 1, and 0—(A) in the table—agree with coefficients in rectangle A in the division formula.[5] Feedback values, in oval F (and F in the table) are subtracted (modulo two) from all except the highest order term during long division. Similarly, the value in the circle labeled J (and J in the table) is the new input in both long division and shift register tabulation.

After each shift, a rectangle in the long division formula shows coefficients agreeing with stages 2, 1, and 0 in the shift register, an oval shows values fed back into stages 2, 1, and 0 in the shift register, and a circle shows the next input to the shift register. Letters label each pertinent result. At the conclusion of seven shifts, the remainder, $X^2 + X + 1$, is given by coefficients in stages 2, 1, and 0. Since this is nonzero, an error has been detected.

---

[5] These coefficients at this point represent the polynomial $X^2 + X$ rather than $X^6 + X^5$, but the following four shifts of the shift register effectively multiply X^4, as each is equivalent to multiplication by X.

Premultiplication by X^3 in the encoder is obtained by feeding the input message bits in at the output of stage 2 rather than at the input of state 0. Further details are examined in Problem 6.15.

☐   ## 6.3.11   Error Detecting Capabilities of Polynomial Codes

Since addition of one modulo two inverts a bit while addition of zero modulo two leaves a bit unchanged (see Eq. (6.2)), the received message after $T(X)$ is transmitted can be written as

$$C(X) = T(X) \oplus E(X), \tag{6.18}$$

with $E(X)$ a polynomial representing transmission errors and containing nonzero terms where errors occur. That is,

$$E(X) = e_0 + e_1 X + e_2 X^2 + \ldots + e_{n-1} X^{n-1}, \tag{6.19}$$

with $n = r + k$ the total number of bits in a code word, and $e_i = 1$ if there is an error in transmitting the ith bit, $i = 0, 1, \ldots, n - 1$, and $e_i = 0$ otherwise.

We know, from Eqs. (6.15b) and (6.16), that $G(X)$ divides $T(X)$.[6] Also, any received polynomial divisible by $G(X)$ is accepted as correct, with an error detected if a non-zero remainder is obtained. Thus the only undetectable errors are those whose error patterns, $E(X)$, are divisible by $G(X)$. This allows us to determine the error detecting capabilities of polynomial codes.

1. *Single bit errors.* If $C(X)$ has a single bit in error, $E(X) = X^i$ where i denotes the position of the error. If $G(X)$ has more than one term, it cannot divide X^i, so this guarantees all single errors will be detected.

2. *Double bit errors.* A double bit error can be represented by the polynomial $E(X) = X^i + X^j$ with $i < j$ and i and j denoting positions of the bits in error. We can rewrite this as $E(X) = X^i(1 + X^{j-i})$. For the error to be detected, neither X^i nor $1 + X^{j-i}$ can be divisible by $G(X)$. If $G(X)$ has a factor with three terms, neither X^i nor $1 + X^{j-i}$ is divisible by $G(X)$, so all double errors will be detected.

3. *Odd numbers of errors.* We will show that all odd numbers of errors are detect-able if $G(X)$ has a factor of $(1 + X)$. That is, if $E(X)$ has an odd number of bits in error, it is not divisible by $(1 + X)$. The proof follows.

    Assume $E(X)$ contains an odd number of ones and is divisible by $(1 + X)$. Then $E(X) = (1 + X) A(X)$, with $A(X)$ a polynomial, so in modulo two arithmetic $E(1) = (1 \oplus 1)A(1) = 0$. If $E(1) = 0$ modulo two, however, it must contain an even number of ones, contradicting our original assumptions. This means that $(1 + X)$ cannot divide $E(X)$ if $E(X)$ represents an odd number of zeros.

4. *Error bursts.* A burst of errors refers to a group of bits within a received message which contains all of the errors.[7] We define the length of a burst, b, to be the number of bits in a group having at least its first and last bits in error. For exam-

---

[6] One polynomial divides another if the remainder after division is zero.

[7] The standard definition of a burst error in coding theory literature allows a burst error in one part of a received word to be coupled with different errors in other parts of the same word. Our analysis here assumes that only a single burst of errors occurs, however.

ple, if $E(X)$ is the polynomial corresponding to the error pattern (0010100000010000), $E(X)$ contains an error burst of length $b = 10$.

In general, we can write an $E(X)$ of this form as $E(X) = X^i E_1(X)$ with $E_1(X)$ a polynomial of b terms and degree $b - 1$. Thus $E(X)$ for the above example is $X^2 + X^4 + X^{11} = X^2(1 + X^2 + X^9)$. Assuming $G(X)$ at least meets our criterion for detecting single errors, X^i is not divisible by $G(X)$ since it has a single term. Hence, the error will be undetected only if $E_1(X)$ is divisible by $G(X)$.

It is impossible for $G(X)$ to divide $E_1(X)$ if the length, b, of the burst is less than the length, $(r + 1)$, of $G(X)$, since this would imply a polynomial of greater degree divided a polynomial of lower degree. This implies that any burst of length $b < (r + 1)$ is detectable if $G(X)$ is of degree r, with r check bits included in each transmitted message.

When b is equal to the number, $r + 1$, of bits in $G(X)$, it will be undetected only if $E_1(X)$ is identical to $G(X)$. The first and last bits in both $G(X)$ and $E_1(X)$ are ones,[8] with the remaining $b - 2 = r - 1$ bits in $E_1(X)$ either 0 or 1, so only a fraction $1/2^{(r-1)}$ of the possible bursts of length $r + 1$ will be undetected.

Analysis of bursts of length $b > r + 1$ is more complex. If a burst error of length $b > r + 1$ occurs, so $E_1(X)$ is of degree greater than r, there are a variety of possible ways in which $G(X)$ could conceivably divide $E_1(X)$. If $G(X)$ does divide $E_1(X)$, we can write $E_1(X) = A_1(X) \cdot G(X)$, with $A_1(X)$ the quotient when $E_1(X)$ is divided by $G(X)$. Since $E(X)$ is a polynomial of degree $b - 1$ and $G(X)$ is a polynomial of degree r, $A_1(X)$ must be of degree $(b - 1 - r)$ and contain a total of $(b - 1 - r) + 1 = b - r$ terms. Furthermore, the first and last terms of $E_1(X)$ and of $G(X)$ are ones, so the first and last terms of $A_1(X)$ must also be ones, and there are $b - r - 2$ terms in $A_1(X)$, which can be either 1 or 0. Thus there are a total of 2^{b-r-2} possible $A_1(X)$ polynomials, and the same number of ways in which $E_1(X)$ could be divisible by $G(X)$.

Since there are 2^{b-2} possible $E_1(X)$ polynomials, the fraction of bursts of length b that are undetectable is

$$\frac{2^{b-r-2}}{2^{b-2}} = \frac{1}{2^r}.$$

summary of error detection capabilities

In summary, if $G(X)$ has $(1 + X)$ as a factor and one factor with three or more terms, the following protection will be given:

Single bit errors	100% protection
Two bits in error (separate or not)	100% protection
An odd number of bits in error	100% protection
Error bursts of length less than $r + 1$ bits	100% protection
Error bursts exactly $r + 1$ bits in length	Fraction $[1 - 1/2^{(r-1)}]$ of bursts detected
Error bursts of length greater than $r + 1$ bits	Fraction $[1 - 1/2^r]$ of bursts detected

example of error detection capabilities

For example, if $G(X)$ is of degree 16, contains $(1 + X)$ as a factor, and has one factor with at least three terms, then it will detect all single or double errors,

---

[8] Recall that $G(X)$ was defined to have a unity coefficient for $X^0 = 1$ and is of degree r. The definition of $E_1(X)$ also implies the first and last coefficients are 1.

all odd numbers of errors, and all bursts of length 16 bits or less. A fraction $1 - 1/2^{15} = 0.999969$ of the bursts of length 17 bits will be detected as well as a fraction $1 - 1/2^{16} = 0.999984$ of the bursts of length greater than 17.

6.3.12   Polynomials in Standards

A few standardized generating polynomials, $G(X)$, include the following:

$$CRC\text{-}12: \; 1 + X + X^2 + X^3 + X^{11} + X^{12}$$
$$CRC\text{-}16: \; 1 + X^2 + X^{15} + X^{16}$$
$$CRC\text{-}CCITT: \; 1 + X^5 + X^{12} + X^{16}$$
$$CRC\text{-}32: \; 1 + X + X^2 + X^4 + X^5 + X^7 + X^8 + X^{10} + X^{11} + X^{12} + X^{16} +$$
$$X^{22} + X^{23} + X^{26} + X^{32}$$

The first three of these each contain $(1 + X)$ as a factor, with $G(X)/(1 + X)$ irreducible in each case (that is, containing no further factors aside from itself).[9] The CRC-32 polynomial is a thirty-second order irreducible polynomial. Data link control protocols discussed in Chapter 7 that use these polynomials include Bisync–CRC-12 for six-bit code words and CRC-16 for eight-bit code words; DDCMP–CRC-16; HDLC and related protocols—CRC-CCITT; IEEE 802 and various U.S. Department of Defense protocols—CRC-32.

6.3.13   FEC versus ARQ

forward error correction

automatic repeat request

Forward error correction (FEC) uses codes to correct (or attempt to correct) errors at the receiver. The major alternative approach to error control is error detection with *Automatic Repeat ReQuest (ARQ)*. For FEC, the receiver attempts to correct errors, but for ARQ it simply detects errors and asks the transmitter to retransmit erroneous blocks.

As was indicated in Subsection 6.3.8, when FEC is used, it is commonly implemented with rate 2/3 or rate 1/2 (or similar rate) convolutional codes, or with BCH or Reed-Solomon codes with similar overhead. On the other hand, typical codes for ARQ error detection use on the order of 16 parity check bits to protect up to several thousand data bits. If a typical code having 16 bits for error protection is used, the probability of an undetected block error in a block of 800 bits has been pessimistically estimated at 10^{-8} [BURT72].

This is an example of a general principle; a relatively small percentage of the transmitted bits need to be error protection overhead bits if ARQ is used. A very simple example was given in the analysis of a Hamming (7,4) code, with tabulation of results in Tables 6.2 and 6.3; the improvement in error

---

[9]No factors exist in modulo two arithmetic. Over the field of complex numbers, any polynomial of degree n factors into the product of n terms, since it has n complex roots (not all necessarily distinct).

detecting capability over error correcting capability is more pronounced for more powerful codes, however.

ARQ requires extra overhead for retransmission requests and retransmitting blocks in error, but total overhead for data links with reasonable error rates is often far less than it would be with FEC. Detailed reasons for this are given in [BURT72], which shows that essentially all standard error correcting codes are poorly matched to error statistics normally observed on data links such as telephone lines.[10] Moreover, standard error detecting codes detect a very high percentage of errors without requiring many parity check bits, and error detection (accomplished by recomputing parity check bits at the receiver and comparing with received bits) is far easier than error correction (which involves actually locating and correcting errors).

ARQ is almost universally used for error control in DLC protocols. The primary exceptions are cases where data links have sufficiently low error rates to allow use of FEC without excessive check bits and where variable delays (experienced with ARQ when retransmissions are needed) are more detrimental than the high overhead of FEC.

6.4   ARQ Protocols

Several variants on ARQ protocols are common. One choice is between using both positive and negative or only positive acknowledgments. Positive acknowledgments indicate correct reception of frames, or more precisely that parity bits recomputed at the receiver are valid. Negative acknowledgments may be returned if recomputed bits are invalid. It is possible to have a correct ARQ protocol that uses only positive acknowledgments, with retransmission of messages not acknowledged within a timeout period, but negative acknowledgments can speed retransmissions since they may be returned before the original transmitter would finish timing out.

The primary advantage of using both positive and negative acknowledgments is greater efficiency, but the pure positive acknowledgment approach is more elegant since it eliminates "superfluous" messages. The most prominent DLC protocol using only positive acknowledgments, ARPANET DLC, originated in the research community. All other DLC protocols discussed in Chapter 7, developed by manufacturers and/or standards committees, use both positive and negative acknowledgments to gain efficiency.

A pure negative acknowledgment protocol is not logically valid. If one were used, the transmitter of a frame would not be able to distinguish be-

---

[10] FEC has been making a comeback recently, especially for situations with long communications delays (for example, deep space probes) or situations with very adverse conditions (such as design conditions for military systems involved in enemy attack). FEC coding is common on communications satellite links, or even for compact disc players. See [BERL87] for a review of the current state of the art.

tween correct reception of a frame and complete loss of the frame; either would result in no response by the receiver. Complete loss is not unusual since it can be caused by a failure to correctly receive synchronizing characters needed for the receiver to lock on to frames (see Problem 6.2).

6.4.1 Stop-and-Wait ARQ

Important DLC protocols use *stop-and-wait ARQ* with separate frames for all positive or negative acknowledgments. Stop-and-wait has been used primarily in DLC protocols designed for half duplex operation. After transmitting a frame, the transmitter waits for a reply before sending another frame. Thus frame transmission alternates with acknowledgments (positive or negative) from the receiver of the original frames.

The protocol will function adequately with only positive acknowledgments plus timeouts or with both positive and negative acknowledgments. (Timeouts are also necessary in the latter case for reasons discussed later.) Both versions are illustrated in Fig. 6.4, with a positive acknowledgment denoted by ACK and a negative acknowledgment by NAK. Each part of the figure shows

Figure 6.4 Stop-and-Wait ARQ. (a) With both positive and negative acknowledgments. (b) With positive acknowledgments plus timeouts.

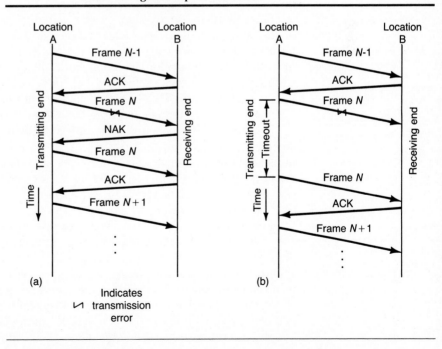

Figure 6.5   Stop-and-Wait ARQ with Lost Frames. Original data frame lost in (a), acknowlegment frame lost in (b). Arrows going off course indicate frames were not received.

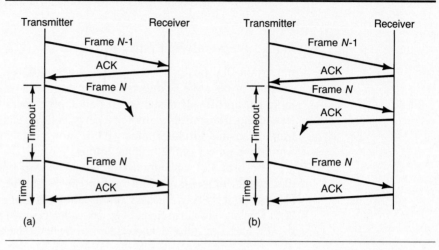

(a)   (b)

transmission of frames numbered $N - 1$, N, and $N + 1$. A transmission error occurs the first time frame N is transmitted, but the second transmission is successful. All other transmissions are error free.

Since we often use diagrams similar to these, it is worthwhile to elaborate on their interpretation. The vertical axis corresponds to time, increasing from top to bottom. Downward slopes of lines indicate time to transmit frames from source to destination, including both time to put them on the data link (lengths of frames divided by data link rates) and propagation delay from source to destination. (When such delays are not important for understanding material, we sometimes use horizontal lines to simplify diagrams.)

The illustrated procedures work as long as all transmissions are received adequately enough to be recognized. That is, acknowledgment frames are transmitted without error and data frames are recognized as such by the receiver (so it knows whether to return an ACK or a NAK). As we have indicated, cases where either data frames or responses (acknowledgments) are lost are not uncommon, however.

In a network using only positive acknowledgments and timeouts, loss of either the original data frame or the acknowledgment frame causes the original transmitter to time out and retransmit; it cannot distinguish either situation from a situation where lack of a response indicates the frame was received erroneously. This leads to the scenarios in Fig. 6.5.

From the point of view of the original transmitting station, the scenarios in the figure are equivalent. In each the transmitter sends a frame, fails to get a reply, and retransmits. The cases differ significantly from the point of view of the receiver, however; in case (b) it accepted and acknowledged frame N after the first transmission; but in (a) it did not accept frame N. Since the DLC protocol does not examine contents of frames, the receiver is not able to distinguish the retransmitted frame N from a new frame unless identification is placed in a frame header. A one-bit sequence number, distinguishing odd numbered from even numbered frames is sufficient to distinguish between original transmissions and retransmissions. For stop-and-wait ARQ, with only positive acknowledgments, at least a one-bit sequence number is needed in the header of all frames. ARPANET DLC, used in the DoD Reference Model and discussed in Chapter 7, uses this approach.

For stop-and-wait ARQ using both positive and negative acknowledgments, a similar problem requires numbering of acknowledgments. Since the receiver is required to respond with either a positive or a negative acknowledgment after each frame, lack of a response implies either the original frame transmission or its acknowledgment was lost. Although timing out and retransmitting if no reply is received is a valid way to handle situations with no reply, a common approach is for the original transmitter to time out and send an enquiry (ENQ) message to the receiver to ascertain its status. When a station using this implementation receives an ENQ, it is required to repeat its last previous frame. In cases where the original frame was received correctly, this can eliminate retransmission of what could be a lengthy data frame, but it can also lead to the scenarios in Fig. 6.6.

In both (a) and (b) the transmitter is sending a sequence of frames. In (a) it sends frame N after receiving an ACK to frame $N - 1$, but frame N never gets to the receiver, which of course does not reply. After a timeout, the transmitter sends ENQ and the receiver repeats the ACK it sent to frame $N - 1$. In

Figure 6.6   Enquiry Sequences. (a) Lost frame. (b) Lost ACK.

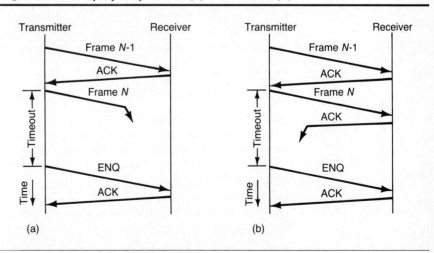

Figure 6.7   Enquiry Sequences with Alternating Acknowlegments to Remove Ambiguities.

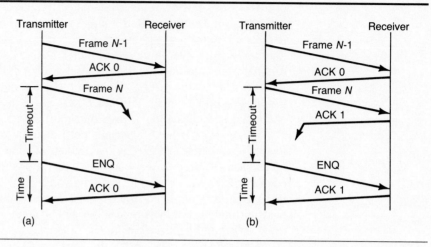

(a)   (b)

(b) frame N was received correctly and acknowledged, but the ACK was lost. Upon receiving the ENQ, the receiver obediently repeats its last ACK. To the transmitter, the two situations are the same since the frames it sends and receives are identical. The protocol must be modified if the transmitter is to know whether frame $N - 1$ or frame N is being acknowledged after the ENQ.

A one-bit acknowledgment count eliminates confusion. This is equivalent to using two ACKs, ACK 0 (ack zero), and ACK 1 (ack one); ACK 0 acknowledges even-numbered data frames and ACK 1 odd-numbered data frames. If N is odd, the sequences in Fig. 6.6 become those in Fig. 6.7. Alternating ACK 0 and ACK 1 responses remove the ambiguity. The Binary Synchronous (Bisync) DLC discussed in Chapter 7 uses this approach, with enquiries and two positive acknowledgments, plus negative acknowledgments.

6.4.2   Continuous ARQ

Continuous ARQ is standard with DLC protocols usable on full duplex communications links. Both frame transmissions and acknowledgments occur simultaneously. This eliminates dead time waiting for acknowledgments.

There are two main variants of continuous ARQ. When a NAK is received at the original transmitter, it may already have transmitted several frames following the one in error. In *go-back-N-ARQ* the erroneous frame and all succeeding frames already transmitted are retransmitted. In *selective reject ARQ*, only the erroneous frame is retransmitted.

go-back-N-ARQ

selective reject ARQ

Figure 6.8   Continuous ARQ. (a) Go-back-*N*. (b) Selective reject.

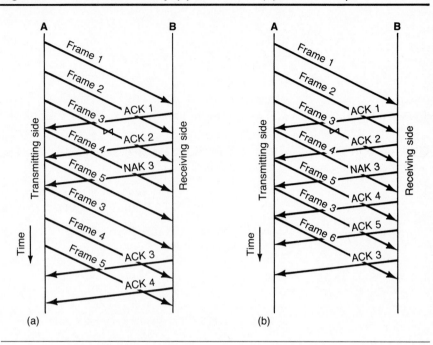

Figure 6.8 illustrates both variants; for simplicity, data transmission from A to B only is illustrated.[11] One error, in frame 3, is illustrated. ACKs and NAKs identify frames acknowledged. Frames 4 and 5 have been transmitted by the time the NAK to frame 3 gets back. Hence, in go-back-*N* ARQ, the transmitter retransmits frames 3, 4, and 5 before going ahead with frame 6. In selective reject ARQ, only frame 3 is retransmitted, followed by frame 6, and so forth.

Although selective reject is more efficient in using the capacity of a data link by requiring fewer retransmissions, go-back-*N* is more widely implemented. It simplifies software by not requiring the receiver to reorder messages. Selective reject also becomes complex when multiple errors occur.

6.4.3   Piggy-backed Acknowledgments and Sequence Counts

piggy-backed acknowledgments

Piggy-backed acknowledgments to data frames from A to B are obtained by using special fields in frame headers of data frames from B to A. A field reserved for acknowledgments and identifying the frames acknowledged is sufficient. Several protocols in Chapter 7 use this approach.

---

[11] Diagrams in Chapter 7 illustrate two-directional transmission.

Acknowledgments that contain sequence count fields (piggy-backed sequence counts) make it possible to acknowledge several frames at a time. A positive acknowledgment (piggy-backed or via a separate frame) for frame N may imply positive acknowledgments for all previously unacknowledged frames with numbers less than N.[12]

☐ ### 6.4.4   Communications Efficiency Calculations

effective rate of trans-
mitting data bits, R_e

Computation of the performance of DLC protocols involves analysis of the effect of retransmissions. We give approximate analyses of one parameter, R_e, *the effective rate of transmitting data bits.*

$$R_e = \frac{\text{Number of data bits accepted by the destination}}{\text{Total time to get those bits accepted}}. \qquad \textbf{(6.20)}$$

An equivalent parameter that differs only by a multiplicative constant is transmission efficiency, η, defined as

$$\eta = \frac{R_e}{R}, \qquad \textbf{(6.21)}$$

with R link transmission rate.[13] Our analyses of R_e can be converted to equivalent analyses of η by dividing numerical values by R.

A first step in computing R_e is to compute the probability distribution for the number of transmissions necessary to get a frame accepted. If P is the probability of an error on any given frame transmission, and the errors on different frames are independent, the probability that j transmissions will be required before success is the probability of $j - 1$ unsuccessful transmissions followed by one successful transmission or

$$P[j \text{ transmissions}] = P^{j-1}(1 - P). \qquad \textbf{(6.22)}$$

This describes a geometric probability distribution with mean

$$\bar{N}_T = \sum_{j=1}^{\infty} j P^{j-1}(1 - P) = \frac{1}{1 - P}. \qquad \textbf{(6.23)}$$

The most common approach to computing P is to use the formula

$$P = 1 - (1 - p)^K \qquad \textbf{(6.24)}$$

---

[12] Two approaches to specifying numbers for frames acknowledged are common: Some protocols specify the number for the last frame acknowledged, but others specify the number for the next frame needed (the number of the last frame acknowledged plus one). The approaches are equivalent, but the differences cause confusion.

[13] R_e is equivalent to what is called TRIB, transmission rate of information bits, by standards committees [AMER80], a mild example of the complex terminology typical of standards bodies.

with p representing the bit error rate, that is probability of an error on an individual bit, and K the number of bits in a frame. This is equivalent to using Eq. (6.3) (from our analysis of error protection capabilities of codes) and has the same limitations; that is, the analysis is based on an assumption that bit errors are independent even though errors normally occur in bursts. Equation (6.24) gives overly conservative (high) values for P since the burst nature of errors implies that errors affect fewer different frames than would be the case with independent errors.

bit versus burst error rates

A modification giving more accurate results is to replace bit error rate, p, by a similar parameter known as burst error rate, defined in [BALK71] as the probability an error burst starts on a given bit.[14] It is more reasonable to assume error bursts are independent than to assume that bit errors are independent; if this assumption is made and p is burst error rate, Eq. (6.3) gives an approximation to the probability k error bursts begin within n bits.[15]

One difficulty with this is the fact that burst errror rates are seldom tabulated, but [BALK71] indicates burst error rate is typically 1/3 to 1/10 bit error rate (that is a typical burst contains 3 to 10 errors). Using a burst error rate of 1/3 of the bit error rate for p in Eq. (6.24) is reasonable, and it gives computations more realistic than those obtained using the original bit error rate.

In subsequent computations we use Eq. (6.24) to compute the probability of one or more transmission errors in a frame. We assume that the error detecting code is powerful enough to catch all transmission errors; normally this is very nearly true and any other assumption would greatly complicate analysis. We also assume that there are no errors on either positive or negative acknowledgments. For reasonable line error rates (burst error rates of 10^{-4} or less) and long data frames (100 characters or more), the probability of error for data frames will be so much larger than that for control frames (for example, acknowledgments) that errors on control frames will not significantly affect results. Furthermore, including control frame errors would greatly complicate analysis.

With these assumptions, analysis of stop-and-wait ARQ systems is a modification of the analysis in Section 4.9. We again assume K bit frames, including n_h overhead bits, acknowledgment frames (positive or negative) of n_a bits, line turnaround time of t_{ta} seconds, and propagation delay of t_p seconds. As in deriving Eq. (4.1) it takes $[(K + n_a)/R] + 2(t_p + t_{ta})$ seconds to send a data block and get a positive or negative response.[16]

---

[14] [BALK71] defined burst error rate as total number of bursts observed divided by total number of bits transmitted, with a heuristic definition of which errors fall within bursts. The burst error rate was found to be insensitive to reasonable variations in the heuristic definition.

[15] A verification (left to readers) is the fact that the formula allows computation of probability distribution functions of frame error rate (probability of one or more errors in a frame), from distribution functions of burst error rates. Both distribution functions are in [BALK71] and [BURT72].

[16] We have assumed equal turnaround times at each end of the communications links; if this is not true, one of each should be included in this delay instead of twice the common value.

Figure 6.9   Variation of Effective Transmission Rate with Frame Length in Characters for Burst Error Rates, p, of 10^{-4} and 10^{-5}. Stop-and-wait ARQ on terrestrial 4800 bps link with 100 msec turnaround time.

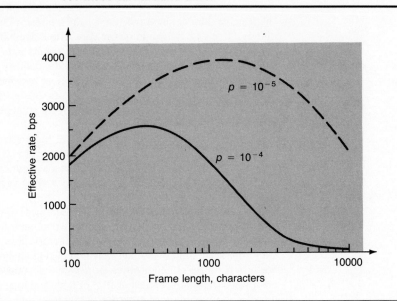

On average, $\bar{N}_T = 1/(1 - P)$ attempts are made before acceptance, so Eq. (6.20) gives

R_e *for stop-and-wait automatic repeat request*

$$R_e = \frac{K - n_h}{\dfrac{1}{1 - P}\left\{\dfrac{K + n_a}{R} + 2(t_p + t_{ta})\right\}} = \frac{(1 - P)(K - n_h)R}{K + n_a + 2(t_p + t_{ta})R}. \tag{6.25}$$

Note that Eq. (4.1), in Section 4.11, is a special case of Eq. (6.25) (for no errors, that is $P = 0$).

Figure 6.9 shows R_e values obtained from this formula for $n_h = n_a = 48$ bits, $t_p = 10$ msec, and $t_{ta} = 100$ msec. R_e is plotted versus frame length (in eight-bit characters) for two burst error rates, 10^{-4} and 10^{-5}. R_e peaks for an optimum frame length. For frame lengths less than optimum, an excessive fraction of the bits are overhead; for lengths greater than optimum, probability of retransmission is overly high. Optimum frame length is analyzed in the next section.

For continuous ARQ without transmission errors, the denominator for Eq. (6.20) is K/R, the time between successive acceptances of data blocks (see Fig. 6.10). Results with transmission errors differ, however.

For go-back-N ARQ with transmission errors, the number of frames transmitted to get a frame accepted is computed as follows. Transmission of one frame (that accepted) is always necessary. With probability $P(1 - P)$ transmission of N extra

Figure 6.10 Continuous ARQ Without Transmission Errors.

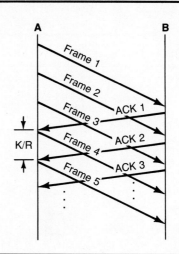

frames is necessary (retransmission of N frames); with probability $P^2(1 - P)$, transmission of $2N$ extra frames is necessary, and so forth.[17] Hence,

$$\bar{N}_T = 1 + NP(1 - P) + 2NP^2(1 - P)$$

$$+ \ldots = 1 + \frac{NP}{1 - P} \tag{6.26}$$

$$= \frac{1 + (N - 1)P}{1 - P}$$

with the denominator of Eq. (6.20) K/R times this. Thus in this case Eq. 6.20 yields

R_e for go-back-N ARQ

$$R_e = \frac{K - n_h}{\left[\dfrac{1 + (N - 1)P}{1 - P}\right]\dfrac{K}{R}} = \frac{(1 - P)(K - n_h)R}{[1 + (N - 1)P]\,K}. \tag{6.27}$$

Values from this formula, under essentially the same conditions assumed for Fig. 6.9 ($n_h = n_a = 48$ bits and burst error rates of 10^{-4} and 10^{-5}) are plotted in Fig. 6.11.[18] Go-back-3 ARQ has been assumed.

A value of $N = 3$ for go-back-N is used with prominent DLC protocols in situations with short propagation delays.[19] Retransmitted frames include the erroneous frame,

---

[17] Our derivation ignores the possibility that the initial transmission of the frame being examined could have been ignored due to an error in an earlier frame causing go-back-N retransmission. It also ignores the fact that whether errors occur in ignored frames or not is immaterial. Both ignored factors are second-order effects, and including them would greatly complicate the analysis.

[18] The other parameters assumed for Fig. 6.9 are not needed.

[19] See the discussion of HDLC and protocols related to it in Chapter 7.

Figure 6.11   Variation of Effective Transmission Rate with Frame Length for Burst Error Rates, p, of 10^{-4} and 10^{-5}. Go-back-3 ARQ on terrestrial 4800 bps link assumed.

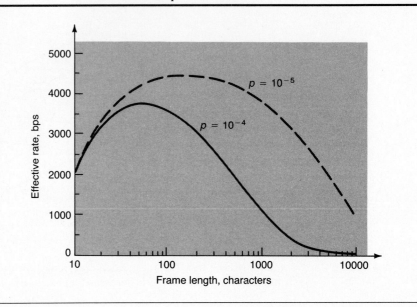

plus the frame being transmitted when the negative acknowledgment came in; these protocols also specify that frames with bad check fields are ignored and no negative acknowledgment transmitted until the next frame is received. For other protocols discussed in Chapter 7, $N = 2$ would be appropriate, since they usually call for return of NAKs immediately after receiving frames with bad check bits.

If delays before returning responses are long, the value of N for go-back-N ARQ depends on frame length. An appropriate value can be computed as follows. Let t_p = propagation delay as before, t_{rr} = reaction time at the receiver (to recognize a frame has come in), t_{rt} = reaction time at the transmitter,[20] and t_a (equal to n_a/R) = time to transmit an acknowledgment. The total time before a response can arrive is then

$$T = 2t_p + t_{rt} + t_{rr} + t_a = 2t_p + t_{rt} + t_{rr} + n_a/R, \qquad \textbf{(6.28)}$$

N for go-back-N ARQ   so the value of N is

$$N = \left\lceil \frac{T}{M/R} \right\rceil + 2 \qquad \textbf{(6.29)}$$

for protocols ignoring erroneous frames. The constant is 1 instead of 2 otherwise; $\lceil x \rceil$ is the smallest integer $\geq x$, that is, x rounded upward if x is not an integer.

[20] For full duplex transmission, as is used for continuous ARQ, these reaction times are normally far less than the turnaround times included in delays for stop-and-wait ARQ. (Reaction times should be included in the turnaround times assumed earlier.)

Figure 6.12 Value of N for Go-Back-N ARQ on 4800 bps Communications Satellite Link Using HDLC-like Protocol.

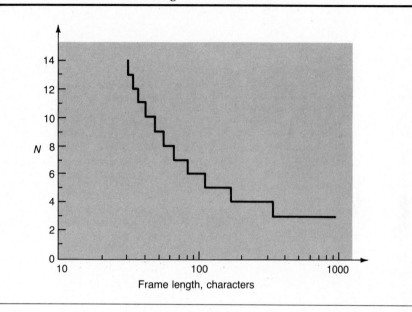

Frame length, characters

Figure 6.12 is a plot of N, from this formula, versus frame length for $n_a = 48$, $t_{rt} = t_{rr} = 2$ msec, $R = 4800$ bps, and $t_p = 270$ msec (typical of communications satellites), so $T = 554$ msec $= 0.554$ sec.[21] All frames are an integral number of eight-bit characters. N is constant over intervals along the horizontal axis, with unit jumps at the ends of these intervals.[22] Thus $N = 8$ for $M/8 = 56$ characters to 66 characters (448 bits to 528 bits) since the $T/(M/R)$ values are $.554 \times 4800/448 = 5.94$ and $.554 \times 4800/528 = 5.04$. Also $N = 7$ for $M/8 = 67$ to 83 characters, and so forth.

Plots of R_e versus frame length are given in Fig. 6.13. They were obtained from Eq. (6.27), using the appropriate N for each frame length. Although the plots are similar to those in Fig. 6.11, they exhibit discontinuities at points where the value of N changes.

Selective reject ARQ is simpler to analyze than go-back-N ARQ.[23] The time required for a frame to be accepted (the denominator of Eq. 6.20) is K/R times the average number of attempts before successful transmission, given by $1/(1 - P)$. (This

---

[21] With a shorter propagation time such as $t_p = 10$ msec for terrestrial propagation, N would equal 3 throughout the entire range of frame length in the figure.

[22] The larger values of N would not necessarily be permissible, since a limit on N may be imposed by the protocol. See Chapter 7 for details.

[23] Our model is an idealization. We assume selective reject is used to recover from all errors, although recovery from multiple errors via selective reject is so complex standard protocols rely on techniques such as go-back-N when multiple errors are known to have occurred. Multiple attempts at selective reject recovery for one frame may also be prohibited.

Figure 6.13   Variation of Effective Transmission Rate with Frame Length for Burst Error Rates, p, of 10^{-4} and 10^{-5}. Go-back-N ARQ for 4800 bps communications satellite link assumed, with N chosen appropriately to implement HDLC-like protocol.

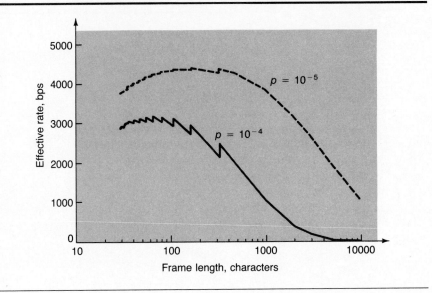

Figure 6.14   Variation of Effective Transmission Rate with Frame Length for Burst Error Rates, p, of 10^{-4} and 10^{-5} and Selective Reject ARQ.

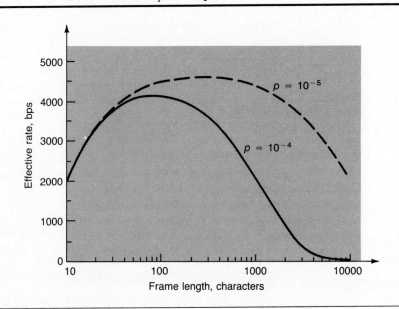

represents time chargeable to transmission of this particular frame rather than total elapsed time.) Thus R_e is given by

R_e for selective reject ARQ

$$R_e = \frac{K - n_h}{\dfrac{1}{1 - P}\dfrac{K}{R}} = \frac{(1 - P)(K - n_h)R}{K}. \tag{6.30}$$

Values from this formula are plotted in Fig. 6.14; the value of n_h has again been assumed to be 48 bits.

□ 6.5 Optimum Frame or Packet Sizes

As is obvious from Figs. 6.9 through 6.14, certain frame sizes maximize effective rate. Before we present techniques for computing optimum frame sizes, we point out that other criteria for optimizing frame sizes can be important and may even outweigh the criteria studied here. For example, in Subsection 3.2.5 we derived the formula (originally given as Eq. 3.6)

optimum packet length from Chapter 3 consideration

$$K_{max}^{opt} \approx n_h + \sqrt{\frac{E[M]n_h}{j - 1}} \tag{6.31}$$

for packet length maximizing efficiency gains due to pipelining effects when transmitting packets over j hops. (Here $E[M]$ is the expected length of a message to be broken up into packets and n_h is header length. K_{max}^{opt} is optimum maximum packet length, with all packets save the last one of this length and the final packet of whatever shorter length is necessary to complete transmission.) Errors were ignored in the previous analysis. In the current analysis all frames are of the same length and only one hop is considered; the optimization maximizes effective transmission rate over one hop in the presence of errors.

Both the criteria leading to Eq. (6.31) and those leading to Figs. 6.9 through 6.14 should be considered in choosing packet or frame size. For multihop communication with low error rates, the pipelining effects leading to Eq. (6.31) may be more important than the techniques for coping with errors analyzed here. Values from Fig. 3.7 should be compared with optimum frame lengths computed here if both factors are important.

Computation of optimum frame lengths when using ARQ error recovery in the presence of transmission errors can be accomplished via techniques like those in Subsection 3.2.5, that is, differentiating the equation for R_e with respect to K (even though K must be an integer) and setting the derivative equal to zero. For the stop and wait case, this yields (see Problem 6.22a)

optimum packet length for stop-and-wait ARQ

$$K_{stop\text{-}and\text{-}wait}^{opt} = \frac{n_h - \Delta n}{2} + \sqrt{\left(\frac{n_h + \Delta n}{2}\right)^2 - \frac{n_h + \Delta n}{\log_e(1 - p)}} \tag{6.32a}$$

where

$$\Delta n = n_a + 2(t_p + t_{ta})R \tag{6.32b}$$

Table 6.5   Optimum Frame Sizes from Eqs (6.32), (6.33), and (6.34) for Parameters Used for Figs. 6.9, 6.11, and 6.14, respectively.

ARQ Type	Burst Error Rate	
	10^{-5}	10^{-4}
Stop-and-wait	1278 characters	364 characters
Go-back-3	279 characters	92 characters
Selective reject	277 characters	90 characters

is the number of bits that can be transmitted during the interval from the time when the frame leaves the transmitter until the acknowledgment has been completely received.

A similar calculation for go-back-N ARQ, with N fixed, requires additional approximations for a simple solution to be found. If we approximate the probability of at least one error in a frame, $1 - (1 - p)^K$, by Kp, the first nonzero term in its expansion (since optimum frame size should yield a small probability of transmission errors), we obtain (Problem 6.22b)

optimum packet length for go-back-N ARQ

$$K^{\text{opt}}_{\text{go-back-}N} \approx \frac{n_h^{\cdot}}{2} + \sqrt{\left(\frac{n_h^{\cdot}}{2}\right)^2 - \frac{n_h}{\log_e(1 - p)}} \tag{6.33a}$$

with

$$n_h^{\cdot} = n_h\left(1 - \frac{N - 1}{N}\frac{p}{\log_e(1 - p)}\right) \tag{6.33b}$$

an altered header length (appearing twice in Eq. 6.33a, the original n_h appearing once). A similar analysis for selective reject ARQ gives (Problem 6.22c)

optimum packet length for selective reject ARQ

$$K^{\text{opt}}_{\text{selective reject}} = \frac{n_h}{2} + \sqrt{\left(\frac{n_h}{2}\right)^2 - \frac{n_h}{\log_e(1 - p)}}. \tag{6.34}$$

If the parameters used in generating Figs. 6.9, 6.11, and 6.14 are used in these formulas, the frame lengths in Table 6.5 are obtained.

All optimum frame sizes in the table agree well with those read from Figs. 6.9, 6.11, and 6.14, except for the optimum for go-back-3 with 10^{-4} burst error rate. In this case the optimum frame size from the figure is approximately 60 characters rather than the 92 characters in the table. This discrepancy results from approximating $1 - (1 - p)^K$ by Kp, an approximation off by approximately 3 percent at the optimum point. Although this results in a 50 percent error in optimum frame length, the reduction in R_e is less than 2 percent. Hence, even under these conditions Eq. (6.33) gives frame lengths almost as good as optimum values. A more precise solution can be obtained by trial and error.

Trial and error is the most realistic approach to optimization when the value of N for go-back-N ARQ varies with frame length since the discontinuities in R_e curves (see Fig. 6.13) make other solution techniques impractical.

Other criteria for defining optimum packet or frame sizes may also be important. For example, the frame size might be chosen to minimize the memory needed for buffering at the destination (see Problem 3.9 for a related type of optimization). Technological trends, which are making main memory less and less expensive, are reducing the importance of this type of optimization, however. An overall optimization including all relevant factors could be useful, but would probably be too complex to be feasible.

6.6 Flow Control for DLCs

Flow control is used at a variety of layers in a network architecture to protect various parts of the network against overloads. It receives major emphasis at higher layers of a network architecture, and is a primary topic of Chapter 8. A limited amount of flow control is performed at the data link layer also, so in this section we briefly discuss techniques used at this layer.

Flow control at the data link layer is exerted, by a station receiving data frames, when it runs into congestion (or other) problems limiting the amount of data it is prepared to receive. Special acknowledgment frames may be used for this purpose, or acknowledgments may be delayed if this means the transmitter must stop transmitting.

WACK   One approach to DLC flow control is through use of what we call a *WACK* (*wait acknowledgment*).[24] If **B**, after receiving a data frame, returns a WACK to **A**, this tells **A** that **B** is busy and temporarily unable to accept additional data frames. **A** consequently stops transmitting data frames but occasionally sends an ENQ frame to **B** to ascertain whether it is still busy. When **B** responds to an ENQ with an ACK, rather than another WACK, data frame transmission can resume. This is illustrated in Fig. 6.15.[25]

A still simpler scheme, applicable in stop-and-wait ARQ since the next data frame transmission from **A** must wait until **B** has acknowledged the previous one, is for **B** to simply withhold its acknowledgment until it is ready to receive. This looks to **A** like situations depicted in Figs. 6.6 and 6.7, with either the data frame or its acknowledgment lost, so **A** would proceed as in these figures. In continuous ARQ there is normally a maximum number of data frames that can be transmitted without an acknowledgment. This creates a window of permissible transmitted data frames, which is advanced by the number of outstanding frames acknowledged each time an acknowledgment comes in. This is called *sliding window flow control.* It is examined in detail in Chapter 8 since it is used at several levels of protocol architectures.

---

[24] The WACK terminology is used in binary synchronous communications and similar protocols; a largely equivalent control frame used in HDLC and related protocols is called RNR, Receive Not Ready. See Chapter 7 for discussions of these protocols.

[25] At least two different acknowledgments would, of course, be used to avoid the types of ambiguities discussed in Subsection 6.4.1.

Figure 6.15   DLC Flow Control Using WACK.

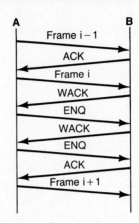

6.7   Summary

Problems addressed by DLC protocols have been presented, plus standard approaches to solving them. Synchronization is handled in several different layers of a network architecture. The primary types handled by DLC protocols are character, frame, and content synchronization, plus synchronization of access. Use of special framing characters, delimiting fields by special characters and use of count fields are the major DLC synchronization techniques.

Transmission errors in modern DLCs are handled either by FEC, or more commonly by ARQ. Good codes for correcting, or detecting, errors are needed. Either stop-and-wait or continuous ARQ can be used, with the latter subdivided into go-back-N and selective reject protocols according to retransmission strategy. Each approach was described, and techniques for computing optimum data frame sizes for each were discussed.

The chapter concluded with an introduction to simple techniques for flow control used by DLC protocols.

Problems

6.1 Briefly describe five types of synchronization needed for data communications and indicate which parts of a telecommunication network are normally responsible for each.

6.2 **a.** Some DLC protocols rely on reception of the flag bit pattern, 01111110, at both the beginning and the end of a frame, to obtain frame synchronization. If the first flag is not received properly, the receiver never locks onto synchronization; if the final flag is not received properly it could search indefinitely for the end of the frame. Compute and plot (using a log scale for each axis) the probability of improper frame synchronization (at least one of two flags not received properly) versus character error rate, P_c (probability of error in transmitting a character), for P_c ranging from 10^{-2} to 10^{-7}. Assume that errors in transmitting characters are independent.

b. Other DLCs rely on receiving at least two successive SYN, 10010110, characters for frame synchronization. Compute and plot (using a log scale for each axis) the probability of improper frame synchronization (not seeing two successive SYNs at the receiver) versus P_c, for P_c ranging from 10^{-2} to 10^{-7} for the following cases: (i) Two SYNs transmitted at the beginning of each frame, (ii) Three SYNs transmitted at the beginning of each frame, (iii) Four SYNs transmitted at the beginning of each frame, and (iv) Five SYNs transmitted at the beginning of each frame.[26]

6.3 Using two-dimensional parity checks (in rows and columns), compute character parity bits and the block check character for the following set of data characters. Use odd parity for both row and column parity bits and compute the final parity bit as row parity on the character formed from column parity bits.

$$
\begin{array}{ccccccc}
1 & 0 & 0 & 1 & 0 & 0 & 0 \\
1 & 1 & 0 & 0 & 0 & 0 & 1 \\
1 & 1 & 0 & 1 & 1 & 0 & 1 \\
1 & 1 & 0 & 1 & 0 & 0 & 1 \\
1 & 1 & 0 & 0 & 1 & 1 & 1 \\
0 & 1 & 0 & 0 & 0 & 1 & 0 \\
1 & 1 & 0 & 0 & 1 & 0 & 1 \\
\end{array}
$$

6.4 Consider the two-dimensional parity check illustrated in Fig. 6.1.

a. Prove that if even parity is used for both row and column parity checks, the value of ϕ_{cr} is the same regardless of whether ϕ_{cr} is computed as a parity check on the last row or a parity check on the last column.

b. Prove that if odd parity is used for both row and column parity checks, the *only* circumstances where the value of ϕ_{cr} is the same regardless of whether it is computed as a parity check on the last row or a parity check on the last column are as follows: (i) The total number of rows and the total number of columns (including parity bits) are both even, and (ii) The total number of rows and the total number of columns (including parity bits) are both odd.

6.5 A natural way to improve on two-dimensional parity checks is to use three-dimensional parity checks. In this approach, data blocks are represented by rectangular parallelepipeds (boxes) in three-dimensional space. Parity checks are then computed in each of three directions (horizontally, vertically, and in depth). The resulting code words can be transmitted as a text book, line by line, page by page.

a. Compute the code word for the data block:

$$d = (d_{111}, d_{112}, d_{113}, d_{121}, \ldots, d_{133}, d_{223}, \ldots, d_{233}, d_{311}, \ldots, d_{333})$$
$$= (1\ 0\ 0\ 1\ 0\ 0\ 0\ 1\ 1\ 0\ 0\ 0\ 1\ 1\ 0\ 1\ 1\ 1\ 0\ 1\ 1\ 1\ 0\ 0\ 1\ 1).$$

---

[26] Some references recommend using at least five successive SYNs as "good programming practice" when using such DLC protocols.

Figure P6.6   Figure for Problem 6.6(c).

```
X   •   •   •   •   •

•   •   X   •   X   •

•   •   •   •   •   •

•   X   •   X   X   •

•   •   •   •   •   •

X   X   X   •   •   •
```

Indicate row, column, and page. Use even parity, insert parity bits in the transmitted stream at the end of each row, column, and page and add a final page of parity bits. Give the code word both in page form (listing the rows and columns for each page) and in vector form. Compute parity checks for corner positions any way you wish; generalization of Problem 6.4(a) shows that it makes no difference.

b.   Discuss the error protection capabilities of such codes.

6.6   Determine the following for a two-dimensional parity check:

    a.   Prove that any error pattern with fewer than four errors is detectable.

    b.   Prove that any error pattern with an odd number of errors is detectable.

    c.   Sketch undetectable error patterns containing 4, 6, 8, and 10 errors. Your sketches should be in the format shown in Fig. P6.6 with X's representing errors.

6.7   Assume that a code has the minimum distance

$$\underset{i,j}{min}\ d_H(i,j) = 2a + 1.$$

If this code is used to correct all combinations of c or fewer errors, prove that for any c in the range of $1 \le c \le a$, the code cannot be used to simultaneously detect all combinations of d errors for $c < d < 2a$.

6.8   Prove that it is possible to simultaneously correct all combinations of c or fewer errors and simultaneously detect all combinations of $c + 1$ to d errors, with $d > c$, if inequality (6.9) is satisifed.

6.9   Assume that a block code is defined by a set of parity check equations. Each equation is of the form

$$\phi_k \oplus v_{k1} \oplus v_{k2} \oplus \ldots \oplus v_{kr_k} = 0,$$

with the v_{kj} referring to particular bit positions. (See the definition of a typical Hamming code in Problem 6.13 for an example; r_k is the number of bit positions checked by the kth parity bit.)

    a.   Prove that the sum (modulo two) of any two valid code words is a code word.

    b.   Prove that if i, j, and k are code words, then $d_H(i,j) = d_H(i \oplus k, j \oplus k) = d_H(k, k \oplus j \oplus i)$. Hence, for any code word, j, at Hamming distance d from i, there exists a code word, $k \oplus j \oplus i$, at distance d from k.

    c. Use these results to prove that the minimum Hamming distance between two code words is equal to the minimum weight of any nonzero code word. (Note that the all zero n-tuple is always a code word for such codes.)

6.10 Prove that if an (n,k) code (with k data bits and $n - k = r$ parity bits in each code word) is able to correct all single errors, the number of parity bits must satisfy

$$2^r - 1 \geq n.$$

Hint: In order for the set of those recomputed parity bits that disagree with received ones to identify the position of the error, the set of "parity violations" must uniquely identify the position of the error.

6.11 An $(8,4)$, single error correcting, double error detecting, Hamming code is constructed as follows. Bits of codewords are numbered from 1 to 8. Bits whose numbers are powers of 2 are check bits; the rest are data bits. (Note that 2^0 is a power of 2.) Each check bit in position 1, 2, or 4 forces the parity of a collection of bits (including itself) to be even; a data bit in position j is checked by each check bit whose position number occurs in the expansion of j as a binary number. For example, bit 7 is checked by bits in positions 1, 2, and 4, bit 5 is checked by bits in positions 1 and 4, and so forth. The check bit in position 8 is an overall parity bit forcing the total number of ones in a code word to be even.

    a. Using the above description, write out the set of parity check equations bits must satisfy.

    b. Construct the complete set of 16 legal code words.

    c. What is the minimum Hamming distance between distinct code words? You may use the results of Problem 6.9 to simplify this computation.

    d. State an algorithm for using comparisons between received and recomputed check bits to determine the position of any single error.

    e. Show that all 28 possible double errors can be detected, while still correcting all single errors.

6.12 A list of all code words in a particular $(15,4)$ code is given below.

(0000 0000 0000 000)	(1000 1111 0101 100)
(0001 1110 1011 001)	(1001 0001 1110 101)
(0010 0011 1101 011)	(1010 1100 1000 111)
(0011 1101 0110 010)	(1011 0010 0011 110)
(0100 0111 1010 110)	(1100 1000 1111 010)
(0101 1001 0001 111)	(1101 0110 0100 011)
(0110 0100 0111 101)	(1110 1011 0010 001)
(0111 1010 1100 100)	(1111 0101 1001 000)

    a. What is the minimum distance for this code? (You may use the results of Problem 6.9 to simplify this computation.)

    b. If this code is used purely for error correction, how many errors can it correct?

    c. If this code is used purely for error detection, how many errors can it detect?

    d. List all feasible combinations of $c \geq 1$ and $d > c$ such that the code can be used to correct c or fewer errors and simultaneously detect $c + 1$ to d errors.

6.13 Consider a Hamming code with block length 7 bits. Parity checks, ϕ_1, ϕ_2, and ϕ_3 are specified to cover data bit positions 1, 4, 6, and 7; 2, 4, 5, and 6; and 3, 5, 6, and 7, respectively. That is,

$$\phi_1 \oplus v_4 \oplus v_6 \oplus v_7 = 0,$$
$$\phi_2 \oplus v_4 \oplus v_5 \oplus v_6 = 0,$$
$$\phi_3 \oplus v_5 \oplus v_6 \oplus v_7 = 0.$$

Bits 1, 2, and 3 are parity bits, while bits 4, 5, 6, and 7 are data bits (which may be chosen arbitrarily).

a. List all 16 legitimate code words.

b. Prove that the minimum distance between any two code words is 3. (You may use the results of Problem 6.9 to simplify this computation.)

c. Show that this code contains one word of weight 0, seven of weight 3, seven of weight 4, and one of weight 7.

d. Show how all seven possible single errors can be detected and corrected. (Identify and consider each one separately.)

e. Show how all 21 double errors can be detected if this is used as an error detecting code (without attempting to correct errors).

6.14 List all 16 legitimate code words for the polynomial code defined by the generator polynomial in Eq. (6.17) and verify that they are identical to the code words found in Problem 6.13. Thus the polynomial code is identical to the Hamming code.

6.15 Show that the encoding circuit in Fig. 6.2 properly encodes a message represented by message polynomial $M(X) = X^3 + X^2 + 1$ by comparing the results of dividing $X^3 \cdot M(X)$ by $G(X)$ with the shift register contents in a manner analogous to construction of Table 6.4.

6.16 Describe the error detecting capabilities of CRC-12, CRC-16, CRC-32, and CRC-CCITT codes as completely as possible. You may use the fact that none of the polynomials describing these codes include factors aside from (possibly) $X + 1$ and an irreducible polynomial.

6.17 Develop a flow chart for ARQ error control in a DLC protocol for two-way simultaneous transmissions over a channel that does not introduce errors but can lose frames. Send and receive sequence numbers, with maximum values of seven (that is, numbers modulo eight), may be used. Use positive acknowledgments sent in separate acknowledgment frames, and allow for the possibility of an acknowledgment acknowledging more than one data frame.

6.18 In a sequence of frame transmissions, data frame D0 has just been positively acknowledged. Four more frames—D1, D2, D3, and D4—are transmitted. The next return transmission, received after D4 has been transmitted, is a negative acknowledgment to D2. Assume that this implicitly acknowledges any previously transmitted but unacknowledged data frames.

a. What action does a go-back-N protocol take?

b. What action does a selective repeat protocol take?

6.19 Compute R_e for stop-and-wait ARQ assuming the following: a 64-kbps link, all characters eight bits, 200 data characters per frame, six characters for either data frame headers or acknowledgment frames, 10 msec propagation time and 20 msec turnaround time, and a burst error rate of 10^{-4}.

6.20 Compute R_e for go-back-N ARQ, using the same parameters as in Problem 6.19, plus $t_{rt} = t_{rr} = 2$ msec. Use an appropriate value of N for a HDLC-like protocol.

6.21 Compute R_e for selective reject ARQ, using the same parameters as in Problem 6.20.

6.22 **a.** Derive Eq. (6.32a).
  b. Derive Eq. (6.33a).
  c. Derive Eq. (6.34).

6.23 Use the expansion

$$\log_e(1 - x) = -\sum_{n=1}^{\infty} \frac{x^n}{n}$$

to show that as $p \to 0$,

$$K^{\text{opt}} \to \sqrt{\frac{n_h}{p}}$$

for either go-back-N or selective reject ARQ and

$$K^{\text{opt}} \to \sqrt{\frac{n_h + \Delta n}{p}}$$

for stop-and-wait ARQ. Compare these approximations, for parameter values used in constructing Table 6.5, with the values in the table.

7 Data Link Layer Protocols

7.1 Introduction

The data link layer handles data transfer over a single communications link without relays or intermediate nodes. The link can be point-to-point, multipoint, broadcast, switched, or nonswitched and may use any of a variety of media including twisted pair cables, coaxial cables, fiber optics, communications satellite, and other radio links. Most local area networks are also treated as single links from the point of view of the data link layer.

During data reception, the data link layer converts a raw bit stream from the physical layer into recognizable characters, frames, and so forth. It strips off the data link header and trailer from each frame and uses these to accomplish its functions. During transmission, the data link layer takes a data unit presented to it by the network layer (including higher layer headers), breaks it into frames, adds a header and trailer to each frame, and presents it to the physical layer in appropriate form for transmission as a raw bit stream. Communication between the data link layer and the physical layer is via an interface.[1]

Modifications to the message by the data link layer at the transmitter are for interpretation by the data link layer at the receiver. *Peer-to-peer communication* takes place between data link layer implementations at the ends of the connection, using data flows via physical layers and the communications link. The peer-to-peer protocol at the data link layer is commonly called a *data link control (DLC) protocol*.

Sophisticated DLC protocols ensure that essentially all transmission errors are corrected, so the primary function of these data link layers is said to be to ensure reliable transfer of user data. However, other common DLC protocols put less emphasis on error handling.

Chapters 5 and 6 discuss medium access control, synchronization, and error control functions of the data link layer. Other functions include activation and termination of data link connections over paths built up by the

---

[1] This is the OSI type of interface, between protocol layers rather than devices.

303

Figure 7.1   DLC Protocols Treated in Chapter.

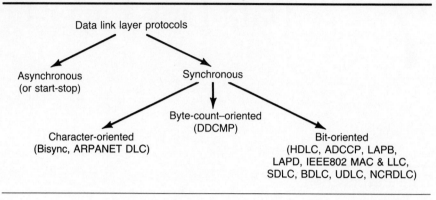

physical layer, identification of sender and/or receiver, abnormal condition recovery, and link management [CONA83].

This chapter discusses the DLC protocols shown in Fig. 7.1. Precise descriptions use state diagrams or other representations similar to those describing physical layer protocols in Chapter 4, but these are very complex.[2] We give detailed descriptions of protocols, but not complex state diagrams.

A choice of DLC protocols to consider should involve the following. Start-stop DLCs are the oldest and most rudimentary, but they are probably still the most common. Bisync was the first really successful synchronous protocol and is still widely used. ARPANET DLC involves interesting and useful modifications of Bisync but is not widely used. DDCMP and the bit-oriented protocols represent the current generation of DLC protocols, with HDLC and variants most prominent in standards. A survey of all protocols treated gives a useful picture of how protocols evolve with experience, however.

7.2   Asynchronous or Start-Stop Data Link Controls

The most rudimentary DLCs in common use are variants of asynchronous or start-stop DLCs. (The two terms are used interchangeably.) Start-Stop DLCs use the transmission format shown in Fig. 7.2.

START bit

In start-stop transmission the line idle state is a voltage level corresponding to binary "one." The line is idle until a *START bit* lasting one bit time and equivalent to a binary "zero" is transmitted. This reverses the state of the line at the beginning of a character; bits encoding the character follow. For com-

---

[2] For example, a finite state machine description [ECMA81] of "selected procedures" in one protocol, HDLC, takes 38 pages, approximately half diagrams and the rest explanatory material.

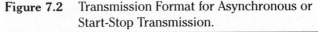

Figure 7.2   Transmission Format for Asynchronous or
Start-Stop Transmission.

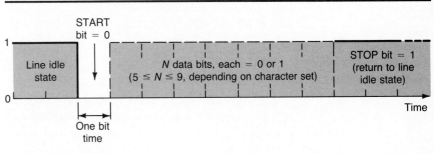

STOP bit

mon character sets, a character can be 5 to 9 bits, each either "1" or "0."[3]
Character transmission concludes by transmission of a *STOP bit*, equivalent
to a return to the idle state (or transmission of a binary "1").

For the STOP bit, a minimum duration (1, 1.5, or 2 bit times for different
implementations), but no maximum duration, is specified. The true duration
is the time the line remains idle before transmission of the next character
begins. Transmissions from data processing equipment normally occur at the
maximum rate allowed, with all STOP bits equal to the minimum allowable
duration; data transmissions originating from a user at a keyboard or other
terminal, on the other hand, are characterized by widely varying STOP bit
durations, depending on typing speed, think times between periods of typing,
and so forth.

*isochronous
transmission*

The start-stop format in Fig. 7.2 is sometimes used when both ends of the
data link are clocked and send data at the maximum rate allowed. This is
called *isochronous transmission*.

7.2.1   Reception of Start-Stop Data

Functions of the transmitter in start-stop transmission consist basically of
appending START and STOP bits, plus parity bits if used, to characters and
sending them on their way. We concentrate on receiver functions.

Start-stop DLCs are often implemented with USARTs (see Subsection
4.9.2). The mode register in a USART can be set for characters of 5, 6, 7, or 8
bits, either with or without parity (as an extra bit). Also, it can be set for
STOP bit durations of 1, 1.5, or 2 bit times. Thus the USART is designed to
handle all normal variants of start-stop transmission.

[3] Typical character lengths are Baudot code (5 bits), Baudot plus parity (6 bits), ASCII (7 bits),
ASCII plus parity or EBCDIC (8 bits), EBCDIC plus parity (9 bits). Even longer character
lengths occasionally occur. For example, characters in Hollerith punched-card format
require 12 bits, or 13 bits if a parity bit is added.

The procedure for extracting timing information is discussed in Subsection 4.9.2. The USART looks for a "1" to "0" transition, indicating the beginning of a START bit, during periods when the line is idle. The voltage is then sampled at approximately the middle of each bit during the time to transmit the character and its associated START and STOP bits (plus parity bit if one is used), with samples decoded as "1s" or "0s." The transmitted character is retrieved by stripping off START, STOP, and parity bits. Appropriate flags are set if parity is not valid or other error conditions are detected.

The primary function of a start-stop DLC protocol is character synchronization. Content synchronization of a limited type is also achieved, since position of a bit indicates whether it is used for synchronization (START or STOP bit), error control (parity bit), or data (all other bits).

7.2.2   Error Control

Rudimentary error control is possible if parity is used.[4] As Chapter 6 indicates, a single parity bit can detect any odd number of errors, but not errors on an even number of bits. Error statistics for data links, such as telephone links, indicate double, quadruple, . . . errors are common so a single parity bit does not give strong error protection. Start-stop DLCs do not prescribe how information about parity check validity is to be used; the information is simply passed to higher protocol layers that may use (or ignore) it.

An alternative form of error control used with start-stop transmission is echoing, also discussed in Chapter 6. This leaves recovery up to the user.

7.2.3   DLC Overhead

Start-stop DLC overhead, or fraction of bits used for synchronization and error handling, is independent of message length. It is typically 20 percent to 44 percent. The former applies for eight-bit characters with no parity and STOP bit durations of one bit time. The latter applies for five-bit characters with parity and STOP bit duration of two bit times. These figures ignore overhead due to retransmissions after errors are detected; such retransmissions may or may not occur, depending on whether invalid parity bits lead to retransmissions or not.

7.2.4   Limitations of Start-Stop DLC Protocols

Start-stop DLC protocols are rudimentary; they are used for limited function terminals or devices, and they provide little more than character synchronization and rudimentary error control. Activation and termination of data link connections, identification of sender and/or receiver, segmenting data into

---

[4]There are systems that attach parity bits at the transmitter but ignore them at the receiver, however, so use of parity bits is no guarantee of error control.

frames for transmission, flow control, abnormal condition recovery, and link management functions are not included or relegated to other layers. Synchronization of messages longer than characters is not treated by such protocols.

Start-stop DLC protocols are the most common DLC protocols for slow-speed links with simple terminals. They are often used at speeds of up to approximately 1200 bps. At higher rates, synchronous transmission is standard.

7.3   Character-Oriented DLC Protocols—Bisync

control characters

Character-oriented DLC protocols are a class of synchronous DLC protocols, in which characters follow each other at clocked intervals. They are based on *control characters* in a character set. The ASCII character set in Fig. 7.3 contains most common control characters; other character sets contain analogous characters. In addition to alphanumeric characters and symbols, it contains "characters" NUL, DLE, SP, and so forth. Some of these are for terminal control. For example, NUL is a null character, SP indicates space, but others are for communications control. Communications control characters are defined in Table 7.1. Supplementary control characters are also used in some character-oriented DLC protocols.

transparent text mode

During normal operation, all characters transmitted are chosen from a character set on which the protocol is based. For example, only ASCII characters are used in a protocol based on ASCII. It is possible to operate in *transparent text mode,* with arbitrary bit patterns transmitted, but this requires more complex control sequences.

Rather than going into detail about character-oriented protocols in general, we present an example: IBM's Binary Synchronous Communications protocol, also known as BSC or Bisync. Prior to development of the bit-oriented protocols discussed in the next section, Bisync was the de facto standard DLC protocol for synchronous communications systems. It is still one of the most common DLC protocols for synchronous communication at rates of approximately 1200 bps or higher.

7.3.1   Normal Text Transmission in Bisync

master and slave stations

Bisync operates only in half duplex mode—that is, a station can transmit or receive but it cannot do both simultaneously. All data transfers are from a station operating as a *master station* to a station operating as a *slave station.* In situations where more than one station on a link may need to transmit data, the master/slave status of stations may be interchanged.

Figure 7.4 illustrates a typical Bisync frame format, for a frame including user data. All frames begin with at least two SYN (synchronous idle) charac-

Figure 7.3 ASCII Character Set, the U.S. National Version of CCITT International Alphabet No. 5. All versions contain the same communications control characters.

| | | Bits $b_7b_6b_5$ | | | | | | |
| | | 000 | 001 | 010 | 011 | 100 | 101 | 110 | 111 |
		0	1	2	3	4	5	6	7
0000	0	NUL	DLE	SP	0	@	P	`	p
0001	1	SOH	DC1	!	1	A	Q	a	q
0010	2	STX	DC2	"	2	B	R	b	r
0011	3	ETX	DC3	#	3	C	S	c	s
0100	4	EOT	DC4	$	4	D	T	d	t
0101	5	ENQ	NAK	%	5	E	U	e	u
0110	6	ACK	SYN	&	6	F	V	f	v
0111	7	BEL	ETB	'	7	G	W	g	w
1000	8	BS	CAN	(	8	H	X	h	x
1001	9	HT	EM	)	9	I	Y	i	y
1010	10	LF	SUB	*	:	J	Z	j	z
1011	11	VT	ESC	+	;	K	[	k	{
1100	12	FF	FS	,	<	L	\	l		
1101	13	CR	GS	-	=	M	]	m	}
1110	14	SO	RS	.	>	N	^	n	~
1111	15	SI	US	/	?	O	_	o	DEL

Bits $b_4b_3b_2b_1$

Table 7.1   ASCII (or International Alphabet No. 5) Communication
Control Characters and Their Meanings.

ACK (Acknowledge): A character transmitted by a receiver as a positive
acknowledgment response to a sender.

DLE (Data Link Escape): A character that changes the meaning of a limited number
of contiguously following characters.

ENQ (Enquiry): A character requesting a response from another station. It may be
used to obtain identification, or status, or both.

EOT (End of Transmission): A character used to conclude a transmission that may
have contained one or more text messages and headings.

ETB (End of Transmission Block): A character indicating the end of a block of data
for communication purposes.

ETX (End of Text): A character terminating a sequence of characters started with
STX and transmitted as an entity.

NAK (Negative Acknowledge): A character transmitted by a receiver as a negative
acknowledgment response to the sender.

SOH (Start of Heading): A character used at the beginning of a sequence of
characters containing an address or routing information and referred to as the
"heading." STX terminates the heading.

STX (Start of Text): A character that precedes a sequence of characters to be treated
as an entity and transmitted through to the destination. Such a sequence is
referred to as "text." STX may also terminate a sequence of characters started by
SOH.

SYN (Synchronous Idle): A character used to provide a signal from which
synchronization may be achieved or retained. It may also be transmitted as a fill
character in the absence of any other character.

ters.[5] (In ASCII code with even parity in the b_8 position, a SYN character is
$b_8b_7b_6b_5b_4b_3b_2b_1$ = 10010110 = 226 octal, as shown in Fig. 7.3.) These characters are used to obtain character synchronization with the aid of circuitry
at the receiver that continually looks for two successive SYN characters. SYN
characters may also be inserted in the middle of a frame if data characters
are not available for transmission. Such SYNs are ignored by the receiver.

Figure 7.4   Format of Typical Bisync Text Frame.

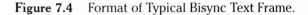

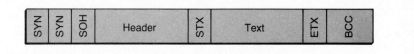

---

[5]Although only two SYN characters are required for the receiver to lock onto character synchronization, more than two are often sent to avoid losing frames if one SYN is received
erroneously. Beginning each frame with five SYN characters is sometimes recommended as
"conservative programming practice" [MCNA82, p. 126].

block check characters

If a heading is used, the two SYN characters are followed by SOH (start of heading), plus a heading terminated by STX (start of text); otherwise STX comes immediately after the two initial SYNs. Text (that is, user data) follows, with ETX (end of text) or ETB (end of transmission block) terminating it. If several frames are used to convey a complete message, ETB terminates all frames except for the last one, where ETX is used. The frame ends with one or two (depending on implementation) *block check characters* (BCC) for error protection. In some implementations one or two *PAD characters* (meaningless characters) are appended at the end of a frame so that no useful data are lost if the transmitting modem turns off its transmitter prematurely.[6]

For ASCII implementations, error checking uses two-dimensional parity checking with a single parity bit on each character plus an overall parity character obtained as described in Subsection 6.3.4 (see Fig. 6.1). The overall parity character forms a one-character BCC. For EBCDIC implementations, the BCC is a 16-bit or two-character block check obtained from a polynomial code generated by CRC-16 (see Subsection 6.3.12). Polynomial code block checks are also used for some other character sets; for example, CRC-12 is used for some six-bit characters. BCC accumulation begins after the first SOH or STX following a line turnaround and continues through the ETX or ETB ending text transmission.

Error control uses stop-and-wait ARQ. Each frame with user data must be acknowledged, either positively (with an ACK) or negatively (with a NAK). An ACK is sent if the format of the frame is satisfactory and the BCC (supplemented by individual parity bits when they are used) indicates no error; a NAK is returned otherwise. An ACK frame is transmitted as SYN SYN ACK[7] while a NAK frame is transmitted as SYN SYN NAK. A typical sequence of data exchanges is similar to that in Fig. 6.4(a).

typical supervisory frames

No Bisync frames besides frames with user data contain BCCs or similar error checking bits. Format checks verify that supervisory frames are in correct format, with branches to error recovery routines if they are not. Lack of a BCC for supervisory frames is a weakness of Bisync, however.

The procedures stated work as long as all transmissions make it through adequately. That is, control frames are received without error and text messages are recognized sufficiently for the receiver to know whether to return an ACK or a NAK. It is possible for no return frame to be received, though; a typical reason is that the receiver did not lock onto character synchronization.

If the master station fails to receive an ACK or a NAK during a specified timeout after transmitting a frame, it sends an ENQ (enquiry) frame. The slave is then required to repeat its most recent transmission. With only one ACK, this leads to ambiguities (see Subsection 6.4.1), so two ACKs are used: ACK 0

---

[6] This type of premature modem turnoff is common with some older modems.

[7] The last "character" is either ACK 0 or ACK 1 since two ACKs are used.

("Ack Zero") and ACK 1 ("Ack One").[8] ACK 0 acknowledges even-numbered data frames and ACK 1 acknowledges odd-numbered data frames.

7.3.2   Transmission of Transparent Text

Transmission of transparent text, or arbitrary bit streams, requires modification of the protocol since such text can contain bit patterns identical to control characters. If these patterns occur in normal text, the protocol dictates control actions should be taken, and such actions initiated when not desired can produce unpredictable results. For example, a bit stream corresponding to ETX or ETB would inform the protocol that this is the end of text, or of a transmission block, with the BCC to follow. The rest of the actual frame would be ignored (and the BCC would probably indicate an erroneous frame).

In Bisync *transparent text mode* is initiated by transmission of the character sequence (DLE STX) instead of simply STX.[9] A DLE (data link escape) character tells the protocol handler succeeding control characters are to be handled in an alternative manner. Any bit stream is then allowed in the text portion of the frame, as long as its length is an integral number of character lengths. Transparent text mode is terminated by the two-character sequence (DLE ETX) or (DLE ETB). Any idling characters inserted must consist of the complete two character sequence (DLE SYN).

character insertion

With this approach, the only bit combination in transparent text that could cause difficulty is that corresponding to DLE. *Character insertion* is used to avoid problems. If a DLE pattern is encountered at the transmitter during the text portion of a frame, it inserts an extra DLE. The receiver scans incoming transparent text for a DLE. If one is detected and is followed by a second DLE, the receiver deletes the second DLE. If the next character is ETX or ETB, transparent text is terminated. If the character is not SYN, DLE, ETX, or ETB, branching to an error recovery routine is appropriate.

This character insertion algorithm causes some increase in error rates since even a single bit error in transmitting a DLE will cause the algorithm to perform incorrectly. In particular, a one-bit error in the DLE preceding an ETX will cause the receiver to fail to locate the end of this text block, possibly concatenating a later transmission together with this one. Such errors will usually be detected through format checking or through the BCC, however.

7.3.3   Control of Access to Data Link

contention mode and supervised mode

Bisync provides two modes of controlling access to a data link: *contention mode,* which is restricted to point-to-point links, and *supervised mode,* which must be used on multipoint links and may be used on point-to-point links.

---

[8] Since standard character sets define only one ACK, Bisync uses two-character sequences for the two ACKs—"DLE 0" for ACK 0 and "DLE 1" for ACK 1.

[9] This assumes that both ends of the data link are prepared to handle transparent text.

We will concentrate on supervised mode but will briefly discuss contention
mode. Both are applicable whenever the data link is in a control state; it is in
a control state whenever it is not in a message transfer state (a state allowing
it to transfer messages). Appendix 7A contains a list of signals that may be
exchanged in a control state.

Contention mode is used between two symmetrical stations on a point-
to-point link, with neither station having control over the link. One station is,
primary and secondary   however, considered *primary* and the other *secondary* to help resolve situa-
stations   tions where both contend for control almost simultaneously. ("Primary" and
"secondary" should not be confused with "master" and "slave" designations.
Either station can become master during user data transmission.)

At any time when the link is not actively used, a station on a link using
contention mode may bid for control by transmitting an ENQ frame. The
receiving station is required to reply with an ACK 0 frame if it is able to
receive and a NAK frame otherwise.[10] If both stations bid essentially simultaneously, the primary ignores all ENQs it receives after sending its own and
persists until it receives a positive or negative reply, transmitting ENQs at
three-second intervals. The secondary, on the other hand, replies either positively or negatively to all ENQs it receives, whether it has itself transmitted
an ENQ or not. This ensures that conflicts are resolved in favor of the primary.

control and tributary   In supervised mode one station is the *control station* and all others are
stations   *tributary stations.* The control station has primary responsibility for controlling access to the link. It is initially the master, but can temporarily assign
master status to a tributary so that it can transmit user data frames. Only one
station can have master status at a given time; all others must have slave
status.

There are two approaches to message transmission in supervised mode:
selection and polling   *selection* and *polling.* Selection is used for transmissions from the control
station to tributaries, and polling enables tributaries to send to the control
station. Polling and selection frames are identical except for their use of a
polling address or a selection address (see Table A7.1 in the appendix to this
chapter); each tributary station must have both addresses, which must be
distinct. Group addresses, recognized by several stations, are acceptable for
selection frames. The address is transmitted twice in many implementations,
since polling and selection frames do not contain BCCs and strange things
happen if an address is changed to another valid one by a transmission error.

ordinary and fast   There are two primary types of selection frames: *ordinary* (or "polite")
selections   *selection* and *fast selection.* In ordinary selection the control station asks a
tributary station if it is able to accept data and does not transmit the data
unless it receives a positive reply. In fast selection data are included with the
selection frame. If the tributary is not able to accept the frame, it is lost—
which means that fast selection is risky; if successful, though, it gets data to

---

[10] Table A7.1 in the appendix to this chapter gives precise formats for these frames.

Figure 7.5   Typical Supervised Mode Bisync Transmissions.

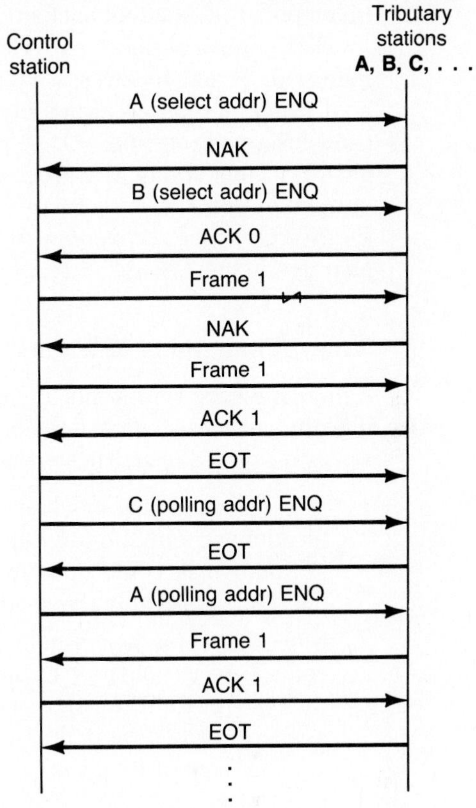

the destination quickly. The control station maintains master status through-out selection and transmission of frames to stations it selects.

Polling enables tributaries to transmit. When the control station polls a tributary, the tributary is allowed to transmit one or more frames. Because data frames are transmitted only from master stations to slave stations, re-ception of a poll also conveys master status on the selected tributary, which remains master until it terminates its transmission by sending an EOT frame.[11] No additional polls are permitted before EOT is returned.

A typical data transfer sequence, including both polling and selection, is given in Fig. 7.5. Ordinary (polite) selection is illustrated.

In the sequence shown, the control station first selects tributary A, but receives a negative reply (a NAK). It then selects B, receives a positive reply

---

[11] An EOT frame transmitted by any station puts the data link back into control state, with the control station again master.

(ACK 0), and transmits a data frame. An error occurs on the first attempt to send the frame, which is NAKed, but retransmission is successful and an EOT frame puts the data link back in control state.

The control station next polls C and receives a negative reply (EOT); it then polls A and receives a data frame. After receiving an ACK 1 from the control station, as confirmation the frame has been received correctly, A (now master), transmits EOT to put the data link back into control state and give up master status. The process can continue with either another selection or another poll.

Normal error recovery sequences are of the stop-and-wait ARQ type illustrated in Fig. 6.4(a).[12]

7.3.4   Limitations of Bisync

Although Bisync represents a major advancement over start-stop, it has significant weaknesses that helped motivate the development of the DLC protocols discussed next. These weaknesses include the following:

1. It can be used only in a half duplex mode and would require major modifications to operate full duplex. Rather than using modified versions of Bisync, which could operate in a full duplex manner, DLC protocols for full duplex operation have adopted distinctly different approaches.

2. It operates in a "select-hold" mode, in which a pair of stations maintain control over (hold) a data link until they have finished communicating; this can include transfer of long messages or long error recovery sequences. Only selection and polling frames contain addresses, so other stations cannot be given access to the data link before data transfer is complete.

3. Incomplete error checking is used, with block check characters only on data frames and supervisory frames unchecked except for format checks. Although transmission errors are more likely on data frames, since they tend to be longer, errors on supervisory frames occur also, and such errors can be difficult to handle.

4. The variety of options and special features make implementation difficult. Even determining which characters are to be included in BCCs requires complex logic.

5. Handling transparent text is clumsy in comparison with its handling in newer protocols.

6. The master/slave status of stations can be poorly defined. Interchanging master status is dangerous since frames passing status may be lost or damaged. Loss of an EOT intended to return the system to control state

---

[12]Sequences for recovery from other types of errors, such as absence of any response to selection or polling frames, are given in various references, for example, [HOUS87].

is difficult to handle; recovery is not defined by Bisync, but it is a user option![13]

7. Individual responses, using separate supervisory frames, are required for each data frame. Newer protocols allow acknowledgment of several data frames with one response, and may use returned data frames for acknowledgments.

8. Ambiguities can result from the multiplicity of ways in which some control characters are used. An example is ENQ, used for selection or polling (employing messages of identical formats), or to request repetition of a previous transmission.

9. Experience has shown that Bisync was not defined with adequate precision. A few details were not clearly specified, allowing implementation options interpreted differently by different implementers. This has caused equipment planned to be compatible to actually be incompatible.

Other DLC protocols were motivated by, and successfully address, some of these deficiencies. Bisync (and start-stop) will be widely used for quite a few more years, though.

7.4 ARPANET DLC

The data link layer of the ARPANET, employed in ARPANET portions of the DoD Reference Model, uses some interesting variations on Bisync. Figure 7.6 gives the format of ARPANET frames; it is essentially that of Bisync transparent text mode.

Aside from using a three-byte CRC (three block check characters obtained from a polynomial code) to replace Bisync's one- or two-byte BCC, the main modifications for ARPANET DLC are the use of eight logically independent full duplex channels multiplexed onto each data link and the use of only positive acknowledgments for error recovery. We do not give the detailed format of packet headers here since it is complex;[14] the headers do not separate out information for different levels as cleanly as do headers in newer architectures, so they contain a good deal of information that is superfluous at this point. Relevant header fields are as follows:

1. Two 16-bit addresses, for source and destination nodes. Inclusion of both in each frame allows the protocol to operate in a full-duplex manner and to interleave transmissions with different sources and destinations.

2. A one-bit sequence number keeping track of the (odd or even) sequence number for this particular frame. Separate odd/even counts are main-

---

[13] One of the authors discovered this when simulating the protocol.

[14] See Chapter 9, Fig. 9.16, for the precise header structure.

Figure 7.6   Format of ARPANET Message Frame.

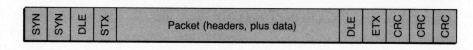

tained for frames flowing over each of eight "logical channels." (See the discussion below of how logical channels operate.)

3. A three-bit logical channel number identifying to which of the eight logical channels the frame belongs.

4. An eight-bit acknowledgment field giving the (odd or even) sequence number of the last frame received successfully on each logical channel.

logical channel operation

Each incoming packet is assigned to the lowest numbered *logical channel* currently free, regardless of the logical channel used by other packets belonging to the same message. Each logical channel uses its own independent stop-and-wait ARQ, with retransmission of any frame not positively acknowledged by the next time its logical channel's transmission turn comes up. Timeouts limit the maximum time before retransmission.

pipelining

Use of eight independent logical channels produces a *pipelining* effect and considerably increases efficiency of the DLC over that of ordinary stop-and-wait protocols. Rather than waiting for an ACK to come back before another frame can be transmitted, a new frame can normally be assigned to an unused logical channel and transmitted almost immediately. The fact that return frames update acknowledgments for all eight logical channels greatly speeds up acknowledgments also. Transmission errors do not greatly slow down acknowledgments, since the next returning frame will contain acknowledgment bits for all channels.

The main weakness of the DLC is that it is very likely to deliver frames to their destination out of order, since transmission over some logical channels (including retransmissions) may take considerably longer than transmissions over others. ARPANET makes no effort to reorder frames on individual links, relying on reordering at the destination; this makes things difficult for higher architecture layers, however. The reordering problem is one reason this protocol is not widely used despite its high efficiency.

7.5   Byte-Count–Oriented Protocols—DDCMP

DDCMP is an acronym for *Digital's Data Communications Message Protocol*, developed by Digital Equipment Corporation. It can be used on synchronous or asynchronous data links, for half duplex or full duplex transmission, and

Figure 7.7 Format of DDCMP Frame.

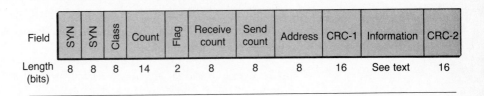

Field	SYN	SYN	Class	Count	Flag	Receive count	Send count	Address	CRC-1	Information	CRC-2
Length (bits)	8	8	8	14	2	8	8	8	16	See text	16

on point-to-point or multipoint links. It is similar to Bisync and ARPANET DLC in using standard communications control characters for control functions and separating frames into header and text portions. There are three types of frames: information, supervisory, or maintenance.

The format of DDCMP frames is given in Fig. 7.7. The first ten bytes are a header, which must be present in all frames and contains a separate block check. Information and maintenance frames contain an information field and a second block check, but supervisory frames contain neither.[15]

7.5.1 Interpretation of Fields in DDCMP Frame

ASCII characters with even parity are used for control characters. Fields in DDCMP frames, and basic protocol operation, are as follows:

- Initial SYNs are used for character synchronization, as in Bisync.

information, supervisory and maintenance frames

- The class field indicates type of frame; for *information frames* it contains SOH; for *supervisory* or *maintenance frames* it contains ENQ or DLE, respectively. Maintenance frames are used for system diagnostic purposes and are not discussed further here.

byte-count oriented

- The count field for information frames gives the length, in eight-bit characters, of the information field (which contains user data). It can vary from 1 to $2^{14} - 1 = 16,383$ characters; a zero count is not allowed. The count is responsible for the name *byte-count oriented* and allows the receiver to locate the end of user data by counting characters. The count is not affected by characters used, so transparent text is handled as readily as text in a character set as long as information field length is a multiple of eight bits. In supervisory frames the count indicates supervisory frame type plus possibly additional information; they have neither information fields nor CRC-2s.[16]

---

[15] Some terminology here has been altered from that in Digital Equipment Corporation literature; we substitute standard terminology for HDLC and similar protocols to simplify comparisons.

[16] For example, a negative response (NAK) frame uses this field to give a reason for the NAK.

- The flag field contains two flag bits for specialized purposes.[17]
- The Receive Count and Send Count fields number received and transmitted frames and allow piggybacking of acknowledgments (see Subsection 6.4.3). The Send Count is the number, modulo 256, of the frame being transmitted. (With eight-bit counts, the count goes back to 0 after it has reached 255.) The Receive Count is the number, also modulo 256, of the last successive frame without detected errors that the transmitter has received from the station it is transmitting to. It is obtained from the value the latter station placed in its Send Count for this last accepted frame.

A Receive Count of N implies frame N and all lower numbered frames (modulo 256) have been received correctly. Hence one Receive Count can acknowledge a number of frames; for example if the last frame B previously received from A contained 253 in the Receive Count field and the current frame contains 3 in the same field, this positively acknowledges B's frame numbers 254, 255, 0, 1, 2, and 3. In theory, one Receive Count field can positively acknowledge up to 255 frames. It cannot unambiguously acknowledge 256 since correct reception of 256 successive frames would return the Receive Count field to its previous value, so it would be impossible to distinguish between 256 and 0 messages received successfully.

Supervisory frames can also be used for acknowledgments. ACK and NAK frames each use a Receive Count (but neither uses a Send Count). An ACK with Receive Count of 3 in the situation above would acknowledge the set of frames listed while a NAK with Receive Count of 2 would positively acknowledge frames 254, 255, 0, 1, and 2 and negatively acknowledge 3.

A REP (Reply) supervisory frame uses a Send Count but no Receive Count and requests a reply to the frame specified by the Send Count. It is sent after no reply to that frame has been received within a predefined timeout. If the identified frame has been received correctly, an ACK is returned; otherwise a NAK is returned. The Receive Count in either the ACK or NAK contains the number of the last successive message received correctly.

NAKs use the last part of the count field to specify the reason for the NAK.[18] Tributary stations on multipoint links do not NAK frames with bad header block checks, since the address could be in error; timeout recovery is used for such errors. NAKs are returned under such circumstances on point-to-point links, though. Stations ignore frames received out of sequence.

---

[17] The two flags are Quick Sync (Q), to tell the receiver this frame will not abut the next frame and will be followed by SYN characters, and Select (S), to control transmission on multipoint and half-duplex data links. The effect of S is similar to that of EOT in Bisync. DDCMP also considers an information frame transmitting station to be a master and a receiving station to be a slave, and S causes interchange of master and slave status.

[18] Reasons that can be specified include header BCC error, data field BCC error, buffer temporarily unavailable, message header format error, and a few others.

Figure 7.8   Typical DDCMP Startup Sequence (from [WECK82]). © 1980
            IEEE. Reprinted by permission.

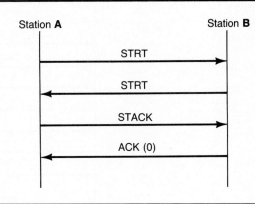

Two other types of supervisory frames are defined. Start (STRT) frames
establish initial contact and synchronization on a link, and Start Acknowledgment (STACK) frames acknowledge STRT frames; they initialize frame timers,
counters, and so forth and put stations in known states. A three-way handshake, illustrated in the next subsection, is used for startup.

For multipoint links the address field gives the address of the tributary
station. Direct transmissions between tributary stations are not allowed;
since the control station is involved in all frame transmissions, the tributary
station address identifies the only station that can vary. For point-to-point
links, an address field is included, but it is ignored by the receiver.

CRC-1 and CRC-2   *CRC-1* is a separate block check for the header portion of the frame. One
reason for a separate check can be appreciated by considering the potential
impact of an error in the most significant bit of the count field. *CRC-2*, in
frames with an information field, checks only the information field.

7.5.2   Typical Frame Sequences

We illustrate two frame sequences, a startup sequence and a typical message
flow sequence. Both are for point-to-point links.

Startup is illustrated in Fig. 7.8. Station **A** first sends a STRT frame to
Station **B**. **B** receives the STRT and sends its own STRT back to **A**, acknowledging the other STRT and initiating a three-way handshake (consisting of
this frame and the two following ones). **A** sends back a STACK, acknowledging
B's STRT and causing **B** to enter a running state. Station **B** completes the
handshake by sending an ACK (rather than a STACK since it is now in a running state). This causes **A** to enter a running state, and completes startup.

Figure 7.9   Typical DDCMP Message Flow Sequence (from [WECK82]).
© 1980 IEEE. Reprinted by permission.

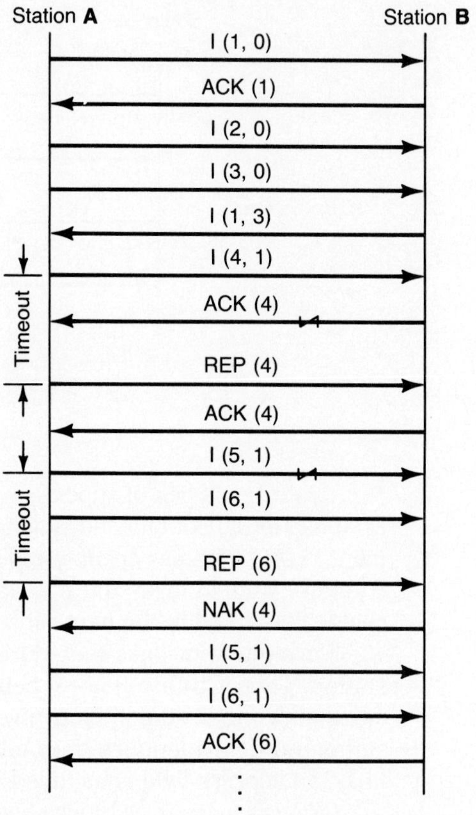

Figure 7.9 shows message flow. Numbers in parentheses for information (I) frames indicate Send and Receive counts, respectively.[19] For supervisory frames, the number is the count used, Receive Count for ACK and NAK frames and Send Count for REP frames.

The sequence starts with Station **A** sending I (1, 0) to Station **B**; this first frame sent by **A** has Send Count of 1 and Receive Count of 0 since no frames have been received from **B**. **B** uses an ACK to acknowledge I (1, 0), as it has no data to return. **A** next sends two more information frames, both acknowledged by I (1, 3) from **B**; the Receive Count of 3 indicates I (3, 0) and all preceding frames from **A** were received correctly. The next information frame

---

[19]They appear in reverse order in Fig. 7.7, but in this order in HDLC frames. For consistency, and to minimize confusion, we use the HDLC order in this type of figure in both cases.

from **A** is acknowledged by **B**, but the ACK contains an error (or is lost). This causes **A** to resort to timeout recovery, sending REP (4) after the timeout. **B** then ACKs frame 4 again.

An alternative error recovery procedure is possible and is illustrated next. I (5, 1) from **A** is received in error (or lost), but **B** does not send back a NAK. The error causes I (6, 1) to be out of sequence so it is discarded. Timeout recovery is used, with a reply to the frame with a Send Count of 6 requested. Since this was not received correctly and the last correctly received frame had a Send Count of 4, NAK 4 is returned. Go-back-*N* retransmission is used, with I (5, 1) and I (6, 1) both retransmitted. The sequence concludes with correct reception of these two frames and verification of reception by ACK(6).

Although the sketched sequences illustrate important aspects of typical operation, they do not illustrate full duplex operation, operation on multipoint links, or other aspects. Modifications for these types of operation should be reasonably obvious after the HDLC material in the next section is studied.

7.5.3   Limitations of DDCMP

Although DDCMP overcomes most of the problems of Bisync, it has not been widely adopted by companies other than Digital Equipment Corporation. The major limitation to its adoption has been use of bit-oriented protocols (see the next section).

One limitation of Bisync that DDCMP has not addressed is interchange of master and slave status. This still occurs in DDCMP, so problems resulting from poorly defined master status are possible.

Although DDCMP allows piggybacking of positive acknowledgments, and acknowledgment of multiple frames with one response, separate negative acknowledgments are required for each frame received in error. On the other hand, some DLC protocols allow the equivalent of negative acknowledgments via information frames (see the discussion of HDLC and related protocols in the next section). This is not possible in DDCMP.

Overall, DDCMP is an excellent data link control protocol, but it had the misfortune of being developed close to the same time when more widely publicized protocols were developed.

7.6   Bit-Oriented Protocols—HDLC

The most widely publicized and implemented current generation DLC protocols are *bit-oriented protocols*. Such protocols transmit bit streams irrespective of subdivision of these streams into characters. They are capable of transmitting bit streams of lengths that are not integral multiples of character lengths (but implementations often require stream lengths to be integral multiples of a character length).

A number of largely equivalent protocols fall into this class. They include *HDLC* (High-level Data Link Control—the ISO standard); *ADCCP* (Advanced Data Communications Control Procedure—an ANSI[20] standard); *SDLC* (Synchronous Data Link Control—IBM's version); *BDLC, NCRDLC, UDLC* (Burroughs', NCR's, and Univac's versions, respectively); *LAPB* (in the CCITT X.25 standard); and *LAPD* (proposed for use in ISDN networks). All are largely equivalent, to the level discussed here, but they can be grouped into two families: vendor standards (SDLC, BDLC, UDLC, NCRDLC, and so forth) and standards committee versions (HDLC, ADCCP, and variants). Vendor standards include limited lists of options, but HDLC and ADCCP have numerous options, chosen so essentially everyone can be satisfied.[21] LAPB and LAPD are basically subsets of HDLC, with specified options and minor additions. IEEE 802 DLC protocols for local area networks are closely related.

The original bit-oriented protocol was IBM's SDLC. HDLC and ADCCP were developed as a result of IBM taking its protocol to standards committees before publicly announcing it. Essentially all bit-oriented DLC protocols are now very nearly compatible except for variations in options. In the following discussion we use the term HDLC, but most of it applies to any protocol listed.

Some possible HDLC link structures are shown in Fig. 7.10. All frames are exchanged, in either direction, between a *primary* and a *secondary* station. Frames from primary to secondary are called *commands* (whether they command the secondary to do anything or not) and frames from secondary to primary are called *responses* (whether they respond to a command or not). Only one primary may be active on a link at a given time, but several secondaries may be active.

primary and secondary stations

commands and responses

In the original protocol the only allowable configurations were the unbalanced point-to-point configuration in (a) and the unbalanced multipoint configuration in (b).[22] Political pressures, however, led to allowing the configuration in (c), with two combined stations (stations containing both a primary and a secondary) on a point-to-point link. The primary motivation for this configuration came from cases where links crossed international boundaries, with neither nation willing to accept secondary status for its station. Although with this configuration either station can operate as a primary or a secondary, a frame is always transmitted between a primary and a secondary, so one station has to accept secondary status during each frame's

---

[20] ANSI is the American National Standards Institute.

[21] As [TANE81, p. 168] states, "The nice thing about standards is that you have so many to choose from; furthermore, if you do not like any of them, you can just wait for next year's model."

[22] The word unbalanced refers to the fact the stations are not equal.

Figure 7.10   HDLC Link Structures. (a) Unbalanced point-to-point link. (b) Unbalanced multipoint link. (c) Balanced point-to-point link with two combined stations.

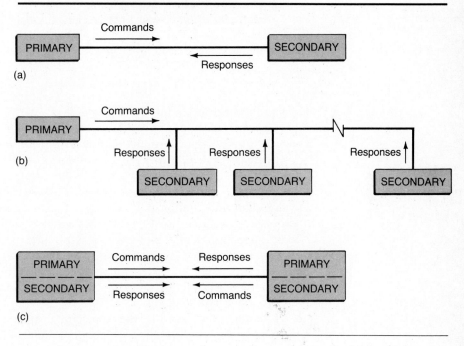

transmission. In particular, when the appropriate frame to be transmitted is a response, the station has to act like a secondary.

7.6.1   HDLC Frame Format

Flag

zero insertion algorithms

The normal format of HDLC frames is given in Fig. 7.11; options allow modified formats. The Flag field contains an eight-bit *Flag* character, 01111110 (that is, a zero, six ones, and another zero).[23] The bit pattern is not allowed to occur in a transmitted frame between initial and trailing Flags; an algorithm discussed below ensures this. If frames abut each other for transmission, the trailing Flag for one can serve as the initial Flag for the next.

A *zero insertion algorithm* inserts a zero at the transmitter after five consecutive ones in the Address, Control, Information, and FCS portion of each frame, so a Flag cannot appear in a transmitted bit stream in this portion of a frame. Circuitry at the receiver counts successive ones; after five ones it

---

[23] This is identical to the Flag character in the I.430 physical layer interface discussed in Chapter 4.

Figure 7.11   HDLC Frame Format with Normal Address and Control Fields.

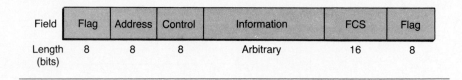

Field	Flag	Address	Control	Information	FCS	Flag
Length (bits)	8	8	8	Arbitrary	16	8

looks at the next bit and throws it out if it is a zero. If it is a sixth one, the following bit is examined to verity whether a Flag has been found. If a zero comes next, a Flag has been found; otherwise the frame is aborted.[24]

The Address Field always contains the address of the *secondary* station. There may be several secondaries on a multipoint link, but only one primary is allowed, with all transmissions between a primary and a secondary, so this identifies the only station that can vary. Group addresses can be used to address several stations, with an all ones address (defined to be a broadcast address) addressing all stations. An option to use *extended address fields* is available. If it is used, the address field is $8 \cdot n$ bits long, with n representing any integer. The first bit of the final address field octet is one, while the first bit of each preceding address field octet is zero, so the extended address field is recursively extendible.

The control field identifies the type of frame and contains frame sequence counts. Extended control fields are optional. Control field formats are given in Fig. 7.12.

The information field contains user data. Its length is arbitrary; it is zero for some frames, including supervisory frames, but may be long for user data frames. Any bit pattern is allowed, so transparent text is handled naturally.

The FCS field contains a frame check sequence (HDLC terminology for what Bisync and DDCMP call a BCC). It is obtained from a polynomial code generated by CCITT-16 (see Subsection 6.3.12) and checks the contents of the address, control, and information fields, excluding zeros inserted by the zero insertion algorithm.

information frames   *Information frames* are used for normal data transfer and are identified by zero in the first bit of the control byte. N(S) is a modulo eight Send Count giving the number of the frame being transmitted and N(R) is a modulo eight Receive Count (see below). (If extended control fields are used, counts are modulo 128.) These counts are used in essentially the same way as send and receive counts in DDCMP. As in DDCMP, piggy-backing of acknowledgments on information frames and acknowledgment of several frames at once are allowed. Differences include the following:

---

[24]An abort sequence consists of at least seven, but not more than 15, successive ones.

Figure 7.12 HDLC Control Field Formats. (a) Basic control field. (b) Extended control field.

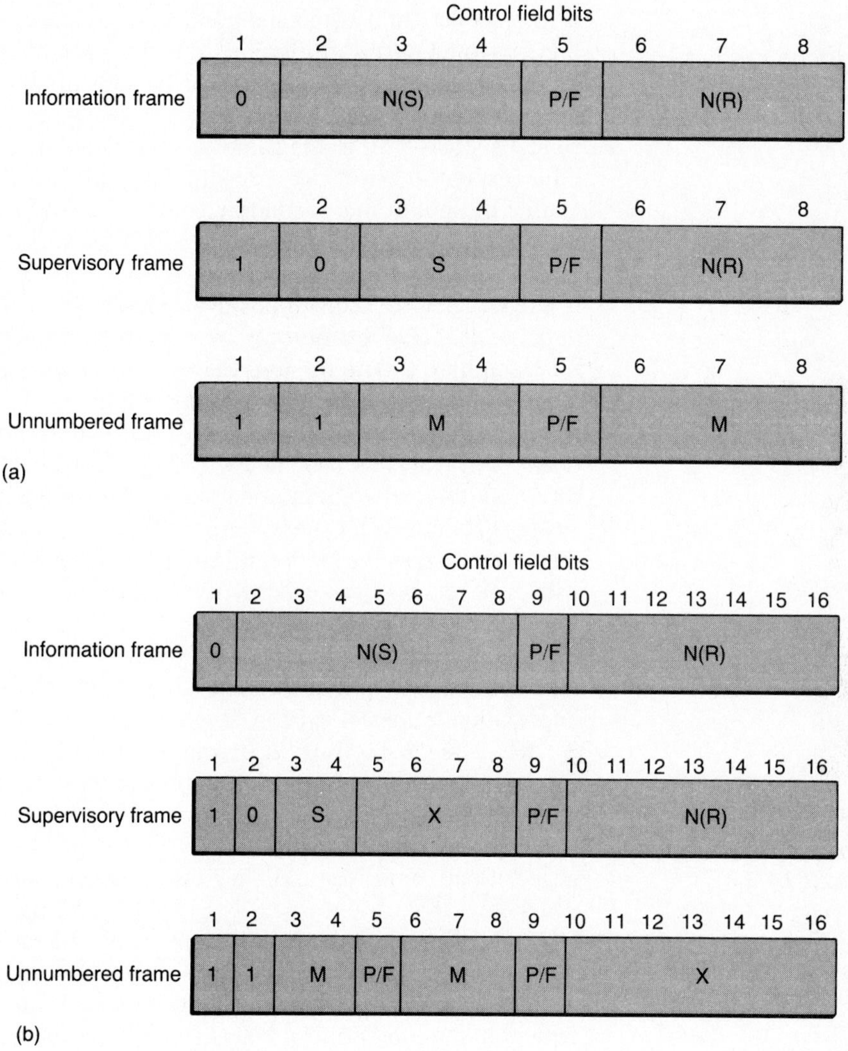

(a)

(b)

1. Counts are modulo eight (or 128) rather than 256; this restricts the maximum number of frames acknowledged at once to seven—or 127—rather than 255.

2. N(R) gives the number of the next frame the transmitter is looking for rather than that of the last good frame seen. This makes the number one

higher than an analogous number for DDCMP, ignoring modulus. Also, the Send Count for the first frame is zero instead of one.

Interpretation of the P/F bit depends on whether the frame is a command or a response. In a command (sent from primary to secondary), it is a poll bit used to request transmission from the secondary. In a response (sent from secondary to primary), it is a final bit indicating the end of a sequence of frames. P and F bits are always exchanged in pairs—that is, a primary cannot send a second P bit to the same secondary before it has received a returned F bit (except in error recovery situations with timeouts) and a second F bit cannot be sent from a secondary before it receives a new P bit. We will soon find that this has interesting implications for ARQ operation.

supervisory frames

Supervisory frames are identified by "10" in the first two bits of the control byte. They are used for positive and negative acknowledgments, if no user data are available for return frames, as well as for other purposes. The two S bits identify the type of supervisory frame. There are four possible combinations and four types of supervisory frames.[25] (The P/F bit and N(R) count have the same interpretation as for information frames.) These are Receive Ready (RR), Receive Not Ready (RNR), Reject (REJ), and Selective Reject (SREJ), but not all are required to be implemented; REJ and SREJ are options.

An RR frame is a general purpose frame most often used for acknowledgment; it indicates the station transmitting it has correctly received all frames with N(S) numbers (as used by the station the RR is sent to) less than N(R) (modulo eight), and is looking for a valid frame containing an N(S) equal to N(R). It also indicates the station is ready to receive information frames. RNR is similar, except for indicating the station is temporarily not ready to receive information frames; it is HDLC's version of a WACK (see Section 6.6).

REJ is similar to NAK in Bisync or DDCMP: it is a negative acknowledgment for the frame with send count equal to N(R), and requests recovery by go-back-*N* retransmission. SREJ is also a negative acknowledgment for frame number N(R) but only requests retransmission of the negatively acknowledged frame. Both REJ and SREJ also positively acknowledge all previously unacknowledged frames with N(S) values less than N(R). Frames with bad FCS values are ignored rather than being rejected, since the FCS checks address and control fields as well as the information field, and negative acknowledgments by the wrong receiver (bad Address Field) or of the wrong type of frame (bad Control Field) could have unpredictable consequences. A REJ or SREJ is triggered by reception of an out of sequence frame (often caused by ignoring a frame) rather than by one with bad FCS.

unnumbered frames

Unnumbered frames are used for housekeeping purposes, including link startup and shutdown, specifying modes of operation (options such as use of extended control fields). They are identified by "11" in the first two bits of the

---

[25] Eight types could be specified by differentiating between commands and responses, but this is not done for supervisory frames. It is done for unnumbered frames, however.

control field and contain no sequence counts (hence their name). The five bits labeled M are modifier bits specifying the type of unnumbered frame. This allows 32 different commands and 32 different responses to be specified, but a smaller number have been identified. "X" bits in the Extended Control Field format are reserved and set to zero. The P/F bit has the same interpretation as before; for the extended control field format, the bits in the fifth (basic control field position for P/F) and the ninth position (extended control field position) are identical.[26]

A few unnumbered frames are discussed below. They are not usually used during normal communication, aside from link startup and deactivation.

7.6.2 HDLC Operational Modes

Three modes of operation for HDLC are defined, with options for each. *Normal response mode* and *asynchronous response mode* are defined for unbalanced configurations with one primary and one or more secondaries. *Asynchronous balanced mode* is defined for balanced point-to-point configurations with two combined stations.

NRM

Normal response mode (*NRM*) is the mode best adapted for use on multidrop links with one primary and several secondaries. Under NRM, a secondary can only transmit after receiving a poll addressed to it by the primary. It may then send a series of responses, but after it sets the F bit in a response, it cannot transmit any more until it receives another poll.

ARM

Although *asynchronous response mode* (*ARM*) is defined for systems with one primary and one or more secondaries, if more than one secondary is present all but one must be in a quiescent (disconnected) mode at any given time. Mode setting commands issued by the primary ensure this. With one primary and one active secondary, either active station may initiate transmission at any time without waiting for a poll or an F bit. This leads to a "freewheeling" mode of operation, which can be more efficient than NRM and reduce response times.

P and F bits must still be paired; a transmission by the primary after sending a P bit, and before receiving an F bit, cannot have the P bit set (save for error recovery situations with timeouts), and a transmission by the secondary after sending an F bit, and before receiving a P bit, cannot have the F bit set. A frame (command) from the primary with P bit set, moreover, requires the secondary to return a response with F bit set as soon as possible.

ABM

Asynchronous balanced mode (*ABM*) is the mode of operation between two combined stations over point-to-point links. Either end can initiate transmission without waiting for a poll from the other end. P and F bits must still

---

[26] Mode setting commands may change the type of control field used, so this allows a receiver to receive and respond to such commands regardless of which format it is looking for. All other bits in the second octet of extended control fields are reserved and set to zero so that receivers looking for the basic format do not lose information.

be paired. That is, a P bit set by a station acting as a primary must be paired with an F bit received from the other station (acting as a secondary), before another P bit can be sent, and vice versa. The only exceptions are for error recovery purposes. These allow the primary to send another P bit after a timeout period, under carefully prescribed conditions. Whether a station is acting as a primary or a secondary in transmitted sequences can be determined by the fact that the address field always contains the secondary's address.

The examples in the next section cover NRM and ABM; ARM is less common and NRM and ABM illustrate protocol operation adequately.

7.6.3   Typical Data Exchanges

Figure 7.13 illustrates data transfer sequences for two-way alternate (half duplex) communication using NRM. Information frames are denoted by (I, n, m, Y), where I stands for information frame, n = N(S), m = N(R), and Y is either P or F, as appropriate; frames with neither P nor F bit set have Y omitted. Supervisory frames are identified by type of supervisory frame and N(R).

The four parts of the figure show similar data transfer sequences with four approaches to error recovery. In each case the primary first sends three information frames to the secondary, with the P bit set in the third to allow the secondary to respond. An error occurs in the second frame, which is ignored by the secondary, but the third frame is received correctly and its P bit allows the secondary to transmit.[27]

In Fig. 7.13 (a) the secondary returns the response (REJ,1,F) to indicate frame 1 was received erroneously (more precisely frame 2 was received out of sequence), so it is looking for a valid frame number 1. It has set the F bit on this response and is prohibited from transmitting again until it receives another command frame with P bit set. The primary retransmits the erroneous (I,1,0) frame plus the other frame it has already transmitted (I,2,0); it then transmits another I frame (I,3,0,P), with new poll bit deferred to this frame.[28]

Part (b) is essentially the same as (a), except for use of SREJ for error recovery. Only the erroneous frame is retransmitted, with the primary going on to transmit two new frames.

checkpointing

Parts (c) and (d) illustrate error recovery by *checkpointing,* a technique based on pairing of P and F bits and allowing the equivalent of a negative acknowledgment without using either REJ or SREJ. A transmission with F bit on from the secondary is not allowed if a P bit, in frame (I,2,0,P) has not been received, so the received F bit verifies reception of (I,2,0,P). A response frame

---

[27] If this frame had been received in error, it would also have been ignored so the secondary would not have responded. Error recovery procedures would then require timeouts.

[28] Retransmitted frames have the same information field and N(S) values as in their previous transmission, but use P or F bits and N(R) values applicable at retransmission times.

Figure 7.13 HDLC Two-Way Alternate Communication Using NRM. (a) REJ error recovery. (b) SREJ recovery. (c) Checkpoint recovery using RR. (d) Checkpoint recovery using I Frame.

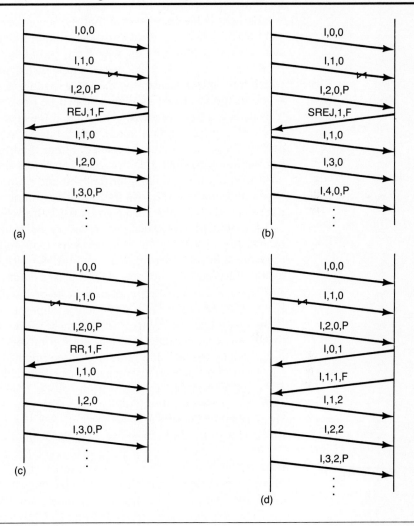

with F bit on but N(R) less than 3, the value applicable with all transmissions through (I,2,0,P) received correctly, allows the primary to determine which frames to retransmit. Go-back-*N* retransmission is used.

In part (c) checkpointing with a Receive Ready frame returned is illustrated. The Receive Ready response with F bit on completes a *checkpoint*

checkpoint cycle

cycle by pairing an F bit with a previous P bit.[29] The N(R) value of 1 indicates the secondary is looking for a good frame number 1, although the F bit indicates frame number 2, with P bit, got through. The primary hence retransmits its frames with N(S) numbers 1 and 2.

Part (d) is the same except that the secondary station has data ready to send when polled and sends two information frames, setting the F bit in the second. The second I frame, with F bit, completes the checkpoint cycle and produces the equivalent of a negative acknowledgment. The primary deduces which frames to retransmit in the same manner as in (c).

Checkpoint error recovery can also be used to recover from errors on transmissions from secondary stations. If the secondary sends a frame with F bit set and later gets a frame with P bit set but N(R) value less than expected, one of two things happened. Either the frame with F bit got through but the frame pointed to by the returned N(R) value did not, or the primary has timed out while waiting for the F bit and retransmitted its P bit. In either case the secondary retransmits frames beginning with the one identified by the returned N(R) value, using go-back-*N* retransmission.

A combination of checkpoint recovery and timeout recovery is adequate, though use of REJ and/or SREJ recovery may improve efficiency. Both REJ recovery and SREJ recovery are options for HDLC; most implementations include REJ recovery but use of SREJ is less common.

There are interesting contrasts in error handling between DDCMP and HDLC. DDCMP ignores out of sequence frames and negatively acknowledges frames with bad check bits.[30] HDLC ignores frames with bad check bits and negatively acknowledges out of sequence frames. Either approach works, as is verified by the fact that both are well-verified DLC protocols.

An example of NRM use on a multipoint link, illustrating polling plus both SREJ and checkpoint recovery from errors, is given in Fig. 7.14. Two-way alternate communication is illustrated. The notation is altered from that in Fig. 7.13 by adding the address field of each frame to the frame description; this is always the address of the secondary station.

The primary first polls **B** via a Receive Ready frame with P bit set. **B** responds with six I frames, (B,I,0,0) through (B,I,5,0,F), the last with F bit set. An error occurs in transmitting (B,I,3,0), but all other frames are received correctly. The erroneous frame is ignored, and SREJ triggered by (B,I,4,0), which is an out of sequence frame (after (B,I,3,0) was ignored). After receiving the F bit, the primary uses (B,SREJ,3) to inform **B** of the error but does not set the P bit to allow retransmission; instead, it polls **C** and **D**. Both return

---

[29] A P bit paired with a previous F bit can also complete a checkpoint cycle.

[30] More precisely, any frame with a bad BCC for text but a good BCC for the header is always NAKed. For point-to-point links any frame with a bad header BCC is also NAKed, but for multipoint links only control stations NAK such frames with tributary stations ignoring them since the error could be in the address.

Figure 7.14 Use of HDLC on Multipoint Link, Including Polling and Error Recovery (adapted from [CARL80]). © 1980 IEEE. Reprinted by permission.

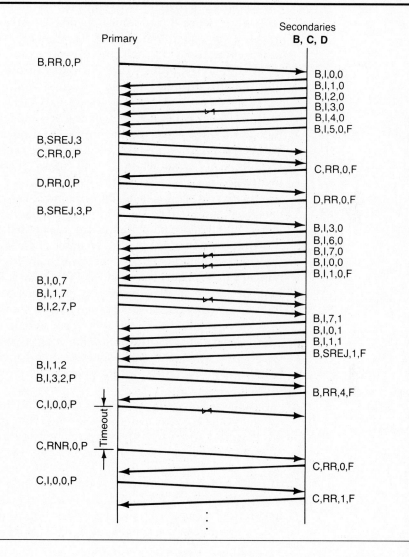

negative responses (Receive Ready frames with F bits); the primary then allows **B** to retransmit by resending the SREJ, this time with P set.

After receiving the P bit, **B** retransmits (B,I,3,0) and sends four additional frames, (B,I,6,0) through (B,I,1,0,F), but those with N(S) equal to 7 and 0 are received in error and ignored. The "out of sequence" frame the primary next receives, (B,I,1,0,F), has N(S) value indicating two missing frames. SREJ recovery is used only for one missing frame,[31] so the primary uses checkpoint recovery; it sends three information frames, the last with P bit completing a checkpoint cycle. There is an error on one, but the one with P bit is received correctly.[32] Hence **B** retransmits the first erroneous frame and all succeeding frames it has sent; it then informs the primary of the new error via SREJ. The primary retransmits to complete recovery from all errors to this point.

The final part of the figure illustrates recovery from a poll frame error. This requires timeout recovery. A second P bit, not paired with an F bit, is sent after timeout; at this time the primary checks to see what happened to its (C,I,0,0) frame, but does not allow **C** to send information frames, by transmitting RNR with P bit set; this prohibits **C** from sending information frames but requires it to return a supervisory frame. The returned (C,RR,0,F) frame completes a checkpoint cycle and allows recovery from the final error illustrated.

Figure 7.15 illustrates full duplex communication using ABM between two combined stations (see Fig. 7.9). REJ is used for recovery from the one error indicated. All information frames are transmitted as commands.

Transmission is full duplex, so frames flow in both directions simultaneously. Relative durations of frames, which are important in full duplex transmission, are explicitly shown in Fig. 7.15. A frame is delineated by two arrows (although the second arrow for one frame may be used as the first arrow for the next). Vertical distance between arrows shows time to transmit a frame (frame length/link speed); this time is shorter for supervisory frames than for information frames, so arrows delineating a supervisory frame are closer together. Horizontal crosshatching indicates **A** to **B** frames and vertical crosshatching **B** to **A** frames, with superimposed crosshatching if transmissions overlap.

Transmission is freewheeling, either station transmitting when ready, except during error recovery. The second frame from **A**, (B,I,1,0), is received erroneously.[33] (An error is represented by two error symbols, ⊢⊣, joined by an arrow parallel to those delineating a frame.) **B** ignores the erroneous frame and sends a response, (B,REJ,1) after the next (out of sequence) frame, (B,I,2,0), arrives (as soon as it has finished the transmission already underway).[34] When the REJ gets to **A**, it is about to finish transmitting an information frame; it completes transmission and retransmits frames with N(S) values of 1–4.[35] Retransmitted frames use updated N(R) counts.

---

[31] HDLC allows only one SREJ that has not been cleared by a retransmission, timeout, or completion of a checkpoint cycle to exist at any time.

[32] Timeout recovery would have to be used if there were an error on this frame.

[33] Note the use of the secondary's address in information frames transmitted as commands.

[34] The REJ is a response, so **B** acts as a secondary and uses its own address.

[35] **A** is allowed to abort this transmission rather than completing it.

Figure 7.15 An Example of Asynchronous Balanced Mode HDLC. This is a modified version of figure in [CARL80]. Shading indicates direction of transmission. © 1980 IEEE. Reprinted by permission.

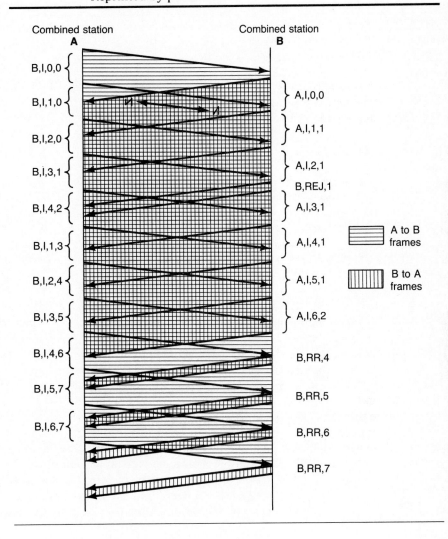

Station **A** transmits information frames continuously throughout the period sketched, sending seven different information frames plus four retransmissions, but **B** runs out of data after transmitting seven information frames, plus the REJ. During the time when it has no information frames to send, **B** simply sends back acknowledgments (as responses).

Table 7.2   HDLC Unnumbered Commands and Responses.

Unnumbered Format Commands		Unnumbered Format Responses	
Mode-Setting Commands		*Mode-Setting Responses*	
SNRM	Set normal response mode	UA	Unnumbered acknowledgment
SARM	Set asynchronous response mode	DM	Disconnected mode
SABM	Set asynchronous balanced mode	RIM	Request initialization mode
SNRME	Set normal response mode extended		
SARME	Set asynchronous response mode extended		
SABME	Set asynchronous balanced mode extended		
SIM	Set initialization mode		
DISC	Disconnect		
Data Transfer Commands		*Data Transfer Responses*	
UI	Unnumbered information	UI	Unnumbered information
UP	Unnumbered poll		
Recovery Commands		*Recovery Responses*	
RSET	Reset	FRMR	Frame reject
Miscellaneous Commands		*Miscellaneous Responses*	
XID	Exchange identification	XID	Exchange identification
		RD	Request disconnect

7.6.4   Initialization Sequences

Unnumbered commands and responses are listed in Table 7.2 and are used for initialization of stations and similar functions. Figure 7.16 shows typical initialization and disconnect sequences. A similar sequence is used for initializing a station to operate in any mode, although each station would have to be a combined station before initialization for ABM would be possible.

7.6.5   HDLC Subsets Used in ISO (X.25) and ISDN

Subsets of HDLC are used as DLC protocols for the X.25 architecture (a subset of the ISO architecture), and for the current version of ISDN Layer 2.

LAPB   The subset of HDLC used in X.25, *LAPB* (Link Access Protocol B), is HDLC asynchronous balanced mode with REJ error recovery and all information frames treated as commands. This is precisely the protocol illustrated in Fig. 7.15.

Figure 7.16 Typical HDLC Initialization and Disconnect Sequences.
[CARL80] © 1980 IEEE. Reprinted by permission.

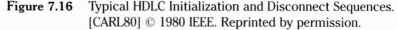

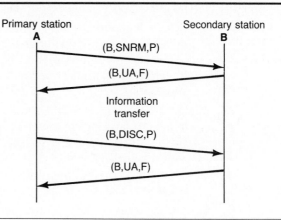

LAPD

A DLC protocol for ISDN in ISO Recommendation I.441 is called *LAPD* (Link Access Protocol D) and defined for operation on the ISDN D channel. It is an extension of LAPB and allows operation in either asynchronous balanced mode or asynchronous balanced mode extended (with extended control field and modulo 128 counts). The extended address field option is used with all addresses two octets long (see Subsection 7.6.1).[36] Information frames are treated as commands. Unnumbered information frames (also treated as commands) can be used for transfer of data without verification of receipt.

The set of allowable unnumbered commands and responses in Table 7.2 is extended to allow two information frame commands (using unnumbered format) and two responses to acknowledge these frames. The command frames and responses alternate, as in transmission of even- and odd-numbered frames with ACK 0 and ACK 1 responses in Bisync; this provides a limited capability for verified delivery while using the unnumbered format.

The ISDN standards also prescribe the LAPB version of HDLC for use on B channels [CCIT84], [CCIT88].

7.6.6 Limitations of HDLC

HDLC and its relatives currently represent the state of the art in DLC protocols. Under most conditions they are more flexible, and more efficient, than the other protocols we have discussed. Nevertheless, HDLC has weaknesses:

---

[36] The extended address field option allows seven bits from each address octet for addressing, with the final bit an address extension bit (see Subsection 7.6.1). Six bits of the first octet are used for a service access point identifier (SAPI), identifying the (OSI type) interface between data link and network layers, with the seventh bit indicating whether this is a command or a response frame. The seven bits available from the second octet are used for a terminal endpoint identifier (TEI), identifying the actual user at the address specified.

1. One limitation results from the fact that HDLC and ADCCP have so many options. Different HDLC devices may not be able to communicate if they do not support the same options.

2. HDLC's limitation to a single address is adequate in point-to-point communications or multipoint communications with a single primary, but it is not adequate in multipoint broadcast environments with all stations treated alike. Both destination and source addresses are needed in such situations.

3. HDLC has also been criticized on the basis that its layering is not clear.[37] In the OSI layering context, its use of flags and of zero insertion are considered to be physical layer features. It is primarily connection-oriented, with acknowledgment of all frames (although it provides limited support for unacknowledged data transfer using unnumbered data frames), so it is not fully suitable for connectionless services.

4. The portions of HDLC frames most susceptible to errors are the initial and trailing flags. The FCS does not check these flags and an error in transmitting either flag destroys framing. An error in the initial flag may mean a complete frame is lost since character synchronization is never obtained; an error in the final flag may mean the initial flag of the following frame is interpreted as the trailing flag of the previous frame.[38] Such erroneous "frames" are, with high probability, detected by the FCS, but error recovery for two frames is necessary under such circumstances.

5. Insertion or deletion of bits is common in HDLC. This can happen if bits in the vicinity of zero insertions are corrupted or bit errors transform bit patterns into patterns that would have called for zero insertion. The polynomial code for the FCS is powerful in detecting errors corresponding to changing zeros into ones or vice versa, but it is less powerful in detecting bit insertions or deletions. (This illustrates a common phenomenon; techniques that are optimum or nearly optimum under certain assumptions may be decidedly suboptimum if the assumptions are invalid. Murphy's Law is confirmed again![39])

6. Ambiguities similar to those for stop-and-wait ARQ with one acknowledgment (discussed in Subsection 7.3.4) occur when the unnumbered format is used with alternating P and F bits in commands and responses. Under these circumstances frames with P bits may be viewed as equivalent to information frames and frames with F bits as ACKs. If an unnumbered command with P bit receives no response and a second unnumbered

---

[37] HDLC predates the OSI model, so the fact it doesn't fully follow OSI layering is not surprising.

[38] If the final flag of this frame is used as initial flag of the next, the final flag of the following frame would be the one interpreted in this manner.

[39] As many readers know, Murphy's Law can be summarized as "If anything can go wrong, it will."

command with P bit is sent after a timeout, there is no way to tell whether an unnumbered response with F bit is a response to the first or second command.[40]

7. There are circumstances under which immediate negative acknowledgments for DDCMP (sent after finding a bad BCC rather than after an out of sequence frame) cause transmission efficiency of DDCMP to be greater than that of HDLC despite more overhead bits in DDCMP frames. If long frames are used and high error rates cause appreciable numbers of retransmissions, the extra delays before retransmission in HDLC can outweigh efficiency losses due to extra overhead bits in DDCMP. Such conditions are not common, though. Usually HDLC is slightly more efficient than DDCMP.

7.7 Local Area Network and MAP/TOP DLC Protocols

MAC and LLC

DLC protocol standards for LANs have been proposed by the IEEE 802 committees (see Subsection 3.9.1). These protocols are divided into two sublayers: *medium access control (MAC)* and *logical link control (LLC)* (see Fig. 7.17). One common LLC sublayer is described in the IEEE 802.2 standard,

Figure 7.17 Subdivision of Local Area Network Data Link Layer.

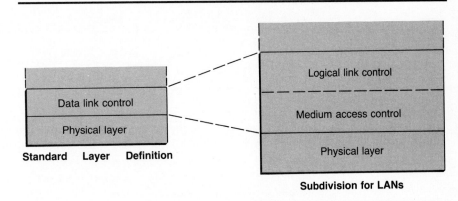

---

[40] The only situation we have discussed where unnumbered commands and responses are used in this manner is in Fig. 7.16 with unnumbered commands and responses for link activation and deactivation. Under the circumstances for this figure, if the primary does not receive a response it repeats exactly the same command sent before (that is, B,SNRM,P or B,DISC,P in Fig. 7.16), so it makes no difference which is acknowledged. In a few circumstances, including use of unnumbered information frames for user data (allowed but not encouraged by the protocol), there are more serious problems. No similar problems occur with information or supervisory frames, though, since they use sequence numbers.

Figure 7.18   Formats of IEEE 802 MAC Frames. (a) CSMA/CD. (b) Token bus. (c) Token ring: (c.i) Token. (c.ii) LLC data frame.

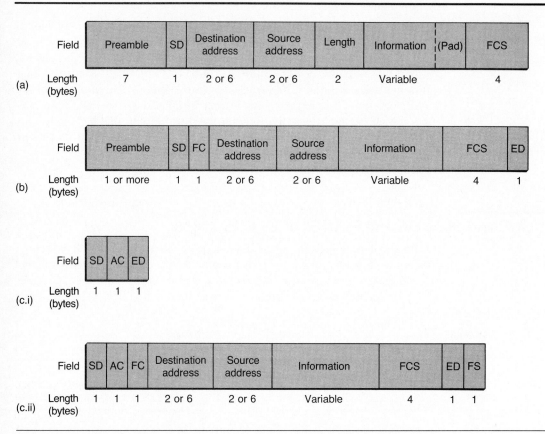

but different MAC sublayers are proposed for different types of LANs. Standards have been adopted for carrier sense multiple access/collision detection (CSMA/CD) networks (802.3 standard), token bus networks (802.4 standard), and token ring networks (802.5 standard). A closely related standard for metropolitan area networks (802.6 standard) is being developed. The medium access protocols are discussed in Chapter 5. We briefly discuss MAC standards, then describe the LLC layer.

MAP and TOP

MAP and *TOP* (see Subsection 3.9.2) use IEEE 802 protocols for their lower layers. MAP uses IEEE 802.4 through the data link layer. Similarly, TOP uses either 802.3 or 802.5 through the data link layer. Thus differences between MAP/TOP and IEEE 802 protocols occur only at higher layers.

Formats of MAC frames are illustrated in Fig. 7.18, with descriptions of fields in Table 7.3. In comparison with frames for other DLC protocols, the

Table 7.3   Interpretation of MAC Frame Fields.

Preamble: Bit pattern for stabilization and synchronization.

SD (Start Delimiter): Bit pattern indicating the start of a frame. For the token ring and token bus it includes non-data bits, that is, signalling patterns not in the normal signal set, so a SD is easy to recognize.

FC (Frame Control): Used in token ring and token bus networks to distinguish between types of frames, including MAC frames and LLC frames.

AC (Access Control): Used in token ring to control access according to priority and reservation algorithms.

Destination Address and Source Address: Address of destination and source, respectively.

Length: Length of data portion of frame in bytes; used only in CSMA/CD.

Information: User data. For LLC frames this is LLC data in Fig. 7.20. For CSMA/CD frames, extra pad characters fill out this subframe if total frame length is less than a specified minimum value (normally 512 bits).

FCS (Frame Check Sequence): A 32-bit frame check sequence derived from a polynomial code based on CRC-32 (see Subsection 6.3.12).

ED (End Delimiter): A sequence of bits indicating the end of a frame. Used in token bus and ring networks, with non-data bits to make ED easy to recognize. (Not needed in CSMA/CD because of length field.)

FS (Frame Status): A byte containing status bits, such as an address recognized and a frame copied bit. Used only in token ring LLC data frames.

frames are complex, with 12 to 26 bytes (96 to 208 bits) of MAC overhead per frame.[41] The high data rates of LANs, normally 1 to 10 Mbps or even higher, make it possible to use this type of overhead with no significant performance penalty under normal loading conditions. Some LANs have been designed with a "bits to burn" design philosophy, trading off transmission of extra bits for simplifications in protocols or architecture.

The frame structures for token bus and token ring networks are similar. A separate format for tokens is used in token rings, but no such format is shown for token buses since their token format is a special case of the format in Fig. 7.18 (b). Each token network uses an end delimiter to locate the end of frame, but CSMA/CD networks use a length field. Thus the token networks use the same approach to framing as HDLC, where the final flag is an end delimiter, and CSMA/CD networks use the same approach as DDCMP.

Either two- or six-byte addresses can be used with any of the LANs. Formats for address fields are given in Fig. 7.19. The number of potential addresses with six-byte addresses is especially generous; with 46 bits used for addresses (see Fig. 7.19) around 16,000 addresses are available for each man, woman, and child on earth!

---

[41] This calculation ignores the Pad field in the CSMA/CD format. If this field is needed, overhead can be slightly higher.

Figure 7.19   Address Fields for IEEE 802 MAC Protocols.
(a) Two-byte address field. (b) Six-byte address field.
I/G bit specifies individual address or group address and
U/L bit whether addresses are assigned on a universal
basis or by a local authority.

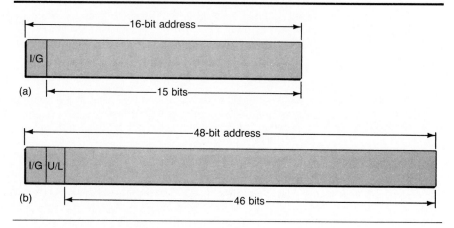

protocol data units

The format of an LLC *protocol data unit* (PDU) is given in Fig. 7.20. A LLC PDU is the information field in a MAC frame in Fig. 7.18.[42] The format is similar to that of HDLC frames, with flags and FCS characters omitted, except for using two address fields. The addresses are in addition to the two- or six-byte addresses in MAC headers; LLC addresses identify application programs or ports at nodes; MAC addresses identify only nodes. If a node contains a file server, a report generator, and a printer driver, they might have separate LLC addresses but a common MAC address.[43] Equivalents of HDLC flags and FCS characters are included in MAC headers and trailers (see Fig. 7.18). LLC functions are also patterned after functions in HDLC.

Figure 7.21 illustrates formats of LLC address fields. The I/G bit in the DSAP (destination service access point) address field tells whether this is an individual address or a group address, but only individual SSAP (source service access point) addresses are allowed since a unique source must be identified. The C/R bit shows whether the frame is a command or response.[44] This bit seems more appropriate for a control rather than an address field, but putting it in this location preserves symmetry between the address fields.

---

[42] As usual, a complete data unit from the higher layer (including higher layer headers) is treated as data (for the information field) by the lower layer.

[43] Addresses are called service access point (SAP) addresses, since they identify locations where services are available.

[44] Since each frame contains both a source address and a destination address, the HDLC convention of letting the address determine whether a station is acting as a primary (sending commands) or a secondary (sending responses) cannot be used.

Figure 7.20   Format of IEEE 802.2 Logical Link Control Protocol
Data Units (PDUs).

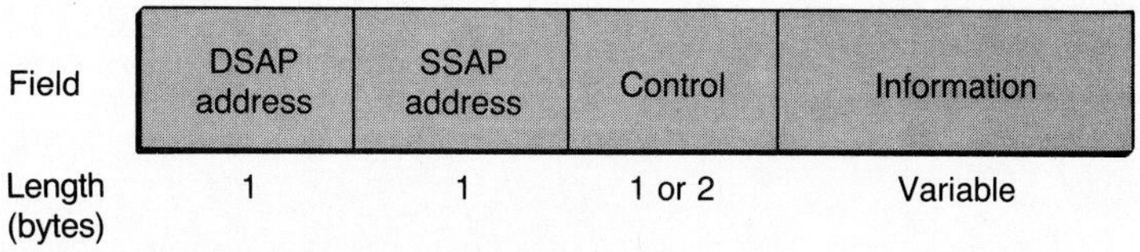

Field	DSAP address	SSAP address	Control	Information
Length (bytes)	1	1	1 or 2	Variable

Figure 7.21   IEEE 802.2 LLC Address Fields.

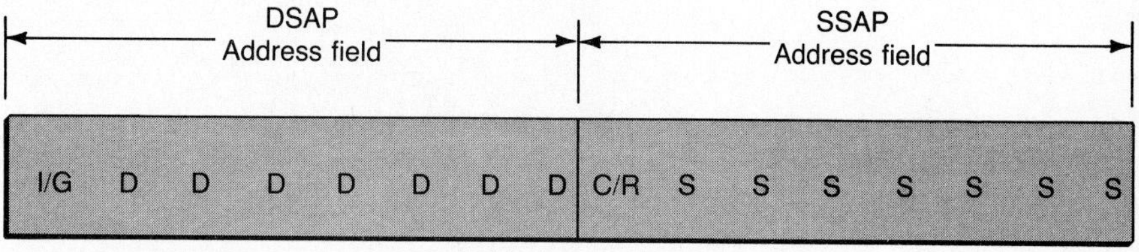

DSAP Address field								SSAP Address field							
I/G	D	D	D	D	D	D	D	C/R	S	S	S	S	S	S	S

Two types of operation are defined:

Type 1 operation

- *Type 1 operation (connectionless).* PDUs are exchanged without establishment of a DLC connection and are not acknowledged within the LLC sublayer, although acknowledgments by higher layers can be used. No LLC flow control or error recovery is used. PDUs are patterned after unnumbered format HDLC frames, with unnumbered information frames for data. Responsibility for ensuring reliable communication is placed on higher layers. Type 1 operation is best adapted to broadcast communication or to use of group addresses (multicast communication), as well as to connectionless point-to-point communication.

Type 2 operation

- *Type 2 operation (connection-oriented).* PDUs are patterned after HDLC asynchronous balanced mode extended, with two-byte control fields. Sequencing, flow control, and error recovery are provided using techniques taken from HDLC. Techniques for establishing, using, resetting and terminating data link layer connections are also patterned after HDLC.

7.8   Overhead in Synchronous DLC Protocols

As discussed in Subsection 7.2.3, start-stop DLC protocols have a percentage DLC overhead that is independent of message length (typically 20 to 44 percent). In contrast, percentage overhead in a frame for synchronous DLC

Figure 7.22   Variation of Percentage DLC Overhead in Data Frame with
Information Field Length for ASCII Implementation of
Bisync, with Character Parity Bits Counted in DLC Overhead
and Six Character Headers.

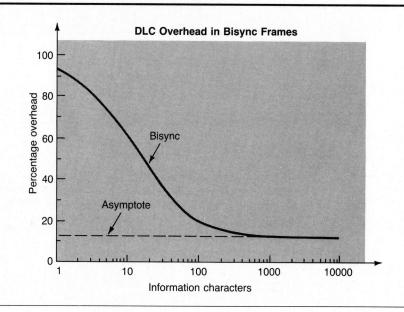

protocols decreases with increasing message length. The percentage DLC
overhead in an individual frame tends to zero with increasing message length
for most synchronous DLC protocols, since a fixed number of bits is used for
overhead. The only exception to this, among synchronous protocols we have
discussed, is for Bisync implementations using character parity plus a one-
character BCC (for example, ASCII implementations); for such cases DLC
overhead never drops below 12.5 percent since one bit of every eight is used
for character parity. (This is not completely fair, however; DDCMP and HDLC
often treat ASCII characters as eight-bit characters but ignore character parity
bits.)

bisync overhead   A typical computation of frame overhead for ASCII transmission in Bisync
is as follows. Assume K is total frame length in bits, including overhead, n_h is
header length in bits, and n_d is the number of data bytes in a frame. There
are then a total of $n_h + n_d$ overhead bits per frame, which means that fractional overhead is

$$O_{\text{frame}} = \frac{n_h + n_d}{K} \tag{7.1}$$

Figure 7.23 Comparison of DLC Overhead in Data Frames for DLC Protocols. Bisync with six byte headers and EBCDIC implementation (or character parity not charged to DLC overhead) has overhead similar to that of DDCMP. Without headers, Bisync overhead is essentially equivalent to that of HDLC under the same conditions.

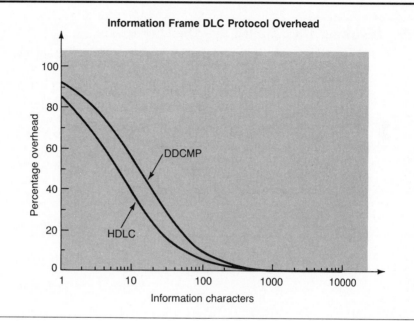

with K given by

$$K = 8n_d + n_h. \tag{7.2}$$

For the frame format in Fig. 7.4, with six character headers, there are a total of 12 characters of DLC overhead (ignoring any PADs). Variation of O_{frame} with data field length is plotted in Fig. 7.22.

DDCMP and HDLC overhead

Computation of overhead for DDCMP and HDLC (or for Bisync if character parity bits are not charged to DLC protocol overhead) is simpler. It is

$$O_{\text{frame}} = \frac{n_h}{K} \tag{7.3}$$

with 96 bits overhead for DDCMP (see Fig. 7.7) and 48 bits overhead for HDLC (see Fig. 7.11). K is still found from Eq. (7.2). Overhead for EBCDIC implementations of Bisync is typically in this range, depending on header length. These values are plotted in Fig. 7.23.

If the effects of errors and retransmissions are included, the appropriate formulas for transmission efficiency are essentially those in Eq. (6.25) for Bisync (stop-and-wait ARQ) and (6.27) for DDCMP and HDLC (go-back-N ARQ) (see Problems 7.22 and 7.23).[45]

7.9   Invoking Data Link Layer Services

The OSI architecture specifies that data link services are invoked by *primitives* exchanged across the boundary with the network layer. OSI data link service primitives are listed in Table 7.4 [CONA83]. A limited subset are illutrated below; the others provide additional power and flexibility. For example, Negotiate primitives available during an Establish Phase provide a basis for negotiation of service parameters to satisfy user requirements.

Figure 7.24 shows an example of the use of primitives for a protocol such as HDLC. This is an expanded version of sequences in Fig. 7.16; it explicitly

Table 7.4   ISO Data Link Service Primitives.

	Types			
Services	*Request*	*Indication*	*Response*	*Confirm*
Establish Phase				
DL-Connect*	X	X	X	X
DL-Activate	X			X
DL-Acquire	X			X
DL-Identify	X	X	X	X
DL-Negotiate	X	X	X	X
Transfer Phase				
DL-Data	X	X	X	X
DL-Expedited Data	X	X	X	X
DL-Flow Control	X	X		
DL-Reset	X	X	X	X
DL-Notify	X	X		X
DL-Abort	X	X		
Terminate Phase				
DL-Disconnect	X	X		X
DL-Deactivate	X			X

*DL prefix indicates layer in which primitive used. © 1983 IEEE. Reprinted by permission.

---

[45] A minor modification needs to be made to Eq. (6.25) if Bisync transmission with ASCII characters is considered and character parity bits are counted as overhead. The term $K - n_h$ in the numerator becomes $K - n_h - n_d$, since one bit per data character needs to be subtracted to give data bits.

Figure 7.24 Typical Application of DLC Service Primitives for Protocol Such as HDLC (adapted from [CONA83]). © 1983 IEEE. Reprinted by permission.

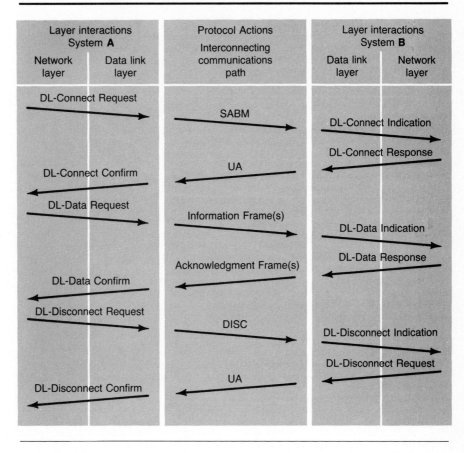

shows how protocol layers and interfaces interact. The figure shows *Establish, Transfer,* and *Terminate phases* for HDLC using ABM.

The Establish Phase is initiated when the network layer in System **A** passes a DL-Connect Request primitive across the interface with its data link layer. The HDLC protocol responds by generating an unnumbered SABM (set asynchronous balanced mode) command and passing it across the communications link to **B**. When the data link layer at **B** receives SABM, it notifies its network layer via a DL-Connect Indicate primitive, and the network layer responds positively by passing a DL-Connect Response primitive to its data link layer. This in turn generates an Unnumbered Acknowledgment (UA) message and passes it across the link to **A**. The Establish Phase handshake is completed when the data link layer at **A** passes a DL-Connect Confirm prim-

Figure 7.25   Use of Primitives with Character-Oriented DLC
Protocol (adapted from [CONA83]). © 1983 IEEE.
Reprinted by permission.

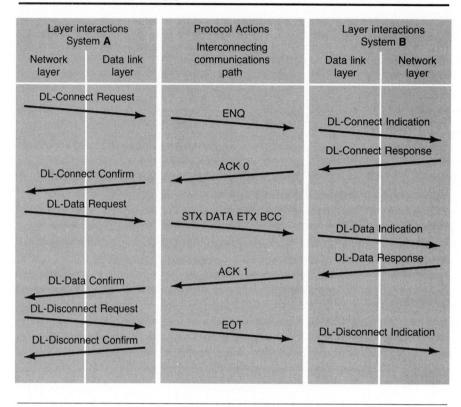

itive to its network layer. The link is then in Transfer Phase for communication from **A** to **B**. A similar Establish Phase with transmissions in the opposite
direction could be used to set up a return path.

During Transfer Phase, illustrated for one directional transfer, transfer is
initiated when the network layer at **A** passes a DL-Data Request primitive,
with data to be transferred, to its data link layer. The data are packaged into
frames and transmitted across the link by HDLC. Each arrival of a correctly
sequenced frame without detected error causes exchange of DL-Data Indication and Response primitives across the interface between the data link
and network layers at **B**. The DL-Data Response primitives cause Acknowledgment Frames with appropriate Receive Counts to be passed across the
link to **A**; these in turn cause the data link layer at **A** to tell its network layer,
via DL-Data Confirm primitives, that the frames were received correctly.

The Terminate Phase is handled similarly. It is initiated by a DL-Disconnect Request primitive passed across the interface at **A**. This causes an Unnumbered DISC (disconnect) frame to be sent to **B** and DL-Disconnect Indication and DL-Disconnect Response primitives to be passed across its interface. DISC is acknowledged by another UA frame, with the Network Layer at **A** informed by a DL-Disconnect Confirm primitive.

Detailed procedures and nomenclature for corresponding functions with other protocols vary, but basic ideas are similar to these. This is illustrated by Fig. 7.25, which shows a corresponding sequence of frames and primitives for character oriented protocols such as Bisync. ISO terminology for primitives is used, but frames use Bisync formats.[46] The sequence is essentially the same as in Fig. 7.24, save for substituting Bisync frames for HDLC frames. The final Terminate Phase handshake is less secure, however, since Bisync does not provide for acknowledgment of the final EOT from **A**. Similar sequences are used with each DLC protocol we have studied.

7.10 Summary

DLC protocols implement functions at the data link layer of the OSI Reference Model and most other networking architectures. These include transferring data between network entities and (in some protocols) recovering from transmission errors; providing for activation, maintenance, and deactivation of data link connections; grouping bits into characters and message frames; character and frame synchronization; and medium access control and flow control.

The simplest DLC protocols in common use are asynchronous or start-stop DLCs, which provide little more than character synchronization and rudimentary error control. More comprehensive functions are provided by synchronous DLC protocols.

Major categories of synchronous DLC protocols include character-oriented, byte-count-oriented and bit-oriented DLCs. Character-oriented DLCs are based on use of communication control characters from standard character sets, and are exemplified by Bisync and ARPANET DLC. Byte-count-oriented DLCs use a count field to indicate the number of bytes in an information (user data) field; DDCMP is the most prominent example. Bit-oriented DLCs use a specified bit pattern called a Flag for synchronization purposes and prohibit appearance of this Flag at any spot within a frame aside from its first and last characters. A wide variety of current DLC protocols use this approach, including vendor protocols such as SDLC and standards such as HDLC and ADCCP. Other protocols such as those used in IEEE 802 local area networks and X.25 and ISDN are closely related.

---

[46] Initial SYN characters have been omitted to simplify the figure.

Appendix 7A   Listing of Bisync Control State Frames

Table A7.1   Bisync Control State Signal Frame Formats (adapted from [EISE67]).
Possible PAD Characters Not Included.

Type	Format	Remarks
Contention mode selection	ϕ ENQ*	Request for permission to transmit
Supervised mode selection	ϕ A (up to 6 noncontrol characters) ENQ*	Basic selection sequence, where A is selection address of station
	ϕ A (up to 6 noncontrol characters) STX text . . .	Fast selection for text transmission
	ϕ A (up to 6 noncontrol characters) DLE STX transparent text . . .	Fast selection for transparent text transmission
	ϕ A (up to 6 noncontrol characters) SOH heading	Fast selection for heading transmission
Selection replies	ϕ ACK 0*	Basic affirmative acknowledgment
	ϕ NAK*	Basic negative acknowledgment
	ϕ (up to 7 noncontrol characters) ACK 0*	Affirmative acknowledgment with end-to-end control characters
	ϕ (up to 7 noncontrol characters) NAK*	Negative acknowledgment with control characters
	ϕ WABT*	Wait before transmit
	ϕ (up to 7 noncontrol characters) WABT*	Wait before transmit with control characters
Supervised mode polling	ϕ A (up to 7 noncontrol characters) ENQ*	Basic polling sequence, with A polling address of station
Polling replies	ϕ EOT*†	Negative reply ending transmission
	ϕ STX text . . .	Affirmative reply followed by text block
	ϕ DLE STX transparent text . . .	Affirmative reply followed by transparent text block
	ϕ SOH heading . . .	Affirmative reply followed by heading block
Disconnect	ϕ DLE EOT†	Disconnection initiated (switched network only)
Circuit assurance	ϕ (up to 15 noncontrol characters) ENQ*	"I am, who are you?" sequence with characters for identification and control (switched network only)
Identification	ϕ (up to 15 noncontrol characters) ACK 0*	"I am" response with characters for identification and control (switched network only)
	ϕ DLE EOT†	Originator's response unsatisfactory and disconnection being initiated
	ϕ (up to 15 noncontrol characters) NAK*	Not ready to receive with up to 15 characters for identification and control

Table A7.1 *continuing*

Type	Format	Remarks
	φ (up to 15 noncontrol characters) WABT*	Temporarily not ready to receive with characters for identification and control

φ Synchronizing pattern (SYN SYN)
*Change in direction of transmission
†No reply expected

Problems

7.1 Start-stop communication is used to transmit ASCII characters over a 1200 bps link. A terminal user types in 10 characters in 4 seconds. This is followed by a 0.5 second delay for computation before a 60-character response comes back from the computer, with characters from the computer spaced together as closely as the DLC protocol will allow. After a 5-second delay for user "thinking time," a comparable sequence is repeated. Stop bits are one-bit time in duration and a parity bit is used for each character.

    a. Sketch a timing diagram for the sequence of events listed, indicating the duration of each relevant interval and the number of characters transmitted during the interval.

    b. Compute the average communication link utilization during the total time period, with link utilization the percentage of time the link is transmitting any type of bits (start, stop, parity, or data bits).

    c. Compute the percentage of time the link is transmitting data bits during the total time period.

7.2 Compute the maximum possible rate of transmitting data bits on a 2400 bps line using a start-stop protocol, with even parity (and parity bits not counted as data). Consider the following cases (giving six answers): (i) Stop bit durations of 1, 1.5, and 2 bits, and (ii) Character lengths of 5 and 7 bits.

7.3 Consider a multidrop network using Bisync, with a control station and 10 subsidiary stations. The average distance of a subsidiary station from the control station is 1000 miles. Eight-bit characters are transmitted, with station addresses (either polling or selection addresses) each two characters. A PAD character is added to each message. Find the maximum number of Bisync polls per second that can be transmitted, assuming no errors and that no tributaries return data. Assume that line speed is 4800 bps, modem turnaround time is 250 msec, and propagation delay is 1 msec per 100 miles.

7.4 A sequence of Bisync transmissions is illustrated in Fig. P7.4. For each transmission, numbers 1 through 8, identify the transmitting and the receiving station and give the format of the frame transmitted. Be as specific as possible, including such parameters as numbers for ACKs, but ignore PADs.

Figure P7.4   Bisync Sequence for Problem 7.4.

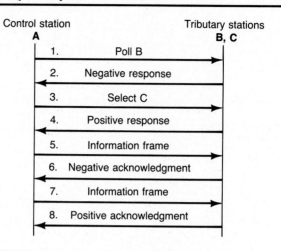

7.5   What is the maximum number of frames that can be transmitted before acknowledg-
        ment in DDCMP? What ambiguities can occur if more are transmitted without
        acknowledgment? Why is it usually desirable to transmit fewer than the maximum
        number before acknowledgment?

7.6   Complete the message flow sequence, between stations 1 and 2, shown in Fig. P7.6
        using DDCMP and operating in a half duplex mode.

7.7   Will a protocol with only positive acknowledgments and timeouts work satisfactorily?
        If so, what is the purpose of using negative acknowledgments?

7.8 a. If a bit string 01111001 10111110 11111101 00000010 (transmitted in left to right
         order) is subjected to zero insertion in HDLC, what is the output string? (Spaces
         have been added after each byte for clarity only; they should be ignored when
         answering the question.)
     b. Describe what happens at the receiver if the transmitted sequence is received
         without errors.
     c. Describe what happens at the receiver if the transmitted sequence is received
         with a single error in the: (i) 6th transmitted bit, (ii) 15th transmitted bit, and (iii)
         16th transmitted bit.

7.9   Figure P7.9 illustrates what is purported to be a sequence of SDLC full duplex trans-
        missions between a primary station and one secondary station.[47] SDLC is identical to
        HDLC operating in normal response mode (save for a few minor distinctions that have
        nothing to do with this problem).
     a. List the errors in this sequence.
     b. Sketch a corrected sequence accomplishing the same data transfers.

---

[47] Except for changing the format to agree with the format in Fig. 7.15, the sequence is identical
     to one in a well-known book on telecommunications, which we prefer not to identify.

Figure P7.6 Figure for Problem 7.6.

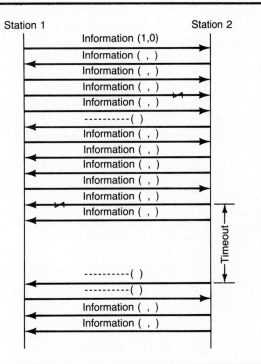

Station 1   Station 2

Information (1,0)

Information (,)

Information (,)

Information (,)

Information (,)

---------- ( )

Information (,)

Information (,)

Information (,)

Information (,)

Information (,)

Information (,)

Timeout

---------- ( )

---------- ( )

Information (,)

Information (,)

Figure P7.9 Figure for Problem 7.9.

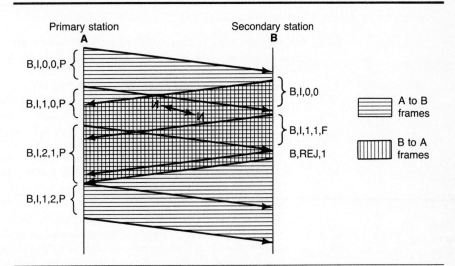

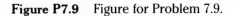

Primary station
A

Secondary station
B

B,I,0,0,P

B,I,1,0,P

B,I,2,1,P

B,I,1,2,P

B,I,0,0

B,I,1,1,F

B,REJ,1

A to B
frames

B to A
frames

7.10   Describe four distinct techniques for handling transparent text transmission. Briefly discuss each and give an example of a DLC protocol implementing it.

7.11   Briefly describe how each DLC protocol in Fig. 7.1 handles the following functions. (All bit-oriented protocols can be grouped together under the name "HDLC.") If a protocol does not handle a particular function, state so.
   a.   Bit synchronization
   b.   Character synchronization
   c.   Content synchronization

7.12   Stations **A** and **B** send data over a 1 Mbps satellite channel with 540 msec round trip propagation delay. Two hundred data blocks, each containing 250 characters (of user data), are to be sent from **A** to **B**. Assume supervisory frames are short enough to require negligible time and no waiting in the receiver. If the first data block contains a transmission error but no other data blocks contain errors, calculate total time to transmit the data blocks using each of the protocols below. In each case assume that the number of data blocks transmitted before acknowledgment is either the maximum number allowed by the protocol or the total number to be transmitted, whichever is smaller.
   a.   DDCMP
   b.   HDLC with extended control field and SREJ option

7.13   Briefly compare the operation of the Bisync, DDCMP and HDLC data link control protocols with respect to the following factors:
   a.   Defining frame boundaries
   b.   Handling received frames with invalid parity check sequences
   c.   Handling out-of-sequence frames
   d.   Operation on half duplex and full duplex channels
   e.   Checking all frames for transmission errors
   f.   Bytes of overhead per frame (ignoring any Bisync headers)

7.14   Figure P7.14 shows data exchanges between stations **A** and **B** using the LAPB (X.25) DLC protocol. Are the exchanges consistent with the protocol? If not, redraw with the necessary corrections.

7.15   Consider the asynchronous balanced mode for HDLC, using standard options for X.25 (save for allowing SREJ in one case below), and use the notation in Fig. 7.14 to describe frames.
   a.   Assume that Stations **A** and **B** communicate over an error-free channel. Give a complete sequence of frames (from initiation to termination) for a dialogue in which **A** transmits 5 information frames to **B** and **B** transmits 3 information frames to **A**. All information frames are of the same time duration, with the first transmission from **A** beginning approximately one-half information frame time before the first transmission from **B**. Propagation delay is short and successive transmissions from each end occur as soon as is permitted by the protocol. Each end includes a P bit in its first information frame, with later polls as soon as permitted by the protocol.
   b.   Indicate how the system would recover from an error in the second frame sent from **A** to **B**, for the sequence of transmissions above using the following: (i) REJ error recovery, (ii) SREJ error recovery, and (iii) Checkpointing with information and RR frames used to convey the information needed.

Figure P7.14   Figure for Problem 7.14.

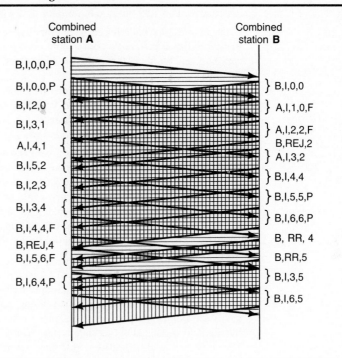

7.16   Both RR and REJ frames positively acknowledge receipt of frames up to (and including) that with send sequence number N(R) − 1, and indicate frame N(R) is expected.
Explain why you should or should not use RR for all positive and negative acknowledgments. Give reasons.

7.17   Two stations, **A** and **B**, are using the LAPB DLC. If station **A** receives the last I (information) frame, in a sequence of I frames from station **B**, in error, how would the
stations recover from this situation?

7.18   In a multipoint link, if a frame sent from the control (either primary or master) station
to a secondary has an error in the text portion, explain how the secondary would
behave when using the following:

a.   Bisync
b.   DDCMP
c.   HDLC-NRM mode

7.19   Compute and plot DLC overhead versus information characters per frame for IEEE
802.3 (CSMA/CD) LANs. Give separate plots for 2-byte and 6-byte MAC address fields,
and include both MAC and LLC overhead in your computations (using 2-byte LLC
control fields). Your plots should resemble Fig. 7.23, using the same types of axes and
spanning the same range of frame lengths. (Ignore any limitations on frame length
imposed by the DLC protocol.)

7.20   Repeat Problem 7.19 but make plots for IEEE 802.4 (token bus) LANs.

7.21   Repeat Problem 7.19 but make plots for IEEE 802.5 (token ring) LANs.

7.22   Compute and plot effective rate (R_e) for Bisync versus frame length in characters for the cases listed below. Put both plots on the same axes, and include frame lengths varying from 100 to 10,000 eight-bit characters. (Your plots should resemble those in Fig. 6.9).

     a.   ASCII transmission with burst error rate of 10^{-4} on a terrestrial 4800 bps link with 10 msec propagation time and 100 msec turnaround time. Headers are 5 bytes, PAD characters are used, and parity bits are considered to be overhead.

     b.   Repeat Problem 7.22 (a) but substitute EBCDIC transmission for ASCII.

7.23   Compute and plot effective rate (R_e) versus frame length in characters for the cases listed below. Put both plots on the same axes and include frame lengths varying from 100 to 10,000 eight-bit characters. (Your plots should resemble those in Fig. 6.11.)

     a.   HDLC transmission with burst error rate of 10^{-4} and terrestrial 4800 bps link with $N = 3$ (the minimum allowed under these circumstances).

     b.   DDCMP transmission with burst error rate of 10^{-4} and terrestrial 4800 bps link with $N = 2$ (the minimum allowed under these circumstances).

7.24   **(Design Problem)** Describe an implementation of the OSI Data Link interface in Fig. 7.24, with X.21 used as a physical layer. List necessary functions and indicate how you would handle them. A partial list includes accepting primitives (with parameters such as connection identification plus user data), constructing primitives, adding characters to data units or removing them, placing signals on circuits, interpreting incoming signals, and recognizing message types. Your list should be more specific (including description of each primitive). You may define primitive formats, as the OSI model leaves this up to implementers. Ignore details of setting up a connection, but include a connection request and recognize connection completion.

7.25   **(Design Problem)** Repeat Problem 7.24 with the following changes: Implement interactions in Fig. 7.25 instead of those in Fig. 7.24, use Bisync as the DLC protocol instead of HDLC, and use an RS-232C physical layer instead of X.21. Make any other changes these alterations require.

8 Routing and Flow Control

8.1 Introduction

Routing algorithms determine paths taken by data flowing from source to destination, while *flow control algorithms* limit traffic allowed on links or into specified portions of a network to avoid excessive congestion. Although they are often studied separately, they are closely related since routing algorithms can introduce congestion, and require flow control, if they route too much traffic into the same areas. We discuss both in this chapter.

Example of routing and flow control relationship

A simple example of the relationship between routing and flow control is based on Fig. 8.1. We represent networks by *nodes* and *links*. Boldface numbers identify nodes; the other numbers indicate capacities.

We assume that all traffic flows either from **1** to **4** or from **3** to **6**. If each flow is less than four units or so, the direct paths, **1-4** and **3-6** are adequate and no flow control is needed with this routing.[1] If either flow equals five units, however, the direct path will saturate if all traffic is routed over this path and delay will become excessive,[2] so flow control is needed. On the other hand, if a routing algorithm diverts one unit or so to go over the link between **2** and **5** under these conditions, excessive congestion is avoided and no flow control is needed. If the two traffic flows total more than about 12 units, congestion occurs regardless of routing, with saturation of all three top to bottom links for total input rate of 15 units or more. Flow control to limit traffic is then essential.

Figure 8.2 illustrates the general relationship between routing and flow control. As the figure indicates, routing affects delay, which affects the amount of traffic allowed into the network by flow control. This in turn affects delay, and so forth.

---

[1] A useful rule of thumb for systems involving queuing is that queuing delays do not normally become excessive for loading less than approximately 80 percent, corresponding to four units of traffic on a link with capacity five units.

[2] According to queuing theory, delay is normally infinite if traffic is at least equal to link capacity and no traffic is rejected due to some mechanism such as flow control.

Figure 8.1   A Simple Telecommunication Network.

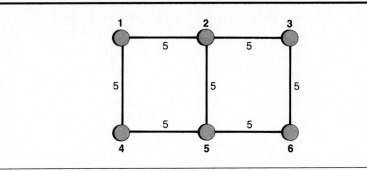

The type of communications session shown in Fig. 8.1, with one source and one destination and a single network, is the most common type, and it is this type of session that we will concentrate on. Other types of communications sessions, such as broadcast sessions, with transmissions destined for all nodes, or multicast sessions, with a subset of nodes receiving each transmission, also are important. Medium access techniques such as those in Chapter 5 are commonly used in such situations. Broadcast or multicast routing in networks without true broadcast capability requires relaying to multiple destinations by intermediate nodes, but fully satisfactory techniques are not available. Routing and flow control across multiple networks often involves use of independently developed intranetwork routing and flow control algorithms during transit through networks, and they are even less well understood.

Figure 8.2   Interaction of Routing and Flow Control (from [BERT87], Bertsekas/Gallager, *Data Networks,* © 1987, p. 299. Reprinted by permission of Prentice-Hall, Inc., Englewood Cliffs, NJ.).

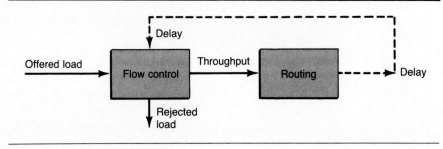

8.2   Classification of Routing Algorithms

Routing algorithms perform two main functions: *selection of routes* between source/destination pairs and *delivery of messages* to their destinations after routes have been selected. The second function is relatively straightforward; it involves use of data structures known as routing tables. The first function often involves a collection of algorithms operating at different nodes, while supporting each other by exchanging services or information. We emphasize the first function but include a discussion of routing tables.

Routing algorithms can be classified in a variety of ways. Table 8.1 lists elements that can be used to classify them. A fundamental element is the decision place. Both *centralized* and *distributed* routing algorithms are used. For centralized algorithms, a central location is in charge of routing, obtaining information from nodes in the network and disseminating routes out to nodes. Distributed algorithms involve cooperation among nodes to share information and perform calculations. A special case is one in which the source node decides how a packet should be routed and includes information describing the route in the packet header. In important networks a subset of nodes, called routing nodes, handle routing.

Table 8.1   Elements of Routing Algorithms.

Decision Place	*Decision Time*
Each node (distributed)	Packet (datagram)
Central node (centralized)	Session (virtual circuit)
Source node	
Subset of nodes	*Network Information Source*
	None
Routing Strategy	Local
Static	Adjacent nodes
Adaptive	Nodes along route
Update time	All nodes
Continuous	
Periodic	*Performance Criterion*
Major load change	Number of hops
Topology change	Cost
	Delay
	Throughput

(Adapted from [STAL88b] W. Stallings, *Data and Computer Communications, 2/e,* © 1988. Reprinted by permission of MacMillan Publishing Company, New York, NY.)

There are tradeoffs between centralized and distributed approaches.[3] Centralized algorithms are simpler and less likely to lead to inconsistent routing tables, since consistency of information at various locations operating asynchronously and interconnected by error-prone links is not required. On the other hand, information for such algorithms must be obtained from locations throughout the network and can be seriously outdated. Distributed algorithms have converse advantages and disadvantages. They may adapt more rapidly and are likely to have more accurate information describing the network in the immediate vicinity of a node making decisions, but it is difficult to coordinate nodes in order to optimize routing and avoid inconsistent routes.

Routing may be *static* or *adaptive.* Static strategies do not attempt to respond to network changes but adaptive strategies emphasize responding to changes.[4] Routing *update times* for adaptive algorithms may be virtually continuous, periodic (with a variety of periods), or may occur only in response to such changes as addition or deletion of nodes or links. Also, decisions may be made for every packet (common for datagrams) or just when a session or connection is established (the usual virtual circuit approach).

Network *information sources* range from none (for static algorithms ignoring network conditions), through local information (available directly at a node doing routing), information from adjacent nodes, information accumulated along their route by previous packets, to information from all nodes. *Performance criteria* for algorithms attempting to optimize performance include number of hops between source and destination, cost of facilities used (defined in a variety of ways), delay to reach the destination, and throughput. With an appropriate definition of cost, most criteria can be defined as types of cost, so we emphasize a cost criterion.

Additional attributes of an ideal routing algorithm are [BELL86][5]:

- *Correctness:* The algorithm must work.

- *Computational simplicity:* The algorithm should use minimum processing capacity at each node. It should also place a minimum burden on bandwidth of links between nodes.

- *Adaptiveness to changing traffic and topologies, or robustness:* The algorithm must adapt to changing levels of traffic flow and find alternate routes when nodes and/or links fail or come back into service.

---

[3]A distinction between centralized and distributed is in some cases more a feature of implementation than of algorithm used. Some algorithms may be implemented in a centralized or distributed fashion, that is, computations and information dissemination can be handled either way.

[4]The distinction between static and adaptive is, in practice, a matter of frequency of changes since major network changes, such as installation of new equipment, are likely to lead to routing changes in any network.

[5]Adapted from Bell/Jabbour, "Review of Point-to-Point Network Routing Algorithms," *IEEE Communications Magazine,* © 1986 IEEE. Reprinted by permission.

- *Stability:* The algorithm must converge without excessive oscillation, while adapting to changing traffic loads and topologies.

- *Fairness:* The algorithm must be equitable to all users, within constraints of assigned priorities.

- *Optimality:* The routing algorithm should be able to provide the "best" routes that minimize mean packet delay and maximize throughput.

No algorithms fully meet all of these criteria. In fact, some are contradictory. For example, maximizing throughput tends to increase delay. The effectiveness of a routing algorithm, however, can have an impact on how rapidly delay increases with increasing throughput, and a good algorithm can yield a high threshold for the throughput where delay becomes excessive and flow control is necessary.

8.3 Routing Tables

Routing tables at each node indicate how that node should route packets. Possible routing tables for the network in Fig. 8.3 are in Table 8.2. Routes are chosen to minimize the number of hops to reach the destination, with arbitrary choices where two or more minimum hop routes exist.

The tables indicate, for each destination, the next node to which a packet should be sent. For example, information from **1** to **5** is sent from **1** to **4**, then directly to **5**. Inconsistent tables may cause unsuccessful routing. Inconsistent tables can lead to ping-ponging (when the table at **i** indicates the best route to **m** is via **j**, and the table at **j** indicates the best route is via **i**, so packets "ping-pong" between **i** and **j**), looping (packets returning to a node traversed earlier, then continually going around this loop), and similar phe-

Figure 8.3   Example of a Telecommunication Network.

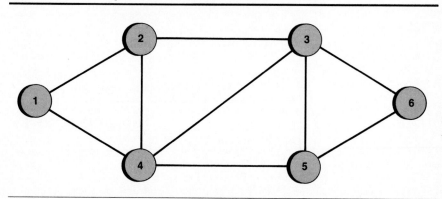

Table 8.2   Examples of Routing Tables for Network in Fig. 8.3. Tables at other nodes will resemble those shown.

Routing Table at Node 1				
Destination node	2	3	4	5	6
Next node	2	2	4	4	2

Routing Table at Node 4				
Destination node	1	2	3	5	6
Next node	1	2	3	5	5

nomena. It is difficult to fully eliminate such glitches when using distributed algorithms in complex networks, especially during adaptation to changes.

After tables have been set up, routing involves simply looking up information in the tables, without computation. Highly adaptive algorithms may, however, require significant computations to maintain routing tables.

Variants on these tables are common. Some are trivial—for example, identification of outgoing link by link name or number rather than by the next node. Information for updating tables may also be included. For example, algorithms based on minimizing cost often include estimated cost to reach the destination by the optimum route in addition to the next node or initial link on the route. This leads to Table 8.3, which is the type of table used in some prominent networks.

Routing tables for virtual circuits identify virtual circuit numbers instead of destinations, but these may change at nodes en route since nodes assign numbers independently. For example, if nodes assign numbers in order of requests and **1**'s virtual circuit number 1 goes through **2**, **2** may not have number 1 available when it sets up this virtual circuit. Number changes are handled by including them in routing tables. Table 8.4 gives an example of virtual circuit routing tables, illustrating portions of tables pertaining to the

Table 8.3   Routing Table for Node **1** in Fig. 8.3 Including Cost Values. Costs are taken as the number of hops necessary to reach destination.

Routing Table at Node 1				
Destination node	2	3	4	5	6
Next node	2	2	4	4	2
Cost	1	2	1	2	3

Table 8.4 Examples of Routing Tables for Virtual Circuits in Fig. 8.4.

Routing Table at Node 1					
Incoming virtual circuit (-,-)	(-,-)	(-,-)	(**2**,1)	(**4**,2)	(**4**,4)
Outgoing virtual circuit (**2**,1)	(**4**,2)	(**4**,4)	(-,-)	(-,-)	(-,-)

Routing Table at Node 4				
Incoming virtual circuit	(**1**,2)	(**1**,4)	(**3**,5)	(**5**,4)
Outgoing virtual circuit	(**5**,4)	(**3**,5)	(**1**,4)	(**1**,2)

three bidirectional virtual circuits between **1** and **6** in Fig. 8.4.[6] The pair (**X**, n) identifies previous node and virtual circuit number for an incoming virtual circuit or next node and virtual circuit number for an outgoing virtual circuit, with blanks (-,-) indicating a virtual circuit originating or terminating at the node with the table. Once routing tables have been established during virtual circuit setup, no further computation is necessary during the duration of the virtual circuit.

Data in routing tables must contain all information needed to identify routes. If routes depend on message class, quality of service requirements, and so forth, additional data may need to be included in tables.

Figure 8.4 Network with Three Bidirectional Virtual Circuits Between **1** and **6**, with Routing Via **1-2-3-6**, **1-4-3-6**, and **1-4-5-6**.

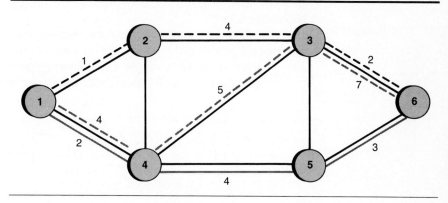

---

[6] Identical virtual circuit numbers for both directions of transmission are used here to simplify the figure, but different numbers can be, and often are, used.

8.4   Shortest-Path Routing

As we indicated earlier, criteria for defining optimum routing algorithms often can be considered to be minimization of some type of cost function. Examples of link cost functions and results of minimizing them include the following; in each case the cost of a single link is specified.

A cost of one for each link passed through en route from source to destination is used to *minimize number of hops* traversed. If the cost equals delay in traversing the link, an optimum route minimizes delay.[7] If cost equals length of a link, the optimum *minimizes distance.* If cost is cost of communications facilities (line usage charges, for example), the optimum *minimizes communications cost.* Another cost function can be defined so the optimum maximizes reliability. The reliability of a path, π, from a to b via $i, j, \ldots,$ m, n is

$$R_\pi(a,b) = p_{ai}p_{ij}\cdots p_{mn}p_{nb}, \tag{8.1}$$

with p_{rs} the probability link rs is operational.[8] This can be converted into a sum by taking logarithms, then into a function to be minimized by taking the negative of the result. Thus if the cost of link ij is $-\log_e p_{ij}$, the optimum *maximizes reliability* (see Problem 8.2).

shortest path algorithms   Algorithms to find optimum routes with link costs given are called *shortest-path algorithms*; a wide variety are treated in the operations research literature [DEO80], [DREY69], [FORD62], [FRAN71], [PIER75], [SYSL83]. We discuss algorithms that are used in major networks. We make a fundamental assumption throughout our discussion of routing algorithms: The load presented to the network is feasible—that is, the network has adequate capacity to handle it.[9] A tree of optimum routes from one source to all other nodes or equivalently to one destination from all other nodes is computed by one class of algorithms; another class computes shortest paths between all sources and all destinations.[10]

We use the network in Fig. 8.5 to illustrate routing algorithms. All links are bidirectional (full duplex). Costs for links are indicated by numbers adjacent to them; costs are assumed equal for each direction of transmission.

We represent the nodes in a network by a set, N, with N elements. Let D_i

---

[7] A cost function based on delay should include the impact of traffic, but this can cause instabilities to occur as traffic is shifted from heavily loaded to lightly loaded links, causing delays on the latter to go up and delays on the former to drop, producing reverse shifts later, and so forth.

[8] Equation (8.1) is based on an assumption that link failures are independent so the probability that the entire path is operational factors into the product of probabilities individual links are operational.

[9] Flow control may be necessary to limit the load to feasible values. We do not attempt to model its effects at this point, however.

[10] It is interesting to note that an algorithm for the "simpler" problem of finding an optimum route between one source node and one destination node, and guaranteed to require fewer computations than are necessary for the most efficient algorithms in the first class, is not known.

Figure 8.5 Example Network for Shortest Path Algorithms. (This network is found in a variety of references, for example [SCHW80b].)

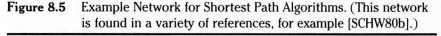

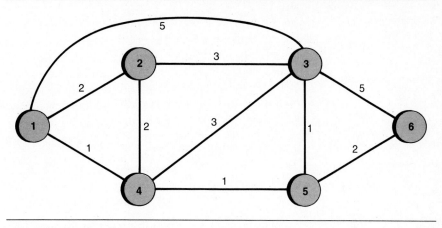

represent the minimum distance for reaching node **i** from a specified origin node (taken as node **1**) and c_{ij} represent distance (cost) between nodes **i** and **j**, with $c_{ii} = 0$ for all i and $c_{ij} = \infty$ if **i** and **j** are not adjacent. For each node **i**, we wish to find a *path,* **1jk . . . mi**, consisting of links **1j, jk, . . . , mi**, such that total distance along this path is the minimum possible. For such a path, we say node **k** is the successor of node **j** and node **j** is the predecessor of node **k**. We assume $c_{ij} \geq 0$ for all i and j; this is consistent with the problem analyzed.[11] The c_{ij}s may be written as a cost matrix **C**. For Fig. 8.5 **C** is

$$C = \begin{bmatrix} 0 & 2 & 5 & 1 & \infty & \infty \\ 2 & 0 & 3 & 2 & \infty & \infty \\ 5 & 3 & 0 & 3 & 1 & 5 \\ 1 & 2 & 3 & 0 & 1 & \infty \\ \infty & \infty & 1 & 1 & 0 & 2 \\ \infty & \infty & 5 & \infty & 2 & 0 \end{bmatrix}.$$

8.4.1 The Bellman-Ford-Moore Algorithm

The *Bellman-Ford-Moore algorithm* [BELL58], [FORD56], [MOOR59] is based on *Bellman's principle of optimality* [BELL57], [BELL62].[12] We present it in terms of finding optimum routes from a given source node to all other nodes

---

[11] The Bellman-Ford-Moore algorithm will find optimum routes with some costs negative (as long as no loops with total distance negative exist), but other algorithms will not allow any negative costs. Since any costs we are interested in are positive, we assume this for all cases.

[12] Various names for this algorithm are found in the literature. For example, [BERT87] calls it the Bellman-Ford algorithm and [SCHW87] the Ford-Fulkerson algorithm (from its presentation in [FORD62]).

in the network, but it can be presented equally well in terms of finding optimum routes to a prescribed destination node from all other nodes (see Problem 8.5).[13]

Bellman's principle of optimality is intuitively reasonable. It states that, for problems in a broad class, if an optimum sequence of actions is decomposed into a subsequence taking a system from its initial state to an intermediate state, followed by another subsequence taking the system to the final state, the subsequence of actions to reach the intermediate state must be the optimum way to reach this state and the subsequence to go from there to the final state must be the optimum way to do this.

Bellman's equation, expressing his principle of optimality, is

$$D_i = \min_j [D_j + c_{ji}], \text{ for all } i \neq 1$$
$$D_1 = 0 \tag{8.2}$$

We prove this is a condition for optimality as follows. The second equation reflects the fact that the source is at zero distance from itself. Any other destination must be reached from an immediately preceding node, which we label j. Total distance to i is thus c_{ji} plus the distance to reach j, but this latter distance must be D_j, distance to j along the optimal route, since otherwise there would be a route with shorter distance found by going to j along the optimal route and then directly to i. Thus distance along the optimal route is of the form $D_j + c_{ji}$, and the optimum j is the one minimizing this sum.

The Bellman-Ford-Moore algorithm *iterates on number of hops in a path.* It is based on first finding optimal paths (routes) from a prescribed source node to all other nodes subject to the constraint no path contains more than one hop, then finding optimal paths subject to the constraint no path contains more than two hops, and so forth. The shortest path subject to the constraint no path contains more than h hops is called the shortest ($\leq h$) path.

Let $D_i^{(h)}$ be the length of the shortest ($\leq h$) path from **1** to i, and define $D_i^{(h)} = 0$ for all h. The Bellman-Ford-Moore algorithm is then as follows:

Let

Bellman's principle of optimality

Bellman's equation

Bellman-Ford-Moore algorithm

$$D_i^{(0)} = \begin{cases} 0 & i = 0 \\ \infty & i \neq 0 \end{cases} \tag{8.3}$$

Then, for each successive $h \geq 0$, let

$$D_i^{(h+1)} = \min_j [D_j^{(h)} + c_{ji}], \text{ for all } i \neq 1 \tag{8.4}$$

---

[13] [STAL88] calls the Bellman-Ford-Moore algorithm the "backward search" algorithm and Dijkstra's algorithm (discussed in the next section) the "forward search" algorithm, but either can be used with either a forward or a backward search. Our presentations use search directions the reverse of his; we also give a homework problem to modify each to reverse its search direction.

and connect i with the predecessor node, j, for which the minimum is obtained (eliminating any connection of i with a different predecessor formed during an earlier iteration). Continue until no more changes occur in the next iteration.

A proof that Eq. (8.4) gives the minimum length of a ($\leq h$) path is essentially identical to the proof given for Bellman's optimality principle, noting that any ($\leq h + 1$) path to a node other than **1** consists of a ($\leq h$) path plus a final link. Upon completion of the algorithm the nodes in the network will always be connected by a tree, called the *shortest-path spanning tree,* with **1** at the root of the tree. (Problem 8.3 proves that the result is always a tree.) The optimum route to i can be found by following the tree from its root to i.

shortest-path spanning tree

Our statement of the Bellman-Ford-Moore algorithm does not specify the order in which nodes should be searched, and several implementations of the algorithm have been developed by varying the order of search. An example will illustrate one technique for ordering the search. The algorithm is readily implemented as a labeling algorithm, which puts a label of the form (P,D) on each node, with P its current predecessor node and D its current estimated distance from the source node. If a node has not been assigned a predecessor node, we use -1 for P; initial distances are given by Eq. (8.3). Labels are updated during later iterations of the algorithm, with the algorithm terminating when no more label changes occur.

Consider the network in Fig. 8.5. (The algorithm can also be applied to networks with directional links; see Problem 8.9.) Figure 8.6 illustrates the Bellman-Ford-Moore algorithm. At each step, current labels are shown and links interconnecting nodes with their predecessors are highlighted. Initialization labels are shown in part (a); the source node is assigned a label of (-1, 0) and all others are assigned labels of (-1, ∞). The value of D for the ith node is $D_i^{(0)}$, with P undefined. The figure also shows a queue (Q) of nodes that are potential predecessor nodes during later steps. Immediately after initialization the only realistic predecessor node is the source node, so $Q = \{1\}$.

The next step is to remove **1** from Q and look at the nodes directly connected to it as possible successor nodes; these nodes are **2**, **3**, and **4**. Since $D_1^{(0)} + c_{1i} < D_i^{(0)} = \infty$ for $i = 2, 3$, and 4, connecting any of these to **1** reduces its distance, so its distance estimate becomes $D_1^{(0)} + c_{1i} = c_{1i}$, with $P = 1$. This yields the labels and connections in part (b). Potential predecessor nodes are now **2**, **3**, and **4**, so Q becomes $\{2,3,4\}$. The remaining steps are of the same form. At each step, the node at the head of the queue is removed and its number becomes j for computation of $D_j^{(h)} + c_{ji}$. If $D_j^{(h)} + c_{ji} < D_i^{(h)}$ (the value of D in node i's label), $D_j^{(h)} + c_{ji}$ replaces D, with j the new P; since i's label has been changed, it is also added to Q. If $D_j^{(h)} + c_{ji} \geq D_i^{(h)}$, i's label is left unchanged.[14]

When **2** is considered as a potential predecessor, no reductions in distance are found. Thus Q is reduced to $\{3,4\}$ with no changes in labels. Remov-

---

[14] In the case of equality, changing the label is optional.

Figure 8.6 Illustration of Bellman-Ford-Moore Algorithm. (a) Initialization. (b) After examining successor nodes for node **1**. (c) After considering successor nodes for nodes **2** and **3**. (d) After considering successor nodes for **4**. (e) After considering successor nodes for **5**. No changes when considering successor nodes for **6** or **3** (again). Since this empties **Q**, the algorithm terminates.

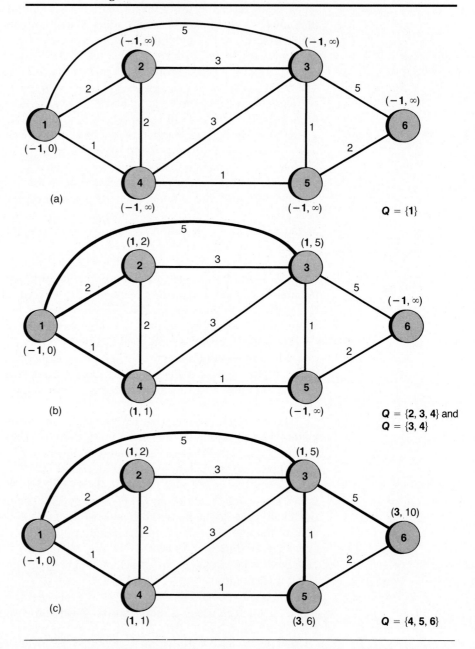

Figure 8.6 *continued*

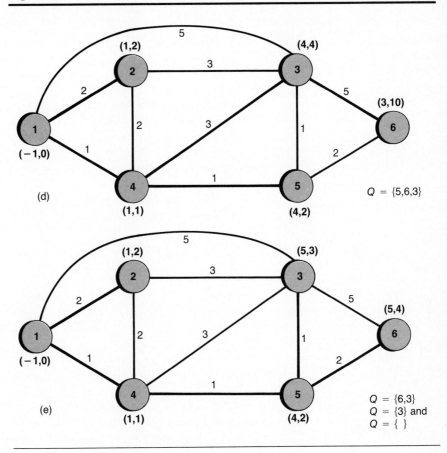

(d) Q = {5,6,3}

(e) Q = {6,3}
Q = {3} and
Q = { }

ing and examining **3**, though, results in labeling **5** and **6**, with results shown in part (c); both nodes are added to **Q**, which becomes {**4,5,6**}. When **4** is removed, labels on **5** and **3** change. As **5** was already in **Q**, it does not need to be added, but **3** comes back in since labels based on its previous value may need to be changed. An example of a needed change is the label for **6**, which is still based on the previous label for **3**; such discrepancies are corrected when the node reentering **Q** is reexamined, if not before. Here the final path to **6** is via **5** instead of **3**, with the label on **6** corrected when **5** is examined.

A pseudo code version of the Bellman-Ford-Moore algorithm (adapted from [SYSL83]) is shown below. We give a version allowing any node, *s*, to be the source node. It uses vectors *d* and *p* to represent distance estimates and current predecessors at nodes, $d(n)$ being node *n*'s distance estimate and $p(n)$ the applicable-predecessor node.

INITIALIZATION
       for $n \leftarrow 1$ to N do
              begin
                     $D(n) \leftarrow \infty$;
                     $P(n) \leftarrow -1$
              end;
       $d(srce) \leftarrow 0$;
       {initialize Q to contain $srce$ only}
       insert $srce$ at the head of Q;

*pseudo code for
Bellman-Ford-Moore
algorithm*

ITERATION
       while Q is not empty do
              begin
                     delete the head node j from Q;
                     for each link jk that starts at j do
                            begin
                                   newdist $\leftarrow D(j) + c_{jk}$;
                                   if newdist $< D(k)$ then
                                          begin
                                                 $D(k) \leftarrow$ newdist;
                                                 $P(k) \leftarrow j$;
                                                 if $k \notin Q$ then insert k at the tail of Q;
                                          end;
                            end
              end.

A variety of ways to order elements in Q have been investigated; some reduce average computational time. A useful modification is to insert nodes at the tail of Q if they have not previously been examined, but insert them at the head of Q if they have previously been in Q and hence examined. The motivation for this is to correct labels based on previously examined nodes *d'Esopo-Pape version* as soon as possible. This is called the *d'Esopo-Pape* version of the algorithm (see Problem 8.6).[15] It modifies the final if in the pseudo code to [SYSL83]:

> if k was never in Q then insert k at the tail of Q
> else if k was in Q but is not currently in Q then
>   insert k at the head of Q.

Extensive experiments [DENA79], [DIAL79], [VANV78] indicate that if the network is large and sparse (that is, if it has a low ratio of the number of links to number of nodes) the d'Esopo-Pape modification of the Bellman-Ford-Moore algorithm is faster than virtually any other known algorithm for finding

---

[15] This modification is commonly attributed to d'Esopo and Pape; it has been examined in [PAPE74], [PAPE80], but [POLL60] attributes the original suggestion to d'Esopo.

Figure 8.7 Maximal Length Path with N Nodes.

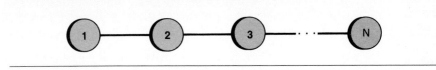

shortest paths from a given source node to all other nodes.[16] Most telecommunication networks are of this form.

The Bellman-Ford-Moore algorithm is one of the most common algorithms in operational networks, but the most commonly implemented version is a distributed version discussed in Subsection 8.4.4. A few networks using it are mentioned at the end of Subsection 8.4.4.

computational requirements

Worst-case computational requirements of the algorithm can be estimated as follows. For a network with N nodes, a path can contain at most $N - 1$ links (see Fig. 8.7). Hence, there can be a maximum of $N - 1$ iterations before convergence. For each iteration, the minimization must be done for $N - 1$ nodes and, for each node, the minimization is over at most $N - 1$ alternatives. Thus the amount of computation is at worst $O(N^3)$ (see Section 2.7). A more careful accounting [PALL84] shows the amount of computation is $O(mL)$, with L representing the number of links and m the number of iterations required for termination. This is significantly less than $O(N^3)$ for many networks.

8.4.2 Dijkstra's Algorithm

Dijkstra's algorithm is another popular shortest-path algorithm. Its worst-case computational requirements are less than those of the Bellman-Ford-Moore algorithm, though there are many situations where the Bellman-Ford-Moore algorithm requires fewer computations.

Rather than iterating on number of hops in a path, as in the Bellman-Ford-Moore algorithm, Dijkstra's algorithm *iterates on the length of a path.* We present it in the form of an algorithm for finding shortest paths to a specified destination from all other nodes, although it can be presented equally well as an algorithm for finding shortest paths from a specified source to all other destinations (see Problem 8.7). It first finds the shortest path from some node to the destination and establishes this path. It then finds the next shortest path and establishes it, and so forth. The detailed algorithm (for destination node **1**) is shown below.

---

[16] It is possible to devise artificial networks, never seen in practice, for which the number of computations with the d'Esopo-Pape version grows exponentially with network size (see [SYSL83]).

Let N_c represent a set of nodes currently connected with the destination and D_j represent the current estimate of distance of node j from the destination. Initially $N_c = \{1\}$, $D_1 = 0$, $D_j = \infty$, $j \neq 1$. Dijkstra's algorithm is as follows:

- *Step 1.* (Updating labels.) For all $j \notin N_c$, set

Dijkstra's algorithm

$$D_j = \min_{i \in N_c} [D_j, D_i + c_{ij}]. \tag{8.5}$$

If the minimum is the same as the previous D_j, leave the tentative connection of j to a successor unchanged; otherwise change it to connect j to the successor, i, for which the minimum is obtained.

- *Step 2.* (Finding next closest node.) Find $i \notin N_c$ for which

$$D_i = \min_{j \notin N_c} D_j \tag{8.6}$$

Set $N_c = N_c \cup \{i\}$ and permanently connect i to the appropriate successor node. If $N_c = N$ stop; the algorithm is complete. Otherwise return to Step 1.

Figure 8.8 illustrates Dijkstra's algorithm. We present it as another labeling algorithm, which assigns labels of the form (S,D) to nodes, with S the successor node and D estimated distance from the destination. We use -1 for S when a successor node has not been assigned. Initial labels are tentative, but they become permanent (and illustrated entirely in boldface type) when connections become permanent. Initially the destination node has label $(-1,0)$. Node i, $i \neq 1$, has initial label $(-1,\infty)$.

Part (b) shows the results of the first iteration. Distances of nodes adjacent to 1 are updated first; the value of D for 4 is minimum, so its connection is made permanent and 4 added to N_c. In the next iteration, distances are again updated. Only distances to nodes adjacent to 4 need to be considered, since distances to 1 were computed earlier. In each iteration, computation can be saved by realizing that only nodes adjacent to the newest member of N_c can require distance updates. The remaining iterations are similar. In each, distances of nodes not in N_c are updated by computing distances via the newest member of N_c and updating if this reduces distance; the node not in N_c with smallest distance is added to N_c and its connection made permanent. The final shortest-path spanning tree is identical to that in Fig. 8.6(e). This should be expected, despite the different search directions, since the network studied is bidirectional with symmetric distances.

A pseudo code version of Dijkstra's algorithm (adapted from [SYSL83]) is shown below. It uses a Boolean vector to distinguish permanently labeled nodes from temporarily labeled ones; when the ith node becomes permanently labeled, the ith element changes from *false* to *true*. A variable **new** identifies the newest member of N_c. The iteration is performed $N - 1$ times, each time adding one node aside from the destination to the permanently labeled set.

INITIALIZATION:
  for $n \leftarrow 1$ to N do
    begin
      $D(n) \leftarrow \infty$;
      final $(n) \leftarrow$ false;
      $S(n) \leftarrow -1$;
    end;
  $D(dest) \leftarrow 0$;
  final$(dest) \leftarrow$ true;
  $new \leftarrow dest$;

pseudo code for
Dijkstra's algorithm

ITERATION
  for $n \leftarrow 1$ to $N-1$ do
    begin
      for every immediate predecessor i of **new** if not final(i) do
        begin
        newdist $\leftarrow D(new) + c_{new,i}$;
        if *newdist* $< D(i)$ then do
          begin
          $D(i) \leftarrow$ *newdist*;
          $S(i) \leftarrow new$;
          end;
      end;
    find the node k with smallest temporary label which is $\neq \infty$
    final(k) $\leftarrow$ true;
    $new \leftarrow k$
  end.

Each step in Dijkstra's algorithm requires a number of computations proportional to the number of nodes, N, and the steps are repeated $N - 1$ times, so the number of computations in the worst case is $O(N^2)$ rather than $O(N^3)$ as in the Bellman-Ford-Moore algorithm. This can be a significant reduction in computations. On the other hand, the number of links, L, in many networks is much smaller than $N(N - 1)$, its worst-case value,[17] and the number, m, of iterations of the Bellman-Ford-Moore algorithm needed for these networks may be small, $m << N$, so $mL << N^3$. Its required number of computations can then be less than the $O(N^2)$ requirement of Dijkstra's algorithm.

computational
requirements

Dijkstra's algorithm is faster than the Bellman-Ford-Moore algorithm if a network is small and dense (with a high ratio of number of links to number of nodes). It is used in the current routing algorithm for the ARPANET. (See Chapter 9 for details.)

---

[17]This is the number of links in a completely connected network with directional links. For nondirectional links, the number is one half this.

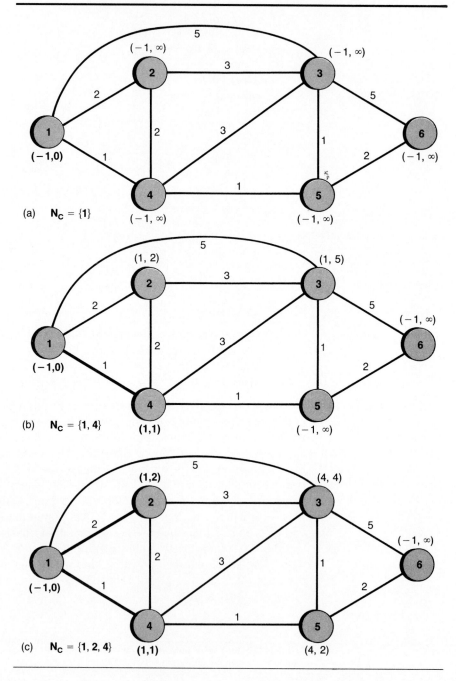

Figure 8.8 Dijkstra's Algorithm. (a) Initialization. (b) After first iteration. (c) After second iteration. (d) After third iteration. (e) After fourth iteration. (f) After fifth iteration, which completes algorithm. Boldface labels and heavy lines indicate permanent connections.

(a) $N_C = \{1\}$

(b) $N_C = \{1, 4\}$

(c) $N_C = \{1, 2, 4\}$

Figure 8.8   *continued*

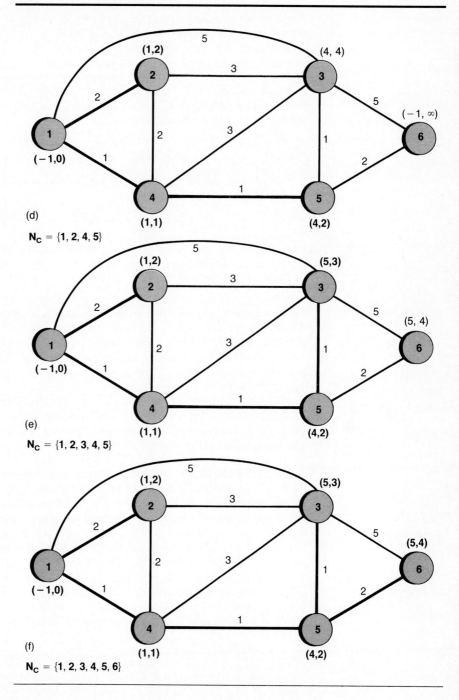

(d)

$N_C = \{1, 2, 4, 5\}$

(e)

$N_C = \{1, 2, 3, 4, 5\}$

(f)

$N_C = \{1, 2, 3, 4, 5, 6\}$

☐ 8.4.3   The Floyd-Warshall Algorithm

The *Floyd-Warshall algorithm* computes shortest paths between all pairs of nodes.[18] It *iterates on nodes allowed as intermediate nodes* in paths. It starts with each pair of nodes "connected" via a direct link between them (of infinite length if no direct link exists). It next calculates shortest paths under the constraint only node **1** can be used as an intermediate node, then with the constraint only **1** and **2** can be used, and so forth. More precisely, let $D_{ij}^{(n)}$ be the shortest path length from node i to node j with the constraint that only **1, 2, . . . , n** can be used as intermediate nodes on paths. The Floyd-Warshall algorithm is as follows:

Floyd-Warshall algorithm

Initially,

$$D_{ij}^{(0)} = c_{ij} \text{ for all } i, j \ i \neq j. \tag{8.7}$$

and all pairs of nodes are connected directly to each other. For $n = 0, 1, . . . , N - 1$, compute for all $i \neq j$

$$D_{ij}^{(n+1)} = \min \left[D_{ij}^{(n)}, D_{i(n+1)}^{(n)} + D_{(n+1)j}^{(n)} \right] \tag{8.8}$$

If the first term gives the minimum, leave the routing between i and j unchanged. Otherwise, alter it to be the concatenation of the previous routing from i to $n+1$ followed by the previous routing from $n+1$ to j. After $n = N - 1$, make all routes fixed as the final solution has been obtained.

The algorithm is illustrated in Fig. 8.9. Indicating shortest-path trees between all pairs of nodes would lead to a figure too complex to be meaningful, so the figure shows current distances between pairs of nodes with paths in parentheses. For example, (**145**) indicates that the path between **1** and **5** follows the route **1→4→5**. Since links are bidirectional, 15 pairs of nodes are considered.[19] Node pairs are listed with nodes in numerical order. Reverse direction routes may be obtained by reading nodes from last to first.

Figure 8.9 is unavoidably complex as it illustrates development of routes between 15 node pairs.[20] Typical computations are as follows. In part (a), $D_{16} = \infty$ but $D_{13} = 5$ and $D_{36} = 5$, so a route via **3** with distance 10 exists (and no other route via **1**, **2**, and **3** as intermediate nodes is shorter); hence in (b) $D_{16} = 10$ with routing **136**. In part (c), $D_{16} = 9$ but $D_{15} = 2$ and $D_{56} = 2$. This means that in (d) $D_{16} = 4$ with routing the concatenation of **145** and **56** routings in (c) or **1456**. Since routes are bidirectional, the **13** route in (d) is found from the **15** and **35** routes in (c) in a similar manner, with **153** the concatenation.

route matrix

A way to represent routes between all pairs of nodes is needed for computer implementation of the algorithm. A suitable representation is in the form of a *route matrix*, $\boldsymbol{R} = [R_{ij}]$, with R_{ij} the number of the next to last node along the shortest path from i to j; the last node has to be j. (An alternative form of this matrix is considered in Problem 8.12.) If this entry is **k**, the principle of optimality says the optimal path to j must be the optimal path from i to k concatenated with the link kj, so the second from last node along the optimal path is R_{ik}, and so forth. Diagonal elements of the matrix may be set to **0**, or any value prohibited for the rest of the entries, since they

---

[18] The Floyd-Warshall algorithm was originally published by Floyd [FLOY62], who based it on an algorithm for transitive closure of Boolean matrices published by Warshall [WARS62].

[19] The algorithm can also be applied with directional links.

[20] For the same size example with directional links, 30 pairs would have to be considered.

do not convey useful routing information. For the example of the Floyd-Warshall algorithm in Figure 8.9, R is

example of
route matrix

$$R = \begin{bmatrix} 0 & 1 & 5 & 1 & 4 & 5 \\ 2 & 0 & 2 & 2 & 4 & 5 \\ 4 & 3 & 0 & 5 & 3 & 5 \\ 4 & 4 & 5 & 0 & 4 & 5 \\ 4 & 4 & 5 & 5 & 0 & 5 \\ 4 & 4 & 5 & 5 & 6 & 0 \end{bmatrix}$$

The optimum route from **1** to **6** can be read as follows. The next to last node is $R_{16} = 5$; the node preceding **5** is $R_{15} = 4$; the node preceding this is $R_{14} = 1$, so the complete route is **1456**.

The pseudo code version of the Floyd-Warshall algorithm below (adapted from [SYSL83]) computes distances and routes, storing routes in the form of an R matrix and distances in the form of a D matrix.

INITIALIZATION
        for $i \leftarrow 1$ to N do
                for $j \leftarrow 1$ to N do
                        begin
                            $D_{ij} \leftarrow c_{ij}$;

pseudo code for Floyd-
Warshall algorithm

                            if $i = j$ then $R_{ij} \leftarrow 0$
                            else if $c_{ij} < \infty$ then $R_{ij} \leftarrow i$
                            else $R_{ij} \leftarrow \infty$;
                        end;

ITERATION
        for $k \leftarrow 1$ to N do
                for $i \leftarrow 1$ to N do
                        if $D_{ik} \neq \infty$ then do
                            begin
                                for $j \leftarrow 1$ to N do
                                    begin
                                      newdist $\leftarrow D_{ik} + D_{kj}$;
                                      if newdist $< D_{ij}$ then do
                                          begin
                                            $D_{ij} \leftarrow$ newdist;
                                          $R_{ij} \leftarrow R_{kj}$;
                                      end;
                              end;
                end.

During initialization, D is set equal to C and R is defined to give all one-step routes with finite distance by inserting i at position ij if c_{ij} is finite. During the iteration, Eq. (8.8) is used to compute distances, with the next to last node on the path between k and j inserted into the path between i and j when routing via k reduces distance.

As each of N steps in the algorithm must be executed for each pair of nodes, and the number of node pairs is $O(N^2)$, the computation involved is $O(N^3)$. This same value would be obtained if the Dijkstra algorithm were repeated for each source node to find the same set of routes.

The Floyd-Warshall algorithm is used in some networks with centralized routing. An example is the TYMNET network discussed in the next chapter.

Figure 8.9   Illustration of Floyd-Warshall Algorithm. (a) Initialization, no intermediate nodes allowed, unchanged for **1** and **2** allowed as intermediate nodes. (b) **1, 2,** and **3** allowed as intermediate nodes. (c) **1, 2, 3,** and **4** allowed. (d) **1, 2, 3, 4,** and **5** allowed as intermediate nodes. No further changes with **6** also allowed, so this is final solution.

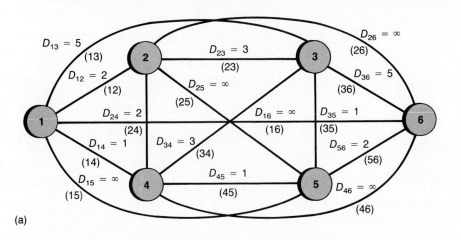

(a)

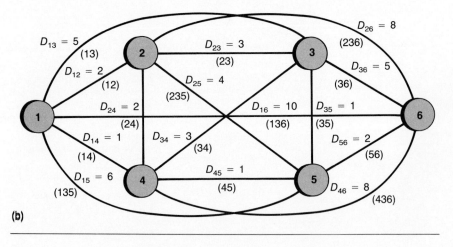

(b)

8.4.4   Distributed Asynchronous Bellman-Ford-Moore Algorithm

One of the main reasons for popularity of the Bellman-Ford-Moore algorithm is the fact it can be implemented in a distributed manner, each node performing calculations independently and exchanging results with other nodes. When implemented in this way, it is convenient to consider it computing the

Figure 8.9   *continued*

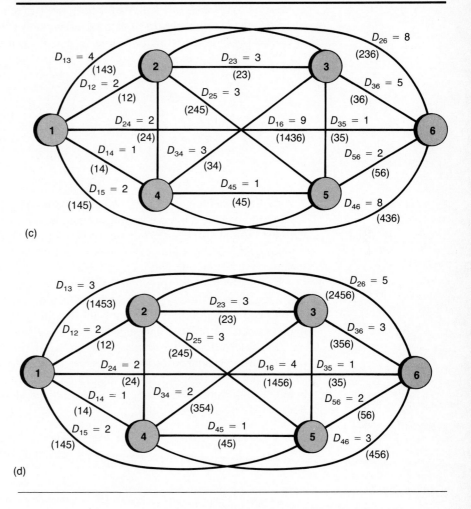

(c)

(d)

shortest path from an arbitrary node, say i, to a specific destination node, labeled 1 in our discussion,[21] rather than the shortest path from 1 to i. We hence rewrite Eq. (8.2) with minimization over initial hop distance plus remaining distance to the destination rather than final hop distance plus distance to the start of the final hop. If we let D_i be the distance from i to the destination (node 1), the appropriate form of Bellman's equation is

modified Bellman's equation

$$D_i = \min_{j \in N(i)} [c_{ij} + D_j], \ i \neq 1$$

$$D_1 = 0$$

(8.9)

---

[21] In practice a separate computation is necessary for each destination node.

with $N(i)$ denoting the current set of neighbors of i, that is, nodes directly connected to i. Each node can readily keep track of distances to its neighbors.

One way of implementing a distributed version of the algorithm is to have each node, i, independently and simultaneously compute

$$D_i^{(h+1)} = \min_{j \in N(i)} [c_{ij} + D_j^{(h)}], \; i \neq 1$$

$$D_1^{(h+1)} = 0$$

(8.10)

with initial conditions

$$D_i^{(0)} = \infty, \; i \neq 1$$

$$D_1^{(0)} = 0.$$

(8.11)

Node i then exchanges results with its neighbors (to update their D_j estimates) and executes the computation again with h incremented by one. With the given initial conditions, the algorithm will terminate after at most $N - 1$ iterations, if N is the number of nodes; after this each node, i, will know both the shortest distance D_i and the outgoing link on the shortest path to node $\mathbf{1}$.

Two problems have prevented this precise approach from being implemented. One is getting all nodes to start the algorithm at the same time. The second is devising a mechanism for aborting the algorithm and restarting it if a link status or length changes while the algorithm is running.

A simpler alternative, implemented in a number of networks, does not require either synchronization or starting with the initial conditions in Eq. (8.11). This eliminates a need for algorithm initiation and restart protocols. It simply operates indefinitely by following the distributed asynchronous Bellman-Ford-Moore algorithm:

At reasonably regular intervals at each node $i \neq \mathbf{1}$ compute,

distributed asynchronous Bellman-Ford-Moore algorithm

$$D_{i\,new} = \min_{j \in N(i)} [c_{ij} + D_{j\,rec}],$$

(8.12)

using the latest distance estimates $D_{j\,rec}$ received from neighbors $\mathbf{j} \in N(\mathbf{i})$ and the latest status and lengths of outgoing links from i to its neighbors. Also transmit the $D_{i\,new}$ estimates to the neighbors at reasonably regular intervals.

There is no need to synchronize either computations or transmission of estimates to neighboring nodes. Furthermore, arbitrary nonnegative initial values of distance estimates are allowed at each node. This means no modifications in estimates are necessary to obtain new initial values for the algorithm after network changes are detected since the estimates previously in use suffice.

The algorithm is still valid under these circumstances. It converges to correct distances in a finite time if no changes occur during this time, regardless of initial conditions. Hence, if changes are infrequent, the algorithm will converge after each change. Even if several changes occur in a short interval, the algorithm will converge if the time before the next change is long enough. A proof of convergence is given in [BERT87].[22]

This distributed version of the Bellman-Ford-Moore algorithm has been used in a variety of networks, with different cost functions and other varying details. It was the original routing algorithm in the ARPANET and is used by the Canadian DATAPAC network and in Digital Equipment Corporations DNA, among others. Each of these networks is discussed in Chapter 9. An example of routing calculations is given in Subsection 9.5.1. Problem 8.14 also involves some typical calculations.

For datagram transmission using this algorithm, successive packets in a message may travel by different routes, possibly arriving out of order. Long delays are possible for packets routed during periods when the algorithm is adapting to changes. This is acceptable for datagrams, but the algorithm is not well adapted for virtual circuit routing. Virtual circuit routes set up during convergence may be far from optimum. During adaptation to changes, in fact, the algorithm is prone to looping; this is intolerable for virtual circuit routing.

8.5   Stability Problems in Shortest-Path Routing

Shortest-path routing algorithms can have significant stability problems if distance is affected by traffic on communications links. A simple example is based on Fig. 8.10, which shows nodes **1** and **2** connected by identical channels, C_1 and C_2. Traffic is initially routed between **1** and **2** via C_1, but shortest-path routing based on expected delay is used. We assume that the routing algorithm bases delay estimates on packets currently in the network, without including the effect of routing changes. At a given time, the expected delay is proportional to the number of packets queued for transmission over the channel, since each must be transmitted before a newly arriving packet. Hence, the routing algorithm will, at its next decision instant, decide to route all traffic over C_2. This will in turn cause C_1 to be selected at the next decision instant, and so forth. Routing will oscillate back and forth between C_1 and C_2.

A worse type of oscillation is possible, as is shown by the example in Fig. 8.11 (a simplified version of an example in [BERT87]), a ring network with eight nodes. Node **8** is the destination node for all traffic, with **1–3** and **5–7** each presenting 1 unit of traffic to the network and **4** presenting $\varepsilon \ll 1$ unit. The measure for length of a link is simply traffic over the link. We assume that each node computes its shortest path to the destination every T seconds and adjusts its routing accordingly.

---

[22] More precisely, convergence is proved under the further assumptions that nodes never stop updating their estimates or receiving messages from all their neighbors, that all initial distance estimates are nonnegative, and that old distance estimates are eventually purged (that is, destroyed if not received by neighbor nodes within a finite time interval after they are generated).

Figure 8.10   Simple Example of Oscillatory Routing.

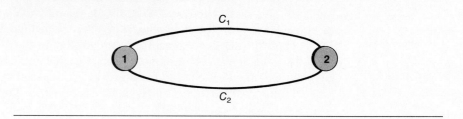

The initial routing routes traffic from **1–3** clockwise and traffic from **4–7** counterclockwise. This is a very good routing, allowing traffic to be as nearly balanced in the two directions as is possible without splitting traffic from one source between two routes. The resulting traffic on links is illustrated in Fig. 8.12 (a). For simplicity, only nonzero values are shown; all other traffic values (including traffic in the reverse direction on labeled links) are zero. Traffic values are identical to lengths of links with the distance measure used.

After T seconds, the nodes compute distances to the destination in both clockwise and counterclockwise directions and adjust routings to give the pattern shown in part (b). The two changes result from the fact that **4** computes the length of the clockwise path as $0 + 1 + 2 + 3 = 6$ and the length of the counterclockwise path as $\varepsilon + 1 + \varepsilon + 2 + \varepsilon + 3 + \varepsilon = 6 + 4\varepsilon$ and

Figure 8.11   Eight-Node Ring Network for Illustration of Shortest-Path Routing Instability.

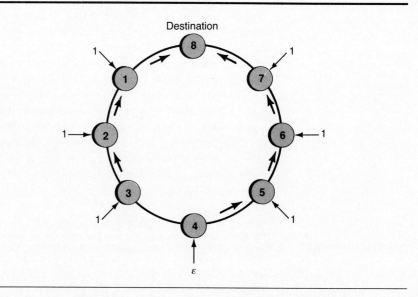

Figure 8.12   Routing Changes for Network in Fig. 8.11. Arrows indicate routing directions, with numbers representing total traffic on links. (a) Initial routing. (b) First rerouting. (c) Second rerouting. (d) Third rerouting. Routing will then alternate between patterns in (c) and (d).

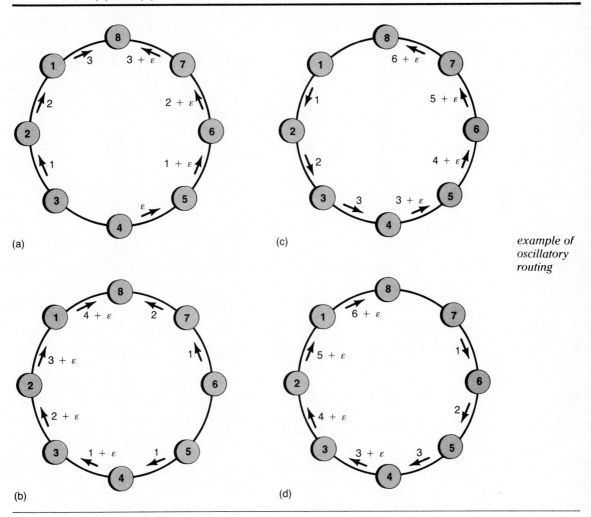

(a)

(b)

(c)

(d)

example of oscillatory routing

switches to the clockwise path, and **5** computes the lengths as 6 and $6 + 3\varepsilon$, respectively, and also switches.

With the second routing, **1** through **5** all compute counterclockwise distance as 3, with clockwise distances ranging from $4 + \varepsilon$ to $11 + 4\varepsilon$. Hence, all reverse their routes for the third routing and all nodes now transmit counterclockwise. The clockwise distance then becomes zero for all nodes, with all counterclockwise distances positive, so all transmit clockwise on the

Figure 8.13   Effect of Staggered Rerouting Times on Stability of Shortest-Path Routing. Horizontal shading indicates nodes transmitting clockwise and vertical shading nodes transmitting counterclockwise; heavy borders indicate nodes eligible to change routing. (a) Initial routing. (b) First rerouting. (c) Second rerouting. (d) Third rerouting. (e) Fourth rerouting. (f) Fifth rerouting. Additional reroutings similar.

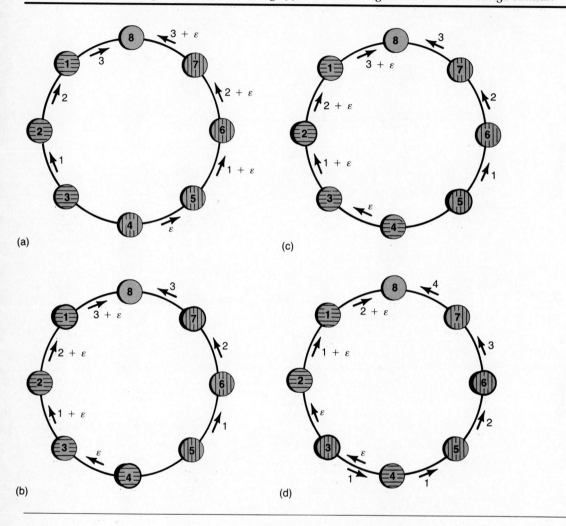

fourth routing. Routing alternates between the latter two patterns indefinitely. This is about the worst type of routing (without looping) that can be visualized.

Oscillatory behavior results from the way in which link arrival rates depend on routing, with routing in turn dependent on arrival rates. This produces a positive feedback mechanism leading to oscillations; it can be ana-

Figure 8.13   *continued*

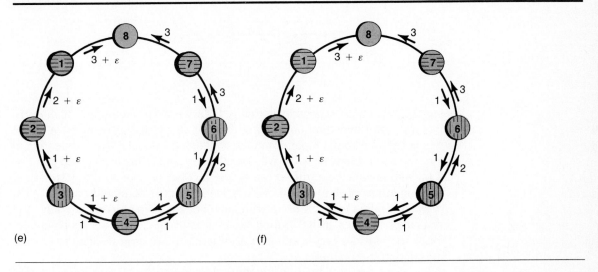

(e)   (f)

lyzed by techniques for studying stability in feedback control theory
[BERT82b]. This type of instability can occur whenever distance increases
monotonically with increasing traffic and is zero for zero traffic. It can be
damped by adding a positive constant, known as a *bias factor,* to each link
length (see Problem 8.15); this type of bias factor was used in the original
ARPANET routing algorithm, which would have shown this type of instability
otherwise. Despite the bias factor, the original ARPANET algorithm occasionally failed to operate satisfactorily, so the entire algorithm was eventually
replaced with one based on Dijkstra's algorithm. Both algorithms are discussed in Chapter 9.

   Another way of stabilizing routing is by staggering times at which nodes
change routings. This occurs in networks using virtual circuit routing, since
each routing remains fixed for the duration of the corresponding virtual circuit and virtual circuits end at different times. A simple example of routing
changes resembling those in virtual circuits is given in Fig. 8.13. This is the
same as Fig. 8.12, except for the fact that each node can change its route only
once every $3T$ seconds, with times staggered so that **1**, **4**, and **7**, then **2** and
5, and finally **3** and **6** change their routings together.[23] The first five reroutings
are shown; the pattern thereafter is similar. Note the decrease in variability
of successive routing patterns.

   Some common routing algorithms, such as the Bellman-Ford-Moore algorithm, also demonstrate unexpected, and possibly undesirable, behavior dur-

---

[23] Although the variability of the routing is decreased in this manner, proving that routing
never becomes unstable with a systematic pattern of virtual circuit durations such as this
is complex.

Figure 8.14   Linear Geometry to Illustrate Update Problems.

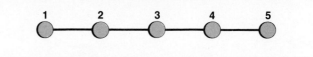

ing their reaction to geometry changes. As we have mentioned, temporary looping is not uncommon during periods of adaptation. A demonstration of this is complex, but simple examples of geometry update give indications of the way such things happen. We consider the simple linear geometry in Fig. 8.14, with distance measure given by the number of hops to the destination. Routing tables of the form in Table 8.3 (listing next node and cost for each possible destination) are used, with each node sending a copy of (only) the distance estimates in its routing table to its neighbors after each update. Since we consider only routing to destination **1**, we omit designation of the destination from tables.

For our first example of updating, we assume that **1** has been disconnected (link **1-2** down) long enough for routing tables to have stabilized. It is reconnected at time 0. We write the entries in routing tables for destination **1** in form "$d, \boldsymbol{X}$," with "d" representing distance and "$\boldsymbol{X}$" next node. ("-" indicates a "don't care" entry). These entries are updated as in Table 8.5.

Thus the tables stabilize to correct new entries after four exchanges. Next, assume **1-2** has been up long enough for table entires to stabilize and then suddenly goes down. Table 8.6 illustrates routing table updates.

The first entry in the second row is found by recalling that **3** has informed **2** it has a distance 2 route to **1**.[24] After link **1-2** goes down, **2** recalls this and decides to route via **3**, with distance found by adding 1, its distance to **3**, to **3**'s value. Other entries are obtained similarly, assuming each node exchanges distance estimates with neighbors between successive reroutings.

Routing tables adapt much more slowly to bad news than they do to good news. This anomaly has motivated a variety of "fixes" in networks using this

Table 8.5   Routing Table Updates When Node **1**
                  Comes Up After Being Down.

		Node		
2	3	4	5	
∞,-	∞,-	∞,-	∞,-	Initially
1,1	∞,-	∞,-	∞,-	After 1 exchange
1,1	2,2	∞,-	∞,-	After 2 exchanges
1,1	2,2	3,3	∞,-	After 3 exchanges
1,1	2,2	3,3	4,4	After 4 exchanges

Table 8.6   Routing Table Updates When Node **1**
Goes Down After Being Up.

		Node		
2	**3**	**4**	**5**	
1,1	2,2	3,3	4,4	Initially
3,3	2,2	3,3	4,4	After 1 exchange
3,3	4,2	3,3	4,4	After 2 exchanges
5,3	4,2	5,3	4,4	After 3 exchanges
5,3	6,2	5,3	6,4	After 4 exchanges
7,3	6,2	7,3	6,4	After 5 exchanges
. . .	. . .	. . .	. . .	
∞,-	∞,-	∞,-	∞,-	After ∞ exchanges

type of algorithm. One technique almost always employed is to declare a
node unreachable as soon as the estimated distance to the node exceeds a
finite threshold, normally somewhat in excess of the maximum expected distance for reaching any node in the network. The time for estimates to reach
this threshold can still be excessive, however.

hold down

A variety of other techniques to improve convergence have been proposed, most based on heuristics. For example, the original ARPANET routing
algorithm used a technique called *"hold down,"* in which normal updating of
routing tables was altered during a period (called hold-down time) after deterioration of a route that had previously been best. During the hold-down time,
distance estimates sent to neighbors were distances to reach the destination
via the route that was previously best, so the distance estimate **2** would send
to **3** immediately after the **1-2** link went down would be ∞ instead of 3. Then
3 would use hold down itself and send this ∞ estimate to **4**, and so forth. For
the example given, hold down would result in quick convergence to the correct routing tables; it also worked in many, but not all, other situations where
the original algorithm did not converge quickly enough.[25]

□ 8.6   Optimal Routing

It is possible to improve performance of routing algorithms, especially under heavy
load conditions, by allowing them to split traffic among multiple paths between a
given source and destination. Routing algorithms for splitting traffic among paths in

---

[24] We assume that nodes only exchange distance estimates, so **3** has not told **2** that its route
is via **2**!

[25] Other approaches to handling the anomaly have also been developed; some are discussed
in [SCHW87], [MCQU74], [MCQU78a], and [HAGO83].

Figure 8.15   Network Used for Example of Notation for Optimal Routing.

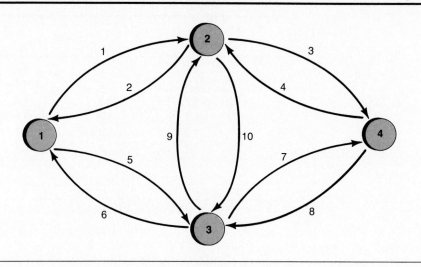

bifurcated routing

this manner to minimize delay are more complex than shortest-path routing algorithms. Routing with traffic split among paths is called *bifurcated routing*. We give an introduction to such techniques but a full treatment is beyond the level of this text.

The notation required is complex; we use Fig. 8.15 to illustrate it. The figure shows four nodes and 10 directed links. Each link is numbered.

Assume two simultaneous data flows, from **1** to **4** and from **2** to **3**. (Normally each would be accompanied by a flow in the opposite direction that would be treated in the same manner but would make the example more complex). We label the **1** to **4** flow by index 1 and the **2** to **3** flow by 2, and assume the rates are r_1 and r_2 (bps), respectively. We consider these to be rates of flow of two commodities to be routed through the network.

There are four paths between **1** and **4**, $p_1 = (1,3)$, $p_2 = (5,7)$, $p_3 = (1,10,7)$, and $p_4 = (5,9,3)$, and three paths between **2** and **3**, $p_5 = (10)$, $p_6 = (2,5)$, and $p_7 = (3,8)$. P_k represents the set of paths available for commodity k, so $P_1 = (p_1, p_2, p_3, p_4)$ and $P_2 = (p_5, p_6, p_7)$. Furthermore, f_{pk} represents rate of flow on path p_k and λ_i the rate of flow on link i (both in bits per second).[26] We must have $f_{p1} + f_{p2} + f_{p3} + f_{p4} = r_1$ and $f_{p5} + f_{p6} + f_{p7} = r_2$ to handle the two commodity flows. Furthermore, $\lambda_1 = f_{p1} + f_{p3}$, $\lambda_2 = f_{p6}$, $\lambda_3 = f_{p1} + f_{p4} + f_{p7}$, and so forth. In general,

$$\sum_{p_i \in P_k} f_{p_i} = r_k ,\qquad(8.13)$$

and

$$\lambda_k = \sum_{\substack{\text{all paths } p_i}} f_{p_i} ,\qquad(8.14)$$

_{containing link k}

---

[26] In order to minimize complexity of notation, we define λ_i in bits per second rather than in packets per second as was the case for related analyses in Chapter 5. The quantity λ_i here corresponds to $\lambda \bar{X}$ in Eq. (5.2).

Figure 8.16 Typical Term in the Summation in Eq. (8.17).

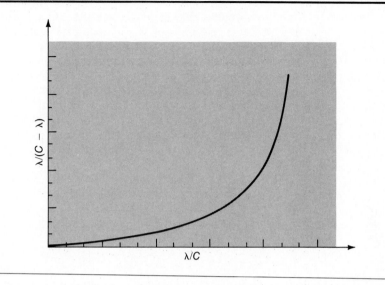

so these quantities are computable from information about paths and rates of flows, given a routing that prescribes how flows are divided among paths.

The goal of optimal routing is to choose flows, f_{p_i}, subject to the constraint in Eq. (8.13), plus a constraint that all flows are nonnegative, in order to satisfy an optimality criterion. The most common criterion is *minimizing average delay*, given (under assumptions of Poisson arrivals and exponential message lengths) by [KLEI75a]

average delay formula

$$T = \frac{1}{\gamma} \sum_{i=1}^{N} \frac{\lambda_i}{C_i - \lambda_i},$$
(8.15)

with N representing the number of links in the network, C_i the capacity of link i, and γ the total external arrival rate to the network in bits per second.[27]

The minimization is simplified by the nature of the quantities involved. As the plot of a typical term, $\lambda/(C - \lambda)$, in Fig. 8.16 indicates, the terms increase monotonically with λ, with their rate of increase also increasing. That is, both first and second derivatives of $\lambda/(C - \lambda)$ are positive.

Some mathematical terms relevant to optimum routing deserve mention. The properties given are equivalent to the mathematical definition of a convex function. Furthermore, if $F_0 = \{f_{p10}, f_{p20}, \ldots\}$ and $F_1 = \{f_{p11}, f_{p21}, \ldots\}$ are any two assignments of flows satisfying the constraints of the problem (nonnegative flows plus Eqs. (8.13) and (8.14)), then $\alpha F_0 + (1 - \alpha)F_1$ also satisfies the constraints for $0 \le \alpha \le 1$. This means that the space of possible solutions satisfies the mathematical definition of a convex space. The problem of finding the optimum routing is thus equivalent to minimizing a convex function over a convex space. A well-known mathematical result is that this

---

[27] Although we treat only the distance measure in Eq. (8.17) here, the solution techniques we give are applicable with other distance measures with appropriate changes in details.

implies any local minimum is a global minimum, greatly simplifying the solution process.

A wide variety of iterative solution techniques are feasible. It is adequate to find a feasible set of flows as a starting point, then iteratively alter flows in a manner ensuring that T decreases for successive iterations. It is even possible to ignore the constraint $\lambda_i \leq C_i$ for all i, since the steep slope of the function in Fig. 8.16 near $\lambda = C$ guarantees any reasonable minimization technique will quickly move away from a routing tending to saturate links, as long as a small step size for adjusting flows is used. Several steepest descent algorithms have been developed. Most are based on the partial derivatives, $\partial T/\partial f_{p_j}$ evaluated for the paths, p_j.

Since

$$\frac{\partial T}{\partial \lambda_k} = \frac{1}{\gamma} \frac{C_k}{(C_k - \lambda_k)^2},$$ (8.16)

and

$$\frac{\partial \lambda_k}{\partial f_{p_j}} = \begin{cases} 1 & \text{if link } k \in \text{ path } p_j, \\ 0 & \text{otherwise} \end{cases}$$ (8.17)

the standard formula for differentiating a function of a function yields

$$\frac{\partial T}{\partial f_{p_j}} = \frac{1}{\gamma} \sum_{\substack{\text{all links } k \\ \text{in path } pj}} \frac{C_k}{(C_k - \lambda_k)^2}.$$ (8.18)

For any given set of flows, $\boldsymbol{F} = \{ f_{p_1}, f_{p_2}, \ldots \}$, the partial derivatives can be evaluated by computing λ_i values from Eq. (8.14) and substituting the results into Eq. (8.18).

Consider the flow of the kth commodity. If $p_i \in \boldsymbol{P}_k$, and $\partial T/\partial f_{p_i}$ is smaller than $\partial T/\partial f_{p_j}$ for any other $p_j \in \boldsymbol{P}_k$, deviating a small increment Δf of flow from path p_j to p_i will increase the contribution of p_i to T by approximately $(\partial T/\partial f_{p_i})\Delta f$ and decrease the contribution of p_j by approximately $(\partial T/f_{p_j})\Delta f$. Our assumption that $\partial T/\partial f_{p_i}$ is the smallest such value, however, indicates the increase is smaller in magnitude than the decrease, so there is a net decrease. Similar calculations can be used to reroute other link flows or commodities. A variety of algorithms differing in details such as selecting step sizes and convergence criteria have been proposed [FRAT77], [GALL77], [BERT87].

An optimum solution is reached when, for each commodity k, the $\partial T/\partial f_{p_i}$ values are equal at the flows being used for all $p_i \in \boldsymbol{P}$ with nonzero flow, and each is less than or equal to $\partial T/\partial f_{p_j}$ for each p_j with zero flow.

A simple example of the calculations needed for routing algorithms of this type is based on the network configuration in Fig. 8.17. The network consists of two nodes, **1** and **2**, with r bps input at **1**, plus links, 1 and 2, with capacities C_1 and C_2, respectively, directed from **1** to **2**.

Assume $r \leq C_1 + C_2$, so it is possible to route the entire input. Also assume an initial routing of αr bps via link 1 and $(1 - \alpha)r$ bps via link 2, with $0 \leq \alpha \leq 1$ and $\alpha r < C_1$, $(1 - \alpha)r < C_2$ (so the initial routing is feasible). We define two paths, $p_1 = (1)$ and $p_2 = (2)$. Initially $f_{p_1} = \alpha r$, $f_{p_2} = (1 - \alpha)r$, and the total input rate is $\gamma = r$. Applying Eq. (8.18) to find derivatives at initial flows gives

Figure 8.17 Simple Network Used for Sample Optimal
Routing Calculations.

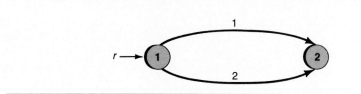

$$\frac{\partial T}{\partial f_{p1}} \bigg|_{\substack{f_{p1} = \alpha r, \\ f_{p2} = (1-\alpha)r}} = \frac{1}{r} \frac{C_1}{[C_1 - \alpha r]^2},$$ (8.19a)

and

$$\frac{\partial T}{\partial f_{p2}} \bigg|_{\substack{f_{p1} = \alpha r, \\ f_{p2} = (1-\alpha)r}} = \frac{1}{r} \frac{C_2}{[C_2 - (1-\alpha)r]^2}.$$ (8.19b)

If the first term is smaller than the second, deviating a small amount of flow from f_{p2} to f_{p1} will reduce T unless $f_{p2} = 0$ and $f_{p1} = r$. The partial derivatives can be recomputed at the new flow values and the process repeated until flows converge to optimum values.

For large networks, an iterative algorithm such as one in the references cited would be used, but it is possible to solve directly for optimum flows in this extremely simple network. Assume $C_1 \geq C_2$. In this case, the optimum solution must have $\alpha \geq 1/2$—that is, it makes no sense to send more traffic on the slower link than on the faster link. There are two cases to consider, one where the optimum solution sends all traffic over p_1 and the other where the optimum solution divides traffic between the links. For the first case, Eq. (8.19a) evaluated at $\alpha = 1$ must be $\leq$ Eq. (8.19b) at $\alpha = 1$, or (after some algebra)

$$(C_1 - r)^2 - C_1 C_2 \geq 0.$$ (8.20)

This is plotted versus r, over the range from $r = 0$ to $r = C_1 + C_2$ in Fig. 8.18. The inequality is valid as long as $r \leq C_1 - \sqrt{C_1 C_2}$.

For r values exceeding this, traffic is routed over both links. The value of α can then be found by setting the right-hand sides of Eqs. (8.19a) and (8.19b) equal. After some algebra (see Problem 8.16), this yields for the rates on the two paths,

$$\alpha r = \frac{\sqrt{C_1} [r - (C_2 - \sqrt{C_1 C_2})]}{\sqrt{C_1} + \sqrt{C_2}},$$ (8.21a)

$$(1 - \alpha)r = \frac{\sqrt{C_2} [r - (C_1 - \sqrt{C_1 C_2})]}{\sqrt{C_1} + \sqrt{C_2}}.$$ (8.21b)

Typical plots of αr and $(1 - \alpha)r$ are given in Fig. 8.19.

Optimal routings found via iterative algorithms in more complex networks are similar. Low capacity paths tend to be used only at higher loads.

Figure 8.18   Plot of Function in Inequality (8.20).

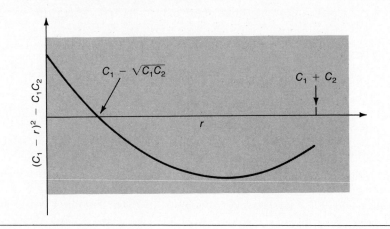

Figure 8.19   Rates on Links in Network in Fig. 8.17 with
Optimal Bifurcated Routing, Bertsekas/Gallagher, *Data
Networks,* © 1987, p. 380. Reprinted by permission of
Prentice-Hall, Inc., Englewood Cliffs, NJ.

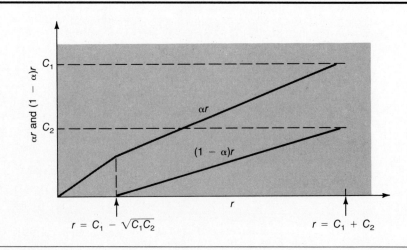

8.7   Other Routing Algorithms

A variety of other routing strategies have been proposed, with some implemented in networks. We briefly discuss the following types of routing: source routing, random routing, flooding, and backward learning.

source routing

Source routing is determined by the source of a packet and implemented by including information describing the route in the header of each packet. Nodes along the route examine the header, stripping off the portion pertaining to them (in some implementations)[28] and using the information to determine routing. No routing tables are needed. A variety of algorithms for choosing information for the header may be used, ranging from fixed routing through use of algorithms such as the Dijkstra algorithm.

random routing

Although *random routing* has received considerable attention in the literature, few if any networks have actually implemented it. In the purest form each node randomly chooses the outgoing link on which it sends or relays a packet destined for another node, choosing each outgoing link with equal probability. Modifications allow variations in probabilities of links, such as giving higher probability to links heading in the general direction of the destination or tending to choose the outgoing link with shortest transmission queue size. This version is sometimes called the *"hot potato" algorithm,* an analogy based on the natural behavior of someone tossed a hot potato—getting rid of it as quickly as possible. None of the variants approach the efficiency of shortest-path algorithms, however. Their virtues are that they require essentially no knowledge of network topology at nodes and they are robust, eventually getting a packet to its destination if there is a path available. However, the time required to get it there can be very long and inefficient routes can generate substantial amounts of surplus traffic.

flooding

Flooding is similar to random routing in requiring essentially no knowledge of network topology, and can generate even more traffic. In flooding, each node that receives a packet destined for another node sends copies out on each outgoing link except (in most implementations) the link on which it was received. Under low traffic conditions, flooding is the most robust routing algorithm known, with highest probability of finding a route if one exists, since multiple copies of packets explore all possible routes to the destination. It has been implemented in networks where a robust routing algorithm is of prime concern; this includes some military networks with extreme survivability requirements. Flooding is also a common technique for disseminating routing information and for broadcasting information to all nodes. If input traffic for such applications is low, the extra packets have little impact.

---

[28] In other implementations the header is left undisturbed, which avoids the need to recompute error-detecting codes protecting the header, but this requires each node to scan the entire address portion of the header to locate its address rather than simply looking at the first part of the header.

Figure 8.20 Packets Generated by Flooding Algorithm Using Maximum
Lifetime Field Described by Hop Count of Three. Source is **1**
and destination is **6**. Shadings in squares depicting packets
identify original "parent" packet. (a) Original packets.
(b) Packets generated after first set received. (c) Packets
generated after second set received (adapted from [STAL88b],
W. Stallings, *Data and Computer Communications, 2/e,*
© 1988. Reprinted by permission of MacMillan Publishing
Company, New York, NY.).

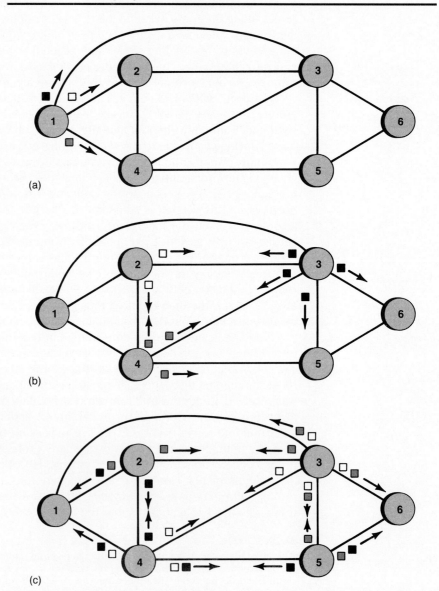

In flooding implementations, a technique for *limiting lifetime* of packets is needed since one packet entering the network would generate an infinite number of copies otherwise. Two approaches have been used. In one, each packet contains a field prescribing its maximum lifetime (in hops, seconds, or other appropriate measure), and this field is decremented at each node, with the packet discarded when the field reaches zero. In the other approach, each node maintains a list of packets it has seen, using a packet identifier, and discards packets it has seen before. Figure 8.20 illustrates traffic using a maximum lifetime approach; hop count defines lifetime and its maximum is three.

Node **1** initially creates three copies of a packet, sending one to each neighbor. Each relays the packet to each of its neighbors aside from the one it was received from, and decrements the hop count to two. This creates nine more packets. One reaches the destination, which keeps it; the other eight are flooded out (with hop count decremented to one), generating 22 more packets. Of these, four more reach the destination. All 22 are discarded by nodes receiving them, after they decrement the hop count to zero. In all, 34 packets are generated. This compares with two copies (or transmissions of the original packet) for shortest-path routing, with distance equal to hop count and path **1→3→6**.

If the second approach of checking packet identification and discarding packets seen previously were used, the first two iterations would be the same, but only **5** would relay on the third iteration, relaying one to four copies—depending on whether it independently relays copies from **3** and **4** (two copies of each) or recognizes it has received copies from both and relays one copy on the link on which no copy has been received. This reduces the number of copies from 34 to a number in the range 13 to 16. Implementation is more complex, however; each node must keep a list of identifications of all packets received within at least the maximum lifetime of packets and compare each received packet with this list.

backward learning

Backward learning is based on symmetric network conditions that ensure a good route in one direction is also a good route in the opposite direction and use of some type of randomized routing (such as flooding) to initially find routes. Refer again to Fig. 8.20. Since **6** first received a flooded packet from **1** via **3**, it may conclude the best way to send a packet to **1** is via **3**. Node **3**, in turn, first received the flooded packet from **1** via the direct connection, so it relays **6**'s packet directly to **1**.

Backward learning has merit in cases with symmetric traffic and reasonably regular transmission of traffic, such as flooded broadcast messages, allowing needed information to be obtained. In some cases, though, it can lead to terrible routing. An example is a military network subject to jamming. Jamming of a node may disrupt its reception of data without impairing its ability to transmit. Backward learning in such a network could easily direct traffic into a "black hole" by sending it directly into a jammed link.

8.8   Introduction to Flow Control

One motivation for *flow control* is familiar to anyone who has observed rush hour traffic in a major metropolitan area. During rush hours, traffic slows to a crawl, with vehicles spending as much time stopped and waiting for traffic to clear as they do moving. The maximum traffic handling capacity of highways occurs at heavy but not peak loads, when they are reasonably full but traffic is moving at close to the speed limit. In theory virtually all motorists might benefit from a technique to limit traffic admitted to highways to values no greater than those giving maximum traffic handling capability. Even those denied immediate entrance might travel enough faster once admitted to make up for lost time. This is not feasible for public highways, however.

Computer networks are even more prone to congestion, since error control and data link control algorithms involve retransmission if no verification of receipt is received within a timeout period. Thus delays cause extra traffic! Figure 8.21 illustrates throughput of a network without flow control. Initially the curve increases linearly, as throughput equals offered load at light loads. Eventually, though, the network becomes congested so throughput increases less rapidly than offered load. It often declines above a certain load, even *deadlock*   dropping to zero under extreme loads. This is known as *deadlock*.

The impact of reasonably good flow control is indicated by Fig. 8.22. The curve labeled "Without Flow Control" is identical to that in Fig. 8.21. A curve labeled "Ideal" in which throughput is equal to arrival rate until the system saturates, then levels off at saturation throughput, is included. This is not achievable but is a case with which performance of flow control algorithms

Figure 8.21   Throughput of Typical Network Without Flow Control.

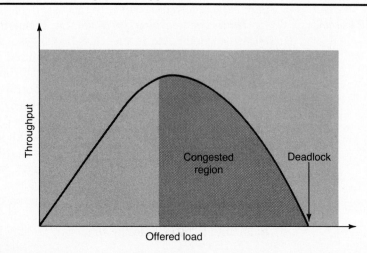

Figure 8.22 Impact of Flow Control on Throughput.

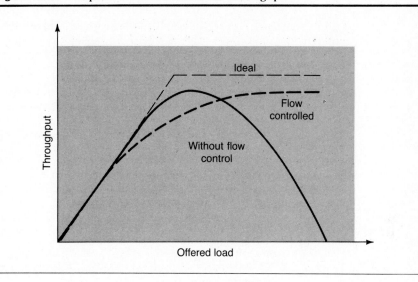

can be compared. The third curve shows the impact of flow control. At light loads in all three cases throughput equals arrival rate. At intermediate loads, rejection of traffic by flow control can decrease throughput below that without flow control, but throughput at heavy loads may be far better.

8.8.1 Functions of Flow Control

The main *functions of flow control* [GERL82] are as follows:

- Prevention of throughput and response time degradation due to network and user overload
- Deadlock avoidance
- Fair allocation of resources among users
- Speed matching between the network and its users

Throughput degradation and deadlocks are illustrated in Figs. 8.21 and 8.22 and are discussed throughout the rest of this chapter.

Fair allocation of resources implies rejection of traffic treats all users equitably. This can conflict with maximizing total throughput. Figure 8.23 indicates $n + 1$ users of an n link network, n requiring the use of one link each and the other requiring use of all n links. Each user has desired data rate of 1 data unit per second and each link has capacity 1. If the n link user is shut out entirely, the others can be fully accommodated giving total throughput of n, but if all users are treated the same, the maximum rate per

Figure 8.23   Example of Conflict Between Fairness and Throughput
Maximization (from [BERT87], Bertsekas/Gallager, *Data
Networks,* © 1987, p. 425. Reprinted by permission of
Prentice-Hall, Inc., Englewood Cliffs, NJ.).

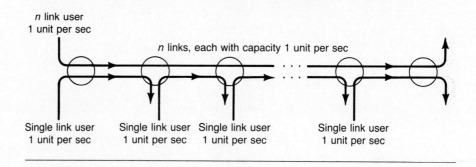

user becomes $1/2$ and total throughput is $(n + 1)/2$. For n large this is about
half the previous value.[29]

Speed matching prevents either network or users from swamping the
other. A simple example is a slow keyboard/printer accessed by a 56 kbps
data link. Unless flow control is used, the data link can completely swamp
the keyboard/printer. Speed mismatches of this general type are common.

Another term used for what we call flow control is *congestion control.*
Some authors make a distinction between the terms, for example, defining
flow control as limiting traffic to prevent congestion and congestion control
as dealing with congestion once it appears, or defining flow control to operate
at certain protocol layers and congestion control to operate at other layers.
Much of the literature treats the terms as synonymous, though, and we follow
this practice.[30]

8.8.2   Levels for Exercising Flow Control

Flow control is commonly exercised at several levels in a network to avoid
congestion of different portions of the network. Common levels are indicated
in Fig. 8.24. The precise subdivision of levels in the figure is not applicable to
all networks due to variability in architectures.

Transport level, or *end-to-end,* flow control is used to protect the destination. Often this involves protecting buffers for a user process from overflow

---

[29] This illustrates some questions about the way fairness should be defined. For example, is it
fair to require n users to sacrifice half the throughput they would be able to obtain otherwise
in order to give a single user throughput equal to theirs? No general answer seems to be
feasible.

[30] Another term occasionally used for the same type of functions is traffic control.

Figure 8.24 Levels for Exercising Flow Control (adapted from [GERL82]). © 1980 IEEE. Reprinted by permission.

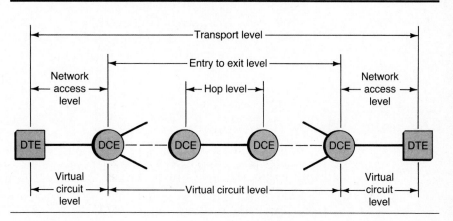

and is accomplished by protocols operating at the equivalent of the OSI transport layer—the first true end-to-end layer in the OSI architecture.

Hop level flow control protects individual communications links, or buffers for accessing communications links, from congestion. As we show below, effects of flow control at the hop level propagate out to other hops through what is called a *"back pressure"* mechanism, so hop level flow control also protects multihop paths. Hop level flow control can be exercised by functions at the OSI data link layer or at the network layer.

The subdivision of the other levels into network access and entry to exit levels depends on the architecture. *Entry to exit* flow control has protection of the exit switch (or DCE) from congestion as its major goal; it is commonly handled by functions at the equivalent of the OSI network layer. The *network access level* can be viewed as a special case of the hop level, with primary function to prevent either the user's DTE or the network entry DCE from swamping the other. Functions at the data link layer or the network layer may be used.

Networks using virtual circuits may also use flow control at the *virtual circuit* level. End-to-end virtual circuits may be implemented as a cascade of three virtual circuits, two at network access level and one at entry to exit level, as shown in Fig. 8.24, or a single end-to-end virtual circuit may be implemented.[31] The virtual circuit level is distinct from the other levels discussed, even when it uses the same equipment as another level and appears at the same location in the figure, since multiple virtual circuits may be mul-

---

[31] The cascaded form of virtual circuits is used in X.25 based networks, discussed in Chapter 11. The cascaded virtual circuits appear to be a single virtual circuit from the user's viewpoint, however.

tiplexed on the same facilities with flow control on individual virtual circuits as well as on composite flows across shared facilities. For example, several virtual circuits may share the same network access path; in this case flow control at the network access level might be used to protect the access path from congestion while flow control on an individual virtual circuit might protect the user process using this virtual circuit.

8.9   Sliding Window Flow Control

One of the most common approaches to flow control is *sliding window flow control.* In this type of flow control, a transmitter is given a permit to transmit a "window" of data—a prescribed number of packets or frames—and is not allowed to transmit more until it receives another permit. The receiver can regulate flow by withholding or granting permits. This type of flow control can be exerted by standard data link layer protocols, but it is also used at a variety of other layers in network architectures. Most current networks use window flow control at one or more layers of their architecture.

Figure 8.25 illustrates window variations for half duplex communication using an HDLC-like protocol.[32] Transmission windows are shown by the shaded portions of the circles, with initial window size at each end equal to three.[33] I n, m indicates an information (user data) packet with $N(S) = n$, $N(R) = m$; supervisory packets are shown by type and $N(R)$ value. Each acknowledgment received (piggybacked or via any control packet aside from RNR) advances the window by the number of packets acknowledged.

Initially **A** transmits three I packets to **B**, decrementing its window size from three to zero. **B** then transmits a packet with $N(R)$ acknowledging all three packets, so **A**'s window size goes back to three (packets 3, 4, and 5) upon receipt of this. **B**'s window size is in turn decremented to two. The figure next shows an alternative way of adjusting window sizes. After receiving one additional packet from **A**, **B** "slams" the window closed by transmitting RNR. Upon receiving RNR, **A** closes its window until it receives the RR shown.

A number of different ways of manipulating windows are in use, or have been suggested. We discuss the most widely used ones. Transmission errors are ignored to simplify the presentation. We use the terminology in Table 8.7. Although throughput often depends on conditions at the receiver not fully reflected in these parameters, such as congestion resulting from all data flows reaching the receiver, the parameters can be used to compute maximum possible throughput.

---

[32] A similar diagram for full duplex communication would be much more complex.

[33] Although the maximum number of frames transmitted before acknowledgment is limited (by the HDLC protocol) to seven, a smaller number is allowed as a user option.

Figure 8.25 Variations in Transmission Windows When Using Sliding Window Flow Control, with Initial Window Sizes of Three, and HDLC-like Data Link Protocol. Shaded portions of circles indicate window openings.

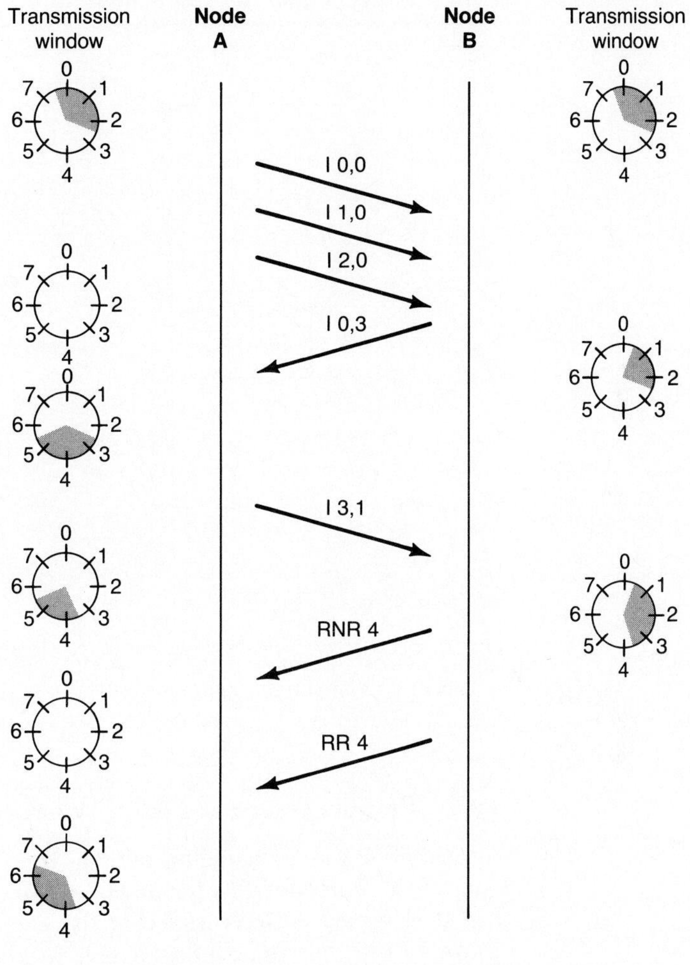

Table 8.7 Definitions of Parameters Used in Analyses of Window Sizes.

T_t	Packet transmission time (packet length in bits/transmission rate in bps)
T_p	Permit transmission time (permit length in bits/transmission rate in bps)
τ	Propagation delay in seconds
W	Window size in packets
d	Time, in seconds, from start of transmission of window until receipt of next permit

Figure 8.26   Window Flow Control with Permit Issued at End of Window or Via Acknowledgments with One Acknowledgment Issued at End of Each Window.

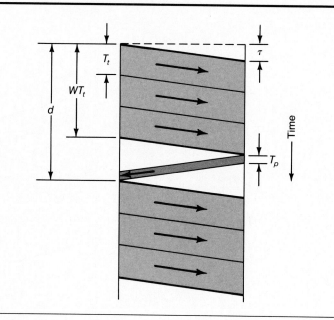

8.9.1   Permits Issued at End of Windows

The simplest way of handling permits is to issue one permit per window. Figure 8.26 illustrates this for a window size, W, of three packets, constant packet size, and permits issued at the end of windows; the illustration is drawn for transmission over a single link. As the figure indicates,

$$d = WT_t + T_p + 2\tau. \tag{8.22}$$

Furthermore, the pattern in the figure repeats periodically, if the receiver does not delay permits and the transmitter transmits whenever it is allowed to do so, so the transmitter transmits for WT_t seconds out of every d seconds. If the data rate of the transmitter is R bps, this gives maximum achievable average throughput, or a maximum effective rate of

$$R_e = \frac{WT_t}{d}R, \tag{8.23}$$

with d given by Eq. 8.22.

Figure 8.27 Maximum Achievable Effective Rate for 2400 Bps Terrestrial Network with Window Flow Control and Permits Issued at End of Windows. $T_p = \tau = 20$ msec.

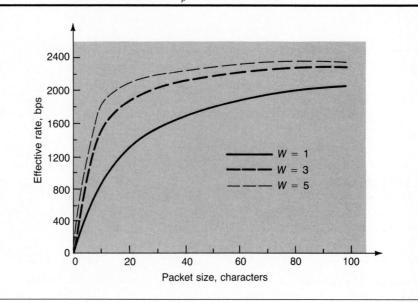

The effective rate is always less than R.[34] Degradation in rate is most severe for long propagation delays (such as those related to communications satellite links) or long permit packets (for example, permits piggybacked on long user data packets). Figure 8.27 contains plots of R_e versus packet size for a terrestrial link with $\tau = 20$ msec, $R = 2400$ bps, and $T_p = 20$ ms (time to transmit a six-character permit). Figure 8.28 shows corresponding plots for a satellite link with $\tau = 270$ msec, $R = 56,000$ bps, and $T_p = 1$ msec (time for a seven-character permit).[35] The reduction in throughput is much more severe for the satellite link.

8.9.2 Permit Issued After Receipt of First Packet in Window

One way to reduce throughput degradation is to return a permit for another window after receiving the first packet in a window, rather than waiting until receipt of the last packet in the window. IBM's SNA flow control uses this

---

[34] The actual transmission rate for user data will be still lower due to overhead for data link protocols, network layer protocols, and so forth.

[35] The extra character allows use of an extended control field for modulo 128 sequence counts.

Figure 8.28   Maximum Achievable Effective Rate for 56 Kbps Satellite Link
Using Window Flow Control and Permits Issued at End of
Windows. $T_p = 1$ Msec, $\tau = 270$ Msec.

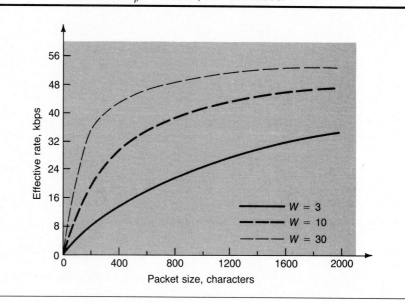

approach; it can allow continuous transmission if window duration is at least
the delay to get the permit back. That is,

$$WT_t \geq d \tag{8.24a}$$

with d given by

$$d = T_t + T_p + 2\tau. \tag{8.24b}$$

This is illustrated in Fig. 8.29. If the delay to get the permit back
approaches T_t—that is, $T_p + 2\tau << T_t$, effective window size can be as high
as $2W - 1$ immediately after a permit is received.

If either propagation delay, τ, or transmission time for the permit, T_p, is
large, the delay to receive the permit can exceed window duration. This is
illustrated in Fig. 8.30.

For the situation in Fig. 8.29, $R_e = R$, while for that in Fig. 8.30, $R_e = (WT_t/d)R$. In general,

$$R_e = \min \left\{ \frac{WT_t}{d}, 1 \right\} R \tag{8.25}$$

with d given by Eq. (8.24b). This is plotted in Fig. 8.31.

Figure 8.29 Window Flow Control with Permit Returned After Receipt of First Packet in Window and $WT_t \geq d$ so Continuous Transmission is Possible. $W = 3$ illustrated.

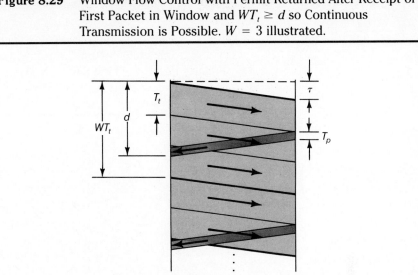

Figure 8.30 Window Flow Control with Permit Returned After Receipt of First Packet in Window and $WT_t < d$ so Continuous Transmission Is Impossible. $W = 3$ illustrated.

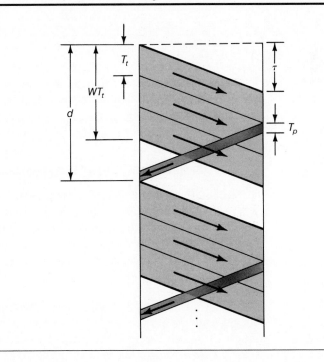

Figure 8.31   Variation of Maximum Possible Effective Rate with Permit
Delay for Window Flow Control with New Permit Issued
After Receipt of First Packet in Window or Via Use of
Acknowledgments with Acknowledgment Issued After
Each Packet.

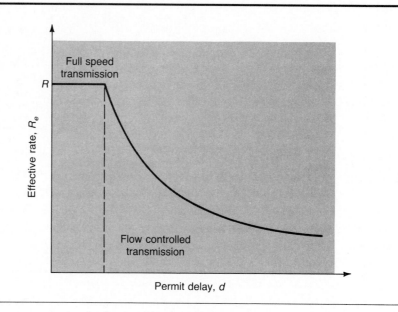

Variation of effective rate with packet size is shown in Figs. 8.32 and 8.33,
using parameters for Figs. 8.27 and 8.28, respectively. Significantly better
throughput is obtained. As the caption indicates, the curves also apply with
an acknowledgment per packet (see the next subsection).

8.9.3   Windows Advanced by Acknowledgments After Packets

A third alternative is the normal window algorithm for DLC protocols, using
acknowledgments returned after packets are received. These acknowledgments serve as permits, though with a slightly different interpretation (see
Fig. 8.25). Window size is now the maximum number of unacknowledged
packets allowed. Hence, if every packet is acknowledged, each acknowledgment advances the window by one (not W). More generally, if an acknowledgment acknowledges k packets, it advances the window by k. Figs. 8.34 and
8.35 illustrate this window mechanism, with an acknowledgment after each
packet, with the parameters used for Figs. 8.29 and 8.30. (T_p now represents
time to transmit an acknowledgment.) Effective transmission rates are the
same as they are for permits returned after the first packets in windows,

Figure 8.32 Maximum Achievable Effective Rate for 2400 Bps Terrestrial Network with Permits Issued After First Packets in Windows or Via Use of Acknowledgments with One Acknowledgment per Packet. $T_p = \tau = 20$ msec.

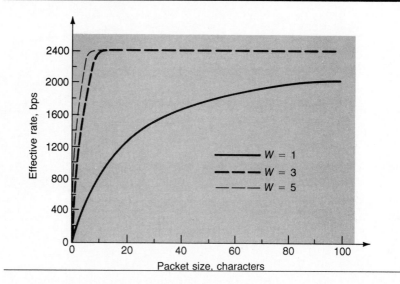

Figure 8.33 Maximum Achievable Effective Rate for 56 Kbps Satellite Network with Permits Issued After First Packets in Windows or Via Use of Acknowledgments with One Acknowledgment Per Packet. $T_p = 1$ msec, $\tau = 270$ msec.

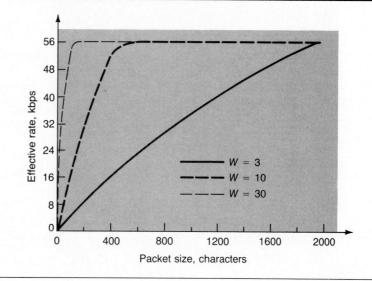

Figure 8.34   Window Flow Control with Acknowledgment Returned After Receipt of Each Packet and $WT_t \geq d$ so Continuous Transmission is Possible. $W = 3$ illustrated.

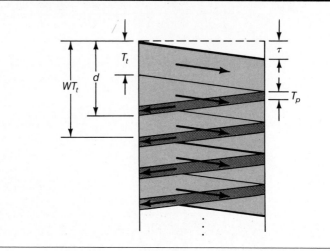

Figure 8.35   Window Flow Control with Acknowledgment Returned After Receipt of Each Packet and $WT_t < d$ so Continuous Transmission Is Impossible. $W = 3$ illustrated.

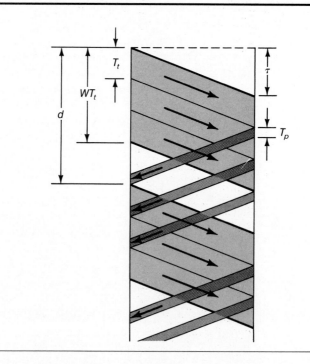

which means that Figs. 8.31–8.33 are applicable. If one acknowledgment per window is used, the protocol becomes identical to that in Fig. 8.26 with Figs. 8.27 and 8.28 giving effective rates. Intermediate frequencies of acknowledgments yield intermediate rates.

8.9.4   Credit Mechanisms Allowing Variable Size Windows

All of our examples so far have assumed fixed window sizes, but more flexibility is obtained if variable window sizes are used and adjusted according to congestion. This can be implemented by including a *"credit"* field in permits to specify the number of additional packets allowed to be sent. An example is shown in Fig. 8.36.

At the beginning of the sequence, **A** has permission to send seven packets to **B**, numbered 0–6. **A** starts with the first window shown and transmits packets 0, 1, and 2, decrementing its window for each. At this point **B** acknowledges packets 0 through 2 via an acknowledgment for 2.[36] It also sends a Credit of 5, allowing **A** to transmit five packets following the last one acknowledged (that is, packets 3–7); this advances the window at **A** by one position since packets 3–6 were already in the window. Before receiving the ACK 2, Credit 5 packet,[37] **A** transmits packets 3 and 4, making corresponding reductions in its window. **A** then exhausts its credit by transmitting packets 5, 6, and 7, and is unable to transmit until it receives more credit. The final transmission of ACK 7, Credit 7 from **B** restores the credit to its original value.

This type of variable credit mechanism is the most common type of window flow control at the ISO transport layer. Both the ISO transport protocol and the DoD transmission control protocol (the two most prominent transport layer protocols discussed in Chapter 11) use this type of credit mechanism. Variable window sizes are also used in some network layer protocols, such as the SNA path control layer discussed in Chapter 9.

8.9.5   End-to-End Versus Hop-by-Hop Window Flow Control

Window flow control can be implemented on either an *end-to-end* or a *hop-by-hop* basis, and both approaches may be used in a single network since they protect against different kinds of congestion.

End-to-end flow control for a three-hop path and permits returned after the first packet in each window is illustrated in Fig. 8.37. The minimum time to return a permit is at least NT_t for an N hop path, even with short propagation delays and short permits. It can easily exceed $2NT_t$ with piggybacked

---

[36] We assume that the numbers in acknowledgments identify the last packet acknowledged, not the next one expected, and that acknowledging one packet implicitly acknowledges all preceding packets with lower sequence numbers.

[37] The long delay for the acknowledgment packet could reflect network congestion.

Figure 8.36   Example of Variable Credit Protocol for Window Flow
Control (adapted from [STAL88b], W. Stallings, *Data and
Computer Communications, 2/e,* © 1988. Reprinted by
permission of MacMillan Publishing Company,
New York, NY.).

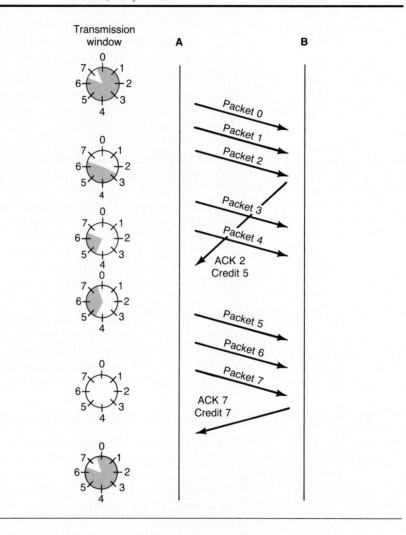

permits on data packets. Since $WT_t \geq d$ is necessary for continuous transmission, a common rule of thumb is to choose window sizes, W, in the N to
$3N$ range for end-to-end flow control across an N hop path.

This rule of thumb for choosing window sizes can lead to a fairness problem when both short path and long path communications occur in the same

Figure 8.37   End-to-End Flow Control Packets for Three-Hop
Communications Path. Window flow control used with
permit returned after receipt of first packet in window.

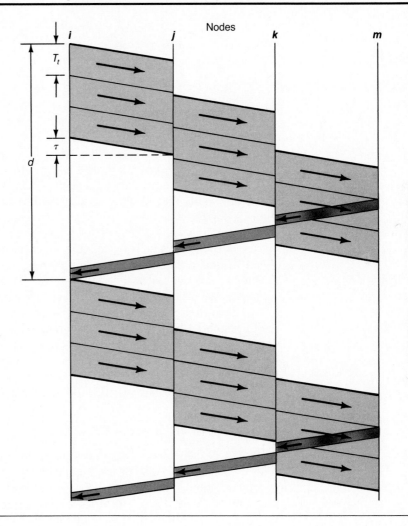

network. Under heavy traffic conditions, the number of packets buffered at a
congested node that originated from one communications session tends to
reach window size for that session. This will be larger for the long path session, if the rule of thumb is employed, so it will gain preferential service if
packets at this node are served in order of arrival.

Flow control for virtual circuits can be exerted on a hop-by-hop basis as
well as on an end-to-end basis. The hop-by-hop alternative is not as suitable

Figure 8.38   Network for Discussion of Hop-by-Hop Flow Control.

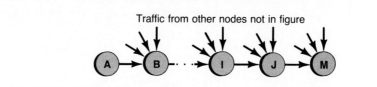

for datagrams, though, since hops taken by datagrams are not predictable. There are advantages to hop-by-hop flow control when it is feasible. Permit delays are shorter, so smaller windows may be used. (If the rule of thumb of $W = N$ to $3N$ is used, window sizes for hop-by-hop flow control are on the order of 1 to 3.) The maximum number of packets buffered along an end-to-end path is still N to $3N$, but packets tend to be evenly distributed among nodes along the path rather than concentrated at a single node.

Hop-by-hop window flow control does not have as rapid an effect on the source of traffic causing congestion as does end-to-end flow control, since it directly limits traffic only over a hop coming into a node where congestion develops. This causes an effect known as *backpressure,* which eventually reduces traffic from the source, however (see Fig. 8.38).

backpressure

Assume that congestion develops at **J**, causing it to use flow control to limit incoming traffic. This will cause buffers at **I** to fill up so it will in turn exert flow control. Similarly, all nodes upstream of **J** will in turn exert flow control until flow control causes traffic from the original source causing congestion, **A**, to be limited. Backpressure causes the effects of hop-by-hop flow control to propagate to other nodes and links in a communications path.

8.9.6   Use of Windows at Receivers

So far our discussion has been focussed on the use of windows at the transmitter. Windows may also be used *at the receiver* to determine which packets may be accepted. If the receiver window size is greater than one, packets may be accepted out of order. This does not necessarily mean they are passed to the user process out of order, though; most implementations require that they be passed to the user process in order. Packets arriving out of order but within the receiver window do not have to be retransmitted, however, if packets filling in gaps are received reasonably promptly.

Acceptance of nonsequential packets can cause problems with sequence counts. An example in [TANE88] illustrates this. We discuss this example and then indicate how standard DLC protocols such as HDLC avoid the problem. Assume modulo eight sequence counts, and that a receiver has a receive window including packets 0–6 (a total of seven packets). It correctly receives packets 0, 1, . . . , 6, acknowledges all seven, sends them to the user process and updates its window to accept packets 7, 0, . . . , 5. A disaster causes all acknowledgments to be lost, however, so the transmitter times out and retransmits packet 0. Unfortunately, 0 is in the current receiver window, so this packet is accepted as a new (not a retransmitted packet).

Receipt of a "new" packet 0 causes the receiver to again acknowledge the last seven packets it "correctly received in sequence" (that is, to send another acknowledgment to packets 0–6), so the transmitter forgets about retransmitting packets and transmits packets 7, 0 (new version) and so forth from the original sequence. As soon as packet 7 has been accepted, the receiver sends it plus the previously accepted packet 0 (retransmitted version) to the user process, since all holes in the received sequence have been filled in. This results in the user process being sent an invalid packet. With this implementation, a receive window size of 7 can allow invalid packets to be accepted.

The difficulty in this situation arises from the fact the old receiver window (before packets were received) and the new window (after acknowledging received packets) overlap. If the window size, W, is chosen so no overlap can occur, the problem will not arise. A value of $W = M/2$ or 4 will suffice.

Receiver windows in a data link control such as HDLC are implemented in a manner that avoids this problem. Receiver windows are of size one unless selective reject (SREJ) is implemented; without SREJ frames are not accepted unless they are in order. When SREJ is used, frames following an erroneous packet but within the receive window can be accepted. SREJ is triggered by receipt of any out of sequence frame, however. Usually this results from ignoring an erroneous frame received just before the frame triggering the reject. In the example, however, frames 0–6 have been acknowledged so SREJ 7 would be triggered by the retransmitted frame 0, which is out of sequence (expected sequence number is seven). When this SREJ is received at the transmitter, it observes that its $N(R)$ value of 7 implies the receiver thinks an untransmitted frame, number 7, has been lost. The correct response to this condition is for the transmitter to send an unnumbered packet called Frame Reject (FRMR) allowing recovery from a variety of anomalous situations such as this.[38] HDLC SREJ implementations do not require that window sizes be limited to $M/2$. The key difference from the example in [TANE88] is the fact that SREJ is used for acknowledgment, rather than using an ordinary acknowledgment, when an out of sequence frame has been received.

8.9.7   Impact of Window Flow Control on Delay

The impact of flow control on delay is at least as important as its impact on achievable throughput, but we can give only rough approximations to its impact on delay without sophisticated mathematical analysis. It is possible

---

[38] FRMR is not discussed in Chapter 9 since it is used only in special situations similar to this. The FRMR packet contains information to aid in diagnosing situations similar to this.

to show, however, that delay is roughly proportional to the number of actively flow-controlled processes for any type of window flow control. To show this, we note that if process i has window size W_i the number of packets from this process allowed in the network at time t is $\leq W_i$. We denote this quantity by $\beta_i(t)W_i$, where $0 \leq \beta_i(t) \leq 1$ and $\beta_i(t)$ may vary with t.[39] The total number of packets in the network with n processes is then equal to a sum of such terms,

$$\sum_{i=1}^{n} \beta_i(t) \, W_i.$$

A standard formula from queuing theory, Little's theorem (see [KLEI75b]), states that the average number in the system is the average arrival rate, λ, times the average time in the system, T, so

$$T = \frac{\sum_{i=1}^{n} \beta_i(t) \, W_i}{\lambda}, \tag{8.26}$$

is average delay, with λ total arrival rate for all processes.

Although each $\beta_i(t)$ may fluctuate with time, their average contributions to the sum will be reasonably constant, especially if n is large. So the value of T from the formula in Eq. (8.26) is roughly proportional to the total number of flow controlled processes. If this number is large, delay may be excessive. It can be reduced by using small windows, but this can unnecessarily throttle users at light loads.

8.10   Buffer Allocation Approaches to Flow Control

Another general category of flow control algorithms is that of *buffer allocation algorithms*. Altering buffer allocations for different classes of traffic can limit traffic under heavy load conditions when congestion occurs. Buffer allocation strategies are often used in conjunction with techniques such as window flow control; the combination is more effective than either alone.

Several buffer allocation strategies have been developed to avoid deadlocks that can occur when buffers are full, but buffer allocation approaches to deadlock prevention are less important with current technology than they were when buffers were more expensive; it is now usually feasible to simply assign extra memory for buffers and thereby prevent deadlocks. Good buffer allocation schemes are helpful in reducing congestion as well as in preventing deadlocks, however, and some strategies developed for deadlock prevention help to reduce congestion. We summarize strategies to prevent deadlock, then examine other buffer allocation strategies for flow control.

---

[39] We make no attempt to analyze how it varies with t.

Figure 8.39   Direct Store and Forward Deadlock.

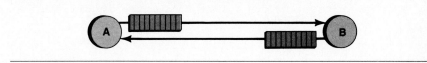

8.10.1   Buffering Strategies for Deadlock Prevention

The simplest type of deadlock in networks is *direct store and forward deadlock*. This is illustrated in Fig. 8.39.

Direct store and forward deadlock may occur when all buffers, both for traffic sent out from a node and for traffic coming into the node, are in a single pool. If all buffers at **A** are full of packets destined for **B** and all buffers at **B** are full of packets for **A**, neither node will be able to accept incoming traffic.[40] Thus neither node can clear traffic from its buffer in order to have space for incoming packets. The result is deadlock!

The straightforward approach to avoiding direct store and forward deadlock is to guarantee the entire buffer pool for a link cannot be filled up by packets destined to be sent out over the link. If the maximum number of buffers for outgoing traffic over link i is no more than $b_i < B_i$, with B_i the total number of buffers for that link, there will always be some space for incoming traffic and direct store and forward deadlock will not occur. This is called a

channel queue limit

channel queue limit strategy and is used in some form in almost all telecommunication networks. Using separate receive and transmit buffers is an alternative approach that simplifies buffer management but is less efficient in using total buffer capacity.

Although the channel queue limit strategy eliminates direct store and forward deadlocks, other types of deadlocks are still possible. A type of deadlock known as *indirect store and forward deadlock,* which can occur for networks using the channel queue limit strategy, is illustrated in Fig. 8.40.

In the situation illustrated the outgoing buffer at **A** for link 1 is full of b_1 packets destined for **C**, the outgoing buffer at **B** for link 2 is full of b_2 packets for **D**, and so forth. Hence each node's outgoing buffer is full of packets destined for a node at least two hops away, so no node can place any more packets in its outgoing queue and no packets can reach their destination.[41] Deadlock!

Indirect store and forward deadlock can be prevented by slightly more elaborate buffer allocation strategies. A structured buffer pool approach that

---

[40] Any packets delivered to either node will be discarded since there are no buffers for them. The data link protocol will not allow packets at the transmitting node to be discarded before acknowledgments are received, and no acknowledgments can be accepted.

[41] The destinations are precisely two hops away in this example, but the example can easily be generalized as long as all destinations are at least two hops away.

Figure 8.40   Indirect Store and Forward Deadlock.

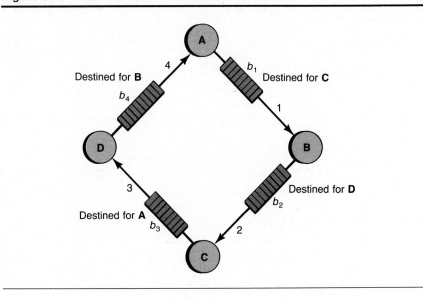

prevents both direct and indirect store and forward deadlock is discussed below. First, though, we mention one other type of deadlock that was observed in the early days of the ARPANET but is applicable only for networks that use datagram transmission at the network layer, with conversion to virtual circuits by reassembling packets at their destinations. This is called *reassembly deadlock* and is illustrated in Fig. 8.41.

The figure shows packets comprising parts of messages labeled A, B, and C en route to the destination host. Each message consists of four packets, and four buffers are available for message reassembly at the destination node (an IMP attached to the destination host). Channel queue limit flow control is used with four buffers for outgoing traffic on each link. Message A has been partly reassembled at the destination, but packet A_2 from this message cannot reach this node since it is at node **1** and blocked by four packets from mes-

Figure 8.41   Reassembly Deadlock for ARPANET and Similar Networks.

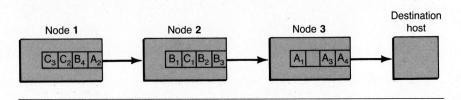

Figure 8.42   Structured Buffer Pool.

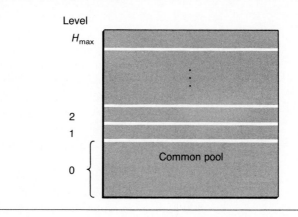

sages B and C filling up outgoing traffic buffers at node **2**. Furthermore, reassembly of either message B or message C, which could release reassembly buffers, is impossible because reassembly buffers are not available. Deadlock!

Reassembly deadlock can be prevented by *buffer reservation*. If enough buffers for reassembly of all messages en route are reserved at the destination before transmission, it will be able to reassemble all incoming packets into messages. This is essentially the same as window flow control, with confirmation of reservations replacing permits to transmit windows, but includes reservation of buffer space. The approach has been adopted for the ARPANET and other DoD Reference Model networks; the message used for buffer reservation is called a Request for Next Message, RFNM (see Subsection 9.5.2).

We now discuss a technique for avoiding both direct and indirect store and forward deadlock. If it is used for a network where reassembly deadlock can occur, we assume the buffer reservation scheme just discussed is also *structured buffer pool* used. The strategy we discuss is called *structured buffer pool* strategy. It uses a buffer pool structure of the type in Fig. 8.42.

The figure shows buffers arranged in levels from 0 through H_{max}. H_{max} is the maximum number of hops on a transmission path in the network (a function of topology and routing algorithm). At level 0, there is a common pool of unrestricted buffers. At level $i, 1 \le i < H_{max}$, the buffers are reserved for packets that have covered at least i hops. Thus a packet that has covered four hops so far would be able to use buffers at levels 0 through 4. The highest level, H_{max}, buffers are reserved for packets that have covered H_{max} hops or have reached their destination.[42] Any packets that have traveled H_{max} hops

---

[42] If reassembly at the destination is required, reservation of reassembly buffers at level H_{max} before transmission of the first packet in messages is assumed, so it is possible to reassemble all messages that have reached their destination.

without reaching their destination are discarded since they have traveled further than should be possible and are assumed to be looping.

Packets that have traveled greater distances have a wider range of buffers to choose from and hence are less likely to be blocked and discarded than packets that have traveled shorter distances. This is reasonable since more network resources have already been expended on the former class of packets. This approach will always prevent both direct and indirect store and forward deadlock, as we now prove. The proof involves showing that buffers at level k cannot fill up permanently. Hence, a nonlooping packet in a level k − 1 buffer is eventually able to move to a level k buffer a hop closer to its destination, and looping packets eventually make it to level H_{max} where they are discarded. We show this by induction, starting with packets in level H_{max} buffers.

Since packets in level H_{max} buffers have either reached their destination (where adequate reassembly buffers are guaranteed) or are due to be discarded, they remain in buffers only briefly before being passed to user processes or discarded. This clears positions in level H_{max} buffers for packets in level H_{max} − 1 buffers, so they will soon be able to advance. Furthermore, removal of these packets from level H_{max} − 1 buffers will allow packets at H_{max} − 2 to advance. Continuing, we see packets in buffers at any level will eventually be able to advance and deadlock is impossible.

input buffer limiting   A simplified version of the structured buffer pool strategy, known as *input buffer limiting,* only divides the transmission buffer pool at a node into two levels. Packets in transit via this node are eligible to use buffers at either level, but (input) packets originating at this node can use only the lower level. This strategy has been studied carefully, both via analysis and simulation, and reasonable guidelines for sizes of the two levels of the buffer pool to avoid congestion have been developed.

8.10.2   Buffering Strategies for Maximizing Throughput

The buffering strategies above avoid deadlocks, but we have not showed that they provide high throughput. The proof that the structured buffer pool strategy prevents store and forward deadlocks just shows that buffers will eventually be available, not that they will be available soon enough to maintain high throughput. However, if it is properly designed (including choice of relative sizes of various portions of the buffer pool), the strategy can significantly improve throughput. Major improvements are possible when a good buffering strategy supplements window flow control.

The primary buffering strategy that has been studied from the point of view of its effect on achievable throughput is the input buffer limit (IBL) strategy. Analytical models have been given by Lam and Schwartz [LAM79] [SCHW87], among others, with a thorough simulation study by Lam and Lien [LAM81]. We give a simple heuristic discussion that provides an intuitively satisfying explanation of the guidelines for choosing relative sizes of buffer pool levels that analysis and simulation have shown to be satisfactory.

Let the total number of buffers at a typical node be N, with N_I of these available for input packets (originating at the node). Transit packets (being relayed between nodes) are allowed to use any of the N buffers, so $N_T = N - N_I$ are available only to transit packets. Assume that the average number of hops a packet traverses between source and destination is H. Then each input packet allowed into the network generates an average of H transit packets relayed between nodes en route to the destination. If the network is to be able to handle all transit packets under heavy traffic conditions, when input buffers tend to be full, there need to be at least H times as many buffers available to transit packets as there are available to input packets. This implies that buffer sizes should satisfy $N_T > N_I H$, or

$$\frac{N_I}{N_T} < \frac{1}{H} \tag{8.27}$$

This is precisely the guideline recommended by the theoretical analyses cited above, but these analyses are based on assumptions of homogeneous traffic. The simulation study in [LAM81] considered a wider variety of traffic patterns, plus other variations, and came up with the slightly altered criterion

$$\frac{N_I}{N_T} < \frac{\alpha}{H}, \tag{8.28}$$

with α a parameter, $0 < \alpha < 1$, reflecting network inhomogeneities. Values of α as low as 0.4 seem advisable for some circumstances, but values close to 1 are normally adequate.

The type of improvement in throughput obtainable with input buffer limiting is indicated by Fig. 8.43, which illustrates the form of curves plotting throughput values versus total buffer size with and without IBL in [LAM81]. All curves in the reference were obtained with window flow control, so throughput improvements with IBL are in addition to those with window flow control. No numerical values are given as the figure has been abstracted from several curves in [LAM81]. Eq. 8.28 was used to compute relative buffer sizes.

As Fig. 8.43 indicates, with large buffers throughput without IBL may equal that obtainable with IBL. However, IBL is an effective and inexpensive way to reduce congestion and increase throughput. It gives high throughput with much smaller numbers of buffers.

IBL by itself does not eliminate store and forward deadlock; a more complex buffering strategy such as that in the previous subsection is necessary for this. IBL can be a part of the more complex strategy, however. If the transit buffer level in IBL is broken up into levels corresponding to the number of hops that transmit packets have traveled, a structured buffer pool strategy is obtained. No guidelines for determining relative sizes of other buffer pool levels that are comparable to Eq. 8.30 have been published, however.

Figure 8.43   Typical Throughput Improvements Obtainable with Input
Buffer Limiting in Network Already Using Window
Flow Control.

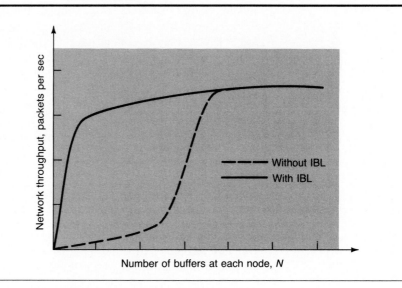

8.11   Other Approaches to Flow Control

The input buffer limiting technique, combined with window flow control,
has been implemented in several networks, including the DATAPAC network
described in Chapter 9.

A variety of other approaches have been proposed and some implemented.
We briefly discuss packet discarding, isarithmic flow control, and use of
choke packets.

packet discarding

Packet discarding—or simply discarding excess packets—is a common
technique for reducing the effective network load. For networks using datagram transmission, this is standard practice. It is effective but imposes extra
burdens on higher protocol layers since any recovery from situations in
which packets are lost must be handled by these higher layers.

isarithmic flow control

Isarithmic flow control [DAVI72], [PRIC77] was developed at the National
Physical Laboratory in England. It limits the total number of packets in the
network by using permits that circulate within the network. Whenever a node
wants to send a packet, it must first capture a permit and destroy it. The
permit is regenerated when the destination node removes the packet from

the network. This ensures that the total number of packets in the network will never exceed the number of permits initially present.

There are problems with this approach, however. Equitable distribution of permits to nodes throughout the network is a problem for which no fully satisfactory solution is known. Nodes needing permits may not be able to obtain them rapidly enough to keep throughput at reasonable levels ("permit starvation"). Although the total number of packets in the network is guaranteed to be reasonable, individual nodes can be overloaded. Circulation of free permits (in some type of random walk fashion) to ensure they can eventually be captured by nodes needing them puts an extra communications load on the network. Finally, permit destruction in any manner (by transmission errors, failure to regenerate permits when packets reach their destinations, and so forth) can cause permanent loss of communications capacity. It is not easy to find how many permits still exist while the network is operational.

Despite these problems, the isarithmic approach has been found to be reasonably effective in networks without hop level flow control. It has never been widely implemented, however.

choke packets

The *choke packet* approach is widely used, though not necessarily by this name. The basic idea is for nodes detecting congestion to send packets, "choke packets," back to the source of any message sent into the congested region. The source is then required to reduce or eliminate this type of traffic, "choking" off its flow. The original proposal for this approach [MAJI79] included specific details, such as approaches to monitoring congestion and using the choke packets. The approach is used in DNA and other networks. Use of Receive Not Ready frames in HDLC or WACK frames in Bisync serves the same purpose. The SNA pacing mechanism for flow control also uses packets that perform essentially this function. Thus the general approach is common. It has been found to work well in reducing congestion.

□ 8.12   Combined Optimal Routing and Flow Control

The approach to optimal routing in Section 8.6 can be generalized to determine jointly optimal routing and flow control. This implies choosing the commodity flows, r_k (data flows between specified sources and destinations) and the flows on paths, f_{p_i}, defined in Section 8.6 to jointly satisfy an optimality criterion.

The basic equations used in Section 8.6 are repeated here for ease of reference. Average delay is given by

$$T = \frac{1}{\gamma} \sum_{i=1}^{N} \frac{\lambda_i}{C_i - \lambda_i},$$
(8.29)

with traffic on the links given by

$$\lambda_k = \sum_{\substack{\text{all paths } p_i \\ \text{containing link } k}} f_{p_i}$$
(8.30)

Figure 8.44   Form Assumed for Penalty Function for Throttling Commodity Flows in Joint Optimization of Routing and Flow Control.

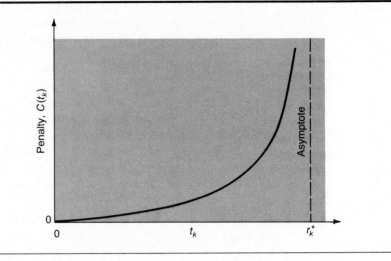

and the path flows, f_{p_i}, satisfying the constraints,

$$\sum_{p_i \in P_k} f_{p_i} = r_k. \tag{8.31}$$

The solution to the problem of choosing the r_k and f_{p_i} values to jointly minimize T is trivial, but unfortunately uninteresting; it is $r_k = 0$ for all k and $f_{p_i} = 0$ for all paths p_i. This is not satisfactory since it chokes off all traffic! A reasonable way to find a more satisfactory solution is available, however.

The difficulty with the approach above is that it does not include any penalty in the optimality criterion for reduced data flows, although a solution with data flows below their desired values is obviously less satisfactory than a comparable solution without throttling of flows. We introduce a penalty factor for reduced data flows to account for this.

We define r_k^* to be the desired flow of commodity k (that is, the flow obtained with no throttling of flows), r_k to be actual flow of this commodity, and

$$t_k = r_k^* - r_k \tag{8.32}$$

penalty for throttling flows

to be the throttled portion (that is, the portion of the desired flow not carried by the network). We choose a penalty (or cost) function, $C(t_k)$, of the form in Fig. 8.44, that is, a nonnegative function of the same general form as the function in Fig. 8.16, and approaching ∞ as t_k approaches r_k^* (implying the flow, r_k, of the kth commodity approaches zero).

Our problem is to choose the t_k and f_{p_i} values that minimize

$$T^* = T + \sum_k C(t_k) \tag{8.33}$$

under the constraints that all t_k and f_{p_i} values are nonnegative, plus

$$\sum_{p_i \in P_k} f_{p_i} + t_k = r_k^*. \tag{8.34}$$

Careful comparison of this with Section 8.6, reveals that the problems are equivalent. If the t_k values are considered equivalent to extra f_{p_i} flows, the solution techniques in Section 8.6 can be applied. For penalty functions of form

$$C(t_k) = \frac{\alpha_k}{r_k^* - t_k}, \tag{8.35}$$

with each α_k a positive constant, derivatives with respect to t_k values are of essentially the same form as derivatives with respect to f_{p_i} values.

Following one of the grand (?) traditions of mathematics—considering a problem solved once it has been reduced to a previously solved problem—we end our treatment of the joint optimization problem at this point. Problem 8.18 involves finding the optimum solution for a simple case for which the optimum can be found analytically. Further details can be found in the literature (for example, [BERT87]).

8.13 Summary

Routing and flow control are essential components of network layer protocols and are closely related. Routing determines the paths through which data flow from source to destination, and flow control limits the traffic allowed into portions of a network to minimize congestion and improve performance.

The routing problem can be divided into two main components: finding appropriate routes and delivery of messages to their destinations once routes have been selected. The latter function is normally performed with the aid of routing tables. Routing algorithms handle the former function.

A variety of routing algorithms are used, but some of the most interesting are based on techniques to find shortest paths in networks. These algorithms use iterative solution techniques, with a variety of possible approaches to iteration. For example, they may iterate on number of nodes in a path, on length of paths, or on nodes allowed to be intermediate nodes in paths. An example of an algorithm based on each iteration criterion is given. Stability problems can develop with these algorithms, though, especially if the distance measure is based at least partially on traffic. Optimal routing techniques allow traffic-based measures to be used without stability problems, and they appropriately divide traffic among routes. However, they are relatively complex.

Flow control limits traffic to minimize congestion. It is especially critical in telecommunication networks, since excessive delays tend to increase traffic in such networks due to retransmissions after timeouts. Sliding window and buffer allocation techniques to flow control are the most common approaches, and work best in combination. Combined optimal routing and flow control is possible but complex.

Problems

8.1   Construct routing tables of the forms shown in Tables 8.2 and 8.3, for minimal hop routing, to be located at all other nodes in the network in Fig. 8.3.

8.2   Prove that if the cost of link ij is $-\log_e p_{ij}$, with p_{ij} the probability link ij is operational, and link failures are independent, then the minimum cost route is the route with maximum reliability. Also show that $-\log_e p_{ij} \geq 0$, so an algorithm valid only for nonnegative costs is suitable for minimization.

8.3   Prove that the set of shortest paths found by the Bellman-Ford-Moore algorithm always forms a tree.

8.4   Assume that our assumption of positive costs for the routing problem is violated. More specifically, assume that it is possible to find a loop with negative net cost along some path between a desired source and a destination node. If a routing algorithm successfully obtained a minimum cost routing for this problem, what would the minimum cost be and how would it be obtained?

8.5   Modify the pseudo code for the Bellman-Ford-Moore algorithm so that it describes an algorithm to find the shortest paths to a specified destination node from all other nodes.

8.6   Apply the d'Esopo-Pape modification of the Bellman-Ford-Moore algorithm to find shortest paths from node **1** to all other nodes for the network in Fig. 8.5. Sketch the results of steps in the algorithm (in the form of Fig. 8.6) and compare convergence rates of the two approaches. Comment on your results.

8.7   Modify the pseudo code for Dijkstra's algorithm so that it describes an algorithm to find the shortest paths from a specified source node to all other nodes.

8.8   **a.**   Use the Bellman-Ford-Moore algorithm to find the shortest paths from Rome to each of the other cities in Fig. P8.8.[43]

       b.   Use the Dijkstra algorithm to find the shortest paths to Rome from each of the other cities.

8.9   Apply the Bellman-Ford-Moore algorithm to find shortest paths from Node **1** to all other nodes for the directional network in Fig. P8.9. Propagation on each link is possible only in the direction of the arrow.

8.10   Apply Dijkstra's algorithm to find shortest paths to node **10** from all other nodes in Fig. P8.9.

8.11   Use the Floyd-Warshall algorithm to find shortest paths between all pairs of cities in Fig. P8.11.

8.12   **a.**   Show that an alternative to the use of matrix R, defined in Subsection 8.4.3 to represent routes obtained by the Floyd-Warshall algorithm, is use of a matrix R' with R_{ij} the second node along the optimal route from i to j.

       b.   Find R' for the example of the Floyd-Warshall algorithm in Subsection 8.4.3.

       c.   What is the general relationship between R and R'?

---

[43] Problems 8.8 and 8.9 adapted from Mischa Schwartz, *Computer Communication Network Design and Analysis* © 1977. Reprinted by permission of Prentice-Hall, Inc., Englewood, NJ.

Figure P8.8 Figure for Problem 8.8.

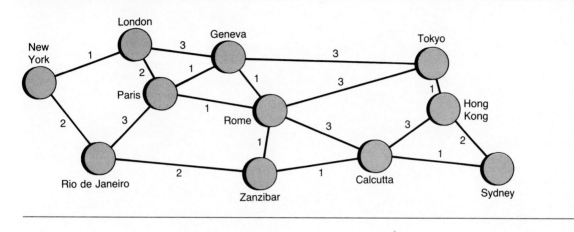

Figure P8.9 Figure for Problems 8.9 and 8.10.

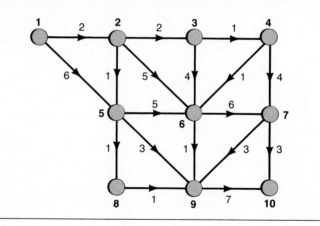

8.13 [BERT87]. The number shown next to each link of the network in Fig. P8.13 is the probability of the link failing during the lifetime of a virtual circuit from node **1** to node **7**. All links shown are directed links, allowing propagation only in the directions of the arrows. Links fail independently of each other. Find the most reliable path from **1** to **7**. What is its reliability?[44]

[44] Problems 8.13 and 8.24 adapted from Bertsekas/Gallagher, *Data Networks,* © 1987, pp. 408 and 446. Reprinted by permission of Prentice-Hall, Inc., Englewood Cliffs, NJ.

Figure P8.11   Figure for Problem 8.11.

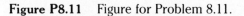

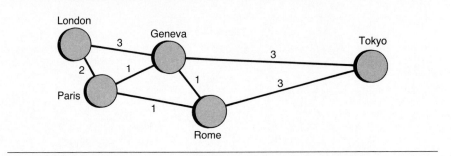

Figure P8.13   Figure for Problem 8.13.

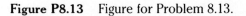

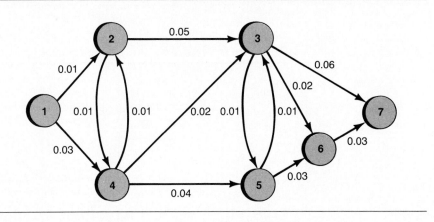

Figure P8.14   Figure for Problem 8.14.

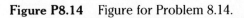

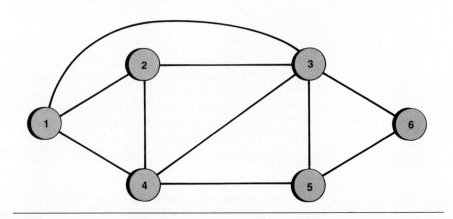

8.14   For the network in Figure P8.14, assume a cost matrix

$$
C = \begin{bmatrix}
0 & 1 & 2 & 2 & \infty & \infty \\
1 & 0 & 2 & 2 & \infty & \infty \\
2 & 2 & 0 & 3 & 2 & 2 \\
2 & 2 & 3 & 0 & 1 & \infty \\
\infty & \infty & 2 & 1 & 0 & 3 \\
\infty & \infty & 2 & \infty & 3 & 0
\end{bmatrix}
$$

with each node only knowing distances to its immediate neighbors. Nodes **2–6** have current estimates of total distance to **1** given by: $D_2 = 1, D_3 = 2, D_4 = 2, D_5 = 4$ and $D_6 = 5$. At time t_0 all nodes simultaneously broadcast their distance estimates to their immediate neighbors. Apply Equation (8.12) (with D_1 known to be zero) to find the new distance estimates, $D_{i\ new}$, computed at nodes **2–6** if the distributed asynchronous version of the Bellman-Ford-Moore algorithm is used and these distance estimate broadcasts happen to occur simultaneously. Also indicate the next nodes that **2–6** would send packets destined to **1** on in implementing what they think is the optimum route to **1**.

8.15   Consider the example of routing instability in Figs. 8.11 and 8.12, but modify the cost of a link to be α plus the flow on the link, with α a positive constant. Thus even a link with no flow has cost of α.

   a.   For $\alpha = 1$ demonstrate that the instability vanishes, with the only oscillation in routing for node **4**.

   b.   Consider an initial routing with all traffic routed clockwise, but with this same type of cost structure. What is the minimum value of α such that the routing eventually converges to that in (a)?

8.16   Verify Eqs. (8.21a) and (8.21b)

8.17   One way to resolve how to allocate capacity in a "fair" way for the network in Fig. 8.23 is to define a penalty function for failing to carry the full one data unit per second generated by each user. Capacity allocation can then be chosen to minimize the total penalty. A possible penalty function is α/r_i for the ith user, with r_i the traffic carried for this user, $0 \le r_i \le 1$ and α a positive constant; since the penalty is infinite at $r_i = 0$, each user is guaranteed some capacity when traffic is optimized. By symmetry, each single link user will be allocated the same traffic, so the total penalty is $(n\alpha/r_1) + (\alpha/r_n)$ with r_1 traffic allotted to each single link user and r_n traffic allotted to the n link user. An obvious constraint for the optimum solution is $r_1 + r_n = 1$. Find the optimum values for r_1 and r_n under this constraint.

   Hint: This problem can be solved readily with the aid of Lagrange multipliers. Look up the Lagrange multiplier technique before attempting to use it if you do not already know it. The constraints $r_1 \ge 0$ and $r_n \ge 0$ are automatically met due to the penalty function used.

8.18   Consider sliding window flow control, with permits issued after receipt of the first packet in a window, operating over a communications satellite link at 1 Mbps. All data packets consist of 500 eight-bit characters, with permit packets short enough that their duration may be neglected. The one-way propagation delay for the link is 250 ms. Find the minimum window size (W) allowing transmission at the full 1 Mbps rate in the absence of errors.

8.19   Briefly discuss the relative advantages and disadvantages of the techniques for issuing permits in sliding window flow control discussed in Section 8.9—that is, permits

Figure P8.21   Figure for Problem 8.21.

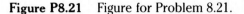

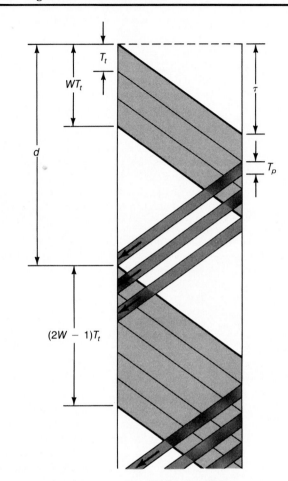

issued at end of windows, after receipt of the first packet in a window, via acknowledgments and via returning credit.

8.20   An intuitively appealing window flow control technique is to issue a permit after receipt of each packet, with each permit advancing the window by W packets.

    a.   Give a formula for permit delay (d) for this situation.

    b.   Show that the effective window size can approach ∞ for any $W \geq 2$.

    c.   Give conditions under which this gives a realistic approach to window flow control.

8.21   Figure P8.21 illustrates window flow control, with a permit issued after receipt of each packet, each permit allowing transmission of W packets after it has been received. The most recently received permit determines the window opening.

Figure P8.22 Figure for Problem 8.22.

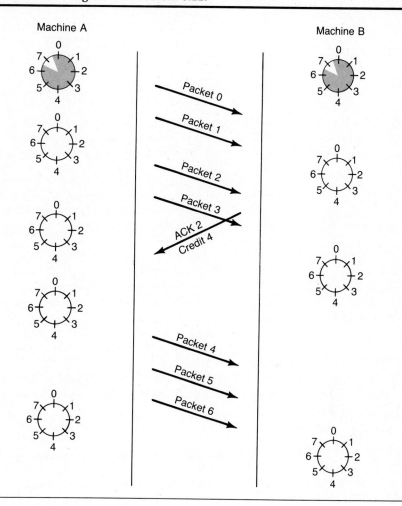

Machine A

Machine B

Packet 0

Packet 1

Packet 2

Packet 3

ACK 2
Credit 4

Packet 4

Packet 5

Packet 6

a. Show that if τ is constant and very large, and $W \geq 2$, periods of continuous transmission alternate with idle periods, with the length of the kth period of continuous transmission $[k(W - 1) + 1]T_t$.

b. Why is this approach to flow control not realistic?

8.22 A credit mechanism is used for flow control, with initial credit of seven packets, as indicated in Fig. P8.22. Shade in all sending machine and receiving machine window sizes, aside from the two already shaded in, to show window sizes at each instant where a circular diagram is given.

8.23 Hop-by-hop window flow control for data transfer from **1** to **3** is used in the three-node network in Fig. P8.23, with $W = 3$ for each of the two links in the path. The

Figure P8.23   Figure for Problem 8.23.

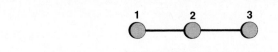

source at **1** has an inexhaustible supply of packets, and processing times at nodes and propagation delays are negligible. Each node returns a permit as soon as possible after receiving the first packet in a window, but each node only has buffers for four packets and does not return a permit unless at least W of these are free. A buffer becomes free as soon as the packet stored in it has been passed to the DLC protocol for transmission. Each packet transmission on link **12** takes one second and each packet transmission on link **23** takes two seconds. Permit transmission on each link takes one second. Find the times, from $t = 0$ to 15 seconds at which a packet transmission starts at node **1** and at node **2**.

8.24   Consider combined optimal routing and flow control for the simple link connecting origin and destination illustrated in Fig. P8.24. Assume a cost function of the form

$$T^* = \frac{r}{C - r} + \frac{\alpha}{r^* - t},$$

where the first term represents a penalty for large delay and the second term represents a penalty for small throughput. C is the capacity of the link and α is a positive weighting factor.

a.   Specify the equivalent routing problem and its constraints.

b.   Show that there will be no flow control ($t = 0, r = r^*$) if

$$r^* < C \frac{\sqrt{\alpha}}{\sqrt{\alpha} + \sqrt{C}}.$$

c.   Show that when the condition in (b) is violated so that there is flow control,

$$r = C \frac{\sqrt{\alpha}}{\sqrt{\alpha} + \sqrt{C}}.$$

Figure P8.24   Figure for Problem 8.24.

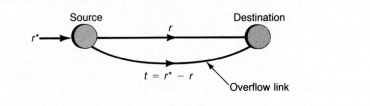

9   Network Layer Protocols

9.1   Introduction

According to the OSI Reference Model the purpose of the network layer is to provide *end-to-end communications capability* to the transport layer. It should allow the transport layer to ignore data transfer technology and relaying and routing considerations, mask peculiarities of the data transfer medium from higher layers, and provide switching and routing functions to establish, maintain, and terminate network layer connections and to transfer data between communicating users. The transport layer should not be concerned with whether circuit switching or packet switching, optical fiber, twisted pair, coaxial cable, microwave or satellite links, or local area networks are used for communications. It should be concerned only with quality of service (measured by such parameters as response time, throughput, error rate, and so forth) and its cost. These are ambitious goals.

Figure 9.1 shows the location of the network layer in the OSI hierarchy for packet-switched networks, with network layer functions performed at each node. Only one intermediate node is shown in the figure, but information flow up through the network layer and back down occurs at all intermediate nodes between source and destination. For circuit-switched networks during data transfer the diagram in Fig. 9.2 is more appropriate; there is a direct path between source and destination, so only physical layer functions are needed at intermediate nodes. During circuit connection and disconnection, though, the diagram in Fig. 9.1 is appropriate, since network layer functions such as routing are needed.[1]

The network layer is the most *conceptually complex* layer in the reference model since peer processes at all nodes must work together to accomplish functions such as routing and flow control (see Chapter 8). The lower two layers—physical and data link—involve cooperation only between adja-

---

[1] An alternative viewpoint is that functions used during connection and disconnection are circuit establishment and disestablishment functions in the physical layer. From this point of view, the network layer in circuit-switched networks completely vanishes.

Figure 9.1   The OSI Reference Model Layer Hierarchy for Packet-Switched Networks.

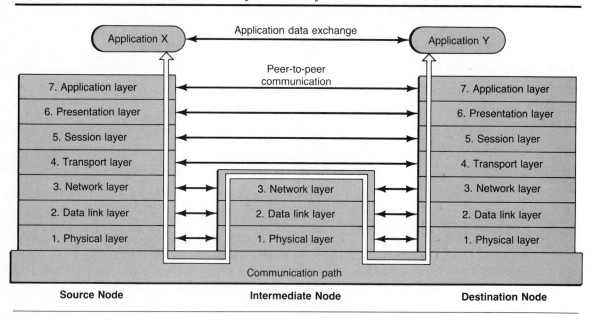

Figure 9.2   The OSI Reference Model Layer Hierarchy for Circuit-Switched Networks During Data Transfer Phase.

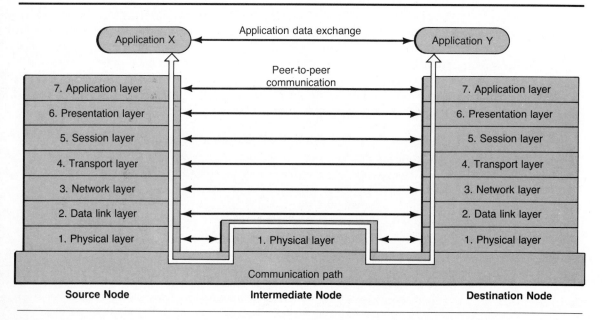

cent nodes[2] while higher layers—transport, session, presentation, and application—involve cooperation only between nodes at either end of the communications link. The layer's complexity is reflected in the fact that network layer standards are less well developed than those for other layers.

X.25

The difficulty of network layer definition has been increased by the fact that developers of the architecture have tried to respond to the needs of a variety of incompatible networks, some predating the OSI model. The CCITT *X.25 standard* discussed below is the most widely implemented OSI network layer standard,[3] but it is not a complete OSI network layer. It is a standard for an interface with a packet network and is normally used to provide a virtual circuit replacement for an ordinary communications circuit, with or without OSI compatibility at higher layers. A variety of techniques are used in X.25-based vendor networks to provide network layer functions not included in X.25.

We illustrate how X.25 fits into the architecture by discussing the X.25 standard and four important X.25 networks that differ widely in implementation; detailed descriptions of the networks are in the appendix to this chapter. This is followed by a discussion of the network layer for circuit-switched networks and additional aspects of the OSI network layer standards. Corresponding ARPANET protocols are presented next, followed by corresponding layers in DNA and SNA. The chapter concludes with presentations of how the network layer fits into the IEEE 802 LAN protocol hierarchy and the MAP and TOP architectures plus the current status of the ISDN network layer.

The OSI Reference Model considers internetworking to be handled by a sublayer of the network layer, but it is important and complex enough to merit separate consideration. Internetworking is the subject of Chapter 10.

9.2   X.25

The *X.25* recommendation was approved by the CCITT in 1976, with revisions in 1980 and 1984. It predates the OSI model, but it now conforms to the lowest three layers of the model (partially implementing the network layer) after cooperation between CCITT and ISO. X.25 defines standards for connection of data terminals, computers, and other systems, or DTEs, to packet-switched networks via DCEs. Figure 9.3 shows an X.25 network with connections from DTE **A** to DTE **B** and DTE **A** to DTE **C**.

X.25 protocols are used *only at the interface* to the packet network, with vendor or operator protocols within the network. These vendor or operator

---

[2] This excludes the medium access control portion of the data link layer, the only other layer or sublayer that involves cooperation among peer processes at all nodes but is less complex.

[3] More precisely, the top layer—the packet layer—of X.25 is the most widely implemented OSI network layer standard. X.25 also contains a physical layer and a data link layer.

Figure 9.3   Typical X.25 Network Configuration.

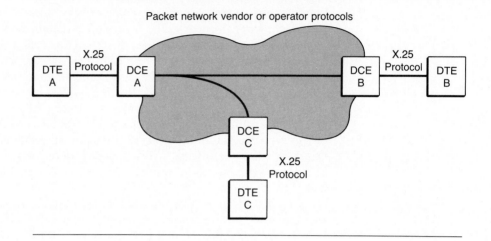

protocols encompass network protocols not specified by X.25 but required for a full network layer implementation.

9.2.1   X.25 Layers

The X.25 standard specifies three layers for the DTE to DCE interface: a *physical layer* (X.21),[4] a *data link layer* (HDLC LAP-B), and a *packet layer*[5] (see Fig. 9.4). The packet layer is the primary subject of our discussion since it corresponds to the reference model network layer. X.21 and LAP-B are discussed in Chapters 4 and 7, respectively.[6]

We have indicated a user process above the packet layer in Fig. 9.4 rather than the upper four layers of the OSI Reference Model. The upper four OSI layers would be used if this were an OSI-compatible implementation, but many X.25 networks are not OSI-compatible.

9.2.2   Virtual Circuit Services

The packet layer provides virtual circuit service to user processes (the transport layer for ISO implementations).[7] Up to 4095 virtual circuits may be assigned to one physical channel (that is, a LAP-B connection). Twelve bits

---

[4]The X.21bis interim standard is allowed in place of X.21. This allows RS-232C or V.24 interfaces to be used, with minor modifications, on networks with analog lines and modems.

[5]The X.25 standard uses the term "level" instead of "layer." We use "layer" for consistency.

[6]An extension of LAP-B, allowing transmission of frames over multiple parallel communications links treated as a common resource to increase throughput or reliability, is an option in X.25.

[7]Earlier versions of the standard included an optional datagram service, but this was never implemented and was dropped in the 1984 revision.

Figure 9.4 X.25 Layers.

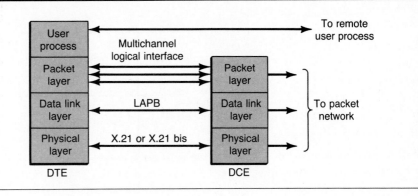

in packet headers identify virtual circuits,[8] using a four-bit logical group number and an eight-bit logical channel number. Both are discussed below.

Two types of virtual circuits are provided: permanent virtual circuits and virtual calls. *Permanent virtual circuits* (PVCs) are comparable to private lines for ordinary telephone service; they are permanently set up, so no connect or disconnect phases are necessary, and are used between locations with traffic justifying their cost. *Virtual calls* (VCs) are comparable to dial telephone calls and require connect and disconnect phases.

permanent virtual circuit

virtual call

Virtual calls at a particular location (DTE) are classified as incoming or outgoing calls. *Logical channel numbers* at each end identify virtual circuits to the DTE and DCE at that end. Logical channel numbers for incoming virtual calls are assigned by the DCE while those for outgoing virtual calls are assigned by the DTE. This raises the potential for conflicts, but conflicts are largely avoided by the numbering scheme in Fig. 9.5.

Logical channel numbers fall into four categories.

1. The lowest numbers are assigned to permanent virtual circuits.

2. Numbers designated for one-way incoming virtual calls are assigned next; such logical channel numbers correspond to numbers for an auto-answer type of service.

3. The highest numbers are for one-way outgoing virtual calls; these correspond to numbers for an auto-dial type of service.

4. An intermediate range of numbers is used for two-way, either incoming or outgoing, virtual calls.

---

[8] Twelve bits allow 4096 virtual circuit numbers, but the all zero address is not a valid address.

Figure 9.5   Assignment of Logical Channel Numbers in X.25.

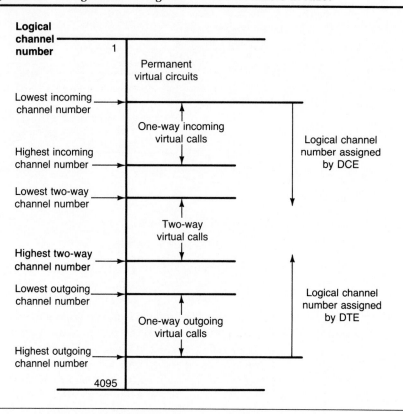

A subscriber (DTE location) subscribes for a predetermined set of numbers in each group—for example, 10 permanent virtual circuits, 20 one-way incoming call numbers, 20 one-way outgoing call numbers, and 15 two-way virtual call numbers. Contiguous sets of logical channel numbers in each category are assigned.

Logical channel numbers have only local significance. Logical channel numbers at each end of a virtual circuit must be mapped into each other so the ends can communicate, but X.25 leaves this to the network administration. An example of the type of mapping required is given below.

search algorithm for assigning logical channel numbers

To assign a logical channel number for a virtual call, a *search algorithm* is used. For an incoming virtual call, the network (DCE) will assign the lowest available logical channel number, in the set allotted (for communication with that DTE via one-way incoming virtual calls), if any is available. If not, it will search the numbers allotted for two-way virtual calls, from lowest to highest number. If all numbers are busy, it will clear the virtual call request with a

*example of search
procedure*

diagnostic code indicating a reason for clearing.[9] Similarly, for outgoing virtual calls, the DTE will search allotted numbers from highest to lowest number. Conflicts can occur only within the two-way virtual call set of numbers where search ranges used by the DCE and the DTE overlap.

As an example of this procedure, a virtual call from DTE **A** to DTE **C** in Fig. 9.3 might be assigned logical channel number 3753 by DTE **A**, if this is the highest available number among those dedicated for outgoing calls; this number is used for communication between DTE **A** and DCE **A**. After the virtual call makes its way through the network, it becomes an incoming call at DCE **C**. DCE **C** might assign it logical channel number 236, if this is the lowest available number among those dedicated for incoming calls. DTE **C** and DCE **C** would use this number. Translations between logical channel numbers at the two ends of a virtual circuit are the responsibility of the network.

9.2.3   X.25 Packet Formats

Transmitted frames in X.25 are *HDLC LAP-B frames* with normal header and trailer; the information part of frames consists of X.25 packets. X.25 data and control packet formats are illustrated in Fig. 9.6. Fields in the packet formats are defined in Table 9.1. Data and control packets are identified by 0 and 1, respectively, in bit 1 of byte 3.

9.2.4   Virtual Calls

VC setup and clearing

Virtual call setup and clearing employ sequences of control packets, with arbitrary sequences of data transfer packets (possibly mixed with some control packets) allowed between setup and clearing. A typical sequence is illustrated in Fig. 9.7. Only the data transfer portion of this sequence would normally be used on permanent virtual circuits. Packet exchanges for setup and clearing are similar to sequences of primitives, but differ in that actual packets are exchanged between a DTE and its associated DCE, so transfers are externally visible. (Primitives are exchanged between protocol layers on one system and are not externally visible.) A frame passing through the network (dotted arrows in figure) does not have to be identical to the frame received by the DCE at the network entrance since the standard does not specify internal network operations, but it must contain all pertinent information.

Call Request and Incoming Call packets contain such information as addresses of calling and called DTEs, assigned virtual call circuit numbers[10]

---

[9] This is equivalent to a busy tone in telephone service, though with a bit more information in the diagnostic code.

[10] The virtual circuit number in the Incoming Call packet will be that assigned to the virtual circuit by DCE **C**, not that originally assigned by DTE **A**.

Figure 9.6   X.25 Packet Formats. (a) Normal (three-bit sequence numbers) data packet format. (b) Control packet format.

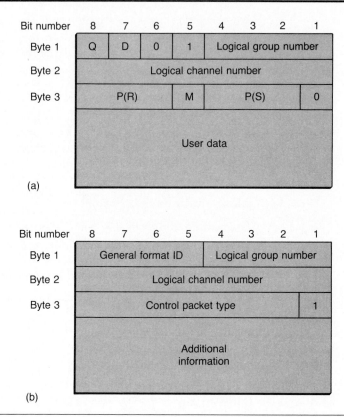

(a)

(b)

and special options or facilities requested. Corresponding parameters, including identical virtual circuit numbers, are in Call Accepted and Call Connected packets, but negotiation of options or facilities is allowed. After completion of setup, the virtual circuit is ready for data transfer. The clearing procedure is similar, but the Clear Confirm packets have only local significance; the Clear Confirm from DCE **A** is returned shortly after DCE **A** received DTE **A**'s Clear Request, and does not indicate the Clear Request has been successfully forwarded to DCE **C** and DTE **C**. Similarly, DTE **C**'s Clear Confirm is not forwarded to DCE **A** and DTE **A**. Any data DTE **C** might have sent after DTE **A** sent the Clear Request is lost.

Descriptions of call setup and clearing procedures are given by the state diagrams in Fig. 9.8, which also define optional sequences. They are interpreted in the same manner as state diagrams in earlier chapters.

Table 9.1 Fields in X.25 Packet Headers.

- The four-bit logical group number and eight-bit logical channel number in each header give the 12-bit virtual circuit number.
- Bits 5–8 in byte 1 comprise a general format identifier (GFI), with the form of the GFI for data packets shown.[a]
- Seven bits in byte 3 of control packets specify the type of packet.
- Use of the Q (qualified data) bit in the data packet header is not defined in the standard; it may distinguish two types of data streams, such as one for user device control and one for user data.
- The D bit indicates whether acknowledgments and flow control have local, $D=0$, or end-to-end, $D=1$, significance.
- The M (more data) bit is one in all except the last packet in a message; hence it is used to locate the end of a message.
- The data field may contain up to 128 bytes unless a different maximum is negotiated. Possible maximums range from 16 to 4096 in powers of 2, but not all are supported.
- P(S) is a send sequence count analogous to N(S) in HDLC.
- P(R) is a receive sequence count analogous to N(R) in HDLC.

[a]An alternative GFI for data packets has bits 6 and 5 set to 1 and 0, respectively. The P(R) and P(S) sequence numbers are then seven bits each, with a fourth header byte added to accommodate the extra bits.

Figure 9.7 Typical X.25 Virtual Call Packet Sequences. Sequences used for permanent virtual circuits omit setup and clearing phases.

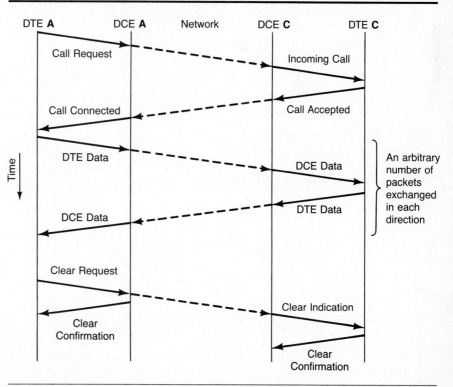

Figure 9.8 State Diagrams for X.25 Call Setup and Clearing Procedures. (a) Call setup. (b) Call clearing. (Reprinted with permission from ITU [International Telecommunications Union], CCITT [International Telegraph and Telephone Consultative Committee], "Interface between Data Terminal Equipment (DTE) and Data Circuit-Terminated Equipment (DCE) for Terminals Operating in the Packet Mode and Connected to Public Data Networks by Dedicated Circuit," *CCITT Red Book,* vol VIII.3 1984. Updated material can be found in the *CCITT Blue Book* and may be obtained from the ITU General Secretariat, Place des Nations, CH-1211 Geneva 20, Switzerland.)

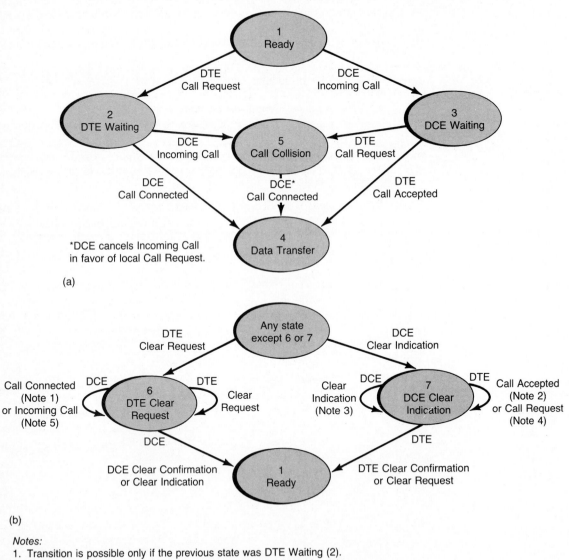

Notes:
1. Transition is possible only if the previous state was DTE Waiting (2).
2. Transition is possible only if the previous state was DCE Waiting (3).
3. Transition may take place after time-out.
4. Transition is possible only if the previous state was Ready (1) or DCE Waiting (3).
5. Transition is possible only if the previous state was Ready (1) or DTE Waiting (2).

9.2.5   Data Transfer

normal data packets

Normal data packets on a virtual circuit, using the headers in Fig. 9.6(a), are numbered with consecutive $P(S)$ values, modulo eight for three-bit $P(S)$ fields and modulo 128 for seven-bit $P(S)$ fields. These sequence counts apply for a particular virtual circuit and are in addition to the similar $N(S)$ counts added by LAP-B at the data link layer. $N(S)$ values are assigned consecutively for successive packets flowing across the link, regardless of which virtual circuit they are assigned to; the data link layer does not even know which virtual circuit is being used. $P(S)$ counts are used in conjunction with $P(R)$ counts for flow control at the packet layer.

Sliding window flow control with window size W is used; the default W is two, but other values (up to the $P(S)$ and $P(R)$ modulus minus one) may be negotiated. $P(R)$ values acknowledge packets as in HDLC—that is, $P(R) = n$ acknowledges all packets with $P(S)$ less than n (modulo the appropriate value).[11] The transmitter can transmit a maximum of W packets, beginning with $P(S)$ equal to the last returned $P(R)$, before acknowledgment. The receiver accepts only data packets next in sequence and within the window; any other data packet causes it to reset the virtual circuit (see the next subsection), so the receive window size is one.

control packets

If data packets are available for return, they may be used to update $P(R)$. *Control packets* can also be used and have the same names as similar HDLC supervisory frames: *Receive Ready (RR), Receive Not Ready (RNR),* and *Reject (REJ).*[12] RR and RNR frames may be transmitted either from DCE to DTE or vice versa, but REJ frames can be sent only from DTE to DCE. Supervisory frames contain $P(R)$ but no $P(S)$ counts. RNR, in addition to updating $P(R)$, indicates the DTE or DCE transmitting it is temporarily unable to accept new data packets; the receiver for this RNR is then prohibited from transmitting data packets until RR clears the RNR condition. Thus a DCE or DTE has two ways of slowing down arrival of packets: postponing transmission of an updated $P(R)$ to reopen the window or transmission of RNR.

Figure 9.9 illustrates flow control between a DTE and its associated DCE. It shows advancing the window by transmitting a data packet, slamming it shut by transmitting RNR, and reopening it with RR.

The definitions of "sender" and "receiver" in Fig. 9.9 depend on the value of the D bit in the packet format (Fig. 9.6a). If $D = 0$, acknowledgments have local significance and flow control is exerted only between a DTE and its DCE. For communication between DTE **A** and DTE **C**, this would mean the sender is DTE **A** and the receiver is DCE **A**, or vice versa, or one is DTE **C** and the other DCE **C**. If $D = 1$, however, acknowledgments have end-to-end significance and flow control is exerted end-to-end, that is, the sender is DTE **A** and

---

[11] $P(R)$ and $P(S)$ fields in Fig. 9.6(a) are in reverse order from corresponding $N(R)$ and $N(S)$ fields in the HDLC control field, Fig. 7.11; this causes a reversal of fields in Fig. 9.9.

[12] REJ is not available in all implementations.

Figure 9.9   Illustration of X.25 flow control for $W = 3$ and modulus of 8. Shaded portions of circles indicate windows applicable at time of sketch.

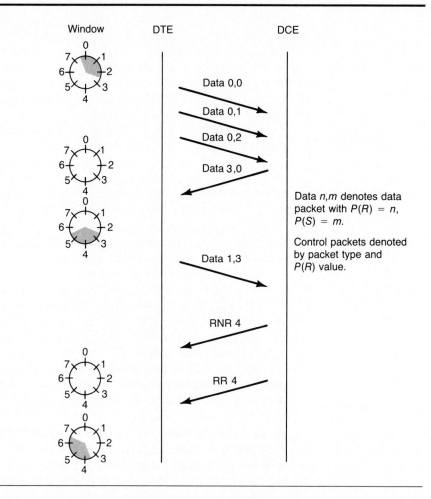

the receiver is DTE **C**, or vice versa. $D = 1$ values are not recognized by some X.25 networks, however.

Both local and end-to-end flow controls are needed. Local flow control can prevent a slow communications link (DCE) from being swamped by a fast data processor (DTE). Other speed mismatches may occur end-to-end, or end-to-end congestion may arise because of multiple virtual circuits active at one time; end-to-end acknowledgments are also used by higher level protocols.

complete packet sequence

Groupings of packets into sequences are defined. An *A* Packet is one in which the *M* bit is one, the *D* bit is zero, and the user data field is full. A *B* Packet is any packet that is not an *A* Packet. A *complete packet sequence* is zero or more *A* Packets followed by a *B* Packet. The network is allowed to split a large packet up into a complete packet sequence for transmission or to combine a complete packet sequence into a single packet before sending it to its destination.

A limited amount of "out of band" data[13] can be transferred by control packets. This is useful when urgent information needs to be delivered quickly; an example might be an "abort" signal from a user who has discovered an infinite loop in a computer program transmitted for execution. *Interrupt packets* are used for this purpose; they are control packets that can carry up to 32 bytes of user data and are delivered to the destination DTE at higher priority than normal data packets. Furthermore, they cannot be blocked by an RNR or a closed window. Acknowledgments of such packets are always end-to-end. A DTE may not send a second interrupt packet on a virtual circuit, though, until acknowledgment of the previous interrupt packet is received.

Another technique for transmitting limited amounts of data, while avoiding the overhead of call setup and clearing, is provided by a *Fast Select* facility. A Call Request packet can request Fast Select by appropriate coding of fields. A maximum of 128 bytes of user data can be included in the packet. The network passes this data to the receiving DTE, which can immediately return a Clear Request with up to 128 bytes of data. This is similar to use of datagrams for short interactive transfers. Alternatively, Fast Select can allow the receiving DTE to return a Call Accepted packet, with up to 128 bytes of user data, after which normal data transfer can begin.

9.2.6   Error Handling

The emphasis of error control at the data link layer is on recovery from transmission errors, but such errors should rarely be seen by the network layer because of the error protection at the data link layer. Error handling at network and higher layers emphasizes protection against other errors.[14]

REJ packets, in implementations that use them, are normally sent to indicate packets have been discarded. A typical situation is one where a DTE runs out of buffer space and does not send RNR in time to keep an overflow packet from arriving. Upon discarding the overflow packet, it should issue a REJ, with

---

[13] This refers to data transferred by procedures not used for normal data to ensure rapid delivery.

[14] Protection against transmission errors in higher layers is sometimes provided for situations in which higher layers are used with lower layer implementations without strong transmission error protection. We give examples of this when discussing the transport layer in Chapter 11. LAP-B's transmission error protection is adequate, though.

appropriate $P(R)$ value; go-back-N retransmission is used. Other errors include incorrect packet formats, invalid entries in packet fields, receipt of packets with $P(R)$ counts outside the allowable range, or expiration of any of several timeouts. Long listings of errors and responses are in the standard.

There are three other responses to errors: *reset, clear,* and *restart,* depending on the type and severity of the error and the type of virtual circuit.

reset

- *Reset* reinitializes a virtual circuit (either virtual call or permanent virtual circuit) by returning the lower edge of the window for each transmission direction to 0 (so the next data packet received must have $P(S) = 0$ to be accepted) and discards all data and interrupt packets currently in the network. The virtual circuit remains connected and in the data transfer state.

clear

- *Clear* is used for more serious errors affecting a single virtual circuit and corresponds to disconnecting that virtual cirtuit. It is used only for virtual calls since permanent virtual circuits are always set up. Figures 9.7 and 9.8(b) illustrate Clears. All packets in transit when a Clear occurs are lost, and the virtual circuit must be set up again before it can be reused.

restart

- *Restart* is used for the most serious error conditions; it affects all virtual circuits, clearing all virtual calls, and resetting all permanent virtual circuits.

Each approach involves potential loss of data. X.25 does not provide a procedure guaranteeing delivery of data in transit when a virtual circuit is cleared or reset. This complicates protocols at the transport layer when they are designed for reliable service with X.25 at the network layer.

9.2.7   Interconnection of Simple Terminals— X.3, X.28, and X.29

Many terminals attached to a typical X.25 based network are simple asynchronous terminals, with minimal function, and unable to handle the synchronous X.25 protocols. Consequently, additional functions and standards have been defined to allow such terminals to be used. The basic function provided is that of a *packet assembler/disassembler (PAD),* which assembles packets from characters generated by a terminal and disassembles packets received from the network to send individual characters to the terminal.

Triple X

Three standards, X.3, X.28, and X.29, known as *Triple X,* have been defined for this purpose. Figure 9.10 shows the relationship of these standards.

X.3 defines the PAD, normally implemented in software at a DCE interfacing a simple Character Mode DTE (C DTE) to the network. X.28 defines the interface between the C DTE and the DCE containing the PAD, including data and control exchanges between them. Finally, X.29 describes protocols allowing parameters in the PAD to be set by an intelligent Packet Mode DTE (P DTE) in the network.

Figure 9.10   X.3, X.28, and X.29 (from [MEIJ82]). From *Computer Network Architectures,* by Anton Meijer and Paul Peeters. © 1982 by Anton Meijer and Paul Peeters. Reprinted by permission of Computer Science Press, an imprint of W.H. Freeman and Company.

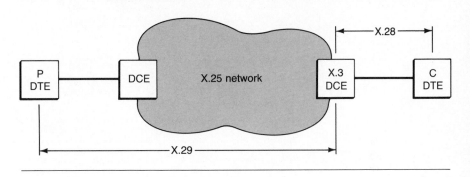

9.2.8   X.25-Based Packet Data Networks

Figure 9.11 illustrates how a typical X.25 packet data network is configured and interfaces with DTEs. Each DCE is connected to a *packet switching node* (*PS*) within the packet data network. The PSs are interconnected in a manner determined by the network vendor or operator. Protocols between PSs and between DCEs and PSs are also determined by the vendor or operator. Protocols at the interfaces are described by the X.25 standard, regardless of the protocols within the packet data network.

Table 9.2 summarizes major differences among four of the most prominent current X.25-based packet data networks—DATAPAC (Canada), TELENET (USA), TRANSPAC (France), and TYMNET (USA)—illustrating that there is little uniformity in the way vendors implement such networks. Detailed descriptions of all four networks are in Appendix 9A.

The first three networks listed have been especially influential in development of X.25. The history of X.25 began in 1974 when Bell Canada announced a decision to install DATAPAC, a network based on packet switching. Simultaneously, they announced a standard protocol (which became the basis of X.25) to provide access to features of DATAPAC. Bell Canada approached other communications suppliers, primarily European PTTs, concerning DATAPAC-like networks. France and England submitted a proposal to the CCITT during 1975, the result of a joint study between DATAPAC, TELENET, and TRANSPAC. This evolved into the X.25 standard.

TYMNET[15] was not as directly involved in the original development, though it was the first to begin service, in 1971, in comparison with 1974 for

---

[15] Acquired by McDonnell Douglas in 1984.

Figure 9.11   Typical X.25-Based Packet Data Network.

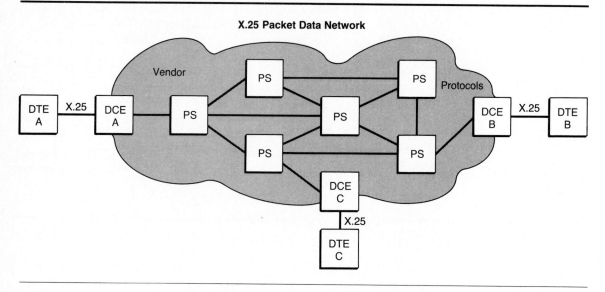

Table 9.2   Comparisons of Four Public Packet Data Networks.

Network	DATAPAC	TELENET	TRANSPAC	TYMNET
Internal packet transmission	Datagrams	Virtual circuits	Virtual circuits	Virtual circuits
Packet assembly	At entry to network	At entry to network	At entry to network	Reassembled at each node
Routing computations	Distributed with exchange of data among nodes	Partially centralized, but with some local data used	Partially centralized, but with some local data used	Centralized, all computation at central location
Flow control	X.25 sliding window, discarding packets, disconnecting VCs	X.25 (both end-to-end and per-hop basis)	X.25 sliding window, prohibiting new VCs, delaying ACKS, disconnecting VCs	X.25 sliding window, sliding window on each link with back pressure

TELENET, 1977 for DATAPAC, and 1978 for TRANSPAC. It has also been influential. TYMNET was originally oriented towards computer timesharing and other types of remote computer access, with X.25 access a relatively recent addition.

These four networks are probably the four largest public packet data networks currently in use, as well as the four most influential ones. All four are linked together and interfaced to a variety of other worldwide networks. It is possible to use any for communication with users around the world.

As the table indicates, one network uses datagrams internally but the others use virtual circuits. Three transmit packets essentially unchanged (aside from possible header changes) over all hops, while the other reassembles packets at each node. One uses almost completely distributed routing computations, two use a combination of centralized and distributed computations,[16] and the fourth uses completely centralized computations. A variety of flow control techniques are also used, the only common denominator being the X.25 sliding window control required by the X.25 standard.

A variety of other comparisons can also be made. For example, the TYMNET approach of using short packets, with reassembly at each node, is better suited for short interactive transfers than for long file transfers. If most messages are short, combining several into one packet, with a single header and trailer, can reduce overhead. Thus TYMNET is best for handling short interactive transfers but the others are better for handling long messages.

The important point in these comparisons is that packet data network implementations are *highly variable.* X.25 puts minimal constraints on packet networks employing X.25 at interfaces with customer equipment.

9.2.9   Limitations of X.25

In comparison with other OSI protocols, X.25 was developed under great time pressure from vendors or operators of networks who wanted their networks to present standardized user interfaces; [SIRB85] gives a survey of its history. It was developed and approved in record-breaking time, with at least one person calling it a "well-engineered political coup" [POUZ80]. It was adopted with known ambiguities under the assumption that an imperfect standard was better than no standard. The networks, and customers implementing the 1976 version, were required to modify their implementations to align with later versions at considerable expense. Nevertheless, these networks have hailed X.25 as a great success since significant delays in adoption would have led to far more variability in interfaces with public packet networks.

X.25 (including its three layers) is often claimed to be *the* OSI Reference Model implementation of the physical, data link, and network layers. It is the implementation of these layers most widely recognized as conforming to the model, but is not a complete implementation of the layers. It was not originally designed to conform to the reference model, but it was modified to make it conform. Specific limitations include the following:

1. Standards bodies have not attempted to make X.25 a complete version of the cited reference model layers. The X.25 packet layer, especially, is an incomplete implementation of the equivalent reference model layer. A complete network layer includes routing, relaying, flow control, and related protocols within the packet network, but X.25 defines only an inter-

---

[16] As Appendix 9A indicates, the approaches used are radically different despite this similarity.

face to a packet network. The undefined aspects are where most of the complexity of the network layer comes in, as all nodes within the packet network must work together to provide these functions.

2.   X.25's lack of a graceful close, and potential loss of data when errors occur, force higher protocol layers to recover from such losses. No alternate routing procedures are built into the protocols; virtual circuits are simply cleared under a variety of conditions and responsibility for reestablishing them left to the user, adding to the complexity of upper layers.

3.   Although X.25 is a standard for interfacing with a packet network, it contains end-to-end features, especially when the D bit is used; interrupt packets are also end-to-end. This makes the distinction between the network layer and the transport layer (supposedly the first end-to-end layer) vague.[17] X.25 contains a few aspects of an ISO transport layer, but it is not (and is not intended to be) a transport layer standard.

4.   Since X.25 was designed to serve as an interface to packet networks, its primary emphasis has been on remote access rather than on resource sharing (that is, sharing of data and programs by users at remote locations). Although resource sharing is handled primarily by upper layers in the OSI hierarchy, capabilities need to be built into lower layers. Such capabilities have not been built into X.25. For example, although transfer of expedited data such as X.25 interrupt packets is at times essential to get urgent information through quickly, there is no requirement that networks with X.25 interfaces provide special handling for interrupt packets or any type of expedited data service.

5.   X.25 is poorly suited for some applications of packet networks. An example is packet voice, which cannot tolerate long or (especially) variable delays. Discarding voice packets that suffer excessive delays is far better than delivering them, but X.25 provides no mechanism for doing this, and the packet formats do not provide for a way to label voice packets as different from other types of packets and subject to special handling. Many persons also feel the datagram service dropped from the standard in 1984 will be needed for important applications, especially those involving short interactive data exchanges.

6.   X.25 was developed for use on public packet data networks using telephone-type facilities. It can be used over other facilities, but cannot necessarily take advantage of their features. For example, it cannot take advantage of possibilities for broadcast communication when geosynchronous communications satellites, packet radio, or local area networks are used.

---

[17] The D bit appears to have aroused more controversy in standards bodies than any other single bit used in any protocol standard!

For the type of operation for which it was designed, X.25 has been successful. The public packet networks for which it was designed are important, but are not the only types of networks that need Layer 3 protocols.

9.3 Circuit-Switched Network Layer

As was stated earlier, a network layer is *not needed* at intermediate nodes of circuit-switched networks during data transfer, since a direct path between source and destination exists. Similarly, a network layer is not needed at end nodes during data transfer; it does not add header or trailer information or perform flow control or routing. Although Fig. 9.2 shows the network layer at end nodes of circuit-switched networks during data transfer, this is for compatibility with other OSI model diagrams; it is null during this time.

During call setup and clearing phases of circuit-switched networks, functions considered to be network layer functions are performed by switching equipment; these involve dialogues between nodes to set up and disconnect circuits. Examples are given in the discussion of the X.21 interface in Chapter 4. Figure 4.6 is a typical example; in the dialogue illustrated two DTEs exchange information to set up a circuit, report on progress in the setup phase, exchange data and clear the circuit. All portions of this dialogue save the data transfer portion involve network layer functions.[18] This type of "network layer" implementation is common to connection and disconnection phases of circuit-switched networks, including those using other interfaces such as RS-232C with auto-dialing equipment, but details differ substantially.

9.4 OSI Network Layer

Aside from X.25, and protocols for internetworking in the next chapter, few protocols for the OSI network layer have been completely defined. There has been no real attempt to define OSI standards for such important network layer functions as routing and flow control, since these are considered to be implementation dependent. Instead, the OSI standards at this level have concentrated on defining functions and services the network layer provides to the transport layer and its interfaces with the transport and data link layers.

Three primary ISO standards have been published. A standard describing the Internal Organization of the Network Layer (IONL), a Network Layer Ser-

---

[18]A data link layer protocol, prescribing formats of the data transmissions (for example, all information characters preceded by two SYNs) is also illustrated in Fig. 4.6.

Figure 9.12   Internal Organization of the OSI Network Layer.

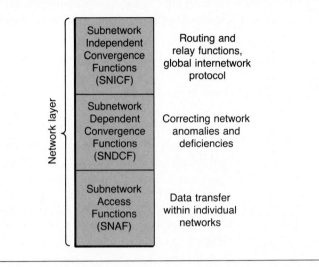

vice Definition, defining services to be provided by the network layer (including exchanges of primitives to invoke services), and a standard for network layer addressing have been adopted. We describe the first two.

9.4.1   Internal Organization of the Network Layer

The IONL Standard (ISO 8648) describes organization of the network layer into the three sublayers illustrated in Fig. 9.12. The terminology for sublayers is typical of that coming from standards committees. As was indicated above, none of the three sublayers specify algorithms such as those for routing or flow control. This subdivision of the network layer into three sublayers causes some persons to claim that the OSI Reference Model actually has at least nine layers instead of seven. According to ISO documentation, however, individual sublayers can be bypassed, but complete layers (or all sublayers in one layer) cannot be bypassed.

SNICF sublayer   Functions in the *SNICF sublayer* do not require accommodation to specific networks. The primary functions are internetwork routing and relaying plus other functions to implement an internetwork protocol. The internetworking standards in the next chapter are defined for this sublayer.

SNDCF sublayer   The *SNDCF sublayer* has been included to allow for situations where some networks do not provide all features assumed by the SNICF sublayer. Relatively little progress has been made on defining this sublayer, which should be highly dependent on individual networks used.

SNAF sublayer

The *SNAF sublayer* is defined to specify how network layer entities make use of functions of a network—for example, operation of a protocol describing an interface to a network. X.25 fits into this sublayer.

9.4.2 Network Layer Services

The Network Layer Service Definition (ISO 8348) describes connection-oriented operation, with an addendum specifying connectionless operation. We discuss connection-oriented service first and then connectionless service.

The *connection-oriented mode* includes *connection establishment, data transfer,* and *connection-release* phases of operation. A connection is defined by the following:

definition of connection

- A path established by a three or more party agreement between end systems and a network or networks

- Parameter values and options determined by negotiation

- Identification of the connection (for example, the virtual circuit number)

- The context within which successive units of data are logically related, sequenced, and controlled.

No relationship of this type between end systems is established for *connectionless operation.* All that is needed is an *association between the communicating entities,* which determines characteristics of data to be transmitted, plus an agreement between each entity and the service provider.

A total of six services are associated with connection-oriented service (one each with the connection-establishment and connection-release phases and four with the data transfer phase), but only one is defined for connectionless service. Each service is invoked by use of one or more primitives (Request, Indication, Response, and Confirmation).[19] Figure 9.13 illustrates use of primitives at the network layer.

Figure 9.14 shows the network services for connection-oriented service and associated primitives during the three phases of Connection Setup, Data Transfer, and Disconnection. There are a total of six network services defined, one each for Connection Setup and Disconnection and four for Data Transfer. Two of the latter—Receipt Confirmation and Expedited Data Transfer—are optional, but Data Transfer and Reset are required. Services during Connection Setup and Disconnection are obvious ones for these phases.

---

[19] These are true primitives, passed across interfaces between protocol layers implemented in one system. They are not externally visible since no information flows across communications links. The OSI standards indicate that conformance to the reference model only requires conformance to its externally visible manifestations, so primitives do not have to be implemented in precisely the manner implied by our discussion.

Figure 9.13   OSI Use of Primitives for Network Layer Services.

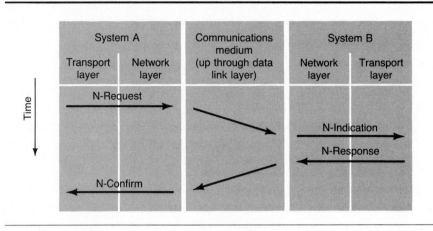

Figure 9.15 is a state transition diagram showing allowed sequences of primitives. The "N-Disconnect primitives" shown correspond to either Requests or Indications, depending upon whether release is initiated by the service user (transport layer) or service provider (network layer), respectively. A detailed state transition diagram for release would resemble Fig. 9.8(b).

A variety of parameters are associated with primitives; these are listed in Table 9.3. For connection-establishment primitives, the list is especially extensive since connection establishment involves negotiation of whether optional facilities such as receipt confirmation or expedited data are to be used.

Limited amounts of user data can be included in primitives. User data is, of course, included with N-Data or N-Expedited-Data Request or Indication primitives. Reset and Disconnect Request and Indication primitives also allow a reason for the Reset or Disconnect to be specified.

There is one connectionless network service, N-Unitdata, with Request and Indication primitives. Each primitive includes source and destination addresses, quality of service parameters and user data. Each data unit is treated independently. The network layer does not handle flow control or keep data units in sequence; data units may follow different routes and arrive in different order from that in which they entered the layer. Resequencing, flow control, error handling, and so forth are left to higher layers.

Quality of Service   Parameters for *Quality of Service* (*QOS*) are also defined and have been adopted by a number of networks. Adequate QOS parameters are essential if users are to be satisfied with network service. Parameters for connection-oriented network service are listed in Table 9.4, and those for connectionless service in Table 9.5.

Figure 9.14 Network Services and Primitives for Connection-Oriented Service.

Phases						
Connection Setup		**Data Transfer**				**Disconnection**	
Service	Primitives	Required service	Primitives	Optional service	Primitives	Service	Primitives
Connection Establishment	N-Connect Request	Data Transfer	N-Data Request	Receipt Confirmation	N-Data Acknowledge Request	Connection Release	N-Disconnect Request
	N-Connect Indication		N-Data Indication		N-Data Acknowledge Indication		N-Disconnect Indication
	N-Connect Response	Reset	N-Reset Request	Expedited Data Transfer	N-Expedited Data Request		
	N-Connect Confirm		N-Reset Indication		N-Expedited Data Indication		
			N-Reset Response				
			N-Reset Confirm				

Figure 9.15   State Transition Diagram for Sequences of Network Service Primitives in OSI Reference Model Architecture. (Reprinted with permission from ITU [International Telecommunications Union], CCITT [International Telegraph and Telephone Consultative Committee], "Network Service Definition for Open Systems Interconnection (OSI) for CCITT Applications," *CCITT Red Book,* vol VIII.5 1984. Updated material can be found in the *CCITT Blue Book* and may be obtained from the ITU General Secretariat, Place des Nations, CH-1211 Geneva 20, Switzerland.)

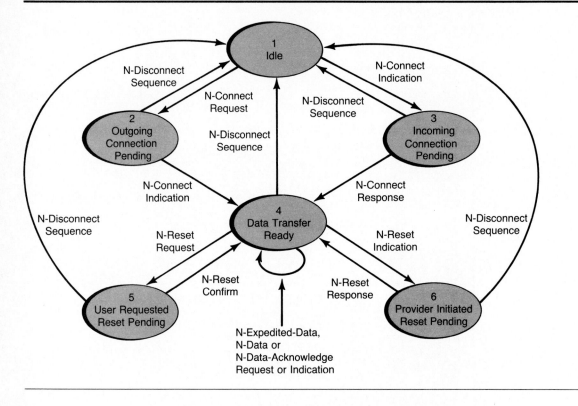

Table 9.3   Parameters Associated with Primitives Used in OSI Connection-Oriented Network Service.

Primitive	Parameters
N-Connect Request or N-Connect Indication	Called address, calling address, receipt confirmation selection, expedited data selection, QOS-parameter set, NS-user data
N-Connect Response or N-Connect Confirm	Responding address, receipt confirmation selection, expedited data selection, QOS-parameter set, NS-user data
N-Data Request or N-Data Indication	NS-user data, confirmation request
N-Data-Acknowledge Request or	

Table 9.3 *continuing*

Primitive	Parameters
N-Data Acknowledge Indication	
N-Expedited-Data Request or N-Expedited-Data Indication	NS-user data
N-Reset Request	Reason
N-Reset Indication	Originator, reason
N-Reset Response or N-Reset Confirm	
N-Disconnect Request	Reason, NS-user data, responding address
N-Disconnect Indication	Originator, reason, NS-user data, responding address

(Reprinted with permission from ITU [International Telecommunications Union], CCITT [International Telegraph and Telephone Consultative Committee], "Network Service Definition for Open Systems Interconnection (OSI) for CCITT Applications," *CCITT Red Book,* vol VIII.5 1984. Updated material can be found in the *CCITT Blue Book* and may be obtained from the ITU General Secretariat, Place des Nations, CH-1211 Geneva 20, Switzerland.)

Table 9.4 QOS Parameters for OSI Connection Oriented Network Service.

- *Network connection establishment delay:* Time delay to establish a connection.
- *Network connection establishment failure probability:* Ratio of establishment failures to attempts.
- *Throughput:* Amount of data transferred per unit time in each direction.
- *Transit delay:* Elapsed time between a request primitive for sending data from a network service user and an indication primitive delivering data to the destination.
- *Residual error rate:* Ratio of number of incorrect, lost, and duplicate data units to total number transferred across the network service boundary.
- *Transfer failure probability:* Ratio of total transfer failures to total transfer attempts.
- *Network connection resilience:* Probability of network service provider invoked release and reset during a network connection.
- *Network connection release delay:* Delay between a network service user invoked disconnect request and successful release signal.
- *Network connection release failure probability:* Ratio of release requests resulting in release failure to total attempts.
- *Network connection protection:* Extent to which network service provider tries to prevent unauthorized masquerading, monitoring, or manipulation of network service user data.
- *Network connection priority:* Relative importance of network connection with respect to other network connections and to data on the network connection.
- *Maximum acceptable cost:* Limit of cost for the network service.

Table 9.5   QOS Parameters for OSI Connectionless Network Service.

- *Transit delay:* Elapsed time between N-Unitdata request and the corresponding N-Unitdata indication.
- *Protection from unauthorized access:* Prevention of unauthorized monitoring or manipulation of network service user originated information.
- *Cost determinants:* Specification of cost considerations for the network service provider to use in selection of a route for the data.
- *Residual error probability:* Likelihood that a particular unit of data will be lost, duplicated or delivered incorrectly.
- *Priority:* Relative priority of a data unit with respect to other data units that may be acted upon by network service provider.
- *Source routing:* Designation by network service user of the path data are to follow to the destination address.

9.5   ARPANET CSNP-CSNP Network Layer Protocols

ARPANET CSNP-CSNP protocols that fit into the OSI network layer have been some of the most influential protocols developed for network layer functions. (Recall that we use CSNP for an IMP or TIP.) The main references for the DoD Reference Model, [CERF83], [ENNI83], and [PADL83], do not call this a separate layer, but the protocols are essential. Current versions are implemented in the ARPANET and Defense Data Network and are the network layer CSNP-CSNP protocols in our discussion of the DoD Reference Model in Chapter 3. Aside from internetwork protocols discussed in Chapter 10, they are the main network layer protocols in ARPANET and the DDN.

CSNP-CSNP protocols

Two categories of *CSNP-CSNP protocols* are defined: adjacent CSNP and source CSNP–destination CSNP protocols. Transmission between adjacent CSNPs is via datagrams, but the source CSNP–destination CSNP protocol provides extra functions to give the equivalent of virtual circuit service; this involves end-to-end (source CSNP–destination CSNP) acknowledgments, recovery from lost or duplicate packets, and resequencing packets. We emphasize routing and flow control in CSNP-CSNP protocols. Primitives are not defined, but primitives for adjacent CSNP protocols would be similar to N-Unitdata requests and indications. A more complete set of primitives would be needed for source CSNP–destination CSNP protocols.

Figure 9.16 shows the format of the 128-bit ARPANET CSNP-CSNP packet header, with definitions of fields in Table 9.6, but it is complex and we will not discuss all of its fields. For data flowing through the ARPANET or DDN, this header is used in addition to the data link layer header and trailer in Fig. 7.6 and, in many situations, headers for the DoD Internet Protocol in Chapter 10 and the DoD Transport Layer Protocol in Chapter 11.

Figure 9.16   ARPANET Data Packet CSNP–CSNP Header (from [TANE88] Andrew Tanenbaum, *Computer Networks, 2/e,* © 1988, p. 260. Reprinted by permission of Prentice-Hall, Inc., Englewood Cliffs, NJ.).

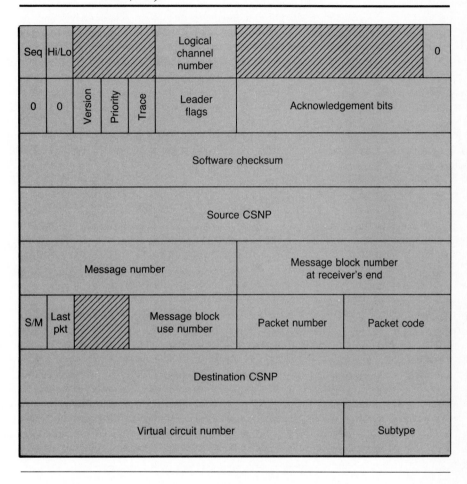

ARPANET does not make the clear distinction between data link and network layers made in more recently designed networks. This is reflected in the header in Fig. 9.16, which contains fields relating to the data link layer, the network layer, the source CSNP–destination CSNP protocol, a host-CSNP protocol, and a host-host protocol. The Seq, Logical channel number, and Acknowledgment bits fields are used by the data link layer protocol, with most of the rest used by the source CSNP–destination CSNP protocol.

Table 9.6　Fields in ARPANET CSNP–CSNP Data Packet Header.

- *Seq* 1-bit sequence number for stop-and-wait protocol between adjacent CSNPs (see Section 7.4.).
- *Hi/Lo* tells end of the line packet came from. Alerts CSNP if line looped back on itself for maintenance purposes, so data arrives back at sender.
- *Logical channel number* tells which of eight stop-and-wait channels packet belongs to (see Section 7.4).
- *Version bit* tells which version of CSNP program is being used. Allows installation of new version of the program while network operating.
- *Priority bit* provides two levels of priority for messages.
- *Trace bit* tells subnet to trace packet so network management can determine how it was routed and perform other measurements.
- *Leader flags* passed by source host to source CSNP, sent across network and passed to destination host, for use by host-CSNP protocol.
- *Acknowledgment bits* provide piggybacked acknowledgments for each of eight logical channels (see Section 7.4); odd/even sequence number count for last packet received correctly for each channel.
- *Software checksum* computed by source CSNP and verified by each CSNP en route to detect transmission errors or CSNP memory failures.
- *Source CSNP* identifies packet's source CSNP.
- *Message number* assigned by source CSNP, with consecutive messages between a pair of hosts numbered consecutively for flow control.
- *Message block number* used by destination CSNP as index into its tables to find variables pertaining to traffic for host-host pair.
- *S/M bit* indicates single-packet or multi-packet message.
- *Last pkt bit* indicates last packet of multipacket message.
- *Message block use number* incremented each time message block used to prevent problems due to reclaiming unused message blocks.
- *Packet number* numbers packets of multipacket message.
- *Packet code* indicates type of message; 0 for normal data, other codes for various control packets.
- *Destination CSNP* tells routing algorithm where to send packet.
- *Virtual circuit number* used by host-host protocol to identify what virtual circuit packet pertains to.
- *Subtype* distinguishes normal packets from datagrams.

9.5.1　ARPANET Routing

The first ARPANET routing algorithm was replaced by a new algorithm in 1979 after "patches" to the original algorithm could not solve its problems. We discuss both algorithms, since both have been influential. It is instructive to consider why problems were encountered with the original algorithm.

Both algorithms are distributed adaptive algorithms; each CSNP computes routing tables using information from other CSNPs in addition to its

Figure 9.17 The ARPANET As It Existed in 1979.

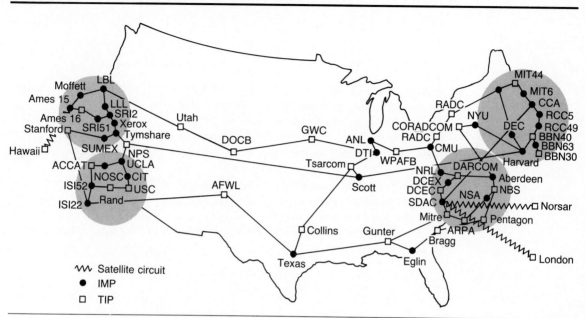

own measurements and attempts to minimize delay to reach the destination. Also both are influenced by congestion variations or other network changes during message transfer. Since routes may vary for successive packets comprising a message, these routing algorithms contribute to the possibility packets may arrive at the destination out of order.[20]

Figure 9.17 shows a configuration of the ARPANET as it existed in 1979. CSNPs are multiply connected by 50 kbps leased lines. (A few lines in various generations of ARPANET have used other speeds.)[21] Even brief inspection of the figure will verify that numerous pairs of nodes are separated by eight to 10, or even more, hops. The algorithms we describe are executed by each CSNP along the path from source to destination.

original ARPANET routing algorithm

The original routing algorithm was based on the *distributed asynchronous Bellman-Ford-Moore algorithm* described in Subsection 8.4.4. Nodes used information they measured to update delay estimates and shared information with neighbors. Each node kept track of two vectors:

---

[20] The ARPANET DLC protocol in Section 9.4 also makes it likely that packets will arrive out of order.

[21] The newer DDN uses 56 kbps lines.

$$
\mathbf{D}^{(i)} = \begin{bmatrix} D_1(i) \\ \cdot \\ \cdot \\ \cdot \\ D_N(i) \end{bmatrix} \text{ and } \mathbf{S}^{(i)} = \begin{bmatrix} S_1(i) \\ \cdot \\ \cdot \\ \cdot \\ S_N(i) \end{bmatrix}, \tag{9.1}
$$

with $\mathbf{D}^{(i)}$ node **i**'s estimates of minimum delays to each of the network nodes, $\mathbf{S}^{(i)}$ the optimum successor nodes for each destination according to node **i**'s view of the network, and N the number of nodes. Each node also kept track of estimated delays to its neighbors. Estimated delay to reach a neighboring node was computed as the number of messages queued for transmission to that node plus a bias factor to avoid stability problems (see Section 8.5), and total delay found by summing delays along the route to the destination.

Periodically (approximately every 2/3 second) the ARPANET CSNPs exchanged delay vectors with their neighbors. This allowed each node to update its delay and successor node tables and adjust its routing. Updated delay estimates were given by

$$
D_j(i) = \min_{k \in N(i)} [D_k(i) + D_j(k)], \tag{9.2}
$$

with $N(i)$ the set of neighbors of node **i**. This is equivalent to Eq. (8.12), but generalized to allow different destination nodes, not just node **1**.

Consider an example based on the simple network in Fig. 9.18, with node **1** initially using the delay and successor vectors

$$
\mathbf{D}^{(1)} = \begin{bmatrix} 0 \\ 1 \\ 2 \\ 4 \\ 3 \\ 6 \end{bmatrix} \text{ and } \mathbf{S}^{(1)} = \begin{bmatrix} 0 \\ 2 \\ 3 \\ 2 \\ 5 \\ 3 \end{bmatrix}.
$$

Also assume that **1**'s estimates of delays to reach its neighboring nodes (**2, 3** and **5**) were 1, 2, and 3, respectively; these values agree with those for nodes **2, 3**, and **5** in the table. Zeros for delays and successor nodes in the position representing node **1** denote the tables are at node **1**. Node **1** then receives the following delay vectors from its neighboring nodes.

$$
\mathbf{D}^{(2)} = \begin{bmatrix} 2 \\ 0 \\ 5 \\ 3 \\ 3 \\ 4 \end{bmatrix}, \mathbf{D}^{(3)} = \begin{bmatrix} 1 \\ 4 \\ 0 \\ 1 \\ 3 \\ 2 \end{bmatrix} \quad \text{and} \quad \mathbf{D}^{(5)} = \begin{bmatrix} 2 \\ 3 \\ 1 \\ 1 \\ 0 \\ 3 \end{bmatrix}.
$$

Figure 9.18   ARPANET-Type Network for Routing Example.

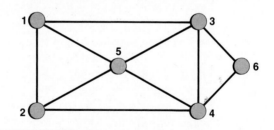

Applying Eq. (9.2) yields the new delay and successor vectors:

$$\mathbf{D}^{(1)}_{new} = \begin{bmatrix} 0 \\ 1 \\ 2 \\ 3 \\ 3 \\ 4 \end{bmatrix} \quad \text{and} \quad \mathbf{S}^{(1)}_{new} = \begin{bmatrix} 0 \\ 2 \\ 3 \\ 3 \\ 5 \\ 3 \end{bmatrix}.$$

(Readers should verify these calculations.)

For this particular example, two of node **1**'s delay estimates changed, both that for delay to node **4** and that for delay to node **6** decreasing. One change in a successor node results, node **3** replacing node **2** as the preferred successor node to reach node **4**.

In addition to these vectors used for routing, each node maintained a vector indicating its estimate of the minimum number of hops to reach each destination. This vector was used for determining connectivity; a node was declared disconnected any time the estimated number of hops to reach it exceeded the number of nodes in the network. The hop count, $\mathbf{H}^{(i)} = [h_j^{(i)}]$ at node **i**, was updated as follows:

$$h_k^{(i)} = \begin{cases} 1 & k \in \mathbf{N}(i), \\[2ex] 1 + \min_{j \in \mathbf{N}(i)} h_j^{(i)} & \text{otherwise} \end{cases} \tag{9.3}$$

that is, the number of hops to a neighbor node was taken as 1 and the number of hops to any other node as 1 plus the smallest number of hops from a neighbor node to the other node.

The strongest point of this old algorithm was simplicity. Nodes did not need to know the topology of the network (that is, which nodes were connected to which). They simply kept track of queue lengths on their outgoing links, received delay estimates and hop counts, and computed next nodes along routes using the formulas above. The algorithm also proved to be inexpensive in terms of use of transmission facilities for routing messages, utilization of CPU cycles at nodes, and required memory space [MCQU78a].

Although this algorithm operated fairly satisfactorily for several years, there were enough problems to motivate installation of a completely new algorithm in 1979. Delay estimates were only approximations to actual delay; delay at a node is strongly influenced by queue length, but factors such as packet length, propagation delay (especially for satellite links), node processing time, and so forth also affect delay. Furthermore, queue lengths change rapidly, and sample sizes were not large enough to be statistically significant, so "optimum" routes changed rapidly, possibly several times while a packet was en route. Updating of tables (every 2/3 second) was rapid in comparison with dissemination of appropriate information, and nodes often updated tables based on inconsistent views of the network. The algorithm for determining node reachability also reacted much more quickly to "good news" than to "bad news" (for reasons explained in Section 8.5). Although the algorithm normally delivered packets rapidly, there were occasions when packets spent extremely long times in the network, apparently due to looping.[22]

new ARPANET routing algorithm

The new routing algorithm uses actual measurements of delays between nodes, rather than queue length, as a delay measure. These measurements are broadcast to all nodes regularly, so each has good estimates of internodal delays throughout the network; each node is also required to know the *geometry* of the entire network. This allows nodes to use a standard shortest-path algorithm to compute truly optimum routes to destination nodes (assuming consistent data). A modified version of *Dijkstra's algorithm* in Chapter 8 is used. The primary modifications have been developed to minimize computations when delays change; for example, no recomputation is necessary when a delay on a link in the shortest-path routing tree decreases or when a delay on a link not in the shortest-path tree increases, since neither change can alter optimum routes.

Delay over a link is measured by time stamping each packet with its arrival time at a node and the time it is sent out from the node (on a final successful transmission if retransmissions are needed). The time to transmit the packet on the link (obtained by table lookup indexed by packet length and line speed) and propagation delay (known for each link) are added and the resulting values averaged over a 10-second period to give an estimate of delay. If this differs from the previous estimate by more than a threshold, the

---

[22]Some spectacular malfunctions are worth mentioning; they resulted from failures never visualized during design. (No fully satisfactory method for avoiding such problems is known, but designers should be aware of the possibilities; techniques used in ARPANET to try to minimize problems are in [MCQU78a].) In one case an IMP malfunctioned every few days, so it forwarded a delay table of all zeros to its neighbors. The other IMPs took this to mean the only IMP that was not operating properly provided the best route to everywhere! Soon a major portion of network traffic was directed toward the malfunctioning IMP and

new estimate is transmitted to all neighbors for flooding out over the entire network. Immediately after an update has been sent out, the threshold for changes is 64 msec, but this is reduced by 12.8 msec each time it is not exceeded, ensuring at least one update per 60 sec. Any time a topology change is detected, however (for example, a line going down or coming up), an update is sent out immediately.

The new algorithm has worked well since its installation, with very few problems encountered.

9.5.2 ARPANET Flow Control

Flow control in ARPANET is based on end-to-end windows, so it is part of the source CSNP–destination CSNP protocol. The entire packet stream between a pair of hosts is viewed as a single data flow. For each flow, a window of eight messages is allowed between origin and destination nodes. Each message may consist of one or more packets, up to a maximum of eight, so window size measured in packets can range from eight to 64. Each transmitted message carries a number indicating its position in the window.

Upon disposing of a complete message, the destination node sends a control message back to the origin to allow it to advance its window by one position. In the ARPANET, this message is called RFNM (ready for next message) and also serves as an end-to-end acknowledgment. If an RFNM is not received within a time-out period, the origin sends a control packet to the destination to inquire whether the message was received. This protects against lost RFNMs and provides for retransmission of lost messages.

An additional mechanism is used to ensure that there is enough memory space to reassemble messages at their destination. Each multipacket message must reserve enough buffer space for reassembly at the receiver before it can be transmitted. A reservation request (REQALL), or request for allocation, is sent from the origin to the destination node. The reservation is granted when the destination sends an allocate (ALL), message back to the origin. In order to avoid delays from requesting and receiving separate allocations for each message comprising a long file transfer, ALL messages are piggybacked on returning RFNMs for multipacket messages so that there is no reservation delay for messages after the first message in a file. If the reserved buffer space is not used by the origin node within a timeout, it is returned to the destination via a "give back" message.

---

normal operations came to a halt. In another case an IMP had an incorrect instruction, due to a memory failure, in its routing program. Consequently, it computed an incorrect pointer to routing data in memory, using whatever data was there as routing data. This caused routing throughout the network to oscillate wildly as the failed IMP sent out incorrect data. In another case the IMP in Aberdeen, Maryland, identified itself as the IMP at UCLA in California (5000 km and many hops away). As a result, no East Coast traffic in the United States was able to reach UCLA, even though everything near there was working properly.

Figure 9.19   Typical Phase IV DNA Network.

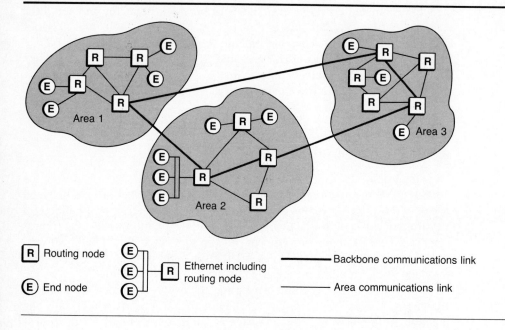

Single packet messages do not need to obtain a reservation before being transmitted. If such a message finds the destination's buffers full, however, it will be discarded. After a timeout, an explicit buffer reservation will be obtained and it will be retransmitted.

Historically, the ARPANET has been influential in the development of flow control algorithms since several deadlock conditions, such as those discussed in Section 8.10, were observed during its operation. This has motivated much of the work that has been done on deadlock avoidance.

9.6   DNA Routing Layer

The *DNA routing layer* implements a *datagram* service, which delivers packets on a best-efforts basis, relying on the next (end communications) layer to convert this into a virtual circuit service. It also incorporates flow control mechanisms to limit congestion. We concentrate on routing and flow control. A typical Phase IV DNA network is illustrated in Fig. 9.19 (which is a duplicate of Fig. 3.18).

The *DNA routing layer header* is illustrated in Fig. 9.20; its fields are described in Table 9.7.

datagram service

Since the DNA routing layer operates a *datagram service,* the major primitives needed are equivalent to the N-Unitdata Request and Indication primi-

Figure 9.20   DNA Routing Layer Header.

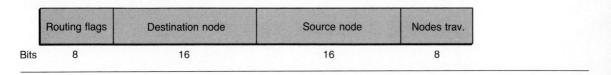

Routing flags	Destination node	Source node	Nodes trav.
Bits 8	16	16	8

tives in the ISO architecture. DNA literature defines the needed functions as procedure calls rather than primitives, however.

9.6.1   DNA Routing

Level I and Level II routes

Routing uses *user-defined* link costs and adapts only to topology changes (such as a node or link coming up or going down), not to congestion. Link costs may be specified by users, but a suggested set of costs is provided, with proposed costs inversely proportional to link capacity. Two levels of routes are computed: *Level I routes* within areas and *Level II routes* between areas. Routes are computed only between *routing nodes;* only one link between an end node and its associated routing node exists, so this link completes all routes involving that end node. Level I routing nodes compute only Level I routes, while Level II routing nodes compute both Level I and Level II routes. No distinction is made between classes of service; all packets are treated alike.

A packet with destination address in its own area is routed to its destination by the routing node serving the source. If the destination is in another area, it is routed to the closest Level II routing node in the source's area, from this node to a Level II routing node in the subarea containing the destination; and finally within this area to the destination node. All routing is done by a variation on the *old ARPANET* (Bellman-Ford-Moore) routing algorithm, adapting only to network topology changes.

Table 9.7   Fields in DNA Routing Layer Header.

- *Routing Flags* (8 bits): Flags used by layer. They include:
  - *Return to sender bit:* indicates packet is on its way back to source.
  - *Return to sender request bit:* Requests packet be returned to sender if it cannot be delivered to destination.
  - *Other bits:* Indicate whether Phase III or Phase IV header formats are being used, whether padding is being used, and so forth.
- *Destination node* (16 bits): Address of destination node.
- *Source node* (16 bits): Address of source node.
- *Nodes traversed* (8 bits): Number of nodes packet has passed through on its way to destination node. Two high order bits of field set to zero so maximum count is 63.

9.6.2   DNA Routing Layer Flow Control

The routing layer implements datagram service, so it does not guarantee message delivery. The primary flow control techniques are discarding transit packets and blocking packets originating at a node from entering the network if the queue for a particular outgoing link exceeds a threshold. The threshold is determined by *Irland's square root rule* [IRLA78] and is equal to

$$Q_{T_i} = \left\lceil \frac{N_{B_i}}{\sqrt{N_{L_i}}} \right\rceil, \tag{9.4}$$

with Q_{T_i} the threshold on queue size at node **i**, N_{B_i} the total number of buffers at the node, N_{L_i} the number of outgoing links from the node, and $\lceil x \rceil$ the smallest integer $\geq x$. This has been shown to be an excellent compromise between allowing a single outgoing link to use up all buffer space and reserving equal fractions of the buffer space for each outgoing link.

DNA also uses a *hop count* to limit maximum lifetime of a packet. This guards against the possibility of a packet looping indefinitely due to inability of the routing mechanism to get it to its destination. A count field in the routing layer header maintains a count of how many nodes the packet has visited and is incremented each time the packet arrives at a node. When this count exceeds a user-defined threshold (in excess of the maximum number of hops on any path between two nodes in the network), the packet is discarded.

9.7   SNA Path Control Layer

session

In SNA a logical connection between two network addressable units (NAUs) is called a *session*. A session can be activated, tailored to provide various protocols, and deactivated upon request. Sessions compete for resources such as links. The *path control layer* is the primary layer handling this competition and is the SNA layer corresponding most closely to the OSI network layer.

virtual circuits

The primary functions of path control are routing and flow control. All communication is via *virtual circuits,* and path control is concerned with establishment, operation, and clearing of virtual circuits. It also resolves addressing for subareas and elements, performs message segmentation and reassembly, and provides for transmission priorities and classes of service such as fast response, secure routes, or reliable connections.

subarea and peripheral nodes

An SNA network is illustrated in Fig. 9.21 (a duplicate of Fig. 3.21b). Host (**H**) and communication processor nodes (**CP**) are relatively full function nodes, called *subarea nodes,* and participate in routing, but terminal (**T**) and communication controller nodes (**CC**) have limited capabilities and do not participate in routing. Only subarea nodes recognize complete network

Figure 9.21   Typical SNA Network.

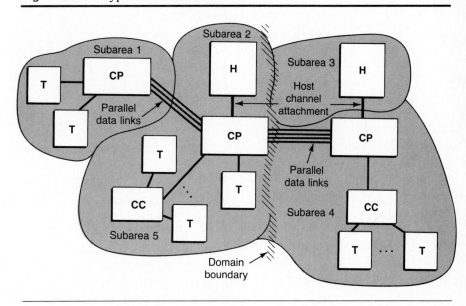

addresses; *peripheral nodes* recognize only local addresses defined within their own subareas.

Path control headers have several formats to allow for varying node capabilities and SNA releases.[23] Figure 9.22 illustrates three of six formats; fields are defined in Table 9.8.[24] FID2 is used between a boundary node (subarea node to which peripheral nodes are attached) and an adjacent communication controller node, FID3 is used between a boundary node and minimal function peripheral node and, FID4 is used between hosts and **CP**s (or subarea nodes).[25] Header lengths are 6, 2, and 26 bytes, respectively.

In the following section we concentrate on SNA's approaches to routing and flow control, plus procedures for transmission group control to enable several parallel communications links to appear to higher layers as equivalent to a single higher speed link. Other functions such as segmentation and reassembly, plus use of transmission priorities and classes of service, are

---

[23] These are called Transmission Headers in SNA documentation, in keeping with SNA's tendency to define as many different terms as possible. We identify them with the name of the layer adding and using them in keeping with terminology used for other architectures.

[24] The formats are complex, so we give only fields we discuss. The term "other control data" indicates fields largely irrelevant for our discussion.

[25] The name "FID" is another example of SNA's unnecessary proliferation of terminology; simply calling these different path control header formats would be more meaningful.

Figure 9.22 SNA Path Control Headers. (a) FID2 format. (b) FID3 format. (c) FID4 format.

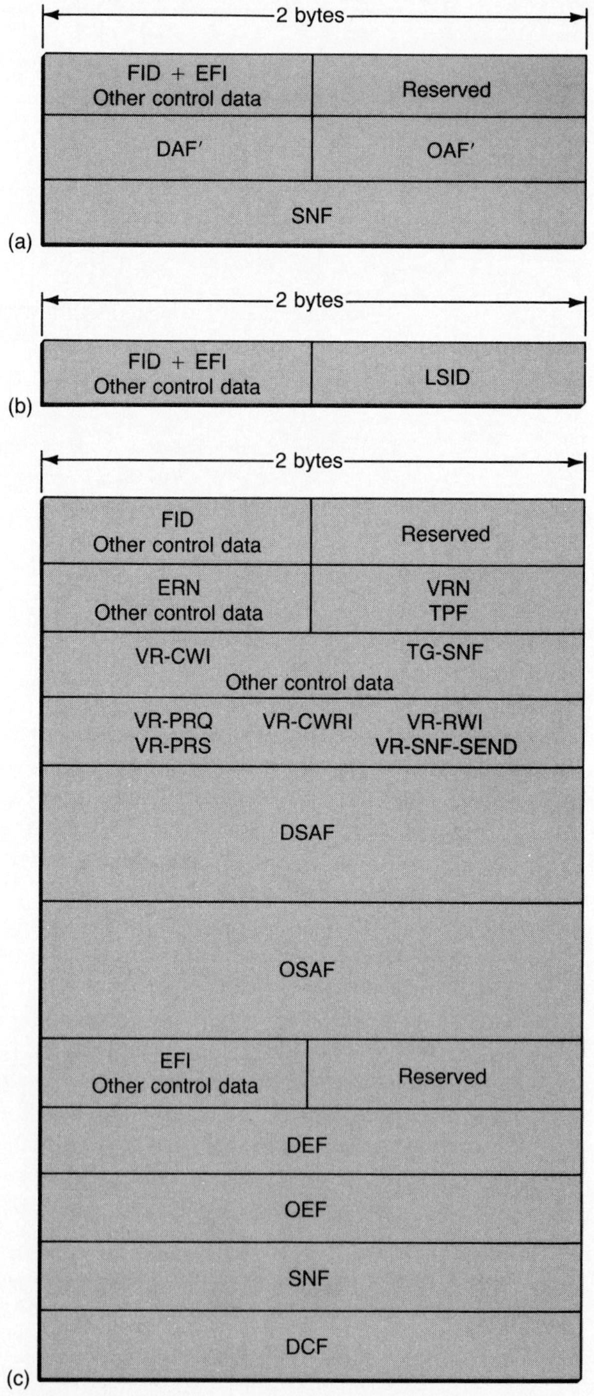

Table 9.8 Fields in SNA Path Control Headers.

- *FID* (4 bits): Format Identifier (indicating type of format)
- *EFI* (1 bit): Expedited Flow Indicator
- *DAF'* (1 byte): Destination Address Field (version used within subarea)
- *OAF'* (1 byte): Origin Address Field (version used within subarea)
- *SNF* (2 bytes): Sequence number field (sequence count for data units)
- *LSID* (1 byte): Local Session Identification (identifies other party involved in session)
- *ERN:* (4 bits): Explicit Route Number (see routing discussion)
- *VRN* (4 bits): Virtual Route Number (see routing discussion)
- *TPF* (2 bits): Transmission Priority Field (data unit priority)
- *TG SNF* (12 bits): Transmission Group Sequence (sequence number for transmissions over transmission group; see description)
- *VR-CWI* (1 bit): Virtual Route Change Window Indicator (see description of flow control for use of this and the following five bits)
- *VR-PRQ* (1 bit): Virtual Route Pacing Request
- *VR-PRS* (1 bit): Virtual Route Pacing Response
- *VR-CWRI* (1 bit): Virtual Route Change Window Reply Indicator
- *VR-RWI* (1 bit): Virtual Route Reset Window Indicator
- *VR-SNF-SEND* (12 bits): Virtual Route Send Sequence Number Field (sequence number for transmissions over virtual route)
- *DSAF* (4 bytes): Destination Subarea Address Field (complete network address of destination subarea)
- *OSAF* (4 bytes): Origin Subarea Address Field (complete network address of source subarea)
- *DEF* (2 bytes): Destination Element Field (identifies destination within subarea)
- *OEF* (2 bytes): Origin Element Field (identifies source within subarea)
- *DCF* (2 bytes): Data Count Field (size of data field)

omitted for brevity and because of similarities to the ways in which these functions occur in other network architectures.[26]

Procedure calls in SNA serve the purpose of ISO network layer primitives. A total of 50 procedures are listed in the description of the path control layer in [IBM80], and multiple procedure calls from higher layers can be used for most. These procedure calls are far too complex to present here.

9.7.1 SNA Routing

The principles of SNA routing are simple but thoroughly disguised by the SNA literature. Routing for a given virtual circuit is essentially *fixed and defined at system generation*. Rather than using adaptive routing to provide enhanced

---

[26] In some circumstances it is also possible to block data units received from the next higher, transmission control, layer together for transmission as a single data link control layer frame. If this is done, deblocking is performed with the aid of count fields in path control headers.

Figure 9.23   SNA Network Showing Subarea Nodes and
Transmission Groups.

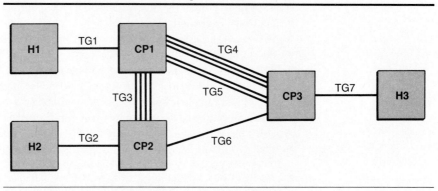

reliability and recovery, SNA uses transmission groups consisting of multiple parallel links, plus multiple virtual circuits between pairs of nodes. SNA terminology for a virtual circuit is *virtual route*, with virtual route definition including priority level and other quality of service parameters.

virtual route

As was noted above, only subarea nodes (**H** or **CP**) participate in routing. There is one possible route between a peripheral node (**T** or **CC**) and the closest subarea node, so this automatically completes routing to or from a peripheral node. Figure 9.23 is a diagram of an SNA network, including only subarea nodes; this is adequate for describing routing. Transmissions between subarea nodes use the FID4 path control header, containing information for routing. The DSAF and OSAF locate subareas, and the DEF and OEF locate network addressable units within subareas.

The figure shows a network with six subarea nodes (**H1–H3**, **CP1–CP3**) and seven transmission groups (TG1–TG7). A *transmission group* consists of one or more parallel communications links treated as one communications facility in routing.[27] TG1, TG2, and TG7 each consist of one link between a host computer and its communications processor; each would normally be a processor channel. TG3 is four parallel links between **CP1** and **CP2**. Two transmission groups, TG4 (three parallel links) and TG5 (two parallel links), connect **CP1** and **CP3**, while one transmission group, TG6 (a single link) connects **CP2** and **CP3**.

explicit route

An SNA route (an *explicit route* in SNA terminology) is a list of subarea nodes and transmission groups connecting source to destination. Three explicit routes between **H1** and **H3** are {**H1**, TG1, **CP1**, TG4, **CP3**, TG7, **H3**}, {**H1**, TG1, **CP1**, TG5, **CP3**, TG7, **H3**} and {**H1**, TG1, **CP1**, TG3, **CP2**, TG6, **CP3**,

---

[27] Transmission groups, including the links comprising them, are defined as a part of system generation. The links in a group should normally be equivalent. For example, a transmission group might consist of four 4800 bps terrestrial links or two 56 kbps satellite links.

class of service

TG7, **H3**}. All routes are reversible, so information can flow in either direction.

Route selection is transparent to a user, who simply supplies a *class of service* for use between two nodes. Class of service involves such things as priority, throughput, cost, delay, security, integrity, the type of communications facilities to be used, and so forth. These may be specified implicitly by indicating that a session should be set up for interactive traffic, file transfers, secure traffic, and so forth. A Class of Service Name (COSNAME) identifies the class of service selected. The network selects a physical path providing the desired class of service without the user being aware of mechanisms involved.

Each COSNAME is translated into a list of virtual routes, with the mapping of COSNAME plus origin/destination subarea pair to virtual route defined at system generation. At most, 24 virtual routes between a pair of subarea nodes may be defined, each identified by one of eight virtual route numbers and one of three priority levels.[28] The COSNAME is used to select a list of virtual routes that provide the desired quality of service; the list is assumed to be in order of preference.[29] The virtual route assigned is the first available one on the list.

Each virtual route is mapped into a predetermined explicit route. A total of 16 explicit routes between a given source/destination pair are allowed. Multiple virtual routes may be multiplexed over the same explicit route and multiple sessions multiplexed on a given virtual route.

Routing in SNA for a given source and destination is, thus, reduced to the *predetermined* mappings

$$Class\ of\ service \rightarrow Virtual\ route \rightarrow Explicit\ route,$$

with mappings by the network. The user can select a preferred class of service or manipulate the order of the list of virtual routes associated with a class of service, but mappings are performed by the system. Subarea nodes share information concerning the status of explicit routes passing through them (inoperative, operative, or active) so nodal routing tables may be updated and the status of each virtual route determined. Only active explicit routes may be used for routing, but operative explicit routes may be activated and then used.

If an explicit route becomes inoperative, due to failure of all links in at least one of its transmission groups or for some other reason, the network will attempt to reestablish sessions using that explicit route. The first available virtual route in the list of virtual routes associated with the desired class of service and source/destination pair will be used to reestablish service, if any such virtual route exists.

---

[28] The VRN is four bits and the TPF is two bits in the FID4 header. This allows up to 16 virtual route numbers and four priorities, but not all are used.

[29] It is possible to manipulate the order of the list when a session is initiated.

9.7.2 Transmission Group Control

One feature of path control is its procedure for controlling a multilink transmission group so it appears to higher layers as a single physical link of higher capacity than the individual links.

Communications links in a multilink transmission group are treated as parallel servers for an outbound message queue.[30] Messages are placed in this queue in order of arrival for each priority, higher priority messages ahead of lower priority ones. Messages may arrive out of order at the destination node because of variable length messages, different link speeds, or error recovery delays. Multiple copies of the same message may also arrive since retransmissions during error recovery may be over both the link where the failure occurred and other links. Resequencing and discarding of duplicate messages are performed at *each intermediate node* using the 12-bit transmission group sequence number field in the FID4 header. Out of sequence messages are not forwarded to the user until all messages preceding them are received and proper sequencing restored.[31] An example from [MEIJ82] illustrates this resequencing:

Expect	Receive	Action
2	2	Pass through No. 2
3	5	Queue No. 5
3	3	Pass through No. 3
4	6	Queue No. 6
4	4	Pass through Nos. 4, 5, and 6
7	7	Pass through No. 7

9.7.3 SNA Path Control Layer Flow Control

pacing

SNA's primary flow control mechanisms are based on a variation of the standard window flow control approach. The SNA term for this type of flow control is *pacing.* Pacing is employed on both a session basis and on a virtual route (virtual circuit) basis, with slight variations in approach. Session level pacing is part of transmission control (SNA's Layer 4) while virtual route pacing is a function of path control. We discuss the latter here; session level pacing is treated in Chapter 11. The 12-bit VR-SNF-SEND field in the FID4 header is used as a sequence number for virtual route pacing.

A window is k PIUs, the data units handled by Path Control. A transmitter is allowed to send a window upon receipt of authorization from the receiver.

---

[30] Messages handled by path control are called Basic Transmission Units (BTUs) in SNA terminology. Each BTU can consist of one or more Path Information Units (PIUs) with a PIU a path control header plus the message unit passed down from transmission control, SNA's next higher layer. Note how the terminology proliferates!

[31] Details of Transmission Group Control illustrate SNA's extremely careful protection against errors. Because of the risk that transmission errors could be data dependent, and repeated if retransmissions are on the same link, a retransmitted frame is always scheduled for re-

Permits are normally returned after receipt of the first PIU in a window, using the approach in Subsection 8.9.2. The first PIU in a window contains a request for authorization to transmit another window, made by setting the VR-PRQ bit in the FID4 header. Authorization to transmit a window—a pacing response—is granted by setting the VR-PRS bit in the header of a returning message, or by a control message if no return message is available. Figure 9.24 shows a typical sequence for window size of 4 and a three-hop virtual route.

Delay variations in the figure are caused by network congestion. It is possible for a pacing response to be received in time to avoid any delay at the transmitter, but a delayed pacing response forces the transmitter to suspend transmissions. The receiver can force this by withholding the pacing response. With short delays, it is possible for a pacing response to return immediately after a transmitter transmits its first PIU. In this case, instantaneous window size at a transmitter can be as great as $2k - 1$.

Windows are dynamically adjusted between predefined minimum and maximum sizes as congestion varies. Normally minimum size is the number of hops on a virtual route and the maximum is three times this (see Subsection 8.9.7). Window sizes are decreased when congestion is detected and increased when there is no congestion if size limits are not violated. Both minor and severe congestion are treated, with definition of each left to implementers. Nodes along explicit routes determine when congestion occurs.

If a node detects minor congestion, it sets the VR-CWI bit in a path control header. The destination notes this and sets the VR-CWRI bit in its next pacing response to indicate that window size should be decreased by one, if this does not reduce it below its minimum; this causes gradual traffic reduction. If severe congestion is detected, adjustment is more rapid and more severe. A node detecting severe congestion does not wait for the delayed notification above; instead it sets the VR-RWI bit in the path control header of the first message it sees sent toward the original transmitter on the explicit route. The transmitter immediately "slams" its window size to the minimum value.

If no network congestion is detected, neither of these adjustment procedures applies. The transmitter's window size adjustment under these conditions depends on whether transmissions have been delayed waiting for a pacing response or not. If transmissions have not been delayed, the window size appears adequate so no change is made. If transmissions have been delayed, however, it increases the window size by one if the maximum size limit is not violated. This allows the network to recover from past congestion.

---

transmission on the next link after that on which the error occurred, but a copy is retained for retransmission on the original link to obtain information about the behavior of the link. Furthermore, when the TG-SNF reaches its maximum possible value of 4095, a "TG Sweep" action is initiated, rather than wrapping the count around; this involves quiescing links until all are free, then transmitting a PIU with TG-SNF of 0 and waiting until all links are again free before proceeding. This avoids the possibility any old PIUs with small sequence numbers could be accepted as new.

Figure 9.24   Typical SNA Transmission Sequence with Virtual Route
Pacing and Initial Window Size of $k = 4$. PIU numbers
indicate values in VR-SNF-SEND field of header if FID4 Path
Control headers used.

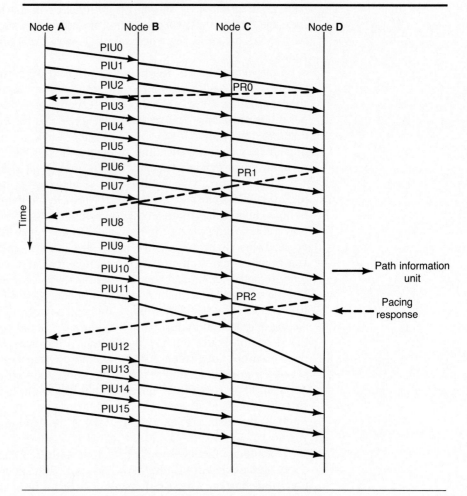

9.8   The Network Layer in IEEE 802 and MAP/TOP

As was mentioned in Chapter 3, no real network layer standards for IEEE 802
are yet available, although they are under development. MAP and TOP both
implement the OSI connectionless network layer service. The Connectionless
Internetwork Sublayer, discussed in Chapter 10, is the primary service
implemented.

9.9   ISDN Network Layer

The ISDN network layer is currently being defined. It is expected to handle two categories of functions. The first category, which has received primary emphasis so far, is that of functions that directly control *connection establishment.* The second category contains functions, beyond those provided by the data link layer, relating to *message transport;* examples include rerouting of signaling messages in the event of D-channel failure, multiplexing, and message segmenting and blocking.

The functions performed include the following:

- Processing primitives for communicating with the data link layer
- Generation and interpretation of network layer messages for peer-level communication
- Administration of timers and other logical entities used in call control
- Administration of resources, including B-channels and packet-layer logical channels such as X.25
- Ensuring that services provided are consistent with user requirements
- Routing and relaying
- Network connection control
- Conveying user-to-network and network-to-user information
- Network connection multiplexing
- Segmentation and reassembly
- Error detection and recovery
- Sequencing
- Flow and congestion control
- Restart

A variety of messages to accomplish these functions have been defined. Messages for circuit-mode connection control are listed in Table 9.9.

Figure 9.25 illustrates use of these messages to set up and take down a circuit-switched connection. The calling terminal first sends a Setup message to the network to initiate call establishment and convey information about the call. The network returns an acknowledgment indicating additional information is needed. The calling terminal supplies needed information and is informed that call setup is underway by a Call Proceeding message.

After acquiring needed information, the network termination at the calling location sends a message (or messages) across the network to the network termination at the called location to set up the call. The precise message is implementation dependent and is not specified in the standard; some typical message exchanges in the standard are based on use of X.25 networks between calling and called terminals and include X.25 setup messages. When

Table 9.9   ISDN Network Layer Messages for Circuit-Mode
Connection Control.

Call Establishment Messages	*Call Information Phase Messages*
Alerting	Resume
Call proceeding	Resume acknowledge
Connect	Resume reject
Connect acknowledge	Suspend
Progress	Suspend acknowledge
Setup	Suspend reject
Setup acknowledge	User information
Call Clearing Messages	*Miscellaneous Messages*
Disconnect	Congestion control
Release	Facility
Release complete	Information
	Notify
	Status
	Status enquiry

Figure 9.25   ISDN Call Control Procedure for Simple Circuit-Switched Calls
(adapted from [KANO86] Sadahiko Kano, "Layers 2 and 3
ISDN Recommendations," *IEEE Journal on Selected Areas in
Communications,* © 1986 IEEE. Reprinted by permission.).

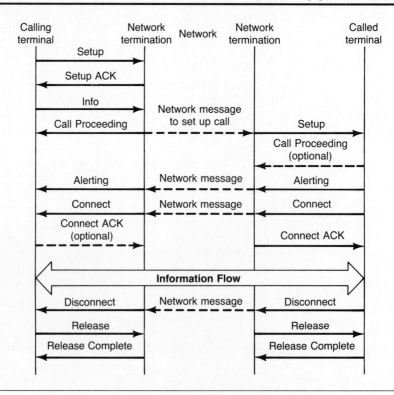

the information reaches the network termination at the called location, a Setup message is passed to the called terminal, which may return a Call Proceeding message or immediately send an Alerting message, indicating that the user is being alerted to the attempted call. For the case illustrated, the user accepts the call and returns a Connect message. After appropriate messages to convey information about the latter two messages to the calling end, the connection is complete and information flow begins.

After information flow is complete, either end may initiate disconnection. In the figure, the called terminal initiates it by sending a Disconnect message to its network termination, which responds with a Release message; the called terminal in turn acknowledges this and completes its disconnection with a Release Complete message. The network termination also passes information about disconnection across the network to the other end, which goes through an analogous disconnection sequence. There is no end-to-end verification of disconnection, however.

Similar sequences for a wide variety of other situations are given in the standard, but it is still being defined.

9.10   Summary

This chapter discusses the layers corresponding to the OSI network layer, exclusive of internetworking aspects, for each networking architecture treated. The network layer is the most conceptually complex of any reference model layer, since it involves cooperation among all nodes in the network to accomplish functions such as routing and flow control.

X.25 is commonly cited as the OSI network layer standard, but it is primarily a standard for interfacing customer equipment to a packet data network and lacks a number of functions commonly associated with the network layer. The OSI network layer standards define most functions needed, but leave implementation details to the network provider. Corresponding protocols for the ARPANET, DNA, and SNA are far more precise and complete, but show little compatibility with each other. ISDN network layer standards are incomplete, but rapid progress is being made on developing them.

Appendix 9A   Descriptions of Four Packet-Switching Networks

9A.1   DATAPAC

DATAPAC is a packet-switched network provided by Telecom Canada (formerly the TransCanada Telephone System). It provides service to all major cities in Canada, plus gateways to major networks in the United States, Europe, and the Far East. As of 1980, there were more than 2000 user connections at speeds of 110 to 9600 bps. Terminals can be X.25 packet mode or character mode, with access via a PAD in the latter case. Packet switches are based on Northern Telecom SL-10 Packet Switching Systems, and internodal trunks are purely digital full duplex 56 kbps links in Telecom Canada's DATAROUTE system. All nodes are at least doubly connected for reliability, and multiple trunks between a pair of nodes may be available. Access can also be via DATAROUTE links where this service is available (in more than 70 cities in Canada).

Two levels of priority are provided. Priority packets are limited to 128 bytes of user data and given nonpreemptive priority over normal packets for transmission across trunks (limited to 256 bytes of user data).

DATAPAC provides *end-to-end virtual circuit service* to users but uses *datagram routing.* Each virtual circuit exists logically end to end, but datagrams are sent through the network independently. Although datagrams can arrive out of sequence, be lost, or be duplicated, hardware and software at network access points resequence them, retrieve lost datagrams, and discard duplicates to provide virtual circuit service to these points.

Initial setup of a Virtual Call (VC) begins when an X.25 Call Request packet is received on a customer line. A source VC process is created to handle the call. It uses the communications subnet to deliver the Call Request, as a datagram, to the destination DTE. When the destination responds with Call Accepted, delivered as a datagram, the acknowledgment is returned to the source VC process, which notifies the source host of VC establishment.

Once this has been accomplished, data transfer can begin. Flow control parameters determine when a data packet from a customer line can be accepted and buffered in the memory of a local line processor. Packets are injected as datagrams into the communications subnet under end-to-end VC flow control. Each datagram is individually acknowledged upon arrival at the destination in order to provide the guaranteed delivery of end-to-end VC service. Any datagrams not acknowledged within a 3-second timeout are retransmitted; if an acknowledgment is not received on four successive attempts, during a total time period of 12 seconds, the VC is cleared.

Datagrams entering the subnet, or already in the subnet, are routed on the basis of information in a routing vector table (RVT) at each node. Separate RVTs are used for priority and normal datagrams, allowing different routing for the two classes. Entries in the RVT indicate, for each possible destination node, the number of the transmission queue for the outgoing link forming the first link in the "best" path to the destination. The "best" path is defined to be a minimum delay path, with delay estimated on the basis of number and speed of links traversed to get to the destination. Loading is not considered, so updates to RVTs are based only on changes in the operational status of links.

Global routing information processes at each node maintain the RVTs using a modification of the *distributed Bellman-Ford-Moore algorithm* in Chapter 8. Rather than simply computing minimum estimated delays to nodes and the outgoing links to be used, they compute estimated delays to all nodes when using each possible first link. Updates sent to a neighboring node consist of minimum delay to reach the destination via all initial links other than the link interconnecting the two nodes. Thus the neighbor at the other end of the preferred initial link gets a higher estimated delay than the others (and should be less likely to send packets back to the originator of the update). This is a routing scheme designed to minimize loops [CEGR75].

Flow control uses both the X.25 end-to-end *sliding window method* and *input buffer limiting.* Either input datagrams or datagrams already within the subnet may be discarded when the memory space for buffers becomes congested. Precedence in access to buffers is given to datagrams already in the subnet rather than datagrams just entering it, since the network has already expended resources on transit datagrams. There is also a limit on buffer size for link queues, with datagrams destined for a link discarded when accepting them would mean exceeding the limit.

The retransmission scheme described becomes effective when datagrams are discarded, since this means no acknowledgments are returned. Especially severe congestion leads to the "ultimate" flow control mechanism of clearing virtual calls. As was indicated above, this occurs when no acknowledgment is returned on four successive attempts to transmit a datagram, each with 3-second timeout.

9A.2   TELENET

TELENET is one of two major public packet-switched networks in the United States (along with TYMNET). It offers service to all 50 states and the District of Columbia, plus connections into more than 40 foreign countries. As of 1983, the network contained more than 200 switches and 40 56 kbps trunks, plus concentrators and access lines. It supports more than 1000 host computers and 100,000 terminals.

Access can be obtained in several ways. A customer may obtain a dedicated channel or use dial-in service. TELENET is available through dial-up local service from over 95 percent of the business phones in the United States. Character mode terminals may interface through an X.25 PAD-like facility; TELENET's version, though, predates the X.3, X.28, and X.29 standards and is more extensive. Packet switches are based on custom computers in a TELENET Processor (TP) family. A current version is TP4000, which can be configured as a packet switch, concentrator, multiplexer, or terminal-handling device.

The network is *hierarchical* with 56 kbps trunks connecting backbone nodes. Level I is a backbone network with nodes in major U.S. cities, each connected to at least two other backbone nodes via 56 kbps trunks. Level I nodes also serve as connection points for Level II nodes. Each of these concentrates traffic from one or more telephone area codes. Access is from Level III nodes over a tree network of concentrators and multiplexers. Traffic is routed first to the appropriate backbone node, then along the backbone network to the appropriate destination backbone node and on to the destination. An end-to-end virtual circuit connection is established by connecting in a series several X.25 segments, one for each intermediate link.[32] *X.25 flow control, error control,* and so forth are used throughout the network.

Routing establishment, at VC setup, follows a centralized-distributed approach similar to a delta routing approach in [RUDI76]. Possible paths are determined centrally by a *Network Control Center* (NCC) and distributed in the form of routing tables to backbone switches; the tables consist of a set of choices for outgoing links for each destination node, divided into two classes: a primary set of preferred choices (corresponding to shortest-path routes) and a secondary set of alternates.

---

[32] More precisely, a series of X.75 links is set up, using the X.75 protocol in the next chapter.

At call setup time, a secondary choice is used only if no primaries are available. A link selection algorithm at each node is invoked to choose the outgoing link to be used. This is chosen, from the set designated by the NCC, as the link with maximum currently unused capacity. If no outgoing link is available at a node, the call is cleared back to the previous node, which tries an alternative link. This continues until a path is found if one accessible by the algorithm exists; otherwise, the call is blocked. Routing loops and indefinite path searches for undeliverable packets are avoided via a hop count and "node visited" list in a utility field of Call Request packets.

Routing tables transmitted to nodes by the NCC are based on topology and estimated flows. Link or node failures cause alarm messages to be sent to the NCC, which distributes the information throughout the network.

9A.3   TRANSPAC

TRANSPAC is the French public packet-switching service. It has 22 switching centers and more than 13,000 dial-up and directly connected subscribers, plus gateways to most other major public packet-switched networks. It supports more than 100,000 daily calls. User access is direct for packet-mode terminals, or via a PAD otherwise.

For reliability, at least two 72 kbps lines, following different paths, connect each node to the network. Each node consists of a control unit (CU), a CII Mitra 125 minicomputer, plus several switching units (SUs). SUs execute data link protocols and access protocols for customers connected to the node. Routing is handled by the CU using information from a *Network Management Center* (NMC). Six local control points gather statistics and perform test and reinitialization procedures after node or link failures, but general supervision, including most routing computation, is handled by the NMC.

There are two classes of nodes. One class is connected in a mesh form with alternate route capability. In the second class each node homes in via a single link to a node of the first class. Messages to nodes of the second type are routed to the "target" node to which they are connected in a manner similar to TELENET hierarchical routing.

Routing is on a *single path per VC* basis. Routing of a Virtual Call follows the path taken by a Call Request Packet as it is forwarded through the network. This path is determined by routing tables at each node; the tables designate a unique preferred outbound link for each destination node.

Routing tables are constructed in an *essentially centralized* fashion by the NMC. The CU at each end of a link computes an estimated cost for that link based on utilization of link capacity and of buffers. Both measured and estimated data (declared by users at connection setup) are used to compute these estimates. The cost is set to infinity if the link is carrying its maximum permissible number of VCs or buffer occupancy has exceeded a preset threshold. Otherwise a coarsely quantized function of utilization is used. Nodes send updated link cost estimates to the NMC whenever a change is noticed. The NMC assigns cost to a link, for use in routing computations, equal to the maximum of the cost estimates from nodes at the ends of the link.

The NMC computes routing tables, including preferred outgoing link and estimated cost to reach the destination, for each source and destination node and distributes this information to individual nodes. To minimize computation, the shortest-path computation is limited to minimization over a prescribed subset of four or five paths joining each pair of nodes. Some local adaptivity is also incorporated into routing. Node **i** can use its own estimates of costs to reach its neighbor nodes to try to improve on cost estimates of the NMC. Mathematically, the algorithm it uses can be expressed as follows.

Let $C(k,d)$ be the NMC's estimate of cost of the minimal cost route to destination d from node **k.** Let $C_N(i,k)$ be the NMC's estimate of the cost of the direct link between node **i** and node **k** and let $C_L(i,k)$ be node **i**'s local estimate of this same cost. Then node **i** chooses the intermediate node, **k,** for which

$$C(k,d) \;+\; \underset{k}{\text{Max}}[C_N(i,k),\, C_L(i,k)] \tag{9A.1}$$

is minimum. Thus the node uses its own estimate of the delay to a neighboring node if this is higher than the NMC's estimate, but uses the NMC's estimate otherwise. Ties are resolved by giving priority to the shortest-hop path.

Flow control is distributed, and exercised by each node independently. One flow control technique is the technique above of *setting link costs to infinity* if a link is carrying its maximum permissible number of VCs or buffer occupancy at a node has exceeded a preset threshold. This prohibits new routes from being established over heavily loaded links or through heavily loaded nodes. If buffer utilization exceeds another threshold, traffic is slowed down on virtual circuits by *delaying acknowledgments* to neighboring nodes. If buffer utilization exceeds a still higher threshold, some virtual circuits are *disconnected,* following predefined priorities.

9A.4   TYMNET

TYMNET was originally developed for computer time-sharing, but later took on a network function as well. It is probably the largest packet-switching network in the world in terms of switching nodes [SCHW87] and now has more than 1000 nodes, almost all connected to at least two other nodes to give a distributed topology with alternate path capability. It also has facilities for international access to more than 40 other countries. The minicomputer nodal processors, called *TYMNET Engines,* are interconnected by leased backbone links at 9.6 and 56 kbps. A variety of other link speeds are also used. Users may access the network with a wide range of asynchronous and synchronous protocols and rates from 110 to 9600 bps.

Routing is done centrally on a virtual circuit, fixed-path basis by a supervisory program running on one of four possible computers. Only one supervisor is active, with the others dormant but capable of taking over if the active one fails. This takes roughly 2 to 2.5 minutes, during which existing routes are maintained but no new ones can be established, and occurs about once a week on average.

A distinction from most virtual circuit networks is that TYMNET uses a *"composite" packet internode protocol.* Data from different VCs traveling on the same trunk can be concatenated in one frame to reduce link overhead. Each frame has a maximum length of 66 bytes, including a two-byte header and a four-byte trailer, and may contain several user logical records. Each logical record is preceded by a two-byte header giving a channel number for that logical record (identifying the virtual circuit) and the length of the logical record in characters. Figure 9A.1 illustrates the frame format.

Logical records can range from a few to 58 characters (66 character maximum frame minus two character frame header, four character frame trailer, and two char-

Figure 9A.1   TYMNET Frame Format.

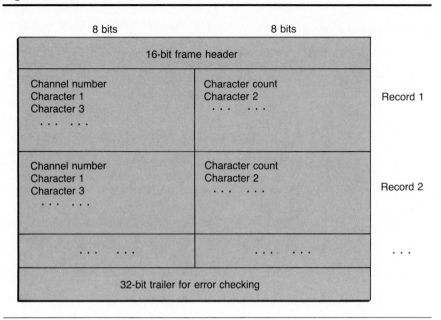

acter logical record header). Packets are transmitted as soon as they are available, without waiting for a logical record of specified size to be assembled. They are disassembled and reassembled at each intermediate node they flow through. Logical records within a packet arriving at a node may be destined for different outgoing links from that node, dictated by routing tables for their virtual circuits, and are reassembled into packets for appropriate links. The routing table at a node matches each virtual circuit's channel number on an incoming link with a channel number on an outgoing link.

The routing computation by the supervisor is based on a minimal cost algorithm reflecting class of service. A modified version of the Floyd-Warshall algorithm in Chapter 8 is used. A typical use of class of service is to steer interactive users away from satellite links by increasing the cost of such links in routing computations for them. Costs of links depend on link speed, link loading at the time a virtual circuit is set up, and class of service. Link loading is incorporated by adding a penalty to cost of a link if either or both ends complain of "overloading."

The supervisor establishes a route by sending a *"needle"* packet to the node originating the Call Request. The needle contains a list of all nodes along the route and threads its way through the network, building the virtual circuit as it goes. Each node sends the needle to the next node on the list, using a currently unassigned channel number, and sets up its routing table appropriately. User data follows behind the needle. If the needle is unable to thread its way through to the destination, due to a recent link failure or some type of error, the circuit is "zapped" back to its origin. At the end of a session a *"Path Zapper" Packet* is used to release channels and buffers in the path.

Flow control is on a per node, per virtual circuit, basis. For each logical channel on each link, a node establishes a *window size* in bytes indicating the maximum number of bytes to be sent. The receiving node must grant permission to transmit after this. It can ask for additional data or withhold permission to reduce congestion. If a window at a node is exhausted, the node withholds permission from the preceding node along the path. The resulting "back pressure" propagates backward until it reaches the originating node and turns off transmission at that point. Periodically, every half second, each node sends a "back pressure vector" to its neighbors, with one bit for each virtual circuit traversing it; the bit indicates whether the applicable window is empty or not.

Problems

9.1 **a.** Sketch the *complete* format of each packet exchanged between DTE **A** and DCE **A** and between DTE **C** and DCE **C** in the X.25 sequence illustrated in Fig. 9.7, including *all* overhead bits or characters. You may sketch a limited number of packet formats and indicate which each packet uses.

b. What can be said about packets exchanged between the two DCEs regarding their format or contents, number of packets exchanged, and so forth?

9.2 Assume that you have an X.25 application requiring end-to-end confirmation of receipt. How can this be obtained? List advantages and disadvantages of your approach.

9.3 Note that the packets exchanged during the clearing sequence in Fig. 9.7 have only local significance. Is it possible to guarantee that both ends of an X.25 connection are always made aware of clearing? Suggest a stronger clearing sequence, and indicate how it might be implemented within X.25.

9.4 **a.** What happens in X.25 if a DTE and its associated DCE both attempt to put a call through at the same time?

b. What happens if a DTE and its associated DCE both attempt to clear a call at the same time?

9.5 A datagram of 1000 bytes is to be transferred across an X.25 network with maximum data field of 128 bytes.

a. How many A and B packets are needed? To the extent possible, give values of M and D fields, and the number of data bytes in each.

b. What is the total number of bytes transferred between each DTE and associated DCE involved in the transmission? Assume normal X.25 fields are used, that is, that there are no extended field options.

c. What can we say about the number of bytes transferred between the DCEs at each end of the path?

9.6 **a.** Assume that the sequence in Fig. 9.9 illustrates the only packets transferred across a LAP-B connection during a particular period, with initial $N(S)$ and $N(R)$ counts of zero at both the DCE and the DTE. Give $N(S)$ and $N(R)$ values for each of the packets transferred between the DTE and its DCE.

b. How are the $N(S)$ and $N(R)$ sequence numbers determined if more than one X.25 logical connection is multiplexed onto the same LAP-B connection?

9.7 Compare the state transition diagram for X.25 in Fig. 9.8 with that for OSI network service in Fig. 9.15 with respect to the following:

a. Nature of data units exchanged.

b. Source and destination of data units exchanged.

c. Requirements for implementations to precisely follow sequences in state diagrams.

d. Types of handshaking procedures used for connection and disconnection.

9.8 Using the OSI network layer primitives in Fig. 9.14 and the state diagram in Fig. 9.15, sketch sequences of primitive exchanges for the following possible disconnect sequences.

a. Connection release initiated by one network service user.

b. Connection release initiated simultaneously by both network service users.

c. Connection release initiated by network service provider.

d. Connection release initiated simultaneously by a network service user and network service provider.

9.9 Indicate how each of the Quality of Service parameters for OSI network service in Tables 9.4 and 9.5 might be measured. Give formulas where possible.

9.10 Assuming an ARPANET CSNP-CSNP data packet contains only the overhead from the ARPANET DLC and the standard data packet CSNP-CSNP header, plot fractional overhead (as defined in Section 7.8) for ARPANET connections, for the number of data characters (on a log scale) ranging from 1 to 10,000 bytes. Your plots should resemble Figs. 7.22 and 7.23.

9.11 Assume that node 1 in Figure 9.18 is using the old ARPANET routing algorithm and has correct values for the minimum number of hops to each of the other nodes.

a. What is $\mathbf{H}^{(1)}$?

b. Assume that link **2–4** fails and is correctly reported to node 1 by its neighbors. What values of $\mathbf{H}^{(2)}$, $\mathbf{H}^{(3)}$, and $\mathbf{H}^{(5)}$ are reported to 1.

c. What is the new $\mathbf{H}^{(1)}$?

9.12 Initial values of $\mathbf{D}_{old}^{(4)}$ and $\mathbf{S}_{old}^{(4)}$, in the network in Fig. 9.18, for the old ARPANET routing algorithm are

$$\mathbf{D}_{old}^{(4)} = \begin{bmatrix} 4 \\ 3 \\ 3 \\ 0 \\ 1 \\ 2 \end{bmatrix} \quad \text{and} \quad \mathbf{S}_{old}^{(4)} = \begin{bmatrix} 3 \\ 2 \\ 3 \\ 0 \\ 5 \\ 6 \end{bmatrix}$$

Node **4** then receives the following distance vectors from its neighbors:

$$\mathbf{D}_{new}^{(2)} = \begin{bmatrix} 2 \\ 0 \\ 5 \\ 3 \\ 3 \\ 4 \end{bmatrix}, \mathbf{D}_{new}^{(3)} = \begin{bmatrix} 1 \\ 4 \\ 0 \\ 1 \\ 3 \\ 2 \end{bmatrix}, \mathbf{D}_{new}^{(5)} = \begin{bmatrix} 2 \\ 3 \\ 1 \\ 1 \\ 0 \\ 3 \end{bmatrix}, \text{ and } \mathbf{D}_{new}^{(6)} = \begin{bmatrix} 3 \\ 4 \\ 3 \\ 4 \\ 3 \\ 0 \end{bmatrix}.$$

Find $\mathbf{D}_{new}^{(4)}$ and $\mathbf{S}_{new}^{(4)}$.

Figure P9.13   Figure for Problem 9.13.

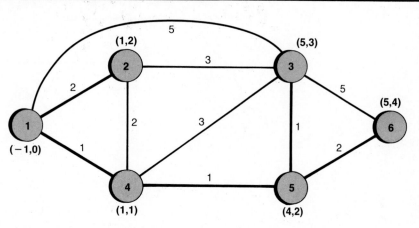

9.13   Figure P9.13 shows the solution (from Fig. 8.8) for the shortest-path spanning tree to node 1 from all other nodes (or from 1 to all other nodes) via Dijkstra's algorithm. Heavy lines indicate the spanning tree, numbers by links indicate their lengths, and (n, d) by a node indicates its predecessor node on the shortest-path spanning tree is n and the distance to 1 is d. Assume that the new ARPANET routing algorithm is used for routing in this network. The following distance changes are reported to node 1 in succession (not all at once). In each case indicate whether or not it is necessary to recompute the shortest-path spanning tree, according to criteria in the text. If recomputation is necessary, give the new shortest path spanning tree and distances to nodes, using the same type of diagram. (Don't forget that changes are cumulative.)

a. c_{34} changes from 3 to 5.
b. c_{12} changes from 2 to 1.5.
c. c_{35} changes from 1 to 3.
d. c_{56} changes from 2 to 1.
e. c_{36} changes from 5 to 3.

9.14   Assuming that a DNA data packet contains only overhead from the DDCMP DLC and the standard routing layer header, plot fractional overhead (as defined in Section 7.8) for DNA connections, for the number of data characters (on a log scale) ranging from 1 to 10,000 bytes. Your plots should resemble Figs. 7.22 and 7.23.

9.15   The link cost assignment to be used for routing suggested in DNA manuals is

$$c_{ij} = \begin{cases} 1 & \text{if } r_{ij} \geq 100,000 \text{ bits/second} \\ \dfrac{100,000}{r_{ij}} & \text{if } 4,000 < r_{ij} < 100,000 \text{ bits/second} \\ 25 & \text{if } r_{ij} \leq 4,000 \text{ bits/second} \end{cases}$$

with r_{ij} the bit rate of link ij. Compute the c_{ij} values for the network in Fig. P9.15 (rounding each up to the next larger integer to simplify calculations), and find the

Figure P9.15   Figure for Problems 9.15–9.17.

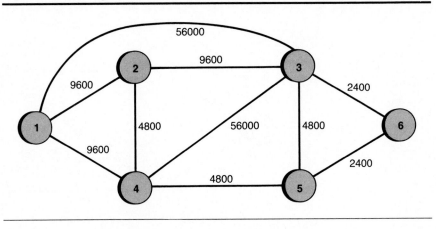

shortest-path spanning tree from node **1** to all other nodes as it would be computed by DNA. Numbers by links correspond to bit rates.

9.16   The hop count threshold (determining when packets should be discarded in DNA) recommended by DNA manuals is twice the hop count for the longest path between any two nodes in the network. Since a path is defined as a sequence of hops that does not visit the same node twice, this is twice the maximum possible number of hops that can be made without coming back to any point visited earlier. Find this hop count threshold for the network in Fig. P9.15.

9.17   Assume that each node in the DNA network in Fig. P9.15 has buffers for 12 messages. Compute, at each node, the threshold on queue length for discarding packets according to Irland's square root rule.

9.18   List all nodes en route between a source at end node E_{11} and a destination at end node E_{33} for the DNA network in Fig. P9.18, for two logical DNA routes, one going through Area 2 and the other going directly from Area 1 to Area 3. For each route indicate whether each routing node along the route is a Level I routing node or a Level II routing node.

9.19   Assuming that an SNA data packet contains only overhead from SDLC and FID4 path control headers, plot fractional overhead (as defined in Section 7.8) for SNA connections, for the number of data characters (on a log scale) ranging from 1 to 10,000 bytes. Your plots should resemble Figs. 7.22 and 7.23.

9.20   **a.**   How many different origin (or destination) subarea address fields are allowed by the SNA FID4 format for path control headers? How many different origin (or destination) element fields are allowed?

   b.   What is the total number of different origins or destinations that can be addressed (including both subarea and element within a subarea) with these fields?

   c.   Current SNA network addresses can consist of an eight-bit subarea address and a 15-bit element address. Compute the fraction of the capacity of each field, and of total addressing capability, that can be used.

Figure P9.18 Figure for Problem 9.18.

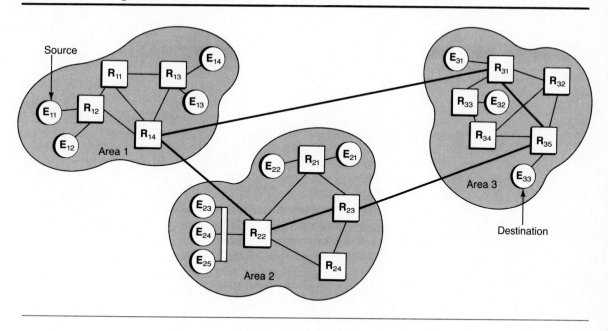

9.21   **a.**   How many different NAUs can a communications controller node address, using the SNA FID2 Path Control header? Where are these addresses translated to full network addresses?

b.   The format of the LSID in FID3 path control headers is given below. The first bit determines whether the SSCP is involved in a session or not, the second bit whether the PU is involved, and the last six bits the local address of an LU if one is addressed. How many LUs can a terminal node using FID3 address via the local address portion of the LSID? Where are the addresses it uses translated to full network addresses?

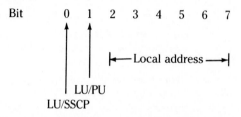

9.22   A simple two-domain SNA network is illustrated in Fig. P9.22. Characteristics of the transmission groups are tabulated as follows:

Figure P9.22   Figure for Problem 9.22.

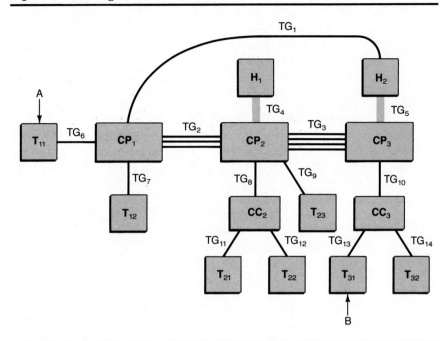

Transmission Group Number	Description	Other Relevant Information
1	One 56 kbps satellite link	270 msec propagation delay
2	Three 4800 bps terrestrial links	Negligible propagation delay
3	Four 9600 bps terrestrial links	15 msec propagation delay
4,5	S/370 channel	500,000 byte/second rate
6,7	One 2400 bps link	Locally attached
8	One 4800 bps link	Locally attached
9	One 2400 bps link	Locally attached
10	One 4800 bps link	Locally attached
11–14	One 2400 bps link	Locally attached

a. Describe two explicit routes between a user at location A, at terminal node T_{11}, and a user at location B, at terminal node T_{31}. To the extent possible, make these explicit routes use distinct facilities.

b. Two sessions are to be set up between A and B, one with Class of Service Name, "FILE_TRANSFER," and the other with Class of Service Name, "REAL _TIME." The former is mapped onto Virtual Route Number 1 and the latter is mapped onto Virtual Route Number 6. Assuming that the Class of Service Names are descriptive, what is an appropriate priority level (low, medium, or high) for each of these Virtual Routes?

c. Choose the appropriate Explicit Route, from among those you described in (a), for each of the Class of Service Names above, and describe the SNA mappings from Class of Service to actual route for each case.

9.23 a. Complete the table below depicting actions of SNA Transmission Group Control during resequencing of packets.

Expect	Receive	Action
2	4	
2	3	
2	2	
4	4	
5	6	
5	5	

9.24 a. What are the normal minimum and maximum window sizes for SNA virtual route pacing over the virtual route for which pacing is illustrated in Fig. 9.24?

b. Assume Node **B** in Fig. 9.24 detects minor congestion just before relaying PIU5. What does it do to combat this congestion? What are the reactions of other nodes in the network? Sketch appropriate changes to the sequence of PIU transmissions.

c. Instead of the situation in (b), assume that node **B** detects severe congestion just before relaying PIU6. Assuming that the transmissions illustrated are the only ones passing through **B**, what does it do? What are the reactions of other nodes? Sketch appropriate changes to the sequence of PIU transmissions.

9.25 Compare the DNA routing layer and SNA path control layer with respect to:
a. Type of packet transmission used
b. Provision for nodes with varying capabilities
c. Approach to error handling
d. Routing
e. Flow control
f. Extra features provided

9.26 Assume that the network in between calling and called terminals in Fig. 9.25 is an X.25 based network, with X.25 connection and disconnection phases used in addition to those shown. Draw an extended message sequence diagram including these X.25 phases. Show all necessary X.25 messages.

9.27 (**Design Problem**) Describe your own implementation of an X.25 based packet data network. Select the X.25 options you plan to implement, list necessary functions, and indicate how you handle them. A partial list of functions includes the following: accepting messages from DCEs (with parameters such as connection identification, plus user data); constructing messages to be sent to DCEs; defining messages to be passed across the packet data network, including techniques for synchronization, segmentation and reassembly, and framing; choosing routing and flow control techniques and how to collect information needed for each. Your implementation needs to provide all essential network layer functions not included in X.25.

10 Internetworking

10.1 Introduction

Thousands of telecommunication networks are in use around the world, and the pace of installation is increasing. Such networks range from LANs serving individual buildings or campuses to networks spanning the globe; from networks devoted to specialized applications to networks primarily used to access functions at remote locations. Many networks specialize in true resource sharing—the sharing of data and programs by users at different locations. Internetworking techniques allow users of different networks to work with each other. The scale of network interconnection is impressive, with internetworks themselves interconnected so that at least some applications can be accomplished across internetwork boundaries. As we mentioned in Chapter 3, these networks form a metanetwork called *Worldnet,* used daily by communities throughout the world.

Worldnet

In this chapter we discuss the motivations for internetworking and some of the major problems involved. We then discuss prominent internetworking standards now in use and being developed. These include two standards for bridges to interconnect LANs at the data link layer plus three standards—X.75 and the DoD and ISO IP standards, for interconnection at the network layer. We then give a brief discussion of Digital Equipment Corporation's DNA/SNA gateway, which is a typical example of interconnection at a higher layer, the transport layer in this case. The chapter concludes with a discussion of some of the strange and unpredictable phenomena that may be observed in complex internetworks.

10.1.1 Motivations for Internetworking

There are a wide variety of situations in which it would be advantageous for a user to have access to another network. Electronic mail is an obvious example; there are numerous occasions when it is necessary to send mail to someone on another network.[1] Banks have found it advantageous to tie together networks so that a customer can use a remote teller machine from another bank, possibly on the other side of the country, to obtain a cash advance or transact other business. Interconnection of automated reservation systems (airline, hotel or motel, car rental, theatre ticket, and so forth) with each other and with credit card verification systems can allow a complete itinerary to be booked and credit check made in one transaction. Corporations may have LANs at a variety of locations plus one or more wide area networks between locations; there are advantages to interconnecting them so a terminal or workstation can access several networks. This can greatly speed up iterating among stages of a process, such as original design, design checking, cost estimation, market forecasting, manufacturing, and maintenance. Different networks may be used for each of these. Interconnection of corporate networks can allow rapid dissemination of pertinent information to subcontractors and make possible the automated entry of orders for such things as supplies.

Internetworking has been motivated by the fact that the value to a user of many networks is strongly correlated with the number of other users that can be reached. This has been sufficient motivation to interconnect networks, including X.25 networks; one standard for internetworking was developed to interconnect X.25 networks.

Prominent organizations pushing internetworking have included military organizations. A wide variety of networks are used for military applications. These include networks to obtain and coordinate information from sensors (such as radars), networks to communicate among battle units, vehicles and command centers, networks spanning small geographical areas or theatres of operations, and networks with global coverage. Several versions of each type of network may be in operation, using different frequency bands, transmission media, protocols and technologies, and so forth. It is also desirable to interconnect networks operated by the various branches of the military and networks operated by the armed forces of allied nations. A prominent internetworking standard originated as a part of the DoD Reference Model.

The success of LANs has led to emphasis on interconnecting LANs. This has resulted in changes in internetworking philosophies, with some successful techniques involving interconnection at levels other than that equivalent to the OSI network layer—the layer where the OSI model says internetworking should occur.

Another impetus toward internetworking has come from the research community. Many internetworking problems are extremely difficult and have aroused research interest. Some of the first networks to be interconnected were networks primarily used by research organizations.

---

[1] Electronic mail between the United States and New Zealand, used by the authors while working on this book, passed through at least three networks en route.

10.1.2 Requirements for Network Interconnection

A few requirements for network interconnection are the following:

- Provide communications facilities between networks. A bare minimum is the provision of a communications path plus the equivalent of the OSI physical and data link layers.
- Provide routing between users on different networks.
- Provide an accounting service to keep track of the use of various networks and facilities and to maintain status information.

goal of internetworking

Each service provided should not require modification to the architecture of any interconnected network; most of the complexity of internetworking arises from this requirement. Different networks are not often compatible, so providing full access from other networks when existing networks cannot be altered is difficult. Limited access, such as provision of electronic mail, is less complex, but even this is not simple. The majority of current applications of internetworking provide access for selected applications, but the *goal* of internetworking is to provide access to all facilities of each network by users of other networks.

10.1.3 Relays, Bridges, Routers, and Gateways

relay

The OSI Reference Model calls a device to interconnect two systems not connected directly to each other a *relay*. If the relay shares a common layer n protocol with other systems, but does not participate in a layer $n + 1$ protocol, it is a layer n relay. Current terminology is as follows, however [PERL88]:

repeater, bridge, router, gateway

- *Repeater:* Physical layer relay
- *Bridge:* Data link layer relay
- *Router:* Network layer relay
- *Gateway:* Any relay at layer higher than network layer

Unfortunately this terminology is far from uniform. The term gateway is often used for any relay operating at the data link layer, network layer, or any higher layer. This terminology is, in fact, used in standards documents to describe devices operating at the network layer or lower.

Figure 10.1 illustrates various types of relays. Networks A and B are so homogeneous they may be considered to be two parts of the same network. The *repeater* joining them passes signals in both directions, amplifying and regenerating them as they pass through. Such repeaters are often used to extend the range of networks. Signals are not changed, except for "cleaning them up," so repeaters have no effect on protocols. We will not treat repeaters further since they do not affect protocols.

Figure 10.1   Interconnection of Networks.

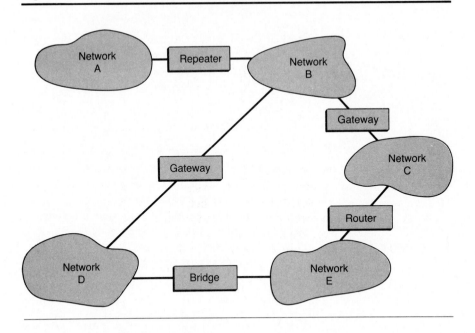

Networks **D** and **E** are very nearly, if not completely, homogeneous, so they are interconnected by a *bridge* operating at the data link layer. Networks **C** and **E** are less homogeneous and are interconnected by a router operating at the network layer. The other interconnections are by gateways, operating at higher layers, and allow for still less homogeneity. Simple routers or gateways can operate with homogeneous networks, but some versions are capable of interconnecting heterogeneous networks.

Gateways may be separated into two halves called half gateways. Half gateways in two networks are illustrated in Fig. 10.2. An almost identical diagram could be drawn for half bridges or half routers.[2]

When half relays are used, each performs half of the translation or mapping necessary, so the net effect of dividing the relay into halves is nil. A common reason for using half relays is political; since each half relay is under the control of one network, jurisdictional problems and resolution of "finger pointing" (when something goes wrong and each administration says the other is at fault) are simplified. This is especially true when the relay is at a national boundary; use of a full relay would require two nations to jointly own and operate a piece of equipment costing a few thousand dollars to inter-

---

[2]Half gateways are regularly discussed in the literature, but similar terminology for bridges or routers is less common. Nevertheless, we use this terminology to simplify our discussions.

Figure 10.2   Replacement of Gateway by Two Half Gateways.

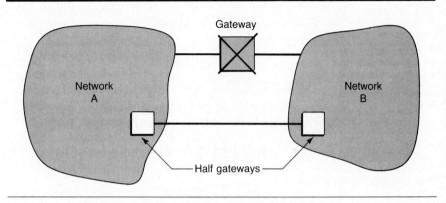

connect networks representing millions of dollars in investment. The resulting political problems can be insurmountable.

When half relays are used, each half normally operates between the protocols used in its network and a set of intermediate protocols. There are significant advantages to standard intermediate protocols. Without a common intermediate representation, designing relays between n different networks could require a distinct mapping between each network and each of the $n-1$ others, or $n(n-1)$ mappings. With a common intermediate protocol, though, $2n$ mappings suffice—one from each network to the common protocol and one from the common protocol to each network. Relays can be manufactured as half relays that fit together in various combinations, due to the common protocol. The difference in number of mappings is significant for n large. For example, for $n = 10$, $n(n-1) = 90$ and $2n = 20$, while for $n = 100$, $n(n-1) = 9,900$ and $2n = 200$.

10.1.4   Network Interconnection Issues

Development of appropriate approaches for internetworking is heavily dependent upon resolution of a variety of issues. The list of questions in Table 10.1 illustrates issues that need to be resolved. Appropriate answers require a clear understanding of user requirements.

10.1.5   Common Network Differences

A wide variety of differences between networks need to be resolved by relays interconnecting them. Some differences are listed in Table 10.2.

Compatibility of protocols does not mean they are identical; it means differences must be of a type that can be handled easily. Use of ASCII at one end and EBCDIC at the other end of a path can be handled through translation

Table 10.1 Internetworking Issues.

- What functions will be provided to users?
- Should those functions be equally accessible by users in all interconnected networks?
- What security constraints must be in effect to prevent network resources from being compromised?
- What level of transparency can be provided to users so access to new network resources is accomplished via existing mechanisms?
- What level of network protocol compatibility will be required to allow interworking between any two arbitrary networks?
- What levels of performance capability will be required?
- What "political" considerations have to be taken into account when interconnecting networks?
- How can the combined network be effectively managed?
- How effectively can fault isolation be accomplished?
- How will resource utilization be cross-charged?
- How effectively can the combined network migrate to new technologies?

(From [MORE86], Digital Equipment Corporation. Reprinted by permission.).

Table 10.2 Network Differences to be Resolved for Internetworking.

Addressing and Naming Schemes

- These vary widely. Some networks use hierarchical naming and addressing, with names of form SYSTEM.ENTITY, extendable to NETWORK.SYSTEM.ENTITY for internetworking; others use flat name structures with no subfields.
- Names and addresses are permanently assigned in some networks, but they are dynamically assigned during times that entities are active in others.
- Some networks use abbreviated local area names while others do not. Broadcast or multicast names can be used in some but not all networks.
- Differences must be reconciled. A global addressing scheme and directory service are often provided.

Routing Techniques

- These also vary widely. Obvious differences are between datagram networks, with routes varying according to congestion and other conditions, and virtual circuit networks with fixed routing during sessions.
- Gateways or internetworking facilities must coordinate to route data between stations on different networks.

Information Quanta

- These are also inconsistent. Units of information in different networks are not just bits, but packets, messages, blocks, and so forth.
- These information units are commonly used in numbering schemes for network control.
- Even basic units for numbering schemes are not consistent; sometimes successive numbers represent byte counts, sometimes packet counts, sometimes message counts.
- Inconsistencies make it complex to translate numbers from one protocol into another.

bridge

The definition of a *bridge* in the literature is not uniform. A definition requiring almost complete network homogeneity is given in [STAL88b]. A bridge between networks **A** and **B** is defined to be a device that does the following:

- Reads all frames on **A** and accepts those addressed to **B**.
- Retransmits accepted frames onto **B**, using the medium access protocol for **B**.
- Does equivalent functions for **B** to **A** traffic.

IEEE 802 standards committees are also developing standards for bridges to interconnect different types of LANs in a single internetwork. Since MAC headers differ for different types, they must be modified by such bridges. Since the FCS checks both headers and data they also have to recompute it.

Some examples of LAN bridges are given in Fig. 10.3. Part (a) shows two half bridges between homogeneous IEEE 802 LANs separated by an HDLC link.[4] The bridge encapsulates the LAN MAC frame (which consists of user data, LLC header, MAC header, and trailer) in an HDLC frame by adding its own data link header and trailer. A more complex configuration, using two half bridges separated by an X.25 link, is given in (b). The point-to-point circuit in (a) is replaced by an X.25 virtual circuit, using all three layers of the X.25 protocol.[5] The X.25 header is added by the X.25 packet layer, with the X.25 data link layer adding the same type header and trailer as the data link layer in (a).

In either Fig. 10.3 (a) or (b) the intervening link is of much lower capacity than the LAN, so flows across the link should be limited by some type of flow control. Unfortunately, few possibilities for flow control exist at the data link layer level, so flow control is difficult [GERL88a]. The most popular type of flow control involves dropping packets when buffers are full, but this can be counterproductive because of the retransmissions it generates. Other techniques include input buffer limits, "choke" packets, and buffer reservations. Either link can also add delays, but this can cause problems in some applications.

The IBM token ring architecture includes specification of a bridge architecture that is the basis for a source routing bridge architecture being developed by the IEEE 802.5 committee. We discuss this architecture next, and then present the architecture proposed by the IEEE 802.1 committee.

---

[4] Refer to IEEE 802 data link protocols in Chapter 7 for descriptions of the LLC and MAC headers and trailers. X.25 headers are discussed in Chapter 9.

[5] The OSI layering structure becomes blurred here; the MAC sublayer of the LAN data link layer operates above the X.25 packet layer (that is, the X.25 packet layer adds its header to a frame containing the MAC header).

Figure 10.3 Bridges Between LANs Separated by: (a) Point-to-point link. (b) X.25 network (adapted from [STAL87a], W. Stallings, *Local Networks: An Introduction, 2/e,* © 1987. Reprinted by permission of MacMillan Publishing Company, New York, NY.).

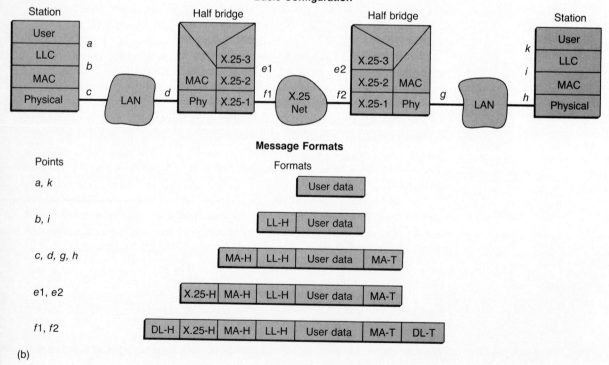

10.2.2 Source Routing Bridges

A bridge interconnection of two rings using token ring bridges, as recommended by the IEEE 802.5 committee, is illustrated in Fig. 10.4. It is also possible to expand ring capacity by using a high-speed backbone ring to which local rings are interconnected via bridges. This is shown in Fig. 10.5.

The bridge operates as in the description above, but it is a single entity, not two half bridges. The bridge reads all frames on each ring, leaving all intraring frames undisturbed but accepting those destined for the other ring. It then retransmits the accepted frames on the appropriate ring.

source routing

Source routing is used for internetwork frames. An optional routing information (RI) field in the frame header is present only when frames are destined for another network. This field occurs immediately after the two address fields in the IEEE 802.5 frame format in Fig. 7.18(c.ii); otherwise the frame format is identical to that for IEEE 802.5. The RI field, if present, contains an ordered list of bridges that must forward the frame in order for it to reach the destination ring. The format of resulting MAC frames is given in Fig. 10.6. Each two-byte route designator contains a 12-bit segment (individual ring) number and four-bit bridge number. Presence of the RI field is indicated by the individual/group address (I/G) bit in the source address field (see Fig. 7.19).[6] The LLC protocol data unit (PDU) format is in Fig. 7.20.

The source routing approach is best adapted to cases where the source normally knows a satisfactory route to the destination, so it can supply the contents of the RI field. If a satisfactory route is unknown, a search is initiated to explore all possible routes between source and destination and to provide the source with a list of routes; the source can then choose the one it desires based on any criteria it is set up to use.

Figure 10.4   Interconnection of Token Rings by a Bridge.

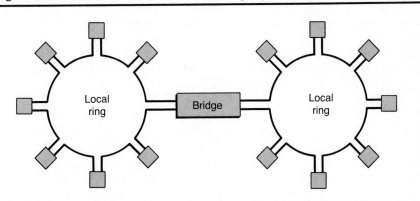

---

[6]Since the source address must be an individual address, this bit would not be used otherwise.

Figure 10.5 Interconnection of Token Rings via Backbone Ring and Bridges (from [STRO87], N. Strole, "The IBM Token-Ring Network—A Functional Overview," *IEEE Network,* © 1987 IEEE. Reprinted by permission.).

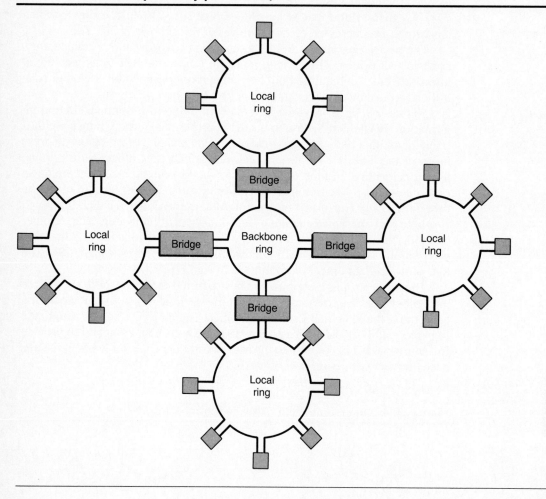

The *search procedure* is as follows [DIXO88]: A "single route broadcast" frame, designated by bits in the routing control field, is sent from the source and addressed to the destination; it is constrained to flow over a spanning tree, so it appears once, and only once, on each ring in the network.[7] This frame will reach the destination, if the spanning tree is valid, since it propagates over each ring in the network.

---

[7] The procedure to determine a spanning tree is not specified, but a variety of procedures are available.

Figure 10.6 Format of Token Ring MAC Frame with Routing Information Field.

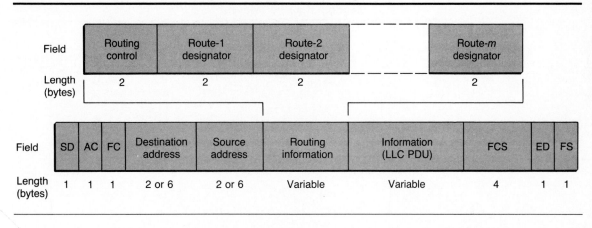

The destination returns an "all routes broadcast" frame to the original source, with an empty route designator field. This frame explores all possible routes back to the source, with route designators added at each bridge it (or any copy) traverses. Each bridge checks to see if the ring number designating its outgoing ring is contained in the current route designator field. If so, it does not forward the frame since this would cause it to flow around a loop. If the ring number is not present, the route designator for this ring number/bridge combination is added and the frame forwarded. This causes frames listing all possible routes to be returned to the original source, which can choose the one it wishes to include in the RI field the next time it transmits a frame to the same destination.

example of search procedure

As an example of this search procedure, consider the interconnected networks in Fig. 10.7, which consist of five LANs and six bridges.[8] Station S1 wants to send a message to station S2, but does not know the route.

Figure 10.8 illustrates the messages exchanged during the learning process, with B*j* denoting Bridge *j*. Part (a) shows how the initial "single route broadcast" message from S1 propagates, assuming Bridges 3 and 5 are not part of the spanning tree used. One message flows via the route **LAN1** → B1 → **LAN2**. Another flows via the route **LAN1** → B2 → **LAN3**, where it splits into two messages, one following the path **LAN3** → B6 → **LAN5** and the other following the path **LAN3** → B4 → **LAN4**. A total of 5 LAN messages are involved.

After receiving the "single route broadcast" message, S2 returns an "all routes broadcast" message. The routing for this message is more complex, since it explores all possible routes. Copies circulate over possible paths until

---

[8]Since the source routing bridge is currently advocated only for ring networks, the LANs shown are rings. Any type of LAN could be used with source routing bridges, however.

Figure 10.7   Interconnected LANs to Illustrate Route Learning in
Source Routing Bridges.

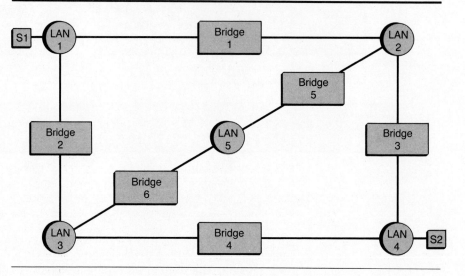

Figure 10.8   Routing of Broadcast Messages During Route Learning for
Network in Fig. 10.7. (a) Initial "single route broadcast"
message, assuming Bridges 3 and 5 not in spanning tree used.
(b) "All routes broadcast" message.

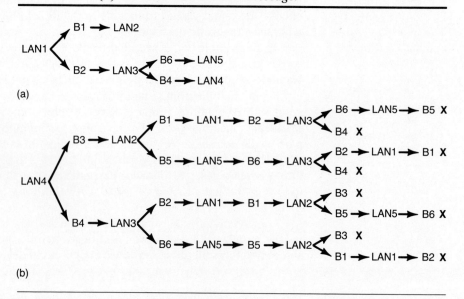

they come to bridges that would take them back to LANs they have already visited. Bridges determine this by examining the routing headers being developed. Part (b) of the figure illustrates the paths followed. A typical path is the top one, **LAN4** → B3 → **LAN2** → B1 → **LAN1** → B2 → **LAN3** → B6 → **LAN5** → B5. At this point, Bridge 5 does not relay the message since this would take it to **LAN2**, which it has already visited. A count of LAN messages during this phase indicates that a total of 15 more messages are needed. Four of these reach S1, which can take its choice of the four routes they indicate. Finding the route, though, takes a total of 20 LAN messages. Putting route finding overhead for this network into perspective is investigated in Problem 10.5.

Once the appropriate route is known, the RI field is used in frames sent to that destination. Routing is then trivial; if a bridge is to forward a frame, it will find its bridge number in the RI field flanked by segment numbers identifying the two rings it interconnects. It forwards the frame only when this is true. This can be done at high speed, as may be necessary for situations with high speed backbone networks, such as that in Fig. 10.5. Frames are buffered in bridges until they have been transferred to the next ring.

10.2.3 Spanning Tree Bridges

The alternative IEEE 802 architecture is a *spanning tree bridge* architecture being developed by the IEEE 802.1 committee. These bridges are fully transparent to end stations, which operate in the same manner regardless of whether the network they are using contains bridges or not.

Spanning tree bridges perform three basic functions:

- Frame forwarding
- Learning station addresses
- Resolving possible loops in the topology by participating in a spanning tree generation algorithm

Figure 10.9 shows a spanning tree bridge. Each bridge receives frames from and transmits frames on the LANs it interconnects. It maintains a forwarding data base containing information used to forward or discard frames in the manner discussed below.

The forwarding data base contains entries for all destinations currently known to the bridge. An entry contains the station address and port number on which a frame should be forwarded to reach the destination. If this is the port on which the frame was received, no forwarding is necessary since the frame is already on the correct LAN, and the frame is discarded.[9] If not, the frame should be forwarded on the port indicated.

---

[9] In the case of a token ring, the frame would be relayed on around the ring on which it was received rather than discarded. It would be discarded for CSMA/CD or token bus LANs since it would have already reached its destination.

Figure 10.9   Spanning Tree Bridge (adapted from [BACK88], F. Backes, "Transparent Bridges for Interconnection of IEEE 802 LANS," *IEEE Network,* © 1988 IEEE. Reprinted by permission.).

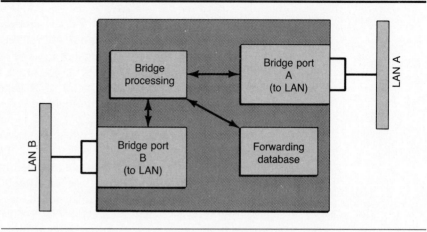

When a frame is received on a bridge port, the destination address in the header is compared to the forwarding data base. If the address is found, the frame is forwarded or discarded as indicated above. If the address is not found, the frame is "flooded" by transmitting it on all ports in the forwarding state (determined by the spanning tree algorithm below) except the one on which it was received. Frames for group addresses are flooded at each bridge unless the forwarding data base contains a special entry to restrict flooding for this group address to certain areas of the bridged LAN.

The bridge *learning process* is based on remembering which port frames are received on. The source address of any frame received without error is compared against the fowarding data base. If the source address is not found, it is added along with the number of the port on which it was received. If the source address is found in the data base, but associated with a different bridge port, the port identifier for that entry is changed. If the source address is associated with the port on which the frame was received, the entry is left untouched save for updating a timer. In all three cases, a timer for the entry is set to indicate that this is a "fresh" entry.

Data is removed from the data base when the associated timer indicates the entry is "stale," using a timeout determined by network management. When the spanning tree algorithm below detects a topology change, the time-out is decreased to ensure that bridges quickly age out stale information.

The spanning tree generation algorithm ensures that only a finite amount of flooding traffic is generated by confining flooded messages to a spanning tree. Many approaches to finding spanning trees are possible, but a specific algorithm has been adopted for the IEEE 802.1 MAC Bridge Draft Standard. Requirements for the algorithm to operate correctly are as follows:

- Each bridge must have a unique bridge identifier. In the standard this identifier is a priority field plus a second field to ensure uniqueness.

- A unique well-known group address that will always be received by all bridges on a particular LAN must exist for each LAN.

- Each bridge port must be identified uniquely within the bridge by a "port identifier."

The spanning tree is calculated as follows. First, a unique root bridge is selected. This is arbitrarily chosen as the bridge with the lowest bridge identifier. Next, each bridge determines which of its bridge ports is "in the direction" of the root, that is, which has the least cost path to the root.[10] This is designated to be its "root port." Finally, a unique "designated bridge" is selected for each LAN; this is the bridge offering the least cost path to the root from that LAN. A bridge places its root port and all of its ports connected to LANs for which it is the designated bridge into a "fowarding" state. The other bridge ports are placed in a "blocking" state.

example of spanning tree calculation

The following example (adapted from [BACK88]) illustrates this procedure. Consider the interconnected LANs in Fig. 10.10. Six LANs, **LAN 1**, **LAN 2**, . . . , **LAN 6** and seven bridges, B1, B2, . . . , B7 are indicated[11] with available ports indicated by dotted lines. Port identifiers are in parentheses near bridge ports. Costs of all links are assumed to be equal.

Figure 10.11 shows results of the algorithm. The first step is selecting B1 as the root bridge as it has the lowest identifier. Each bridge (aside from B1) then selects the port with lowest cost path to the root bridge as its root port, using the lower numbered port in case of ties; the resulting root ports are indicated. Next, the bridge offering the lowest cost route to the root bridge is chosen as the designated bridge for each LAN; these bridges are indicated in square brackets. Finally, each bridge puts its root port, and all other ports connected to LANs for which it is the designated bridge, into a forwarding state. Such ports are indicated by solid lines in the figure; all other ports are in the blocking state and indicated by dotted lines. The interconnection of LANs and bridges by forwarding ports (solid lines) forms the spanning tree (the solid lines in the figure interconnect bridges and LANs in a tree configuration).

Solution for the spanning tree in this example was based on an assumption that all information needed for the solution was available at each bridge. The information needed for the algorithm to operate is exchanged between bridges using messages called bridge protocol data units (BPDUs). These contain information about which bridge is believed to be the root and how far away (in terms of path cost) the root bridge is believed to be; the bridge

---

[10] Costs are associated with each LAN or other link in the internetwork. These costs are normally determined so that high speed links have lower costs than low speed links.

[11] These are simplified identifiers. A typical bridge identifier might be 123, 08007C000942 with the first three digits indicating priority and the rest the globally assigned bridge address. Only relative magnitudes of identifiers are needed for the algorithm, however.

Figure 10.10   Geometry for Example of Spanning Tree Algorithm. (From F. Backes, "Transparent Bridges for Interconnection of IEEE 802 LANS," *IEEE Network,* © 1988 IEEE. Reprinted by permission).

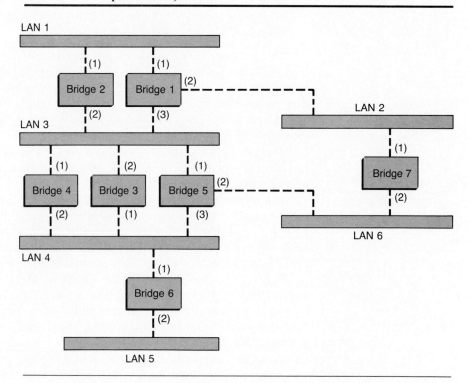

and port identifier of the sending bridge and the age of the information in the BPDU are also included. Algorithms for using this information to modify the spanning tree as the network changes are in the draft standard.

The state of a bridge port is never changed directly from blocking to forwarding, since this could introduce temporary loops. Instead, a port is changed from a blocking to a listening state for a period of time. It is next changed to a learning state while it acquires the information it needs for its forwarding data base. It enters the forwarding state only after both preliminary phases have been completed successfully.

10.2.4   Comparison of IEEE 802 Bridge Approaches

The debate over which approach to bridges is better is vociferous, with proponents of both approaches vigorously defending their approach. A valiant attempt to present all sides to the arguments, as well as of a similar debate

Figure 10.11 Results of Applying Spanning Tree Algorithm to Topology in Fig. 10.10. (From F. Backes, "Transparent Bridges for Interconnection of IEEE 802 LANS," *IEEE Network,* © 1988 IEEE. Reprinted by permission).

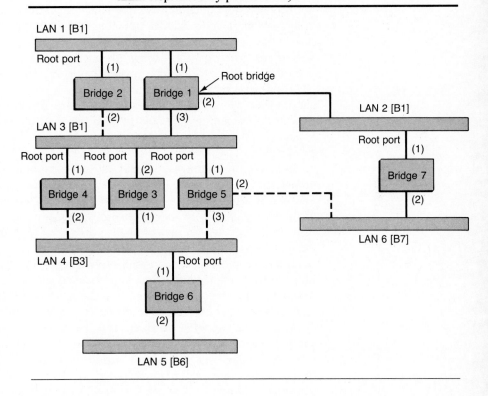

about relative merits of bridges versus routers, is contained in the January 1988 issue of *IEEE Network.*[12] Readers who study this issue and other papers on the topic should realize that some arguments presented are true only in limited contexts, and defend positions of the authors or of the companies they work for. An excellent critique of such arguments is the slogan, "Optimality Differs According to Context," found in the closing paper [PADL88] in the issue.

The most relevant point in favor of the spanning tree approach appears to be transparency. No modifications to existing or installed equipment are needed if compatible LANs are interconnected by spanning tree bridges. A

---

[12] In order to present all viewpoints, roughly half a dozen papers "judged to be unworthy of publication by some of the reviewers (mostly, those with different views)" were included [GERL88b].

significant base of installed equipment that would need to be modified to fully utilize source routing is already in use.

The most significant advantage of source routing bridges seems to be applicability at high speeds. Prominent candidates for high-speed backbone networks interconnecting token rings include networks (such as FDDI) operating over fiber optic links at 100 Mbps or higher speeds;[13] since fiber optic links now operate at gigabit-per-second speeds, future LANs operating at such speeds are likely. Simply reading a route from a header and routing packets according to this information is faster than consulting a forwarding database. High-speed searches of tables are possible—for example, a scan of an 8192 station table of 6-byte addresses in 4 μsec [HAWE86]—but this is inevitably slower than scanning a bit stream for a fixed pattern. Whether it is possible to search tables rapidly enough to keep up with FDDI, or faster, rates is problematical.

Additional advantages claimed for source routing include flexibility and controllability of routes. Hosts have control over communications paths, allowing splitting of loads between routes, and so forth. Whether this extra flexibility is significant is uncertain, however. Most LANs have enough excess capacity and operate at high enough speeds that impacts of suboptimal routing and no use of parallel paths are minimal.

Source routing has been criticized because of the possibility of excessive times to learn unknown routes, but published examples of slow learning are based on very unrealistic geometries. The learning process can require substantial amounts of time with more realistic geometries and can impact performance if it happens too often, but under normal operating conditions, where the great majority of the routes needed are already known,[14] the performance impact will be minimal (see Problem 10.5). Thus source routing is best-suited to cases where most routes are already known.

How successful either approach will be at interconnecting dissimilar LANs is uncertain. The IEEE 802 LAN protocols *differ significantly* at the MAC layer. For example, the token bus and token ring standards include use of priorities at the MAC layer while the CSMA/CD standard does not. Only the token bus standard allows distinction at the MAC layer between frames requiring a response and those not requiring one. There are also incompatible implementations, including two speeds for token rings, two speeds and four different physical layer specifications for CSMA/CD, and three speeds and eight modulation schemes for token busses.

Source routing bridges are considered by some persons to be routers, which operate at the network layer rather than the data link layer [PERL88].

---

[13] The 100 Mbps FDDI protocol is very similar to the IEEE 802.5, token ring protocol and ease of bridging between 802.5 token rings and FDDI rings was given "the highest priority" in development of the FDDI protocol [ROSS87].

[14] The great majority of user conversations over networks are, for most users, with a limited number of other users. The routes needed for these should be known.

From this viewpoint the extra RI field added to frames is a network layer header[15] and the procedure for learning routes is a network layer protocol. This is reasonable, but one difficulty with it is political. The standards community, to avoid turf wars, has divided responsibility for LAN-related OSI standards between IEEE and ANSI at the boundary between the data link layer and the network layer [CHAP88]; hence, IEEE 802 committees are not supposed to be working on source routing bridges if they are indeed routers!

The only thing that can be predicted with confidence is that both types of bridges will soon be widely implemented; advocates of each type are developing products.

10.3 Routers and Gateways

The most common alternative to bridge interconnection of LANs is use of routers, which operate at the equivalent of the OSI network layer. The most prominent current approaches to interconnecting other types of networks, such as wide area networks, also involve interconnection at the network layer, so their relays fit the definition of routers. However, the term "gateways" is used in the literature describing most internetworking techniques. Since the terms "router" and "gateway" are used interchangeably in much of the literature, we discuss both together. We cover important techniques, especially those exemplified in standards, and concentrate on approaches allowing reasonably full sharing of facilities of one network by users of another. This does not mean the majority of implementations have been set up to handle full resource sharing, however. Remote access and sharing of a small number of resources are more common.

We discuss three standards for internetworking: *X.75,* the *DoD Internet protocol,* and the *ISO Internet protocol.* Each relay provides the equivalent of a network layer interface, and hence fits the definition of a router, but in each case the term "gateway" is used in the literature. We then discuss a true gateway, the DNA/SNA gateway offered by Digital Equipment Corporation for interconnecting these two types of networks. Other routers and gateways are offered by vendors.

If each network has the same network access interface, a simple relay can be used, providing either a datagram or virtual circuit interface, depending upon which is provided by the individual networks. The major example

X.75

of this approach is *X.75,* which provides internetworking of X.25 networks. Concatenated virtual circuits are set up between source and destination (keeping the X.25 bias toward virtual circuits). The "gateways" fit our defini-

---

[15] Even if this is viewed as a network layer header, the protocol still alters one bit in the MAC header, the I/G bit in the source address field.

tion of routers, since they operate at the network layer.[16] They are sometimes called DCE level relays since special DCEs are often used to implement them.

DoD internet protocol, ISO internet protocol

More complex relays are implemented with special purpose computers. Prominent standards are the *DoD Internet Protocol (IP)* and a similar *ISO IP* standard. Both IP protocols provide datagram service, relying on higher level protocols to supplement services as desired; the supplemented service can provide the equivalent of virtual circuit service if so desired. Only minimal assumptions about services provided by underlying networks are needed to provide datagram service. This is the primary reason for choosing datagram service for the IP protocols.

10.3.1   X.75

X.75 has been developed to provide internetworking of X.25-based networks. X.75 interconnection of X.25 packet data networks is common; most public X.25 networks are linked together via X.75.[17] Figure 10.12 illustrates X.25 networks interconnected via X.75. The figure shows four X.25 networks in a ring configuration; the geometry is arbitrary, however. Only two DTE-DCE pairs have been illustrated for simplicity, but thousands of pairs could use the four networks.

Figure 10.12   Interconnection of X.25 Networks via X.75.

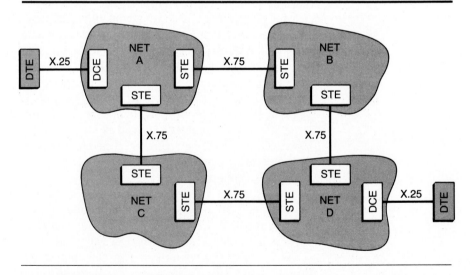

---

[16] Alternative terminology, specialized to the X.75 case, is introduced below.

[17] A reminder of the impact of political and legal factors on the field is given by the fact that X.75 is not likely to be allowed as an interface between public and private networks.

signaling terminal exchange

X.75 relays are implemented as half "gateways" (half routers) with each half a special type of DCE called a *signaling terminal exchange* (STE). Each user interface is an X.25 interface with X.75 protocols used between networks. As was indicated in Chapter 9, protocols within each network are not specified in the standards, and a variety of implementations that preserve the X.25 and X.75 interfaces may be used.

The basic function of X.75 is the interconnection of X.25 networks by setting up DTE-DTE virtual circuits between users. Each internetwork DTE-DTE virtual circuit is spliced together from a source DTE-DCE segment, an intranetwork DCE-STE segment to the first half gateway, an internetwork STE-STE segment to the other half of this first gateway, as many additional intra-network and internetwork STE-STE segments as are necessary to reach the destination network, and finally intranetwork STE-DCE and DCE-DTE segments to reach the destination. Seven segments would be spliced together, via either of two routes, to interconnect the two DTEs illustrated. Each segment is a separate virtual circuit, with its own flow control and error handling.[18]

The STE uses the X.25 packet format, but modifies the logical channel number to obtain an STE to STE dialogue.[19] It also adds a utilities field to the X.25 header of control packets to specify quality of service and similar parameters for STE to STE segments. Aside from this, STEs perform no encapsulation or other modification of the X.25 header. X.75 packet formats are illustrated in Fig. 10.13.

multilink procedure

The X.25 and X.75 packet levels also support a *multilink procedure* (*MLP*), which provides for use of multiple links between STEs. This is analogous to use of multilink transmission groups in SNA (see Subsection 9.7.2). MLP establishes rules for link transmission and resequencing for delivery to and from the links, making multiple channels between STEs appear as one channel with greater capacity than any individual channel.

X.75 does not offer permanent virtual circuits, so call setup and clearing are necessary. Standard X.25 control packets (with the added fields mentioned above) are used for setup and clearing. Rather than being sent directly from origin to destination DCE as in X.25, however, packets flow intranetwork from the origin DCE to the first STE on the path, from STE to STE as many times as necessary to reach the destination network, and finally to the destination DCE, setting up or clearing virtual circuits on each segment en route. The data transfer phase is identical to that in X.25, except for packets flowing through virtual circuits in cascade. X.75 is invisible to X.25 users during data exchanges.

---

[18] Except for end-to-end flow control obtained through use of the D bit.

[19] Recall that different logical channel numbers are used at the two ends of an X.25 virtual circuit.

Figure 10.13   X.75 Packet Formats. (a) Data packet. (b) Control packet.

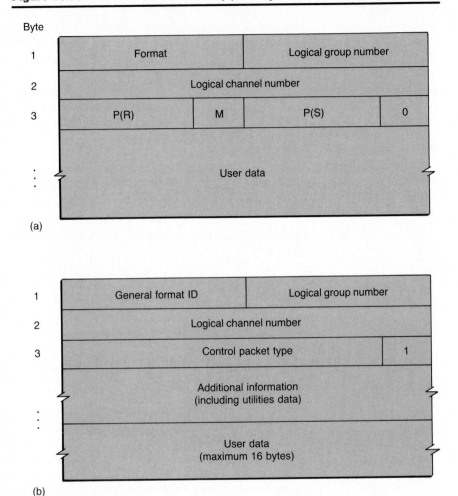

(a)

(b)

10.3.2   DoD's Internet Protocol (IP)

DARPA Internet

The largest single internetwork today in terms of number of interconnected networks is the *DARPA Internet.* As of late 1987 the DARPA Internet interconnected several thousand individual networks in the United States and Europe and connected over 20,000 computers. The growth rate in late 1987 was estimated at 15 percent per month [COME88]! One year's growth at this rate would more than quintuple the size of the network!

Figure 10.14 depicts the Internet as it existed in 1983. The figure illustrates the wide variety of types of networks interconnected without attempt-

Figure 10.14 The DARPA Internet in 1983 (from [HIND83], Hinden/Haverty/Sheltzer, "The DARPA Internet: Interconnecting Heterogeneous Computer Networks with Gateways," *Computer,* © 1983 IEEE. Reprinted by permission.).

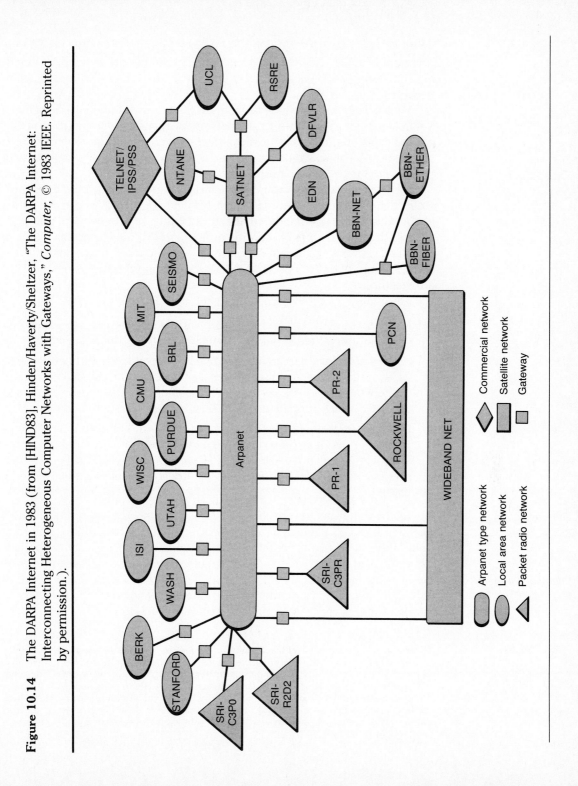

ing to show a few thousand networks in one diagram. The networks include a variety of LANs, ARPANET-type networks, commercial networks (including X.25 networks), packet radio networks, and communications satellite networks.

Because of the diversity of networks in the Internet, minimal assumptions are made about capabilities of any network. The primary assumption is that each is capable of accepting a packet and delivering it with reasonably high probability to a specified destination on that network. Guaranteed delivery of packets by a network is not assumed.

Gateways attach to networks as hosts, with an internet protocol (IP) providing datagram service between hosts. In keeping with minimal assumptions about network service and the definition of datagrams, delivery is not guaranteed. DoD's transmission control protocol (TCP), which is discussed in Chapter 11, provides guaranteed delivery, resequencing, flow control, error control, and so forth. TCP is the primary transport protocol in the DoD architecture.

Advantages for the datagram approach to IP include the following:

- *Flexibility.* DoD's IP can deal with a variety of networks, some offering connection-oriented service and some only connectionless service.

- *Robustness.* Since datagrams are routed independently through the network, with no predefined paths, they can be routed around affected areas if a network or gateway becomes unavailable. Also, the network can react quickly to congestion by making decisions for individual datagrams.

- *Connectionless application support.* Applications that need internet routing but for which a connection-oriented protocol would introduce excessive overhead can be handled. TCP, on top of DoD's IP, provides such services as reliable delivery and flow control for applications needing them.

gateway-gateway protocol

Additional protocols are also used. The *gateway-gateway protocol* (GGP) operates between gateways to exchange reachability and routing information.[20] It helps construct routing tables which give, for each destination network, the next gateway to which datagrams should be sent. These protocols are invisible to ordinary hosts and users. An *Internet Control Message Protocol* (ICMP) allows hosts to interact with gateways, and hosts and gateways to interact with internet monitoring and control centers.

internet control message protocol

---

[20]Recent growth in the Internet has caused the number of gateways to become too large for all to communicate with each other. Hence networks are now grouped into *autonomous systems* (for example, all systems on one university campus). Gateways are divided into exterior gateways (at least one per autonomous system), which communicate with other exterior gateways via an exterior gateway protocol (EGP) and interior gateways, which communicate only with other gateways in their autonomous system via an interior gateway protocol (IGP). This is being extended to group autonomous systems into *autonomous confederations* [COME88].

Figure 10.15 Operation of DoD IP Protocol (adapted from [STAL88b], W. Stallings, *Data and Computer Communications, 2/e,* © 1983. Reprinted by permission of MacMillan Publishing Company, New York, NY.).

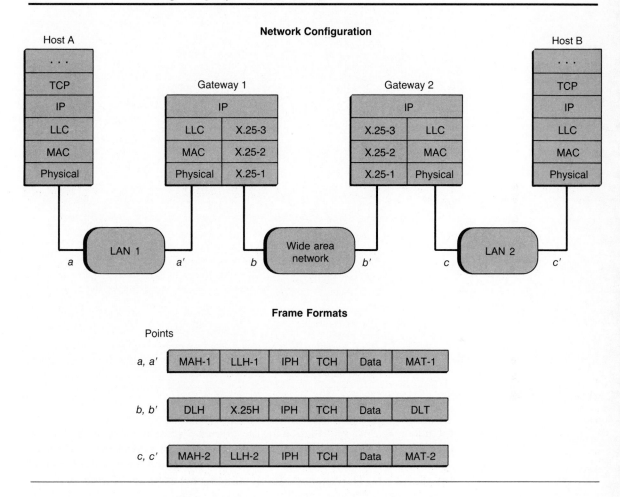

Figure 10.15 shows operation of DoD's IP in a situation where information passes through two LANs and one wide area network. The information IP receives at host **A** (data plus headers added by higher layers) is *encapsulated* with an IP header containing information for internetwork operation. LLC and MAC in **LAN 1** add headers, with MAC also adding a trailer. This is passed in the form of a datagram across **LAN 1** to gateway 1, which strips off the MAC and LLC headers and trailer, and analyzes the datagram to determine whether

Figure 10.16   DoD IP Header Format.

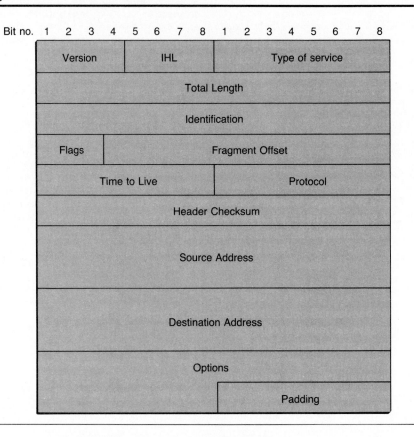

it contains control information for the gateway[21] or data for a host further on. It decides it is to route the frame across the wide area network to gateway 2, adds X.25 packet and data link headers and trailer, and sends the datagram to the network. The datagram eventually reaches gateway 2, which strips off the headers and trailer added by gateway 1, and decides to add the LLC and MAC headers and trailer for **LAN 2** and send the datagram to its destination at host **B**.

This approach of encapsulating datagrams in IP headers and sending them on their way avoids protocol translations in the process of sending data across network boundaries. The data will, with high probability, reach their destinations if the IP header contains appropriate information.

Two primitives are defined at the user-IP interface. *Send* is used to request transmission of a data unit, and *Deliver* is used to notify a user of

---

[21] Control information would be sent via ICMP.

Table 10.3   Fields in DoD IP Header.

- *Version* (4 bits): Version number, allows protocol evolution.
- *Internet Header Length, IHL* (4 bits): Length of header in 32-bit (4-byte) words. Minimum is five, so header is at least $5 \times 4 = 20$ bytes.
- *Type of Service* (8 bits): Specifies parameters such as precedence, desired reliability, delay and throughput values.
- *Total Length* (16 bits): Total data unit (datagram or fragment) length including header, in bytes. Maximum is 65,535.
- *Identification* (16 bits): Sequence number which, with source and destination addresses and user protocol, uniquely identifies a datagram.
- *Flags* (3 bits): One bit, a More flag, is used for fragmentation and reassembly.[a] Another, if set, prohibits fragmentation. (Datagram may then be discarded if too long to pass through a network.) Third bit not used.
- *Fragment Offset* (13 bits): Indicates where in datagram a fragment belongs. Measured in 64-bit or 8-byte units, considering data only. All fragments other than the last must contain a data field that is a multiple of 8 bytes long.
- *Time to Live* (8 bits): Measured in gateway hops and decremented at each gateway. Ensures that datagrams or fragments do not loop indefinitely.
- *Protocol* (8 bits): Indicates the higher level protocol to receive the data field at the destination.
- *Header Checksum* (16 bits): Checksum for header only. Reverified and recomputed at each gateway, since some fields may change.[b]
- *Source Address* (32 bits): Allows a variable allocation of bits to specify network and host within network. A network number is assigned by the internetwork administrator and host number by the network. Mappings of large address fields, such as 48-bit IEEE 802 LAN addresses, cause problems, but algorithms exist.
- *Destination Address* (32 bits): Same type coding as Source Address, using format appropriate for destination.
- *Options* (variable): Encodes options requested by sender, including security level, source routing, record routing, stream identification (allowing special handling for some traffic types), and timestamp (request to add timestamp as datagram passes gateway).
- *Padding* (variable): Ensures that header ends on 32-bit boundary.
- *Data* (variable): Data field must be a multiple of eight bits in length. Its length, plus the header length, must be that stated in Total Length field.

---

[a]Although we have used the term segmentation for this elsewhere, the term fragmentation is thoroughly embedded in the DARPA Internet literature, so we use it here.
[b]It is computed by first taking the 16-bit ones complement sum (that is, sum modulo $2^{16} - 1$) of all 16-bit words in the header, then taking the ones complement of the result. The checksum is assumed to be zero during the computation.

arrival of a data unit. A variety of parameters are encoded into the *DoD IP header,* which has the format shown in Fig. 10.16 and the fields in Table 10.3.

fragmentation example   An example will illustrate *fragmentation* and header fields. Assume that a datagram with 636 data bytes arrives at a network with a maximum length restriction of 256 bytes (including headers). Also assume that the IP header length is the minimum possible of 20 bytes. We consider two approaches to fragmentation.

In the first approach, each fragment except for the last is the maximum length possible. With 20-byte IP headers, this allows 236 data bytes per fragment, but the number in all except for the final fragment must be divisible by eight. Hence this is adjusted downward to $29 \times 8 = 232$ and fragments with 232, 232, and 172 data bytes used. The header length is four times the IHL value, total length is data field length plus header length, and the fragment offset is the number of data bytes (from the fragmented datagram) in fragments preceding the one under consideration. Thus the values for pertinent IP header fields are as follows:

> First fragment:
> IHL = 5
> Total Length = 252
> Fragment Offset = 0
> More = 1
>
> Second fragment:
> IHL = 5
> Total Length = 252
> Fragment Offset = 29
> More = 1
>
> Third fragment:
> IHL = 5
> Total Length = 192
> Fragment Offset = 58
> More = 0

For the second approach, all fragments are as nearly equal in length as possible. The average number of data bytes per fragment is $636/3 = 212$. This is not a multiple of 8, so lengths of the first two must be modified. The most nearly equal lengths are found by increasing one length and decreasing the other to yield $27 \times 8 = 216$, $26 \times 8 = 208$ and 212 bytes. This gives the following:

> First fragment:
> IHL = 5
> Total Length = 236
> Fragment Offset = 0
> More = 1
>
> Second fragment:
> IHL = 5
> Total Length = 228
> Fragment Offset = 27
> More = 1
>
> Third fragment:
> IHL = 5
> Total Length = 232

Table 10.4   ICMP Messages.

- *Destination Unreachable:* There are several possibilities for this. The gateway might not know how to reach the destination network, the destination host might not be reachable within the network, or the desired user protocol or service access point at the destination might not be available. If source routing is used, the specified route might not be available; if *Don't fragment* has been set, it might have been impossible to reach the destination without fragmentation. The returned message includes the entire IP header plus the first 64 bits of the original datagram.

- *Time Exceeded:* The time to live value of the datagram has expired. The original header plus 64 bits are returned.

- *Parameter Problem:* A syntactic or semantic error in the IP header has been detected. The original header plus 64 bits are returned, and a field in the message points to the byte where the error was detected.

- *Source Quench:* This provides a rudimentary form of flow control. Either a gateway or a host may use this message to request a source to reduce the rate at which it is sending traffic. The original header plus 64 bits of the datagram triggering the complaint are returned.

- *Redirect:* This message is returned if a gateway determines a shorter route to the destination would be used if the datagram had been sent to a different gateway. It advises the source to send future traffic via the other gateway. The original header plus 64 bits of the datagram are returned, but the gateway forwards the datagram toward its destination.

- *Echo:* This message instructs the recipient to return the same message to the source to provide a mechanism for verifying that communication is possible between source and destination.

- *Echo Reply:* Response to an echo message.

- *Timestamp:* This, along with the next message, provides a mechanism for sampling delay characteristics of the internet. The sender includes a message identifier plus the time the message is sent (originate timestamp).

- *Timestamp Reply:* Response to Timestamp message. The receiver appends a receive Timestamp and a Transmit Timestamp to the original message and returns it.

$$\text{Fragment Offset} = 53$$
$$\text{More} = 0$$

ICMP

The *Internet control message protocol (ICMP)* is a required companion to IP. Its function is to provide feedback from gateways to hosts about problems in the communications environment. It also allows hosts and gateways to interact with internet monitoring and control centers. Although it is basically at the same level as IP, ICMP uses IP to transfer its messages; it constructs an ICMP message and passes it to IP, which encapsulates it with an IP header and transmits it to the destination gateway or host. Nine types of ICMP messages have been defined; they are listed in Table 10.4.

DoD internet protocol operation

The DARPA Internet operates as follows. A host originating an Internet message passes a Send primitive across the user-IP interface, with data and parameters for the IP header. IP builds the IP header and passes it and the data to the lower layers in the originating network. They encapsulate the data unit in appropriate headers and trailers and transmit the resulting frame to the first gateway en route to the destination. This gateway strips off lower layer headers and trailers and examines the IP header to determine what to do. It makes appropriate modifications in the IP header, such as decrementing the time to live field, and fragments the datagram if necessary, adds lower layer headers, and sends it toward the destination. This continues until the datagram reaches the destination, which strips off headers and trailers and accepts the data or takes other appropriate actions.

ICMP reports back problems encountered during this process, and the GGP protocols maintain network reachability and routing information to aid in routing datagrams to their destinations. Extra error control, flow control, guaranteed delivery, or similar requirements are handled by TCP (or an alternative transport layer protocol).

The DoD internet protocol has been used successfully for over a decade. It is employed by a variety of networks aside from those comprising the DARPA Internet, and it is the most successful internetworking standard to date.

10.3.3   ISO's Internet Protocol

ISO's Internet Protocol, the connectionless internetworking protocol developed by ISO, is based on the DoD IP standard and is similar to it. The two primitives for connectionless network service listed in Subsection 9.4.2— N-UNITDATA.request and N-UNITDATA.indication—are used at the IP interface with the next higher protocol layer (normally the ISO transport layer). The header format is shown in Fig. 10.17, with fields defined in Table 10.5. The header is divided into four parts, with the first two always present, unless source and destination are connected to the same network; if source and destination are connected to the same network, only the eight-bit protocol identifier field is used and the IP protocol is null.

An Optional Part may also be included in the header. Each option is encoded in three fields: Parameter Code, Parameter Length, and Parameter Value. Some options are one byte long, but others are of variable length. Parameters include: Security, defined by the user, Source Routing, with a list of gateways to be visited, Recording of Route, used to trace a route, Quality of Service parameters such as desired reliability and delay values and Priority.

operation of ISO Internet Protocol   Operation of IP is as follows. A host originating an Internet message passes an N-UNITDATA.request across the user-IP interface, with data and parameters for an IP header. IP builds its header and passes data and header to lower layers in the originating network, which encapsulate them in headers and trailers and transmit the frame to the first gateway. This gateway strips off lower layer headers and trailers and examines the IP header, makes appropriate modifications in the header (that is, decrementing the time-to-live field), segments the datagram if necessary, adds lower layer headers, and

Figure 10.17 ISO IP Header Format.

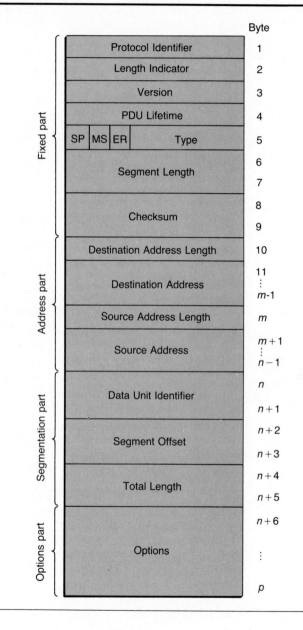

Table 10.5   Fields in ISO IP Header.

Fixed Part Fields (Always Present)

- *Protocol Identifier* (8 bits): Indicates whether ISO's IP is being used. Equivalent to the Protocol field in the DoD IP header.
- *Length Indicator* (8 bits): Specifies header length in bytes.
- *Version* (8 bits): Included to allow evolution of the protocol. Specifies version being used.
- *PDU Lifetime* (8 bits): Specifies lifetime of Data Unit as a multiple of 500 msec. It is determined and set by the source station, then decremented by each gateway the data unit visits by each 500 msec of estimated delay for that hop (transmission time to gateway plus processing time).
- *Flags* (3 bits): The SP flag is one if segmentation is permitted and zero otherwise; it is set by the originator of a datagram and cannot be changed. MS is one in all except the last segment in a datagram. ER indicates whether the source station desires an error report if an IP data unit is discarded.
- *Type* (5 bits): Currently indicates whether this is a Data PDU (with user data) or an Error PDU (containing an error report).
- *Segment Length* (16 bits): Total data unit length, including header, in bytes.
- *Checksum* (16 bits): Checksum computed on entire header. Since header fields may change at each gateway, it is reverified and recomputed at each. The algorithm to compute the checksum is given in Appendix 10A.

Address Part Fields (Always Present)

- *Destination Address* (variable): Address of destination.
- *Source Address* (variable): Address of source.
- *Address Length* (8 bits): Precedes each address.

Segmentation Part Fields (Only Present if SP Flag in the Fixed Part Set to One)

- *Data Unit Identifier* (16 bits): Uniquely identifies Protocol Data Unit. This requires the station IP entity to assign values for the PDU's destination, which are unique during the maximum time a PDU may stay in the internet.
- *Segment Offset* (16 bits): Indicates where in PDU this segment belongs. Measured in 8-byte units, considering data only; all segments other than the last must contain a data field that is a multiple of 8 bytes long.
- *Total Length* (16 bits): Total length of original PDU in bytes, including one header and data.

---

sends it toward the destination. This continues until the datagram reaches the destination, which strips off headers and trailers and accepts the data or takes indicated actions. Extra error control, flow control, guaranteed delivery, or similar requirements are handled by the transport layer.

10.3.4   DNA/SNA Gateway

The *DNA/SNA Gateway,* developed by Digital Equipment Corporation for interconnecting DNA and SNA networks, allows users of a DNA network to access and use an SNA network. Development was complex because of dissimilarities between the two architectures [MORE86]. Despite superficial similarities, their implementations include radical differences:

- At the equivalent of the OSI network layer, DNA provides connectionless service with adaptive routing but SNA provides connection-oriented service with quasi-fixed routing.

- At the OSI transport layer level, DNA uses a symmetric, three-way handshake for connection, but SNA uses an asymmetric, three-party negotiation.

- At the session layer, DNA provides process-binding and access-control functions, while SNA provides data-phase services such as chains, brackets, and multiple acknowledgment schemes.

- At the application layer, radically different applications are included in the architectures.

- A central DNA application service is file transfer and access, which SNA does not support.

- Conversely, a widely used SNA application service is remote job entry, which DNA does not support.

Figure 10.18 shows the gateway. It is a true gateway, which operates at the equivalent of the ISO transport layer. A one-way encapsulation approach is used. SNA sessions are mapped onto DNA logical links, with all SNA session data (including connection establishment messages) carried over to the data phase of the DNA logical link. Higher layer SNA protocols are emulated on the DNA system containing the SNA gateway, or in the user's DNA system, so DNA users can run SNA applications.

Two modules implement the SNA gateway service: an *SNA access module* in any DNA system providing access to SNA and the *SNA gateway module* in the DNA system containing the gateway. The two modules communicate by means of an *SNA gateway access protocol* (*GAP*), which operates over a DNA link interconnecting the two DNA systems. GAP is a DNA network application layer protocol that uses functions provided by the DNA session control and lower layers, including flow control, error control, and message segmentation and reassembly.

Two models for distribution of gateway functions between the user's DNA system and the DNA system containing the gateway have been implemented. In the *gateway access model,* higher layers are contained in a process that executes in the user's system, while in the *server model* all major gateway functions execute in the gateway node. The layers of the architecture are illustrated in Fig. 10.19. For the gateway access model, the two top layers execute in the user's system and the lower three in the gateway node, while for the server model all five layers execute in the gateway node. Descriptions of the layers (which do not precisely follow SNA layers) are given in Table 10.6.

Figure 10.18   Structure of DNA/SNA Gateway (from [MORE86], Digital
Equipment Corporation. Reprinted by permission.).

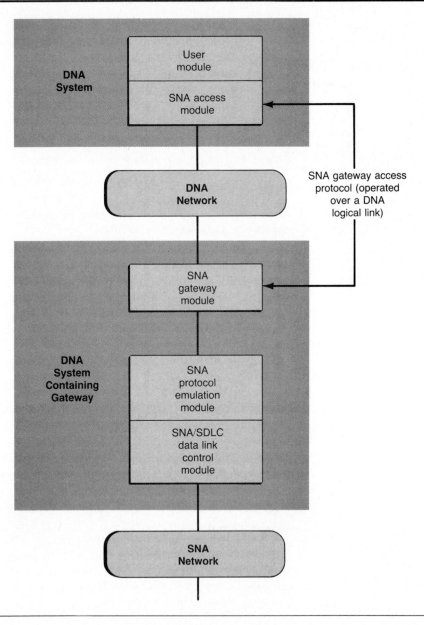

Figure 10.19 Layers of DEC's DNA/SNA Gateway Product Architecture (from [MORE86], Digital Equipment Corporation. Reprinted by permission.).

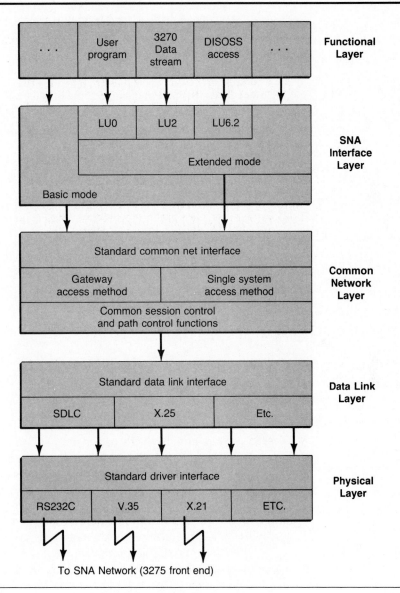

Table 10.6   Layers of DNA/SNA Gateway Architecture.

- *Functional Layer :* This is the highest layer and implements user functions. Translation from SNA presentation protocols to DNA presentation media and formats is done in this layer. The layer can also contain programs supplied by customers or by DEC, as well as entities that are part of DNA/SNA products. Two such products are indicated in Fig. 10.19, IBM 3270 data stream and DISOSS access.

- *SNA Interface Layer :* The SNA interface layer provides access into the SNA network. Three access modes are offered: a basic mode, which is very close to that offered by the common network layer; an extended mode, which offers generic support for SNA data flow control and transmission control protocols; and several LU-mode interfaces, each implementing a particular logical unit type. Higher modes offer specialized, easier-to-use, interfaces at the expense of flexibility.

- *Common Network Layer :* This provides routing and multiplexing functions between the data link layer and SNA interface layer. The data units it receives are routed to the entity in the SNA interface layer owning the SNA session to which the data units belong. The data units it sends are routed to the appropriate data link layer. The layer implements SNA path control protocols and some transmission control protocols. SNA physical unit and logical unit services management functions are also part of the layer.

- *Data Link Layer :* This implements data link protocols, such as SDLC or DDCMP, and corresponds exactly to the data link layer in either architecture.

- *Physical Layer :* This layer corresponds exactly to the physical layer in either architecture. It can support a variety of device types and interface standards and can be implemented in various hardware and software mixtures.

example of gateway message exchanges

An example of message exchanges to establish an SNA session and transfer data illustrates the gateway. Typical message exchanges are illustrated in Fig. 10.20. Dashed two-way arrows indicate sequences of messages to set up and disconnect the DNA logical link.

The first step in establishing a session is for the DNA user program to issue a connect call,[22] with parameters such as name of the gateway node, secondary logical unit (SLU) address to be used (the DNA user's port into the SNA network), name of the primary logical unit (PLU) (network port used by the IBM application accessed), and a variety of other SNA parameters. The SNA access module allocates internal resources and establishes a DNA logical link to the SNA gateway module in the specified gateway node; this SNA gateway node in turn allocates resources for the session and waits.

After these preliminaries, a sequence of messages to set up the session begins. The SNA access module transmits a GAP Connect message to the SNA gateway module. The gateway module allocates the requested SLU (DNA port) address and transmits an SNA Initiate-Self message to the system ser-

---

[22]Establishment of an SNA Session and of a DECNET Logical Link are covered in Chapter 11. The gateway uses standard messages to establish connections.

Figure 10.20   Typical Message Flows Across DNA/SNA Gateway (adapted from [MORE86], Digital
Equipment Corporation. Reprinted by permission.).

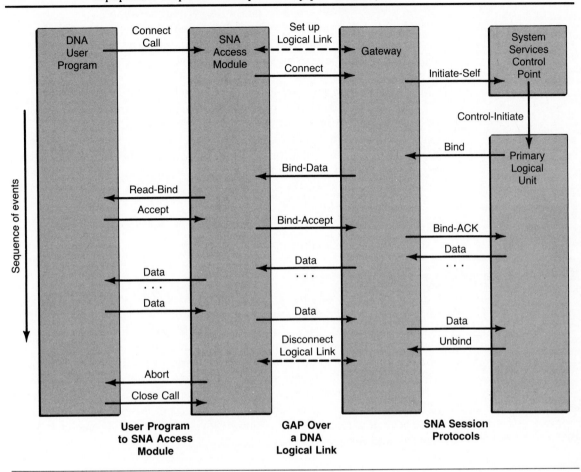

vices control point (SSCP) in the SNA network. The SSCP informs the selected
PLU (SNA port) of this via an SNA Control-Initiate message.

Eventually the PLU transmits an SNA Bind message to the gateway. This
contains information for establishment of the session, and it is forwarded to
the SNA access module as a GAP Bind-Data message, then ultimately to the
user program in response to a Read-Bind call. The user program agrees to
the session by issuing an Accept message, which causes the SNA access module to send a Bind-Accept message to the SNA gateway module. The gateway
module acknowledges the bind, and the LUs are in session.

Following session establishment, the DNA user program can exchange
data messages with the IBM application. Transmit-Calls and Receive-Calls

functions of the SNA access module are used for data exchanges.[23] During data transfer, the SNA gateway module acts as a message switch, passing data to and from the SNA access module without interpretation.

At some point the PLU terminates the session by sending an SNA Unbind message to the gateway. The SNA gateway module then disconnects the DNA logical link with the SNA access module, supplying a reason code in a disconnect message. The user program reads the reason code and issues a Close-Call to cause the SNA access module to deallocate its resources.

The approach does not require DNA specific software or hardware components to be introduced into the SNA environment. SNA-oriented applications on DNA nodes can be written as though they directly use services of the transmission control layer of SNA, and a variety of SNA applications have been implemented successfully in this manner. Each new SNA application protocol, though, requires a complete implementation on the user's DNA node before it can run in the DNA universe.

10.4   Problems in Large Internetworks

Experience with large internetworks, especially those with minimal standards for member networks, such as the DARPA Internet, has shown that unexpected interactions can occur at many levels, some causing serious problems. Some of the most significant have resulted from bugs in implementations. Such bugs are inevitable in internetworks as large and diverse as the DARPA Internet, which inevitably interconnect some networks using network software that has not been fully debugged.[24] Satisfactory explanations of some observed phenomena are not available, but a discussion of observed phenomena points out that precautions are advisable in designing such networks. A discussion of observed problems is given in [BOSA88].[25]

10.4.1   Disagreement over Broadcast Addresses

The most consistent class of problems in DoD IP-based networks is caused by disagreement over broadcast addresses. After a change in the standard for forming broadcast addresses in the DARPA Internet a few years ago and the

---

[23] Higher-level protocol initialization may also be needed before true user data exchanges can begin, but this is not shown. The SNA access and gateway modules are not aware of such details; they simply pass information needed for protocol initialization as if it were any other data.

[24] Other problems have included the case where a "worm" programmed by a graduate student in the United States slowed or halted over 6000 computers attached to the DARPA Internet in November 1988.

[25] Some persons view these problems as "network folklore" rather than real science, but we include a discussion of them since strange things do happen in large networks. Whether the diagnosis of problems is accurate or not, the problems cited have occurred.

subdivision of networks into "subnets" for use by organizational subdivisions, there are normally six possible broadcast addresses on a network! If systems disagree about what the broadcast address is, one system may send a broadcast that another may not recognize as a broadcast. Rather than accepting the broadcasting packet, as each system should do, the system may attempt to forward it. Unfortunately, this process of trying to find the destination often involves generating another broadcast. This generates a barrage of secondary broadcasts that can cause the network to saturate for a few seconds every 30 seconds or so. This is referred to as a *broadcast storm*.

broadcast storm

Every system on a network is supposed to process a broadcast packet, so processing power is exhausted before transmission bandwidth during most broadcast storms. A statewide network in the United States reportedly has so much broadcast traffic that some microcomputer TCP/IP implementations cannot be used; all CPU power would be taken up processing broadcasts!

10.4.2   Chernobyl Packets and Network Meltdown

In some situations, systems have tried to return error messages such as "destination unreachable" rather than attempting to erroneously forward broadcast packets. This is usually better than attempting to forward such packets, but there have been situations in which it caused far more damage than would be expected. A bug in some early IP implementations caused such error messages to be sent to the original destination address rather than back to the original source address! This sometimes resulted in a new broadcast being sent to the destination address in the original packet. If two or more hosts had this problem on the same network, the result was an endless sequence of error messages, saturating all network machines. This has been called *meltdown* of the network, with packets initiating meltdown called *Chernobyl packets*.

meltdown
Chernobyl packets

Networks interconnected by bridges are especially susceptible to broadcast storms, meltdown, and so forth. This is because bridges are less capable of filtering out unnecessary broadcast messages than are routers or other higher level gateways. Bridges normally pass on all broadcast packets. However, routing is a basic function of higher level relays, so such relays can be selective in the packets they send to other networks.

10.4.3   Synchronization Loading Peaks

Overloading problems on a variety of networks have also been caused by synchronization effects that cause a number of systems to produce *loading peaks* at certain times. In [BOSA88] a case is cited where all computers on a network ran a process to resynchronize their clocks at midnight each night. This involved network activity, especially since most computers on the network were diskless. Thus at midnight each of these computers had to read the command file and load all programs and data files needed for its execu-

self-synchronization

tion, with all these operations taking place over the network. The affected network had adequate capacity for normal use, but excessive delays were encountered for a few minutes near midnight every night.

There is a conjectural phenomenon called *self-synchronization,* which may cause processes to synchronize with each other. Generally, processes that should occur at fixed intervals do not schedule their next execution using an absolute clock. Instead, after each activation they schedule wakeup a fixed interval of time later. The operating system, though, may not wake up a process as soon as it becomes eligible to run, and the program usually does some computation between the time it wakes up and the time it schedules the next wakeup. Self-synchronization is thought to occur if there is a feedback mechanism that tends to adjust delays to keep processes running on different systems in synchronization. Such mechanisms appear to have been observed, but no adequate theoretical explanations are available.

Routing processes are especially vulnerable to introducing synchronized activity, since routing updates are initiated when a route changes (typically because a link has gone down). For common algorithms, on the order of ten updates per gateway are required before convergence to new routes, with updates often distributed via broadcasts. This can saturate the input processes of other systems on networks.

10.4.4   Black Holes and Other Routing Problems

A variety of additional problems are due to routing difficulties. These include routes that do not reach their destination (known as black holes), excessive bandwidth wasted due to routing, sustained oscillations, and local and regional overloads.

black holes

Black holes may result from any of several phenomena. During recovery from link failure temporary circular routes can arise. Once a packet enters such a route, it circulates until either the loop is corrected or a maximum time to live is exceeded. On the other hand, links can fail in such a way that packets can pass in only one direction, with no other overt indication of trouble. Routing algorithms not designed to detect this are prone to sending traffic into oblivion.[26]

Some dynamic routing algorithms require significant fractions of network resources. It is easy to find examples where the required bandwidth for routing messages is $O(n^2)$, with n representing the number of routable entities (network nodes or subnetworks). In addition to consuming bandwidth, this can be ruinously expensive when traffic flows over pay-per-packet networks. Some networks are perilously close to a point where the slowest links are always saturated by routing traffic.

---

[26] An example of this mentioned in Section 8.7 is a military network subject to jamming. Since jamming often affects only one direction of transmission on a link, routing algorithms that do not take into account asymmetric links are likely to send packets into black holes.

Sustained oscillations may occur due to unstable feedback of routing decisions into utilization measures that influence routing decisions. Examples are discussed in Section 8.5.

As a link approaches capacity, queuing delays become significant and highly variable. If a network is using a routing algorithm based on measured delay, instability can be induced by a single link approaching capacity. When several links in a region are near capacity, variability in delay can be far greater. As links actually have their capacity exceeded, messages will have to be discarded, leading to variation of both delay and reliability with number of congested links traversed by a route. Under these circumstances, host behavior may conspire with network overload characteristics to prolong the saturation. As more distant hosts become difficult to communicate with, a backlog of requests will develop. When hosts try to provide requested services, they maintain the network in the congested state. It can become impossible to recover except by draconian "load shedding" measures.

10.4.5 General Comments on Large Internetworks

These are only a few of the problems observed with large scale internetworks. Many problems can be, and are being, solved by careful redesign once they have been observed, and techniques for writing software containing fewer bugs are being developed. Nevertheless, it is extremely difficult to predict the impact of network changes when undetected bugs are present. At any given time, the best possible situation seems to be that all detected errors have been corrected. Guaranteeing that no bugs are present is well beyond the current state of the art.

Some weaknesses of current hardware and software validation techniques for ensuring reliability of large scale networks became painfully obvious to many persons in the United States on January 15, 1990, the day that every phone in AT&T's long distance network "seemed to be off the hook."[27] A flaw in software in AT&T switching offices caused a minor problem, originally experienced in New York, to largely disable switching offices throughout the country. More details are given in Section 12.2.

This incident has shaken the confidence of many persons who felt that it was possible to design complex telecommunication networks to have very high reliability. AT&T has led the way in developing techniques for reliable design, and had subjected the faulty software to rigorous testing before it was installed. Although this particular problem was fixed within a few days of its first occurrence, there is currently no way to guarantee similar problems will

---

[27] This is an exaggeration; approximately half the long distance calls attempted during that day were completed. Nevertheless, on the order of 65 million attempted calls failed to be completed, and many switches suffered total down time equal to that anticipated during several hundred years, according to standard design criteria for AT&T switching offices of no more than two hours down time in forty years.

never recur in the AT&T network, or in other systems of comparable complexity.

10.5   Summary

A major emphasis of telecommunication network studies today is the development and implementation of techniques for internetworking. The major problems encountered in internetworking stem from the fact that different networks are seldom compatible and that there is a standard assumption in internetworking that the architectures of individual networks cannot be altered.

Some of the most successful internetworking techniques for LANs involve the use of bridges operating at the data link layer. Both source routing bridges and spanning tree bridges are being standardized. The most prominent internetworking standards involving interconnection at the network layer are X.75, developed to internetwork X.25 networks by concatenating virtual circuits, and the DoD and ISO IP protocols, which use datagram transmission. The DNA/SNA gateway is a typical example of relays for interconnection at higher layers.

Strange phenomena are sometimes observed in large and diverse internetworks, with many of them attributable to bugs in the software of some networks. No satisfactory explanations of such phenomena are available, but it is important to be aware of the potential for them to occur.

Appendix 10A   The ISO Checksum Algorithm

The ISO internet protocol checksum is computed by using an algorithm originally given in [FLET82]. It is much easier to implement in software than a polynomial code CRC, yet has almost equivalent error-detection properties. Efficient computational procedures are given in [COCK87].

The algorithm is used to compute a 16-bit checksum included in the ISO IP header. The checksum is originally calculated by the sender, with the field it is to occupy assumed to be all zeros, and the results placed in the appropriate field in the header in such a manner that when the receiver applies the same algorithm (to the header, including the 16-bit checksum), it will get a zero result if there are no errors.

All computations are performed by using eight-bit ones complement arithmetic, which is equivalent to modulo 255 arithmetic. In order to present the algorithm, we define the following terms:

$$L = \text{Length of header in bytes}$$
$$B_i = \text{Value of } i\text{th byte}$$

$$k, k+1 = \text{Byte positions for checksum}[28]$$

B_k and B_{k+1} are set to zero throughout initial computation. Two checksum bytes, R_i, and S_i are calculated by processing the B_i one at a time, using the formulas (modulo 255)

$$R_i = R_{i-1} + B_i \qquad \text{and} \qquad \text{(10A.1)}$$

$$S_i = S_{i-1} + R_i, \qquad \text{(10A.2)}$$

with R_0 and S_0 both set to 0. Upon completion,

$$R = R_L = \sum_{i=1}^{L} B_i, \qquad \text{(10A.3)}$$

$$S = S_L = \sum_{i=1}^{L} (L - i + 1)B_i. \qquad \text{(10A.4)}$$

Thus R is the modulo 255 sum of all bytes in the header, while S is the modulo 255 sum of all bytes weighted (inversely) by position in the header.

The final step in the algorithm is to determine appropriate values to be placed in B_k and B_{k+1}, so that when the receiver performs the same computation, including the bytes in these positions, it will get a result of zero if no errors occur. Since the receiver will compute the same values as the sender plus the results of including B_k and B_{k+1}, we find

$$R + B_k + B_{k+1} = 0, \qquad \text{(10A.5)}$$

$$S + (L - k + 1)B_k + (L - k)B_{k+1} = 0. \qquad \text{(10A.6)}$$

with arithmetic still modulo 255. Solving,

$$B_k = (L - k)R - S, \qquad \text{(10A.7)}$$

$$B_{k+1} = -(L - k + 1)R + S, \qquad \text{(10A.8)}$$

modulo 255.

Problems

🔶 **10.1**   (**Design Problem**). Assume that your university campus has a variety of independent networks operated by different colleges or departments, and that internetworking them is being considered. Give realistic answers to the questions in Table 10.1 and indicate which issues in Table 10.2 need to be addressed. You will need to hypothesize (or preferably learn) a variety of features of your campus networks in order to solve this problem.

---

[28] As Fig. 12.14 indicates, $k = 8$ for the ISO IP header.

Figure P10.4   Figure for Problem 10.4.

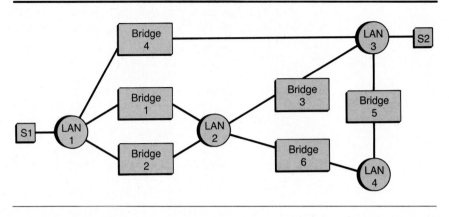

10.2   **a.**   Sketch the complete message format at point *e* in Fig. 10.3(a), showing contents and lengths of all fields. Assume IEEE 802.3 (CSMA/CD) LANs with six-byte address fields, and normal formats (with no extended fields) for each header or trailer.

b.   Sketch the complete formats of messages at points *f*1 and *f*2 in Fig. 10.3(b). Use the same assumptions as in (a).

10.3   Show that the number of LAN messages generated during the "single route broadcast" portion of the route learning process for source routing bridges is independent of the spanning tree used, and give a formula for determining this number for an internet with N LANs and M bridges.

10.4   **a.**   Using the format of Fig. 10.8, diagram the routes followed by broadcast messages exchanged, during route learning for the interconnected source routing LANs in Fig. P10.4, when **S1** learns a route to **S2**. Assume that bridges 2, 3, and 5 are not part of the initial spanning tree.

b.   How many LAN messages are required to learn the possible routes?

10.5   Follow the steps below to put route learning overhead in interconnected LANs with source routing bridges approximately into perspective. Assume that the examples in Figs. 10.7 and 10.8 are typical. The steps involve analyzing an arbitrary, but not unreasonable, traffic mix under the assumption that all routes are known, then examining the impact of route learning.

a.   Define LAN traffic as the total number of bytes transmitted on LANs in the network; a byte confined to one LAN gets counted once, a byte flowing through two LANs gets counted twice, and so forth. Assume the following: no route discovery, MAC addresses of six bytes, LLC control fields of two bytes, and average LLC information field content of a user data message is 100 bytes, with token ring LLC and MAC overhead (including needed routing information) added.

With probability 0.50, a user data message is destined for the same ring, with probabilities 0.20, 0.15, 0.10, and 0.05 it passes through 1, 2, 3, and 4 bridges, respectively. Compute the average LAN traffic, $\bar{N}_{message}$, generated by one user data message.

b. Assume that an average of one supervisory frame is generated for each user data frame and that supervisory frames have the same formats as user data frames (including distribution of routing information field lengths), except for having an empty LLC information field. Compute the average number of bytes in supervisory frames $\bar{N}_{\text{supervisory}}$.

c. Under our assumption of one supervisory message per user data message, total LAN traffic per user data message is given by $\bar{N}_{\text{no learning}} = \bar{N}_{\text{message}} + \bar{N}_{\text{supervisory}}$. Compute this value.

d. Assume that each of the route learning messages, either "single route broadcast" or "all routes broadcast," has the format in Fig. 10.6, with the LLC Information field empty. When either type broadcast message leaves its source, only a routing control subfield is present in the routing information field, but each bridge it passes through adds a two-byte route designator. Compute the average length in bytes, $\bar{N}_{\text{broadcast}}$, of the broadcast frames used in the search process.

e. If a fraction p of the user data frames require route learning, the average amount of LAN traffic per user data frame becomes $\bar{N}_{\text{learning}} = \bar{N}_{\text{no learning}} + 20\, p\, \bar{N}_{\text{broadcast}}$. Plot the percentage increase in LAN traffic, $100(\bar{N}_{\text{learning}} - \bar{N}_{\text{no learning}})/\bar{N}_{\text{no learning}}$ versus percentage of user data frames requiring route learning (on a log scale) as this percentage varies from 0.01 percent to 10 percent. Comment on your results.

10.6 Find the spanning tree computed by the IEEE 802 spanning tree bridge algorithm for the interconnected LANs in Fig. P10.6. Assume that all link costs are equal. Use the lower-numbered bridge or port in any case where there is a tie in the criterion for a choice.

10.7 Compare the internetworking approach used in X.75 with that used in the DoD and ISO IP protocols. Give the major advantages and disadvantages of each approach.

10.8 Compare the format of data packets between two STEs in an X.75 network with the format between DTE and DCE at either end. What fields are changed and/or added? Compare these differences with differences at the entry and the exit of an ordinary X.25 network.

10.9 Assuming that a DoD IP data packet contains only overhead from the ARPANET DLC, the standard data packet CSNP-CSNP header, and a minimal length DoD IP header, plot fractional overhead (as defined in Section 7.8) for internet connections, for the number of data characters (on a log scale) ranging from 1 to 10,000 bytes. Ignore possible problems with maximal packet lengths allowed in some networks. Your plots should resemble Figs. 7.22 and 7.23.

10.10 Sketch the precise format for all headers and trailers, except for the TCH (covered in Chapter 11) in Fig. 10.15 assuming IEEE 802.3 and 802.5 for LANs 1 and 2.

10.11 Assume that a DoD IP packet containing 586 data bytes arrives at a gateway to a network with maximum length restriction (including headers) of 128 bytes and that IP headers are each 28 bytes. Give the length of each fragment and the values of the IHL, Total Length, Fragment Offset, and More fields for each fragment. Assume the following:

a. Each fragment except for the last is the maximum feasible length, with the final fragment whatever length is necessary to complete packet transmission.

b. All fragments are of as nearly equal length as possible.

10.12 Assuming that an ISO IP data packet contains only overhead from HDLC (with no extended fields), the standard X.25 data packet header, and an ISO IP header with no

Figure P10.6   Figure for Problem 10.6.

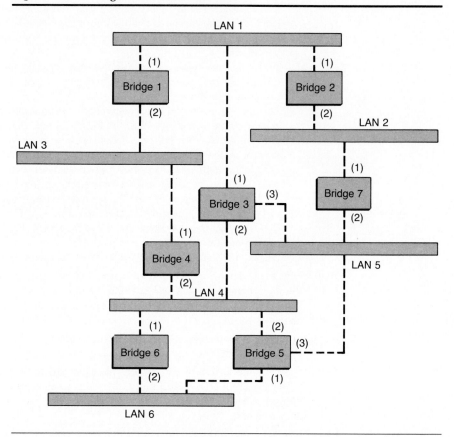

optional or segmentation fields and six-byte addresses, plot fractional overhead (as defined in Section 7.8) for ISO IP, for the number of data characters (on a log scale) ranging from 1 to 10,000 bytes. Ignore possible problems with maximal packet lengths allowed in some networks. Your plots should resemble Figs. 7.22 and 7.23.

10.13   Repeat the fragmentation (or segmentation) example in Subsection 10.3.2 but use the ISO IP protocol instead of the DoD IP protocol. The original datagram length and network maximum length restriction are again 636 bytes and 256 bytes (including headers), respectively. ISO IP headers with no optional part and four-byte addresses are used. Consider the same two cases as in the example, and in each case give the values of the Length Indicator field, the SP and MS flags, and the Segment Length, Segment Offset and Total Length fields for each segment.

10.14   (**Design Problem**) Assume that the only significant difference between two networks is that one uses Bisync as a DLC protocol and the other uses HDLC Normal Response Mode. Develop the basic design of a relay to interconnect these two networks. Include in your design any restrictions that must be placed on operation of one or the other

Figure P10.16 Figure for Problem 10.16.

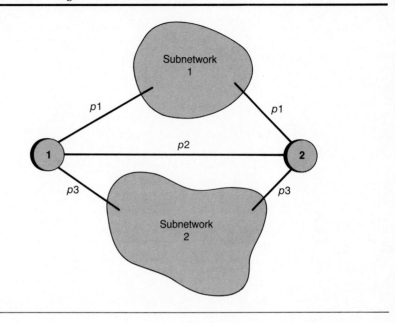

of the DLC protocols, procedures for converting frames from one format to the other, and examples of sequences of transmissions. Also indicate where special problems come in, including any you are not able to fully solve. (If you do a good job, you should find a few examples of such problems; this is a much more complex problem than it may appear to be at first glance.)

10.15 Which of the two models of operation of the DNA/SNA gateway—the gateway access model or the server model—relies more heavily on the user's system? Briefly discuss the advantages and disadvantages of the two models.

10.16 The type of network behavior leading to "network meltdown" is illustrated in Fig. P10.16. Assume node **1** and node **2** are interconnected by three paths, $p1$ via subnetwork 1, $p2$ as a direct path, and $p3$ via subnetwork 2. For simplicity, assume that delays along these three paths are essentially the same, so copies of the same packet sent between the two nodes on different paths arrive at essentially the same time. Assume that **1** initially sends **2** a packet, on the direct path, that **2** misinterprets so that it sends out a broadcast packet on all paths except the one that the packet arrived on. Each of these is misinterpreted by **1** so it responds to each by sending out broadcast packets on all paths except the one each incoming packet arrived on, and so forth. Letting the kth iteration at **1** represent the kth time that **1** receives packets, compute a general formula for the number of packets **1** receives on the kth iteration and evaluate this formula for $k = 10$. Is it reasonable to call the packet starting this sequence a Chernobyl packet?

11 Transport Layer Protocols

11.1 Introduction

Protocols at the equivalent of the *OSI transport layer* provide the basic end-to-end service of transferring data between users. The transport layer interfaces the lower layers—physical, data link, and network—which provide transparent connections between users, with the upper layers—session, presentation, and application—which ensure that information is delivered in correct and understandable form. Protocols at this level are in some ways the *keystone* of a network architecture. They have the task of providing reliable, cost-effective data transmission from source host to destination host, regardless of physical networks and transmission technologies used.

Transport layer protocols are in some ways less complex than those at the OSI network layer, since they involve cooperation only between two nodes at each end of a communications path rather than cooperation among all nodes, as is required for network layer routing and flow control. The fact that nodes may be separated by a widely varying number of hops (and types of communications facilities), with storage at intermediate nodes, produces problems not encountered at other layers, though. Some of the most difficult are due to the fact that good bounds on maximum network delay are not known. Providing for the possibility that delayed packets, including some for which retransmitted duplicates have already been accepted, may turn up at inopportune times adds to complexity of the protocols. The possibility of long and widely varying delays also makes it difficult to design flow control and buffering schemes. Figure 11.1 illustrates the type of variation in delays commonly observed. Mechanisms to enforce bounds on network packet lifetimes are often incorporated in transport layer protocols.

Another problem is the many types of errors such protocols must deal with. Packet-switching networks not only damage or lose packets but also duplicate packets (via retransmissions) and deliver them out of order. These anomalies are especially likely in networks providing datagram service at the network layer. In addition, nodes may be subject to failures, some wiping out their memory of system states before failure.

Figure 11.1   Plot of Round Trip Times as Measured for 100 Successive Datagrams Traveling Across the DARPA Internet (from [COME88], Douglas Comer, *Internetworking with TCP/IP: Principles, Protocols, and Architecture,* © 1988 p. 144. Reprinted by permission of Prentice-Hall, Inc., Englewood Cliffs, NJ.).

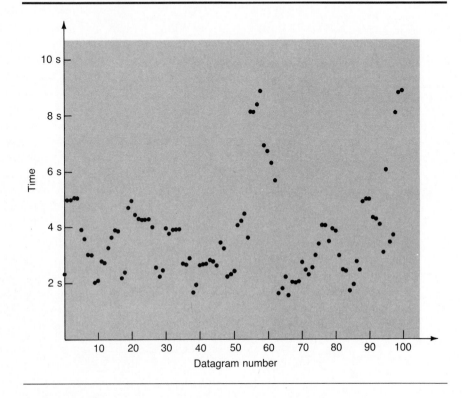

Even networks providing virtual circuit service at the network layer may not provide adequate reliability at this layer, so extra techniques to ensure reliable data transfers may be needed at the transport layer. For example, the possibilities of X.25 resets, clears, or restarts, with possible packet loss, mean that if transport layer protocols are to provide reliable data transfers across X.25 networks they must provide for recovery of lost packets after signaled failures.[1]

---

[1] "Signaled" failures are specified since the network notifies the user of resets, clears, or restarts.

Transport layer protocols also put more emphasis on naming and addressing than protocols at other layers.[2] Techniques for assigning names and addresses, and for efficiently mapping human-readable names onto machine-readable addresses, strongly influence the nature of transport protocols.

Many of the problems addressed by transport layer protocols are very similar to those addressed by lower layer protocols. We do not explicitly address these here, aside from mentioning how specific protocols treat them. Some areas in which techniques needed at the transport layer differ from those used at lower layers include end-to-end error control, connection management, and approaches to determining appropriate timeouts and to congestion control. We briefly discuss these aspects before treating specific protocols, which include techniques for handling these functions.

11.1.1   End-to-End Error Control

We have discussed a variety of techniques for error control as it is applied in various protocol layers. Although error control on individual communications links and at lower protocol layers can eliminate the great majority of the errors that occur, it cannot eliminate all errors. Some end-to-end error checks are recommended no matter how reliable the communications system is. There are always some types of errors that cannot be caught by lower layer checks. An example in [SALT84] based on a problem encountered at M.I.T. indicates how problems can occur. A network system there involved several local networks connected by relays and using a packet checksum on each hop from one relay to the next. Since this appeared to be adequate transmission error protection, application programmers assumed that the network was providing reliable data transmission. Unfortunately, data were not protected while stored in each relay. One relay developed a transient error: While copying data from an input to an output buffer, a byte pair was interchanged, with frequency of interchange about once in every million bytes passed. After source files of an operating system were repeatedly transferred through the defective gateway, some source files were contaminated by the byte exchanges. The only way found to recover was a programmer's nightmare—manual comparison with and correction from old listings!

A listing of threats to a file transfer, from disk at host **A** to disk at host **B**, indicates types of errors that should be considered in a careful design [SALT84]:

---

[2] Protocols for internetworking are the only real exceptions to this. In virtually all cases, however, they must treat names and addresses within networks as fixed and not part of the design of internetworking protocols.

- Even though the file was originally written correctly onto the disk at host **A**, it may contain incorrect data if read now, perhaps because of hardware faults in the disk storage system.

- The hardware processor or its local memory might have a transient error while doing the buffering and copying, either at host **A** or host **B**.

- The software of the file system, the file transfer program, or the data communications system might make a mistake in buffering and copying the data of the file, either at host **A** or host **B**.

- The communications system might drop or change the bits in a packet or deliver a packet more than once.

- Either of the hosts may crash partway through the transaction after performing an unknown amount (perhaps all) of the transaction.

The first two types of errors would normally be handled by protocol layers above the transport layer, but a major emphasis of the transport layer in some protocol architectures is to handle the latter two types of errors. Eliminating the third type is the most difficult, since eliminating it involves writing correct programs and guaranteeing that programs are correct is not always feasible. Both the transport layer and the layers above it are involved in protection against such software errors.

Optimal division of error protection between different layers is a very difficult problem. Strong emphasis on error protection at lower layers may seem superfluous if higher layers are going to redo the job, but it is often much easier to do error correction before the effects of errors have had a chance to propagate in strange and wonderful ways, and this greatly simplifies the job of higher layers. Standard ARQ techniques for protection against transmission errors are relatively inexpensive and catch almost all such errors that may occur (and are also almost universally applied). Regardless of the degree of error protection at lower layers, though, the transport layer must be prepared to handle errors that the lower layers cannot catch. Thus a major emphasis of some standard transport layer protocols is protection against possible data loss in X.25 networks after Resets, Clears, or Restarts.

11.1.2   Transport Layer Connection Management

Connection management involves setting up, using, and taking down connections. Although some type of connection management is handled by almost every layer in a protocol architecture, it is especially complex at the transport layer due to the unpredictability of network delay. In some networks, such as those on which DoD's TCP is used, the maximum time a packet is in the network may occasionally be several hours [WATS81], although the probability it will be more than 30 to 60 seconds is low. Precautions against delayed packets coming in at inopportune times are necessary.

handshake-based
timer-based

Two basic approaches to transport layer connection management, which are often combined, are *handshake-based* and *timer-based*. Handshake-based mechanisms use explicit connection opening and closing packet exchanges. The opening packet exchange guards against openings caused by duplicate packets and may allow negotiation of resources to be used. The closing exchange can be designed to assure that all data have been received and both parties are prepared to close. Timer-based mechanisms are based on sender and receiver keeping track of the system state (for example, sequence numbers, connection or transaction identifiers or responses) long enough to ensure that all old duplicate packets (including retransmissions) have left the system. Combined mechanisms use handshakes coupled with a timer-based mechanism to ensure that some type of packet or connection identifier is unique during the maximum time packets may remain in the system; this unique identifier is used to guard against treating old packets like new ones.

A problem that can occur without careful connection management is illustrated in Fig. 11.2. We use OPEN and CLOSE for packets requesting that a connection be opened and closed, respectively. Send sequence numbers and acknowledgment numbers (probably receive sequence numbers) are shown. The figure shows a two-way handshake to open a connection. A connection between **A** and **B** is opened, used, closed, and later reopened. Unfortunately, a delayed packet transmitted during the first connection shows up during the second connection just in time to be accepted as a new packet, so the true new packet is rejected as a duplicate.

three-way handshake

The standard solution to this type of problem is known as a *three-way handshake,* originally proposed by Tomlinson [TOML75] and elaborated by Sunshine and Dalal [SUNS78b], among others. It is illustrated in Fig. 11.3.

As the figure shows, each end of the connection chooses a unique identifier to identify the connection. **A** includes its ID# in its original OPEN packet, and **B** includes **A**'s ID# in its OPEN packet, plus **B**'s choice of ID#.[3] A deadlock can occur if separate packets are used for **B**'s ACK and OPEN (see Problem 11.1). The connection is not fully open until **A** returns an acknowledgment with the ID# chosen by **B**. After this, each end knows the other's ID# and uses it to validate received packets and identify the source of its acknowledgments. This eliminates the difficulty in Fig. 11.2.

Protocols using three-way handshakes—TCP (Section 11.2), the OSI transport layer (Section 11.3), and the DNA end communications layer (Section 11.4)[4]—use different forms of ID#s. TCP uses an initial sequence number

---

[3] Each of the packets transmitted during the three-way handshake can contain some user data, but these data are not passed on to the user until the connection is fully established. For simplicity, we have not included any data in packets.

[4] SNA uses a three-party negotiation for connection establishment instead of a three-way handshake.

Figure 11.2   Example of Problems with Two-Way Handshake for Connection Establishment. Delayed packet treated like new one.

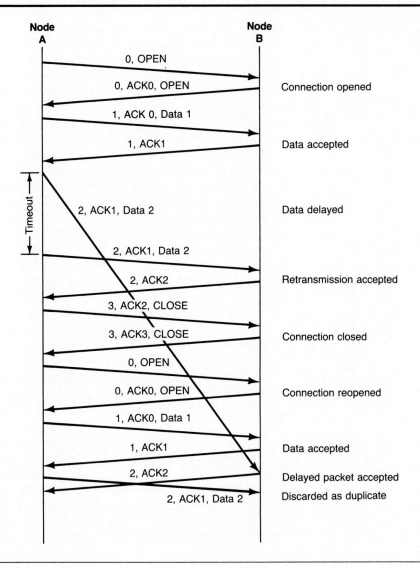

to identify the connection. OSI transport layer protocols use reference numbers chosen by the source and destination of packets; DNA uses a logical link number. In each case, reuse of an ID# is delayed long enough to ensure that

Figure 11.3 Three-Way Handshake for Opening Connection.

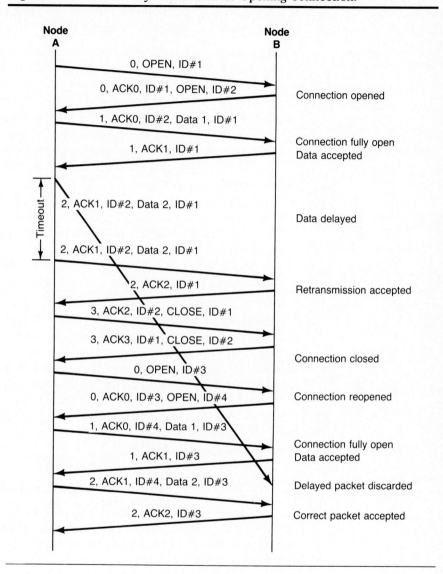

all packets associated with the old number are no longer in the system, combining the three-way handshake with a timer-based mechanism.

If an old OPEN packet from **A** shows up at **B** when no connection is active, it could cause **B** to return an OPEN packet referencing the ID# in this old packet. When **A** receives this, the ID# is invalid, so **A** closes the connection. A variety of other possibilities can also be handled if the following rules [SUNS87b] are followed (see Problem 11.3).[5]

- If a connection has not yet been established (or does not exist), return a CLOSE for any packet acknowledging something that the receiver has not sent earlier. The CLOSE should take its ID# from the acknowledgment field of the offending packet and its acknowledgment field should acknowledge all data and control in the offending packet.

- If the connection has been established, an unacceptable packet should elicit only an empty acknowledgment packet containing the current ID# for the acknowledging node and indicating the ID# expected from the other node.

- A CLOSE packet is validated by checking its ID# as in other packets. If this is valid and the CLOSE acknowledges something the receiver sent but has not yet received acknowledgment for, the CLOSE must be valid. If the connection has not been established, the transport layer protocol then returns to an opening or listening state. If the connection has been established, it is aborted, placed in a nonactive state, and the local user process is notified.

Connection management also involves closing connections after they have been used. Both abrupt and graceful closing procedures are used. Abrupt closing does not guarantee delivery of packets en route when closing is requested, but graceful closing does guarantee delivery. A graceful close requires that a node request close after it has sent all its data, but that it continue to accept data from the other end of a connection until it receives a close request from that end. Graceful closing is illustrated in our discussion of TCP in Section 11.2, and abrupt closing in our discussion of the OSI transport layer protocols in Section 11.3.

11.1.3   Transport Layer Timeouts and Congestion Control

The long and highly variable delays encountered in end-to-end transmission across a network cause special problems in congestion control and in determining appropriate timeouts before retransmissions. It is difficult to get congestion control algorithms to operate rapidly enough to keep congestion from building up to large peaks. The resulting long delays then interact with retransmission algorithms to maintain congestion unless these retransmission algorithms themselves adapt to congestion. This has led to such phenomena as a drop in DARPA Internet throughput between Lawrence Berkeley Laboratory and the University of California at Berkeley from 32 kbps to 40 bps (almost a factor of a thousand) in October of 1986 [JACO88]! The loca-

---

[5] Sunshine and Dalal formulate their rules in terms of TCP terminology. We have modified their statements to fit a wider variety of protocols without changing the basic rules.

tions are separated by three CSNPs and about 400 meters. Recent work has led to major progress on improving these algorithms, but this has not yet been reflected in the standards. We briefly summarize some of the recent results.

Timeouts before retransmission at the transport layer are normally based on estimated round trip transmission time, with the estimated times adjusted continually as packets are transmitted. A common estimate is a weighted average of the most recent round trip time and the previous estimated value. This may be computed from

$$W_n = \alpha W_{n-1} + (1 - \alpha)\tau_n, \tag{11.1}$$

with W_n the nth estimate of average round trip time, τ_n the most recent round trip time, and α a constant weighting factor, $0 \le \alpha < 1$, which determines relative weights given to old average and new round trip time. Choosing α close to 1 makes the weighted average relatively immune to changes that last a short time, while a value of α close to 0 makes the weighted average respond to changes in delay quickly. Normally α near 1 would be used to keep single delays from affecting the average dramatically; a value of 0.9 is normally recommended for TCP. The timeout before retransmission of a packet not previously retransmitted is often chosen to be βW_n, with $\beta > 1$ (a value of 2 is often recommended). Increasing timeouts before retransmission of previously retransmitted packets according to a backoff algorithm (see Chapter 5) gives improved performance in the presence of congestion; exponential backoff with timeout doubled for each new retransmission is usually recommended.

A number of problems with this approach have become apparent. One is that it is difficult to properly account for the effect of retransmitted packets in measuring τ_n values. Two approaches have been used, measuring the time from the first attempt to send a packet until it is acknowledged, including times for all retransmissions, or measuring only the time from the last transmission until acknowledgment. However, both have proved to be less than fully satisfactory. The former can introduce unrealistically large terms into the sum in Eq. (11.1) (possibly infinite terms for lost packets), and the second may be distorted by the fact that the final acknowledgment is sometimes a delayed acknowledgment sent after a previous transmission. An algorithm developed by Karn and described in [KARN88] seems to have solved the estimation problem. It ignores all times for transmitting retransmitted packets in making the W_n calculation, but also keeps a backed off retransmission timeout value until the next packet that does not require retransmission is received. Recomputation of W_n and of the timeout then resumes.

A second problem with the standard algorithm is that the value of β does not depend on congestion, though the variability of round trip transmission times is greatly influenced by congestion. Some recent work by Jacobson [JACO88], which includes an estimate of the variance of round trip transmission time as well as its mean, and bases timeouts on both, has resulted in major improvements in networks in which it has been used. Several other related algorithms developed by Jacobson [JACO88] have also resulted in significant improvements—for example, slowly opening a window for a new connection to avoid immediate saturation when a high-speed system such as a LAN rapidly dumps a full window of packets into a network that cannot handle this much traffic, and some techniques for dynamically changing window sizes when congestion appears. Several of these techniques should be incorporated in standards in the near future.

11.2   DoD Transport Layer Protocols

transmission control protocol (TCP)

user datagram protocol (UDP)

Two transport layer protocols—the *transmission control protocol* (*TCP*) and the *user datagram protocol* (*UDP*) are standards in the DoD Reference Model suite of protocols. Since TCP is by far more common and important, we concentrate on it and give only a brief description of UDP.

11.2.1   Transmission Control Protocol

The transmission control protocol developed by the U.S. Department of Defense, and included in the DoD Reference Model suite of protocols, is the most widely implemented transport layer protocol, with implementations in both DoD and non-DoD networks. Until standardization of the OSI transport layer, it was the only transport layer protocol available for networks interconnecting heterogeneous equipment, such as various LANs. Although it is generally assumed that the OSI transport protocols will eventually replace TCP, this will not happen soon. Recent explosive growth in the DARPA Internet and other networks using TCP implies that TCP will be widely used for quite some time.

TCP is usually implemented in conjunction with DoD's IP, and the protocols are so closely connected that it is sometimes claimed that TCP is designed specifically and exclusively to work with IP. A more accurate assessment is given by Comer in the best current reference on TCP and IP [COME88, p. 136]: "Because the TCP protocol assumes little about the underlying communications system, TCP can be used with a variety of packet delivery systems including the Internet IP datagram delivery service. For example, TCP can be implemented to use dialup telephone lines, a local area network, a high speed fiber optic network, or a lower speed long haul network. In fact, the large variety of delivery systems TCP can use is one of its strengths."[6]

---

[6] However, some details of TCP, such as passing parameters to IP for inclusion in an IP header and incorporating a "pseudo header" containing complete IP addresses into computation of a checksum, would have to be altered or accommodated by an alternative network layer.

TCP provides *reliable virtual circuit service* that "guarantees" to deliver a stream of data from one machine to another without duplication or data loss despite unreliable packet delivery service at the network layer.[7] It treats data as a *stream of eight-bit octets* or bytes, and delivers to the receiver the sequence of bytes passed to it by the source. A data unit handled by the protocol is a segment in TCP documentation, but we call it a TCP data unit for uniformity with OSI transport layer terminology.

11.2.2 TCP Primitives

Tables 11.1 and 11.2 list TCP primitives (user commands in TCP documentation), grouped into Request, Indication, and Confirm primitives according to definitions in Section 3.6.3, to aid comparison with OSI primitives.[8]

The two passive open primitives indicate that the user is willing to accept a connection request from another site; the active open actually requests a TCP connection. Data may be included with OPEN TCP data units, but it will not be delivered to the user application before connection establishment is completed.

11.2.3 TCP Data Unit Format

The format of a TCP Data Unit is given in Fig. 11.4, with definitions of parameters in Table 11.3.

The *checksum* is interesting since it is computed over information not in the data unit as well as information in the data unit. To compute it, a pseudo header of the form in Fig. 11.5 is placed at the beginning of the data unit,[9] enough bytes containing zeros to make total length an integral multiple of 16 bits are appended to the end, and the checksum is computed over the result. The receiver must obtain information to reconstruct the pseudo header (available in IP headers) and include it before verifying the checksum.

The purpose of the pseudo header is to confirm that the data unit has reached its destination. Port addresses in the TCP header identify the port only within source and destination nodes. Hence the additional information in the pseudo header verifies receipt at the correct host within the correct network.

---

[7] It is more accurate to state that TCP delivers the stream of data correctly with high probability, but the TCP literature states that correct delivery is guaranteed. Problem 11.7, on error patterns not detected by the TCP checksum, illustrates one reason why correct delivery is not guaranteed.

[8] None fit the definition of Response primitives, although Request primitives with appropriate choices of parameters at times serve the purpose of Response primitives.

[9] The two addresses are the standard 32-bit IP addresses found in the IP header. The field identified by Zero is all zeros, and is included to make the pseudo header length an integral multiple of 16 bits. The Protocol field identifies the protocol being used as TCP. The TCP Length field specifies the total length of the TCP data unit.

Table 11.1   TCP Service Request Primitives.[a] Parameters are in
parentheses, (   ), and optional parameters in brackets, [   ].

UNSPECIFIED-PASSIVE-OPEN. request (source-port, [timeout], [timeout-action],
[precedence], [security-range])

Listen for connection attempts at specified security and precedence levels from
any remote user.

FULL-PASSIVE-OPEN. request (source-port, destination-port, destination-address,
[timeout], [timeout-action], [precedence], [security-range])

Listen for connection attempts at specified security and precedence levels from
specified user.

ACTIVE-OPEN. request (source-port, destination-port, destination-address, [timeout],
[timeout-action], [precedence], [security])

Request connection to specified user at specified security and precedence levels.

ACTIVE-OPEN-W/DATA. request (source-port, destination-port, destination-address,
[timeout], [timeout-action], [precedence], [security], data, data-length, push-flag,
urgent-flag)

Request connection to specified user at specified security and precedence levels
and transmit data with the request.

SEND. request (local-connection-name, data, data length, push-flag, urgent-flag,
[timeout], [timeout-action])

Send data on specified connection.

ALLOCATE. request (local-connection-name, data length)

Issue incremental buffer allocation for data to be received.

CLOSE. request (local-connection-name)

Close connection gracefully.

ABORT. request (local-connection-name)

Close connection abruptly.

STATUS. request (local-connection-name)

Report connection status.

---

[a]That is, user to TPC primitives.

11.2.4   TCP Initial Connection Establishment

TCP Initial connection establishment is by a three-way handshake with each
end specifying an initial sequence number, ISN, for bytes transmitted from
that end (the ISN serves the purpose of a connection ID#). A simple handshake is illustrated in Fig. 11.6, showing primitives exchanged and data units
transmitted across the network.[10]

---

[10] For simplicity, the slopes of lines to indicate transmission and propagation delays have been
made the same in all three sections of the figure, although delays for transmitting data units
across the network are normally far larger than those for exchanging primitives within a
node.

Table 11.2 TCP Service Indication and Confirm Primitives.[a] Notation same as in Table 11.1.

***

OPEN-ID.confirm (local-connection-name, source-port, destination-port, destination-address)

Informs user of connection name assigned to pending connection requested in an earlier OPEN primitive.

OPEN-FAILURE.confirm (local-connection-name)

Report failure of an active OPEN request.

OPEN-SUCCESS.confirm (local-connection-name)

Reports completion of an active OPEN request.

DELIVER.indication (local-connection-name, data, data-length, urgent-flag)

Reports arrival of data.

CLOSING.indication (local-connection-name)

Reports that remote TCP user has issued a CLOSE.

TERMINATE.confirm (local-connection-name, description)

Reports that connection has been terminated and no longer exists.

STATUS-RESPONSE.confirm (local-connection-name, source-port, source-address, destination-port, destination-address, connection-state, receive window, send-window, amount-waiting-ack, amount-waiting-receipt, urgent-state, precedence, security, timeout)

Reports current status of connection.

ERROR.indication (local-connection-name, description)

Reports service-request or internal error.

***

[a]That is, TCP to user primitives.

Initially a service user at system **A** issues an UNSPECIFIED-PASSIVE-OPEN.request, indicating it is willing to accept a connection from any other port. The service provider (TCP entity at **A**) returns an OPEN-ID.confirm, with local connection name for use when a connection is set up. When another service user, at system **B**, later desires a connection with **A** (now listening for requests), it issues an ACTIVE-OPEN.request containing data needed to set up the connection. The returned OPEN-ID.confirm conveys the local connection name at **B**.

A TCP data unit with SYN bit set (equivalent to a Request for Connection) is sent to **A** with **B**'s choice, x, of ISN. This begins the three-way handshake. **A** returns a Data Unit acknowledging the SYN by setting the ACK bit and returning the correct sequence number, $x+1$, in its acknowledgment number field. **A** also sets the SYN bit, making this its Request for Connection, and sends its choice, y, of ISN. **B** completes the handshake by acknowledging this data unit. The connection has then been established and both **A** and **B** may

Figure 11.4   TCP Data Unit Format.

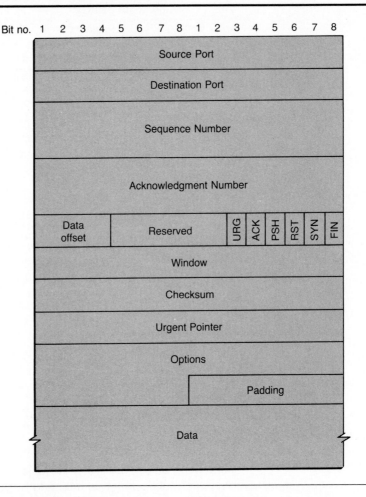

send data units to each other.[11] The first data bytes transmitted are numbered $y+1$ and $x+1$, respectively.

The connection establishment protocol is designed to deal with problems such as reception of delayed duplicate Data Units. Simultaneous requests for connection establishment can also be handled, though a total of four transmissions are then needed (two simultaneous SYNs plus an ACK to each). See Problem 11.3 for a treatment of this in a more general context.

---

[11] Data can be included in each of the data units used to set up the connection, but it cannot be delivered to the user processes until completion of connection establishment.

Table 11.3   Interpretation of TCP Header Fields.

- *Source Port* (16 bits): Identifies source service access point (SAP). Typically the address of an application within the source node or host, and used in conjunction with node or host address to locate the source.
- *Destination Port* (16 bits): Identifies destination SAP, interpreted in the same manner.
- *Sequence Number* (32 bits): Sequence number of first data byte in data unit, except when SYN bit is set (implying the data unit is a request to set up a connection). If SYN bit is set, field indicates the initial sequence number (ISN) to be used over the connection. The sequence number of the first data byte is then ISN + 1.
- *Acknowledgment Number* (32 bits): The sequence number of the next byte that the TCP entity expects, allowing piggybacked acknowledgments.
- *Data Offset* (4 bits): Indicates where data begins in this data unit by specifying number of 32-bit words in the header.
- *Control Bits* (6 bits):
  - *URG (Urgent):* If bit is set, data field contains data considered to be urgent and meriting immediate consideration by the destination TCP user. (Urgent data are then found from the beginning of the field up to a position indicated by the urgent pointer.)
  - *ACK (Acknowledgment):* Indicates acknowledgment field is significant.
  - *PSH (Push):* Set after a push operation, causing immediate delivery of a data unit without waiting for a full-sized one to be assembled, and implying that when reception through this data unit is complete it should be reassembled and passed to higher layers.[a]
  - *RST (Reset):* Indication from transmitting transport entity that receiving entity should break and reset the transport connection.
  - *SYN (Synchronize):* Set in one or more transmissions in each direction to request a connection between two ports and help synchronize sequence numbers.
  - *FIN (Final):* A flag to signal there will be no more transmissions from this port along this channel as sender has reached the end of its byte stream.
- *Window* (16 bits): Specifies the number of bytes, beginning with the one specified in the ACK, that the receiving port is willing to accept without further authorization. Used for credit flow control.
- *Checksum* (16 bits): Checksum field.[b] Used for error detection.
- *Urgent Pointer* (16 bits): Points to the byte following urgent data. Indicates how many bytes, starting with the one to which current sequence number points, are urgent.
- *Options* (variable length):
  - *Buffer size:* Only true option specified at present. Communicates receive buffer size, that is maximum acceptable data unit size, at TCP implementation that sends data unit. Only used in initial call setup.
  - *No operation:* Option code which may be used between options.
  - *End of option list:* Indicates end of the option list.
- *Padding* (variable length): Used to ensure that the TCP header ends on a 32-bit boundary.

---

[a]The initial version of the protocol called this an EOL (end of letter) bit that signified the end of a letter or complete record. Its function was essentially the same as that of the PSH bit.

[b]The checksum is computed by taking the 16-bit ones complement sum (sum modulo $2^{16} - 1$) of all 16-bit words in the header and text (plus a pseudo header, see Fig. 11.5) then taking the ones complement (inverting all bits) of the result.

Figure 11.5   Format of Pseudo Header Used in
TCP Checksum Computations.

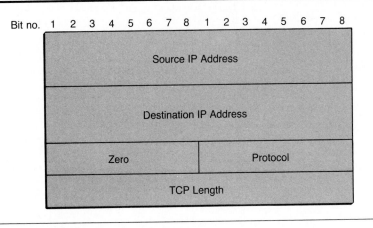

Figure 11.6   Sequence of Messages in Simple Successful
TCP Three-Way Handshake.

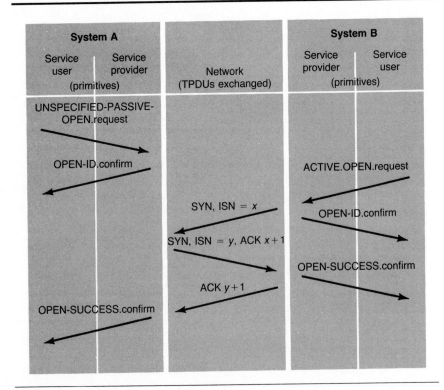

As long as ISNs for new connections are unique within the maximum time a data unit may remain in the network, this handles problems such as a delayed SYN or ACK arriving in the middle of connection. A recommended procedure to ensure unique ISNs is to use a 32-bit counter, updated approximately every 4 microseconds (so numbers cycle approximately every 4.5 hours)[12] to generate ISNs. If the processor at a site crashes in a manner causing it to forget its sequence number, it is required to remain quiet, for a period at least equal to the maximum time a data unit can remain in the network, before attempting to reestablish connections to guarantee it will not reuse the sequence number of a data unit still in the network. If strictly enforced, this could require it to remain silent for several hours, since data units occasionally persist in the network that long, but much shorter silent periods (a few minutes) are normally satisfactory and are commonly used.

11.2.5 TCP Data Transmission

Figure 11.7 illustrates half-duplex data exchange in TCP. It shows successful transmission of 30 data bytes from port **A** to port **B** followed by successful transmission of 100 data bytes from port **B** to port **A**. ISNs of 55 and 202 have been assumed for the two transmissions, with each followed by an ACK before the next data transmission.

TCP flow control implements a credit mechanism using the window field in the header, which specifies the number of additional bytes the receiver is prepared to receive. This type of credit flow control was discussed in Chapter 8; it operates on a byte-by-byte basis rather than on the packet or frame basis used in most flow control. Independent credit adjustments for the two directions of transmission are possible using piggybacked credit fields present in all TCP data units. (Acknowledgments are also piggybacked.) A sequence of transmissions using flow control is illustrated in Fig. 11.8.

For the sequence illustrated, each end initially has permission to send 256 data bytes, with initial sequence numbers 1000 and 500. Port **A** sends its initial window in two data units, then receives permission to send 384 bytes, beginning with byte 1001; this allows it to send 128 more bytes beyond the number already sent, so it continues. Another credit, for 512 bytes, beginning with number 1129, comes in while it sends the next data unit, so it sends bytes through byte 1640. It must then pause until it receives more credit. Port **B** is impacted more severely by flow control, since **A** cuts credit to 128 bytes, beginning with byte 629, in its third data unit. After exhausting its initial credit, **B** receives this and finds its window is still empty. It has to wait for the next data unit before it is allowed to send more data.

A TCP implementation is allowed to collect data from a sending user and to send that data in data units at its own convenience (as allowed by protocol mechanisms), until the end of a letter (TCP's term for a complete record) is

---

[12] $2^{32} \times 4$ microseconds ≈ 4.55 hours.

Figure 11.7   Use of TCP Primitives and Data Units for Half Duplex
Data Exchange (adapted from [SCHW87]).

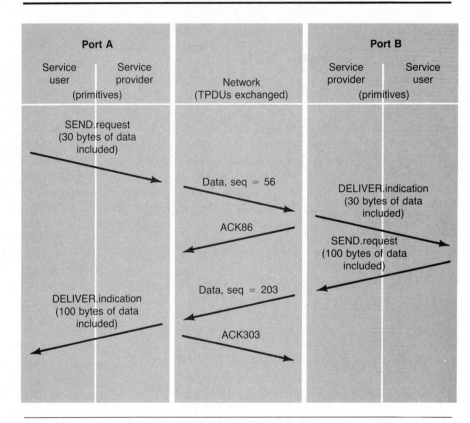

reached. A data unit may contain all or part of a letter, but never contains
parts of more than one letter. To make transfers efficient and to minimize
network traffic, implementations normally collect enough data from a stream
to fill a reasonably large data unit before transmitting it. A push operation
allows an application to force delivery of bytes currently in the stream without waiting for a data unit to fill. The PSH bit in the header is then set so data
will be delivered to the application program at the receiving end.

TCP uses positive acknowledgment ARQ to ensure reliable transmission,
with data units retransmitted if an acknowledgment is not returned within a
timeout. Computing an appropriate timeout interval is difficult, however, because of the wide variability of round trip times before acknowledgments are
received (see Subsection 11.1.3).

The receiver is allowed to keep data units received out of sequence, as
long as they are within the current window, but it only acknowledges bytes

Figure 11.8   Typical TCP Flow Control Sequence. Notation indicates
{Source Port, Destination Port, Sequence Number,
Acknowledgment Number, Credit}. All Data Units contain
128 data bytes.

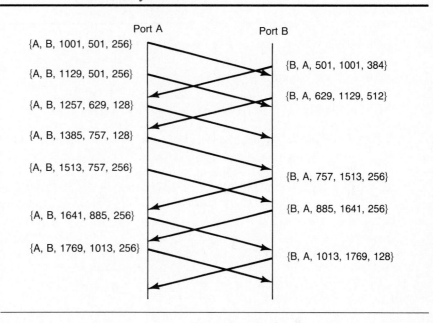

received in sequence with no gaps. If a delayed data unit filling in a gap in
sequence numbers is received, the acknowledgment number is advanced to
include both it and the earlier data units. A site that sends several data units
and does not receive an acknowledgment within its timeout period has the
choice of resending all unacknowledged data units or only sending the first
unacknowledged one and hoping only this one failed to get through successfully. The latter approach is recommended, but not required.

11.2.6   Closing a TCP Connection

The normal means of closing a connection is a *graceful close*. The two halves
of a connection (for transmission in each direction) are closed independently.
When an application program has no more data to send, TCP sets the FIN bit
in the last data unit containing data transmitted from that site. The receiving
TCP acknowledges this data unit and informs the application program that
no more data are expected. TCP then refuses to accept more data for transmission over the connection in that direction, but data can continue to flow
in the opposite direction until the TCP at the other end closes it. When both
directions have been closed, the connection is terminated.

Figure 11.9   Example of TCP Graceful Close (adapted from [SCHW87], M. Schwartz, *Telecommunication Networks: Protocols, Modeling and Analysis,* © 1987 by Addison-Wesley. Reprinted by permission of Addison-Wesley Publishing Co., Inc. Reading, MA.).

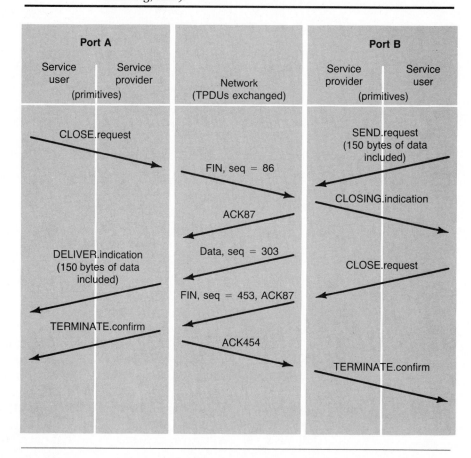

An example of graceful close is given in Fig. 11.9-a continuation of Fig. 11.7. Port **A** decides to close the connection immediately after the sequence in Fig. 11.7, but **B** still has 150 bytes of data to send. **A**'s user process issues a CLOSE.request, while **B**'s user process issues a SEND.request with 150 bytes of data. **A**'s CLOSE.request results in transmission of a FIN data unit (equivalent to Close Request) to **B**. No data are allowed in this data unit, so the sequence count is 86, the value applicable before its transmission. **B** is notified of the FIN via a CLOSING.indication primitive. **A** is then prohibited from sending any more data, but **B** is allowed to finish transmission. After completion **B** issues its own CLOSE.request. When **A** gets **B**'s FIN, it acknowledges it and issues a TERMINATE.confirm. When **B** gets the ACK, it in turn issues a TERMINATE.confirm and the connection is fully terminated.

Under abnormal conditions, such as unrecoverable errors, node crashes, and so forth, a connection may be aborted by transmission of a data unit with RST bit set. This is an abrupt termination with all attempts to send or receive data abandoned. Data in transmission and reception buffers are discarded and the buffers released. Any recovery is the responsibility of higher layers.

11.2.7 User Datagram Protocol

The *user datagram protocol* (*UDP*) provides a transport-level datagram service. It is designed to work with the DoD IP protocol and adds a mechanism that distinguishes among multiple destinations within a given host so multiple application programs executing on the host can send and receive datagrams independently. The UDP header is illustrated in Fig. 11.10.

The Source and Destination Port fields contain 16-bit port numbers used to demultiplex datagrams among processes to receive them at a host. The Source Port field is optional; if used, it indicates the port to which replies should be sent.[13] The Length field gives the number of bytes in the datagram, including UDP header. The Checksum is evaluated in the same manner as the TCP Checksum, including addition of the pseudo header (with UDP instead of TCP in the protocol byte), but checks only the header and pseudo header. It verifies that a datagram has arrived at the correct user port, but there is no error reporting. There is no mechanism for connection establishment.

UDP simply assembles a datagram and hands it to IP for transmission. The checksum is used to check incoming datagrams, with valid ones passed up to the specified port or application program and invalid ones discarded.

Figure 11.10   UDP Datagram (UDP Data Unit) Header.

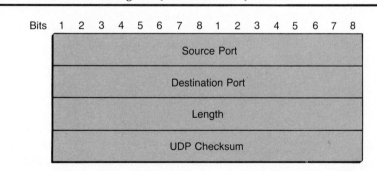

---

[13] If not used, it is set to zero.

Pertinent primitives are simplified versions of the SEND and DELIVER primitives in Tables 11.1 and 11.2 with only the data and data-length parameters.

11.2.8   Limitations of DoD Transport Layer Protocols

The DoD transport layer protocols, especially TCP, have proved to be successful and have found wide acceptance. Nevertheless, the protocols represent an early stage in the development of transport layer protocols, and lessons have been learned from experience with them. We concentrate in this section on the limitations of TCP since it is much more common.

1. An obvious weakness of TCP is its long headers. All data units have the same format, except for optional fields, with minimum header length 20 bytes. When TCP is used in conjunction with IP, at least 20 more bytes of header are added, giving total TCP/IP header overhead of at least 40 bytes. For short messages, as are often used for interactive applications, this can lead to very poor data transmission efficiency.

2. The checksum for TCP is weak, failing to detect a substantial number of error patterns. How serious this is depends on the network layer and data link layer protocols underneath TCP. In many cases HDLC or an equivalent protocol, with strong error detecting capability, will be used at the data link layer. When this is true, the TCP checksum is largely superfluous. When transport layer checksums are really needed, though, it would be desirable to have a more powerful one.

3. The urgent data feature in TCP is marginally adequate for delivering out of band or urgent data, since it does not cause normal flow control to be bypassed or otherwise greatly speed up delivery. Its primary function is to inform the user that some data are urgent and should be processed quickly.

4. The stream orientation of TCP, treating data as a stream of bytes rather than as a sequence of message units, is unusual though it works reasonably well. It required the addition of the Push function to ensure timely delivery of short interactive messages; otherwise enough data to fill a normal size data unit would be accumulated before transmission.

5. All in all, TCP has proved to be remarkably successful, especially when considering that it was the first general purpose transport layer protocol developed. With the addition of UDP for applications better adapted to datagram service, the DoD architecture satisfies most requirements for transport layer protocols reasonably well. At least two other types of transport layer protocols would be desirable, however; a speech protocol guaranteeing sequenced, timely delivery but without high reliability requirements and a real-time protocol guaranteeing high reliability and timeliness [MCFA79].

11.3   OSI Transport Layer Protocols

transport service data units (TSDUs)

transport protocol data units (TPDUs)

In OSI terminology, the session layer accesses the transport layer at transport service access points (TSAPs) and the transport layer accesses the network layer at network service access points (NSAPs). Data units presented to the transport layer by the session layer are *transport service data units (TSDUs)*; those presented to the network layer by the transport layer are network service data units (NSDUs). Data units used by the transport layer for communication of data or control information are *transport protocol data units (TPDUs)*. The transport layer may segment a TSDU into multiple TPDUs at the transmitter and reassemble corresponding TSDUs at the receiver. It may also generate additional TPDUs for control purposes. Also, before presenting NSDUs to the network layer for transmission, the transport layer may concatenate several TPDUs into one NSDU, with separation of the NSDU into corresponding TPDUs at the receiver.[14] This gives flexibility in size of data units.

In the remainder of our treatment, we minimize our use of this terminology, except for using the term TPDU for OSI transport protocol data unit.

11.3.1   OSI Network Service Types

The ISO has defined a variety of transport layer protocols to fit different user requirements and for networks of varying quality [INTE86a–c]. Three types of network service have been defined, with different classes of transport layer protocols for different service types:

network service types

- *Type A:* Network connections with acceptable residual error rate and acceptable rate of signaled failures.
- *Type B:* Network connections with acceptable residual error rate but unacceptable rate of signaled failures.
- *Type C:* Network connections with residual error rate not acceptable to the transport service user.

Type A networks are the simplest to design transport layer protocols for, since the underlying network layer service is assumed to be fully adequate. This can be a function of user requirements as much as of true network quality. Transport layer protocols for such networks tend to have minimal server quality requirements. It is assumed that the basic connection will not introduce errors, loss, or misordering of packets, so there is no need to provide failure recovery services, services to handle loss of data or resequencing, and so forth.

---

[14]Similarly the session layer may concatenate several session protocol data units (SPDUs) into one TSDU at the transmitter and separate them at the receiver.

The class of Type B networks was defined to include X.25 or similar networks used for applications with high reliability requirements. In such situations, X.25 resets, clears, and restarts, with consequent data loss, can occur at an unacceptable rate. Since users are notified of resets, clears, and restarts, they are classed as signaled failures.[15] A transport layer protocol for this environment must provide for recovery from signaled failures.

Type C networks are most difficult to design for, since network service is unreliable. The transport layer must detect and recover from network failures, detect and correct out-of-sequence, duplicate, or misdirected messages, and shield users from problems that may occur due to these anomalies.

11.3.2   Classes of OSI Transport Layer Protocols

A total of five classes of connection-oriented transport layer protocols have been defined [INTE86a,b].

- *Class 0:* Simple class
- *Class 1:* Basic error recovery class
- *Class 2:* Multiplexing class
- *Class 3:* Error recovery and multiplexing class
- *Class 4:* Error detection and recovery class

Protocols corresponding to these classes are called *TP0, . . . , TP4,* respectively. In addition, a connectionless protocol has been defined.

Classes 0 and 2 assume a Type A network with error control, sequencing, and flow control. *Class 0* provides minimal connection establishment, data transfer, and connection termination facilities; it was developed for teletex applications where user requirements are not stringent. *Class 2* also provides for multiplexing multiple transport layer connections over a network layer virtual circuit and for transport layer flow control using a credit mechanism.

Classes 1 and 3 are designed for Type B networks. *Class 1* is similar to Class 0 but includes capability to recover from network layer resets or disconnects by adding sequence numbers to TPDUs (modulo 2^7 or 2^{31}) so lost TPDUs can be detected and retransmitted; expedited data capability is included. *Class 3* is essentially a union of Classes 1 and 2, adding the multiplexing and flow control functions of Class 2 to the error-handling functions of Class 1.

Class 4 is designed for Type C networks. It assumes a worst-case network and provides comprehensive error detection and recovery features, including checksums, timeouts, and retransmissions, recovery from loss of data or control TPDUs, expedited data, and user notification of unrecoverable errors.

---

[15] For a particular virtual circuit, these either cause a network layer reset or a network layer disconnection. N-RESET.indication and N-DISCONNECT.indication primitives indicate such problems to the transport layer.

Multiplexing and flow control for each connection are provided. In terms of function, TP4 is largely equivalent to TCP, but it is not compatible. It is applicable to networks using datagrams at the network layer, some types of local area networks, urban networks with mobile nodes, packet radio networks with radio transmission over fading channels and similar networks.

The connectionless OSI transport protocol, [INTE86c], is a relatively recent addition. There are enough important applications involving transmission of limited amounts of data, where establishment and termination of a connection would cause excessive overhead, to warrant its development.

The U.S. National Bureau of Standards (NBS)[16] has published a series of six volumes [NATI83a–f] describing the U.S. Federal Information Processing Standard (FIPS) for the transport protocol. The FIPS documents give detailed and formal specifications, but only describe Class 2 and Class 4 protocols and an optional connectionless protocol. They also contain a few optional features that are not part of the OSI transport protocol. Aside from options, the FIPS protocols are identical to corresponding OSI protocols.

11.3.3   OSI Transport Layer Primitives and TPDUs

The transport layer uses all four primitive types. Primitives are listed in Table 11.4. The NBS standard adds T-CLOSE.request and T-CLOSE.indication primitives, used for graceful close, and a T-UNIT-DATA.confirm primitive to confirm delivery of UNIT-DATA TPDUs in the connectionless service.

The first four primitives are used in transport connection establishment via handshakes discussed in the next subsection. The next two are used during connection termination, and the rest during data transfer. Only the two primitives indicated by asterisks are used for connectionless service.

A limited number of primitives may be legitimately sent or received in any given state at the transport layer entity. Figure 11.11 indicates which primitives are allowed in various states and the state transitions they cause. Four states are shown:

state definitions

1. *Idle:* No transport connection established and no attempt to establish a connection.

2. *Outgoing connection pending:* T-CONNECT.request has been sent, but no response has been received.

3. *Incoming connection pending:* T-CONNECT.indication has been received, but no response has been sent.

4. *Data transfer ready:* Transport connection has been established and transport layer entity is in data transfer state.

---

[16] The bureau has recently been renamed the National Institute of Standards and Technology (NIST).

Table 11.4	Transport Service Primitives.

Primitive	Parameters
T-CONNECT.request	Called Address, Calling Address, Expedited Data Option, Quality of Service, Data
T-CONNECT.indication	Called Address, Calling Address, Expedited Data Option, Quality of Service, Data
T-CONNECT.response	Quality of Service, Responding Address, Expedited Data Option, Data
T-CONNECT.confirm	Quality of Service, Responding Address, Expedited Data Option, Data
T-DISCONNECT.request	Data
T-DISCONNECT.indication	Disconnect Reason, Data
T-DATA.request	Data
T-DATA.indication	Data
*T-UNIT-DATA.request[a]	Called Address, Calling Address, Quality of Service, Security Parameters, Data
*T-UNIT-DATA.indication[a]	Called Address, Calling Address, Quality of Service, Security Parameters, Data
T-EXPEDITED-DATA.request	Data
T-EXPEDITED-DATA.indication	Data

[a]The two primitives indicated by an asterisk are used for connectionless service and are the only two primitives for this service.

There are two ways to leave the Idle state: sending a T-CONNECT.request (causing transition to Outgoing connection pending) or receiving a T-CON-NECT.indication (causing transition to Incoming connection pending). If the transition is to Outgoing connection pending, the next transition can be back to Idle, if T-DISCONNECT.indication is received (indicating connection was unsuccessful) or the initiating party changes its mind and sends a T-DISCON-NECT.request. Similarly, the next transition from Incoming connection pending can be back to Idle if the request is declined or the calling party changes its mind. Otherwise, legitimate transitions are to Data transfer ready by receiving a T-CONNECT.confirm primitive while in Outgoing connection pending or sending a T-CONNECT.request primitive while in Incoming connection pending. Transitions into Idle from Data transfer ready occur when a T-DIS-CONNECT.request primitive is sent or a T-DISCONNECT.indication primitive is received.

quality of service parameters

Quality of service parameters, plus possible use of an expedited service option, are negotiated during transport connection establishment. Quality of service parameters include the following:

- Expected connection establishment delay
- Connection establishment failure probability
- Throughput

Figure 11.11   State Transition Diagram Showing Allowed Sequences of OSI Transport Layer Primitives. (Reprinted with permission from ITU [International Telecommunications Union], CCITT [International Telegraph and Telephone Consultative Committee], Transport Service Definition for Open Systems Interconnection (OSI) for CCITT Applications," *CCITT Red Book,* vol VIII.5 1984. Updated material can be found in the *CCITT Blue Book* and may be obtained from the ITU General Secretariat, Place des Nations, CH-1211 Geneva 20, Switzerland.)

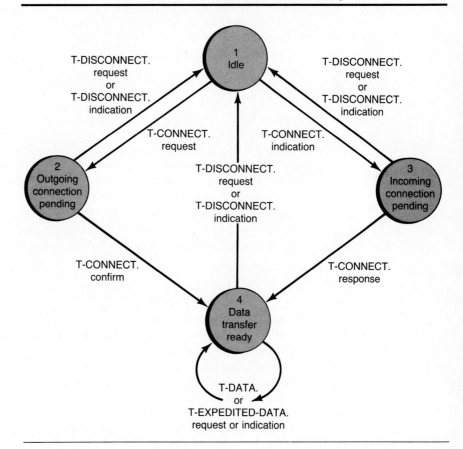

- Transit delay
- Residual error rate
- Probability of failure to transfer TPDUs
- Connection release delay
- Connection release failure probability

- Connection resilience (probability of provider initiated disconnection with no request for disconnection)
- Transport connection protection (extent of attempts to prevent unauthorized monitoring or manipulation of user information)

Quality of service is not guaranteed, since the parameters describing it vary randomly and cannot be predicted precisely. The negotiation allowed is spelled out in detail in standards; in almost all instances quality of service parameters requested by the calling user can only be made poorer by the service provider or by the called user. The expedited data option may be requested by the calling service user and either accepted or declined by the called user; the called user may not request it if it has not first been requested by the calling user.

A total of ten types of TPDUs are defined in the OSI connection-oriented standard, and one type is defined in the connectionless standard. They are listed in Table 11.5. Uses of these TPDUs should be reasonably obvious by now.

Each TPDU consists of three parts: fixed header, variable header, and data field. The latter two parts are not always present; in particular no data field is allowed in DC, ER, EA, AK, and ED TPDUs. DT or UD TPDUs, used for normal data, carry data fields up to a negotiated or network defined length. ED TPDUs, intended for short urgent messages, are limited to 16 bytes of data. Limited amounts of data are allowed in other TPDU types.[17]

Formats of the fixed portions of TPDU headers are given in Fig. 11.12, with definitions of fields in Table 11.6. The variable part includes such parameters as calling and called transport service access point IDs or addresses,[18]

Table 11.5   OSI TPDUs.

- *Connection request, CR* (classes 0–4)
- *Connection confirm, CC* (classes 0–4)
- *Disconnect request, DR* (classes 0–4)
- *Disconnect confirm, DC* (classes 1–4)
- *Data, DT* (classes 0–4 but two formats)
- *Expedited data, ED* (classes 1–4)
- *Acknowledgment, AK* (classes 1–4)
- *Expedited acknowledgment, EA* (classes 1–4)
- *Reject, RJ* (classes 1 and 3)
- *TPDU error, ER* (classes 0–4)
- *Unit data, UD* (connectionless mode only)

---

[17] CR and CC TPDUs may each carry up to 32 octets of data and DR TPDUs may carry up to 64 octets.

[18] These are always included in connectionless mode TPDUs.

Figure 11.12   Fixed Parts of TPDU Headers. (From W. Stallings, *Data and Computer Communications, 2/e,* © 1988. Reprinted by permission of MacMillan Publishing Company, New York, NY).

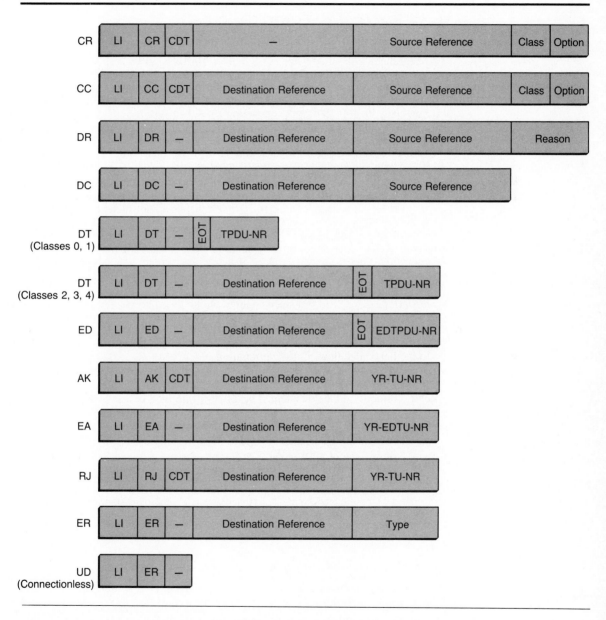

Table 11.6   Definitions of Fields in OSI TPDU Headers.

- *Length indicator (LI)* (8 bits): Length of header, fixed plus variable parts but excluding LI field, in bytes.
- *TPDU code* (4 bits): Type of TPDU, abbreviations for types shown.
- *Credit (CDT)* (4 bits): Credit allocation for flow control. Number of TPDUs, beginning with one pointed to by YR-TPDU-NR in last AK or RJ received, allowed to be transmitted; initial value in CR and CC TPDUs. Ignored if blank. 16-bit CDT is allowed in AK TPDU if 31-bit sequence numbers are used (see TPDU-NR and YR-TU-NR).[a]
- *Source Reference* (16 bits): 16-bit identification number used by transport entity to uniquely identify transport connection in its own system.
- *Destination Reference* (16 bits): Corresponding identification number used by peer entity to identify transport connection uniquely. All zeros in CR TPDU, since reference number assigned by destination. CC TPDU puts source reference from CR in field and fills in its source reference field.
- *Class* (4 bits): Protocol class (0, 1, 2, 3 or 4).
- *Option* (4 bits): Specifies normal (7-bit sequence number, 4-bit credit) or extended (31-bit sequence number, 16-bit credit) flow control fields. Also specifies such options as use of expedited data, alternative protocol classes, explicit flow control, or checksums for appropriate protocol classes.
- *Reason* (8 bits): Reason for requesting disconnect or rejecting transport connection request.
- *EOT* (1 bit): Used when session layer data unit has been segmented into multiple TPDUs. It is set to 1 on last segment.
- *TPDU-NR* (7 bits): Send sequence count, modulo $2^7 = 128$, of a DT TPDU. May be extended to 31 bits, for modulo 2^{31} count, if extended flow control option used, with field extended by 3 bytes.
- *EDTPDU-NR* (7 bits): Send sequence count of ED TPDU. May also be extended to 31 bits if extended flow control option used.
- *YR-TU-NR* (8 bits, 7 bits significant): Next expected DT sequence number ("your TPDU number"). Bit 8 is 0, with 7 bits significant. If extended flow control option used, field extended by 3 bytes, with bit 8 of first byte zero.[b]
- *YR-EDTU-NR* (8 bits, 7 bits significant): The next expected ED sequence number ("your ED TPDU number"). Same conventions on bits and possible extended field as for YR-TU-NR.
- *Type* (8 bits): Type of TPDU error.

---

[a]Then field in figure is 0000 and 16-bit CDT follows after YR-TU-NR.
[b]The order of bytes and bits within bytes in diagrams is a compromise between Big Endian style, with most significant byte or bit first, and Little Endian style, with least significant first (see [COHE81]). Low numbers are on the left for bytes, but on the right for bits within bytes. Thus the EOT bit in Class 2 - 4 DT TPDU is bit 8 of byte 5.

TPDU size, protocol version used, security parameters, checksums (only in Class 4 and connectionless service),[19] quality of service parameters, and so forth. Each parameter uses three subfields, parameter code (eight bits), parameter length (eight bits) and parameter value (one or more bytes).

---

[19]The OSI checksum algorithm described in the appendix to Chapter 10 is used to compute checksums.

Piggybacking of acknowledgments and of flow control credit parameters is not possible with the OSI transport layer protocols.[20] Separate AKs are necessary; after the initial credit is granted in CR and CC TPDUs, further grants or changes can be made only via AK (or RJ) TPDUs.

transport-connection management

Services provided by the transport layer are grouped into two types: *transport-connection management,* responsible for establishing and later terminating a connection, and *data transfer.*

data transfer

11.3.4   OSI Transport Layer Connection Establishment

Transport connection establishment uses a two-way handshake for classes 0–3 and a three-way handshake for class 4. These handshakes are illustrated in Fig. 11.13.

In each case, the service user (session layer) at system **A** passes a T-CONNECT.request primitive across to its service provider (transport layer), with appropriate parameters for setting up the connection. The transport layer entity at **A** generates a Connection Request (CR) TPDU containing the parameter values and sends it across the network[21] to its peer transport layer entity at **B**, which in turn generates a T-CONNECT.indication primitive and passes it to its user, session layer, entity. The connection is accepted, with a T-CONNECT.response primitive, connection confirm (CC) TPDU and T-CONNECT.confirm primitive flowing back across the network from **B** and up to the user at **A**. For Classes 0–3, this concludes Transport Connection establishment since network service is considered to be reliable and confirmation of receipt of the CC TPDU is not required.[22] For class 4, however, the network is unreliable so a Transport Connection is not fully established until a TPDU acknowledging receipt of the CC by **A** is received; an acknowledgment (AK) TPDU transmitted from **A** to **B** is shown serving this purpose, but a data (DT) or expedited data (ED) TPDU could be used.

Either the transport service user at **B** or the transport service provider at **A** can reject the Connection Request, leading to an unsuccessful connection attempt. The two possibilities are illustrated in Fig. 11.14.

The source and destination reference numbers are used in conjunction with TPDU numbers to uniquely identify DT TPDUs in classes 2–4. For those classes where errors are possible (for example, 1, 3 and 4) reference numbers are "frozen" as long as there is a possibility TPDUs previously transmitted but not yet received or acknowledged are still in the network after transport connections are disconnected; they are released for reuse only after the maximum time a TPDU might remain in the network. This eliminates problems

---

[20] Both may be piggybacked in TCP.

[21] This involves flowing down through the network, data link, and physical layers at the source node, across the communications channel, and up through the physical, data link, and network layers at the destination node.

[22] Possible loss is not considered for classes 0 and 2; the TPDU could be lost in classes 1 and 3 but the sender would be notified of this possibility.

Figure 11.13   OSI Transport Connection Establishment Sequences for Successful Connection. (a) Class 0–3 transport protocols. (b) Class 4 protocol (adapted from [SCHW87], M. Schwartz, *Telecommunication Networks: Protocols, Modeling and Analysis,* © 1987 by Addison-Wesley. Reprinted by permission of Addison-Wesley Publishing Co., Inc. Reading, MA.).

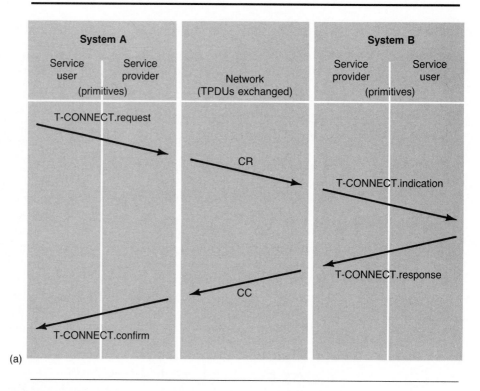

(a)

that might occur because of old TPDUs arriving at inopportune times and being erroneously accepted. Examples of how this resolves potential problems are found in Problem 11.13.

Transport Connection establishment may fail because of problems at either end of the path. If the problems are at the responding end, a Disconnect Request (DR) TPDU is returned by the responding transport layer entity in response to a T-DISCONNECT.request primitive from its session layer entity. If the problems are at the initiating end, its transport layer entity generates a T-DISCONNECT.indication primitive without contacting the other end. In either case, the T-DISCONNECT.indication contains a Reason parameter which is passed to the initiator session layer entity.

Figure 11.13   *continued*

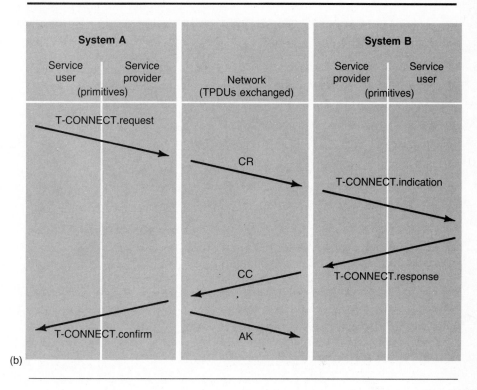

(b)

11.3.5   OSI Transport Layer Data Transfer

Procedures for *data transfer* differ for various classes of service.

connectionless mode transport service

   Connectionless mode transport service operates over either connection-less or connection-oriented network service.[23] With connectionless network service, data units received from the session layer are converted into UD TPDUs, by adding headers, and passed to the network service for transmission. No segmenting into smaller TPDUs is permitted; overly large data units are discarded. Transmission over connection-oriented network service is similar, but may require setting up a network layer connection. If a checksum is used, it is generated at the transmitter and verified at the receiver. TPDUs failing verification are discarded. Receipt verification is unavailable, so any recovery is by a higher layer.[24]

   All *connection-oriented data transfers* use sequence numbers to identify

---

[23] A connectionless network service is more natural, but only connection-oriented network service (such as X.25) may be available; thus the protocol has been defined to work over either.

[24] As is indicated in Subsection 11.3.2, the NBS version of the connectionless service includes a receipt confirmation option.

Figure 11.14   Rejection of Transport Connection Request. (a) Rejection by
user at responding entity. (b) Rejection by service provider.

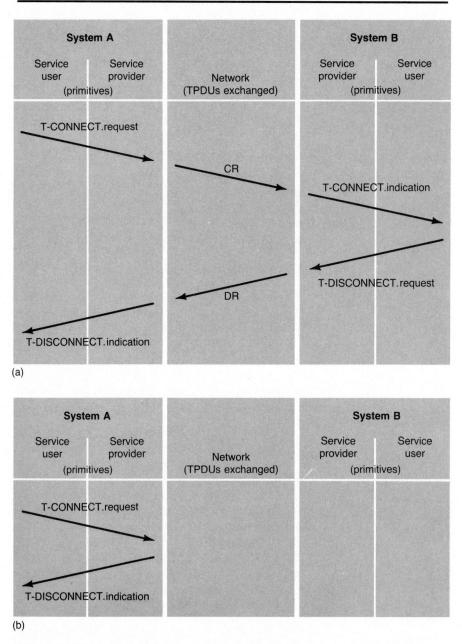

(a)

(b)

TPDUs. The sequence number for the first DT (or ED when allowed) TPDU sent is always 0, with additional DT or ED TPDUs numbered consecutively modulo 2^7 or 2^{31}, depending on the sequence count option.

In classes 0 and 1, session layer data units are simply segmented into TPDUs and transmitted, with reassembly by the receiver. The EOT bit is set in the last TPDU from a session layer data unit. Class 0 does not use AKs, since TPDUs are assumed not to be lost but class 1 provides for possible loss. AKs are returned to verify receipt and indicate that the original sender can release copies of TPDUs.[25] If Expedited Data is used, at most one unacknowledged ED TPDU is allowed in each direction of transmission at any time.

Classes 2 and 3 differ from 0 and 1, respectively, primarily in use of multiplexing. The destination reference field distinguishes transport layer connections multiplexed over one network layer virtual circuit.

Flow control uses the CDT field in CR, CC, AK, and RJ TPDUs. Initial credit is granted in CR and CC TPDUs and subsequent credit in AKs.[26] Flow control is illustrated in Fig. 11.15.

The example shows use of both primitives and TPDUs for data transfer from **A** to **B**.[27] It applies for the first data transfer after a transport connection is opened, since the initial TPDU number is 0. An initial credit of 1 is assumed, so after **A** transmits the first DT TPDU its transport entity waits for an AK with more credit. **B** returns an AK with CDT of 2, allowing transmission of two more TPDUs. This type of sequence continues until the entire session layer data unit has been transmitted, with EOT bit set in the final DT TPDU. The receiving transport entity at **B** then sends the collected session layer data unit to its user via a T-DATA.indication primitive.[28]

Only one unacknowledged expedited data (ED) TPDU is allowed to be outstanding at one time, so expedited data flow takes the form in Fig. 11.16.

Class 4 data transmission operates in the same manner as long as the network operates reliably. The primary differences for class 4 are in its capabilities for error detection and recovery. We discuss these below.

11.3.6 OSI Transport Connection Termination

abrupt disconnection

The OSI protocol provides only for an *abrupt disconnection* procedure that discards data already transmitted but not yet delivered. It is illustrated in Fig. 11.17.

---

[25] An alternative confirmation of receipt procedure is only available in this class. It uses a network layer confirmation service and can only be available if this network layer service is available.

[26] In Class 3, RJ can also be used to update credit.

[27] Data transfers in only one direction and over one transport connection are illustrated. Data transfers can be in both directions and over different transport connections multiplexed over the same virtual circuit. Flow control operates on each transport connection independently.

[28] Alternatively, the receiving entity could send individual TPDUs to its user, each with a T-DATA.indication primitive. This is an implementation decision, not spelled out in the standard.

Figure 11.15   Normal Data Transfer, Classes 2, 3, and 4, Multi-TPDU
Message (adapted from [SCHW87]).

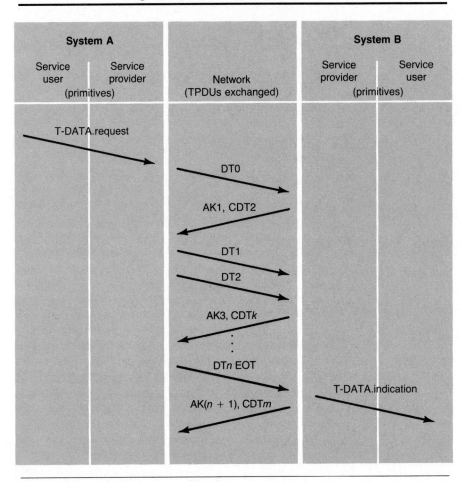

11.3.7   OSI Transport Layer Error Recovery

Since the network service for class 0 and class 2 is assumed to be error-free,
these protocols make no real provision for error recovery.[29] Classes 1 and 3
provide only for reassignment and resynchronization after the transport layer
is notified of network layer resets or disconnects. Class 4 provides *extensive*

---

[29] The class 0 and class 2 protocols acknowledge the possibility of errors to the extent of
releasing the connection if they are informed of a disconnect or reset by the network layer,
however.

Figure 11.16   Expedited Data Flow.

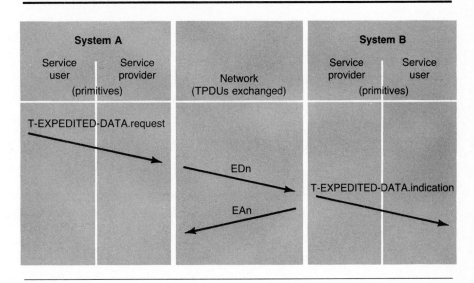

Figure 11.17   OSI Transport Layer Connection Termination.

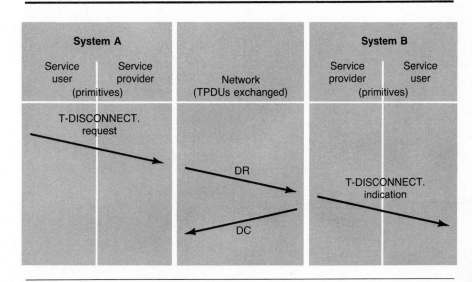

error detection and recovery capabilities, including checksums, timeouts and retransmissions, recovery from loss of data or control TPDUs, expedited data, and user notification of unrecoverable errors. Connectionless service allows checksums as an option, but only uses checksums to detect errors and discard TPDUs. It does not attempt recovery from errors.

In this section we summarize class 1 and class 3 reassignment and resynchronization and then discuss class 4 error-recovery techniques.

Class 1 and class 3 reassignment and resynchronization techniques use two timers: *TTR* (*time to try reassignment/resynchronization*), used by the initiator of the Transport Connection, and *TWR* (*time to wait for reassignment/ resynchronization*), used by the responder.[30] When a transport layer entity is informed of a network layer reset or disconnection, it starts the appropriate timer. If the timer runs out before resynchronization is completed (after reset) or both reassignment and resynchronization are completed (after disconnection), the transport connection is released and reference numbers frozen to prevent reuse while a TPDU from the connection might remain in the network.

After disconnection by the network layer, the initiator's transport layer attempts to reestablish (or reassign) the network layer connection using standard network layer protocols. If it succeeds before its TTR timer runs out, and the responder learns of the new connection before its TWR timer runs out, resynchronization is attempted. If the problem was manifested only by a network layer reset (see Subsection 9.2.6), only resynchronization is attempted.

resynchronization

Resynchronization ensures that both ends of a transport connection retransmit lost, or unacknowledged, TPDUs and are up to date on sequence numbers. The initiator first retransmits any unacknowledged CR or DR TPDUs and then moves to a data resynchronization procedure involving first the retransmission of any unacknowledged ED TPDU, then the transmission of a RJ (reject) TPDU with YR-TU-NR field indicating the next expected TPDU number. This is interpreted in the same manner as an HDLC REJ; it requires retransmission of all TPDUs previously sent beginning with that pointed to by the RJ. The responder uses a passive resynchroniztion procedure. It waits for a valid TPDU and carries out appropriate procedures for this TPDU, then it uses the same data resynchronization procedure.

class 4 error handling

Class 4 provides by far the most comprehensive error-handling procedures of any class. Use of checksums to detect transmission errors is an option selected during transport connection establishment. If a checksum is used, it uses the Fletcher algorithm in the appendix to Chapter 10. TPDUs with invalid checksums are ignored,[31] retransmission occurs after timeouts

---

[30] TWR is typically around two minutes unless otherwise negotiated. It exceeds TTR by at least the sum of maximum disconnect propagation delay plus maximum transit delay.

[31] Note that RJ (reject) TPDUs are available only in classes 1 and 3.

Table 11.7 Timer Parameters Related to Operation of Class 4 Protocol.

Symbol	Name	Definition
E	*Expected maximum transit delay*	A bound on the maximum transit delay for all but a small proportion of data units propagated across the network (for example, a 90th percentile delay).
M	*Maximum transit delay*	A bound on the maximum transit delay for any data units propagated across the network.
A	*Acknowledge time*	A bound for the maximum time between receipt of a TPDU by a transport entity and the transmission of the corresponding acknowledgment.
T_1	*Local retransmission time*	A bound for the maximum time the local transport entity will wait for an acknowledgment before retransmitting a TPDU.
R	*Persistence time*	A bound for the maximum time the local transport entity will continue to transmit a TPDU that requires acknowledgment.
N	*Maximum number of transmissions*	A bound for the number of times the local transport entity will continue to transmit a TPDU that requires acknowledgment.
L	*Bound on references and sequence numbers*	A bound for the maximum time between transmission of a TPDU and receipt of any response relating to it.
I	*Inactivity time*	A bound for the time a transport entity will wait without receiving a TPDU before initiating the release procedure to terminate the Transport Connection.
W	*Window time*	A bound for the maximum time a transport entity will wait before retransmitting up-to-date window information.

discussed below. The protocol uses several timer parameters, listed in Table 11.7, to handle possibilities of lost TPDUs, broken connections, and so forth.[32]

Formulas for a number of these parameters are given in the standard, but their use is not mandatory. Each parameter is a bound, with precise value often difficult to compute. For example, good bounds on maximum network delay are not known, but some parameters listed are such bounds. In practice reasonable estimates for such delays are used even if they are not mathe-

---

[32] The standard includes parameters for delays between transmission by local and reception by remote transport entity (E_{LR} and M_{LR}) and between transmission by the remote and reception by the local (E_{RL} and M_{RL}), but we give only one delay for each, assuming they should be nearly equal. We also list only one acknowledge time, instead of separate A_R and A_L times at the remote and local entities, respectively.

matically rigorous. It is always possible to overbound such delays by using extremely large values, but this can greatly reduce transmission efficiency.

The relationship between E (expected maximum transit delay) and M (maximum transit delay) deserves comment. Both are bounds on maximum transit delay, but M is normally considerably larger than E. M is supposed to be an absolute upper bound, including the possibility the data unit goes by the longest path allowed by the protocols and encounters unusual queuing delays.[33] E is a value that most (for example, 90 percent) of transit delays do not exceed.

The most fundamental timer parameter is T_1 (local retransmission time), which determines when TPDUs should be retransmitted. Five types of TPDUs (CR, CC, DT, ED, and DR) require some type of acknowledgment in class 4. A CR is acknowledged by a CC, a CC by an AK (or DT or ED), a DT by an AK, an ED by an EA, and a DR by a DC. In class 4, however, TPDUs may be lost, so acknowledgments are not guaranteed to arrive (either because the original TPDU did not get through or the acknowledgment did not get back). If T_1 expires without receipt of an acknowledgment after transmission of one of the five TPDU types above, the TPDU is retransmitted. A recommended value for T_1 (in terms of the simplified parameter list in Table 11.4) is

$$T_1 = 2E + A + X, \qquad \textbf{(11.2)}$$

with E and A expected maximum transit delay and acknowledge time, respectively, and X local processing time for a TPDU. Most acknowledgments should be received within time T_1 so this is a reasonable value for the retransmission timer.[34]

The values of N (maximum number of transmissions) and R (persistence time) are closely related. N is the maximum number of times the transport entity will try to retransmit a TPDU that has not been acknowledged, while R is the maximum time during which it will try to retransmit. Since T_1 is the interval between retransmissions, the value of R is stated to be

$$R = NT_1 + \varepsilon \qquad \textbf{(11.3)}$$

with ε representing a small quantity to allow for internal delays, granularity of the mechanism used to implement T_1, and so forth. After time R (or slightly longer) has elapsed with no acknowledgment, the transport layer releases the connection.[35]

---

[33] As has been commented earlier, a good estimate for a delay such as M is extremely difficult to obtain.

[34] Obtaining realistic estimates for the terms in the equation is not necessarily simple. The value of A at each transport entity, though, may be included in the variable part of the CR or CC, making good estimates available after connection establishment. Note that if excessive retransmissions are to be avoided, E should be greater than the mean transit delay; a 90th (or even higher) percentile for transit delay would be a more realistic value for E.

[35] The standard recommends that connection release be delayed for a time $W + M$ after the N transmissions, each with timeout T_1. This is larger than the value of ε that would normally be used in Eq. (11.3), though a small value of ε should suffice for the R used in Eq. (11.4).

Another significant timer value is L, bounding the maximum time between transmission of a TPDU and receipt of any response relating to it. This is used to determine how long a reference number should be "frozen" after a transport connection is closed, prohibiting its use for a new connection to avoid any possibility that a TPDU relating to the earlier connection could be accepted as relating to the later one. The formula recommended for L is

$$L = 2M + R + A. \tag{11.4}$$

With the interpretations for M, R, and A above, this gives a realistic upper bound for the time any response could remain in the network, and hence is a reasonable time for freezing references to prevent their reuse.

The inactivity time (I) provides for the possibility of unsignalled failures of the transport connection. If the connection simply "goes dead" for this length of time, a failure is assumed and the connection broken.

The window time (W) ensures that AKs are returned within fixed time periods after TPDUs are sent and thereby guarantees that an acknowledgment reopening a closed window will be transmitted within a reasonable period of time after congestion clears up. When DT TPDUs are sent, it is not necessary to acknowledge all of them (see Fig. 11.15), so W may allow for reception of multiple TPDUs before generating the acknowledgment.[36]

Since DT TPDUs may arrive out of order, one job of the receiver is to resequence them. It uses the TPDU-NR field for this, with source and destination reference numbers used to demultiplex TPDUs before resequencing. Any DT TPDU within the current window (as determined by sequence number and the credit field for the last credit granting or adjusting TPDU received) may be accepted, but DT TPDUs may only be delivered to the user in sequence. Hence DT TPDUs received out of sequence but within the window must be buffered until all intervening TPDUs are received.

If several TPDUs have not been acknowledged when the T_1 timer expires, the choice between retransmitting only the first or retransmitting all of them is left as an option in the standard. If it turns out that only the first has been lost, retransmitting only this one is the proper choice. However, the transmitter has no way of determining whether this is the case or not.

ED TPDUs are guaranteed to be delivered before any DT TPDUs transmitted after them, but only one unacknowledged ED TPDU may be outstanding at a time. The protocol takes a simple but drastic approach to guaranteeing that no DT TPDU transmitted after the ED is delivered before the ED. It refuses to transmit any more DT TPDUs until the EA comes back. Hence, ED TPDUs can significantly impact delays for DT TPDUs.

The possibility of errors affecting TPDUs used for flow control causes special problems. The protocol explicitly provides for the possibility of using an AK at any time to reduce credit below values previously granted (or to

---

[36] Similarly, the acknowledge time (A) may reflect the possibility of receiving several DT TPDUs before acknowledgment.

increase credit). This includes the possibility of reducing the credit to zero. The problems cited result from the fact that AKs may arrive out of order. For example, assume that two AKs with the same YR-TU-NR field (that is, acknowledging the same DT TPDU) are issued, the second having a smaller CDT value than the first. If the AKs are received out of order, the effect of the credit reduction (probably issued because of congestion) would be lost without modification of the protocol explained so far. The receiver would think the credit had been increased from the smaller value to the larger value instead of vice versa. This problem is handled by including a subsequence number in the variable part of the header field in any AK acknowledging a previously acknowledged TPDU but reducing credit. The credit field from the AK with highest subsequence number among those received is treated as current.

The techniques described handle most error situations that may occur with the unreliable type C networks for which the class 4 protocol was designed; some details have been omitted, however.

11.3.8   Limitations of OSI Transport Layer Protocols

Although the OSI transport layer protocols show advances over earlier protocols, they contain significant weaknesses:

1. The different connection-oriented protocols are a hodgepodge, not fitting together in a systematic structure. The protocols show definite signs of being developed by different groups then packaged together to form a partially consistent set. Although class 3 is basically a union of classes 1 and 2, it does not include all options in the latter two classes.[37] Class 4 is by far the most comprehensive class, but it also omits features in other classes; thus RJ (Reject) TPDUs are available only in classes 1 and 3. Judicious use of this TPDU in class 4, in a manner analogous to use of REJ in HDLC, could improve efficiency.

2. A more serious factor impacting efficiency is lack of piggybacking of acknowledgments and credit fields in data packets. Separate AKs are used, with updating of credit only in AKs (plus RJs in classes 1 and 3). Large sequence number fields, 7 or 31 bits, imply substantial numbers of TPDUs could be transmitted before AKs were required, but implementations use AKs more frequently.[38] Substantial fractions of the traffic devoted to TPDUs could be avoided with piggybacking.

---

[37] Examples include a network expedited variant of expedited data transfer and a confirmation of receipt procedure, both available only in class 1.

[38] The A and W parameters in class 4 will probably require this.

3. Another weakness of the OSI protocols is their lack of a graceful close. The NBS variant incorporating a graceful close is a definite improvement that could eventually find its way into the international standards.

4. TCP's use of PASSIVE-OPEN (that is, Listen) and ACTIVE-OPEN primitives allows entities willing to accept calls to be placed into a listening state, simplifying software for Transport Connection establishment. No PASSIVE-OPEN.request type primitives are available in OSI protocols, implying reliance on interrupts to inform a user process of a connection attempt it may or may not welcome. Partially as a consequence, simultaneous OPEN.requests from both ends always result in one connection in TCP (if negotiation is successful), but may result in two, one, or zero connections with OSI protocols.

5. Neither the OSI nor the DoD protocols include all the types of transport layer protocols it would be desirable to have. At least two other types of transport layer protocols would be desirable: a speech protocol guaranteeing sequenced timely delivery but without high reliability requirements and a real-time protocol guaranteeing high reliability and timeliness [MCFA79].

11.4   The End Communications Layer in DNA

The DNA layer corresponding to the OSI transport layer is the *end communications layer,* which provides reliable, sequential, connection-oriented, end-to-end communications servcie and isolates higher layers from transient errors or reordering of data introduced by lower layers. It also multiplexes multiple End Communications Connections, called logical links, between pairs of nodes or between one node and multiple nodes. Logical links correspond to ports in TCP or reference numbers in the OSI transport layer. End communications layer protocols are called network services protocols (NSPs), terminology carried over from earlier phases of DNA when the end communications layer was the network services layer and the routing layer was the transport layer, a direct interchange of OSI terms.

Since the DNA routing layer provides datagram service, the end communications layer must deal with the same problems as TCP or TP4. It is designed to operate over an unreliable routing layer service that may lose, reorder, or duplicate messages. It also deals with host computers crashing at one or both ends of a communications path or logical link. It uses logical link identifiers and sequence numbers to identify end communications data units in a manner analogous to the use of sequence numbers and port numbers in TCP or sequence numbers and reference numbers in TP4.

11.4.1   End Communications Layer Primitives and Data Units

A partial list of end communications layer *primitives* is given in Table 11.8, with DNA notation altered to identify them as Request, Indication, Response, or Confirm primitives in keeping with OSI terminology.[39] The list includes all primitives relevant to our discussion, but there are a number of others used for housekeeping purposes.

The OPEN primitive corresponds to TCP's PASSIVE-OPEN primitives, with CLOSE available to revoke OPEN. The next eight primitives are used for End Communications Connection (successful or unsuccessful) and disconnecting or aborting connections; the last four primitives are used for transmission and reception of ordinary data messages or interrupt messages.

Table 11.8   Partial Listing of DNA Primitives.

Primitive	Parameters
OPEN.request	Buffer for address of node later requesting connect, buffer for incoming connect data.
CLOSE.indication	Logical link identification (id).
CONNECT-XMT.request	Destination node address, buffer containing connect data.
ACCEPT.response	Logical link id, buffer for accept data.
REJECT.response	Logical link id, buffer for reject data.
CONNECT-STATUS.confirm	Logical link id, buffer for accept or reject data.
DISCONNECT-XMT.request	Logical link id, two bytes of disconnect data, buffer containing up to 16 more bytes of disconnect data.
ABORT-XMT.request	Logical link id, two bytes of abort data, buffer containing up to 16 more bytes of abort data.
DISCONNECT-RCV.indication	Logical link id, two bytes of disconnect data, buffer containing up to 16 more bytes of disconnect data.
DATA-XMT.request	Logical link id, buffer for transmit data, flag indicating whether last byte in buffer is last byte of message.
DATA-RCV.indication	Logical link id, buffer for receive data, flag indicating whether data truncation allowed.
INTERRUPT-XMT.request	Logical link id, buffer for transmitted interrupt data.
INTERRUPT-RCV.indication	Logical link id, buffer for received interrupt data.

---

[39] In DECNET literature, these are described as calls to subroutines, but they serve essentially the same purpose as OSI primitives.

Table 11.9 DNA End Communications Data Units.

Type	Message	Description
Data	Data Segment	A portion of a data message, as received from the session control layer.
Other Data	Interrupt	Urgent data originating from higher layers.
	Data Request	Data flow control information.
	Interrupt Request	Interrupt flow control information.
Acknowledgment	Data Acnowledgment	Acknowledges receipt of either a Connect Confirm or one or more Data Segment messages, plus optionally an Other Data message.
	Other Data Acknowledgment	Acknowledges receipt of one or more Interrupt, Data Request, or Interrupt Request messages.
Control	Connect Initiate	Carries a connect request from session control layer.
	Connect Confirm	Carries a connect acceptance from session, control layer.
	Disconnect Initiate	Carries a connect rejection or disconnect request from session control layer.
	No Resources	Sent when Connect Initiate is received and there are no resources to establish a new logical link.
	Disconnect Complete	Acknowledges receipt of a Disconnect Initiate.
	No Link	Sent when a message is received for a nonexisting link.
	No Operation	Does nothing.

A list of End Communications Messages, or data units, is given in Table 11.9;[40] the format of a DNA Data Segment data unit is given in Fig. 11.18 with fields defined in Table 11.10. It illustrates the general format of End Communications data units. Other data units have a similar format.

11.4.2 End Communications Connection Establishment

The manner in which various DNA commands are invoked, corresponding to use of primitives in the OSI model, is implementation dependent. System calls are used on some processors and library routine calls on others. In the discussion following we use common versions of such calls and describe them in terms of the primitives and data units introduced above.

End Communications connection establishment is via a three-way handshake similar to that in TCP or TP4 and indicated in the diagram in Fig. 11.19; we use slight modifications of DEC terminology to simplify comparison with

---

[40] These are called NSP messages in DECNET manuals.

Figure 11.18   Format of DNA Data Segment End Communications Data Unit.

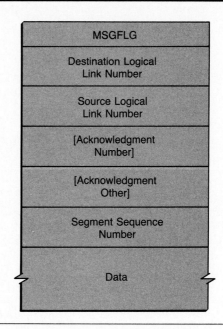

Table 11.10   Fields in DNA Data End Communications Data Units.

- *MSGFLG* (8 bits): Identifies this as a Data Segment End Communications Data Unit and contains end-of-message and beginning-of-message flags.
- *Destination Logical Link Number* (16 bits): Identifies destination logical link.
- *Source Logical Link Number* (16 bits): Identifies source logical link.
- *Acknowledgment Number* (optional, 16 bits if present with 12 bits available for numbers): Piggybacked data message acknowledgment—sequence number of last data message segment received in sequence. Presence indicated by high order bit being 1, and identified as Acnowledgment Number (and either positive or negative acknowledgment) by next 3 bits.
- *Acknowledgment Other* (optional, 16 bits if present with 12 bits available for numbers): Piggybacked acknowledgment for other messages moving in data subchannel. These include data request, interrupt request, and interrupt messages. Presence indicated by high order bit being 1, and identified as Acnowledgment Other by next 3 bits.
- *Segment Sequence Number* (16 bits with 12 bits available for actual numbers): Sequence number for message segment being transmitted.

Figure 11.19   DNA End Communications Layer Connection Establishment.

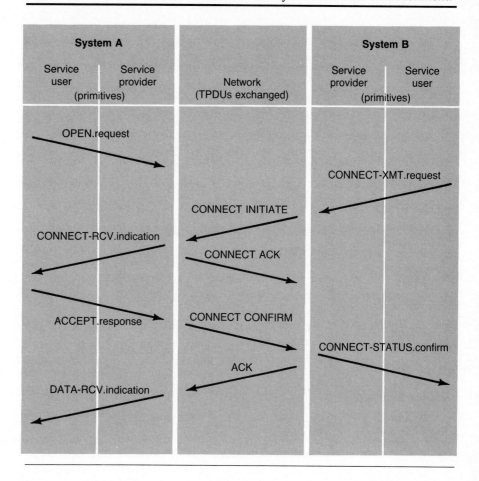

TCP and TP4. The handshake is closer to the TCP handshake than to that for TP4, since a process that wishes to listen for incoming calls must issue a request equivalent to the TCP Passive Open primitives (called OPEN in DNA documentation) and provide a buffer for the incoming message.

The Connect Initiate and Connect Confirm data units shown contain Source and Destination Logical Link Numbers[41] plus additional information, such as segmentation parameters and flow control options, needed to set up the connection. To avoid using the same number to identify two different logical links, DNA avoids reusing the same logical link number used for an

---

[41] The Destination Logical Link Number is set to zero in the Connect Initiate Data Unit, since this must be assigned by the destination.

earlier connection for as long as possible. The Connect Acknowledgment message is used across the data link to confirm receipt of the Connect Initiate,[42] while the final Acknowledgment uses the standard data message acknowledgment format. If the destination process did not wish to accept the connection, a Disconnect Initiate would be returned.

11.4.3   End Communications Layer Data Transfer

The end communications layer is designed specifically for operation above the DNA routing layer, which uses datagram transmission. It handles the error control and flow control functions necessary to provide reliable, sequential, connection-oriented, end-to-end communications service and to isolate higher layers from transient errors or reordering of data introduced by lower layers.

In order to provide error protection, all Data Segment data units are numbered in sequence modulo 4096, using 12-bit segment sequence numbers (see Fig. 11.18), with equivalent numbering for Interrupt, Data Request, and Interrupt Request data units. Each must be either positively or negatively acknowledged within a timeout, using either a piggybacked acknowledgment or a separate Acknowledgment data unit, or the data unit will be retransmitted. Multiple data units may be acknowledged at once since an acknowledgment implies all previous data units of the class acknowledged were received correctly. Any data unit with segment number equal to the current receive acknowledgment number plus one is accepted, and the acknowledgment number incremented. Data units with segment numbers less than or equal to the current receive acknowledgment number are discarded as assumed retransmissions of previously acknowledged data units; if data units with segment numbers greater than the current receive acknowledgment number plus one are received, the receiver has the option of holding them until preceding data units are received or discarding them.

segmentation and reassembly

As in TCP and the OSI transport layer protocols, long messages received from higher layers (session control here) may undergo *segmentation*—the breaking up of long messages into shorter segments before transmission. The maximum segment size is the smaller of the size of a transmit buffer in the source node and the maximum length the destination end communications layer can receive, with the latter number included in Connect Initiate and Connect Confirm messages exchanged when the logical link is set up. The segments are then transmitted as Data Segment data units, with the receiver using the sequence numbers and end-of-message flags for *reassembly* of the messages.

flow control

There are three options for *flow control* for data segment messages, with the choice made during logical link set up: no flow control, segment flow

---

[42] This is an addition to the normal three-way handshake.

control, and complete message flow control. In addition, each may be supplemented by on/off control, equivalent to opening a transmission window and slamming it shut. Interrupt messages use separate flow control, employing the message flow control approach; on/off control is not allowed for interrupt messages since they are short, high-priority messages that must be delivered quickly.

Data Request or Interrupt Request data units are used to implement flow control. Each contains an eight-bit field used to implement variable credit flow control. If flow control is used, the contents of this field indicate the number of data segments, complete messages before segmentation, or interrupt messages (whichever is appropriate) the transmitter is allowed to send, in addition to any credit previously given. The credit field is allowed to be negative, thus revoking previous credit, for data segment flow control, but must be nonnegative for complete message or interrupt message flow control. Another field in Data Request data units contains bits either allowing data to be sent, prohibiting data to be sent, or keeping the status unchanged; this field can also be used when the No Flow Control option is used, allowing rudimentary flow control even in this case. Sequence numbers in Data Request and Interrupt Request data units allow the most recent one to be determined if there are any ambiguities.

Data transfer in the End Communications Layer is similar to data transfer in TCP and the OSI transport layer; a detailed sketch of data transfer has been included in homework (see Problem 11.21).

11.4.4 End Communications Layer Connection Termination

The end communications layer uses graceful close for normal connection termination, with abrupt disconnection used only when it is unavoidable. A graceful close is initiated by transmission of a Disconnect Initiate data unit, which can only be transmitted after the session control layer requests a disconnection and there are no outstanding unacknowledged Data Segment messages (and no previously transmitted Disconnect Initiate data unit is being timed-out). An adequate number of receive buffers for any data transmitted from the other session control module must also be reserved. Use of Abort data units for disconnection, on the other hand, results in abrupt disconnection with consequent loss of data.

Problem 11.22 involves sketching the normal termination process in terms of the primitives and data units given in Subsection 11.4.2.

11.4.5 Critique of DNA End Communications Layer

Since this is the fourth version of DNA, the algorithms used have undergone considerable refinement and do a good job of meeting their requirements. Most of the limitations listed for TCP and the OSI transport layer are not applicable here, so we give a brief critique of the layer instead of a section on limitations.

Since the end communications layer was developed specifically for use above the DNA routing layer, it does not have to provide for the variations in lower layer architectures accommodated by TCP and the OSI transport layer. This has led to a reasonably efficient architecture, with data segment message header lengths varying from seven to eleven bytes (depending on whether the Acknowledgment Number and Acknowledgment Other fields are used or not). No checksum or block check fields are needed since the DDCMP fields are adequate.

Although the end communications layer uses a graceful close under normal circumstances, it puts more reliance on disconnection being initiated at times appropriate for both ends of a connection than does TCP. The protocol, as stated in DEC manuals (for example, in [DIGI83b]), does not appear to guarantee that the end of a logical link not initiating disconnect will be able to finish transmission before the link is shut down.

The end communications layer comes no closer than TCP or the OSI transport layer to satisfying the need for a speech protocol guaranteeing sequenced, timely delivery but without high reliability requirements or a real-time protocol guaranteeing high reliability and timeliness.

Overall, the end communications layer is a carefully thought out and successful version of a transport layer protocol for use in its intended environment.

11.5   Transmission Control Layer in SNA

The transmission control layer in SNA is the layer at a level closest to that of the transport layer in the OSI architecture. It is responsible for establishing, maintaining, and terminating SNA sessions, sequencing data messages, and session level flow control. It also encapsulates messages with transmission control headers[43] and routes data to appropriate points within network addressable units (NAUs).[44]

sessions

The transmission control layer supports *sessions* or logical relationships between communicating NAUs. A fundamental rule in SNA is that no communication can take place if there is no session between the communicating NAUs.[45] Sessions are implemented in SNA as half sessions in a manner anal-

---

[43] Request/response headers (RHs) in SNA's standard terminology; we maintain our use of terminology associating headers with the layers attaching them.

[44] Recall that in SNA all information flow is to or from network addressable units (NAUs). These consist of system services control points (SSCPs), logical units (LUs), and physical units (PUs).

[45] Limited types of communication needed to set up sessions are, of course, allowed.

ogous to implementation of half gateways described in Chapter 10; the logical functions required at one end of a session constitute that end's half session.

Two major groups of transmission control functions[46] are provided for each half session, session control functions, and connection point manager functions. Session control provides support for starting, clearing, and resynchronizing session-related data flows. The connection point manager controls sequence number checking, flow control, and other support functions related to such data flows.[47]

As was indicated in Subsection 3.8.2, the upper layers in SNA—transmission control, data flow control, presentation services, and transaction services—define LUs, or ports into the network. SNA's subsetting of functions according to capabilities of different equipment, functions available, and so forth has led to definition of several types of PUs and LUs but to only one type of SSCP. Table 11.11 lists the different types of SNA NAUs according to current definitions. The amount of detail in this table is a small sample of the type of fine detail in SNA manuals.[48]

logical units

Logical units perform all those functions which are necessary for communication between users during sessions. They are the primary NAUs of interest during normal communication. LU to LU sessions fulfill the purpose of the network, enabling actual transfer of data between users to take place. SSCPs and physical units handle details of setting up sessions and operation of the physical devices used in the network, but are not involved in LU to LU data transfer after LU to LU sessions are established.

Table 11.11   SNA Network Addressable Units.

Network Addressable Unit	Description
SSCP	Functions within a host subarea node to activate, control, and deactivate network resources.
Physical Unit Type 1[a]	Functions within a very simple terminal node controlling attached links and other resources.
Physical Unit Type 2.0	Functions within a cluster controller node or more complex terminal node controlling attached links and other resources.

continued

---

[46] Protocol machines in SNA documentation.

[47] One of the most important such support functions is encryption and decryption of information for security purposes. Encryption is an important and interesting topic, but we omit discussion of it because of space limitations and the fact that few systems currently use it.

[48] Still finer subdivision of logical units is provided by Profiles that define subsets of the functions in the higher layer protocols defining each type, which are selected during session establishment. See SNA manuals for this additional level of definition.

Table 11.11   *continued*

Network Addressable Unit	Description
Physical Unit Type 2.1	Functions in more capable version of Physical Unit Type 2, which is also capable of performing control functions for an adjacent SNA node.
Physical Unit Type 4	Functions within a communications controller node controlling attached links and other resources.
Physical Unit Type 5	Functions within a host computer node controlling attached links and other resources.
Logical Unit Type 0	Functions supporting sessions that use SNA-defined protocols for Transmission Control and Data Flow Control, but end-user or product defined protocols at higher layers.
Logical Unit Type 1	Functions supporting an application program that communicates with single- or multiple-device data processing workstations in an interactive, batch data transfer or distributed data processing environment. The SNA character string or Document Control Architecture is used.
Logical Unit Type 2	Functions supporting an application program that communicates with a single display workstation in an interactive environment, using the SNA 3270 data stream.
Logical Unit Type 3	Functions supporting an application program that communicates with a single printer, using the SNA 3270 data stream.
Logical Unit Type 4	Functions supporting either (1) an application program that communicates with a single- or multiple-device data processing workstation in an interactive, batch data transfer, or distributed data processing environment, or (2) logical units in peripheral nodes that communicate with each other. The SNA character string or Office Information Interchange data stream is used.
Logical Unit Type 6.1	Functions supporting an application subsystem that communicates with another application subsystem in a distributed data processing environment.
Logical Unit Type 6.2	Functions supporting sessions between two applications in a distributed data processing environment. LU Type 6.2 communication is also called Advanced Program-to-Program Communication. This is supposed to be the *only* Logical Unit Type supported in the future.

[a]No further Physical Unit Type 1 implementations are being developed, although existing implementations are still being supported.

Figure 11.20   Format of SNA Transmission Control Data Unit.

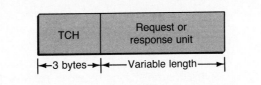

11.5.1   SNA Transmission Control Data Units and Primitives

There are two basic types of data units handled by transmission control, *requests,* and *responses.* Requests contain user data or network commands. Responses either positively or negatively acknowledge receipt of requests. Each data unit has a three-byte transmission control header (TCH).[49] A data unit is depicted in Fig. 11.20.

The first bit in the header indicates whether the data unit is a request or a response. The next two bits indicate to which protocol component the data unit belongs. This can be function management,[50] data flow control, session control, or Network Control. (Upper layers pass parameters down to the transmission control layer for inclusion in its header rather than constructing a header of their own.)[51] The remaining bits are indicators for the protocols.

A total of 45 *SNA procedures* for the transmission control layer are listed in [IBM80], with multiple calls to many of these. These procedure calls serve the purpose of primitives in SNA. We make no attempt to list primitives (nothing in SNA is simple), but examples of how a few are used are included in discussions below.

11.5.2   Transmission Control LU-LU Connection Establishment and Termination

In keeping with the complexity of SNA, a wide variety of types of connections may be established and terminated, with precisely described procedures for each. These include LU-LU, SSCP-LU, SSCP-PU, and SSCP-SSCP connections, with the NAUs at the two ends of the connections either in the same domain (under control of the same SSCP) or in different domains. We consider only establishment and termination of LU-LU connections, or sessions. The LU-LU

---

[49] Request/response header (RH) in SNA documentation.

[50] As was mentioned in Chapter 3, the function management layer was used in earlier versions of the SNA architecture for what was essentially a combination of the transaction services layer and the presentation services layer.

[51] As was mentioned in Chapter 3, a function management header is occasionally used, but it is considered to be part of the request/response unit.

data transfer during these sessions is the primary reason for the existence of SNA.

SSCP-PU and SSCP-LU connections for the physical units and logical units involved in a session must exist before LU-LU session activation, so we assume that such connections have already been activated. Also, we consider the case where the two LUs involved are in separate domains, with SSCP-SSCP connections between the domains already active. All LU-LU sessions must be between a *primary LU* and a *secondary LU,* and we make the additional assumption that the secondary is the one initially requesting the session.[52] This might correspond to the common situation where a terminal user wants to establish a session with some type of processor in the SNA network; the initiator would then normally be the secondary. Under these circumstances, session establishment is handled by the session control component of transmission control in the manner indicated in Fig. 11.21. We have illustrated both transmission control data unit exchanges, and the primitives exchanged between LU services (or the higher SNA layers) and transmission control at the end points, using OSI-like notation for primitives. A more complete figure would include primitives, or equivalents, exchanged between layers at each SSCP location, but these have been omitted to keep the figure from being even more complex.

The sequence sketched begins when LU services at **A** (the secondary) issues a SESSION-INIT.Request primitive to the session control component of transmission control at **A**. This could be the result of a user at a terminal typing something like "Logon **B**." Session control then generates an Initiate Self with **B**, INITSELF(B), data unit to be sent to its SSCP, **SSCP 1** in the figure. This indicates it wishes to activate a LU-LU session between itself and **LU B**[53] and contains the name of the requested LU and information about the requesting **LU A**.[54] The information received is interpreted by **SSCP 1** and various checks made. For the situation in the figure, **SSCP 1** also determines that **LU B** is not in its domain, but it is in the domain of **SSCP 2**.

If all checks are satisfactory, **SSCP 1** then sends a Cross Domain Initiate (CDINIT) data unit to **SSCP 2** specifying a limited number of details about the proposed session. **SSCP 2** makes similar checks and then, for the successful connection establishment in the figure, returns a Response to CDINIT, RSP(CDINIT), agreeing to assist in setting up the session. A considerably longer Cross Domain Control Initiate (CDCINIT) data unit is next used to con-

---

[52] SNA, of course, allows for all conceivable possibilities here, including the primary requesting the session, a third LU (not at either end of the connection to be established) requesting the session, the system operator requesting the session, and so forth.

[53] Initiate Other (INITOTHER) would be used if the initiating LU were a third party, not to be involved in the final session.

[54] The format of INIT-SELF may request later notification, via a NOTIFY Data Unit, of the results of the connection establishment attempt. The figure assumes that no such request is made.

vey details of the proposed session, with a positive response, RSP(CDCINIT), returned by **SSCP 2**.[55] **SSCP 2** also sends a Control Initiate (CINIT) data unit to Session Control for **LU B**, asking it to set up a session via a BIND data unit, and including suggested parameters for the BIND.

After an appropriate primitive, labeled SESSION-INIT.Indication, is sent to LU services at **B**, a positive response primitive is returned and session establishment can begin. This is accomplished by the BIND data unit transmitted directly from **LU B** to **LU A**. BIND data units can be long and complex, and contain the parameters and protocols **LU B** suggests for the session—for example, pacing counts to be used.[56] Under some conditions, the BIND may be negotiable. If this is the case, the RSP(BIND) from **LU A** may suggest alternative parameters. If **LU B** accepts them, the session can then start, with requests and responses flowing back and forth over the two-way connection that has been established; otherwise the session cannot start and **LU B** sends an UNBIND data unit (see procedures for connection termination below). Negotiation of this type is only allowed for certain types of sessions, however.

Since the session has now been established, normal session data flow via an exchange of requests and responses can now take place.[57] This is indicated in the figure. **LU B** also informs its SSCP of successful session establishment, via a Session Started (SESSST) data unit, with **SSCP 2** passing this information on to **SSCP 1** via a Cross Domain Session Started (CDSESSST) data unit.[58]

Although Fig. 11.21 shows (some of) the primitives and data units exchanged during LU-LU connection establishment, it does not clearly indicate which data units flow through which portions of an SNA network. Figure 11.22 is an alternative way of conveying most of the information in Fig. 11.21, but more clearly showing which data units flow through SSCPs and which flow directly between LU locations. Primitives are not shown in Fig. 11.22, however. Each of the paths in Fig. 11.22 could pass through several intermediate nodes between source and destination. Data units are numbered approximately in chronological sequence, but 10 and 11 would normally occur simultaneously with the first data units in 9; 5 and 6 are also essentially simultaneous.

---

[55] Although [IBM80] indicates that this last response is necessary, it is the only data unit in the figure for which it does not give a description and format.

[56] A large part of the information needed for BIND data units is conveyed by specifying transmission control and data flow control profiles to be used. These profiles are described in manuals such as [IBM80].

[57] Another of SNA's numerous options is to separate the start of data traffic from session establishment. If this is selected, an extra Start Data Traffic, SDT data unit is required to start traffic.

[58] The SESSST and CDSESSST data unit exchanges would actually occur simultaneously with exchange of the first Requests and Responses during the session, but it is difficult to show this on the figure.

Figure 11.21   Primitives and Transmission Control Data Units Exchanged During SNA LU-LU Session Establishment, with LUs in Two Domains.

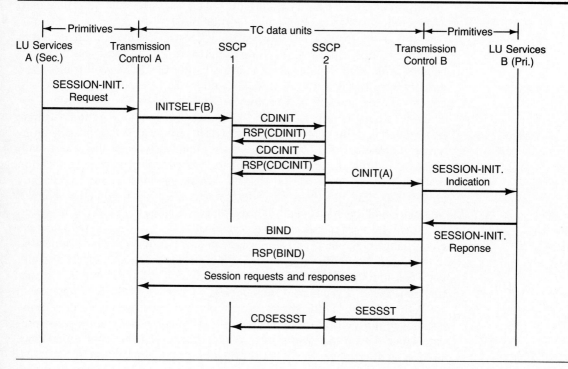

If **LU A** and **LU B** were in the same domain, the session would be established similarly. The five TC data units shown being exchanged between SSCPs would be omitted, but the rest of session establishment would be the same. The LUs would not see any difference.

Data unit exchange for LU-LU connection (session) termination is sketched in Fig. 11.23, for the same situation. **LU A**, the secondary LU, has been assumed to be the LU requesting termination. It does so by sending a Terminate Self, TERM-SELF(B), data unit to **SSCP 1**, which in turn sends a Cross Domain Terminate (CDTERM) to **SSCP 2** and receives a response, RSP(CDTERM). **SSCP 2**, in turn, sends a Control Terminate, CTERM(A), to **LU B**. If **LU B** (the primary) still has information it wants to send on the LU-LU connection, it completes the session before issuing an UNBIND to terminate the session. After it gets a response, RSP(UNBIND) and the two SSCPs are notified via Session Ended (SESSEND) and Cross Domain Session Ended (CDSESSEND) data units, the session is terminated and any facilities it has used are released. As was the case for session establishment, the procedure is identical, from the point of view of the LUs, if both LUs are in the same domain, but the SSCP-SSCP exchanges are omitted.

Figure 11.22 Transmission Control Data Units Exchanged During LU-LU Connection Establishment.

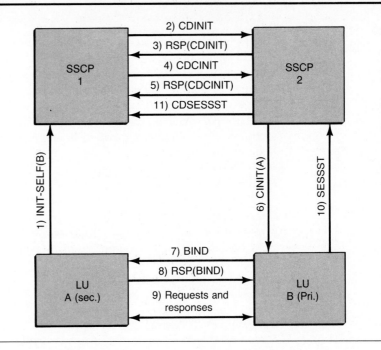

Figure 11.23 Transmission Control Data Units Exchanged During LU-LU Connection Termination.

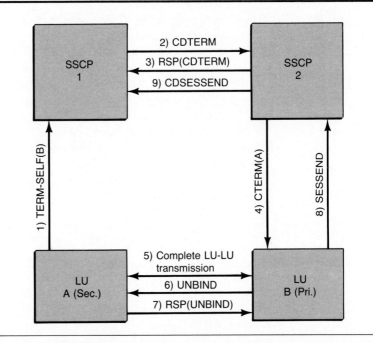

11.5.3   Transmission Control Data Transfer

Data transfer during a session is handled by the connection point manager component of transmission control. Primary functions of the connection point manager are to handle expedited delivery of messages, flow control, sequence number checking, and other support functions.[59]

expedited flow

Expedited flow is used for urgent messages and allows them to bypass the normal queuing and flow constraints. They are also unaffected by normal sequence numbering. Expedited flow is used for recovery mechanisms and similar functions.[60]

The *flow control* mechanism used is the *pacing* mechanism discussed in Subsection 9.7.3 for the path control layer, though with a fixed window size rather than a variable window size. Pacing in the transmission control layer is on a per session basis. The first request in each window must have a pacing request indicator bit on in the transmission control header. A response coming back has a pacing response indicator bit on when the receiver wishes to reopen (or further open) the window.[61]

The sender of a transmission control request data unit can also specify what type of response is desired. There are three possibilities:

1. *No response* desired.

2. Response only in case of error (*exception response*).

3. Always a response (*definite response*).

11.5.4   Critique of SNA Transmission Control Layer

The transmission control layer in SNA, as is the case with the rest of SNA, is the most carefully designed and complete "transport layer protocol" in existence. It can do all jobs it was designed to do, and do them well if users are able to decipher SNA enough to take advantage of its features.

The major single limitation of SNA is its *daunting complexity*. In order to take real advantage of its "embarrassment of riches," a user needs to be expert in both communications and applications programming—and able to decipher almost incomprehensible SNA documentation. One of the best examples of the complexity of SNA is LU6.2, the version of logical unit architecture, which is claimed to be, since its announcement in 1983, IBM's most important networking protocol. LU6.2 is just beginning to find wide adoption

---

[59] This includes enciphering/deciphering if these are implemented.

[60] A bit in the path control header (transmission header in SNA documentation) indicates whether expedited flow is being used or not. Logically, this bit should be in the transmission control header since transmission control handles expedited flow.

[61] The same header bit is used for both the pacing request indicator and the pacing response indicator, with its interpretation depending on whether the data unit is a request or a response.

[KORZ89], with the six-year delay from its announcement due primarily to its complexity. (This is also why we have barely touched on LU6.2.)

The second major weakness of SNA is its requirement for large processors for efficient implementation. This is being alleviated as products suitable for microcomputer applications come out. These products may still use more than their share of resources, though; for example, LU6.2 implementations on a microcomputer often require more than half the available memory [KORZ89].

Of all the networking architectures that we have studied, SNA is the least compatible with the OSI Reference Model. There have been significant efforts to make SNA more compatible with the OSI model, but its basic orientation and structure are so different that achieving compatibility will be difficult.

Despite their limitations, properly implemented SNA installations work, and work well. SNA should continue to be the most widely implemented vendor networking architecture (along with DNA) for the foreseeable future.

11.6 Transport Layer in IEEE 802 Networks and MAP/TOP

The IEEE 802 architecture does not contain a layer equivalent to the transport layer. A large percentage of the LANs that have been implemented use TCP as a transport layer protocol if this layer is needed. The newer OSI transport layer protocols are also being adopted for a number of LANs.

Since MAP and TOP specify full networking architectures, they specify the transport layer to be used. In both cases, the transport layer specified is TP4, the Class 4 OSI transport layer protocol. Since MAP and TOP both use OSI connectionless network layer protocols, they were forced to select the Class 4 service to convert connectionless network layer protocols to connection-oriented (the equivalent of virtual circuit) service at the transport layer.

11.7 Summary

The transport layer interfaces the three lower layers of the OSI Reference Model architecture, which are concerned primarily with data communication, with the three upper layers, which are primarily concerned with ensuring that data are in usable form, and is hence a keystone of the architecture. TCP, TP4, and the DNA end communications layer all assume a datagram or similar network layer, while SNA assumes a virtual circuit network layer. (Other variants of OSI transport layers assume intermediate types of network layers or do not put stringent requirements on the network.) This results in major shifts between relative complexity of the network and transport layers.

The transport layer is the highest layer we treat in this text, since it is the highest layer primarily concerned with communications. The next chapter discusses recent developments in the ISDN architecture.

Problems

11.1 Assuming that the ACK and the OPEN from **B** in the three-way handshake described in Subsection 11.1.2 are transmitted in separate packets, consider the following situation. The packet initiating **A**'s first attempt to establish a connection is delayed and retransmitted, with the retransmission leading to a successful session. Some time later, the delayed packet arrives at **B**, which considers it to be a request to open a new connection and treats it accordingly. Show that this quickly leads to a deadlock condition, assuming that each end discards acknowledgments for old packets and ignores new requests to open a connection when it believes one has already been established.

11.2 Can a version of the three-way handshake solve the three-army problem discussed in Chapter 1? Briefly discuss the relationship between the three-army problem and that solved by the three-way handshake.

11.3 Verify that the three-way handshake, with the extra rules specified by Sunshine and Dalal, will satisfactorily handle each of the following situations. In each case, sketch the sequence of packets that would be exchanged.
  a. A delayed OPEN packet arrives when no connection has been established.
  b. A delayed OPEN packet arrives when a connection has already been established.
  c. A node crashes during a connection, and then comes back up and tries to reestablish the connection. It has forgotten the connection ID#s, though, so uses the wrong ones. (In this case, the "half-open" connection—that is open at the non-crashed end but not at the other—needs to be closed before the connection can be reestablished.)
  d. A node crashes but does not try to reestablish the connection when it comes back up. The other node then tries to send data on what it thinks is an established connection. (The recovery process should again involve closing a "half-open" connection.)

11.4 Assume that the round trip times (τ_n) used in the weighted average in Eq. (11.1) all have the same mean ($\bar{\tau}$) and that a finite W_0 is used for the first calculation (to get W_1). Show that as $n \to \infty$, the expected value of $W_n \to \bar{\tau}$, for any α in the range $0 < \alpha < 1$.
Hint: You may find the following summation to be helpful:

$$\sum_{n=0}^{N} \alpha^n = \frac{1 - \alpha^{N+1}}{1 - \alpha}.$$

11.5 The manner in which the weighted average in Eq. (11.1) adapts to variations in measured round trip times is best visualized by working out an example. Consider the successive round trip time measurements in Fig. P11.5 (a rough approximation to the type of behavior in Fig. 11.1). The first ten values of τ_n sketched are 3, the next five are 8, and the following 10 are back at 3.

Figure P11.5   Figure for Problem 11.5.

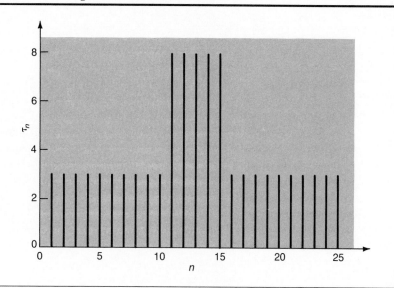

a.   Compute and sketch the values of W_n, $n = 0, 1, \ldots, 25$, assuming $W_0 = 4$ and $\alpha = 0.9$.

b.   If a retransmission timeout of the form βW_n is used, with $\beta = 2$, how many packets would be retransmitted for the situation analyzed?

11.6   Verify that TCP connection establishment is able to handle each of the situations listed in Problem 11.3, with SYN substituted for OPEN, sequence numbers substituted for ID#s, and any other changes necessary to adapt the problem to TCP terminology.

11.7   Briefly describe the types of bit error patterns that are not detected by the TCP checksum.

11.8   Figure P11.8 illustrates a requirement for occasional resynchronization of sequence numbers in TCP. As is discussed in Subsection 11.2.4, the normal method for select-ing initial sequence numbers (ISNs) is to use a clock to generate them. Assuming that the clock generates new ISNs at the rate B per second, the heavy line in the figure indicates the ISNs generated by the clock. (We have approximated a step function by a continuous curve to simplify analysis.) The ISNs are modulo ISN_{max}, so the curve drops back down to zero at time $T_{ISN\ clock}$, the period of the clock. To avoid problems with sequence numbers for old packets overlapping the sequence numbers, no sequence numbers that were generated by this clock during a period t_{max} preceding the present time (the maximum time packets remain in the network) are allowed in the system, leading to the forbidden region in the figure. Under most circumstances, the sequence numbers used during a connection will increase at a rate $b < B$, leading to the more slowly increasing line in the figure. Note that this curve eventually hits the boundary of the region of the forbidden region from the right. At this point sequence numbers must be resynchronized. Compute t_0, the time before resynchron-

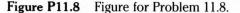

Figure P11.8   Figure for Problem 11.8.

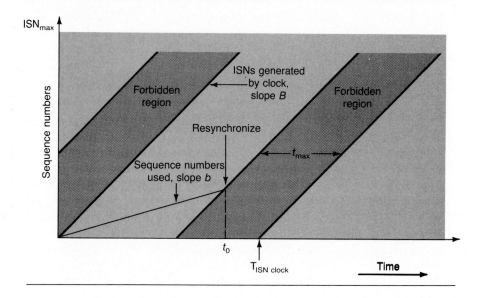

ization is needed, in terms of the illustrated parameters. Under what circumstances does $t_0 \to \infty$, so resynchronization is never needed?

11.9 Assuming that a DoD data packet contains only overhead from the ARPANET DLC, the standard data packet CSNP-CSNP header, a minimal length DoD IP header, and a minimal length TCP header, plot fractional overhead (as defined in Section 7.8) for packets, for the number of data characters (on a log scale) ranging from 1 to 10,000 bytes. Ignore possible problems with maximal packet lengths allowed. Your plot should resemble Figs. 7.22 and 7.23.

11.10 Does the error recovery procedure in TCP use go-back-N retransmission, selective reject transmission, or some other variant of a retransmission strategy? What type of flexibility on choosing a retransmission strategy is given to the implementer?

11.11 What is the maximum possible window size for TCP? Assuming that the round-trip time of a network is at least 2 seconds, what transmission rate is needed in order to transmit a full maximum size window before it would be possible to receive an acknowledgment to the first bytes in the window?

11.12 Assume that the initial credit field in the first transmission in each direction in Fig. 11.8 is increased to 512. Each credit field is reduced to 384 on the second transmission and then 256 on all succeeding transmissions, with each port transmitting as soon as it is allowed to. Sketch the resulting modification of Fig. 11.8.

11.13 Verify that TP4 connection establishment is able to handle each of the situations listed in Problem 11.3, with CR or CC substituted for OPEN, reference numbers substituted for ID#s, and any other changes necessary to adapt the problem to OSI terminology.

11.14 Assume an ISO data packet contains only overhead from HDLC (with no extended fields), the standard X.25 data packet header, an ISO IP header (with no optional or segmentation fields and six byte addresses), and a Class 4 DT TPDU header (with 7-bit TPDU-NR). Plot fractional overhead (as defined in Section 7.8) for ISO IP, for the number of data characters (on a log scale) ranging from 1 to 10,000 bytes. Ignore possible problems with maximal packet lengths allowed in some networks. Your plots should resemble Figs. 7.22 and 7.23.

11.15 Sketch an example of TP4 TPDU transmission in which the credit **A** gives to **B** is reduced below its previous value (that is, fewer packets, including any already transmitted, will be accepted after the second credit than before it). Specifically, assume that **A** has issued credit of 6, beginning with TPDU Number 102, to **B**, and then issues an AK pointing to TPDU Number 102 and with credit of 3. The two credit values arrive out of order, however. Indicate how the protocol handles possible ambiguities in this situation.

11.16 Briefly discuss each of the OSI class 4 timer values described by Eqs. (11.2)–(11.4), and indicate why the result of applying each equation is a realistic estimate of the indicated quantity.

11.17 What is the maximum possible window size in TP4? Why is it difficult to compare this with the maximum possible window size with TCP?

11.18 (**Design Problem**). Develop the basic design for a gateway to convert back and forth between TCP and TP4. Include in your design the mappings of various packet formats into each other, routines for connection establishment, data transfer and disconnection, flow control, and so forth. Indicate where you run into special problems, including problems you cannot solve fully.

11.19 Verify that DNA end communications layer connection establishment is able to handle each of the situations listed in Problem 11.3, with Connect Initiate or Connect Confirm substituted for OPEN, logical link numbers substituted for ID#s, and any other changes necessary to adapt the problem to DNA terminology.

11.20 Assuming that a DNA data packet contains only overhead from the DDCMP DLC, the standard routing layer header, and a standard end communications layer header (with an acknowledgment number field but no acknowledgment other field), plot fractional overhead (as defined in Section 7.8) for DNA connections, for the number of data characters (on a log scale) ranging from 1 to 10,000 bytes. Your plot should resemble Figs. 7.22 and 7.23.

11.21 Draw a diagram, similar to Fig. 11.15 and indicating transfer of the same number of user data packets, for the data transfer phase of the DNA end communications layer.

11.22 Sketch the normal termination process for the DNA end communications layer in terms of the primitives and data units given in Subsection 11.4.2. Your diagram should resemble Fig. 11.9.

11.23 Assuming that a SNA data packet contains only overhead from SDLC, FID4 path control headers, and a transmission control header, plot fractional overhead (as defined in Section 7.8) for SNA connections, for the number of data characters (on a log scale) ranging from 1 to 10,000 bytes. Your plots should resemble Figs. 7.22 and 7.23.

11.24 In contrast with other architecture layers, where the SNA protocols, headers, and so forth seem to be more complex than essentially any of the corresponding elements for other architectures, the transmission control header for SNA is by far the simplest of any transport layer headers we have discussed. Why is SNA able to get away with a relatively simple transport layer? Briefly discuss the major differences in architectures that make this possible.

11.25   Draw diagrams similar to Figs. 11.21 and 11.22, but showing SNA transmission control connection establishment for a case where **LU A** and **LU B** are both in the same domain. Briefly describe each of the primitives and packets illustrated and indicate its functions.

11.26   Transmission control pacing in SNA allows for the following variants:

- From the Primary to the Secondary in one stage.
- From the Primary to the Secondary in two stages:
    - From the Primary Logical Unit to the Boundary Function.
    - From the Boundary Function to the Secondary Logical Unit.
- From the Secondary to the Primary in one stage.
- From the Secondary to the Primary in two stages:
    - From the Secondary Logical Unit to the Boundary Function.
    - From the Boundary Function to the Primary Logical Unit.

Pacing can be used for both directions independently, with independent choices of window sizes. Different window sizes can also be used for the two stages in the second and fourth variant.

a.   What are the advantages to considering these different variants?

b.   What is the maximum possible window size between a terminal node and a boundary function node?

Hint: Pacing windows must use sequence counts available in path control headers of the appropriate FID format.

12 Integrated Services Digital Networks

12.1 Introduction

Integrated Services Digital Networks are expected to unify many of the approaches to telecommunications networking we have been discussing. *ISDN* is viewed, by telephone network management at least, as the goal toward which their efforts to digitize their networks is leading. ISDN standards committees are developing international standards to ensure compatibility of ISDN networks around the globe and eventually lead to what can be viewed as a single world-wide ISDN network.

A wide variety of services are expected to be provided by ISDNs. Table 12.1 lists some services currently planned, along with their data rates and the types of ISDN channel types and facilities expected to be used to provide them. Standard user interfaces are being developed for access to all of these services; users will also be provided more control over their access to communications facilities than they are currently given. Numerous special features that are expected to be available are not listed in the table.

Development of ISDN standards and networks is a continuation of the *digitization of telephone networks,* which is proceeding at a rapid pace. An example is digitization of Southern Bell's facilities. Southern Bell's progress towards ISDN is described as follows [SNEL87]. "A move to stored program switching is happening; electromechanical switch replacement will be complete in Southern Bell by the end of 1989. Fiber must be placed in the trunk and feeder portions of the loop—these routes are single mode (fibers) now in Southern Bell. Fiber in the distribution portion of the loop—this will be economically feasible in certain areas based on POTS[1] only by 1988. Fourth generation switches will be deployed beginning 1989–1990 in the third generation switch's footprint."

POTS

Figure 12.1 is a nomogram, developed by Dr. Gordon Bell of the National Science Foundation. The nomogram shows the relationship among transmission time, transmission data rate, and data set size. It can be used to estimate the bandwidths needed for various services such as those in Table 12.1. To

---

[1] POTS is an acronym for "plain old telephone service."

Table 12.1 ISDN Service Requirements for Home and Business.

Requirements for Home

Service	Data Rate Requirement	Channel Type		Facilities			
		B	D	Circuit Switched	Packet Switched	Channel Switched	Overlay
Telephone	8,16,32,64 kbps	X		X			

Alarms
  Smoke
  Fire
  Police
  Medical
	10–100 bps		X				

Utility metering	0.1–1.0 kbps		X	X	X		
Energy management	0.1–1.0 kbps		X	X	X		

Interactive information services
  Electronic banking
  Electronic yellow pages
  Opinion polling
	4.8–64 kbps	X			X		

Electronic mail	4.8–64 kbps	X			X		
High quality audio	~ 300–700 kbps					X	X
Compressed video	~ 30 Mbps					X	X
Broadcast video	~ 100 Mbps					X	X
Switched video	~ 100 Mbps					X	X
Interactive video	~ 100 Mbps			X		X	X

Requirements for Business

Service	Data Rate Requirement	Channel Type		Facilities			
		B	D	Circuit Switched	Packet Switched	Channel Switched	Overlay
Telephone	8,16,32,64 kbps	X		X			
Interactive data communications	4.8–64 kbps	X	X		X		
Electronic mail	4.8–64 kbps	X			X		
Bulk data transfer	4.8–64 kbps	X		X			
Facsimile graphics	4.8–64 kbps	X		X			
Slow scan/freeze frame TV	56–64 kbps	X		X			
Compressed video conference	~ 1.5 Mbps					X	X

(Adapted from [BHUS84], © 1984 IEEE. Reprinted by permission.). B and D channels are defined in earlier chapters. Channel switched and overlay facilities to be provided in earlier stages of ISDN evolution, while standard networks are not yet capable of handling services, if these facilities are checked.

Figure 12.1   Data Transmission Nomogram (from [ROBE87], attributed
to Dr. Gordon Bell. © 1987 IEEE. Reprinted by permission.).

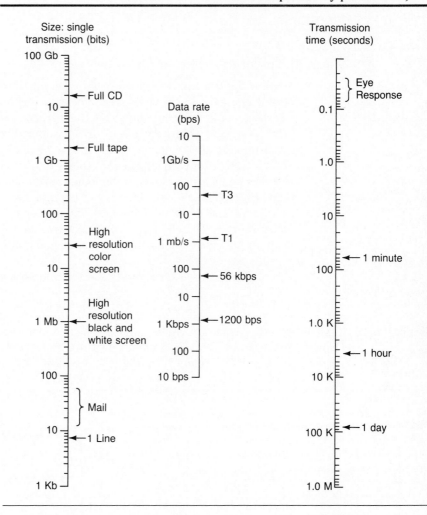

read the nomogram, place a straight edge between data set size at the left
and desired transmission time at the right. The required transmission data
rate is indicated in the middle column. Thus transmission of a high resolution
black and white workstation screen in 0.1 second will require approximately
10 Mbps data transmission rate.

　　We begin our discussion with an overview of current implementations of
digital networks in telephony. This is followed by discussion of the ISDN
architecture (with reference to portions in earlier chapters). A brief discussion of common channel signalling systems is given next. Strategies for migra-

tion toward ISDN are then reviewed. The chapter, and the text, conclude with a presentation of possible future extensions of ISDN.

12.2   Evolution of Telephone Networks

integrated digital networks

During the last two decades, worldwide architectures and standards for digital telephone network transmission of voice and data have been established. Such networks are commonly called *integrated digital networks* (IDNs) with the term integrated referring to use of standardized digital techniques for transmission and switching systems. The telephony IDN is based on *64 kbps PCM* coding of speech signals. This has led to an ISDN architecture in which 64 kbps channels are the basic building blocks. Thus voice transmission techniques in telephone networks, rather than the basic requirements of the new services to be provided, have been the primary factor deciding basic transmission rates for ISDN. Basic components of the IDN, as used for voice applications, include PCM multiplexing and digital multiplexing equipment. These are illustrated in Fig. 12.2.

A *PCM multiplexer/demultiplexer*[2] samples voice band signals, encodes

Figure 12.2   Typical Digital Transmission System, Illustrating PCM and Digital Multiplexer/ Demultiplexers (from [DECI87], Kamilo Feher, *Advanced Digital Communications: Systems and Signal Processing Techniques,* © 1987, pp. 123, 124. Reprinted by permission of Prentice-Hall, Inc., Engelwood Cliffs, NJ.).

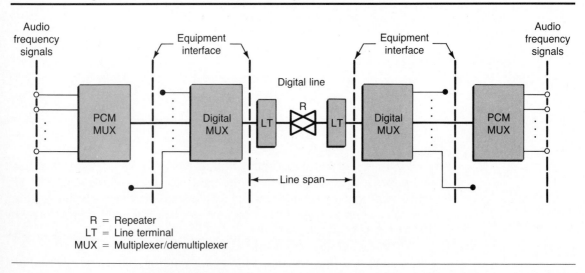

R = Repeater
LT = Line terminal
MUX = Multiplexer/demultiplexer

---

[2]In the United States, terminology for a PCM multiplexer/demultiplexer is "digital channel bank."

them into PCM signals, and multiplexes 24 or 30 of them together (in formats illustrated in Fig. 4.11) to form the first level in the digital multiplexing hierarchy; it also performs the reverse transformations at the receiving end. A *digital multiplexer/demultiplexer* then converts from this format to one of the other, higher speed formats in the digital multiplexing hierarchy used. PCM multiplexers use *byte multiplexing* (a byte of data in each slot, as shown in Fig. 4.11), but *bit multiplexing* is more common at higher levels in multiplexing hierarchies.

The digital hierarchies differ for different areas of the world, considerably complicating ISDN efforts to develop worldwide standards. The major levels of the digital hierarchies currently implemented in different areas are listed in Table 12.2. Interconnection of high speed optical facilities (cross connects) is almost always at Level 3 (44.736 Mbps) in North America, Level 4 (139.264 Mbps) in Europe, and Level 5 (397.20 Mbps) in Japan.[3]

Figure 12.3 illustrates a typical digital transmission chain to provide purely digital connectivity between two PCM switching offices.

Analog switching offices and transmission facilities are still common, so the digital switching and transmission facilities illustrated in Figs. 12.2 and 12.3 comprise only part of the telephone facilities currently available. Current generation stored program control switching offices employ the type of PCM switching indicated in Fig. 12.3, switching signals in digital form.[4] Analog

Table 12.2 Incompatible Digital Multiplexing Hierarchies.

		North American	European	Japanese
Level 1	Bit rate (Mbps)	1.544	2.048	1.544
	Voice circuits	24	30	24
Level 2	Bit rate (Mbps)	6.312	8.448	6.312
	Voice circuits	96	120	96
Level 3	Bit rate (Mbps)	44.736	34.368	32.064
	Voice circuits	672	480	480
Level 4	Bit rate (Mbps)	274.176	139.264	97.728
	Voice circuits	4032	1920	1440
Level 5	Bit rate (Mbps)	Not defined	565.148	397.200
	Voice circuits		7680	5760

(From [JACO86], © 1986 IEEE. Reprinted by permission.).

---

[3] In addition to these discrepancies in basic multiplexing rates, there are discrepancies in the digital encoding used (briefly summarized in [JACO86]). The bit rates for higher order multiplexers are not integer multiples of those for lower order ones. The extra bits allow for incompatibilities in clock rates for lower order multiplex signals feeding a higher order multiplexer; "stuffed" bits are added to fill out frames, along with bits identifying which bits are stuffed and which represent real data. This complicates finding individual data frames within the high rate data stream, however.

[4] The earliest stored program control switching offices, or electronic switching systems, were sophisticated electromechanical switches employing computer control.

Figure 12.3   Typical Digital Transmission Chain (adapted from [DECI87], Kamilo Feher, *Advanced Digital Communications: Systems and Signal Processing Techniques,* © 1987, pp. 123, 124. Reprinted by permission of Prentice-Hall, Inc., Englewood Cliffs, NJ.). Digital transmission rates at levels indicated vary according to geographical area, as indicated in Table 12.2.

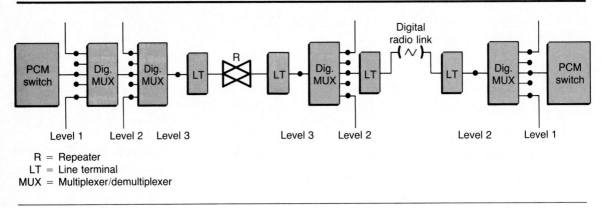

    R = Repeater
    LT = Line terminal
MUX = Multiplexer/demultiplexer

technologies are also advancing and, because of recent advances, are still more cost efficient than digital technologies in a few areas [KOST84].

Until recently the majority of the circuit kilometers in telephone system networks in the United States were analog microwave circuits, but the more than three million kilometers of fiber optic links recently installed have more than doubled the capacity of the networks, even though they comprise only a small percentage of the mileage of copper circuits [SHUM89]; this implies the majority of circuit kilometers are now optical fiber circuits. Digital microwave facilities are also being installed rapidly. Almost *complete digitization* of telephone networks is expected within 10 years.

common channel signaling

Another essential component of integrated digital networks is *common channel signaling* (CCS).[5] This gives complete separation of signaling information from user data by transmitting signaling over a separate network. It is an example of out-of-band signaling (sending signaling over separate facilities) in contrast to the in-band signaling (signaling sent over the same channels as voice) used previously. This greatly increases the flexibility of signaling, makes it possible to offer a wide variety of additional functions, and has helped to significantly reduce average circuit connection times below those achieved without common channel signaling.

STPs

Figure 12.4 is a sketch of part of a typical network employing CCS. Individual *signal transfer points* (*STPs*) have been illustrated to simplify the diagram, but all STPs occur in duplicated pairs, each pair fully connected with

---
[5] Also called common channel interoffice signaling, especially in the U.S. Bell System.

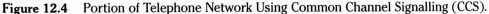

Figure 12.4 Portion of Telephone Network Using Common Channel Signalling (CCS).

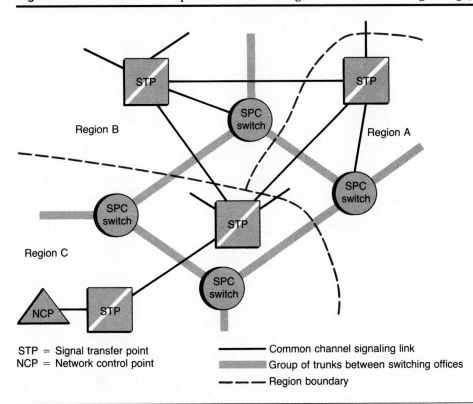

STP = Signal transfer point
NCP = Network control point

——————— Common channel signaling link

████████ Group of trunks between switching offices

– – – – – Region boundary

NCPs

another pair to form a "quad," to achieve high reliability.[6] A typical quad, one pair serving a pair of *Network Control Points* (*NCPs*) as well as a stored program controlled (SPC), or digital, switch,[7] is illustrated in Fig. 12.5. SPC switches are also highly reliable, with reliability design goals of no more than two hours down time in 40 years of operation.

Although the type of network configuration illustrated in Figs. 12.4 and 12.5 was designed to ensure extremely high reliability, a flaw in its implementation led to the AT&T network outage on January 15, 1990 (see Subsection 10.4.5). The problems on this date began after an SPC switch in New York (which we will call switch **A**) took itself out of service during normal recovery

---

[6] This configuration is used in the United States and some other countries with large telephone networks. More ad hoc configurations are used in countries having fewer telephones.

[7] SPC switches use special purpose computers to control switching and include some of the largest and most sophisticated computers ever built. They are much more powerful than STP computers, so it makes sense to have them in charge of the network with the STPs used only to route messages to appropriate SPCs.

Figure 12.5   Typical Quad of STPs Used to Enhance Reliability
of CCS Network.

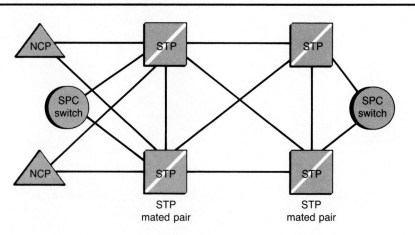

from a minor problem, announcing this to adjacent SPC switches so they could take over its call processing load during the four to six second period normally required for recovery. The adjacent switch processors made notations in their programs to indicate that switch **A** was temporarily out of service.

Upon completion of fault recovery, switch **A** announced its recovery to adjacent switches by resuming call processing, sending call setup messages to these switches; each then noted recovery of **A** by an appropriate update in its program. The software flaw leading to network failure resulted in these switches being vulnerable to certain types of disruption for a few seconds during the period when such updates were being made. The fatal type of disruption was provided when switch **A** sent two very closely spaced call setup messages (within an interval of .01 sec) to an adjacent switch, which we will call switch **B**.[8] This caused some data at **B** to be damaged so the processor at **B** tried to execute an instruction that didn't make sense. The SPC software then told **B** that its processor was insane, so switch **B** took itself out of service.

Unfortunately, when switch **B** returned to service, it repeated switch **A**'s actions, disabling additional SPC switches for brief periods of time. A chain reaction was initiated, eventually disabling switches throughout the AT&T network until AT&T engineers managed to temporarily suspend signalling traffic on backup links so the processors could recover. On the following day,

---

[8]Although we discuss problems experienced by one adjacent switch, all switches adjacent to **A** were susceptible to the same type of problems.

the faulty program was replaced by an earlier version without this bug; a new version of the update causing the problem, with bug removed, was installed a few days afterwards. Although this particular problem was eliminated in this manner, there is currently no way to guarantee that similar problems will never recur.

The CCS network interconnects STPs with each STP serving a particular region.[9] Each STP is a packet data switch that provides concentration of traffic and directs messages to appropriate SPC switches. NCPs contain databases shared by multiple SPC switches and handle such functions as use of "calling cards." When a calling card is used to charge a call, processing is interrupted while a message is sent over the common channel network to verify the card number; this takes only a fraction of a second from the time the number has been entered, allowing call processing to resume without undue time constraints. CCS also allows time-of-day routing[10] and other special features for individual subscribers.

common channel

signaling network

The *common channel signaling network* is a packet-switching network employing datagram transmission to transfer signaling messages among the STPs, NCPs, and SPC switches. Details of the version adopted for ISDN are covered in Section 12.5.

Most current digital networks are asynchronous, with no common reference frequency or clock that forces all transmissions to remain in phase. However, *synchronous networks* with common clocks (using atomic clocks with extremely high stability) are emerging. This will greatly simplify multiplexing digital bit streams from several sources.

12.3 ISDN Overview

Although IDNs are regularly used to transmit data as well as voice signals, they represent the current culmination of networks originally designed to transmit voice. However, ISDNs are being designed with their capability to transmit data emphasized at least as much as their capability to transmit voice.

integrated access

A key element of ISDN will be *integrated access* to all communications services needed by the customer. Rather than separate connections for voice, personal computers or workstations, telemetry data, and so forth, a single

---

[9] There are currently 10 such regions in the United States.

[10] Time-of-day routing can reduce congestion by taking advantage of time zone variations. For example, U.S. traffic between two East Coast locations may be routed via the West Coast during much of the morning since the three-hour time difference ensures minimal West Coast traffic. Similar routing for toll free calls (800-line calls in the United States) can be implemented so that the location where a toll free call is received depends on time of day; handling of 800-line calls in the United States normally involves use of NCP databases, so a variety of special features can be made available.

connection should serve all these needs. ISDN focuses on information move-
ment as well as management. It is designed to give users much greater flexibility and more control over their communications facilities, along with an
ability to tailor communications applications to specific needs and to change
applications rapidly.

distributed processing

out-of-band signaling

Two key aspects of ISDN are *distributed processing* and *out-of-band signaling.* SPC switches, STPs, and NCPs in IDNs already form distributed processing networks, including shared data bases and software-based call
processing routines. Separate signaling channels are also to be made available to the ISDN subscriber (as separate TDM channels); this is out-of-band
signaling since signaling information flows are independent of flows on data
channels. It gives individual subscribers access to shared network interfaces
with signaling and communications links to supplement the signaling and
communications links within telephone networks and helps make a distributed communications architecture possible.

Fundamental roles of ISDN are to provide fully digital communications
facilities for sophisticated voice and data applications and to support and
advance distributed processing capabilities. Eventually, support for such
facilities and capabilities will be an integral part of telephone networks, but
during the period before this is accomplished it is impractical to replace current facilities, valued at hundreds of billions of dollars. From the point of view
of the user, however, it is immaterial how facilities and capabilities are provided, as long as they have adequate availability and reasonable cost. Hence,
initial plans for providing ISDN are to develop independent networks overlay-
ing basic telephone networks.

Figure 12.6 presents a conceptual view of ISDN. ISDN developers are
focusing on defining and standardizing the ISDN interface so network providers and end users can implement and evolve subnetworks independently but
still ensure compatibility. A single interface[11] will access any facilities indicated, including any network listed as well as a wide variety of communications speeds and channel types. The channels to be provided are still
evolving, but the primary ones currently planned are those shown in Table
12.3.

SONET

Higher rates have recently been adopted for internal network use in the
SONET (Synchronous Optical Network) hierarchy [BALL89]. These rates have
not been standardized for the user-network interface, but some will almost
certainly be made available to users in order to provide high rate services,
such as the video-based services in Table 12.1. SONET rates are given in Table
12.4.

---

[11] Calling this a single interface is somewhat exaggerated since two distinct interfaces, basic
rate and primary rate, have already been defined and new broadband interfaces will be
needed soon. A few additional interfaces, distinguished largely by network access point, are
defined below. This is a great reduction in interface types from the current situation,
though.

Figure 12.6 Conceptual View of the ISDN (from [ROCA86], R. Roca, "ISDN Architecture," *AT&T Technical Journal,* vol. 65, Issue 1 January/February 1986, pp. 4–17. Reprinted with permission © 1978 AT&T.).

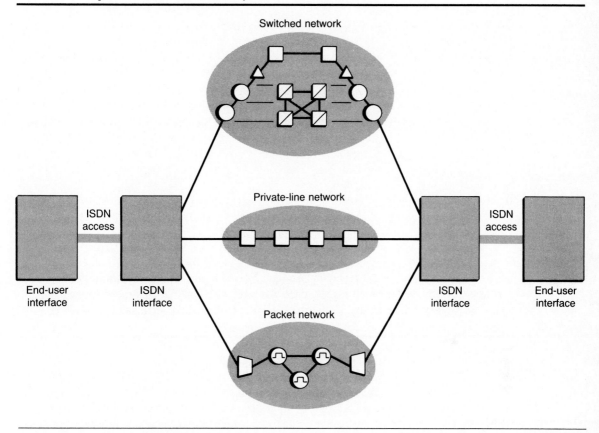

Table 12.3 ISDN Channel Types.

B channel	64 kbps, for user data.
D channel	16 or 64 kbps, primarily for signaling, but also used for telemetry or packet-switched data.
E channel	64 kbps variant of D channel, signaling for circuit switching. Only used with multiple access configurations.[a]
H channels	H0: 384 kbps H1: H11—1536 kbps for 1544 kbps primary rate H12—1920 kbps for 2048 kbps primary rate Used for a variety of user information streams, but not for signaling.

[a]This should not be confused with the E-echo channel of the basic rate interface, despite the fact that both are called E channels.

Table 12.4   Optical Channel N (OC-N) Rates for SONET Transmission.

OC-1	51.84 Mbps
OC-3	155.52 Mbps
OC-9	466.56 Mbps
OC-12	622.08 Mbps
OC-18	933.12 Mbps
OC-24	1244.16 Mbps
OC-36	1866.24 Mbps
OC-48	2488.32 Mbps

Figure 12.7   ISDN Data and Signaling Interfaces Between Network and Customer Equipment (adapted from [ROCA86], R. Roca, "ISDN Architecture," *AT&T Technical Journal,* vol. 65, Issue 1 January/February 1986, pp. 4–17. Reprinted with permission © 1978 AT&T.).

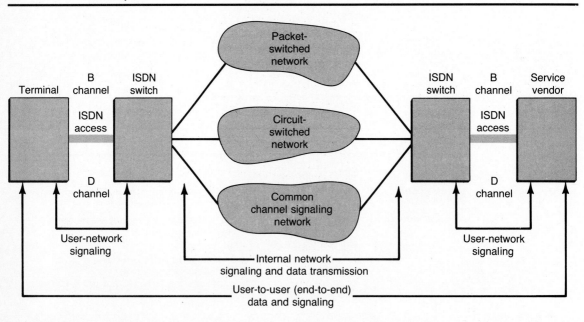

Figure 12.8   ISDN Reference Points and Functional Groupings.

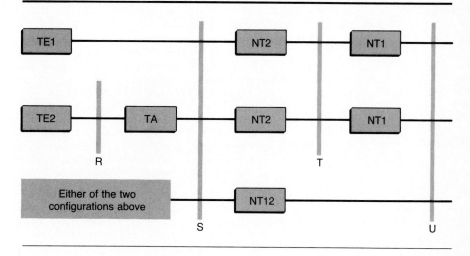

As we have already mentioned, the user will be given more control over the network by being given access to signaling functions. The types of data and signaling to be made available to the user and to be used by the network are indicated in Fig. 12.7.

As the figure indicates, signaling may be transmitted between the user and the network (at either end of a connection) and both signaling and data may be transmitted end-to-end between users. User signaling will be transmitted via the D or E channels, with data usually transmitted via B (or H) channels. The common channel signaling facilities in the figure are available only to the network provider.

ISDN reference points, with associated functional groupings, are indicated in Fig. 12.8. *Reference points* are conceptual points used to separate groups of functions. Functional groupings refer to particular arrangements of physical equipment or equipment combinations. Definitions of functional groupings are given in Table 12.5 and the reference points in Fig. 12.8 are defined in Table 12.6.

12.4   ISDN Reference Model

User, control, and management planes

The ISDN Reference Model is basically the same as the OSI Reference Model, except for defining separate *user, control, and management Information Planes* (or Functional groupings). This is illustrated in Fig. 12.9. The func-

Table 12.5   Definitions of ISDN Functional Groupings.

TE   *Terminal equipment.*

TE1   *Terminal Equipment Type 1* (ISDN Terminal). Such terminal equipment complies with ISDN standards and can be accessed at interface S.

TE2   *Terminal Equipment Type 2* (non-ISDN Terminal). Such terminal equipment supports functions similar to those supported by TE1 equipment but does not comply with ISDN interface standards. It must be accessed at interface R, which in turn is interfaced with S via a TA.

TA   *Terminal Adaptor.* This converts the non-ISDN interface functions at interface R into ISDN acceptable form.

NT1   *Network Termination 1* includes functions belonging to OSI layer 1. In most countries, NT1 equipment is expected to be controlled by the ISDN provider and form a boundary to the ISDN network.[a] In the United States, however, FCC regulations state that provision of NT1 equipment must be competitive, with multiple providers allowed.

NT2   *Network Termination 2* is an intelligent device that may include up through OSI Layer 3 functionality. Possible types of NT2 equipment include a digital PBX, a terminal controller, or a LAN.

NT12   *Network Termination 1,2* is a single device that contains the combined functions of the NT1 and NT2.

(From J. Ronayne, *The Integrated Services Digital Network: From Concept to Application,*
    © 1988 Wiley. Reprinted by permission of John Wiley & Sons, Inc., New York, NY).
[a]The CCITT ISDN standards, in Recommendation I.411, state that access at the U reference
    point is "not envisaged at this time," implying NT1 will be available only from the ISDN
    provider.

Table 12.6   Definition of the Reference Points in Fig. 12.8.

Reference Point R	(R for Rate) Provides a non-ISDN interface between non-ISDN compatible user equipment and adaptor equipment.
Reference Point S	(S for System) Separates user terminal equipment from the network user functions at the interface to NT2.
Reference Point T	(T for Terminal) Separates NT1 equipment from NT2 equipment. Where the ISDN provider provides NT1 equipment, this separates the network provider's equipment from user equipment, but this distinction breaks down in situations (as in the United States) where provision of NT1 equipment is competitive.
Reference Point U	(U for User) Provides an interface to the actual communications line. Although this interface is not recognized in official ISDN documentation, FCC regulations require it to be available in the United States.

© 1987 IEEE. Reprinted by permission.

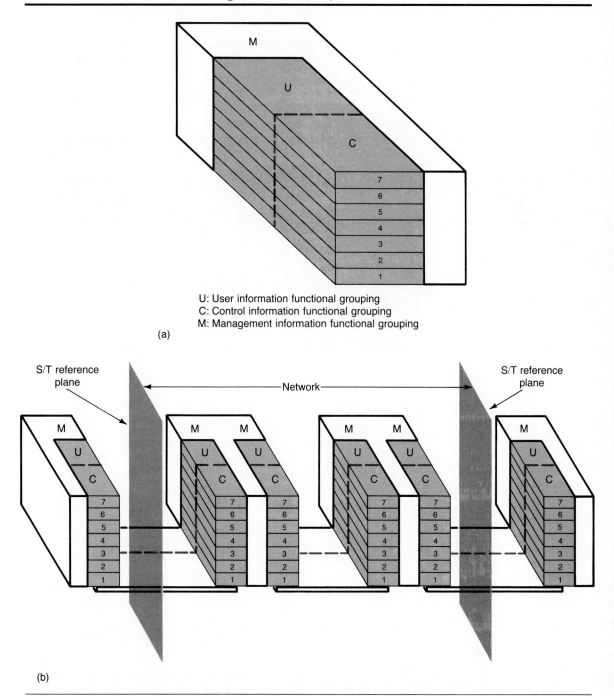

U: User information functional grouping
C: Control information functional grouping
M: Management information functional grouping

(a)

(b)

Figure 12.10   Assignment of Major Functions to Layers in ISDN Architecture (adapted from [BLAC87a]).

Layer		Functions					
7	Application	Application-related functions					
6	Presentation	Encryption/ decryption	Compression/ expansion				
5	Session	Session connection establishment	Session connection release	Session to transport connection mapping	Session connection synchronization	Session management	
4	Transport	Transport connection multiplexing	Transport connection establishment	Transport connection release	Error detection/ recovery	Flow control	Segmenting blocking
3	Network	Routing relaying	Network connection establishment	Network connection release	Network connection multiplexing	Congestion control	Addressing
2	Data link	Data link connection establishment	Data link connection release	Flow control	Error control	Sequence control	Framing synchronization
1	Physical	Physical layer connection activation	Physical layer connection deactivation	Bit transmission	Channel structure multiplex		

High layer functions (or teleservices) — Layers 7–4

Low layer functions (or bearer services) — Layers 3–1

tional groupings interact with each other to perform ISDN functions. We discuss user and control interactions below, but interactions with the management information functional grouping are complex and specialized and will not be discussed. As the figure indicates, both user and control information functional groupings are divided into the standard seven layers in the OSI architecture.[12] Functions in layers are indicated in Fig. 12.10, which is identical to Fig. 3.25.

Since a variety of types of communication, including circuit-switched communication via B channels, packet-switched communication via B channels, and packet-switched communication via D channels, are included in the *ISDN Reference Model* ISDN architecture, the *ISDN Reference Model* needs to account for each of these possibilities. Interactions between user and control functional group-

---

[12] The figure shows seven layers each in the user information and the control information functional groupings, but the ISDN standard control protocol, Common Channel Signaling System Number 7 (CCSS No. 7), currently contains four layers, the lower three corresponding reasonably well to Layers 1–3 of the ISO model and the fourth layer corresponding roughly to ISO Layers 5–7. CCSS No. 7 does not have a layer corresponding to Layer 4, the transport layer, of the ISO Reference Model. See Section 12.5 for details.

Figure 12.11   Circuit-Switched Connection in ISDN (from [DUC85], © 1985 IEEE. Reprinted by permission.).*

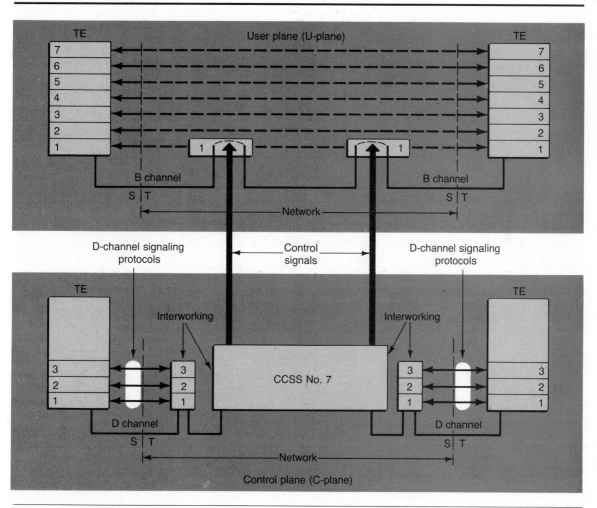

*Some authors, for example, [DECI88], consider all layers of the control plane to be embedded within layer 1 of the user plane at user access points for this configuration.

ings or planes can be visualized via representations such as those in Figs. 12.11–12.13, which illustrate one configuration each for circuit-switched communication via B channels, packet-switched communication via B channels, and packet-switched communication via D channels. The ISDN Reference Model includes protocols for each, plus others.

The user plane portions of these figures should be compared with Fig. 9.2 for circuit-switched communication or Fig. 9.1 for packet-switched commu-

Figure 12.12   ISDN Packet-Switched Communication via a B Channel Over an ISDN–Packet-Switched Network Interconnection (from [DUC85], © 1985 IEEE. Reprinted by permission.).*

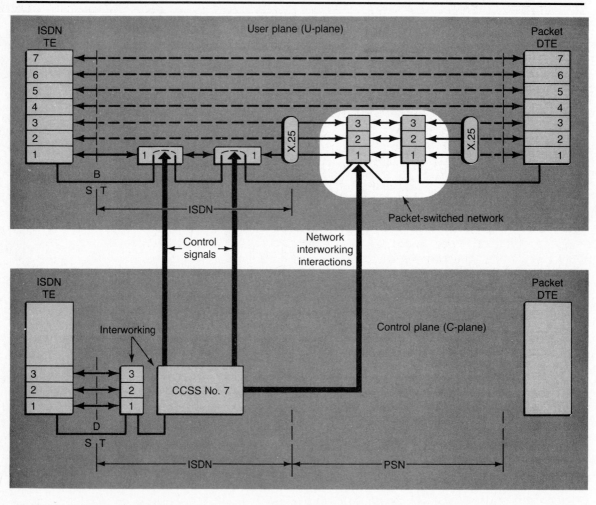

*[DECI88] considers the control plane layers to be embedded in user plane layer 3 in this situation.

nication; the earlier figures show the same structures for layers traversed at intermediate nodes. The control plane portions of the figures here, however, give more details about how connections are set up and taken down.

Descriptions of ISDN protocols at various layers have been given in earlier chapters. The major exception is protocols used by CCSS No. 7; these protocols are discussed in the next section.

Figure 12.13   ISDN packet-switched communication via D channel, with information routed via packet-switched network (from [DUC85], © 1985 IEEE. Reprinted by permission.).

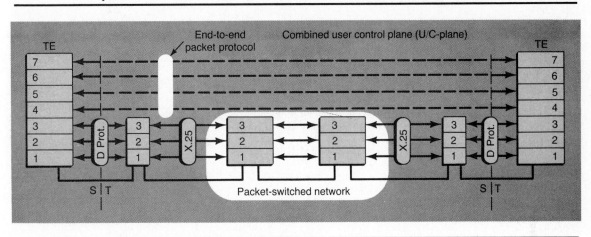

12.5   Common Channel Signaling

The basic motiviations for *common channel signaling* are discussed in Section 12.2. Even before the advent of ISDN, telephone companies were moving toward use of common channel signaling equipment to replace older in-band signaling equipment, with the change motivated by the greater flexibility and other features of common channel signaling. The initial common channel signaling equipment, based on CCITT recommendations for Common Channel Signaling System Number 6 (CCSS#6), used analog trunks with 2400 bps modems. CCSS#6 is widely used, with the AT&T CCSS#6 network in the United States claimed to be the largest private data network in the world [DONO86],[13] but it is being supplanted by *common channel signaling system #7 (CCSS#7)*, using purely digital trunks at 56 or 64 kbps. CCSS#7 is the ISDN version; we thus discuss it here.

CCSS#7

CCSS#7 does not fit the OSI model very well, but considerable effort has been put into improving the match between the protocols. Much of the difficulty in achieving a good fit comes from differences in purposes of the protocols; OSI is designed to enable users to interconnect in a standard, flexible manner, but signaling is typically performed to create a communications subnetwork for users. From this point of view, the entire CCSS#7 structure might be viewed as embedded within the lower three layers of the OSI structure.

---

[13] The AT&T CCSS#6 network was based on the CCITT standard, but with some enhancements to accommodate the special needs of the U.S. network [DONO86].

Figure 12.14   CCSS#7 Architecture as defined by CCITT in 1984 (from [SCHL86]).

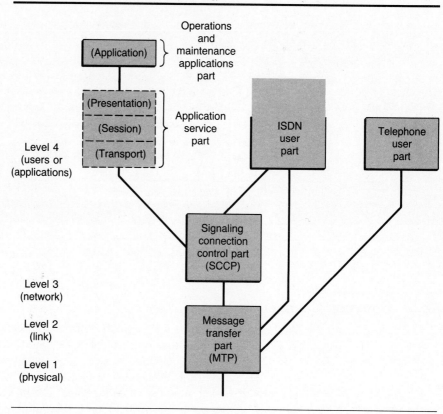

On the other hand, CCSS#7 and other signaling protocols allow application processes (including operations, administration, and maintenance) to communicate with each other in essentially the same way as OSI users, and this implies it contains full OSI functionality.

The best way to represent these distinctions seems to be by separate user and control planes, as shown in Figs. 12.11–12.13. The layered structure for CCSS#7 discussed here can be considered to be a description of the control plane, as illustrated in Figs. 12.11 and 12.12.

Figure 12.14 illustrates the layered structure of the CCSS#7 structure. It basically consists of four layers, the lower three roughly corresponding to X.25 layers and the fourth containing all other functions of the protocol. Originally, the fourth layer had little structure, but more structure is now being introduced.

message transfer part (MTP)

signaling connection control part (SCCP)

The *message transfer part* (*MTP*) provides a highly reliable, connection-less, sequenced data communication service. It consists of three layers: physical, data link, and network. The MTP network layer is, in some situations, supplemented by the *signaling connection control part* (*SCCP*), which becomes an upper portion of the network layer if it is used.

The physical layer implements standard telephone digital communications. The data link layer has similarities to HDLC, using flag characters to delineate frames, along with bit-stuffing to avoid long strings of ones, standard HDLC extended mode sequence numbering, and CCITT CRC-16 checksums. The rest of the frame structure is specialized for the application, however; frame structures are given in [ROEH85].

The network layer portion of MTP contains common transfer functions to provide the data communication service. Its functions include transfer of messages from source to destination on the signaling network with sequencing provided by consistently used signaling link selection codes, reconfiguration of routes after failures, flow control to handle congestion, and sending management information about abnormal conditions in the signaling network. The SCCP modifies the connectionless sequenced service for users requiring enriched connectionless or connection-oriented service. It is to provide five *classes of service,* though not all have been fully specified:

classes of service

- *Class 0:* Basic unsequenced connectionless service
- *Class 1:* Sequenced (MTP-like) connectionless service
- *Class 2:* Basic connection-oriented service
- *Class 3:* Flow control connection-oriented service
- *Class 4:* Error recovery and flow control connection-oriented service

The *telephone user part* provides circuit-related signaling for telephone call control of both analog and digital circuits. The operations and maintenance applications part provides measurements for the MTP and limited applications procedures. An incompletely defined *application service part* is also included in the figure. The transport, session, and presentation sublayers illustrated could some day become part of CCSS#7, probably for connection-oriented applications.

The *ISDN user part* defines functions, procedures and signaling information flows to set up, supervise, and release both voice and nonvoice calls over circuit-switched connections between ISDN access points. Its basic service is setup and release of calls involving single circuit-switched connections between subscriber access lines. It also provides *supplementary ser-vices,* making possible features promoted in ISDN trials. These include the following [APPE86]:

supplementary ISDN services

- Closed user groups
- User access to calling party address identification

Figure 12.15   Basic Call Setup and Release by ISDN User Part of CCSS#7 (adapted from [APPE86]).

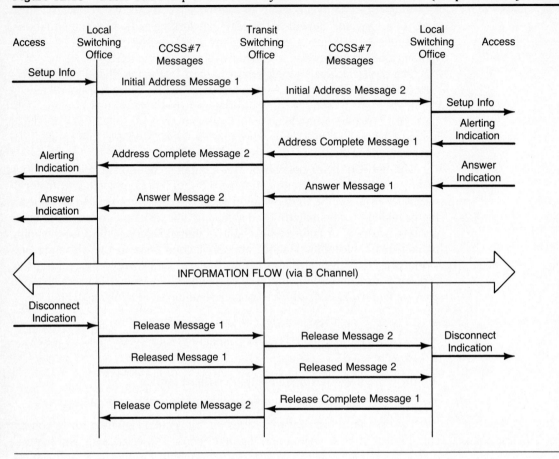

- User access to called party address identification
- Redirection of calls
- Connect when free, waiting allowed
- Call completion to busy subscribers
- Malicious call identification
- In-call modification

Basic call control procedures are illustrated in Fig. 12.15. This should be compared with Fig. 9.25, which illustrates the same procedures from the point of view of the ISDN network layer; Fig. 12.15 illustrates telephone signaling needed for the procedures in the earlier figure. Six (out of 40 standardized so far) message types are illustrated in the figure.

The first Initial Address Message contains information from the D channel SETUP message (see Fig. 9.25); this includes the ISDN number of the called party, type of connection required, identity of selected physical circuit to the succeeding switching office and its characteristics, plus additional parameters. This is modified by the transit switching office on which the selected outgoing circuit terminates to include identity of the circuit selected by the transit switching office and other information relevant to operation of that switching office. On receipt of the latter message, the destination switching office extracts pertinent information to include in a D channel SETUP message to the destination. The two Address Complete Messages serve to acknowledge the Initial Address Messages and indicate the called party has been found to be idle and is being alerted. The Answer Messages indicate that the called party has been connected and cause the B channel circuit to be set up. They may also contain information related to charging or similar administrative functions.

Although only one possibility for call release has been indicated, the protocol is symmetrical; either the calling party or the called party may initiate release. Release may also be initiated by one of the switching offices if it becomes congested. Three types of messages associated with call release are illustrated. Release Messages inform the nodes involved in the call that circuit connections used by the call may be released. Released Messages and Release Complete Messages are exchanged between adjacent switching offices to indicate the portion of the circuit between them has been released and this information has reached the other switching office; after reception of the Release Complete Message, this portion of the circuit is idled.

This has been a bare introduction to CCSS#7. Additional details are given in [ROEHB85], [SCHL86], and [APPE86].

12.6 Evolution Toward ISDN

In order to implement ISDN, techniques for transmitting information in the ISDN format must be developed and appropriate equipment installed. In addition, appropriate ISDN interfaces must be implemented, a combination of D-channel and CCSS#7 signaling must be used, and other details not fully covered in current ISDN standards must be handled in a realistic manner. Major progress is being made in each of these areas and initial ISDN implementations have been in use for several years. In this section we discuss approaches to ISDN implementation.

Implementation is following a progressive, *multiphase* approach. Current plans for implementation published by phone companies seldom mention communications speeds beyond the primary rate (1.544 or 2.048 Mbps), but a great deal of effort is being put into developing higher speed techniques. In this section we concentrate on plans for evolution of systems using basic rate and primary rate access. Higher speeds are discussed in Section 12.8.

The *first phase* is implementation of full-service single ISDN nodes (switching offices implementing ISDN) to verify that ISDN access concepts and standards can deliver promised functionality and services. This has led to a series of trials. A list of early trials, along with brief details of each, is given in Table 12.7. Most have concentrated on implementation of basic mode access, but a few have also implemented primary mode access and other features. A few also include more than one node.

Since definition of ISDN is not complete, it has been impossible for early implementations to fully follow ISDN specifications, with this problem especially severe for the earliest implementations. Even current standards do not specify such things as procedures for terminal maintenance or special features such as conference, transfer, hold, or multiline operation. Thus each implementation uses what the implementers feel are reasonable procedures in such areas. Nevertheless, the early implementations have been faithful to many ISDN details and have provided valuable background and experience.

The *second phase* is creating an ISDN network from a set of nodes to demonstrate the practicality of extending ISDN globally and to verify its cost effectiveness. Such implementations should include a wider variety of features, including both basic mode and primary mode access, a combination of D-channel and CCSS#7 signaling and so-forth. Techniques to provide distributed intelligence, network control, and high-capacity transport switching to allow appropriate capabilities at access points are being developed.

The *final phase* is expected to be network evolution to exploit ISDN's capabilities and develop a wide variety of applications. If the earlier phases are successful, this phase should be characterized by an industry-wide synergy with vendors continually adding value to developments of other vendors. Many of the most important applications developed during this phase may well be applications that have not yet been conceived.

12.7   Growth in ISDN Applications

Although telephone networks are moving rapidly toward becoming IDNs, as was indicated in Section 12.2, the *market* for new applications and services is developing less rapidly. Only a limited number of the services in Table 12.1 have significant markets so far. Growth is being limited by standard factors limiting growth of new services or technologies; the cost of offering new features tends to be high until a critical mass of installations is reached, but this high cost discourages customers from buying needed equipment or subscribing to services. The incremental cost of telephone portions of ISDN will not be high, however. For example Southern Bell's cost targets for ISDN equipment in the near term are 1.5 times the cost of plain old telephone system

Table 12.7   Summary of ISDN Trials.

City/Area	Country	Administration/ Supplier	Date	Brief Details
	Sweden	Televerket/Ericsson	1981	Local network transmission
Wisconsin	USA	Wisconsin Bell/Siemens	1985	Customer acceptance trials, mobile unit
Munich/ Berlin	W. Germany	DBP/various	1984	BIGFON; local wideband ISDN distribution
Tokyo	Japan	NTT	1984	INS trial; 64/16/4/4, B/B/D/D access
Venice	Italy	SIP/Ericsson	1984	Trial and demonstration; I.412 access
London	England	BT/various	1985–86	IDA trial and commercial 64/8/8, B/B/D access
Chicago	USA	Illinois Bell/AT&T	1986	I.412 access; basic access only; customer McDonalds
Phoenix	USA	Mountain Bell/NT	1986–87	I.412 access; basic 1986, primary 1987; DMS100 switch; 3 customers
Phoenix	USA	Mountain Bell/GTE	1986–87	GTD5 EAX
Phoenix	USA	Mountain Bell/AT&T	1986–87	No. 5 ESS
Phoenix	USA	Mountain Bell/NEC	1986–87	Digital adjunct to 1A ESS
Portland	USA	Pacific NW Bell/NT	1987	DMS 100; customer US National Bank; 32 kbps voice channels
Atlanta	USA	Southern Bell/AT&T	1987	No. 5 ESS
Atlanta	USA	Southern Bell/NT	1987	DMS 100
Boca Raton	USA	Southern Bell/Siemens	1987	EWSD
St. Louis	USA	Southwestern Bell/NT	1987	DMS 100; 3 customers
St. Louis	USA	Southwestern Bell/ AT&T	1987	No. 5 ESS
Ottawa	Canada	Bell Canada/NT	1986	DMS 100; CCSS#7 trials
Ottawa	Canada	Bell Canada/NT	1987	DMS 100; basic and primary access
	Belgium	RTT/BTMC	?	System 12; details not known
Mannheim/ Stuttgart	W. Germany	DBP/Siemens/SEL	1986–88	EWSD System 12; comprehensive phased trials
Lannion	France	CNET	1987	E10, MT25, "Renan" project
Heathrow, Florida	USA	Southern Bell/NT	1988	Fiber to home, POTS, ISDN, CATV transport

(Adapted from [RONA88], © 1987 IEEE. Reprinted by permission.).

(POTS) or 1.5 times centrex basic switch costs; the target becomes 1.2 times POTS equipment costs at one million volume level, expected to be reached in 1990 [SNEL87]. For these relatively small increases in cost, customers will gain great increases in capability. Costs of terminals, adaptors, and so forth will be high for a while, however.

A number of ISDN services for business are already making significant inroads. One of the more prominent is electronic mail, strongly preferred over ordinary mail by many users. Use of facsimile graphics is growing rapidly and appears to be the most rapidly expanding of the applications in Table 12.1. ISDN-like services for homes are expanding more rapidly in Europe than in the United States and other areas, due largely to success of the French Teletel system (see Chapter 1). Many applications have been implemented independently of ISDN, but should become much more popular as ISDN drives costs down.

initial offerings   Initial offerings of ISDN-like facilities to the public have tended to generate useful improvements in ordinary telephone service. For example, Ronayne [RONA88,Ch. 7] lists the following features advertised by Southern Bell as part of an ISDN trial in the Orlando, Florida, area: special ringing signals indicating calls from "special people"; blocking phone calls from certain numbers; tracing annoying phone calls; and automatic return of phone calls to a caller who has hung up just before the phone is answered. Though these do not exploit the full power of ISDN, they may be more immediately marketable to the general public than more powerful features.

A much more general list of applications, analyzed in US West ISDN trials, is given in Table 12.8. Virtually all are oriented toward business rather than individual users. User acceptance of the applications is not discussed in the reference.

Unfortunately, some charges from telephone companies in the United States, such as special access charges, have increased so significantly recently that some large customers have shrunk the size of their ISDN-like networks [OTOO87]. This could limit the rate of adoption of ISDN.

Another limitation on immediate acceptance discussed by Ronayne [RONA88] could be a major factor; this is *charging mechanisms* for ISDN services. He illustrates the problem with his experiences during 1981 in using Prestel, British Telephone's viewdata system, to display some 200 pages of information. Although Prestel is available over the ordinary analog telephone network, the billing problems are similar to some that can be anticipated with ISDN. Ronayne summarizes the results as follows: "The task of consolidating the various bills received—some monthly, some quarterly—and from different sources—BT telephones, BT Prestel, Information Provider, rental company—in order to determine the final cost was almost impossible." If the job of reconciling such charges is passed on to the subscriber (the easy solution from the telephone companies' viewpoint), he feels this could kill any interest the subscriber might have in ISDN, which could appear to be an uncontrollable financial drain.

Table 12.8   ISDN Applications Analyzed in US West ISDN Trials
(from [ANDE87b]).

- Video conferencing
- Asynchronous terminal networking
- Wide area networking
- LAN interconnection
- High-speed facsimile
- Coax elimination
- SNA emulation
- Flexible data base access
- Network integration
- Wiring simplification
- PC networking
- VAN/private network interconnection
- Enhanced call management
- Facsimile distribution
- High-speed screen transfer
- Multiple terminals per access line

- Primary rate to PBX interface
- Work at home
- Security, systems monitoring
- Integrated voice, data, image
- Packet-switching B/D channels
- Private line duplication
- D channel to B channel concentration
- Office automation with ISDN
- Enhanced ISDN terminals
- High-speed digital circuit switch
- Network analysis
- Videotex
- PC to facsimile transfer

In summary, technologies and standards for ISDN currently are advancing much more rapidly than ISDN applications. The primary motivation for ISDN so far has been the telephone companies' realization that a unified approach to handling all types of traffic is feasible and will provide a cost effective way of handling varied types of traffic. Much of the ISDN development is still based on conventional techniques in the telephone industry, the prime example of this being the 64 kbps (digital voice band) rate being used as a basic building block for essentially all currently defined portions of ISDN. Relatively little emphasis has been placed on developing applications or on developing suitable mechanisms for charging users for ISDN services. Although we expect ISDN to be successful, these factors could significantly slow down the timetable for its universal adoption.

12.8 Evolution Toward Broadband ISDN

BISDN

A variety of extensions to ISDN are being discussed. Some of the most important of these are extensions to provide true broadband capabilities. In this section we concentrate on *BISDN* (*broadband ISDN*) protocols and proposals. The BISDN proposals assume that optical transmission facilities are used,

Table 12.9   Volumes of Data for Current Data Files and Images
Without Compression (from [ARMB87]).

Data File or Image	Data Volume in Mbits
Relatively large data file	Several × 100
High-resolution computer graphics	20–100
Newspaper page	200–600
A4 facsimile (black/white)	1–4
A4 facsimile (gray)	9–16
A4 facsimile (color)	30–60
Color television image	4–6
High-definition color television image	16–24

though it should be possible to handle some with facilities such as digital microwave.

The most spectacular advances for ISDN applications will probably occur when true broadband facilities become readily available. This will be the first time when true integration of communications facilities for everything from low-speed data to voice to full motion video becomes available. Examples of some types of data files or images containing large volumes of data are given in Table 12.9. In some cases, there are stringent time restraints on the length of time to transmit these files. For example, video images are refreshed 25 or 30 times per second,[14] so the data rate required for full motion video without compression is on the order of 100 Mbps or more. By using data compression techniques, it is possible to compress most of these data sources by a factor of roughly 3–15 at the expense of more complex terminal equipment.[15]

Possibly the ultimate requirement for data transmission speed, based on the numbers in Table 12.9, is a system allowing a user to "browse" through a newspaper as effectively as we often do with a printed version. This can easily involve flipping through the pages rapidly enough for the required data rate to exceed that for high-definition TV; whether there is a sufficient market for this to justify developing it is questionable, however. A similar application, with more probability of a realistic market, is scanning high-resolution pages at several pages per second in a word processing system.

---

[14] The standard number of frames per second is 30 in North America, South America, and Japan; in most of the rest of the world it is 25.

[15] By using interframe coding, transmitting differences between video frames rather than full frames, it is even possible to transmit video teleconference quality TV at approximately 1.5 Mbps [KANE87]. Broadcast quality TV, though, requires 20 to 30 Mbps with the same technique.

Table 12.10   Broadband Services for Commercial and Private
Applications of ISDN.

Broadband Services	Applications	
	Commercial	*Private*
Communication of data, text, graphics		
▪ Data transfer (burst, stream)	X	
▪ Document transfer	X	
▪ Document filing and retrieval	X	
Person-to-person video communication		
▪ Video telephony (including showing documents)	X	X
▪ Video conference	X	
▪ Broadband message handling	X	X
Access to video information		
▪ Broadband videotex	X	X
▪ Video on demand	X	X
Broadcast of programs and data		
▪ Common TV	X	X
▪ Pay TV (pay per channel, pay per view)	X	X
▪ High-definition television (HDTV)	X	X
▪ Cabletext	X	X

BISDN standards are being developed to handle the type of applications indicated in Table 12.9. Specific applications under consideration are listed in Table 12.10.

synchronous transfer mode (STM)

asynchronous transfer mode (ATM)

During development of standards for BISDN, there has been a continuing debate between proponents of *synchronous transfer mode (STM)* and *asynchronous transfer mode (ATM)*. STM is the conventional telephone approach, using ordinary time division multiplexing (see Subsection 2.4.5). Each user is allocated preassigned time slots to transmit information, whether any information is available to send or not. ATM is essentially the same as asynchronous time division multiplexing (see Subsection 2.4.5), but with techniques to speed up its operation. Time slots are not preassigned, but are available on demand to users needing them, with information identifying the source incorporated in headers in each slot. This gives more flexibility for handling dynamically varying loads, and ATM is currently favored for this reason. STM is still under consideration, however.[16] Figure 12.16 illustrates the differences between STM and ATM.

Since ATM is the favored approach, and STM techniques proposed do not differ significantly from those in Subsection 2.4.5, we concentrate on ATM. We shall also discuss *SONET,* a synchronous protocol, for transmission of ATM frames that has been adopted as an international standard.

---

[16] See [GECH89] for a discussion of some of the problems with ATM that still have not been satisfactorily resolved.

Figure 12.16   Synchronous Transfer Mode and Asynchronous Transfer Mode Multiplexing.

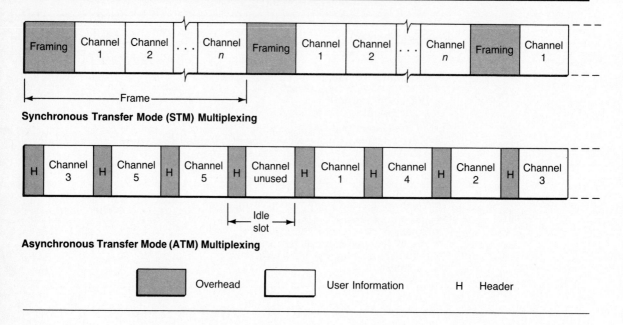

Synchronous Transfer Mode (STM) Multiplexing

Asynchronous Transfer Mode (ATM) Multiplexing

ATM can be considered to be a multiplexing and switching technique confined to layer 1 and basic functions of layer 2 of the OSI model. It packs information to be transferred into fixed-size slots called "cells," each with a header used to identify and switch cells. The term "asynchronous" refers to the fact that cells are not preassigned, so those assigned to a particular connection may occur at irregular intervals. An important difference between ATM and ordinary asynchronous time division multiplexing is the use of fast packet-switching techniques with ATM in order to maintain the desired data rate. Information in the header is used to determine how a cell is routed through the network, without necessarily storing the full cell at a node before sending it on toward its destination.

A recent agreement between the United States and the CCITT sets the ATM *cell size* at 53 bytes.[17] The structure of a cell, as it has been tentatively defined at the time this is being written, is sketched in Figure 12.17. It consists of a five byte ATM cell header, with fields tentatively defined as shown, plus a 48 byte user part, which will normally consist of a two byte adaption layer header, 44 bytes of user data and a two byte adaption layer trailer.

---

[17] This cell size replaces an earlier cell size of 69 bytes which had been adopted by standards committees in the United States.

Figure 12.17 ATM Cell Fields (based on [HÄND89], [SHUM89]).

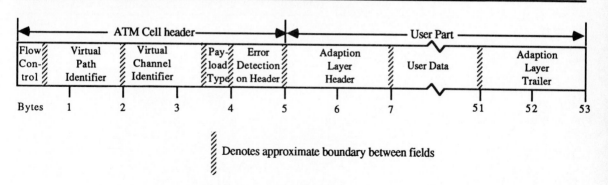

Denotes approximate boundary between fields

The ATM header contains a field containing flow control information, two fields giving a virtual path identifier and a virtual channel identifier,[18] a field indicating the type of payload carried and an error detection field for header data only; a second error detection field for the user part may be contained in the adaption layer trailer. In order to achieve fast switching times for broadband ISDN, only functions such as frame delimiting, multiplexing, and error detection associated with the cell structure are performed at individual network nodes; all other functions (including the upper portions of Layer 2 protocols for the OSI model) are done on an end-to-end basis. Details of one approach to this, frame relay, are given in [CHER89], [LAI89] and [LAMO89].

The current version of a *protocol model for BISDN* using ATM is illustrated in Fig. 12.18. It portrays the way in which all layers of the user plane protocols above the ATM layer (and an adaption layer discussed below) operate on an end-to-end basis. All transmissions, for either the control plane or user plane functions, use the lower three BISDN layers (corresponding to the physical layer and part of the data link layer in the OSI model), however. These all operate on a link-by-link basis.

The *adaption layer* compensates for the fact that data field size in an ATM cell is unlikely to correspond to size of a user message. It may indicate the type of data unit, its beginning or ending, its length, and so forth. Its basic function is to provide alignment of data structures with the ATM cell user data field.

Two bit rates for transmission of BISDN are currently being emphasized internationally—one around 150 Mbps and the other around 600 Mbps. A lower rate (50 Mbps) is also being studied, primarily in the United States. Transmission at each of these rates can be accomplished through the use of *SONET,* the Synchronous Optical Network standard approved as an international standard in 1988 [BALL89]. A list of SONET rates is given in Section 12.3.

---

[18] BISDN standards currently emphasize connection-oriented service, as this terminology indicates, but connectionless service is also being studied.

Figure 12.18   Protocol Model for BISDN Using ATM (adapted from [HÄND89], © 1989 IEEE. Reprinted by permission.).

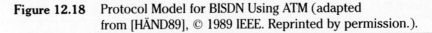
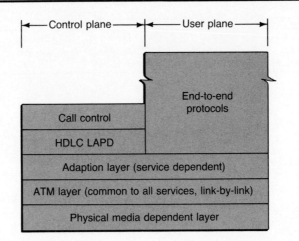

Figure 12.19   Format of SONET STS-1 Frame.

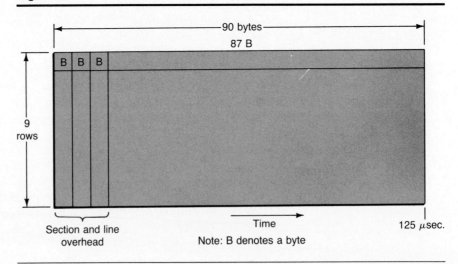

Figure 12.20   Placement of ATM Cells in SONET Frames.

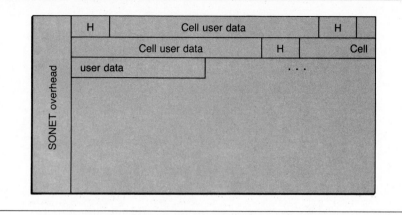

The format of an STS-1 frame, the lowest speed version of SONET, is illustrated in Fig. 12.19. A frame contains $9 \times 90 = 810$ bytes, with one frame transmitted every 125 μsec., or 8000 frames transmitted per second. This corresponds to a transmission rate of $810 \times 8 \times 8000$ bps or 51.84 Mbps. Of this, $36 \times 8 \times 8000$ bps or 2.304 Mbps are overhead bits (including 27 bytes of overhead per frame illustrated plus another 9 bytes discussed below, and illustrated in Fig. 12.21); this leaves 49.536 Mbps available for data (plus ATM headers if ATM is used). The 44.736 Mbps rate for optical digital cross connects in North America (see Section 12.2) is readily accommodated, but the 139.264 Mbps rate in Europe and the 397.20 Mbps rate used in Japan require higher order versions of SONET. These are discussed below.

The SONET frame structure helps maintain synchronization of the transmitted data stream. If it is used to transmit ATM data, successive ATM cells are placed in the data portion of SONET frames in the manner indicated in Fig. 12.20. (This is an idealized version of placement, ignoring offset of the STS-1 payload discussed below.) If the ATM format is used with a 69-byte cell including a 5-byte header, this allows slightly less than 46 Mbps of user data in an STS-1 frame.

In order to accommodate minor variations in the rate of the STS-1 payload with respect to the rate of SONET transmission, the STS-1 payload can begin at a position that floats with respect to a SONET frame. This is illustrated in Fig. 12.21. The location of the first byte of STS-1 path overhead is given by a pointer in the line overhead part of a frame.[19]

---

[19] Techniques for adjusting the start position as required by minor rate variations are given in [BALL89].

Figure 12.21   Location of STS-1 Payload in Interior of
SONET STS-1 Frames.

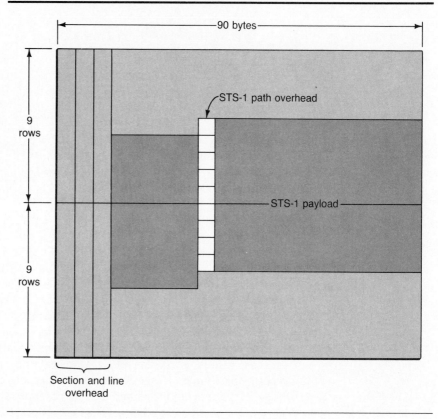

Higher rate SONET, STS-N, signals are obtained by byte interleaving N STS-1 signals to give the frame format in Fig. 12.22. All of the section and line overhead bytes in STS-1 #1 in this format are used, but only part of the other N-1 section and line overhead channels are actually used.

The STS-N frame is the signalling format for the OC-N signals listed in Table 12.4. OC-3, at 155.52 Mbps, and OC-12, at 622.08 Mbps, are receiving primary emphasis since they provide the approximate 150 Mbps and 600 Mbps rates emphasized for BISDN. OC-3 is also adequate for the European 139.264 optical digital cross connect signals, while OC-9, at 466.56 Mbps, can be used for the Japanese 397.20 Mbps rate.

Initial availability of SONET-based transmission systems may be in a switched multimegabit data service being planned by Bell companies in the United States [HEMR88]. Attempts are also being made to integrate SONET into standards for metropolitan area networks [MOLL88], [NEWM88b].

Figure 12.22   Format of SONET STS-N Frame.

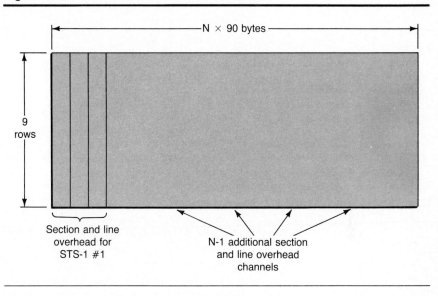

9 rows

N × 90 bytes

Section and line
overhead for
STS-1 #1

N-1 additional section
and line overhead
channels

12.9 Summary

ISDN is a logical outgrowth of the digitization of telephone networks. The integrated digital networks being developed for telephone networks form the basic framework for ISDN, but initial installations will use extra facilities overlaid on existing telephone networks since the tremendous investments represented by current facilities cannot be simply written off.

A major part of work done on ISDN architectures to date has been devoted to the development of interface standards and deciding what functions should be provided. Although much of the literature talks about having a single interface to ISDN, it would be more accurate to say a small number of interfaces will be used since different interfaces are at different speeds and a few other minor variations are planned. This will greatly reduce the number of different communications lines a user will need, though; in addition, the customer will be given much more control over the network, including access to some types of signalling facilities.

The architecture of ISDN basically follows that of the OSI Reference Model, with separate user, control, and management planes. The control plane is expected to be primarily handled by common channel switching system No. 7. Broadband ISDN capabilities are also being developed and are expected to be based on ATM switching and SONET transmission.

Problems

12.1 Use the nomogram in Fig. 12.1 to estimate the data rate needed for each of the following data transfers:
    a. Electronic mail in 1 minute (give range of answers)
    b. Full tape in 2 minutes
    c. Full CD in 1 hour
    d. High-resolution color screen in 30 milliseconds

12.2 For the digital transmission chain in Fig. 12.3, indicate what transmission rate would be used at each point if the telephone network with this transmission chain were located in the following areas:
    a. North America
    b. Europe
    c. Japan

12.3 For each of the levels in the North American, European, and Japanese digital multiplexing hierarchies in Table 12.2, compute the number of overhead bits transmitted per second. Assume that each voice-grade channel uses 64 kbps and count all other bits transmitted as overhead bits.

12.4 One possible approach to allowing the various digital multiplexing hierarchies in Table 12.2 to operate together might be to design a high order multiplexer such that the number of voice-grade channels it carries is an integral multiple of each of the numbers of voice-grade channels in Table 12.2.
    a. What is the minimum possible number of voice-grade channels carried by this type of multiplexer?
    b. What would be the approximate bit rate of this multiplexer? Assume 64 kbps for each voice-grade channel plus 10 percent overhead for framing bits and so forth.

12.5 Describe how common channel signalling makes it possible for time-of-day routing to be provided to subscribers to toll free (800 line in the United States) services. Describe the way in which the appropriate routing to be used (for example, which location should receive calls) can be determined. Describe the sequence of messages exchanged among different facilities in a telephone network in order to set up a call with time-of-day routing.

12.6 A number of CCITT speech coding standards are listed below [MINZ89]. Describe how each might be handled in an ISDN network. Indicate the type of channel that might be used, the most reasonable type of switching, how signaling might be handled, and so forth.

Standard	Analog bandwidth	Transfer Rate	Coding Algorithm
CCITT Rec. G.711	3.2 kHz	64 kbps	PCM
CCITT Rec. G.721	3.2 kHz	32 kbps	ADPCM
CCITT Rec. J.41	15 kHz	384 kbps	PCM
CCITT Rec. J.42	7 kHz	192 kbps	PCM

12.7 A number of CCITT Study Group XVIII video quality proposals are listed below [MINZ89]. In each case indicate whether or not current versions of ISDN can handle the signals. If the signals can be handled, indicate how. If they cannot, indicate as precisely as possible how future versions of ISDN may be able to handle them.

Quality	Transfer Rate	Description
A	92–200 Mbps	High-definition television (HDTV)
B	30–145 Mbps	Digital component—coding signal
C	20–40 Mbps	Digitally-coded NTSC, PAL, SECAM
D	.384–1.92 Mbps	Reduced spatial resolution and movement portrayal
E	64 kbps	Highly reduced spatial resolution and movement portrayal

12.8 Compare the relative advantages and disadvantages of synchronous transfer mode (STM) and asynchronous transfer mode (ATM) for the different classes of traffic expected to be handled by ISDN.

12.9 How many full ATM cells can be accommodated in one SONET STS-1 frame if the standard ATM format is used? How many data byte positions are left over?

12.10 Sketch the manner in which ATM cells would fit into the SONET payload in Fig. 12.21, giving a more precise version of cell locations than that shown in Fig. 12.20.

Glossary

ABM *See* Asynchronous Balanced Mode.

Abrupt Close Close of a connection without attempt to prevent loss of data.

Access Control (AC) Field used in token ring MAC to control access according to priority and reservation algorithms.

Acknowledgment (ACK) Message returned from receiver to transmitter to indicate successful reception of transmission. Also communications control character used for acknowledgment.

ACK0 Bisync acknowledgment for even-numbered message.

ACK1 Bisync acknowledgment for odd-numbered message.

Active Open Used in TCP to request connection with another node.

Address Data structure used to identify the destination of a message.

Address Field Field containing an address.

Advanced Data Communications Control Protocol ANSI version of standard bit-oriented data link control protocol.

ALOHA Algorithm for multiple access in which any station transmits when ready, determines whether collision has occurred, and then retransmits if necessary.

AM *See* Amplitude Modulation.

American National Standards Institute (ANSI) Standards body in United States responsible for a number of telecommunications standards.

American Standard Code for Information Interchange (ASCII) Standard alpha-numeric code common in United States. U.S. version of IA5.

Amplitude Modulation (AM) Modulation technique conveying information via amplitude of carrier signal.

Amplitude Shift Keying Digital version of AM.

ANSI See American National Standards Institute.

Application Layer Highest layer of OSI Reference Model.

ARM *See* Asynchronous Response Mode.

ARPANET DLC Data link control protocol used in ARPANET.

ARPA Network (ARPANET) One of the earliest and most influential computer networks in the United States. Phased out in 1990.

ARPANET Reference Model *See* DoD Reference Model.

ARQ *See* Automatic Repeat Request.

ASCII *See* American Standard Code for Information Interchange.

ASK *See* Amplitude Shift Keying.

Asynchronous Operating without precise clocking.

Asynchronous Balanced Mode (ABM) HDLC mode defined for balanced point-to-point configurations with two combined stations. Either end can initiate transmission without waiting for a poll from the other end.

Asynchronous Data Link Control *See* Start-Stop.

Asynchronous Response Mode (ARM) HDLC mode for systems with one primary and one or more secondaries; if more than one secondary is present, all but one must be quiescent. With one primary and one active secondary, either active station may initiate transmission at any time without waiting for a poll or an F bit.

Asynchronous Transfer Mode (ATM) Broadband ISDN mode similar to asynchronous time division multiplexing except for techniques to speed up operation.

Asynchronous Transmission Transmission of individual characters without prescribed clocking between transmission times of different characters.

Automatic Repeat Request (ARQ) Error control technique in which receiver detects errors and asks for retransmissions.

Average Transfer Delay (T) Average time between the arrival of a packet to a station interface and its complete delivery to the destination station.

Backoff Random delay before another attempt at transmitting in random access medium access control protocols.

Backpressure Propagation of effects of hop-by-hop flow control to upstream nodes.

Backward Learning Routing algorithm based on assumed symmetric network conditions. Source node assumes best route to given destination is via neighbor node that was on best route from destination to source.

Balanced Configuration Point-to-point network configuration in HDLC with two combined stations.

Balanced Mode Transmission Data transmission with information conveyed by differences in voltages on two circuits to minimize effects of induced voltages.

Baud Rate Number of signal changes per second used to convey information. Often misused to denote bit rate of digital signal.

BCC See Block Check Characters.

B Channel Full duplex 64 kbps ISDN channel used for user data.

Bellman-Ford-Moore Algorithm Shortest-path routing algorithm that iterates on number of hops in a route to find shortest-path spanning tree.

Bifurcated Routing Routing that may split one traffic flow among multiple routes.

Binary Signal Encoding Technique for encoding digital information to improve data transmission performance.

Binary Synchronous Communications See Bisync.

BISDN See Broadband ISDN.

Bisync (Binary Synchronous Communication) Character-oriented DLC protocol for half duplex applications; de facto standard before advent of bit-oriented protocols.

Bit Error Rate Average fraction of transmitted bits received erroneously.

Bit Oriented Protocol Data link control protocol based on use of Flag character to begin and end frame, with Flag bit pattern prohibited in rest of frame. Any bit pattern may then occur in body of frame.

Bit Rate Number of bits per second transmitted or received.

Block Check Characters (BCC) Bisync terminology for extra characters added to frames for error protection.

Block Code Error protection code containing fixed number of bits, say n, in each code word, usually with k data bits and $n-k$ parity bits.

Boundary Function Capability in SNA subarea node to handle some functions that nearby peripheral nodes are not capable of handling.

Bridge Data link layer relay.

Broadband ISDN (BISDN) Standards being developed for ISDN to handle applications such as video requiring high bandwidth.

Broadcast Transmission Transmission mode in which each station is assumed to hear every other station.

BSC See Bisync.

Burst Error Rate Probability error burst begins on specified transmitted bit.

Bus Single communications link to which stations are attached so that each station can hear all other stations.

Byte-Count–Oriented Class of data link control protocols using byte count field to determine length of data field.

Byte Timing Circuit Optional X.21 circuit used to maintain byte or character synchronization.

C See Control Circuit.

Carrier Sense Multiple Access (CSMA) Medium access technique involving first sensing medium to see if other signals are already present, then transmitting if no signal is present. Several algorithms handle cases where signal is already present.

Carrier Sense Multiple Access/Collision Detection (CSMA/CD) Combination of CSMA with sensing to detect collisions during transmission. Standard medium access technique in some LANS, including Ethernet.

CCITT See Consultative Committee for International Telegraph and Telephone.

CCS See Common Channel Signaling.

CCSS#7 See Common Channel Signaling System #7.

Channel Queue Limit Limit on number of transmit buffers used by a station to guarantee that some receive buffers are always available.

Character Oriented DLC Protocol Data link control protocol based on use of communications control characters in standard alpha-numeric character set.

Checkpoint Cycle HDLC error recovery cycle formed by pairing an F bit with a previous P bit or vice versa.

Checkpointing HDLC error recovery based on pairing of P and F bits and giving the equivalent of a negative acknowledgment without using either REJ or SREJ.

Choke Packet Packet used for flow control. Node detecting congestion generates choke packet and sends it toward source of congestion, which is required to reduce input rate.

Circuit Switching Type of switching used in ordinary telephone service, with dedicated path between source and destination set up for duration of call.

Class 0 Transport Service OSI transport service assuming a Type A network with error control, sequencing, and flow control; provides minimal connection establishment, data transfer, and connection termination facilities.

Class 1 Transport Service OSI transport service designed for Type B networks. Similar to Class 0 but includes capability to recover from network layer resets or disconnects by adding sequence numbers to TPDUs so that lost TPDUs can be detected and retransmitted; expedited data capability is included.

Class 2 Transport Service OSI transport service assuming a Type A network with error control, sequencing, and flow control; provides connection establishment, data transfer, and connection termination facilities, for multiplexing multiple transport layer connections over a network layer virtual circuit; and for transport layer flow control using a credit mechanism.

Class 3 Transport Service OSI transport service designed for Type B networks; essentially a union of Classes 1 and 2, adding multiplexing and flow control functions of Class 2 to the error handling functions of Class 1.

Class 4 Transport Service OSI transport service designed for Type C networks. Assumes worst-case network and provides comprehensive error detection and recovery features, including checksums, timeouts, and retransmissions, recovery from loss of data or control TPDUS, expedited data, and user notification of unrecoverable errors. Multiplexing and flow control for each connection provided.

Class of Service Information for setting up route supplied by SNA user. Consists of such things as priority, throughput, cost, delay, security, integrity, type of communications facilities to be used, and so forth.

Clear X.25 error control procedure. Corresponds to disconnecting virtual circuit. All packets in transit when a Clear occurs are lost, and virtual circuit must be set up again before it can be reused.

Clear to Send (CTS) The communications equipment (DCE) activates this circuit in RS-232C interface when it is ready to accept data from the DTE.

Coaxial Cable Electrical transmission medium consisting of insulated inner and outer conductors plus shield; used for wide bandwidth electrical signals.

Coding Theory Mathematical theory describing how to encode data into streams of digital symbols at transmitter and decode it at receiver to maximize accuracy of data presented to user.

Collision Result of multiple attempts to transmit at same time on multiple access medium. Usually all colliding transmissions wipe each other out and require retransmission.

Combined Station HDLC station containing both a primary and a secondary and used in asynchronous balanced mode.

Command Any frame sent from a primary to a secondary on an HDLC link.

Common Channel Signaling (CCS) Signaling systems being used and installed in many telephone networks. Completely separates signaling information from user data by sending it over separate signaling network.

Common Channel Signaling System #7 (CCSS#7) Version of CCS used in ISDN architecture.

Common Network Layer Provides routing and multiplexing functions between data link layer and SNA interface layer in DNA/SNA gateway.

Communications Channel Interface Interface between modem or communications adapter and communication link.

Communications for Computers Use of communication facilties to transfer information to or between computers.

Computational Complexity Measure of number of computations necessary to implement algorithm.

Computer Equipment Interface Interface between DTE side of physical layer interface and computer equipment, which converts signals from forms on DTE side of interface to forms accepted by computer equipment.

Computers for Communications Use of computers and computer techniques in communication facilities.

Concentration Sharing of facilities via a demand assignment algorithm.

Confirm A primitive returned to the requesting $(N+1)$st layer by the (N)th layer to acknowledge or complete a procedure previously invoked by a request primitive.

Connectionless Transmission Data transmission without prior establishment of a connection.

Connection-Oriented Transmission Data transmission technique involving setting up connection before transmission and disconnecting it afterward.

Consultative Committee for International Telegraph and Telephone (CCITT) International standards committee responsible for substantial portion of telecommunications network standards.

Contention Mode Mode of operation in which station transmits when ready, then provides for recovery from collisions.

Continuous ARQ Standard approach to ARQ for DLC protocols that are usable on full duplex communications links. Both frame transmissions and acknowledgments occur simultaneously.

Control Character A character (or bit pattern) in standard character set that is used to convey device or communication control information rather than user data.

Control Circuit (C) X.21 interface circuit used to send control information from DTE to DCE.

Control Field Field in frame containing control information.

Control Station On multiaccess link, station that is in charge of such functions as selection and polling.

Convolutional Code Error protection code encoding data bits in a continuous stream. After each k input bits are fed into encoder, a total of n output bits are obtained with each output bit depending on current group of k input bits as well as $N - 1$ previous groups of k input bits.

COSNAME Identifies class of service in SNA.

CRC Check characters based on polynomial code and added to frames for error protection.

CRC-12 Generator polynomial used for error protection in Bisync versions with six-bit alphanumeric code words.

CRC-16 Generator polynomial used for error protection in Bisync versions with eight-bit alphanumeric code words and in DDCMP.

CRC-32 Generator polynomial used for error protection in IEEE 802 and various U.S. Department of Defense protocols.

CRC-CCITT Generator polynomial for error protection used in HDLC and related DLC protocols.

Credit Permit to transmit number of messages specified in permit itself; allows use of variable size windows.

CSMA *See* Carrier Sense Multiple Access.

CSMA/CD *See* Carrier Sense Multiple Access/Collision Detection.

CSNP Communications subnet processor interfacing ARPANET host with network.

CTS *See* Clear to Send.

DARPA Internet World's largest internetwork, linking together thousands of networks around world; sponsored by U.S. Defense Advanced Research Projects Agency.

Data Circuit-Terminating Equipment (DCE) Standards committee terminology for modem or similar data communications equipment.

Data Flow Control Layer Layer 5 in SNA.

Datagram Packet or short message transmitted through network without previously setting up a connection. Different datagrams are handled independently of each other.

Data Link Control (DLC) Protocol Protocol implementing functions of data link layer.

Data Link Escape (DLE) A communications control character that changes the meaning of a limited number of contiguously following characters in Bisync.

Data Link Layer Layer 2 of OSI Reference Model and most other networking architectures.

DATAPAC Major packet data network in Canada.

Data Set Ready (DSR) The modem (data set) activates this circuit in RS-232C interface when it is powered up and ready for use.

Data Terminal Equipment (DTE) Standards committee terminology for data processing equipment interfaced to communications link.

Data Terminal Ready (DTR) The DTE (terminal or computer) activates this circuit in RS-232C interface to let the modem know it is ready to send and receive data.

D Bit Bit in X.25 packet header determining whether acknowledgments have local or end-to-end significance.

DCE *See* Data Circuit-Terminating Equipment.

D Channel Full duplex 16 kbps or 64 kbps ISDN channel (in basic rate and primary rate interfaces, respectively) used to transmit control data or user data.

DDCMP *See* Digital's Data Communications Message Protocol.

DDN *See* Defense Data Network.

Deadlock Condition under which throughput of network, or part of network, goes to zero due to congestion.

DECNET *See* Digital Network Architecture.

Defense Data Network (DDN) Network split off from ARPANET (see definition) to handle U.S. military needs. Also called MILNET.

Demand Assignment Allocation of part of capacity of medium or other equipment to users only when they request to use it.

Demodulation Process of returning modulated signal to its original form.

Destination Service Access Point (DSAP) Address of service at destination.

Destination Service Access Point (DSAP) Address Field Field in IEEE 802 LLC header identifying DSAP.

Differential Modulation Modulation in which information is encoded into differences of successive signals.

Digital's Data Communications Message Protocol (DDCMP) Data link control protocol developed by Digital Equipment Corporation and used in DNA.

Digital Network Architecture (DNA) Digital Equipment Corporation's layered network architecture for telecommunications. Also called DECNET.

Dijkstra's Algorithm Shortest-path routing algorithm that iterates on length of path to find shortest-path spanning tree.

Distributed Asynchronous Bellman-Ford-Moore Algorithm Version of Bellman-Ford-Moore algorithm that can be performed in distributed manner by nodes without common clocking.

DLC *See* Data Link Control Protocol.

DLE *See* Data Link Escape.

DNA *See* Digital Network Architecture.

DNA/SNA Gateway Gateway developed by Digital to interconnect DNA and SNA networks.

DoD Reference Model Networking architecture developed for the U.S. Department of Defense as an outgrowth of the ARPANET project.

Domain Portion of SNA network under control of one SSCP.

DSAP *See* Destination Service Access Point.

DSR *See* Data Set Ready.

DTE *See* Data Terminal Equipment.

DTR *See* Data Terminal Ready.

EBCDIC (Extended Binary Coded Decimal Interchange Code) Eight-bit alphanumeric code used in many IBM products.

E Channel 64 kbps channel used for control of circuit switching in primary rate ISDN interface.

Echoing Approach to error control with simple terminals in which character printed at terminal is result of transmitting character to receiver and then echoing it back to transmitter. Operator is supposed to enter any erroneous characters again.

ED *See* End Delimiter.

E (Echo) Channel 16 kbps ISDN basic rate channel echoing contents of D channel back from DCE to DTEs. Used in bidding for access to multipoint link.

Effective Rate of Transmitting Data Bits Total number of data bits accepted divided by the time to get them accepted.

EIA-232D New version of RS-232C physical layer interface adopted in 1987.

Electrical Characteristics Include voltages or current levels, timings of signals, and their interpretation as zeros or ones. Part of physical layer description.

End Communications Layer Layer 4 of DNA.

End Delimiter (ED) Sequence of bits used by IEEE 802 MAC to indicate the end of a frame. Used in token bus and ring networks, with nondata bits making ED easy to recognize.

End Node Node in DNA network that has one attachment to network and does not handle routing.

End of Text (ETX) A communications control character terminating a sequence of characters started with STX and transmitted as an entity in Bisync.

End of Transmission (EOT) A communications control character used to conclude a Bisync transmission that may have contained one or more text messages and headings.

End of Transmission Block (ETB) A communications control character indicating the end of a block of Bisync data for communication purposes.

Enquiry (ENQ) Communications control character in standard alphanumeric codes used for some type of enquiry. Used in Bisync in messages for polling and selection, and to ascertain status of receiver.

EOT *See* End of Transmission.

Equalization Adjustment of characteristics of receiver and/or transmitter to match characteristics of communications channel.

ETB *See* End of Transmission Block.

Ethernet Local area network using CSMA/CD medium access; developed by Xerox.

ETX *See* End of Text.

Expedited Data Data given preferential treatment, via priorities, and so forth. Normally limited to short packets.

Explicit Route SNA route consisting of list of subarea nodes and transmission groups connecting source subarea to destination subarea.

Extended Address Field HDLC address field $8 \times n$ bits long, with n representing any integer. The first bit of the final address field octet is one, while the first bit of each preceding address field octet is zero.

Extended Control Field Two byte control field in HDLC allowing modulo 128 sequence counts.

F *See* Flag.

Fairness Equitable treatment of all users, especially in limiting traffic by flow control.

FC *See* Frame Control.

FCS *See* Frame Check Sequence.

FDDI High-speed (100 Mbps) communications architecture standardized by ANSI, primarily for fiber optic links.

FDM *See* Frequency Division Multiplexing.

FEC *See* Forward Error Correction.

Fiber Optics Glass or similar fibers and associated circuitry that transmit optical signals; used for high rate digital communications systems.

FID Used in SNA to denote different path control header formats (transmission header formats in SNA documentation).

FID2 SNA FID used between a boundary node and an adjacent communications controller node.

FID3 SNA FID used between a boundary node and minimal function peripheral node.

FID4 SNA FID used between hosts and CPs (or subarea nodes).

Fixed Assignment Predetermined allocation of parts of capacity of medium or other equipment to users.

Fixed-Priority-Oriented-Demand Assignment (FPODA) Medium access technique in which one station acts as master and controls channel based on requests from stations. Each station may request an allocation of data slots for future traffic.

Flag (F) Special character used to obtain byte and frame synchronization in HDLC and other protocols.

Flooding Routing algorithm in which each node receiving a packet destined for another node sends copies on each outgoing link except (in most implementations) the link on which it was received.

Flow Control Regulation of traffic allowed into specific portions of a network to avoid excessive congestion.

Floyd-Warshall Algorithm Shortest-path routing algorithm that computes shortest paths between all pairs of nodes. It iterates on nodes allowed as intermediate nodes.

FM *See* Frequency modulation.

Forward Error Correction (FEC) Approach to error control in which redundancy is included in transmitted messages to allow correction of errors at receiver without retransmission.

Forwarding Data Base Data base containing entries for all destinations currently known by spanning tree bridge. Entry contains station address and port number on which a frame should be forwarded to reach the destination.

Fourier Transform Mathematical formula giving frequency representation of time signal.

FPODA *See* Fixed-Priority-Oriented-Demand Assignment.

Fragmentation *See* Segmentation.

Frame Data unit generated by data link layer, consisting of user data plus all pertinent headers and trailers.

Frame Check Sequence (FCS) HDLC terminology for extra characters added to frame for error protection.

Frame Control (FC) Field used in IEEE 802 token ring and token bus MAC to distinguish between types of frames, including MAC frames and LLC frames.

Frequency Division Multiplexing (FDM) Communications channel sharing technique with different signals transmitted in different frequency bands.

Frequency Modulation (FM) Modulation technique conveying information via instantaneous frequency of carrier signal.

Frequency Shift Keying (FSK) Digital version of FM.

Full Duplex Transmission Data transmission in both directions simultaneously.

Functional Layer Highest layer of DNA/SNA gateway architecture.

Functional Specifications Assign meanings to circuits (or pins) for physical layer. It is common to classify circuits into categories of data, control, timing, and grounds.

G *See* Normalized Offered Traffic.

Gateway Relay operating at any layer higher than network layer. Also often used for network layer relays.

Gateway-Gateway Protocol (GGP) DARPA Internet protocol operating between gateways to exchange reachability and routing information.

General Format Identifier (GFI) Field in X.25 packet header identifying type of packet.

Generating Polynomial Polynomial used to generate a particular polynomial code.

Geometry Geometrical configuration of links and nodes in network. Sometimes called topology.

Geosynchronous Communications Satellite Communications satellite positioned at elevation over equator causing it to have 24-hour rotational period so it appears to be stationary from position on earth.

GGP See Gateway-Gateway Protocol.

Go-Ahead Message Message or bit pattern, sent by station in network using polling to indicate the end of its transmission, allowing the next station to be polled or to transmit.

Go-Back-*N* ARQ Variant of continuous ARQ in which the erroneous frame and all succeeding frames already transmitted are retransmitted after an error has been detected.

Graceful Close Technique for closing a connection that attempts to ensure no data are lost. Side requesting Close accepts data from other side until that side also requests Close.

Half Duplex Transmission Data transmission in one direction at a time, with turnaround to reverse direction.

Half Gateway Half of a gateway. Gateways and other relays are often divided into two halves, one in each network, to simplify design and maintenance.

Handshake Sequence of messages exchanged to both convey information and verify its receipt.

Handshaking Sequence Sequence of messages sent back and forth to both convey information and verify its receipt.

H Channel Full duplex channel in ISDN primary rate interface operating at 384 kbps (H0), 1536 kbps (H11), or 1920 kbps (H12).

HDLC *See* High-Level Data Link Control.

Header Information attached to front of packet or message by Layer N at transmitter for interpretation by Layer N at receiver.

High-Level Data Link Control (HDLC) The ISO standard bit-oriented data link control protocol.

Host-Host Layer Layer III of DoD Reference Model.

Hub Polling Variant of polling in which polled stations are in a configuration in which each station can pass a poll on to the next station after completing its transmission, rather than requiring additional poll by central station.

I *See* Indication Circuit.

I.430 Basic rate physical layer interface defined for ISDN.

I.431 Primary rate physical layer interface defined for ISDN.

IA5 *See* International Alphabet No. 5.

ICMP *See* Internet Control Message Protocol.

IDN *See* Integrated Digital Networks.

IEEE 802 Set of standards for LANs being developed by IEEE 802 Committee.

IEEE 802.2 Logical link control protocol developed by IEEE 802 Committee.

IEEE 802.3 CSMA/CD-MAC protocol developed by IEEE 802 Committee.

IEEE 802.4 Token bus MAC protocol developed by IEEE 802 Committee.

IEEE 802.5 Token ring MAC protocol developed by IEEE 802 Committee.

I/G Bit Bit in IEEE 802 MAC address field distinguishing between individual and group addresses.

IMP *See* CSNP.

Indication A primitive returned to layer $(N+1)$ from layer N to advise of activation of a requested service or of an action initiated by the layer N service provider.

Indication Circuit (I) X.21 circuit used to send control information from DCE to DTE.

Information Field A field containing user data.

Information Frame Frame in HDLC, DDCMP, or related protocols containing user data.

Initial Sequence Number (ISN) Generated at each end of TCP connection to help to uniquely identify that connection.

Input Buffer Limiting Buffering strategy that divides buffer at a node into two classes, both available to transit packets but only one available to packets input at node.

Integrated Digital Networks (IDN) Integrated networks for transmission of voice and data for which standards have been developed by telephone companies in past two decades.

Integrated Services Digital Network (ISDN) Integrated network being developed by telephone companies to serve all data communications needs of users.

Interface Boundary between equipment or between protocol layers, across which information flows.

International Alphabet No. 5 (IA5) Internationally standardized alphanumeric code with national options. ASCII is United States version.

International Organization for Standardization (ISO) International standards organization responsible for telecommunications networking standards including OSI Reference Model.

Internet Control Message Protocol (ICMP) DARPA Internet Protocol that allows hosts to interact with gateways, and hosts and gateways to interact with internet monitoring and control centers.

Internet Layer Layer II of DoD Reference Model.

Internet Protocol (IP) Protocol for datagram transmission across interconnected networks. Both DoD Reference Model and OSI Reference Model include versions of IP.

Internetworking Interconnection of various networks.

Interrupt Packet High priority X.25 packet that can contain up to 32 bytes of user data.

Intersymbol Interference Distortion of signals due to preceding or following pulses affecting desired pulse amplitude at time of sampling.

Inverse Fourier Transform Inversion of Fourier transform to convert frequency representation of signal to time representation.

IP *See* Internet Protocol.

Irland's Square Root Rule Rule used in DNA to define threshold for discarding packets and blocking input packets because of congestion.

Isarithmic Flow Control Approach to flow control in which transmission permits circulate throughout network. Node wishing to transmit must first capture permit and destroy it, then recreate permit after transmission finished.

ISDN *See* Integrated Services Digital Network.

ISDN Reference Model Overall architecture for ISDN; basically similar to OSI Reference Model except for separating functions into user, control, and management planes.

ISN *See* Initial Sequence Number.

ISO *See* International Organization for Standardization.

Isochronous Transmission Transmission of characters in start-stop format but with both ends of communications link clocked.

Labeling Algorithm Algorithm for shortest path routing or similar problems which labels individual nodes, updating labels as appropriate to reach a solution.

LAN *See* Local Area Network.

LAPB CCITT X.25 version of bit-oriented data link control protocol.

LAPD ISDN D Channel version of bit-oriented data link control protocol.

Layered Network Architectures Currently the basis of all telecommunication network architecture standards, with functions allocated to different layers and standardized interfaces between layers.

Level I Route DNA route within subarea.

Level II Route DNA route between subareas.

Line Turnaround Time Time required to reverse direction of transmission on communication line.

Link Communication facilities interconnecting a pair of
   nodes in a telecommunication network.

LLC See Logical Link Control.

Local Area Network (LAN) Network spanning an area
   of not more than a few kilometers in diameter.

Logical Channel Number Identification of X.25 virtual
   circuit used by DCE and DTE pair at one end of
   virtual circuit.

Logical Link Control (LLC) Upper half of data link
   layer in IEEE 802 architecture.

Logical Unit (LU) User port into SNA network. Comprises the upper layers of the architecture.

Looping Routing which sends packets around loop
   returning to starting point.

LU See Logical Unit.

MAC See Medium Access Control.

MAN See Metropolitan Area Network.

Manufacturing Automation Protocol (MAP)
   Networking architecture, based on token bus protocol, developed by General Motors for automating
   manufacturing processes.

Master Station Station in charge of error recovery, and
   so forth, on Bisync link.

Mechanical Characteristics Include physical dimensions of plugs or connectors, assignment of circuits to pins, connector latching and mounting
   arrangements in physical layer.

Medium Access Control (MAC) Lower half of data link
   layer in IEEE 802 architecture; handles medium
   access functions. Used to keep stations sharing
   common communications link from interfering
   with each other.

Message Switching Switching technique involving
   transmission of messages from node to node, with
   storage at intermediate nodes until next portion of
   path available.

Metropolitan Area Network (MAN) Network spanning
   area up to a few dozen kilometers in diameter.

MILNET See Defense Data Network.

Modem Modulator/demodulor used to translate digital
   data into form suitable for transmission over analog communication facilities and vice versa.

Modulation Technique to transform frequency characteristics of signal to fit it into passband of communications channel or otherwise improve its
   characteristics for transmission.

Multidrop Line Communications line geometry with all
   stations attached to branches off main communications link. Each station can communicate with a
   central station, but stations do not necessarily
   hear each other otherwise.

Multiple Token Operation Variant of token passing for
   rings in which a free token is transmitted immediately after the last bit of the data packet, allowing
   multiple tokens on ring (but only one free token)
   simultaneously.

Multiplexing Technique for sharing facilities employing
   fixed assignment of portions of capacity.

NAK See Negative Acknowledge.

Name Symbol identifying some type of resource.

NAU See Network Addressable Unit.

NCP See Network Control Point.

Negative Acknowledge (NAK) A communications control character transmitted by a receiver as a negative acknowledgment response to the sender.

Negative Acknowledgment Message returned from
   receiver to transmitter to indicate unsuccessful
   reception of transmission.

Network Access Layer Layer I of DoD Reference Model.

Network Addressable Unit (NAU) Addressable entity
   in SNA: SSCP, LU, or PU.

Network Applications Layer Layer 6 of DNA.

Network Control Point (NCP) Control data base in
   common channel signalling network.

Network Layer Layer 3 of OSI Reference Model.

Network Management Layer Provides decentralized
   management for a DNA network.

Network Termination 1 (NT1) ISDN equipment that
   includes functions belonging to OSI Layer 1.

Network Termination 2 (NT2) Intelligent ISDN device
   that may include up through OSI Layer 3
   functionality.

Network Termination 1,2 (NT12) Single ISDN device
   that contains combined functions of NT1 and NT2.

Node Data processing and other equipment at one geographical location in a telecommunication
   network.

Non-Data Bit Bit with encoding violating normal format;
   used for special control purposes.

Nonpersistent CSMA Variant of CSMA in which station
   sensing carrier waits random backoff time and
   repeats carrier sensing.

Normalized Average Transfer Delay ($\hat{T}$) Average
   transfer delay (T) divided by packet transmission
   time at the clock rate of the medium.

Normalized Network Throughput (S) Network
   throughput in packets per second divided by maximum throughput possible at clock rate of medium.
   Less than one.

Normalized Offered Traffic (G) The average number
   of attempted packet transmissions per second
   divided by the average number of packet transmis-
   sions/second possible at the clock rate of the
   medium. May exceed one.

Normal Response Mode (NRM) HDLC mode for use on links with one primary and one or more secondaries. Under NRM, a secondary can transmit only after receiving a poll addressed to it by a primary; it may then send a series of responses, but after it sets the F bit in a response, it cannot transmit any more until it receives another poll.

NPDU *See* Protocol Data Unit.

N(R) HDLC Receive sequence count. Used for piggy-backed acknowledgments.

NRM *See* Normal Response Mode.

N(S) HDLC Send sequence count.

NT1 *See* Network Termination 1.

NT2 *See* Network Termination 2.

NT12 *See* Network Termination 1,2.

Null Modem Appropriately wired RS-232C cable allowing two DTEs to talk directly to each other.

Nyquist Intersymbol Interference Theorem Theorem proved by Nyquist showing that transmission at Nyquist transmission rate is possible with zero intersymbol interference.

Nyquist Sampling Rate Minimum rate of sampling of analog waveform needed to allow perfect reconstruction of waveform from samples.

Nyquist Sampling Theorem Theorem proved by Nyquist and showing it is possible to precisely reconstruct analog signals from samples if an adequate number of samples is taken.

Nyquist Transmission Rate Maximum number of pulses per second that can be transmitted over channel without intersymbol interference; equal to twice bandwidth in Hz.

Optical Fiber Communications medium used for fiber optics.

OSI Reference Model Telecommunication networking architecture developed by International Organization for Standardization (ISO) and adopted as an international standard; also known as Reference Model for Open Systems Interconnection.

Pacing IBM terminology for type of window flow control used in SNA.

Packet Assembler/Disassembler (PAD) Used to connect simple, character mode terminal, which is not capable of assembling and disassembling packets, to X.25 network. Performs these functions for terminal.

Packet Discarding Approach to flow control that simply discards packets when congestion occurs.

Packet Layer Layer of X.25 corresponding to partial OSI network layer. (Packet level in X.25 standards.)

Packet Radio Network Telecommunication network using radio links and packet switching.

Packet Switching Switching technique involving breaking long messages up into shorter packets transmitted individually through network, with storage at intermediate nodes until next portion of route available.

PAD *See* Packet Assembler/Disassembler.

Parity Bit Extra bit added to transmitted data to help in error control. Bit ensures number of ones in data, or subset of data, is even (for even parity) or odd (for odd parity).

Passive Open Used in TCP to put node into state in which it is listening for open requests from other nodes.

Path Control Layer Layer 3 in SNA.

PDU *See* Protocol Data Unit.

Peer-to-Peer Communication Standard communications process in layered network architectures, with entities at same layer talking to each other.

Peripheral Node SNA node that does not recognize full network addresses. Classified as terminal node or cluster controller node.

Permanent Virtual Circuit (PVC) X.25 virtual circuit that is permanently set up so no connection or disconnection phases are needed.

P/F *See* Poll/Final bit.

Phase Modulation (PM) Modulation technique conveying information via variations in phase of carrier signal.

Phase Shift Keying (PSK) Digital version of PM.

Physical Control Layer SNA terminology for OSI physical layer.

Physical Layer The lowest layer of the OSI Reference Model and most other networking architectures.

Physical Unit (PU) SNA logical interface for device control functions.

Piggy-Backed Acknowledgments or Sequence Counts Acknowledgments or sequence counts implemented by using special fields in ordinary data messages, so separate transmissions are not needed.

Ping-Ponging Routing that causes packet to bounce back and forth between two nodes.

Pipelining Simultaneous use of different facilities to accomplish parts of a job, giving speed up due to simultaneity.

PM *See* Phase Modulation.

Poll/Final Bit Bit in HDLC frame control field. If frame is a command, bit is a poll bit asking station to reply. If frame is a response, bit is a final bit identifying last frame in message.

Polling Technique for controlling access to communications medium by central station. Central station sends out message (poll) to other stations to enable them to transmit.

Polynomial Code Class of error protection codes based on properties of polynomials over a field of two elements.

Positive Acknowledgment *See* Acknowledgment.

***p*-persistent CSMA** Variant of CSMA in which station sensing carrier continues to sense until no carrier is present, then immediately transmits with probability *p* and backs off with probability 1 - *p*.

P(R) X.25 receive sequence count.

P(S) X.25 send sequence count.

Preamble Bit pattern used in IEEE 802 MAC for stabilization and synchronization.

Presentation Layer Layer 6 of OSI Reference Model.

Presentation Services Layer Layer 6 in SNA.

Primary Station Station in charge of HDLC link.

Primitive Message passed between adjacent layers in layered networking architecture. Not externally visible.

Procedural Specifications Specify sequences of control and data messages to set up, use, and deactivate physical layer connections.

Process/Applications Layer Layer IV (highest layer) of DoD Reference Model.

Propagation Time Time required for signal from transmitter to reach receiver at propagation speed of communications medium.

Protocol Set of mutually agreed upon rules of procedure stating how two or more parties are to interact to exchange information.

Protocol Data Unit (PDU) Type of data unit used by a particular protocol. Often preceded by letter to indicate protocol layer—for example, NPDU and TPDU for network and transport layers.

PSK *See* Phase Shift Keying.

PU *See* Physical Unit.

Pulse Code Modulation (PCM) Transmission of analog information in digital form via first sampling it and then encoding each sample with a fixed number of bits.

PVC *See* Permanent Virtual Circuit.

Quality of Service (QOS) Parameters such as error rates and time delays that describe quality of network service as perceived by user.

Quantization Encoding samples of an analog waveform into one of a finite number of possible values in order to allow digital encoding of samples.

R *See* Receive Circuit and Ring Indicator.

Raised Cosine Spectrum Pulse A pulse with spectrum proportional to a constant plus a portion of a cosine function and producing zero intersymbol interference when transmitted at appropriate rate.

R-ALOHA *See* Reservation ALOHA.

Random Access Unscheduled access to communications medium in which stations transmit when ready (possibly after sensing medium), and later resolve any conflicts that arise.

Random Routing Routing in which each node en route from source to destination forwards packet on randomly chosen outgoing link.

RD *See* Received Data.

R_e *See* Effective Rate of Transmitting Data Bits.

Ready for Next Message (RFNM) ARPANET message returned by destination of message to allow source to advance its window.

Reassembly Reconstituting messages or packets from shorter packets, fragments, or segments.

Reassembly Deadlock Type of deadlock due to shortage of reassembly buffers.

Reassignment OSI terminology for reestablishment of connection.

Receive Circuit (R) Circuit in X.21 interface used to send data from DCE to DTE.

Receive Count Field in header giving count of successive frames received correctly and allowing piggybacked acknowledgments.

Received Data (RD) The sequence of pulses representing data received by the DCE is placed on this circuit in RS-232C interface.

Received Line Signal Detector (RLSD) The circuit that is activated by the modem (data set) in RS-232C interface when it is detecting a carrier signal from another modem.

Receive Not Ready Frame (RNR) HDLC frame similar to RR, except for indicating the station is temporarily not ready to receive information frames.

Receive Ready Frame (RR) General purpose HDLC frame often used for acknowledgment; it indicates station has correctly received all frames with N(S) numbers (used by station RR is sent to) less than N(R), and is looking for valid frame with N(S) equal to N(R).

Receive Window Set of packets or data units receiver is allowed to accept.

Reference Model for Open Systems Interconnection *See* OSI Reference Model.

Reference Point Conceptual point in ISDN architecture used to separate groups of functions.

Reference Point R ISDN reference point providing a non-ISDN interface between non-ISDN compatible user equipment and adaptor equipment.

Reference Point S ISDN reference point separating user terminal equipment from the network user functions at the interface to NT2.

Reference Point T ISDN reference point separating NT1 equipment from NT2 equipment.

Reference Point U ISDN reference point providing an interface to communications line. Not internationally recognized.

Reject Frame (REJ) HDLC frame used as negative acknowledgment for frame with send count equal to N(R), and requesting recovery by Go-Back-N retransmission.

Relay OSI terminology for a device used to interconnect two systems not directly connected to each other.

REP *See* Reply Supervisory Frame.

Repeater Physical layer relay.

Reply Supervisory Frame (REP) DDCMP supervisory frame requesting a reply to the frame specified by a send count.

Request A primitive sent by layer $(N+1)$ to layer N to request a service.

Request to Send (RTS) Data processing equipment (DTE) activates this circuit in RS-232C interface when it is ready to send data.

Reservation ALOHA (R-ALOHA) Modification of ALOHA medium access technique in which stations are allowed to reserve part of capacity of medium before using it.

Reservations Technique for multiple access in which stations reserve part of capacity of medium before using it.

Reset X.25 error control procedure that reinitializes a virtual circuit by returning lower edge of the window for each transmission direction to 0 and discarding all data and interrupt packets in the network. Virtual circuit remains connected and in the data transfer state.

Response A primitive provided by layer $(N+1)$ in reply to an indication primitive. Also a frame sent from secondary to primary on HDLC link.

Restart Used for serious X.25 error conditions; affects all virtual circuits, clearing all virtual calls and resetting all permanent virtual circuits.

Resynchronization Used in OSI to ensure that both ends of a transport connection retransmit lost, or unacknowledged, TPDUs and are up to date on sequence numbers.

RFNM *See* Ready for Next Message.

Ring Geometry for LAN or other network with stations attached to link in shape of ring (that is, closing on itself after passing through all stations). All communication normally flows in one direction around ring.

Ring Indicator (R) The modem (DCE) activates this circuit in RS-232C interface to tell the DTE that a ringing signal due to an incoming call is being received.

RLSD *See* Received Line Signal Detector.

RNR *See* Receive Not Ready Frame.

Roll Call Polling Polling technique in which the central station maintains list of stations to be polled and periodically polls each station in turn.

Route Information needed to send a message to a specified address.

Route Matrix Matrix used to indicate all routes, with ij entry the next to last node along the path from i to j.

Router Network layer relay.

Routing Technique for sending message to a specified address.

Routing Layer Layer 3 of DNA.

Routing Node Node in DNA network that handles routing functions.

Routing Table Table containing information used by node and maintained at that node.

RR *See* Receive Ready Frame.

RS-232C Most common physical layer interface in United States. Virtually identical to EIA-232D interface and international V.24 interface.

RTS *See* Request to Send.

S *See* Normalized Network Throughput.

Sampling Determining values of an analog waveform at predetermined instants of time.

SD *See* Start Delimiter.

Scheduling Setting up orderly sequence of access to communications medium to avoid conflicts.

Secondary Station Any station on HDLC link that is not the primary station.

Segmentation The breaking up of a packet into shorter packets for transmission across network with short maximum packet length. Sometimes called fragmentation.

Selection Approach to transmission from control station on multiaccess link to another station in which control station appends address to message and sends it, possibly after first verifying that the station is ready to receive.

Selective Reject ARQ Variant of continuous ARQ in which only the erroneous frame is retransmitted after error has been detected.

Selective Reject Frame (SREJ) HDLC frame used as negative acknowledgment for frame, but only asking for retransmission of this one frame.

Semantics Meanings of signals exchanged, including control information for coordination and error handling.

Send Count Field in header giving sequence number of frame being transmitted.

Service Access Point (SAP) Point at which a service is provided.

Session Control Layer Layer 5 of DNA.

Session Layer Layer 5 of OSI Reference Model.

Shannon Channel Capacity Maximum rate at which it is theoretically possible to transmit information across communications channel with probability of error approaching zero.

Shannon Channel Capacity Theorem Theorem describing theoretical limits on transmitting information with error probability approaching zero.

Shortest-Path Routing Routing that minimizes cost or distance function.

Shortest-Path Spanning Tree Tree of all shortest paths from one source node to all possible destination nodes or to one destination node from all possible source nodes.

Signal Constellation Set of signals used at transmitter in communications system.

Signal Element Timing Circuit X.21 circuit used to help maintain bit synchronization.

Signaling Terminal Exchange (STE) Relay or gateway used by X.75.

Signal Transfer Point (STP) Node in signalling network for common channel signalling.

Simplex Transmission Data transmission in only one direction.

Single Packet Operation Variant of token passing for rings in which the free token is transmitted after the transmitting station has received the last bit of its transmitted packet.

Single Token Operation Variant of token passing for rings in which the free token is transmitted after the transmitting station receives the last bit of its busy token (and has also transmitted the last bit of its data packet).

Slave Station Subservient station on Bisync link.

Sliding Window Flow Control Type of flow control in which transmitter is given permit to transmit "window" of packets or other data units, and not allowed to transmit more until it receives another permit.

Slotted-ALOHA Modification of ALOHA medium access technique in which time is divided into slots of duration equal to packet duration and each transmission is required to start at the beginning of a slot.

SNA *See* Systems Network Architecture.

SNAF *See* Subnetwork Access Functions Sublayer.

SNA Interface Layer DNA/SNA gateway layer providing access into the SNA network.

SNDCF *See* Subnetwork Dependent Convergence Functions Sublayer.

SNICF *See* Subnetwork Independent Convergence Functions Sublayer.

Source Routing Routing determined by source of packet and implemented by including information describing route in packet header.

Source Routing Bridge Bridge being developed by IEEE 802.5 committee to interconnect IEEE 802.5 token ring LANs.

Source Service Access Point (SSAP) Address Field Field in IEEE 802 LLC header identifying SSAP address.

SOH *See* Start of Heading.

SONET Synchronous Optical Network standard recently approved as international standard.

Spanning Tree Bridge Bridge being developed by IEEE 802.1 committee to interconnect IEEE 802 LANs.

SPC *See* Stored Program Controlled Switching System.

SREJ *See* Selective Reject Frame.

SSCP *See* System Services Control Point.

Start Acnowledgment Frame (STACK) DDCMP frame to acknowledge STRT Frame.

START Bit Extra bit appended at the beginning of each character in start-stop transmission to help receiver synchronize to beginning of character.

Start Delimiter (SD) Bit pattern indicating start of IEEE 802 MAC frame. For token ring and token bus it includes nondata bits, so SD is easy to recognize.

Start Frame (STRT) DDCMP frame to establish initial contact and synchronization on link.

Start of Heading (SOH) Communications control character used in Bisync at beginning of sequence of characters that contain address or routing information; referred to as the "heading." STX terminates the heading.

Start of Text (STX) Communications control character that precedes a sequence of Bisync characters to be treated as entity and transmitted through to destination. Such a sequence is referred to as "text"; may also terminate sequence of characters begun by SOH.

Start-Stop Data link control protocol for asynchronous transmission of individual characters, each with appended start and stop bits.

State Diagram Diagram illustrating possible states and allowable transitions between states for protocol.

STE *See* Signalling Terminal Exchange.

STM *See* Synchronous Transfer Mode.

STP *See* Signal Transfer Point.

Stop-and-Wait ARQ Version of ARQ with separate frames for all positive or negative acknowledgments. After transmitting frame, transmitter waits for reply before sending another frame.

STOP Bit Extra bit appended at end of each character in start-stop transmission to help receiver synchronize to breaks between characters.

Store and Forward Deadlock Deadlock due to shortage of transmit or receive buffers.

Stored Program Controlled Switching System (SPC) Digital telephone switching office using special purpose computer for control.

STRT *See* Start Frame.

Structured Buffer Pool Buffer pool structured in manner that guarantees store and forward deadlocks cannot occur.

STX *See* Start of Text.

Subarea Node SNA node handling complete network

addresses. Classified as communication controller node or host node.

Subnetwork Access Functions Sublayer (SNAF) Sublayer of OSI network layer; specifies how network layer entities make use of functions of network (for example, operation of protocol describing interface to network). X.25 fits into this sublayer.

Subnetwork Dependent Convergence Functions Sublayer (SNDCF) Sublayer of OSI network layer; allows for situations where networks do not provide all features assumed by SNICF sublayer.

Subnetwork Independent Convergence Functions Sublayer (SNICF) Sublayer of OSI network layer; contains functions that do not require accommodation to networks. Primary functions are internetwork routing and relaying plus other functions to implement internetwork protocol. OSI IP protocol fits into this sublayer.

Supervised Mode Mode of operation in which one station controls access to link.

Supervisory Frame Frame in HDLC, DDCMP, or related protocols used for link supervisory purposes.

SYN *See* Synchronous Idle.

Synchronization Establishing common timing between transmitter and receiver.

Synchronous Precisely clocked oepration.

Synchronous Data Link Control (SDLC) IBM's version of bit-oriented data link control protocol.

Synchronous Idle (SYN) Communications control character used in Bisync to provide signal from which synchronization may be achieved or retained; may also be transmitted as fill character.

Synchronous Transfer Mode (STM) Broadband ISDN multiplexing mode based on ordinary time division multiplexing.

Synchronous Transmission Transmission of character streams with successive characters occurring at precisely clocked intervals.

Syntax Structure of information communicated, including such things as data format, coding, representation in terms of signal levels, and so forth.

System Services Control Point (SSCP) Functions in some SNA host nodes to control operation of network or domain within network.

Systems Network Architecture (SNA) IBM's layered architecture for telecommunication networks.

T *See* Average Transfer Delay or Transmit Circuit.

T̂ *See* Normalized Average Transfer Delay.

T-1 Carrier System Digital telephone carrier system developed by Bell Labs in United States as first purely digital carrier system.

TA *See* Terminal Adaptor.

TCP *See* Transmission Control Protocol.

TD *See* Transmitted Data.

TDM *See* Time Division Multiplexing.

TDMA *See* Time Division Multiple Access.

TE *See* Terminal Equipment.

TE1 *See* Terminal Equipment Type 1.

TE2 *See* Terminal Equipment Type 2.

Technical and Office Products System (TOP) Networking architecture, based on CSMA/CD bus or token ring, developed by Boeing Computer Services to automate office operations.

Telecommunications Term for technology involving use of both computers and communications to expand capabilities of both.

TELENET Major packet data network in United States.

Terminal Adaptor (TA) Converts non-ISDN interface functions of TE2 into ISDN acceptable form.

Terminal Equipment (TE) ISDN terminology for DTE.

Terminal Equipment Type 1 (TE1) ISDN terminal; complies with ISDN standards.

Terminal Equipment Type 2 (TE2) Non-ISDN terminal; supports functions similar to those supported by TE1 equipment, but does not comply with ISDN interface standards. Must be accessed via terminal adaptor.

Throughput Rate of correctly received traffic passing network boundary.

Time Division Multiple Access (TDMA) Multiple access protocol based on TDM.

Time Division Multiplexing (TDM) Communication channel sharing technique carrying different signals in different time slots.

Timing Diagram Diagram illustrating typical sequence of transitions between states over time.

Timings Times at which data should be transmitted or looked for by a receiver, sequencing of information, speed matching, and so forth.

Token Bus LAN architecture using bus geometry for medium and passing token around to stations attached to bus. Owner of token has temporary control of network.

Token Passing Medium access technique used by token bus and token ring.

Token Ring LAN architecture using ring geometry for medium and passing token around ring. Owner of token has temporary control of network.

TOP *See* Technical and Office Products System.

Topology *See* Geometry.

TPDU *See* Protocol Data Unit.

Trailer Information attached to end of packet or message by Layer *N* at transmitter for interpretation by Layer *N* at receiver.

Transaction Services Layer Layer 7 in SNA.

Transmission Control Layer Layer 4 in SNA.

Transmission Control Protocol (TCP) DoD Reference Model protocol at equivalent of OSI transport layer. Converts datagram service to equivalent of virtual circuit service.

Transmission Group One or more parallel communications links treated as one communications facility in SNA routing.

Transmit Circuit (T) Circuit in X.21 interface used to send data from DTE to DCE.

Transmit Window Set of successive packets or data units transmitter is allowed to send at given time.

Transmitted Data (TD) Circuit on which sequence of pulses representing data sent out from DTE is placed at times dictated by procedural specifications in RS-232C protocol.

TRANSPAC Major packet data network in France.

Transparent Text Mode Transmission mode allowing arbitrary bit patterns to be transmitted.

Transport Layer Layer 4 of OSI Reference Model.

Tree Communications geometry in which single path exists between any two stations, but no loops or multiple paths exist.

Trellis Coding Error protection coding technique that uses densely packed signal constellation but restricts sequence of permissible signals.

Tributary Station Any station on a multi-access link that is not the control station.

Triple X X.3, X.28, and X.29.

Turnaround Time Time to reverse direction of transmission on communications channel or other equipment.

Twisted Pair Most common type of communications link between telephone office and subscriber; consists of pair of copper wires with twists to minimize noise pickup.

Two Dimensional Parity Check Error protection encoding with parity bit added to each character and overall parity character checking corresponding bits in each character.

TYMNET Major packet data network in United States.

Type A Network Service OSI terminology for network connections with acceptable residual error rate and rate of signaled failures.

Type B Network Service OSI terminology for network connections with acceptable residual error rate but unacceptable rate of signaled failures.

Type C Network Service OSI terminology for network connections with residual error rate not acceptable to the transport service user.

Type 1 Operation Connectionless operation for IEEE 802 LLC.

Type 2 Operation Connection-oriented operation for IEEE 802 LLC.

UDP *See* User Datagram Protocol.

U/L Bit Bit in IEEE 802 MAC 48-bit address field distinguishing addresses assigned by universal and local authorities.

Unbalanced Configuration Network configuration in HDLC with one primary station and one or more secondary stations.

Unbalanced Mode Transmission Data transmission with information conveyed by voltages on circuits, all measured with respect to common ground.

Unitdata Data unit used in connectionless mode transmission.

Universal Synchronous/Asynchronous Receiver Transmitter (USART) Standardized component of computer equipment interface produced by semiconductor companies. Handles many functions of interface.

Unnumbered Frames HDLC frames for housekeeping purposes, including link startup and shutdown, specifying modes, and so forth.

USART *See* Universal Synchronous-Asynchronous Receiver Transmitter.

User Datagram Protocol (UDP) Datagram-oriented protocol in DoD Reference Model at equivalent of OSI transport layer.

User Layer Highest layer of DNA.

V.24 Most common physical layer interface in many countries. Largely equivalent to RS-232C and EIA-232D.

Virtual Call (VC) X.25 virtual circuit that is set up when needed, then disconnected afterward.

Virtual Circuit Transmission path set up, end to end, by connection protocol before transmission.

Virtual Route SNA terminology for virtual circuit.

Voice-Grade Channel Standard telephone channel for voice communications; useful bandwidth around 3 kHz.

WACK (wait acknowledgment) Response returned in some DLC protocols to acknowledge received frame and also indicate that receiver is temporarily unable to accept more frames.

Wide Area Network (WAN) Network spanning large geographical area, possibly world wide.

Worldnet Internetwork consisting of interconnected networks around the globe; used daily by various communities throughout world.

X.3 Defines the PAD; normally implemented in software in DCE interfacing simple character mode DTE to X.25 network.

X.21 Physical layer interface included in OSI Reference Model architecture.

X.21bis Interim physical layer interface to ease migration from RS-232C or similar interfaces to X.21.

X.25 Standard defining user interface to public data network. Partial implementation of three lower OSI layers.

X.28 Defines interface in X.25 network between character mode DTE and DCE containing PAD, including data and control exchanges between them.

X.29 Describes protocols allowing parameters in PAD to be set by intelligent packet mode DTE in X.25 network.

X.75 Protocol to interconnect X.25 networks. Sets up concatenated virtual circuits between two ends of connection.

Zero Insertion Algorithm Algorithm used in HDLC to avoid occurrence of flag in main body of frame. Transmitter inserts zero in transmitted bit stream any time five successive ones are observed in this part of a frame; zero is removed by receiver.

Bibliography

[AARO62] Aaron, M. R. "PCM Transmission in the Exchange Plant." *Bell System Technical Journal,* 41 (January 1962):99–141.

[ABAT89] Abate, J. E.; Butterline, E. W.; Carley, R. A.; Greendyk, P.; Montenegro, A. M.; Near, C. D.; Richman, S. H.; and Zampetti, G. P. "AT&T's New Approach to the Synchronization of Telecommunication Networks." *IEEE Communications Magazine,* 27, no. 4 (April 1989):35–45.

[ABRA63] Abramson, N. *Information Theory and Coding.* New York: McGraw-Hill, 1963.

[ABRA70] Abramson, N. M. "The Aloha System, Another Alternative for Computer Communications." *AFIPS Conference Proceedings, 1970 Fall Joint Computer Conference* 37 (1970):281–85.

[ABRA73] Abramson, N. "Packet Switching with Satellites." *Proceedings of Compcon 73* (1973):695–702.

[ABRA80] Abrams, M.; Blanc, R. P.; and Cotton, I. W., eds. *Computer Networks: Text and References for a Tutorial,* 1980. Long Beach, Calif.: IEEE Computer Society, 1980.

[ABRA87] Abrams, M. D., and Jeng, A. B. "Network Security: Protocol Reference Model and the Trusted Computer System Evaluation Criteria." *IEEE Network* 1, no. 2 (April 1987):24–33.

[ACAM87] Acampora, A. S.; Karol, M. J.; and Hluchyj, M. G. "Terabit Lightwave Networks: The Multihop Approach." *AT&T Technical Journal* 66, no. 6 (November/December 1987):21–34.

[ACAM89] Acampora, A. S., and Karol, M. J. "An Overview of Lightwave Packet Networks." *IEEE Network* 3, no. 1 (January 1989):29–41.

[ADAM81] Adams, M. J. *An Introduction to Optical Waveguides.* New York: Wiley, 1981.

[ADOU87] Adoul, J. P. "Speech-Coding Algorithms and Vector Quantization," in [FEHE87a], pp. 133–81.

[AHUJ82] Ahuja, V. *Design and Analysis of Computer Communication Networks.* New York: McGraw-Hill, 1982.

[ALDE86] Aldermeshian, H. "ISDN Standards Evolution." *AT&T Technical Journal* 65, no. 1 (January/February 1986):19–26.

[ALIS85] Alisouskas, V. F., and Tomasi, W. *Digital and Data Communications.* Englewood Cliffs, N.J.: Prentice-Hall, 1985.

[AMER79] American National Standards Institute (ANSI), *Advanced Data Communication Control Procedures.* New York, 1979.

[AMER80] American National Standards Institute (ANSI), *Determination of Performance of Data Communication Systems that Use Bit-Oriented Control Procedures.* New York, 1980.

[ANDE79] Anderson, R. R.; Foschini, G. J.; and Gopinath, B. "A Queueing Model for a Hybrid Data Multiplexer." *Bell System Technical Journal* 58, no. 2 (February 1979):279–301.

[ANDE86] Andersson, J. O.; Bauer, A.; and Carlqvist, B. "An LSI Implementation of an ISDN Echo Canceller: Design and Network Aspects." *IEEE Journal on Selected Areas in Communications* SAC-4, no. 8 (November 1986):1350–58.

[ANDE87a] Anderson, J. M.; Frey, D. R.; and Miller, C. M. "Lightwave Splicing and Connector Technology." *AT&T Technical Journal* 66, no. 1 (January/February 1987):45–64.

[ANDE87b] Anderson, C. P. "ISDN Market Opportunity." *IEEE Communications Magazine* 25, no. 12 (December 1987):55–59.

[ANDR82] Andrews, D. W., and Schultz, G. D. "A Token-Ring Architecture for Local-Area Networks: An Update." *Proceedings of COMPCON Fall '82* (1982):615–24.

[ANDR84] Andrews, F. T., Jr. "ISDN '83." *IEEE Communications Magazine* 22, no. 1 (January 1984):6–10.

[APPE86] Appenzeller, H. R. "Signaling System No. 7, ISDN User Part." *IEEE Journal on Selected Areas in Communications* SAC-4, no. 3 (May 1986):366–71.

[ARMB87] Armbrüster, H., and Arndt, G. "Broadband Communication and Its Realization with Broadband ISDN." *IEEE Communications Magazine* 25, no. 11 (November 1987):8–19.

[ARNO86] Arnon, E.; Chomik, W.; and Elder, M. "Transmission System for ISDN Loops." *Telesis* 13, no. 3 (1986):34–45.

[ATKI80] Atkins, J. D. "Path Control—The Transport Network of SNA." *IEEE Transactions on Communications* COM-28, no. 4 (April 1980):527–38.

[ATKI82] Atkins, J. D. "Path Control—The Network Layer of System Network Architecture," in [GREE82a], pp. 297–326.

[AUER86] Auerbach, J. "File Request Transparency between Heterogeneous Systems." *Proceedings of IEEE International Conference on Communications* (June 1986).

[BACK88] Backes, F. "Transparent Bridges for Interconnection of IEEE 802 LANs." *IEEE Network* 2, no. 1 (January 1988):5–9.

[BALK71] Balkovic, M. D.; Klancer, H. W.; Klare, S. W.; and McGruther, W. G. "High Speed Voiceband Data Transmission Performance on the Switched Telecommunications Network." *Bell System Technical Journal* 50 (April 1971):1349–84.

[BALL89] Ballart, R., and Ching, Y.-C. "SONET: Now it's the Standard Optical Network." *IEEE Communications Magazine* 27, no. 3 (March 1989):8–15.

[BARA86] Baratz, A. E., and Jaffe, J. M. "Establishing Virtual Circuits in Large Computer Networks." *Computer Networks and ISDN Systems* 12, no. 1 (August 1986):23–37.

[BARB83] Barberis, G.; Calabrese, M.; Lambarelli, L.; and Roffinella, D. "Coded Speech in Packet-Switched Networks: Models and Experiments." *IEEE Journal on Selected Areas in Communications* SAC-1, no. 6 (December 1983):1028–38.

[BARK86] Barksdale, W. J. *Practical Computer Data Communications.* New York: Plenum Press, 1986.

[BART83] Bartoli, P. D. "The Application Layer of the Reference Model of Open Systems Interconnection." *Proceedings of the IEEE* 71, no. 12 (December 1983):1404–7.

[BASS84] Bass, P. F., and Brennan, W. "Characteristics of Good Neighbors in a Packet Radio Network." *Proceedings of IEEE Globecom '84.* (1984):498–503.

[BECK86] Beck, P. R., and Krycka, J. A. "The DECnet-VAX Product—An Integrated Approach to Networking." *Digital Technical Journal* (Digital Equipment Corporation, Hudson, Mass.), no. 3 (September 1986):88–99.

[BEER72] Beere, M. P., and Sullivan, N. C. "Tymnet, A Serendipitous Evolution." *IEEE Transactions on Communications* COM-20, no. 3, Part 2 (June 1972):511–15.

[BELL57] Bellman, R. *Dynamic Programming.* Princeton: Princeton University Press, 1957.

[BELL58] Bellman, R. E. "On a Routing Problem." *Quarterly of Applied Mathematics* 16 (1958):87–90.

[BELL62] Bellman, R., and Dreyfus, S. *Applied Dynamic Programming.* Princeton: Princeton University Press, 1962.

[BELL82a] Bellamy, J. *Digital Telephony.* New York: Wiley, 1982.

[BELL82b] Bell Telephone Laboratories, *Transmission Systems for Communications,* 5th ed. Holmdel, N.J.: Bell Telephone Laboratories, 1982.

[BELL84] Bellchambers, W. H.; Francis, J.; Hummel, E.; and Nickelson, R. L. "The International Telecommunication Union and Development of Worldwide Telecommunications." *IEEE Communications Magazine* 22, no. 5 (May 1984):72–83.

[BELL86] Bell, P. R., and Jabbour, K. "Review of Point-to-Point Network Routing Algorithms." *IEEE Communications Magazine* 24, no. 1 (January 1986):34–38.

[BELL89] Bell, T. E. "Technology '89, Telecommunications." *IEEE Spectrum* 26, no. 1 (January 1989):41–43.

[BEND87] Bendel, J. E. "ISDN: The Learning Continues." *IEEE Communications Magazine* 25, no. 12 (December 1987):60–63.

[BENH83] Benhamou, E., and Estrin, J. "Multilevel Internetworking Gateways: Architecture and Applications." *Computer* 16, no. 9 (September 1983):27–37.

[BENH88] Benhamou, E. "Integrating Bridges and Routers in a Large Internetwork." *IEEE Network* 2, no. 1 (January 1988):65–71.

[BENJ83] Benjamin, J. H.; Hess, M. L.; Weingarten, R. A.; and Wheeler, W. R. "Interconnecting SNA Networks." *IBM Systems Journal* 22, no. 4 (1983):344–66.

[BERL68] Berlekamp, E. R. *Algebraic Coding Theory.* New York: McGraw-Hill, 1968.

[BERL87] Berlekamp, E. R.; Peile, R. E.; and Pope, S. P. "The Application of Error Control to Communications." *IEEE Communications Magazine* 25, no. 4 (April 1987):44–57.

[BERT80] Bertine, H. V. "Physical Level Protocols." *IEEE Transactions on Communications* COM-28, no. 4 (April 1980):433–44.

[BERT82a] Bertine, H. V. "Physical Interfaces and Protocols." in [GREE82a], pp. 57–84.

[BERT82b] Bertsekas, D. P. "Dynamic Behavior of Shortest Path Routing Algorithms for Communication Networks." *IEEE Transactions on Automatic Control* AC-27, no. 1 (1982):60–74.

[BERT87] Bertsekas, D., and Gallager, R. *Data Networks.* Englewood Cliffs, N.J.: Prentice-Hall, 1987.

[BHUS83] Bhusri, G. "Optimum Implementation of Common Channel Signaling in Local Networks." *Proceedings of IEEE INFOCOM '83* (1983):129–136.

[BHUS84] Bhusri, G. "Considerations for ISDN Planning and Implementation." *IEEE Communications Magazine* 22, no. 1 (January 1984):18–32.

[BIGL86] Biglieri, Ezio, and Prati, Giancarlo, eds., *Digital Communications: Proceedings of the Second Tirrenia International Workshop on Digital Communications.* Tirrenia, Italy, Sept. 1985, North Holland, Amsterdam, 1986.

[BIND75] Binder, R. "A Dynamic Packet Switching System for Satellite Broadcast Channels." *Proceedings of International Conference on Communications* (1975): 41.1–41.5

[BIND81] Binder, R. "Packet Protocols for Broadcast Satellites," in [KUO81], pp. 175–201.

[BIND87] Binder, R.; Huffman, S. D.; Gurantz, I.; and Vena, P. A. "Crosslink Architectures for a Multiple Satellite System." *Proceedings of the IEEE* 75, no. 1 (January 1987):74–82.

[BIRR84] Birrell, A., and Nelson, B. "Implementing Remote Procedure Calls." *ACM Transactions on Computer Systems* 2, no. 1 (February 1984):39–59.

[BLAC84] Blackshaw, R. (Open Systems Data Transfer), "Integrated Services Digital Networks." OMNICOM (February 1984):1–12.

[BLAC87a] Black, U. *Computer Networks: Protocols, Standards, and Interfaces.* Englewood Cliffs, N.J.: Prentice-Hall, 1987.

[BLAC87b] Black, U. *Data Communications and Distributed Networks,* 2d ed. Englewood Cliffs, N.J.: Prentice-Hall, 1987.

[BLAC89] Black, U. *Data Networks: Concepts, Theory and Practice.* Englewood Cliffs, N.J.: Prentice-Hall, 1989.

[BLAN76] Blanc, R. P., and Cotton, I. W., eds. *Computer Networking.* New York: IEEE Press, 1976.

[BOCH77] Bochmann, G. V., and Chung, R. J. "A Formalized Specification of HDLC Classes of Procedures." *Proceedings of NTC 77,* Los Angeles, Calif., 3A2:1–11.

[BOCH80a] Bochmann, G. V. "A General Transition Model for Protocols and Communication Services." *IEEE Transactions on Communications* COM-28, no. 4 (April 1980):643–50.

[BOCH80b] Bochmann, G. V., and Sunshine, C. "Formal Methods in Communication Protocol Design." *IEEE Transactions on Communications* COM-28, no. 4 (April 1980):624–31.

[BOCH82a] Bochmann, G. V. "A Hybrid Model and the Representation of Communication Services," in [GREE82a], pp. 625–44.

[BOCH82b] Bochmann, G. V., and Sunshine, C. A. "A Survey of Formal Methods," in [GREE82a], pp. 561–78.

[BOGG80] Boggs, D. R.; Shoch, J. F.; Taft, E. A.; and Metcalfe, R. M. "Pup: An Internetwork Architecture." *IEEE Transactions on Communications* COM-28, no. 4 (April 1980):612–24.

[BOGG82] Boggs, D. R.; Shoch, J. F.; Taft, E. A.; and Metcalfe, R. M. "A Specific Internetwork Architecture (Pup)," in [GREE82a], pp. 527–56.

[BOOT81] Booth, G. M. *The Distributed System Environment: Some Practical Approaches.* New York: McGraw-Hill, 1981.

[BORG87] Borgonovo, F. "ExpressMAN: Exploiting Traffic Locality in Expressnet." *IEEE Journal on Selected Areas in Communications* SAC-5, no. 9 (December 1987):1436–43.

[BOSA88] Bosack, L., and Hedrick, C. "Problems in Large LANs." *IEEE Network* 2, no. 1 (January 1988):49–56.

[BRAD86] Brady, P. T. "Relation of User-Perceived Response Time to Error Measurements on ISDN Data Links." *IEEE Journal on Selected Areas in Communications* SAC-4, no. 8 (November 1986):1210–17.

[BRAN83] Branscomb, L. M. "Networks for the Nineties." *IEEE Communications Magazine* 21, no. 5 (1983):38–43.

[BRAN87] Branstad, D. K. "Considerations for Security in the OSI Architecture." *IEEE Network* 1, no. 2 (April 1987):34–39.

[BRAU87] Braunstein, M. R.; Burton, C. L.; and McNabb, S. D. "ASQIC 800 Call Data Master." *AT&T Technical Journal* 66, no. 3 (May/June 1987):21–31.

[BROP86] Brophy, S. G., and Falconer, D. D. "Investigation of Synchronization Parameters in a Digital Subscriber Loop Transmission System." *IEEE Journal on Selected Areas in Communications* SAC-4, no. 8 (November 1986):1312–16.

[BROW87] Brown, H. R. "British Telecom ISDN Experience." *IEEE Communications Magazine* 25, no. 12 (December 1987):70–73.

[BUFF87] Buffer, A. F. "AT&T's Pay-Per-View Television Trial." *AT&T Technical Journal* 66, no. 3, (May/June 1987):54–63.

[BUNN86] Bunner, B.; deHoog, R.; Unsoy, M.; and Wiebe, C. "DMS-100: Switching Architecture for ISDN." *Telesis* 13, no. 3 (1986):14–23.

[BURT72] Burton, H. O., and Sullivan, D. D. "Errors and Error Control." *Proceedings of the IEEE* 60, no. 11 (November 1972):1293–1301.

[BUX81] Bux, W. "Local-Area Subnetworks: A Performance Comparison." *IEEE Transactions on Communications* COM-29, no. 10 (October 1981):1465–73.

[BUX84] Bux, W., and Grillo, D. "End-to-End Performance in Local Area Networks of Interconnected Token Rings." *Proceedings of IEEE INFOCOM* (1984):60–68.

[BUX87] Bux, W.; Grillo, D.; and Maxemchuk, N. F., eds. "Special Issue on Interconnection of Local Area Networks." *IEEE Journal on Selected Areas in Communications* SAC-5, no. 9 (December 1987).

[BUX89] Bux, W. "Token Ring Local-Area Networks and Their Performance." *Proceedings of the IEEE* 77, no. 2 (February 1989):238–56.

[BYRN89] Byrne, W. R.; Kilm, T. A.; Nelson, B. L.; and Soneru, M. D. "Broadband ISDN Technology and Architecture." *IEEE Network* 3, no. 1 (January 1989):23–28.

[CALL83] Callon, R. "Internetwork Protocol." *Proceedings of the IEEE* 71, no. 12 (December 1983):1388–93.

[CAMP87] Campanella, S. J. "Digital Speech Interpolation Systems," in [FEHE87a], pp. 237–81.

[CAMP88] Campbell, V. "Protocol Politics." *LAN Times* V, no. X (November 1988):1, 112–15.

[CAPE79] Capetenakis, J. J. "Generalized TDMA: The Multi-Accessing Tree Protocol." *IEEE Transactions on Communications* COM-27, no. 11 (October 1979):1476–84.

[CAPP86] Cappello, P. R. "Transforming Systolic Arrays in Space-Time (Tutorial)." *Digital Communications: Proceedings of the Second Tirrenia International Workshop on Digital Communications,* Tirrenia, Italy, Sept. 1985, North Holland, Amsterdam (1986):219–34.

[CARG88] Cargill, C., and Soha, M. "Standards and Their Influence on MAC Bridges." *IEEE Network* 2, no. 1 (January 1988):87–89.

[CARL80] Carlson, D. E. "Bit-Oriented Data Link Control Procedures." *IEEE Transactions on Communications* COM-28, no. 4 (April 1980):455–467.

[CARL82] Carlson, D. E. "Bit-Oriented Data Link Control," in [GREE82a], pp. 111–44.

[CARN86] Carney, D. L., and Prell, E. M. "Planning for ISDN in the 5ESS™ Switch." *AT&T Technical Journal* 65, no. 1 (January/February 1986):35–44.

[CARR70] Carr, C. S.; Crocker, S. D.; and Cerf, V. G. "Host/Host Communication Protocol in the ARPA Network." *Proceedings AFIPS Spring Joint Computer Conference* (May 1970):589–97.

[CASO85] Casoria, A. "Interconnections and Services Integration in Public and Private Networks for Office Automation." *Proceedings of IEEE INFOCOM '85* (1985): 56–69.

[CCIT81a] CCITT (International Telegraph and Telephone Consultative Committee). "CCITT V-Series Recommendations." *CCITT Yellow Book,* vol. VIII.1 (Data Communication Over the Telephone Network). Geneva: 1981.

[CCIT81b] CCITT (International Telegraph and Telephone Consultative Committee). "CCITT X-Series Recommendations." *CCITT Yellow Books,* vols. VIII.2 and VIII.3 (Data Communication Networks). Geneva: 1981.

[CCIT84] CCITT (International Telegraph and Telephone Consultative Committee). "CCITT I-Series Recommendations." *CCITT Red Book,* vol. III.5, Geneva: 1984.

[CCIT88] CCITT (International Telegraph and Telephone Consultative Committee). *CCITT Blue Book.* Geneva: 1988.

[CEGR75] Cegrell, T. "A Routing Procedure for the TIDAS Message Switching Network." *IEEE Transactions on Communications* COM-23 (June 1975).

[CERF74] Cerf, V. G., and Kahn, R. E. "A Protocol for Packet Network Intercommunication." *IEEE Transactions on Communications* COM-22 (May 1974):637–48.

[CERF78] Cerf, V. G., and Kirstein, P. T. "Issues in Packet-Network Interconnection." *Proceedings of the IEEE* 66, no. 11 (November 1978):1386–1408.

[CERF81] Cerf, V. G. "Packet Communication Technology," in [KUO81], pp. 1–34.

[CERF83] Cerf, V., and Cain, E., "The DOD Internet Architecture Model." *Computer Networks* (October 1983).

[CHAK88] Chakraborty, D. "VSAT Communications Networks—An Overview." *IEEE Communications Magazine* 26, no. 5 (May 1988):10–24.

[CHAP83] Chapin, A. L. "Connections and Connectionless Data Transmission." *Proceedings of the IEEE* 71, no. 12 (December 1983):1356–71.

[CHAP88] Chapin, A. L. "Standards for Bridges and Gateways." *IEEE Network* 2, no. 1 (January 1988):90–91.

[CHEN85] Chen, P. "Use Hybrid Switches for Voice and Data." *Computer Design* (October 1983):149–53.

[CHEN87] Cheng, Y.-C., and Robertazzi, T. G. "Annotated Bibliography of Local Communication System Interconnection." *IEEE Journal on Selected Areas in Communications* SAC-5, no. 9 (December 1987):1492–99.

[CHER88] Cheriton, D. R. "The V Distributed System." *Communications of the ACM* 31, no. 3 (March 1988):314–33.

[CHES88] Chesson, G. "Protocol Engine Design." *Proceedings 1987 Summer USENIX Conference,* Phoenix, Ariz. (June 1987):209–15. Also in [PART88b].

[CHOR84] Chorafas, D. N. *Telephony, Today and Tomorrow.* Englewood Cliffs, N.J.: Prentice-Hall, 1984.

[CHOU73] Chou, W., and Kershenbaum, A. "A Unified Algorithm for Designing Multidrop Teleprocessing Networks." *Proceedings of Third Data Communications Symposium,* St. Petersburg, Fla. (November 1973):148–56.

[CHOU83a] Chou, W. "Analysis of Data/Computer Networks," in [CHOU83b], pp. 410–41.

[CHOU83b] Chou, W., ed. *Computer Communications Volume I—Principles.* Englewood Cliffs, N.J.: Prentice-Hall, 1983.

[CHOU83c] Chou, W. "Data/Computer Communications Network Structures," in [CHOU83b], pp. 1–36.

[CHOU85a] Chou, W., ed. *Computer Communications, Volume II—Systems and Applications.* Englewood Cliffs, N.J.: Prentice-Hall, 1985.

[CHOU85b] Chou, W. "Optimization of Data/Computer Networks," in [CHOU85a], pp. 1–49.

[CHOW84] Chow, C. H.; Gouda, M. G.; and Lam, S. S. "An Exercise in Constructing Multi-Phase Communication Protocols." *Proceedings of the ACM SIGCOMM '84 Conference* (June 1984).

[CHU79] Chu, W. W., ed. *Advances in Computer Communications,* 3d ed. Dedham, Mass.: Artech House, 1976.

[CHU85] Chu, W. W. "Principles of Distributed Data-Base Design," in [CHOU85a], pp. 434–80.

[CIME86] Cimet, I. A., and Kumar, P. R. S. "A Resilient Distributed Protocol for Network Synchronization." *Proceedings ACM SIGCOMM '86 Symposium on Communication Architectures and Protocols.* Stowe, Vt. (August 1986):358–69.

[CIO88] CIO Publishing. "Global Communications and Computer Strategies for the 90's." *Forbes* 142, no. 6, special advertising supplement (Sept. 19, 1988):139–163.

[CLAR45] Clarke, A. C. "Extra-terrestrial Relays: Can Rocket Stations Give Worldwide Radio Coverage?" *Wireless World* (1945).

[CLAR76] Clark, A. P. *Principles of Digital Data Transmission.* New York: Wiley, 1976.

[CLAR77] Clark, A. P. *Advanced Data-Transmission Systems.* New York: Wiley, 1977.

[CLAR78] Clark, D.; Pogran, K.; and Reed, D. "An Introduction to Local Area Networks." *Proceedings of the IEEE* 66, no. 11 (November 1978):1497–1517.

[CLAR81] Clark, G. C., Jr., and Cain, J. B. *Error-Correction Coding for Digital Communications.* New York: Plenum Press, 1981.

[CLAR87] Clark, D. D.; Zhang, L.; and Lambert, M. "NETBLT: A High Throughput Transport Protocol." *Proceedings ACM SIGCOMM '87,* Stowe, Vt. (August 1987):353–59.

[CLAU86] Claus, J. "ISDN Implementation Strategy of the Deutsche Bundespost—From the ISDN Pilot Project to Commercial Services." *IEEE Journal on Selected Areas in Communications* SAC-4, no. 3 (May 1986):390–97.

[COCK87] Cockburn, A. A. R. "Efficient Implementation of the OSI Transport Protocol Checksum Algorithm Using 8/16-Bit Arithmetic." *Computer Communication Review* 17, no. 3 (July/August 1987):13–20.

[COHE81] Cohen, D. "On Holy Wars and a Plea for Peace." *IEEE Computer Magazine* 14, no. 10 (October 1981):48–54.

[COLE86] Cole, R. *Computer Communications,* 2d ed. New York: Springer-Verlag, 1986.

[COLL83] Collie, B. "Looking at the ISDN Interfaces: Issues and Answers." *Data Communications* (June 1983):125–36.

[COME88] Comer, D. *Internetworking with TCP/IP: Principles, Protocols, and Architecture.* Englewood Cliffs, N.J.: Prentice-Hall, 1988.

[COMR84] Comroe, R. A., and Costello, D. J. "ARQ Schemes for Data Transmission in Mobile Radio Systems." *IEEE Journal on Selected Areas in Communications* SAC-2, no. 4 (July 1984):472–81.

[CONA80] Conard, J. W. "Character-Oriented Data Link Control Protocols." *IEEE Transactions on Communications* COM-28, no. 4 (April 1980):445–54.

[CONA82] Conard, J. W. "Character-Oriented Link Control," in [GREE82a], pp. 87–110.

[CONA83] Conard, J. W. "Services and Protocols of the Data Link Layer." *Proceedings of the IEEE* 71, no. 12 (December 1983):1378–83.

[COOK84] Cooke, R. "Intercity Limits: Looking Ahead to All-Digital Networks and No Bottlenecks." *Data Communications* (March 1984):167–75.

[COOP87] Cooper, R. W. "The Moving Target—Marketing ISDN to Businesses." *IEEE Communications Magazine* 25, no. 12 (December 1987):21–24.

[COUC87] Couch, L. W., II, *Digital and Analog Communication Systems,* 2d ed. New York: Macmillan, 1987.

[CRAV63] Cravis, H., and Crater, T. V. "Engineering of T1 Carrier Systems Repeatered Lines." *Bell System Technical Journal* 42 (March 1963):431–86.

[CRAV81] Cravis, H. *Communication Network Analysis.* Lexington, Mass.: Lexington Books, 1981.

[CROC70] Crocker, S. D.; Heafner, J. F.; Metcalfe, R. M.; and Postel, J.B. "Function Oriented Protocols for the ARPA Computer Network." *Proceedings AFIPS Spring Joint Computer Conference* (May 1972):271–79.

[CROW73] Crowther, W.; Rettberg, R.; Walden, D.; Ornstein, S.; and Heart, F. "A System for Broadcast Communication: Reservation-Aloha." *Proceedings Sixth Hawaii International Conference on System Sciences* (1973):371–74.

[CUMM87] Cummings, J. L.; Hickey, K. R.; and Kinney, B. D. "AT&T Network Architecture Evolution." *AT&T Technical Journal* 66, no. 3 (May/June 1987):2–12.

[CUNN83] Cunningham, I. "Message-Handling Systems and Protocols." *Proceedings of the IEEE* 71, no. 12 (December 1983):1425–30.

[DADD89] Daddis, G. E., and Torng, H. C. "A Taxonomy of Broadband Integrated Switching Architectures." *IEEE Communications Magazine* 27, no. 5 (May 1989):32–42.

[DAM88] Dam, K. W. "The Global Electronic Market." *Pacific Telecommunications* 9, no. 3 (December 1988):23–25.

[DANT67] Dantzig, G. B. "All Shortest Routes in a Graph," in *Theory of Graphs,* P. Rosenstiehl, ed. New York: Gordon and Breach, 1967, pp. 91–93.

[DANT80] Danthine, A. A. S. "Protocol Representation with Finite State Models." *IEEE Transactions on Communications* COM-28, no. 4 (April 1980):632–44.

[DANT82] Danthine, A. A. S. "Protocol Representation with Finite State Models," in [GREE82a], pp. 579–606.

[DAVE72] Davey, J. R. "Modems." *Proceedings of the IEEE* 70, no. 11 (November 1972):1284–92.

[DAVI71] Davies, D. W. "The Control of Congestion in Packet Switching Networks." *Proceedings ACM/IEEE Second Symposium on Problems in the Optimization of Data Communications Systems* (October 1971):46–49.

[DAVI72] Davies, D. W. "The Control of Congestion in Packet Switching Networks." *IEEE Transactions on Communications* COM-20, no. 3 (June 1972):546–50.

[DAVI73] Davies, D. W., and Barber, D. L. A. *Communication Networks for Computers.* London: Wiley, 1973.

[DAVI83] Davidson, J. A. "OSI Model Layering of a Military Local Network." *Proceedings of the IEEE* 71, no. 12 (December 1983):1435–41.

[DAVI88] Davidson, J.; Hathaway, W.; Postel, J.; Mimno, N.; Thomas, R.; and Walden, D. "The ARPAnet Telnet Protocol: Its Purpose, Principles of Implementation, and Impact on Host Operating System Design." *Proceedings Fifth Data Communications Symposium,* Snowbird, Utah (September 1977).

[DAY80] Day, J. D. "Terminal Protocols." *IEEE Transactions on Communications* COM-28, no. 4 (April 1980):585–93.

[DAY81] Day, J. D. "Terminal, File Transfer and Remote Job Protocols for Heterogeneous Computer Networks," in [KUO81], pp. 78–121.

[DAY82] Day, J. D. "Terminal Support Protocols," in [GREE82a], pp. 437–58.

[DAY83] Day, J. D., and Zimmermann, H. "The OSI Reference Model." *Proceedings of the IEEE* 71, no. 12 (December 1983):1334–40.

[DEAS86] Deasington, R. J. *X.25 Explained: Protocols for Packet Switching Networks,* 2d ed. New York: Halsted Press, 1986.

[DEAT83] Deaton, G. A., Jr., and Hippert, R. O., Jr. "X.25 and Related Recommendations in IBM Products." *IBM Systems Journal* 22, nos. 1/2 (1983):11–29.

[DECI82] Dècina, M. "Progress Towards User Access Arrangements in Integrated Services Digital Networks." *IEEE Transactions on Communications* COM-30, no. 9 (September 1982):2117–30.

[DECI83] Dècina, M., and Vlack, D. "Voice by the Packet." *IEEE Journal on Selected Areas in Communications* SAC-1, no. 6 (December 1983):961–62.

[DECI86a] Dècina, M., and Scace, E. L. "CCITT Recommendations on the ISDN: A Review." *IEEE Journal on Selected Areas in Communications* SAC-4, no. 3 (May 1986):320–25.

[DECI86b] Dècina, M.; Gifford, W. S.; Potter, R.; and Robrock, A. A. "Guest Editorial—ISDN: Coming of Age Now!" *IEEE Journal on Selected Areas in Communications* SAC-4, no. 3 (May 1986):313–15.

[DECI86c] Dècina, M.; Gifford, W. S.; Potter, R.; and Robrock, A. A. "Guest Editorial—ISDN: Technology and Implementation." *IEEE Journal on Selected Areas in Communications* SAC-4, no. 8 (November 1986):1186–87.

[DECI87] Dècina, M., and Roveri, A. "ISDN: Integrated Services Digital Network: Architectures and Protocols," in [FEHE87a], pp. 40–132.

[DEJU86] de Julio, U., and Pellegrini, J. "Layer 1 ISDN Recommendations." *IEEE Journal on Selected Areas in Communications* SAC-4, no. 3 (May 1986):349–54.

[DELA86] Delatore, J. P.; Osehring, H.; and Stecher, L. C. "Implementation of ISDN on the 5ESS Switch." *IEEE Journal on Selected Areas in Communications* SAC-4, no. 8 (November 1986):1262–67.

[DENA79] Denardo, E. V., and Fox, B. L. "Shortest-Route Methods: 1. Reaching, Pruning, and Buckets." *Operations Research* 27 (1979):161–186.

[DEO74] Deo, N. *Graph Theory with Applications to Engineering and Computer Science.* Englewood Cliffs, N.J.: Prentice-Hall, 1974.

[DEO80] Deo, N., and Pang, C. Y. "Shortest Path Algorithms: Taxonomy and Annotation." *Networks* 14 (1984):275–323.

[DESJ83] desJardins, R. "Afterword: Evolving Towards OSI." *Proceedings of the IEEE* 71, no. 12 (December 1983):1446–48.

[DETR83] DeTreville, J., and Sincoskie, W. D. "A Distributed Experimental Communications System." *IEEE Journal on Selected Areas in Communications* SAC-1, no. 6 (December 1983):1070–75.

[DIAL79] Dial, R. B.; Glover, F.; Karney, D.; and Klingman, D. "A Computational Analysis of Alternative Algorithms and Labeling Techniques for Finding Shortest Path Trees." *Networks* 9 (1979):215–48.

[DICK83] Dickson, G. J., and de Chazal, P. E. "Status of CCITT Description Techniques and Application to Protocol Specification." *Proceedings of the IEEE* 71, no. 12 (December 1983):1346–55.

[DIFF88] Diffie, W. "The First Ten Years of Public Key Cryptography." *Proceedings of the IEEE* 76, no. 5 (May 1988):560–77.

[DIGI82] Digital Equipment Corporation (DEC). *DECnet Digital Network Architecture (Phase IV) General Description.* Bedford, Mass., Order No. AA-149A-TC (1982).

[DIGI83a] Digital Equipment Corporation (DEC). *DECnet Digital Network Architecture (Phase IV) Network Management Functional Specification.* Bedford, Mass., Order No. AA-X437A-TK (1983).

[DIGI83b] Digital Equipment Corporation (DEC). *DECnet Digital Network Architecture (Phase IV) NSP Functional Specification.* Bedford, Mass., Order No. AA-X439A-TK (1983).

[DIGI83c] Digital Equipment Corporation (DEC). *DECnet Digital Network Architecture (Phase IV) Routing Layer Functional Specification.* Bedford, Mass., Order No. AA-X435A-TK (1983).

[DIGI84] Digital Equipment Corporation (DEC). *DECnet Digital Network Architecture Digital Data Communications Message Protocol, DDCMP, Functional Specification,* Version 4.1.0. Bedford, Mass., Order No. AA-K175A-TK (1984).

[DIJK59] Dijkstra, E. W. "A Note on Two Problems in Connexion with Graphs." *Numerische Mathematik* 1, (1969):269–71.

[DIXO87] Dixon, R. C. "Lore of the Token Ring." *IEEE Network* 1, no. 1 (January 1987):11–18.

[DIXO88] Dixon, R. C., and Pitt, D. A. "Addressing, Bridging, and Source Routing." *IEEE Network* 2, no. 1 (January 1988):25–32.

[DOLL72] Doll, D. R. "Multiplexing and Concentration." *Proceedings of the IEEE* 60, no. 11 (November 1972):1313–21.

[DOLL78] Doll, D. R. *Data Communications: Facilities, Networks and Systems Design.* New York: Wiley, 1978.

[DONN74] Donnan, R. A., and Kersey, J. R. "Synchronous Data Link Control: A Perspective." *IBM Systems Journal* (May 1974):140–162.

[DONO86] Donohoe, D. C.; Johannessen, G. H.; and Stone, R. E. "Realization of a Signaling System No. 7 Network for AT&T." *IEEE Journal on Selected Areas in Communications* SAC-4, no. 8 (November 1986):1257–61.

[DOWD87] Dowd, P. W., and Jabbour, K. "A Unified Approach to Local Area Network Interconnection." *IEEE Journal on Selected Areas in Communications* SAC-5, no. 9 (December 1987):1418–25.

[DREY69] Dreyfus, S. E. "An Appraisal of Some Shortest-Path Algorithms." *Operations Research* 17 (1969):395–412.

[DUC85] Duc, N. Q. "ISDN Protocol Architecture." *IEEE Communications Magazine* 23, no. 3 (March 1985):15–22.

[DUC86] Duc, N. Q. "ISDN Terminals and Integrated Services Delivery." *IEEE Journal on Selected Areas in Communications* SAC-4, no. 8 (November 1986):1188–92.

[DVOR88] Dvorak, C. A.; Hatori, M.; and Redman, T. C. "Guest Editorial: The Evolution of Quality." *IEEE Communications Magazine* 26, no. 10 (October 1988):7–8.

[ECMA81] ECMA (European Computer Manufacturers Association). "Standard ECMA-71, HDLC Selected Procedures." (January 1981).

[EINE88] Einert, D., and Glas, G. "The SNATCH Gateway: Translation of Higher Level Protocols." *Journal of Telecommunication Networks* (Spring 1983):83–102.

[EISE67] Eisenbies, J. L. "Conventions for Digital Data Communication Link Design." *IBM Systems Journal* 6, no. 1 (1967):267–302.

[ELEC69] Electronic Industries Association (EIA). "EIA Standard RS-232-C Interface Between Data Terminal Equipment and Data Communication Equipment Employing Serial Binary Data Interchange." Washington, D.C. (October 1969).

[ELEC77] Electronic Industries Association (EIA). "EIA Standard RS-449, General-Purpose 37-Position and 9-Position Interface for Data Terminal Equipment and Data Circuit-Terminating Equipment Employing Serial Binary Interchange." Washington, D.C. (November 1977) and "Addendum 1 to RS-449." (February 1980).

[ELEC79] Electronic Industries Association (EIA). "EIA Standard RS-366-A, Interface Between Data Terminal Equipment and Automatic Calling Equipment for Data Communication." Washington, D.C. (March 1979).

[ELHA83] Elhakeem, A. K.; Hafez, H. M.; and Mahmoud, S. A. "Spread-Spectrum Access to Mixed Voice-Data Local Area Networks." *IEEE Journal on Selected Areas in Communications* SAC-1, no. 6 (December 1983):1054–69.

[ELLI86] Ellis, R. L. *Designing Data Networks.* Englewood Cliffs, N.J.: Prentice-Hall, 1986.

[ELOV74] Elovitz, H. S., and Heitmeyer, C. L. "What is a Computer Network?" *National Telecommunications Conference 1974 Record* (1974):1007–14.

[EMMO83] Emmons, W. F., and Chandler, A. S. "OSI Session Layer: Services and Protocols." *Proceedings of the IEEE* 71, no. 12 (December 1983):1397–1400.

[ENG87] Eng, K. Y.; Hluchyj, M. G.; and Yeh, Y. S. "A Knockout Switch for Variable-Length Packets." *IEEE Journal on Selected Areas in Communications* SAC-5, no. 9 (December 1987):1426–35.

[ENNI83] Ennis, G. "Development of the DoD Protocol Reference Model." *Proceedings, SIGCOMM '83 Symposium* (1983).

[EPHE87] Ephremides, A.; Wieselthier, J. E.; and Baker, D. J. "A Design Concept for Reliable Mobile Radio Networks with Frequency Hopping Signaling." *Proceedings of the IEEE* 75, no. 1 (January 1987):56–73.

[ESTR87] Estrin, D. "Interconnection Protocols for Interorganization Networks." *IEEE Journal on Selected Areas in Communications* SAC-5, no. 9 (December 1987):1480–91.

[EVER83] Everett, R. R., ed. "Special Issue—SAGE System." *Annals of the History of Computing* 5, no. 4 (October 1983).

[EXLE87] Exley, G. M., and Merakos, L. F. "Throughput-Delay Performance of Interconnected CSMA Local Area Networks." *IEEE Journal on Selected Areas in Communications* SAC-5, no. 9 (December 1987):1380–1390.

[FALC86] Falconer, D. D. "Bandlimited Digital Communications: Recent Trends and Applications to Voiceband Modems and Digital Radio (State-of-the-Art)." *Digital Communications: Proceedings of the Second Tirrenia International Workshop on Digital Communications,* Tirrenia, Italy (September 1985) North Holland, Amsterdam (1986):25–40.

[FALK77] Falk, G., and McQuillan, J. M. "Alternatives for Data Network Architectures." *Computer* 10, no. 11 (November 1977):22–31.

[FALK83a] Falk, G.; Groff, J. S.; Milliken, W. C.; Nodine, M.; Blumenthal, S.; and Edmond, W. "Integration of Voice and Data in the Wideband Packet Satellite Network." *IEEE Journal on Selected Areas in Communications* SAC-1, no. 6 (December 1983):1076–83.

[FALK83b] Falk, G. "The Structure and Function of Network Protocols," in [CHOU83b], pp. 37–80.

[FANO72] Fano, R. M. "On the Social Role of Computer Communications." *Proceedings of the IEEE* 60, no. 11 (1972):1249–53.

[FARO86] Farowich, S. A. "Computers: Communicating in the Technical Office." *IEEE Spectrum* 23, no. 4 (April 1986):63–67.

[FEDE83] Federal Communications Commission (Notice of Inquiry, Docket Number 83-841), "In the Matter of Integrated Services Digital Networks." (August 10, 1983).

[FEHE87a] Feher, K., ed. *Advanced Digital Communications: Systems and Signal Processing Techniques.* Englewood Cliffs, N.J.: Prentice-Hall, 1987.

[FEHE87b] Feher, K. "DIGCOM and DSP: Digital Communications and Digital Signal Processing Overview," in [FEHE87a], pp. 1–39.

[FEHE87c] Feher, K. "Digital Modem (Modulation-Demodulation) Techniques," in [FEHE87a], pp. 316–428.

[FERG85] Ferguson, J. J., and Aminetzah, Y. J. "Exact Results for Nonsymmetric Token Ring Systems." *IEEE Transactions on Communications* COM-33, no. 3 (March 1985):223–31.

[FIEL86] Field, J. A. "Logical Link Control." *Proceedings IEEE INFOCOM '86,* Miami, Fla. (April 1986):331–36.

[FILI89] Filipiak, J. "M-Architecture: A Structural Model of Traffic Management and Control in Broadband ISDNs." *IEEE Communications Magazine* 27, no. 5 (May 1989):25–31.

[FINE84] Fine, M., and Tobagi, F. A. "Demand Assignment Multiple Access Schemes in Broadcast Bus Local Area Networks." *IEEE Transactions on Computers* C-33, no. 12 (December 1984):1130–59.

[FISH84] Fisher, D. E. "Integrated Digital Communications Networking." *IEEE Communications Magazine* 22, no. 1 (January 1984):42–48.

[FISH86] Fishman, D. A.; Lumish, G.; Denkin, N. M.; Schultz, R. R.; Chai, S. Y.; and Ogawa, K. "1.7 Gb/s Lightwave Transmission Field Experiment." *Technical Digest, OFC '86* PDP-11 (February 1986).

[FLET82] Fletcher, J. "An Arithmetic Checksum for Serial Transmission." *IEEE Transactions on Communications* COM-30, no. 1 (January 1982):247–52.

[FLET85] Fletcher, C. "Videotex: Return Engagement." *IEEE Spectrum* 22, no. 10 (October 1985):34–38.

[FLOY62] Floyd, R. W. "Algorithm 97: Shortest Path." *Communications of the ACM* 5 (1962):345.

[FOLE83] Foley, J. S. "Business Data Usage of OSI." *Proceedings of the IEEE* 71, no. 12 (December 1983):1442–45.

[FOLT80a] Folts, H. C. "Procedures for Circuit-Switched Service in Synchronous Public Data Networks." *IEEE Transactions on Communications* COM-28, no. 4 (April 1980):489–95.

[FOLT80b] Folts, H. C. "X.25 Transaction-Oriented Features—Datagram and Fast Select." *IEEE Transactions on Communications* COM-28, no. 4 (April 1980):496–99.

[FOLT82a] Folts, H. C. "Circuit-Switched Network Layer," in [GREE82a], pp. 195–212.

[FOLT82b] Folts, H. C. "Packet-Switched Network Layer for Short Messages," in [GREE82a], pp. 239–48.

[FOLT83a] Folts, H. C., and desJardins, R., eds. "Special Issue on Open Systems Interconnection (OSI)—Standard Architecture and Protocols." *Proceedings of the IEEE* 71, no. 12 (December 1983).

[FOLT83b] Folts, H. C. "The Special Issue on Open Systems Interconnection (OSI)—New International Standards Architecture and Protocols for Distributed Information Systems." *Proceedings of the IEEE* 71, no. 12 (December 1983):1331–33.

[FOLT86] Folts, H. C., ed. *McGraw-Hill's Compilation of Data Communications Standards,* Edition III, New York: McGraw-Hill Information Systems, 1986.

[FORD56] Ford, L. R., Jr. "Network Flow Theory." *Report P-923,* Rand Corporation, Santa Monica, Calif. (August 1956).

[FORD62] Ford, L. R., Jr., and Fulkerson, D. R. *Flows in Networks.* Princeton: Princeton University Press, 1962.

[FORE86] Forecast, J.; Jackson, J. L.; and Schriesheim, J. A. "The DECnet-ULTRIX Software." *Digital Technical Journal,* Digital Equipment Corporation, Hudson, Mass., no. 3 (September 1986):100–107.

[FORN70] Forney, G. D., Jr. "Coding and Its Application in Space Communications." *IEEE Spectrum* 7, no. 6 (June 1970).

[FORN84] Forney, G. D., Jr.; Gallager, R. G.; Lang, G. R.; Longstaff, F. M.; and Qureshi, S. U. "Efficient Modulation for Band-Limited Channels." *IEEE Journal on Selected Areas in Communications* SAC-2, no. 5 (September 1984):632–47.

[FOSC74] Foschini, G. J.; Gitlin, R. D.; and Weinstein, S. B. "Optimization of Two-Dimensional Signal Constellations in the Presence of Gaussian Noise." *IEEE Transactions on Communications* COM-22, no. 1 (January 1974):28–38.

[FRAN71] Frank, H., and Frisch, I. T. *Communication, Transmission, and Transportation Networks.* Reading, Mass.: Addison-Wesley, 1971.

[FRAN81] Franta, W., and Chlamtac, I. *Local Networks.* Lexington, Mass.: Lexington Books, 1981.

[FRAN83] François, P., and Potocki, A. "Some Methods for Providing OSI Transport in SNA." *IBM Journal of Research and Development* 27, no. 5 (September 1983): 452–63.

[FRAN84a] Frank, C. "Legal and Policy Ramifications of the Emerging Integrated Services Digital Network." *Journal of Telecommunication Networks* 3, no. 1 (Spring 1984):47–56.

[FRAN84b] Frankel, M. S. "Telecommunications and Processing for Military Command and Control: Meeting User Needs in the Twenty-First Century." *IEEE Communications Magazine* 22, no. 7 (July 1984):18–25.

[FRAN86] Franks, L. E. "Digital Filters and Equalizers for Data Communication (Survey)." *Digital Communications: Proceedings of the Second Tirrenia International Workshop on Digital Communications,* Tirrenia, Italy, September 1985, North Holland, Amsterdam (1986):205–18.

[FRAT73] Fratta, L.; Gerla, M.; and Kleinrock, L. "The Flow Deviation Method: An Approach to Store-and-Forward Communication Network Design." *Networks* 3 (1973):97–133.

[FREE75] Freeman, R. L. *Telecommunication Transmission Handbook.* New York: Wiley, 1975.

[FREE88] Freer, J. R. *Computer Communications and Networks.* New York: Plenum, 1988.

[FROS78] Frosdick, H. C.; Schantz, R. E.; and Thomas, R. H. "Operating Systems for Computer Networks." *Computer* 11, no. 1 (January 1978):48–57.

[FUCH70] Fuchs, E., and Jackson, P. E. "Estimates of Distributions of Random Variables for Certain Computer Communications Traffic Models." *Communications of the ACM* 13 (December 1970):727–57.

[FUNG86] Fung, K.; Luetchford, J.; and Scales, I. "ISDN Standards Issues." *Telesis* 13, no. 3 (1986):24–33.

[GAGG87] Gaggioni, H. P. "The Evolution of Video Technologies." *IEEE Communications Magazine* 25, no. 11 (November 1987):20–36.

[GAGL78] Gagliardi, R. *Introduction to Communications Engineering.* New York: Wiley, 1978.

[GAGL86] Gagliardi, R. M. "Optical Communications: From Theory to Systems (Survey)." *Digital Communications: Proceedings of the Second Tirrenia International Workshop on Digital Communications,* Tirrenia, Italy, September 1985, North Holland, Amsterdam (1986):159–70.

[GALL68] Gallager, R. G. *Information Theory and Reliable Communication.* New York: Wiley, 1968.

[GALL77] Gallager, R. G. "A Minimum Delay Routing Algorithm Using Distributed Computation." *IEEE Transactions on Communications* COM-23, no. 1 (January 1977):73–85.

[GARE79] Garey, M. R., and Johnson, D. S. *Computers and Intractability.* San Francisco: Calif., W. H. Freeman, 1979.

[GARL77] Garlick, L. L.; Rom, R.; and Postel, J. B. "Reliable Host-to-Host Protocols: Problems and Techniques." *Proceedings of the Fifth Data Communications Symposium* (1977):4.58-65; also in [LAM84c].

[GART87] Gartside, C. H., III; Panuska, A. J.; and Patel, P. D. "Single-Mode Cable for Long-Haul, Trunk, and Loop Networks." *AT&T Technical Journal* 66, no. 1 (January/February 1987):84–94.

[GECH89] Gechter, J., and O'Reilly, P. "Conceptual Issues for ATM." *IEEE Network* 3, no. 1 (January 1989):14–16.

[GEIG86] Geiger, G., and Lerach, L. "ISDN-Oriented Modular VLSI Chip Set for Central-Office and PABX Applications." *IEEE Journal on Selected Areas in Communications* SAC-4, no. 8 (November 1986):1268–74.

[GERL73] Gerla, M. "Deterministic and Adaptive Routing Policies in Packet-Switched Computer Networks." *Data Networks: Analysis and Design—Proceedings of Third IEEE Data Communications Symposium* (November 1973):23–28.

[GERL81] Gerla, M. "Routing and Flow Control," in [KUO81], pp. 122–74.

[GERL82] Gerla, M., and Kleinrock, L. "Flow Control Protocols," in [GREE82a], pp. 361–412.

[GERL84] Gerla, M. "Controlling Routes, Traffic Rates, and Buffer Allocation in Packet Networks." *IEEE Communications Magazine* 22, no. 11 (November 1984):11–23.

[GERL85] Gerla, M. "Packet, Circuit and Virtual Circuit Switching," in [CHOU85a], pp. 222–67.

[GERL88a] Gerla, M., and Kleinrock, L. "Congestion Control in Interconnected LANs." *IEEE Network* 2, no. 1 (January 1988):72–76.

[GERL88b] Gerla, M.; Green, L.; and Rutledge, R. "Guest Editorial." *IEEE Network* 2, no. 1 (January 1988):3–4.

[GERS84] Gersho, A., and Lawrence, V. B. "Multidimensional Signal Constellations for Voiceband Data Transmission." *IEEE Journal on Selected Areas in Communications* SAC-2, no. 5 (September 1984):687–702.

[GERS86] Gersho, A. "Vector Quantization: A New Direction in Source Coding (Tutorial)." *Digital Communications: Proceedings of the Second Tirrenia International Workshop on Digital Communications,* Tirrenia, Italy, September 1985, North Holland, Amsterdam (1986):267–82.

[GIEN79] Gien, M., and Zimmermann, H. "Design Principles for Network Interconnection." *Proceedings of Sixth IEEE-ACM Data Communications Symposium* (1979):109–19.

[GIFF86] Gifford, W. S. "ISDN User-Network Interfaces." *IEEE Journal on Selected Areas in Communications* SAC-4, no. 3 (May 1986):343–48.

[GIFF87] Gifford, W. S. "ISDN Performance Tradeoffs." *IEEE Communications Magazine* 25, no. 12 (December 1987):25–29.

[GITM76] Gitman, I.; Van Slyke, R. M.; and Frank, H. "Routing in Packet Switched Broadcast Radio Networks." *IEEE Transactions on Communications* COM–24 (August 1976).

[GITM78] Gitman, I., and Frank, H. "Economic Analysis of Integrated Voice and Data Networks: A Case Study." *Proceedings of the IEEE* 66, no. 11 (November 1978):1549–70.

[GLAS77] Glaser, A. B., and Subak-Sharpe, G. E. *Integrated Circuit Engineering.* Reading, Mass.: Addison-Wesley, 1977.

[GOUL88] Gould, E. P. "Advanced Traffic Routing as Part of the USA Intelligent Telecommunications Network." *Proceedings of 12th International Teletraffic Congress,* Torino, Italy (June 1988) 6.2iA.2.1–2.7.

[GRAN79] Grangé, J. L., and Gien, M., eds. *Flow Control in Computer Networks.* Amsterdam: North Holland Publishing, 1979.

[GRAY72] Gray, J. P. "Line Control Procedures." *Proceedings of the IEEE* 60, no. 11 (November 1972):1301–12.

[GRAY81] Gray, J. P. "Synchronization in SNA Networks," in [KUO81], pp. 319–68.

[GREE75] Green, P. E., Jr., and Lucky, R. W., eds. *Computer Communications.* New York: IEEE Press, 1975.

[GREE77] Greene, W., and Pooch, U. W. "A Review of Classification Schemes for Computer Communication Networks." *Computer* 10, no. 11 (November 1977): 12–21.

[GREE80a] Green, P. E., Jr. "An Introduction to Network Architectures and Protocols." *IEEE Transactions on Communications* COM-28, no. 4 (April 1980):413–24.

[GREE80b] Green, P. E., Jr., ed. "Special Issue on Computer Network Architectures and Protocols." *IEEE Transactions on Communications* COM-28, no. 4 (April 1980).

[GREE82a] Green, P. E., Jr., ed. *Computer Network Architectures and Protocols.* New York: Plenum Press, 1982.

[GREE82b] Green, P. E., Jr. "The Structure of Computer Networks," in [GREE82a], pp. 3–32.

[GREE82c] Green, P. E., Jr. "Videotex Terminal Protocols," in [GREE82a], pp. 483–508.

[GREE84] Green, P. E., Jr. "Computer Communications: Milestones and Prophecies." *IEEE Communications Magazine* 22, no. 5 (May 1984):49–63.

[GREE86] Green, P. E., Jr. "Protocol Conversion." *IEEE Transactions on Communications* COM-34, no. 3 (March 1986):257–68.

[GREE88] Green, P. E., Jr. ed. *Network Interconnection and Protocol Conversion.* New York: IEEE Press, 1988.

[GREG77] Gregg, W. D. *Analog and Digital Communication: Concepts, Systems, Applications, and Services.* New York: Wiley, 1977.

[GROE88] Groenbaek, I. "Conversion Between the TCP and ISO Transport Protocols as a Method of Achieving Interoperability Between Data Communications Systems." *IEEE Journal on Selected Areas in Communications* SAC-4, no. 2 (March 1986):288–96.

[GRUB75] Grubb, D. S., and Cotton, I. W. "Rating Performance." *Data Communications* (September/October 1975):41–47.

[GRUB81] Gruber, J. G. "Delay Related Issues in Integrated Voice and Data Networks." *IEEE Transactions on Communications* COM-29, no. 6 (June 1981):787–99.

[GRUB83] Gruber, J. G., and Le, N. "Performance Requirements for Integrated Voice/Data Networks." *IEEE Journal on Selected Areas in Communications* SAC-1, no. 6 (December 1983):981–1005.

[GUIN87] Guinn, D. E. "ISDN—Is the Technology on Target?" *IEEE Communications Magazine* 25, no. 12 (December 1987):10–13.

[GUNT81] Günther, K. D. "Prevention of Deadlocks in Packet-Switched Data Transport Systems." *IEEE Transactions on Communications* COM-29, no. 4 (April 1981): 512–24.

[HAEN81] Haenschke, D. G.; Kettler, D. A.; and Oberer, E. "Network Management and Congestion in the U.S. Telecommunications Network." *IEEE Transactions on Communications* COM-29, no. 4 (April 1981):376–85.

[HAGO83] Hagouel, J. "Issues in Routing for Large and Dynamic Networks," Ph.D. diss. Columbia University, 1983.

[HAHN84] Hahn, J. J., and Stolle, D. M. "Packet Radio Network Routing Algorithms: A Survey." *IEEE Communications Magazine* 22, no. 11 (November 1984):41–47.

[HAIL82] Hailpern, B. T. "Specifying and Verifying Protocols Represented as Abstract Programs," in [GREE82a], pp. 607–24.

[HALS85] Halsall, F. *Introduction to Data Communications and Computer Networks.* Workingham, England: Addison-Wesley, 1985.

[HAMI75] Hamilton, D. J., and Howard, W. G. *Basic Integrated Circuit Engineering.* New York: McGraw-Hill, 1975.

[HAMM86] Hammond, J. L., and O'Reilly, P. J. P. *Performance Analysis of Local Computer Networks.* Reading, Mass.: Addison-Wesley, 1986.

[HAMM87] Hammond, J. L., and Spragins, J. D. "Rapidly Reconfiguring Computer Communication Networks: Definition and Major Issues." *Proceedings of IEEE INFO-COM,* San Francisco, Calif. (April 1987).

[HAMM88] Hammond, J. L.; Leathrum, J. F.; Lovegrove, W. P.; Tipper, D. W.; and Spragins, J. D. "Issues in Simulation of Rapidly Reconfigureable Networks." *Proceedings of IMACS International Symposium,* Paris, France (July 1988).

[HAMN88] Hamner, M. C., and Samsen, G. R. "Source Routing Bridge Implementation." *IEEE Network* 2, no. 1 (January 1988):33–36.

[HÄND88] Händel, R. "Evolution of ISDN Towards Broadband ISDN." *Proceedings of IEEE INFOCOM '88* (March 1988):5A.1.1–1.8.

[HÄND89] Händel, R. "Evolution of ISDN Towards Broadband ISDN." *IEEE Network* 3, no. 1 (January 1989):7–13.

[HART88] Hart, J. "Extending the IEEE 802.1 MAC Bridge Standard to Remote Bridges." *IEEE Network* 2, no. 1 (January 1988):10–15.

[HAWE84] Hawe, B.; Kirby, A.; and Stewart, B. "Transparent Interconnection of Local Area Networks with Bridges." *Journal of Telecommunication Networks* 3, no. 2 (Summer 1984):116–30.

[HAWE86] Hawe, W. R.; Kempf, M. F.; and Kirby, A. J. "The Extended Local Area Network Architecture and LANBridge 100." *Digital Technical Journal,* Digital Equipment Corporation, Hudson, Mass., no. 3 (September 1986):54–72.

[HAYE78] Hayes, J. F. "An Adaptive Technique for Local Distribution." *IEEE Transactions on Communications* COM-26, no. 8 (August 1978):1178–86.

[HAYE84] Hayes, J. F. *Modeling and Analysis of Computer Communications Networks.* New York: Plenum Press, 1984.

[HAYK88a] Haykin, S. *An Introduction to Analog and Digital Communication.* New York: Wiley, 1988.

[HAYK88b] Haykin, S. *Digital Communications.* New York: Wiley, 1988.

[HEAR70] Heart, F. E.; Kahn, R. E.; Ornstein, S. M.; Crowther, W. R.; and Walden, D. C. "The Interface Message Processor for the ARPA Computer Network." *Proceedings AFIPS Spring Joint Computer Conference* (May 1970):551–67.

[HEBU88] Hebuterne, D. "STD Switching in an ATD Environment." *Proceedings of IEEE INFOCOM '88* (March 1988):5A.3.1–3.10.

[HELL87] Hellman, M. E. "Commercial Encryption." *IEEE Network* 1, no. 2 (April 1987):6–10.

[HEMR88] Hemrick, C. F.; Klessig, R. W.; and McRoberts, J. M. "Switched Multi-Magabit Data Service and Early Availability Via MAN Technology." *IEEE Communications Magazine* 26, no. 4 (April 1988):9–14.

[HERR86] Herr, T. J., and Plevyak, T. J. "ISDN: The Opportunity Begins." *IEEE Communications Magazine* 24, no. 11 (November 1986):6–10.

[HIGD86] Higdon, M.; Page, J. T.; and Stuntebeck, P. "AT&T Communications ISDN Architecture." *AT&T Technical Journal* 65, no. 1 (January/February 1986):27–34.

[HIND83] Hinden, R.; Haverty, J.; and Sheltzer, A. "The DARPA Internet: Interconnecting Heterogenous Computer Networks with Gateways." *Computer* 16, no. 9 (September 1983):38–49.

[HINT87] Hinton, H. S. "Photonic Switching Technology Applications." *AT&T Technical Journal* 66, no. 3 (May/June 1987):41–53.

[HIRA87] Hirade, K. "Mobile-Radio Communications," in [FEHE87a], pp. 488–572.

[HOBB72] Hobbs, L. C. "Terminals." *Proceedings of the IEEE* 60, no. 11 (November 1972):1273–83.

[HOBE80] Hoberecht, V. L. "SNA Function Management." *IEEE Transactions on Communications* COM-28, no. 4 (April 1980):594–603.

[HOBE82] Hoberecht, V. L. "SNA Higher Layer Protocols," in [GREE82a], pp. 459–82.

[HOBE83] Hoberecht, V. L. "A Layered Network Protocol for Packet Voice and Data Integration." *IEEE Journal on Selected Areas in Communications* SAC-1, no. 6 (December 1983):1006–13.

[HOLL83] Hollis, L. L. "OSI Presentation Layer Activities." *Proceedings of the IEEE* 71, no. 12 (December 1983):1401–3.

[HOLS87] Holsinger, J. L., and Pahlavan, K. "A Historical Review of Voice-Band Modems." *Proceedings IEEE International Conference on Communications '87* Seattle, Wash. (June 1987):12.1.1–1.6.

[HOPP86] Hoppitt, C. "ISDN Evolution: From Copper to Fiber in Easy Stages." *IEEE Communications Magazine* 24, no. 11 (November 1986):17–22.

[HORO78] Horowitz, E., and Sahni, S. *Fundamentals of Computer Algorithms.* Potomac, Md.: Computer Science Press, 1978.

[HOUS87] Housley, T. *Data Communications & Teleprocessing Systems.* Englewood Cliffs, N.J.: Prentice-Hall, 1987.

[HSIE84] Hsieh, W. N., and Gitman, I. "Routing Strategies in Computer Networks." *Computer* 17, no. 6 (June 1984):46–56.

[HU82] Hu, T. C. *Combinatorial Algorithms.* Reading, Mass.: Addison-Wesley, 1982.

[HUNT86] Hunt, L. R. *Data Communications and Networks,* Part I and Part II. Christchurch, New Zealand: Department of Computer Science, University of Canterbury, March 1986.

[HUYN81] Huynh, D., and Kuo, F. F. "Mixed-Media Packet Networks," in [KUO81], pp. 202–39.

[IBM70] IBM Corporation. *General Information—Binary Synchronous Communications.* Research Triangle Park, N.C.: Order No. GA27-3004-2, 1970.

[IBM75] IBM Corporation. *IBM Synchronous Data Link Control: General Information.* Research Triangle Park, N.C.: Order No. GA27-3093-1, 1975.

[IBM80] IBM Corporation. *Systems Network Architecture Format and Protocol Reference Manual: Architectural Logic.* Research Triangle Park, N.C.: Order No. SC30-3112-2, 1980.

[IBM86a] IBM Corporation. *Systems Network Architecture: Concepts and Products.* Research Triangle Park, N.C.: Order No. GC30-3072-3, 1986.

[IBM86b] IBM Corporation. *Systems Network Architecture: Technical Overview.* Research Triangle Park, N.C.: Order No. GC30-3073-2, 1986.

[IDC87a] International Data Corporation (IDC). "Computer Systems and Software for Business and Industry." *Forbes* 139, no. 14 special advertising supplement (June 29, 1987).

[IDC87b] International Data Corporation (IDC). "Office Automation & Desktop Computing." *Forbes* 140, no. 9 special advertising supplement (October 19, 1987).

[IDC87c] International Data Corporation (IDC). "Trends in Communications: Networks for Tomorrow." *Forbes* 140, no. 6 special advertising supplement (September 21, 1987).

[ILYA85] Ilyas, M., and Mouftah, H. T. "Performance Evaluation of Computer Communications Networks." *IEEE Communications Magazine* 23, no. 4 (April 1985): 18–29.

[INOS78] Inose, H., and Saito, T. "Theoretical Aspects in the Analysis and Synthesis of Packet Communication Networks." *Proceedings of the IEEE* 66, no. 11 (November 1978):1409–22.

[INST85a] Institute of Electrical and Electronics Engineers. *IEEE 802.2: IEEE Standards for Local Area Networks—Logical Link Control.* New York: Institute of Electrical and Electronics Engineers, 1985.

[INST85b] Institute of Electrical and Electronics Engineers. *IEEE 802.3: IEEE Standards for Local Area Networks—Carrier Sense Multiple Access with Collison Detection (CSMA/CD) Access Method and Physical Layer Specifications.* New York: Institute of Electrical and Electronics Engineers, 1985.

[INST85c] Institute of Electrical and Electronics Engineers. *IEEE 802.4: IEEE Standards for Local Area Networks—Token-Passing Bus Access Method and Physical Layer Specifications.* New York: Institute of Electrical and Electronics Engineers, 1985.

[INST85d] Institute of Electrical and Electronics Engineers. *IEEE 802.5: IEEE Standard for Local Area Networks—Token Ring Access Method and Physical Layer Specifications.* New York: Institute of Electrical and Electronics Engineers, 1985.

[INTE84a] International Organization for Standardization (ISO) 3309. *Data Communication: High-Level Data Link Control Procedures—Frame Structure,* 3d ed. Geneva: 1984.

[INTE84b] International Organization for Standardization (ISO) 4335. *Data Communication: High-Level Data Link Control Procedures—Consolidation of Elements of Procedures,* 2d ed. Geneva: 1984.

[INTE84c] International Organization for Standardization (ISO) 7809. *Data Communication: High-Level Data Link Control Procedures—Consolidation of Classes of Procedures.* Geneva: 1984.

[INTE84d] International Organization for Standardization (ISO) 7498. *Information Processing Systems: Open Systems Interconnection—Basic Reference Model.* Geneva: 1984.

[INTE86a] International Organization for Standardization (ISO) 8072. *Information Processing Systems: Open Systems Interconnection—Transport Service Definition.* Geneva: 1986.

[INTE86b] International Organization for Standardization (ISO) 8073. *Information Processing Systems—Open Systems Interconnection—Connection Oriented Transport Protocol Specification.* Geneva: 1986.

[INTE86c] International Organization for Standardization (ISO) 8602. *Information Processing Systems: Open Systems Interconnection—Protocol for Providing the Connectionless-Mode Transport Service.* Geneva: 1986.

[INTE87a] International Organization for Standardization (ISO) 8348. *Information Processing Systems: Data Communications—Network Service Definition.* Geneva: 1987.

[INTE87b] International Organization for Standardization (ISO) 8473. *Information Processing Systems: Data Communications—Protocol for Providing the Connectionless-Mode Network Service.* Geneva: 1987.

[INTE87c] International Organization for Standardization (ISO) 8648. *Information Processing Systems: Data Communications—Internal Organization of the Network Layer.* Geneva: 1987.

[IRLA78] Irland, M. I. "Buffer Management in a Packet Switch." *IEEE Transactions on Communications* COM-26, no. 3 (March 1978):328–37.

[IRME86] Irmer, T. "An Idea Turns Into Reality—CCITT Activities on the Way to ISDN." *IEEE Journal on Selected Areas in Communications* SAC-4, no. 3 (May 1986): 316–19.

[JABL87] Jablonowski, D. P.; Paek, U. C.; and Watkins, L. S. "Optical Fiber Manufacturing Techniques." *AT&T Technical Journal* 66, no. 1 (January/February 1987): 33–44.

[JACO78] Jacobs, I. M.; Binder, R.; and Hoversten, E. V. "General Purpose Packet Satellite Networks." *Proceedings of the IEEE* 66, no. 11 (November 1978):1448–67.

[JACO79] Jacobs, I. M.; Binder, R.; Bressler, R. D.; Edmond, W. B.; and Killian, E. A. "Packet Satellite Network Design Issues." *Proceedings National Telecommunications Conference* (November 1979):45.2.1–2.12.

[JACO86] Jacobs, I. "Design Considerations for Long-Haul Lightwave Systems." *IEEE Journal on Selected Areas in Communications* SAC-4, no. 9 (December 1986):1389–95.

[JACO88] Jacobson, V. "Congestion Avoidance and Control." *Proceedings ACM SIGCOMM '88* Stanford, Calif. (August 1988).

[JAFA80] Jafari, H.; Lewis, T.; and Spragins, J. "Simulation of a Class of Ring-Structured Networks." *IEEE Transactions on Computers* C-29, no. 5 (May 1980):385–92.

[JAIN86a] Jain, R. "A Timeout-Based Congestion Control Scheme for Window Flow-Controlled Networks." *IEEE Journal on Selected Areas of Communications* SAC-4, no. 7 (October 1986):1162–67.

[JAIN86b] Jain, R., and Hawe, W. R. "Performance Analysis and Modeling of Digital's Networking Architecture." *Digital Technical Journal* Digital Equipment Corporation, Hudson, Mass., no. 3 (September 1986):25–34.

[JAIN87] Jain, R.; Ramkrishnan, K. K.; and Chiu, D. M. *Congestion Avoidance in Computer Networks with a Connectionless Network Layer.* Digital Equipment Corporation Technical Report TR 506 (1987).

[JAME72] James, R. T., and Muench, P. E. "AT&T Facilities and Services." *Proceedings of the IEEE* 60, no. 11 (November 1972):1342–49.

[JANS84] Janson, P. A., and Mumprecht, E. "Addressing and Routing in a Hierarchy of Token Rings." *Ring Technology Local Area Networks,* Elsevier Science Publishers B. V. (1984):97–109.

[JANS87] Janson, P. A., and Cockburn, A. A. R. "Adding Transparent Internetworking to a LAN Application Interface." *IEEE Journal on Selected Areas in Communications* SAC-5, no. 9 (December 1987):1471–79.

[JARE81] Jarema, D. R., and Sussenguth, E. H. "IBM Data Communications: A Quarter Century of Evolution and Progress." *IBM Journal of Research and Development* 25, no. 5 (September 1981).

[JENQ83] Jenq, Y. C. "Performance Analysis of a Packet Switch Based on Single-Buffered Banyan Network." *IEEE Journal on Selected Areas in Communications* SAC-1, no. 6 (December 1983):1014–21.

[JOEL84] Joel, A. E., Jr. "The Past 100 Years in Telecommunications Switching." *IEEE Communications Magazine* 22, no. 5 (May 1984):64–71.

[JOHN86] Johnson, W. R., Jr. "Foreword to Special Issue on Networking Products." *Digital Technical Journal,* Digital Equipment Corporation, Hudson, Mass., no. 3 (September 1986):8–9.

[JUBI87] Jubin, J., and Tornow, J. D. "The DARPA Packet Radio Network Protocols." *Proceedings of the IEEE* 75, no. 1 (January 1987):21–32.

[JUDI88] Judice, C. N. "4th IEEE International Workshop on Telematics." *IEEE Communications Magazine* 26, no. 10 (October 1988):87–88.

[JUEN76] Jueneman, R. R., and Kerr, G. S. "Explicit Path Routing in Communications Networks." *Proceedings of the International Conference on Computer Communication* (1976):340–42; also in [LAM84c].

[JUEN87] Jueneman, R. R. "Electronic Document Authentication." *IEEE Network* 1, no. 2 (April 1987):17–23.

[KADE81] Kaderali, F., and Weston, J. "Digital Subscriber Loops." *Electrical Communications* no. 1 (1981):71–79.

[KAHL86] Kahl, P. "A Review of CCITT Standardization to Date." *IEEE Journal on Selected Areas in Communications* SAC-4, no. 3 (May 1986):326–33.

[KAHN71] Kahn, R. E., and Crowther, W. R. "Flow Control in a Resource-Sharing Computer Network." *Proceedings ACM/IEEE Second Symposium on Problems in the Optimization of Data Communication Systems* (October 1971):108–16.

[KAHN72] Kahn, R. E. "Resource-Sharing Computer Communications Networks." *Proceedings of the IEEE* 60, no. 11 (November 1972):1397–1407.

[KAHN73] Kahn, D. *The Code Breakers.* New York: Macmillan, 1973.

[KAHN78a] Kahn, R. E.; Gronemeyer, S. A.; Burchfiel, J.; and Kunzelman, R. C. "Advances in Packet Radio Technology." *Proceedings of the IEEE* 66, no. 11 (November 1978):1468–96.

[KAHN78b] Kahn, R. E.; Uncapher, K. W.; and Van Trees, H. L., eds. "Special Issue on Packet Communication Networks." *Proceedings of the IEEE* 66, no. 11 (November 1978).

[KAIS87] Kaiser, P.; Midwinter, J.; and Shimada, S. "Status and Future Trends in Terrestrial Optical Fiber Systems in North America, Europe and Japan." *IEEE Communications Magazine* 25, no. 10 (October 1987):8–21.

[KAJI82] Kajiwara, T.; Ohara, Y.; and Ohta, K. "A Study on the Protocol Conversion Method." *Review of the Electrical Communication Laboratories* 30, no. 6 (1982):1066–75.

[KALI87] Kalish, D., and Cohen, L. G. "Single-Mode Fiber: From Research and Development to Manufacturing." *AT&T Technical Journal* 66, no. 1 (January/February 1987):19–33.

[KAMI86] Kaminski, M. A., Jr. "Computers: Protocols for Communicating in the Factory." *IEEE Spectrum* 23, no. 4 (April 1986):56–62.

[KANE86] Kanemasa, A.; Sugiyama, A.; Koike, S.; and Koyama, T. "An ISDN Subscriber Loop Transmission System Based on Echo Cancellation." *IEEE Journal on Selected Areas in Communications* SAC-4, no. 8 (November 1986):1359–66.

[KANE87] Kaneko, H., and Ishiguro, T. "Digital Television-Processing Techniques," in [FEHE87a], pp. 282–315.

[KANG88] Kang, Y. J.; Herzog, J. H.; and Spragins, J. "FISHNET: A Distributed Architecture for High-Performance Local Computer Networks." *IEEE Transactions on Computers* 37, no. 1 (January 1988):119–23.

[KANO86] Kano, S. "Layers 2 and 3 ISDN Recommendations." *IEEE Journal on Selected Areas in Communications* SAC-4, no. 3 (May 1986):355–59.

[KANY88] Kanyuh, D. "An Integrated Network Management Product." *IBM Systems Journal* 27, no. 1 (1988):45–59.

[KAO66] Kao, K. C., and Hockham, G. A. "Dielectric-Fiber Surface Waveguides for Optical Frequencies." *Proceedings of the IEEE* 113 (July 1966):1151–58.

[KAPR70] Kapron, F. P.; Keck, D. B.; and Maurer, R. D. "Radiation Losses in Glass Optical Waveguides." *Applied Physics Letters* 17 (November 1970):423–25.

[KARN88] Karn, P., and Partridge, C. "Improving Round-Trip Time Estimates in Reliable Transport Protocols." *Proceedings ACM SIGCOMM '87,* Stowe, Vt. (August 1987): 2–7.

[KARP86] Karp, R. M. "Combinatorics, Complexity and Randomness." *Communications of the ACM* 29, no. 2 (February 1986):98–109.

[KAWA87] Kawa, C., and Bochmann, G. "Hierarchical Multi-Network Interconnection Using Public Data Networks." *Proceedings of IEEE INFOCOM* (1987):426–35.

[KELL78] Kelly, P. T. F. "Public Packet Switched Data Networks, International Plans and Standards." *Proceedings of the IEEE* 66, no. 11 (November 1978):1539–49.

[KENT81] Kent, S. T. "Security in Computer Networks," in [KUO81], pp. 369–432.

[KIHA89] Kihara, M. "Performance Aspects of Reference Clock Distribution for Evolving Digital Networks." *IEEE Communications Magazine* 27, no. 4 (April 1989): 24–34.

[KIM83] Kim, B. G. "Characterization of Arrival Statistics of Multiplexed Voice Packets." *IEEE Journal on Selected Areas in Communications* SAC-1, no. 6 (December 1983):1133–38.

[KITA88] Kitawaki, N., and Nagabuchi, H. "Quality Assessment of Speech Coding and Speech Synthesis Systems." *IEEE Communications Magazine* 26, no. 10 (October 1988):36–44.

[KLEI73] Kleinrock, L., and Lam, S. S. "Packet-Switching in a Slotted Satellite Channel." *AFIPS Conference Proceedings* 42 (1973):703–10.

[KLEI75a] Kleinrock, L., and Lam, S. S. "Packet Switching in a Multiaccess Broadcast Channel: Performance Evaluation." *IEEE Transactions on Communications* COM-23, no. 4 (April 1975):410–23.

[KLEI75b] Kleinrock, L. *Queueing Systems, Volume I: Theory.* New York: Wiley, 1975.

[KLEI75c] Kleinrock, L., and Tobagi, F. A. "Packet Switching in Radio Channels: Part I - Carrier Sense Multiple Access Modes and Their Throughput-Delay Characteristics." *IEEE Transactions on Communications* COM-23 (December 1975):1400–16.

[KLEI76a] Kleinrock, L. *Queueing Systems, Volume II: Computer Applications.* New York: Wiley, 1976.

[KLEI76b] Kleinrock, L.; Taylor, W. E.; and Opderbeck, H. "A Study of Line Overhead in the ARPAnet." *Communications of the ACM* 19, no. 1 (January 1976):3–12.

[KLEI78a] Kleinrock, L., and Yemini, Y. "An Optimal Adaptive Scheme for Multiple Access Broadcast Communication." *Conference Record of IEEE International Conference on Communications* (June 1978):7.2.1–2.5

[KLEI78b] Kleinrock, L. "Principles and Lessons in Packet Communications." *Proceedings of the IEEE* 66, no. 11 (November 1978):1320–29.

[KLEI80] Kleinrock, L., and Gerla, M. "Flow Control: A Comparative Survey." *IEEE Transactions on Communications* COM-28, no. 4 (April 1980):553–74.

[KLEI82] Kleinrock, L. "A Decade of Network Development." *Journal of Telecommunication Networks* (Spring, 1982):1–11.

[KLEI88] Kleinrock, L., and Kamoun, F. "Hierarchical Routing for Large Networks: Performance Evaluation and Optimization." *Computer Networks* 1, (January 1977):155–74.

[KNIG72] Knight, J. R. "A Case Study: Airlines Reservation Systems." *Proceedings of the IEEE* 60, no. 11 (November 1972):1423–31.

[KNIG83] Knightson, K. G. "The Transport Layer Standardization." *Proceedings of the IEEE* 71, no. 12 (December 1983):1394–96.

[KOBA84] Kobayashi, K. "The Past, Present, and Future of Telecommunications in Japan." *IEEE Communications Magazine* 22, no. 5 (May 1984):96–103.

[KOBA86] Kobayashi, K. *Computers and Communications.* Cambridge, Mass.: MIT Press, 1986.

[KOKJ86] Kokjer, K. J., and Roberts, T. D. "Networked Meteor-Burst Data Communications." *IEEE Communications Magazine* 24, no. 11 (November 1986):23–29.

[KONH74] Konheim, A. G., and Meister, B. "Waiting Lines and Times in a System with Polling." *Journal of the ACM* 21, no. 3 (July 1974):470–90.

[KORZ89] Korzeniowski, P. "LU6.2 is Finally Finding Fans." *Communications Week, 230* (January 9, 1989):1, 34–35.

[KOST84] Kostas, D. J. "Transition to ISDN—An Overview." *IEEE Communications Magazine* 22, no. 1 (January 1984):11–17.

[KUHN84] Kuhn, M. "Telecommunications in Canada—A Century of Symbiotic Development." *IEEE Communications Magazine* 22, no. 5 (May 1984):104–14.

[KUN86a] Kun, R. "An ISDN Network D-Channel VLSI Architecture." *IEEE Journal on Selected Areas in Communications* SAC-4, no. 8 (November 1986):1275–80.

[KUN86b] Kun, R. "Network VLSI D-Channel Architecture." *Telesis* 13, no. 3 (1986): 46–52.

[KUO81] Kuo, F. F., ed. *Protocols & Techniques for Data Communication Networks.* Englewood Cliffs, N.J.: Prentice-Hall, 1981.

[KURI87] Kurita, O., and Yoshida, T. "Current Subscriber Radio and Recent Trends." *IEEE Communications Magazine* 25, no. 11 (November 1987):44–50.

[KUWA86] Kuwahara, H.; Amemiya, S.; and Murano, K. "Phase Aligned Passive Bus

(PAB) Scheme for ISDN User-Network Interface." *IEEE Journal on Selected Areas in Communications* SAC-4, no. 8 (November 1986):1367–72.

[LAM75] Lam, S. S., and Kleinrock, L. "Dynamic Control Schemes for a Packet Switched Multi-Access Broadcast Channel." *Proceedings of National Computer Conference* 44 (1975):143–53.

[LAM77] Lam, S. S. "Delay Analysis of a Time Division Multiple Access (TDMA) Channel." *IEEE Transactions on Communications* COM-25, no. 12 (December 1977):1489–94.

[LAM79] Lam, S. S., and Reiser, M. "Congestion Control of Store-and-Forward Networks by Input Buffer Limits: An Analysis." *IEEE Transactions on Communications* COM-27, no. 1 (January 1979):127–34.

[LAM81] Lam, S. S., and Lien, Y. C. L. "Congestion Control of Packet Communication Networks by Input Buffer Limits—A Simulation Study." *IEEE Transactions on Computers* C-30, no. 10 (October 1981):733–42.

[LAM83a] Lam, S. S. "Data Link Control Procedures, in [CHOU83b], pp. 81–113.

[LAM83b] Lam, S. S. "Multiple-Access Protocols," in [CHOU83b], pp. 114–55.

[LAM84a] Lam, S. S. "Fundamentals of Computer Communication Networks," in [LAM84c], pp. 1–42.

[LAM84b] Lam, S. S., and Shankar, A. U. "Protocol Verification via Projections." *IEEE Transactions on Software Engineering* (July 1984):325–42.

[LAM84c] Lam, S. S., ed. *Tutorial: Principles of Communication and Networking Protocols.* Silver Spring, Md.: IEEE Computer Society Press, 1984.

[LAM88] Lam, S. S. "Protocol Conversion." *IEEE Transactions on Software Engineering* 14, no. 3 (March 1988):353–62.

[LAND86] Landweber, L. H.; Jennings, D. M.; and Fuchs, I. "Research Computer Networks and Their Interconnection." *IEEE Communications Magazine* 24, no. 6 (June 1986):5–17.

[LANG83] Langsford, A.; Naemura, K.; and Speth, R. "OSI Management and Job Transfer Services." *Proceedings of the IEEE* 71, no. 12 (December 1983):1420–24.

[LAPE86] La Pelle, N. R.; Seger, M. J.; and Sylor, M. W. "The Evolution of Network Management Products." *Digital Technical Journal* Digital Equipment Corporation, Hudson, Mass., no. 3 (September 1986):117–28.

[LATH83] Lathi, B. P. *Modern Digital and Analog Communication Systems.* New York: Holt, Rinehart and Winston, 1983.

[LAUC86] Lauck, A. G.; Oran, D. R.; and Perlman, R. J. "A Digital Network Architecture Overview." *Digital Technical Journal* Digital Equipment Corporation, Hudson, Mass., no. 3 (September 1986):10–24.

[LAWS87] Lawser, J. J., and Oxley, P. L. "Common Channel Signaling Network Evolution." *AT&T Technical Journal* 66, no. 3 (May/June 1987):13–21.

[LECH86] Lechleider, J. W. "Loop Transmission Aspects of ISDN Basic Access." *IEEE Journal on Selected Areas in Communications* SAC-4, no. 8 (November 1986):1294–1301.

[LEDE78] Lederberg, J. "Digital Communications and the Conduct of Science: The New Literacy." *Proceedings of the IEEE* 66, no. 11 (November 1978):1314–19.

[LEE83] Lee, J. "Symbiosis Between a Terrestrial-Based Integrated Services Digital Network and a Digital Satellite Network." *IEEE Journal on Selected Areas in Communications* SAC-1, no. 1 (January 1983):103–9.

[LEE88] Lee, E. A., and Messerschmitt, D. G. *Digital Communication.* Boston: Kluwer Academic Publishers, 1988.

[LEIN85] Leiner, B. M.; Cole, R.; Postel, J.; and Mills, D. "The DARPA Internet Protocol Suite." *IEEE Communications Magazine* 23, no. 3 (March 1985):29–34.

[LEI87] Leiner, B. M.; Nielson, D. L., and Tobagi, F. A. "Issues in Packet Radio Network Design." *Proceedings of the IEEE* 75, no. 1 (January 1987):6–20.

[LEMI81] Lemieux, C. "Theory of Flow Control in Shared Networks and Its Application in the Canadian Telephone Network." *IEEE Transactions on Communications* COM-29, no. 4 (April 1981):399–412.

[LEON89] Leon-Garcia, A., ed. "Special Issue on Broadband Networks." *IEEE Network* 3, no. 1 (January 1989).

[LEWA83] Lewan, D., and Long, H. G. "The OSI File Service." *Proceedings of the IEEE* 71, no. 12 (December 1983):1414–19.

[LI87] Li, T. "Advances in Lightwave Systems Research." *AT&T Technical Journal* 66, no. 1 (January/February 1987):5–18.

[LI88] Li, T., and Linke, R. A. "Multigigabit-Per-Second Lightwave Systems Research for Long-Haul Applications." *IEEE Communications Magazine* 26, no. 4 (April 1988):29–35.

[LICK78] Licklider, J. C. R., and Vezza, A. "Applications of Information Networks." *Proceedings of the IEEE* 66, no. 11 (November 1978):1330–46.

[LIEB85] Liebowitz, B. H., and Carson, J. H. *Multiple Processor Systems for Real-Time Applications.* Englewood Cliffs, N.J.: Prentice-Hall, 1985.

[LIN81] Lin, S., and Costello, D. J., Jr. "Coding for Reliable Data Transmission and Storage," in [KUO81], pp. 240–318.

[LIN83] Lin, S., and Costello, D. J., Jr. *Error Control Coding: Fundamentals and Applications.* Englewood Cliffs, N.J.: Prentice-Hall, 1983.

[LIN84] Lin, S.; Costello, D. J.; and Miller, M. J. "Automatic Repeat Request Error Control Schemes." *IEEE Communications Magazine* 22 (December 1984):5–17.

[LIND73] Lindsey, W. C., and Simon, M. K. *Telecommunication Systems Engineering.* Englewood Cliffs, N.J.: Prentice-Hall, 1973.

[LINI83] Linington, P. F. "Fundamentals of the Layer Service Definitions and Protocol Specifications." *Proceedings of the IEEE* 71, no. 12 (December 1983):1341–45.

[LINN83] Linn, R. J., and Nightingale, J. S. "Testing OSI Protocols at the National Bureau of Standards." *Proceedings of the IEEE* 71, no. 12 (December 1983):1431–34.

[LISS71] Lissandrello, G. J. "World Data Communications as Seen by the Data Processing Systems Designer." *Proceedings of ACM/IEEE Second Symposium on Problems in the Optimization of Data Communications Systems* (October 1971): 130–36.

[LIU86] Liu, H. S.; Thomas, E. P.; and Walters, S. M. "A Working Research Prototype of an ISDN Central Office." *IEEE Journal on Selected Areas in Communications* SAC-4, no. 8 (November 1986):1241–50.

[LOWE83] Lowe, H. "OSI Virtual Terminal Service." *Proceedings of the IEEE* 71, no. 12 (December 1983):1408–13.

[LUCK65] Lucky, R. W. "Automatic Equalization for Digital Communications." *Bell System Technical Journal* 44, no. 4 (April 1965):547–88.

[LUCK68] Lucky, R. W.; Salz, J.; and Weldon, E. J., Jr. *Principles of Data Communication.* New York: McGraw-Hill, 1968.

[LUCZ78] Luczak, E. C. "Global Bus Computer Communication Techniques." *Proceedings of Computer Networking Symposium* (December 1978):58–71.

[LUET86] Luetchford, J. C. "CCITT Recommendations—Network Aspects of the ISDN." *IEEE Journal on Selected Areas in Communications* SAC-4, no. 3 (May 1986): 334–43.

[MACC86] Macchi, O. "Advances in Adaptive Filtering (State-of-the-Art)." *Digital Communications: Proceedings of the Second Tirrenia International Workshop on Digital Communications,* Tirrenia, Italy, September 1985, North Holland, Amsterdam (1986):41–58.

[MACG86] MacGillivary, D. L., and Markvorsen, K. "Bell Canada ISDN Trial Plans." *IEEE Journal on Selected Areas in Communications* SAC-4, no. 3 (May 1986):421–23.

[MAHM83] Mahmoud, S. A.; Chan, W.-Y.; Riordon, J. S.; and Aidarous, S. E. "An Integrated Voice/Data System for VHF/UHF Mobile Radio." *IEEE Journal on Selected Areas in Communications* SAC-1, no. 6 (December 1983):1098–11.

[MAIM60] Maiman, T. H. "Stimulated Optical Radiation in Ruby." *Nature, London* 187 (1960):493–94.

[MAJI79] Majithia, J.; Irland, M.; Grangé, J. L.; Cohen, N.; and O'Donnell, C. "Experiments in Congestion Control Techniques," in [GRAN79], pp. 211–34.

[MANN86] Mann, B. E. "Terminal Servers on Ethernet Local Area Networks." *Digital Technical Journal,* Digital Equipment Corporation, Hudson, Mass., no. 3 (September 1986):73–87.

[MART76] Martin, J. *Telecommunications and the Computer,* 2d ed. Englewood Cliffs, N.J.: Prentice-Hall, 1976.

[MART77] Martin, J. *Future Developments in Telecommunications,* 2d ed. Englewood Cliffs, N.J.: Prentice-Hall, 1977.

[MART82] Martin, J. *Viewdata and the Information Society.* Englewood Cliffs, N.J.: Prentice-Hall, 1982.

[MASS74] Massey, J. L. "Coding and Modulation in Digital Communications." *Proceedings 1974 International Zurich Seminar on Digital Communications,* Zurich, Switzerland (March 1974):E.2.1–2.4.

[MASS86] Massey, J. L. "Cryptography—A Selective Survey (Tutorial)." *Digital Communications: Proceedings of the Second Tirrenia International Workshop on Digital Communications,* Tirrenia, Italy, September 1985, North Holland, Amsterdam (1986):3–21.

[MATH72] Mathison, S. L., and Walker, P. M. "Regulatory and Economic Issues in Computer Communications." *Proceedings of the IEEE* 60, no. 11 (November 1972):1254–72.

[MATH78] Mathison, S. "Commercial, Legal and International Aspects of Packet Communications." *Proceedings of the IEEE* 66, no. 11 (November 1978):1527–38.

[MCCL83] McClelland, F. M. "Services and Protocols of the Physical Layer." *Proceedings of the IEEE* 71, no. 12 (December 1983):1372–77.

[MCDO86] McDonald, J. C. "A Proactive ISDN Implementation Strategy." *IEEE Journal on Selected Areas in Communications* SAC-4, no. 8 (November 1986):1218–21.

[MCFA79] McFarland, R. "Protocols in a Computer Internetworking Environment." *Proceedings of EASCON 79* (1979).

[MCGR83] McGregor, P. V. "Functions and Characteristics of Devices Used in Computer Communications," in [CHOU83b], pp. 314–68.

[MCLE88] McLeod, S. "TELNET," in [STALL88a], pp. 145–76.

[MCNA82] McNamara, J. E. *Technical Aspects of Data Communications,* 2d ed. Burlington, Mass.: Digital Press, 1982.

[MCNA85] McNamara, J. E. *Local Area Networks: An Introduction to the Technology.* Burlington, Mass.: Digital Press, 1985.

[MCNI87] McNinch, B. "ISDN: The Man-Machine Interface." *IEEE Communications Magazine* 25, no. 12 (December 1987):50–54.

[MCQU74] McQuillan, J. M. "Design Considerations for Routing Algorithms in Computer Networks." *Proceedings Seventh Hawaii International Conference on System Sciences* 24 (January 1974):22–24.

[MCQU78a] McQuillan, J. M.; Falk, G.; and Richer, I. "A Review of the Development and Performance of the ARPANET Routing Algorithm." *IEEE Transactions on Communications* COM-26, no. 12 (December 1978):1802–11.

[MCQU78b] McQuillan, J. M. "Enhanced Message Addressing Capabilities for Computer Networks." *Proceedings of the IEEE* 66, no. 11 (November 1978):1517–27.

[MCQU80] McQuillan, J. M.; Richer, I.; and Rosen, E. C. "The New Routing Algorithm for the ARPANET." *IEEE Transactions on Communications* COM-28, no. 5 (May 1980):711–19.

[MEIJ82] Meijer, A., and Peeters, P. *Computer Network Architectures.* London: Computer Science Press, 1982.

[MEIJ88] Meijer, A. *Systems Network Architecture: a Tutorial.* New York: Wiley, 1988.

[MELI86] Melindo, F., and Valabonesi, G. "Network and System Architecture: The Integrated Approach to ISDN in the UT Line." *IEEE Journal on Selected Areas in Communications* SAC-4, no. 8 (November 1986):1251–56.

[MERS87] Merski, R., and Parrish, D. M. "Operations Systems Technology for New AT&T Network and Service Capabilities." *AT&T Technical Journal* 66, no. 3 (May/June 1987):64–72.

[MESS86a] Messerschmitt, D. G. "Design Issues in the ISDN Basic Customer Access (Survey)." *Digital Communications: Proceedings of the Second Tirrenia International Workshop on Digital Communications,* Tirrenia, Italy, September 1985, North Holland, Amsterdam (1986):133–44.

[MESS86b] Messerschmitt, D. G. "Design Issues in the ISDN U-Interface Transceiver." *IEEE Journal on Selected Areas in Communications* SAC-4, no. 8 (November 1986):1281–93.

[MESS87] Messerschmitt, D. G. "Echo Cancellation in Speech and Data Transmission," in [FEHE87a], pp. 182–236.

[METC76] Metcalfe, R. M., and Boggs, D. R. "Ethernet: Distributed Packet Switching for Local Computer Networks." *Communications of the ACM* 19, no. 7 (July 1976): 395–404.

[MICH88] Michel, T. "File Transfer Protocol", in [STALL88a], pp. 92–119.

[MIER86] Mierswa, P. O.; Mitton, D. J.; and Spence, M. L. "The DECnet-DOS System." *Digital Technical Journal,* Digital Equipment Corporation, Hudson, Mass., no. 3 (September 1986):108–16.

[MILL72a] Mills, D. L. "Communication Software." *Proceedings of the IEEE* 60, no. 11 (November 1972):1333–41.

[MILL72b] Mills, N. "NASDAQ—A User-Driven, Real Time Transaction System." *Proceedings Spring Joint Computer Conference, AFIPS Conference Proceedings,* Washington, D.C. (1972):1197–1206.

[MINS67] Minsky, M. L. *Computation: Finite and Infinite Machines.* Englewood Cliffs, N.J.: Prentice-Hall, 1967.

[MINZ89] Minzer, S. E., and Spears, D. R. "New Directions in Signaling for Broadband ISDN." *IEEE Communications Magazine* 27, no. 2 (February 1989):6–14.

[MITC86] Mitchell, O. M. M. "Implementing ISDN in the United States." *IEEE Journal on Selected Areas in Communications* SAC-4, no. 3 (May 1986):398–406.

[MIYA88] Miyahara, M. "Quality Assessments for Visual Service." *IEEE Communications Magazine* 26, no. 10 (October 1988):51–60.

[MOCK88a] Mockapetris, P. V., and Dunlap, K. J. "Development of the Domain Name System." *Proceedings ACM SIGCOMM '88* Stanford, Calif. (August 1988).

[MOCK88b] Mockapetris, P. "Simple Mail Transfer Protocol," in [STALL88a], pp. 120–44.

[MODE86] Modestino, J. W.; Massey, C. S.; Bollen, R. E.; and Prabhu, R. P. "Modeling and Analysis of Error Probability Performance for Digital Transmission Over the Two-Wire Loop Plant." *IEEE Journal on Selected Areas in Communications* SAC-4, no. 8 (November 1986):1317–30.

[MOLL86] Mollenauer, J. F. "Standards for Metropolitan Area Networks." *IEEE Communications Magazine* 26, no. 4 (April 1988).

[MONC86] Moncalvo, A.; Rizzotto, G.; and Valabonesi, G. "ECBM: Low-Cost Echo Canceller in a Silicon Boutique for Terminals, PBX's and CO Access." *IEEE Journal on Selected Areas in Communications* SAC-4, no. 8 (November 1986):1331–36.

[MONT83] Montgomery, W. A. "Techniques for Packet Voice Synchronization." *IEEE Journal on Selected Areas in Communications* SAC-1, no. 6 (December 1983):1022–27.

[MONT87] Montgomery, R. H. "Services and Tariffs with ISDN." *IEEE Communications Magazine* 25, no. 12 (December 1987):17–20.

[MOOR59] Moore, E. F. "The Shortest Path Through a Maze." *Proceedings of International Symposium on the Theory of Switching,* Part II. Cambridge: Harvard University Press, 1959, 285–92.

[MORE84] Moreau, R. *The Computer Comes of Age.* Cambridge: M. I. T. Press, 1984.

[MORE86] Morency, J. P.; Porter, D.; Pitkin, R. P.; and Oran, D. R. "The DECnet/SNA Gateway Product—A Case Study in Cross Vendor Networking." *Digital Technical*

Journal, Digital Equipment Corporation, Hudson, Mass., no. 3 (September 1986):35–53.

[MORR79] Morris, J. M. "Optimal Block Lengths for ARQ Error Control Schemes." *IEEE Transactions on Communications* COM-27 (1979):488–93.

[MOSS86] Mossotto, C.; Perucca, G.; and Romagnoli, M. "ISDN Activities in Italy." *IEEE Journal on Selected Areas in Communications* SAC-4, no. 3 (May 1986):413–20.

[MURA86] Muralidhar, K., and Sundareshan, M. "On the Decomposition of Large Communication Networks for Hierarchical Control Implementation." *IEEE Transactions on Communications* COM-34, no. 10 (October 1986).

[MURA88] Murakami, H.; Hashimoto, H.; and Hatori, Y. "Quality of Band-Compressed TV Services." *IEEE Communications Magazine* 26, no. 10 (October 1988):61–69.

[MURT88] Murthy, K. M. S. "Guest Editorial, Special Series on 'VSAT Communications Networks: Technology and Applications'." *IEEE Communications Magazine* 26, no. 5 (May 1988):8–9, 60.

[MUSS83] Musser, J. M.; Liu, T. T.; Li, L.; and Boggs, G. J. "A Local Area Network as a Telephone Local Subscriber Loop." *IEEE Journal on Selected Areas in Communications* SAC-1, no. 6 (December 1983):1046–53.

[NAGE87] Nagel, S. R. "Optical Fiber—The Expanding Medium." *IEEE Communications Magazine* 25, no. 4 (April 1987):33–43.

[NAGL87] Nagle, J. "On Packet Switches with Infinite Storage." *IEEE Transactions on Communications* COM-35, no. 4 (April 1987):435–38.

[NANT87] Nantz, T. D., and Shenk, W. J. "Lightguide Applications in the Loop." *AT&T Technical Journal* 66, no. 1 (January/February 1987):108–18.

[NATI83a] National Bureau of Standards (NBS), *Specification of a Transport Protocol for Computer Communications, Vol. 1: Overview and Services,* Institute for Computer Sciences and Technology, Gaithersburg, Md. (January 1983).

[NATI83b] National Bureau of Standards (NBS), *Specification of a Transport Protocol for Computer Communications, Vol. 2: Class 2 Protocol,* Institute for Computer Sciences and Technology, Gaithersburg, Md. (February 1983).

[NATI83c] National Bureau of Standards (NBS), *Specification of a Transport Protocol for Computer Communications, Vol. 3: Class 4 Protocol,* Institute for Computer Sciences and Technology, Gaithersburg, Md. (February 1983).

[NATI83d] National Bureau of Standards (NBS), *Specification of a Transport Protocol for Computer Communications, Vol. 4: Service Specifications,* Institute for Computer Sciences and Technology, Gaithersburg, Md. (January 1983).

[NATI83e] National Bureau of Standards (NBS), *Specification of a Transport Protocol for Computer Communications, Vol. 5: Guidance for the Implementer,* Institute for Computer Sciences and Technology, Gaithersburg, Md. (January 1983).

[NATI83f] National Bureau of Standards (NBS), *Specification of a Transport Protocol for Computer Communications, Vol. 6: Guidance for Implementation Selection,* Institute for Computer Sciences and Technology, Gaithersburg, Md. (January 1983).

[NATI83g] National Telecommunications and Information Administration "Primer on Integrated Services Digital Networks (ISDN): Implications for Future Global Communication, CCITT and ISDN" (September 1983).

[NEED78] Needham, R. M., and Schroeder, M. D. "Using Encryption for Authentication in Large Networks of Computers." *Communications of the ACM* 21, no. 12 (December 1978):993–99.

[NEIG86] Neigh, J. L., and Spindel, L. A. "ISDN Evolution in Information Systems Architecture." *AT&T Technical Journal* 65, no. 1 (January/February 1986):45–55.

[NEWM86] Newman, R. "Introduction of ISDN Into the BT Network." *IEEE Journal on Selected Areas in Communications* SAC-4, no. 3 (May 1986):385–89.

[NEWM87a] Newman, D. B., Jr.; Omura, J. K.; and Pickholtz, R. L. "Public Key Management for Network Security." *IEEE Network* 1, no. 2 (April 1987):11–16.

[NEWM87b] Newman, D. B., Jr., ed. "Special Issue on Telecommunications Regulation." *IEEE Communications Magazine* 25, no. 1 (January 1987).

[NEWM88a] Newman, S., Jr. "The Communications Highway of the Future." *IEEE Communications Magazine* 26; no. 10 (October 1988):45–50.

[NEWM88b] Newman, R. M.; Budrikis, Z. L.; and Hullett, J. L. "The QPSX MAN." *IEEE Communications Magazine* 26, no. 4 (April 1988):20–28.

[NEWP72] Newport, C. B., and Ryzlak, J. "Communication Processors." *Proceedings of the IEEE* 60, no. 11 (November 1972):1321–32.

[NIEL85] Nielson, D. "Packet Radio: An Area-Coverage Digital Radio Network," in [CHOU85a], pp. 50–100.

[NIZN86] Niznik, C. A.; Chatterjee, A.; and Walter, D. H. "Software Protocol for Handling ISDN in the Packet-Switched Network." *IEEE Journal on Selected Areas in Communications* SAC-4, no. 8 (November 1986):1230–40.

[NOKE84] Nokes, C. S. "Data Communications and the Large User: The Need for Bridging Systems." *IEEE Communications Magazine* 22, no. 7 (July 1984):11–17.

[NYQU28] Nyquist, H. "Certain Topics in Telegraph Transmission Theory." *Transactions of the AIEE* 47 (April 1928):617–44.

[OBUC87] Obuchowski, E. J. "Access Charge and Revenue Architecture." *AT&T Technical Journal* 66, no. 3 (May/June 1987):73–81.

[OHAR87] Ohara, Y.; Yoshitake, S.; and Kawaoka, T. "Protocol Conversion Method for Heterogenous Systems Interconnection in Multi-Profile Environment." *Proceedings of Seventh IFIP Symposium on Protocol Specification, Testing, and Verification* (1987):405–18.

[OHKO86] Ohkoshi, S.; Matsumoto, K.; and Nakano, S. "A Digital Telephone Set for ISDN." *IEEE Journal on Selected Areas in Communications* SAC-4, no. 8 (November 1986):1193–1201.

[OKUM86] Okumura, K. "A Formal Protocol Conversion Method." *Proceedings of ACM SIGCOMM* (1986):30–37.

[OMNI84] Omnicom, Inc. (Open Systems Communication), "CCITT Completes Work on Integrated Services Digital Network." (July 1984).

[OPDE74] Opderbeck, H., and Kleinrock, L. "The Influence of Control Procedures on the Performance of Packet-Switched Networks." *National Telecommunications Conference 1974 Record* (1974):810–17.

[OPPE83] Oppenheim, A. V.; Willsky, A. S.; and Young, I. T. *Signals and Systems.* Englewood Cliffs, N.J.: Prentice-Hall, 1983.

[OPPE88] Oppen, D. C., and Dalal, Y. K. "The Clearinghouse: A Decentralized Agent for Locating Named Objects in a Distributed Environment." *ACM Transactions on Office Information Systems* 1, no. 3 (July 1983):230–53.

[OREI88] O'Reilly, P. "Implication of Very Wide Fiber Bandwidth for Network Architecture." *Fiber and Integrated Optics* (1988):159–71.

[ORNS72] Ornstein, S. M.; Heart, F. E.; Crowther, W. R.; Rising, H. K.; Russell, S. B.; and Michel, A. "The Terminal IMP for the ARPA Computer Network." *Proceedings AFIPS Spring Joint Computer Conference* 40 (1972):243–54.

[OTOO87] O'Toole, T. J. "ISDN: A Large User's Perspective." *IEEE Communications Magazine* 25, no. 12 (December 1987):40–43.

[PADL83] Padlipsky, M. A. "A Perspective on the ARPANET Reference Model." *Proceedings of IEEE Infocom '83* (1983):242–53; also reprinted in [PADL85].

[PADL85] Padlipsky, M. A. *The Elements of Networking Style and Other Essays and Animadversions on the Art of Intercomputer Networking.* Englewood Cliffs, N.J.: Prentice-Hall, 1985.

[PADL88] Padlipsky, M. A. "At Last, The Last Word." *IEEE Network* 2, no. 1 (January 1988):92–93.

[PAHL88] Pahlavan, K., and Holsinger, J. L. "Voice-Band Data Communication Modems—A Historical Review: 1919–1988." *IEEE Communications Magazine* 26, no. 1 (January 1988):16–27.

[PALA88] Palais, J. C. *Fiber Optic Communications,* 2d ed. Englewood Cliffs, N.J.: Prentice-Hall, 1988.

[PALL84] Pallottino, S. "Shortest-Path Methods: Complexity, Interrelations and New Propositions." *Networks* 14 (1984):257–67.

[PAND87] Pandhi, S. N. "Communications: The Universal Data Connection." *IEEE Spectrum* 24, no. 7 (July 1987):31–37.

[PAPE74] Pape, U. "Implementation and Efficiency of Moore-Algorithms for the Shortest Route Problem." *Mathematical Programming* 7 (1974):212–22.

[PAPE80] Pape, U. "Algorithm 562: Shortest Path Length." *ACM Transactions on Mathematical Software* 5 (1980):450–55.

[PART88a] Partridge, C., and Rose, M. T. "A Comparison of External Data Formats." *Proceedings IFIP TC6 Working Symposium,* Irvine, Calif. (October 1988); also in [PART88b].

[PART88b] Partridge, C., ed. *Innovations in Internetworking.* Norwood, Mass.: Artech House, 1988.

[PART88c] Partridge, C., and Trewitt, G. "The High-Level Entity Management System (HEMS)." *IEEE Network* 2, no. 2 (March 1988):37–42.

[PASU87] Pasupathy, S. "Correlative Coding: Baseband and Modulation Applications," in [FEHE87a], pp. 429–58.

[PAWL81] Pawlita, P. "Traffic Measurements in Data Networks, Recent Measurement Results, and Some Implications." *IEEE Transactions on Communications* COM-29, no. 4 (April 1981):525–35.

[PEEB87] Peebles, P. Z., Jr. *Digital Communication Systems.* Engelwood Cliffs, N.J.: Prentice-Hall, 1987.

[PERL88] Perlman, R.; Harvey, A.; and Varghese, G. "Choosing the Appropriate ISO Layer for LAN Interconnection." *IEEE Network* 2, no. 1 (January 1988):81–86.

[PETE72] Peterson, W. W., and Weldon, E. J. *Error Correcting Codes,* 2d ed. Cambridge, Mass.: MIT Press, 1972.

[PHIL82] Phillips, T. R. "The Information Age: The Future Is Here, Ready or Not," paper presented at Honors Colloquium, Clemson University, Clemson, S.C. (January 1982).

[PHIL88] Phillips, D. "Picking the Right Strategy for Protocol Conversion." *Data Communications* (March 1985):193–205.

[PHIN83] Phinney, T., and Jelatis, G. "Error Handling in the IEEE Token-Passing Bus LAN." *IEEE Journal on Selected Areas in Communications* SAC-1, no. 5 (November 1983):785–89.

[PIAT77] Piatkowski, T. F.; Hull, D. E.; and Sundstrom, R. J. "Special Report: Inside IBM's Systems Network Architecture." *Data Communications* (February 1977):33–48.

[PICK83a] Pickens, R. A. "Wideband Transmission Media I: Radio Communication," in [CHOU83b], pp. 156–203.

[PICK83b] Pickens, R. A. "Wideband Transmission Media II: Satellite Communication," in [CHOU83b], pp. 204–52.

[PICK83c] Pickens, R. A. "Wideband Transmission Media III: Wireline, Coaxial Cable and Fiber Optics," in [CHOU83b], pp. 253–313.

[PICK85a] Pickens, R. A., and Hanson, K. W. "Digitization Techniques," in [CHOU85a], pp. 357–433.

[PICK85b] Pickens, R. A., and Hanson, K. W. "Integrating Data, Voice and Image," in [CHOU85a], pp. 268–356.

[PIER75] Pierce, A. R. "Bibliography on Algorithms for Shortest Path, Shortest Spanning Tree and Related Circuit Routing Problems (1956–1974). *Networks* 5 (1975):129–49.

[PIER84] Pierce, J. R. "Telephony—A Personal View." *IEEE Communications Magazine* 22, no. 5 (May 1984):116–20.

[PISC88] Piscitello, D. M.; Weissberger, A. J.; Stein, S. A.; and Chapin, A. L. "Internetworking in an OSI Environment." *Data Communications* (May 1986):118, 136.

[PITT87] Pitt, D. "Standards for the Token Ring." *IEEE Network* 1, no. 1 (January 1987):19–22.

[PITT87] Pitt, D. A., and Winkler, J. L. "Table-Free Bridging." *IEEE Journal on Selected Areas in Communications* SAC-5, no. 9 (December 1987):1454–62.

[PITT88] Pitt, D. A. "Bridging—the Double Standard." *IEEE Network* 2, no. 1 (January 1988):94–95.

[POKR84] Pokress, R., ed. "Special Issue on Integrated Services Digital Networks." *IEEE Communications Magazine* 22, no. 1 (January 1984).

[POLL60] Pollack, M., and Wiebenson, W. "Solutions of the Shortest-Route Problem— A Review." *Operations Research* 8 (1960):224–30.

[POSN84] Posner, E. C., and Stevens, R. "Deep Space Communication—Past, Present, and Future." *IEEE Communications Magazine* 22, no. 5 (May 1984):8–21.

[POST80] Postel, J. B. "Internetwork Protocol Approaches," in [GREE80], pp. 604–11.

[POST81a] Postel, J., ed. "Internet Protocol, DARPA Internet Program Protocol Specification," RFC 791, Information Sciences Institute, University of Southern California, Marina del Rey, Calif. (September 1981).

[POST81b] Postel, J., ed. "Internet Control Message Protocol, DARPA Internet Program Protocol Specification," RFC 792, Information Sciences Institute, University of Southern California, Marina del Rey, Calif. (September 1981).

[POST81c] Postel, J. B.; Sunshine, C. A.; and Cohen, D. "The ARPA Internet Protocol." *Computer Networks* 5 (1981):261–71.

[POST81d] Postel, J., ed. "Transmission Control Protocol, DARPA Internet Program Protocol Specification," RFC 793, Information Sciences Institute, University of Southern California, Marina del Rey, Calif. (September 1981).

[POST82] Postel, J. B. "Internetwork Protocol Approaches," in [GREE82a], pp. 511–26.

[POUZ78] Pouzin, L., and Zimmermann, H. "A Tutorial on Protocols." *Proceedings of the IEEE* 66, no. 11 (November 1978):1346–70.

[POUZ80] Pouzin, L. "The Seven Curses of the 80s." Presented at INFOTECH, London, November 1980.

[POUZ81] Pouzin, L. "Methods, Tools, and Observations on Flow Control in Packet-Switched Data Networks." *IEEE Transactions on Communications* COM-29, no. 4 (April 1981):413–26.

[POUZ85] Pouzin, L. "Internetworking," in [CHOU85a], pp. 180–221.

[PRAB87] Prabhu, V. K. "Interference Analysis and Performance of Linear Digital Communication Systems," in [FEHE87a], pp. 459–87.

[PRIC77] Price, W. L. "Data Network Simulation Experiments at the National Physical Laboratory, 1968–1976." *Computer Networks* 1 (1977):199–208.

[PRIT84] Pritchard, W. L. "The History and Future of Commercial Satellite Communications." *IEEE Communications Magazine* 22, no. 5 (May 1984):22–37.

[PRUS87] Prussog, A.; Blohm, W.; and Romahn, G. "Multi-Service Terminals—Human Factors Studies with an Experimental System." *IEEE Communications Magazine* 25, no. 11 (November 1987):37–43.

[PUJO88a] Pujolle, G.; Seret, D.; Dromard, D.; and Horlait, E. *Integrated Digital Communications Networks,* vol. 1. Chichester: John Wiley, 1988.

[PUJO88b] Pujolle, G.; Seret, D.; Dromard, D.; and Horlait, E. *Integrated Digital Communications Networks,* vol. 2. Chichester, England: John Wiley, 1988.

[PURS86] Purser, M. *Data Communications for Programmers.* Wokingham, England: Addison-Wesley, 1986.

[QUAR86] Quarterman, J. S., and Hoskins, J. C. "Notable Computer Networks." *Communications of the ACM* 29, no. 10 (October 1986):932–71.

[QURE82] Qureshi, S. "Adaptive Equalization." *IEEE Communications Magazine* 20, no. 2 (March 1982):9–17.

[QURE87] Qureshi, S. U. H. "Adaptive Equalization," in [FEHE87a], pp. 640–714.

[RAHM84] Rahm, M. W. "Broadband Networks: A User Perspective." *IEEE Communications Magazine* 22, no. 7 (July 1984):7–10.

[RANS84] Ransom, M. "Local Area Data Transport Service Overview." *AT&T Bell Laboratories Technical Journal* 63, no. 6 (July-August 1984):1113–34.

[REDE83] Redell, D. D., and White, J. E. "Interconnecting Electronic Mail Systems." *Computer* 16, no. 9 (September 1983):55–63.

[REIN77] Reingold, E. M.; Nievergelt, J.; and Deo, N. *Combinatorial Algorithms: Theory and Practice.* Englewood Cliffs, N.J.: Prentice-Hall, 1977.

[REUD87] Reudink, D. "Advanced Concepts and Technologies for Communications Satellites," in [FEHE87a], pp. 573–639.

[RIDE89] Rider, M. J. "Protocols for ATM Access Networks." *IEEE Network* 3, no. 1 (January 1989):17–22.

[RICH88] Richters, J. S., and Dvorak, C. A. "A Framework for Defining the Quality of Communications Services." *IEEE Communications Magazine* 26, no. 10 (October 1988):17–23.

[ROBE72] Roberts, L. G. "ALOHA Packet System with and without Slots and Capture." ARPANET Satellite System Note 8, (NIC 11290), June 1972; reprinted in *Computer Communications Review* 5 (April 1975).

[ROBE73] Roberts, L. "Dynamic Allocation of Satellite Capacity through Packet Reservation." *Proceedings of National Computer Conference* (1973):711–16.

[ROBE74] Roberts, L. G. "Data by the Packet." *IEEE Spectrum* 11, no. 2 (February 1974):46–51.

[ROBE78] Roberts, L. G. "The Evolution of Packet Switching." *Proceedings of the IEEE* 66, no. 11 (November 1978):1307–13.

[ROBE87] Roberts, M. M. "ISDN in University Networks." *IEEE Communications Magazine* 25, no. 12 (December 1987):36–39.

[ROBI84] Robin, G. "Consumer Installations for the ISDN." *IEEE Communications Magazine* 22, no. 4 (1984):18–23.

[ROCA86] Roca, R. T. "ISDN Architecture." *AT&T Technical Journal* 65, no. 1 (January/February 1986):4–18.

[RODE82] Roden, M.S. *Digital and Data Communication Systems.* Englewood Cliffs, N.J.: Prentice-Hall, 1982.

[RODE83] Rodell, D. D., and White, J. E. "Interconnecting Electronic Mail Systems." *Computer* 16, no. 9 (September 1983):55–63.

[RODE85] Roden, M.S. *Analog and Digital Communication Systems,* 2d ed. Englewood Cliffs, N.J.: Prentice-Hall, 1985.

[ROEH85] Roehr, W. "Signalling System Number 7." Omnicom Inc. (Open Systems Data Transfer) (February 1985):1–16.

[ROFF87] Roffinella, D.; Trinchero, C.; and Freschi, G. "Interworking Solutions for a Two-Level Integrated Services Local Area Network." *IEEE Journal on Selected Areas in Communications* SAC-5, no. 9, (December 1987):1444–53.

[ROGA87] Rogalski, J. E. "Evolution of Gigabit Lightwave Transmission Systems." *AT&T Technical Journal* 66, no. 3 (May/June 1987):32–41.

[RONA88] Ronayne, J. *The Integrated Services Digital Network: From Concept to Application.* New York: John Wiley, 1988.

[ROSN82a] Rosner, R. D. *Distributed Telecommunications Networks via Satellites and Packet Switching.* Belmont, Calif.: Lifetime Learning Publications, 1982.

[ROSN82b] Rosner, R. D. *Packet Switching, Tomorrow's Communications Today.* Belmont, Calif.: Lifetime Learning Publications, 1982.

[ROSS77] Ross, M.; Tabbot, A.; and Waite, J. "Design Approaches and Performance Criteria for Integrated Voice/Data Switching." *Proceedings of the IEEE* 65, no. 9 (September 1977):1283–95.

[ROSS86] Ross, F. E. "FDDI—A Tutorial." *IEEE Communications Magazine* 24, no. 5 (May 1986):10–17.

[ROSS87] Ross, F. E. "Rings are 'Round for Good'!" *IEEE Network* 1, no. 1 (January 1987):31–38.

[ROTH81] Rothnie, J. B., Jr.; Goodman, N.; and Marrill, T. "Database Management in Distributed Networks," in [KUO81], pp. 433–62.

[RUBI83] Rubin, I., and de Moraes, L. M. "Message Delay Analysis for Polling and Token Multiple-Access Schemes for Local Communication Networks." *IEEE Journal on Selected Areas in Communications* SAC-1, no. 5 (November 1983):935–47.

[RUBI87] Rubin, I., and Tsai, Z.-H. "Performance of Double-Tier Access-Control Schemes Using a Polling Backbone for Metropolitan and Interconnected Communications Networks." *IEEE Journal on Selected Areas in Communications* SAC-5, no. 9 (December 1987):1403–17.

[RUDI76] Rudin, H. H. "On Routing and 'Delta Routing': A Taxonomy and Performance Comparison of Techniques for Packet-Switched Networks." *IEEE Transactions on Communications* COM-24, no. 1 (January 1976):43–59.

[RUDI78] Rudin, H.; West, C. H.; and Zafiropulo, P. "Automated Protocol Validation: One Chain of Development." *Computer Networks* 2 (1978):373–80.

[RUDI85] Rudin, H. "An Informal Overview of Formal Protocol Specification." *IEEE Communications Magazine* 23, no. 3 (March 1985):46–52.

[RUDI87] Rudigier, J. J. "Army Implementation of ISDN." *IEEE Communications Magazine* 25, no. 12 (December 1987):44–49.

[RUDO85] Rudov, M. "Marketing ISDNs: Reach Out and Touch Someone's Pocketbook." *Data Communications* (June 1984):239–45.

[RUMS86] Rumsey, D. C. "Support of Existing Data Interfaces by the ISDN." *IEEE Journal on Selected Areas in Communications* SAC-4, no. 3 (May 1986):372–75.

[RUTK82] Rutkowski, A. M., and Marcus, M. J. "The Integrated Services Digital Network: Developments and Regulatory Issues." *Computer Communication Review* (July/October 1982):68–82.

[RYBC80] Rybczynski, A. "X.25 Interface and End-to-End Virtual Circuit Service Characteristics." *IEEE Transactions on Communications* COM-28, no. 4 (April 1980):500–509.

[RYBC82] Rybczynski, A. "Packet-Switched Network Layer," in [GREE82a], pp. 213–38.

[RYDE83] Ryder, K. D. "An Experimental Address Space Isolation Technique for SNA Networks." *IBM Systems Journal* 22, no. 4 (1983):367–86.

[SALT81] Saltzer, J. H.; Clark, D. D.; and Pogran, K. T. "Why a Ring?" *Proceedings of the Seventh Data Communications Symposium* (1981):211–17.

[SALT84] Saltzer, J. H.; Reed, D. P.; and Clark, D. D. "End-to-End Arguments in System Design." *ACM Transactions on Computer Systems* 2, no. 4 (November 1984):277–88; also in [PART88b], pp. 195–206.

[SALW88] Salwen, H.; Boule, R.; and Chiappa, J. N. "Examination of the Applicability of Router and Bridging Techniques." *IEEE Network* 2, no. 1 (January 1988):77–80.

[SAND85] Sandberg, R.; Goldberg, D.; Kleiman, S.; Walsh, D.; and Lyons, B. "Design and Implementation of the Sun Network System." *Proceedings 1985 Summer USENIX Conference,* Portland, Oreg. (June 1985):119–30; also in [PART88b].

[SANF87] Sanferrare, R. J. "Terrestrial Lightwave Systems." *AT&T Technical Journal* 66, no. 1 (January/February 1987):95–107.

[SAST84] Sastry, A. R. K. "Performance Objectives for ISDNs." *IEEE Communications Magazine* 22, no. 1 (January 1984):49–55.

[SCHL86] Schlanger, G. G. "An Overview of Signaling System No. 7." *IEEE Journal on Selected Areas in Communications* SAC-4, no. 3 (May 1986):360–65.

[SCHN83] Schneidewind, N. "Interconnecting Local Networks to Long-Distance Networks." *Computer* 16, no. 9 (September 1983):15–26.

[SCHR84] Schroeder, M. D.; Birrell, A. D.; and Needham, R. M. "Experience with Grapevine: The Growth of a Distributed System." *ACM Transactions on Computer Systems* 2, no. 1 (February 1984):3–23.

[SCHU80] Schultz, G. D.; Rose, D. B.; West, C. H.; and Gray, J. P. "Executable Description and Validation of SNA." *IEEE Transactions on Communications* COM-28, no. 4 (April 1980):661–77.

[SCHU82] Schultz, G. D.; Rose, D. B.; Gray, J. P.; and West, C. H. "Executable Representation and Validation of SNA," in [GREE82a], pp. 671–706.

[SCHU84] Schulke, H. "User Needs for ISDN as Seen by the Banking Community." *Proceedings of the International Conference on Communications* (1984):564–67.

[SCHW72] Schwartz, M.; Boorstyn, R. R.; and Pickholtz, R. L. "Terminal-Oriented Computer Communications Networks." *Proceedings of the IEEE* 60, no. 11 (November 1972):1408–23.

[SCHW77] Schwartz, M. *Computer-Communication Network Design and Analysis.* Englewood Cliffs, N.J.: Prentice-Hall, 1977.

[SCHW80a] Schwartz, M. *Information Transmission, Modulation and Noise,* 3rd ed. New York: McGraw-Hill, 1980.

[SCHW80b] Schwartz, M., and Stern, T. E. "Routing Techniques Used in Computer Communication Networks." *IEEE Transactions on Communications* COM-28, no. 4 (April 1980):539–52.

[SCHW82] Schwartz, M., and Stern, T. E. "Routing Protocols," in [GREE82a], pp. 327–60.

[SCHW84] Schwartz, M. I. "Optical Fiber Transmission—From Conception to Prominence in 20 Years." *IEEE Communications Magazine* 22, no. 5 (May 1984): 38–48.

[SCHW87] Schwartz, M. *Telecommunication Networks: Protocols, Modeling and Analysis.* Reading, Mass.: Addison-Wesley, 1987.

[SEDG88] Sedgewick, R. *Algorithms,* 2nd ed. Reading, Mass.: Addison-Wesley, 1988.

[SEID83] Seidler, J. *Principles of Computer Communication Network Design.* Translated by P. Senn. Chichester, England: Halsted Press, 1983.

[SEIF88] Seifert, W. M. "Bridges and Routers." *IEEE Network* 2, no. 1 (January 1988): 57–64.

[SEMI89] Semilof, M. "SABRE Recovers from Network Crash." *Communications Week* (May 22, 1989):249.

[SENI85] Senior, J. *Optical Fiber Communications: Principles and Practice.* Englewood Cliffs, N.J.: Prentice-Hall, 1985.

[SEVE84] Severson, A. "AT&T's Proposed PBX-to-Computer Interface Standard." *Data Communications* (April 1984):157–62.

[SHAC83] Shacham, N.; Craighill, E. J.; and Poggio, A. A. "Speech Transport in Packet-Radio Networks with Mobile Nodes." *IEEE Journal on Selected Areas in Communications* SAC-1, no. 6 (December 1983):1084–97.

[SHAC87] Shacham, N., and Westcott, J. "Future Directions in Packet Radio Architectures and Protocols." *Proceedings of the IEEE* 75, no. 1 (January 1987):83–98.

[SHAN83] Shankar, A. U., and Lam, S. S. "Specification and Verification of an HDLC Protocol with ARM Connection Management and Full-Duplex Data Transfer." *Proceedings of the ACM SIGCOMM '83 Symposium* (March 1983):38–48.

[SHAN84] Shankar, A. U., and Lam, S. S. "Time Dependent Communication Protocols," in [LAM84c], pp. 504–20.

[SHIE79] Shier, D. R. "On Algorithms for Finding the k Shortest Paths in a Network." *Networks* 9 (1979):195–214.

[SHIE85] Shier, D. R., and Spragins, J. D. "Exact and Approximate Dependent Failure Reliability Models for Telecommunications Networks." *Proceedings of IEEE INFOCOM,* Washington, D.C. (March 1985).

[SHOC78] Shoch, J. F. "Inter-Network Naming, Addressing, and Routing." *Proceedings of COMPCON Fall 78 Conference* (1978):72–79.

[SHOC79] Shoch, J. "Packet Fragmentation in Inter-Network Protocols." *Computer Networks* (February 1979):3–8.

[SHOC80] Shoch, J. F.; Cohen, D.; and Taft, E. A. "Mutual Encapsulation of Internetwork Protocols." *Proceedings of IEEE Symposium on Trends and Applications: Computer Network Protocols* (1980):1–11.

[SHOC82] Shoch, J. F.; Dalal, Y. K.; Redell, D. D.; and Crane, R. C. "Evolution of the Ethernet Local Computer Network." *Computer* 15, no. 8 (August 1982):10–26.

[SHUM89] Shumate, P. W., Jr. "Optical Fibers Reach into Homes." *IEEE Spectrum* 26, no. 2 (1989):43–47.

[SIEW82] Siewiorek, D. P.; Bell, C. G.; and Newell, A. *Computer Structures: Principles and Examples.* New York: McGraw-Hill, 1982.

[SINC88] Sincoskie, W. D., and Cotton, C. J. "Extended Bridge Algorithms for Large Networks." *IEEE Network* 2, no. 1 (January 1988):16–24.

[SINH84] Sinha, R., and Gupta, S. C. "Performance Evaluation of a Protocol for Packet Radio Network in Mobile Computer Communications." *IEEE Transactions on Vehicular Technology* VT-33, no. 3 (August 1984):250–58.

[SINH85] Sinha, R., and Gupta, S. C. "Mobile Packet Radio Networks: State-of-the-Art." *IEEE Communications Magazine* 23, no. 3 (March 1985):53–61.

[SIRB85] Sirbu, M. A., and Zwimpler, L. E. "Standards Setting for Computer Communication: The Case of X.25." *IEEE Communications Magazine* 23, no. 3 (March 1985):35–45.

[SKIL87] Skillen, R. P., ed. "Special Issue on ISDN: A Means Towards a Global Information Society." *IEEE Communications Magazine* 25, no. 12 (December 1987).

[SKLA88] Sklar, B. *Digital Communications: Fundamentals and Applications.* Englewood Cliffs, N.J.: Prentice-Hall, 1988.

[SKRZ87] Skrzypczaak, C. "Intelligent Home of 2010 AD." *IEEE Communications Magazine* 25, no. 12 (December 1987):81–84.

[SLUT88] Slutsker, G. "Good-Bye Cable TV, Hello Fiber Optics." *Forbes* 142, no. 6 (September 19, 1988):174–79.

[SMIT81] Smith, E. A.; Walsh, W. A. G.; and Wilson, M. J. "Impact of Non-Voice Services on Network Evolution." *Electrical Communications* no. 1 (1981):17–30.

[SNEL87] Snelling, R. K. "Environmental Aspects of ISDN." *IEEE Communications Magazine* 25, no. 12 (December 1987):14–16.

[SNYD83] Snyder, A. W., and Love, J. D. *Optical Waveguide Theory.* London: Chapman Hall, 1983.

[SODO85] Sodolski, J., and Plevyak, T., eds. "Divestiture: Two Years Later." *IEEE Communications Magazine* (Special Issue) 23, no. 12 (December 1985).

[SOHA88] Soha, M., and Perlman, R. "Comparison of Two LAN Bridge Approaches." *IEEE Network* 2, no. 1 (January 1988):37–43.

[SOLO88] Solomon, M.; Landweber, L. H.; and Neuhengen, D. "The CSNET Name Server." *Computer Networks* 6 (July 1982):161–172.

[SPIL77] Spilker, J. J., Jr. *Digital Communications by Satellite.* Englewood Cliffs, N.J.: Prentice-Hall, 1977.

[SPRA72] Spragins, J. "Loops Used for Data Collection," in *Computer Communications Networks and Teletraffic.* Brooklyn: Polytechnic Press, Polytechnic Institute of Brooklyn, 1972.

[SPRA77a] Spragins, J. "Dependent Failures in Data Communication Systems." *IEEE Transactions on Communications* COM-25, no. 12 (December 1977):1494–99.

[SPRA77b] Spragins, J. "Simple Derivation of Queueing Formulas for Loop Systems." *IEEE Transactions on Communications* COM-25, no. 5 (April 1977):446–48.

[SPRA79] Spragins, J. "Approximate Techniques for Modeling the Performance of Complex Systems." *Computer Languages* 4 (1979):99–129.

[SPRA81] Spragins, J. D.; Markov, J. D.; Doss, M. W.; Mitchell, S. A.; and Squire, D. C. "Communication Network Availability Predictions Based on Measurement Data." *IEEE Transactions on Communications* COM-29, no. 10 (October 1981):1482–91.

[SPRA86] Spragins, J. D.; Sinclair, J. C.; Kang, Y. J.; and Jafari, H. "Current Telecommunication Network Reliability Models: A Critical Assessment." *IEEE Journal on Selected Areas in Communications* SAC-4, no. 6 (October 1986):1168–73.

[SPRO78] Sproull, R. F., and Cohen, D. "High-Level Protocols." *Proceedings of the IEEE* 66, no. 11 (November 1978):1371–86.

[SPRO81] Sproule, D. E., and Mellor, F. "Routing, Flow, and Congestion Control in the Datapac Network." *IEEE Transactions on Communications* COM-29, no. 4 (April 1981):386–91.

[SRIS83] Sriram, K.; Varshney, P. K.; and Shanthikumar, J. G. "Discrete-Time Analysis of Integrated Voice/Data Multiplexers With and Without Speech Activity Detectors." *IEEE Journal on Selected Areas in Communications* SAC-1, no. 6 (December 1983):1124–32.

[STAL85a] Stallings, W. "Integrated Services Digital Network," in [STAL85b], pp. 6–24.

[STAL85b] Stallings, W., ed. *Tutorial: Integrated Services Digital Networks,* IEEE Computer Society Press, Los Angeles, CA, 1985.

[STAL87a] Stallings, W. *Local Networks: An Introduction,* 2d ed. New York: Macmillan, 1987.

[STAL87b] Stallings, W. *Handbook of Computer-Communications Standards, Volume 1: The Open Systems Interconnection (OSI) Model and OSI-Related Standards.* New York: Macmillan, 1987.

[STAL87c] Stallings, W. *Handbook of Computer-Communications Standards, Volume 2: Local Network Standards.* New York: Macmillan, 1987.

[STAL88a] Stallings, W.; Mockapetris, P.; McLeod, S.; and Michel, T. *Handbook of Computer-Communications Standards, Volume 3: Department of Defense (DoD) Protocol Standards.* New York: Macmillan, 1988.

[STAL88b] Stallings, W. *Data and Computer Communications,* 2d ed. New York: Macmillan, 1988.

[STAR79] Stark, H., and Tuteur, F. B. *Modern Electrical Communications, Theory and Systems.* Englewood Cliffs, N.J.: Prentice-Hall, 1979.

[STAU89] Stauffer, K. E. "DS-1 Extended Superframe Format and Related Performance Issues." *IEEE Communications Magazine* 27, no. 4 (April 1989):19–23.

[STIX89] Stix, G. "Technology '89, Data Communications." *IEEE Spectrum* 26, no. 1 (January 1989):43–46.

[STRA87] Strathmeyer, C. "Voice/Data Integration: An Applications Perspective." *IEEE Communications Magazine* 25, no. 12 (December 1987):30–35.

[STRE82] Stremier, F. G. *Introduction to Communication Systems,* 2d ed. Reading, Mass.: Addison-Wesley, 1982.

[STRO87] Strole, N. C. "The IBM Token-Ring Network—A Functional Overview." *IEEE Network* 1, no. 1 (January 1987):23–30.

[STUM84] Stumpers, F. "The History, Development, and Future of Telecommunications in Europe." *IEEE Communications Magazine* 22, no. 5 (May 1984):84–95.

[STUT72] Stutzman, B. W. "Data Communication Control Procedures." *Computing Surveys* 4, no. 4 (December 1972):197–220.

[SUNS78a] Sunshine, C. "Survey of Protocol Definition and Verification Techniques." *Computer Networks* 2, nos. 4/5 (September/October, 1978):346–50.

[SUNS78b] Sunshine, C. A., and Dalal, Y. K. "Connection Management in Transport Protocols." *Computer Networks* 2 (1978):454–73.

[SUNS79] Sunshine, C. "Formal Techniques for Protocol Specification and Verification." *Computer* 12, no. 9 (September 1979):20–26.

[SUNS81] Sunshine, C. A. "Transport Protocols for Computer Networks," in [KUO81], pp. 35–77.

[SUNS82] Sunshine, C. A. "Addressing Problems in Multi-Network Systems." *Proceedings of IEEE INFOCOM* (1982):12–18.

[SUNS83] Sunshine, C.; Kaufman, D.; Ennis, G.; and Biba, K. "Interconnection of Broadband Local Area Networks." *Proceedings of IEEE-ACM Eighth Data Communications Symposium* (1983).

[SY87] Sy, K. K.; Shiobara, M. O.; Yamaguchi, M.; Kobayashi, Y.; Shukuya, S.; and Tomatsu, T. "OSI-SNA Interconnections." *IBM Systems Journal* 26, no. 2 (1987): 157–73.

[SYLO86] Sylor, M. W. "The NMCC/DECnet Monitor Design." *Digital Technical Journal,* Digital Equipment Corporation, Hudson, Mass., no. 3 (September 1986):129–41.

[SYSL83] Syslo, M. M.; Deo, N.; and Kowalik, J. S. *Discrete Optimization Algorithms with Pascal Programs.* Englewood Cliffs, N.J.: Prentice-Hall, 1983.

[SZEC86] Szechenyi, K.; Zapf, F.; and Sallaerts, D. "Integrated Full-Digital U-Interface Circuit for ISDN Subscriber Loops." *IEEE Journal on Selected Areas in Communications* SAC-4, no. 8 (November 1986):1337–49.

[TAKA86a] Takagi, H. *Analysis of Polling Systems.* Cambridge, Mass.: MIT Press, 1986.

[TAKA86b] Takahashi, T., and Okimi, K. "A Report from Mitake—INS Model System Experiment Site." *IEEE Journal on Selected Areas in Communications* SAC-4, no. 3 (May 1986):376–84.

[TAKA88] Takahashi, K. "Transmission Quality of Evolving Telephone Services." *IEEE Communications Magazine* 26, no. 10 (October 1988):24–35.

[TANE81] Tanenbaum, A. S. *Computer Networks.* Englewood Cliffs, N.J.: Prentice-Hall, 1981.

[TANE88] Tanenbaum, A. S. *Computer Networks,* 2d ed. Englewood Cliffs, N.J.: Prentice-Hall, 1988.

[TANG86] Tang, W. V. "ISDN—New Vistas in Information Processing." *IEEE Communications Magazine* 24, no. 11 (November 1986):11–16.

[TECH80] Techo, R. *Data Communications: An Introduction to Concepts and Design.* New York: Plenum Press, 1980.

[TERP87] Terplan, K. *Communication Networks Management.* Englewood Cliffs, N.J.: Prentice-Hall, 1987.

[THAP84] Thapar, H. K. "Real-Time Application of Trellis Coding to High-Speed Voiceband Data Transmission." *IEEE Journal on Selected Areas in Communications* SAC-2, no. 5 (September 1984):648–58.

[THOM87] Thomas, D. G. "Extending the Boundaries of Lightwave Technology." *AT&T Technical Journal* 66, no. 1 (January/February 1987):2–4.

[THOM88] Thomas, R. H.; Forsdick, H. C.; Crowley, T. R.; Schaaf, R. W.; Tomlinson, R. S.; Travers, V. M.; and Robertson, G. G. "Diamond: A Multimedia Message System Built on a Distributed Architecture." *IEEE Computer* 18, no. 12 (December 1985):65–78.

[THOR83] Thornton, J. E., and Christensen, G. S. "Hyperchannel Network Links." *Computer* 16, no. 9 (September 1983):50–54.

[TOBA75] Tobagi, F., and Kleinrock, L. "Packet Switching in Radio Channels: Part II—The Hidden Terminal Problem in Carrier Sense Multiple Access and the Busy Tone

Solution." *IEEE Transactions on Communications* COM-23 (December 1975): 1417–33.

[TOBA76] Tobagi, F. A., and Kleinrock, F. A. "Packet Switching in Radio Channels: Part III—Polling and (Dynamic) Split Channel Reservation Multiple Access." *IEEE Transactions on Communications* COM-24 (August 1976):832–45.

[TOBA77] Tobagi, F., and Kleinrock, L. "Packet Switching in Radio Channels: Part IV—Stability Considerations and Dynamic Control in Carrier Sense Multiple Access." *IEEE Transactions on Communications* COM-25 (October 1977):1103–20.

[TOBA78] Tobagi, F. A.; Gerla, M.; Peebles, R. W.; and Manning, E. G. "Modeling and Measurement Techniques in Packet Communication Networks." *Proceedings of the IEEE* 66, no. 11 (November 1978):1423–47.

[TOBA80a] Tobagi, F. A., and Hunt, V. B. "Performance Analysis of Carrier Sense Multiple Access with Collison Detection." *Computer Networks* 5, no. 5 (November 1980):245–59.

[TOBA80b] Tobagi, F. A. Multiaccess Protocols in Packet Communication Systems." *IEEE Transactions on Communications* COM-28, no. 4 (April 1980):468–88.

[TOBA82] Tobagi, F. A. "Multiaccess Link Control," in [GREE82a], pp. 145–90.

[TOBA84] Tobagi, F. A.; Binder, R.; and Leiner, B. "Packet Radio and Satellite Networks." *IEEE Communications Magazine* 22, no. 11 (November 1984):24–40.

[TOBA87] Tobagi, F. A. "Modeling and Performance Analysis of Multihop Packet Radio Networks." *Proceedings of the IEEE* 75, no. 1 (January 1987):135–55.

[TODA80] Toda, I. "DCNA Higher-Layer Protocols." *IEEE Transactions on Communications* COM-28, no. 4 (April 1980):575–84.

[TODA82] Toda, I. "DCNA Higher-Layer Protocols," in [GREE82a], pp. 415–36.

[TODD87] Todd, T. D. "A Traffic Scheduling Technique for Metropolitan Area Gateways." *IEEE Journal on Selected Areas in Communications* SAC-5, no. 9 (December 1987):1391–1402.

[TOML75] Tomlinson, R. S. "Selecting Sequence Numbers." *Proceedings ACM SIGCOMM/SIGOPS Interprocess Communications Workshop,* Santa Monica, Calif. (March 1975); also in [PART88b], pp. 232–44.

[TROU86] Trouvat, M. "RENAN: An Experimental ISDN Project in France." *IEEE Journal on Selected Areas in Communications* SAC-4, no. 3 (May 1986):407–12.

[TROU87] Trouvat, M. "The RENAN Project: Opening Up ISDN in France." *IEEE Communications Magazine* 25, no. 12 (December 1987):64–69.

[TSUC88] Tsuchiya, P. "The Landmark Hierarchy: A New Hierarchy for Routing in Very Large Networks." *Proceedings ACM SIGCOMM '88,* Stanford, Calif. (August 1988).

[TURN86] Turner, J. S. "Design of an Integrated Services Packet Network." *IEEE Journal on Selected Areas in Communications* SAC-4, no. 8 (November 1986):1373–80.

[TYME81] Tymes, L. "Routing and Flow Control in TYMNET." *IEEE Transactions on Communications* COM-29, no. 4 (April 1981):392–98.

[TZEN86] Tzeng, C. P. J.; Hodges, D. A.; and Messerschmitt, D. G. "Timing Recovery in Digital Subscriber Loops Using Baud-Rate Sampling." *IEEE Journal on Selected Areas in Communications* SAC-4, no. 8 (November 1986):1302–11.

[UEDA83] Ueda, H.; Tokizawa, I.; and Aoyama, T. "Evaluation of an Experimental Pack-etized Speech and Data Transmission System." *IEEE Journal on Selected Areas in Communications* SAC-1, no. 6 (December 1983):1039–45.

[UNGE87a] Ungerboeck, G. "Trellis-Coded Modulation with Redundant Signal Sets Part I: Introduction." *IEEE Communications Magazine* 25, no. 2 (February 1987):5–11.

[UNGE87b] Ungerboeck, G. "Trellis-Coded Modulation with Redundant Signal Sets Part II: State of the Art." *IEEE Communications Magazine* 25, no. 2 (February 1987): 12–21.

[UNSO81] Unsoy, M. S., and Shanahan, T. A. "X.75 Internetworking of Datapac and Telenet." *Proceedings of the Seventh Data Communications Symposium* (1981):232–39.

[VANV78] Van Vliet, D. "Improved Shortest Path Algorithm for Transportation Net-works." *Transportation Research* 12 (1978):7–20.

[VISS83] Vissers, C. A.; Tenney, R. L.; and Bochmann, G. V. "Formal Description Tech-niques." *Proceedings of the IEEE* 71, no. 12 (December 1983):1356–65.

[VITE79] Viterbi, A. J., and Omura, J. K. *Principles of Digital Communication and Cod-ing.* New York: McGraw-Hill, 1979.

[VOYD83] Voydock, V. L., and Kent, S. T. "Security Mechanisms in High-Level Network Protocols." *ACM Computing Surveys* 15, no. 2 (June 1983):135–71.

[WAKA88] Wakayama, H.; Ohara, Y.; Kobayashi, Y.; and Shiobara, M. O. "Application of OSI Protocols as Intermediary for DCNA-SNA Networks Interconnection." *Pro-ceedings of IEEE III International Conference on Introduction of Open Systems Con-nections Standards* (September 1985).

[WARE83] Ware, C. "The OSI Network Layer: Standards to Cope with the Real World." *Proceedings of the IEEE* 71, no. 12 (December 1983):1384–87.

[WARN80] Warner, C. "Connecting Local Networks to Long Haul Networks: Issues in Protocol Design." *Proceedings of Fifth Conference on Local Computer Networks* (1980).

[WARN86] Warner, C., and Keune, C. M. "New Protocols for Tactical Data." *Proceedings of IEEE MILCOM '86,* Monterey, Calif. (1986).

[WARS62] Warshall, S. "A Theorem on Boolen Matrices." *Journal of the ACM* 9 (1962):11–12.

[WATA87] Watanabe, H. "Integrated Office Systems: 1995 and Beyond." *IEEE Commu-nications Magazine* 25, no. 12 (December 1987):74–80.

[WATE84] Waters, A. G., and Adams, C. J. "The Satellite Transmission Protocol of the Universe Project." *Proceedings of SIGCOMM '84 Symposium on Communications Architectures & Protocols* (June 1984):18–24.

[WATS81] Watson, R. W. "Timer-Based Mechanisms in Reliable Transport Protocol Connection Management." *Computer Networks* 5, no. 1 (February 1981):47–56.

[WATS87] Watson, R. W., and Mamrak, S. A. "Gaining Efficiency in Transport Services by Appropriate Design and Implementation Choices." *ACM Transactions on Com-puter Systems* 5, no. 2 (May 1987):97–120.

[WEBE85] Weber, K. "Considerations on Customer Access to the ISDN." *IEEE Trans-actions on Communications* COM-30, no. 9 (September 1972):2131–36.

[WECK80] Wecker, S. "DNA—The Digital Network Architecture." *IEEE Transactions on Communications* COM-28, no. 4 (April 1980):510–26.

[WECK82] Wecker, S. "DNA—The Digital Network Architecture," in [GREE82a], pp. 249–96.

[WEI84a] Wei, L. F. "Rotationally Invariant Convolutional Channel Coding with Expanded Signal Space—Part I: 180°." *IEEE Journal on Selected Areas in Communications* SAC-2, no. 5 (September 1984):659–71.

[WEI84b] Wei, L. F. "Rotationally Invariant Convolutional Channel Coding with Expanded Signal Space—Part II: Nonlinear Codes." *IEEE Journal on Selected Areas in Communications* SAC-2, no. 5 (September 1984):672–86.

[WEIN83] Weinstein, C. J., and Forgie, J. W. "Experience with Speech Communication in Packet Networks." *IEEE Journal on Selected Areas in Communications* SAC-1, no. 6 (December 1983):963–80.

[WEIN84] Weinstein, S. B. "Emerging Telecommunications Needs of the Card Industry." *IEEE Communications Magazine* 22, no. 7 (July 1984):26–31.

[WEIS87] Weissberger, A. J., and Israel, J. E. "What the New Internetworking Standards Provide." *Data Communications* (February 1987):141–56.

[WELS87] Welsh, F. S. "Lightwave Data Links and Interfaces." *AT&T Technical Journal* 66, no. 1 (January/February 1987):65–72.

[WEST82a] Westcott, J., and Jubin, J. "A Distributed Routing Design for a Broadcast Environment." *Proceedings of IEEE Milcom '82* (October 1982):10.4.1–4.5.

[WEST82b] Westcott, J. "Issues in Distributed Routing for Mobile Packet Radio Networks." *Proceedings of COMPCON '82,* Washington, D.C. (September 1982).

[WIEN84] Wienski, R. M. "Evolution to ISDN with the Bell Operating Companies." *IEEE Communications Magazine* 22, no. 1 (January 1984):33–41.

[WILK72] Wilkov, R. S. "Analysis and Design of Reliable Computer Networks." *IEEE Transactions on Communications* COM-20 (June 1972):660–78.

[WILL84] Williams, R., and Gillman, R. "ISDN Access Protocols—Status and Applications." *Proceedings of National Communications Forum* (1984):181–90.

[WILL87] Williams, R. A. *Communication Systems Analysis and Design: A Systems Approach.* Englewood Cliffs, N.J.: Prentice-Hall, 1987.

[WITT84] Wittke, P. H.; Penstone, S. R.; and Keightley, R. J. "Measurements of Echo Parameters Pertinent to High-Speed Full-Duplex Data Transmission on Telephone Circuits." *IEEE Journal on Selected Areas in Communications* SAC-2, no. 5 (September 1984):703–10.

[WOIN88] Woinsky, M. N. "National Performance Standards for Telecommunications Services." *IEEE Communications Magazine* 26, no. 10 (October 1988):70–81.

[WOLF84] Wolf, J. K. "Statistical Communication Theory—A Worker's View." *IEEE Communications Magazine* 22, no. 5 (1984):121–22.

[WONG82a] Wong, J. W., and Lam, S. S. "Queuing Network Models of Packet Switching Networks: Part 1: Open Networks." *Performance Evaluation* 2 (1982):9–21.

[WONG82b] Wong, J. W., and Lam, S. S. "Queuing Network Models of Packet Switching Networks: Part 2: Networks with Population Size Constraints," in *Performance Evaluation,* vol. 2 (1982):161–80.

[WONG86] Wong, C. L., and Wood, R. "Implementation of ISDN." *Telesis* 13, no. 3 (1986):4–13.

[WONG87] Wong, J. W.; Vernon, A. J.; and Field, J. A. "Evaluation of a Path-Finding Algorithm for Interconnected Local Area Networks." *IEEE Journal on Selected Areas in Communications* SAC-5, no. 9 (December 1987):1463–70.

[WOOD83] Wood, H. M., and Cotton, I. W. "Security in Computer Communications Systems," in [CHOU83b], pp. 369–409.

[WOOD85a] Wood, D. C. "Local Networks," in [CHOU85a], pp. 101–31.

[WOOD85b] Wood, D. C. "Computer Networks: A Survey," in [CHOU85a], pp. 132–79.

[YAJI86] Yajima, T., and Suda, K. "Switching Software Design for ISDN." *IEEE Journal on Selected Areas in Communications* SAC-4, no. 8 (November 1986):1222–29.

[YAMA86] Yamaguchi, H.; Wada, M.; and Yamamoto, H. "A 64 kbit/s Integrated Visual Communication Systems—New Communication Medium for the ISDN." *IEEE Journal on Selected Areas in Communications* SAC-4, no. 8 (November 1986):1202–09.

[YAMA89] Yamamoto, U., and Wright, T. "Error Performance in Evolving Digital Networks Including ISDNs." *IEEE Communications Magazine* 27, no. 4 (April 1989):12–18.

[YOKO89] Yokoi, T., and Kodaira, K. "Grade of Service in the ISDN Era." *IEEE Communications Magazine* 27, no. 4 (April 1989):46–50.

[YU81] Yu, P. S., and Lin, S. "An Efficient Selective-Repeat ARQ Scheme for Satellite Channels and its Throughput Analysis." *IEEE Transactions on Communications* COM-29, no. 3 (March 1981):353–63.

[ZAFI80] Zafiropulo, P.; West, C. H.; Rudin, H.; Cowan, D. D.; and Brand, D. "Towards Analyzing and Synthesizing Protocols," in [GREE80a], pp. 651–60.

[ZAFI82] Zafiropulo, P.; West, C. H.; Rudin, H.; Cowan, D. D.; and Brand, D. "Protocol Analysis and Synthesis Using a State Transition Model," in [GREE82a], pp. 645–70.

[ZHAN88] Zhang, L. "Comparison of Two Bridge Routing Approaches." *IEEE Network* 2, no. 1 (January 1988):44–48.

[ZIEM85] Ziemer, R. E., and Tranter, W. H. *Principles of Communications: Systems, Modulation, and Noise,* 2d ed. Boston: Houghton Mifflin, 1985.

[ZIMM80] Zimmermann, H. "OSI Reference Model—The ISO Model of Architecture for Open Systems Interconnection." *IEEE Transactions on Communications* COM-28, no. 4 (April 1980):425–32.

[ZIMM82] Zimmermann, H. "A Standard Layer Model," in [GREE82a], pp. 33–54.

[ZOLI85] Zoline, K. O., and Lidinsky, W. P. "An Approach for Interconnecting SNA and XNS Networks." *Proceedings of IEEE-ACM Ninth Data Communications Symposium* (1985):184–98.

Index

a. See Normalized time delay

a'. See Normalized ring latency

ABM. *See* Asynchronous balanced mode

Abrupt close, 436, 446, 496, 546, 559, 573, 587, 641

AC. *See* Access control field

Access control, 129, 222, 496
control field (AC) 338–39, 641

ACK 0 and ACK 1, 284, 310–14, 641

ACK bit, 552–53
character, 281–86, 308–14, 316, 319–21, 348
packet, 543–46

Acknowledge. *See* ACK character

Acknowledgment, 189–90, 280–86, 404–7, 439, 444, 456, 473–74, 543, 552–53, 578–80, 641

Acknowledgment (AK) TPDU, 566–74, 578–80

Active Open, 550–51, 581, 641

A/D. *See* Analog to digital conversion

Adaptive differential PCM (ADPCM), 80
routing algorithms, 357–58

ADCCP. *See* Advanced Data Communications Control Protocol

Address, 107, 112–15, 131, 176–77, 315, 464, 494, 503, 541, 549–53, 559, 641
field 262, 315, 323–24, 328–33, 338–41, 641

ADPCM. *See* Adaptive differential PCM

Advanced Data Communications Control Protocol, 304, 322, 336, 347

AH. *See* Application header

AK. *See* Acknowledgment (AK) TPDU

ALL. *See* Allocate

Allocate (ALL), 461

All routes broadcast, 501–3

ALOHA, 12, 207, 224–30, 243–48, 251–53, 641

Alternate mark inversion (AMI) encoding. *See* Bipolar RZ encoding
routing, 446, 469, 473

AM. *See* Amplitude modulation

American National Standards Institute (ANSI), 322, 641
Standard Code for Information Interchange (ASCII), 167, 305–7, 310, 342, 641

AMI. *See* Alternate mark inversion encoding; Bipolar RZ encoding

Amplitude modulation (AM), 49–50, 62, 641
shift keying (ASK), 49, 50, 53, 641

Analog-to-digital (A/D) conversion, 77–79, 114

ANSI. *See* American National Standards Institute

Application header (AH), 121–22
layer, 15, 120–22, 125, 147–48, 430, 539, 641

ARM. *See* Asynchronous response mode

ARPANET, 10–11, 106, 109–10, 280, 283, 304, 315–16, 347, 371, 379, 385, 414–15, 431, 454–62, 641
DLC, 280, 283, 304, 315–16, 347, 414–15, 641
Reference Model. *See* DoD Reference Model

ARQ. *See* Automatic repeat request

ASCII. *See* American Standard Code for Information Interchange

ASK. *See* Amplitude shift keying

Asynchronous balanced mode (ABM), 327–28, 332–33, 641
DLC. *See* Start-stop DLC
response mode (ARM), 327–28, 641
time division multiplexing (ATDM), 63, 631
transfer mode (ATM), 63, 631–33, 637, 641
transmission, 46, 641

ATDM. *See* Asynchronous time division multiplexing

ATM. *See* Asynchronous transfer mode

AT&T divestiture, 10

Automatic equalization, 9
repeat request (ARQ), 7, 100, 279–95, 542, 556, 641

Autonomous systems and confederations, 514

Average cycle time (T_c), 238
delay formula, 387
transfer delay (T), 232, 237–39, 246, 255, 641

B channel, 172–75, 178–79, 335, 473, 612–14, 618–20, 642

Backbone network, 500, 508

Backoff, 225, 228, 246–47, 547, 641

Backpressure, 397, 410, 641

Backwards learning, 391–93, 641

Balanced configuration, 322–23, 642
mode transmission, 157, 164–65, 642

Bandwidth, 31, 48, 60
/data rate tradeoff, 60
/time duration tradeoff, 48

Basic rate access. *See* B channel
Basic transmission Unit (BTU), 470
Baudot code, 305
Baud rate, 49, 641
BCC. *See* Block check characters
BCH. *See* Bose-Chaudhuri-Hocquengham codes
Bellman-Ford-Moore algorithm, 363–69, 371, 379, 457–59, 462, 642
    d'Esopo-Pape version, 368–69
    See also Distributed asynchronous Bellman-Ford-Moore algorithm
Bellman's equation, 364, 377
    principle of optimality, 363–64
BF. *See* Boundary function
Bidding for D channel, 175–78
Bifurcated routing, 386, 642
Binary signal encoding, 31, 57–60, 642
    synchronous communications. *See* Bisync
Bind, 526, 593–95
Biphase encoding. *See* Manchester encoding
Bipolar RZ encoding, 57–60, 174
BISDN. *See* Broadband ISDN
Bisync, 307–15, 321, 334, 342–43, 348–49, 419, 496, 642
Bit error rate, 38, 263, 286–87, 642
    -oriented DLC protocols, 304, 321–37, 347
    rate, 49
    stuffing, 218
    synchronization, 261–62
Black hole, 393, 530
Block check characters (BCC), 310–11, 642
    code, 267–79, 642
Bose-Chaudhuri-Hocquengham (BCH) codes 271, 279
Boundary function (BF), 141, 642
    node, 465
Bridge, 489–92, 496–509, 642
Broadband ISDN (BISDN), 629–36, 642
    protocol model, 633–34
Broadcast transmission, 107, 206, 262, 303, 356, 446, 460, 494, 642
BSC. *See* Bisync
BTU. *See* Basic Transmission Unit
Buffering, 114–15, 412–18, 441, 461–64, 503, 539, 542, 582, 586
    buffer reservation, 415, 461–62
    channel queue limit, 413
    input buffer limiting, 416–418, 464
    Irland's square root rule, 464
    structured buffer pool, 415–16
Burst error rate, 287–93, 642
    errors, 263
Burstiness of data sources, 81
Bus geometry, 111, 206, 216, 223–24, 642
Busy periods, 248–50

Bypass switches, 223, 254
Byte count oriented DLC protocols, 304, 316–21, 347, 642
    synchronization, 262
    timing circuit, 166–67, 642

C. *See* Control circuit
Call Accepted packet, 436–38, 441
    Connected packet, 436–38
    Request packet, 435–38, 441
    setup time, 191–92
Carrier-sense-multiple-access (CSMA), 207, 226–30, 243, 247–52, 642
    /collision detection (CSMA/CD), 106, 145, 207, 229–31, 243, 251–55, 339, 503, 508, 642
CC. *See* Connection confirm TPDU
CCITT. *See* Consultative Committee for International Telegraph and Telephone
    channel carrier system, 37, 178–79
    International Alphabet No. 5 (IA5), 167–71, 647
CCS. *See* Common channel signaling
    CCSS#6. *See* Common channel signaling system #6
    CCSS#7. *See* Common channel signaling system #7
Cdinit, 592–95
Cdsessend, 594–95
Cdsessst, 593–95
Cdterm, 594–95
Cellular mobile radio, 41–42
Central control, 213–16, 224, 357, 444
Centralized reservations, 212
Channel capacity (Shannon), 74–75
    capacity (S_{max}), 234–35, 241, 245–46, 253–56
    characteristics, 33–47, 263
    group, 61
    queue limit, 413–14, 642
Character insertion, 311
    -oriented DLC protocols, 304, 307–16, 347, 642
    synchronization, 262, 309, 317, 347
Checkpointing, 328–30, 642
    checkpoint cycle, 329–30, 642
Checksum, 516–17, 549, 559–62, 568–71, 576
Chernobyl packets, 529
Choke packet, 419, 497, 642
Circuit switching, 95–97, 172, 207, 429, 473–74, 608–14, 619, 642
Class 0 *through* class 4 Transport Service. *See* TP0 *through* TP4
    NP and class P problems, 83
    of service, 469, 643
Clear 442, 464, 540–42, 562, 643
    to send (CTS) circuit, 157–63, 171, 193–95, 643
Clear Confirm packet, 436–38
    Request packet, 436–38, 441
Close packet, 543–46

Cluster controller peripheral node, 140
Coaxial cable, 34, 44, 208, 429, 643
Coding theory, 75–6, 264, 643
Collision, 207, 224–25, 643
Combined station, 322–23, 643
Command frame, 322, 332–35, 337, 643
Common channel signaling (CCS), 605, 608–14, 619–25, 643
  channel signaling system #6 (CCSS#6), 621
  channel signaling system #7 (CCSS#7), 619–25, 643
  channel signaling network, 611
  network layer, 525–26, 643
Communication channels, 31, 33–47
  satellites, 34, 42, 107, 429
  channel interface, 154, 188, 643
  for computers, 2, 643
Complete packet sequence, 441
Completely connected geometry. *See* Fully connected geometry
Computational complexity, 81–83, 356, 369–71, 460, 643
Computer equipment interface, 154, 183–88, 643
  port, 183
Computerized branch exchange (PBX), 207
Computers for communications, 2, 643
Concentration, 33, 60–66, 643
Concentrator capacity, 63–66
Confirm primitive, 124, 643
Congestion control, 108, 396, 421, 440, 473, 546–48
Connect Ack, 585–86
  Confirm, 583–85
  Initiate, 583–86
Connection confirm (CC) TPDU, 566–73, 578
  definition, 449
  identifier (ID#), 543
  oriented transmission, 100, 136, 336, 341, 449–53, 495, 562, 566, 571–73, 581–86, 591–97, 623, 643
  establishment, 96, 125, 306, 344–47, 429, 435–38, 447–53, 464, 473–75, 543–46, 562, 564–65, 569–72, 618, 624–25
  management, 542–46, 569, 618
    handshake-based, 543–46
    rules, 546
    timer-based, 543
  point manager, 589
  request (CR) TPDU, 566–78
Connectionless transmission, 100, 127–28, 336, 341, 449–50, 454, 495, 512–22, 562–66, 571, 597, 623, 643
Consultative Committee for International Telegraph and Telephone, 16, 37, 167, 178–79, 322, 431, 443, 643
Contention mode, 311–12, 348, 496, 643
Content synchronization, 262
Continuous ARQ, 284–85, 288–94, 643

Control character, 262, 307–15, 643
  circuit (C), 165–71, 643
  field, 262, 323–33, 644
  packets, (X.25), 439–40
  plane, 615–21, 637
  state, 312, 348–49
  station, 312–14, 644
Conversions between voltage levels, 183–84
Convolutional code, 267, 271, 644
Coordination of sender and receiver, 18–21, 153, 261
Cosname, 469, 644
Cost, 23–5, 358–64, 370–78, 453–54, 462, 505, 626–28
  See also Distance
Count field, 262, 317, 347, 467
  See also Length field
CR. *See* Connection request TPDU
C/R bit, 340–41
CRC-12, 279, 310, 644
CRC-16, 279, 310, 623, 644
CRC-32, 279, 315, 644
CRC-CCITT (or CCITT-16), 279, 324, 644
Credit, 407–8, 555, 568, 573, 579–80, 587, 644
  See also Permit
Cross connects, 607
CSMA. *See* Carrier-sense-multiple-access
  /CD. *See* Carrier-sense-multiple-access/collision detection
CSNP (communications subnet processor), 129–33, 454–62, 644
  -CSNP protocol, 130–33, 454–62
  -CSNP header, 132, 455
CTS. *See* Clear to send circuit
D bit, 436, 439–41, 446, 644
D channel, 172–79, 335, 473, 613–14, 618–21, 625–26, 644
D/A. *See* Digital to analog conversion
DARPA Internet, 1, 10, 512–20, 540, 548, 644
Data Acknowledgment, 583–86
Data-circuit-terminating equipment (DCE), 155, 159–63, 167–78, 188, 431–43, 644
Data (DT) TPDU, 566–69, 573–80
  field. *See* Information field
  flow control layer, 142–44, 598, 644
  link control (DLC), 114–15, 261
    protocols 261, 280, 303–49, 644
  link escape character. *See* DLE
  link header (DH or DLH) and trailer (DLT or DT), 121–22, 144, 309, 316–17, 324, 454
  link layer, 15, 119–26, 135–37, 142–48, 261–62, 280, 303–49, 429–31, 445, 454, 489–91, 525–26, 539, 618, 622–23, 644
    ARPANET DLC, 315–16
    Bisync, 307–15

Data link layer—Continued
    DDCMP, 316–21
    HDLC, 321–37
    SDLC, 142, 304, 347
    Start-stop, 304–7
  packet (X.25). *See* Normal data packet
  set ready (DSR) circuit, 157–63, 193–95, 644
  terminal equipment (DTE), 155, 159–63, 167–78, 188–
      89, 431–43, 644
  terminal ready (DTR) circuit, 157–63, 194–95, 644
  transfer, 125, 215–23, 313, 344–47, 429, 435–41, 447–53,
      464, 474, 511, 544–45, 562–65, 569–75, 586–87,
      592–96
Data Request, 583, 586–87
  Segment, 583, 586–87
Datagram, 99, 136, 379, 432, 444–46, 494, 514–20, 539–40,
    559, 563, 581, 644
DATAPAC, 379, 443–44, 475–77, 644
DC. *See* Disconnect confirm TPDU
DCE. *See* Data-circuit-terminating equipment
DDCMP. *See* Digital's Data Communications Message
    Protocol
DDN. *See* Defense Data Network
Deadlock, 394–95, 413–16
  direct store-and-forward, 413, 652
  indirect store-and-forward, 413–14, 652
  reassembly, 414, 650
DECNET. *See* Digital Network Architecture
Defense Data Network, 10, 454
Delay, 411–12, 444–46, 453–54, 458–61, 531, 539, 564–65,
    579
  encoding. *See* Miller encoding
  requirements, 80–81, 359
Delayed packets, 539, 543–45, 552
Demand assignment, 207, 254
d'Esopo-Pape algorithm. *See* Bellman-Ford-Moore
    algorithm
Destination address, 101, 222, 315, 338–41, 455–56, 463–
    67, 516–17, 521–22, 552–53, 559, 566, 582
  service access point (DSAP) field, 340–41, 644
Deterministic algorithms, 82
DH. *See* Data link header
Dialing delays, 96
Dialogue, 114–16, 120
Differential modulation, 49, 54
  phase shift keying (DPSK), 54
Digital adapters, 114–15, 154
  multiplexer/demultiplexer, 606–8
  multiplexing hierarchies, 607
  -to-analog (D/A) conversion, 77–79, 114
  switching offices, 2, 13
  transmission, 2, 13, 31
Digital Network Architecture (DNA), 12, 107, 133–38, 148,
    180, 379, 419, 431, 581–88, 597, 644

  data link layer, 135–37, 316–21
  end communications layer, 135–37, 581–88
  network application layer, 135–37
  network management layer, 135–37
  physical layer, 135–37, 180
  routing layer, 135–37, 462–64
  session control layer, 135–37
  user layer, 135–37
Digital's Data Communications Message Protocol
    (DDCMP), 136, 304, 316–21, 330, 337–39, 342–47,
    644
Digitization of telephone networks, 603–11, 637
Dijkstra's algorithm, 364, 369–73, 460–61, 644
Directory service, 494
Disconnect confirm (DC) TPDU, 566–67, 575–78
  request (DR) TPDU, 566–67, 570–72, 575–78
Disconnect Complete, 583
  Initiate, 583
Disconnection, 96–97, 126, 306, 344–47, 429, 435–38, 447–
    53, 464, 474–75, 544–46, 557–59, 562–65, 573–75,
    587, 594–95, 618, 624–25
Distance, 362–65, 368–71, 374–83, 421
  See also Cost
Distributed asynchronous Bellman-Ford-Moore
    algorithm, 376–79, 457–59, 644
  control, 216–24, 357, 444
  reservations, 212
DLC. *See* Data link control
DLE, 307–11, 316–17, 348, 644
DLT. *See* Data link trailer
DNA. *See* Digital Network Architecture
DNA/SNA gateway, 489, 522–28, 645
DoD Reference Model, 127–32, 148, 280–83, 304, 315–16,
    347, 414–15, 454–56, 548–60, 645
  ARPANET DLC, 315–16
  CSNP-CSNP protocol, 130–33, 454–56
  host-host layer, 130–33, 548–60
  internet layer, 130–33, 512–20
  network access layer, 130–33
  process/application layer, 130–33
Domain, 140–41, 465, 591–95, 645
DPSK. *See* Differential phase shift keying
DR. *See* Disconnect request TPDU
DSAP. *See* Destination service access point address field
DSR. *See* Data set ready circuit
DT. *See* Data link trailer or Data TPDU
DTE. *See* Data terminal equipment
DTR. *See* Data terminal ready circuit
Duobinary signaling, 71
Duplicate packets 539, 543, 552, 581

E. See Effective packet transmission time
EA. *See* Expedited acknowledgment TPDU
EBCDIC, 305, 310, 343, 645

E channel, 178–79, 613, 645
Echo suppressors, 188–89
Echoing, 264, 306, 645
E (echo) channel, 174–78, 613, 645
ED. *See* End delimiter or Expedited data TPDU
Effective network throughput (*S'*), 233
  packet transmission time (*E*), 232
  transmission rate (*R_e*) (or Effective rate of transmitting
    data bits), 103, 190, 286–93, 400–7, 645
EGP. *See* Exterior gateway protocol
EIA-232D. *See* RS-232C
Electrical characteristics, 118, 155, 165, 645
Electronic mail, 5, 132, 489
Encapsulation, 515, 520–23, 588
End communications layer, 135–37, 543, 581–83, 645
  delimiter (ED), 222, 338–39, 645
  node, 134, 645
  of text character. *See* ETX
  of transmission block character. *See* ETB
  of transmission character. *See* EOT
End-to-end error control, 541–42
  flow control, 396, 407–10, 439–40
ENQ, 283–84, 308–17, 348, 645
Enquiry character. *See* ENQ
Entry to exit flow control, 397
EOT, 308–9, 313–18, 348, 496, 645
  bit, 568, 573
Equalization, 9, 72–73, 188–89, 263, 645
ER. *See* TPDU error TPDU
Error control, 76, 101, 114–15, 119, 131, 183, 261–95, 303–
    10, 314–15, 326–33, 347, 441–42, 473, 495, 522,
    539–42, 562, 574–80, 618, 623
  correcting codes, 76, 268–70
  detecting codes, 76, 265–71, 277–79
  protection field, 262
  rate, 31, 38, 80–81, 263, 286–93, 429, 453–54, 565
    bit, 38, 263, 286–87
    burst, 287–93
    requirements, 80–81
Establish phase. *See* Connection phase
ETB, 308–9, 645
Ethernet, 12, 135–36, 146, 463, 645
ETX, 262, 316, 645
Expedited acknowledgment (EA) TPDU, 566–67, 575–79
  data, 446, 449–53, 467, 562, 566–69, 573–79, 596
  data (ED) TPDU, 566–69, 573–79, 645
Explicit reservations, 210–11
  route, 467–69, 645
Extended address field, 324, 645
  control field, 324–25, 645
Exterior gateway protocol (EGP), 514

F. *See* Flag character
Fairness, 22, 359, 395–96, 408–9, 645

Fast circuit switching, 101
  select, 441
FC. *See* Frame control field
FCS. *See* Frame check sequence
FDDI, 216, 508, 645
FDM. *See* Frequency division multiplexing
FEC. *See* Forward error correction
Federal Information Processing Standards (FIPS), 563, 581
Fiber optics, 13, 34, 44–45, 256, 429, 603, 631–37, 645
FID (FIDs)
  FID2, 465–67, 645
  FID3, 465–67, 645
  FID4, 465–67, 470–72, 645
File transfer errors, 541–42
FIN bit, 552–53, 558–59
FIPS. *See* Federal Information Processing Standards
Fixed assignment, 207–13, 254, 645
  –priority-oriented-demand assignment (FPODA), 212,
    645
Flag character (F), 176–77, 262, 323–24, 340, 623, 646
Flat name space, 112, 494
Flooding, 391–93, 504, 646
Flow control, 10, 101, 106, 114–19, 131, 142, 295–96, 347,
    355–56, 394–421, 429, 439–40, 445, 454, 464–65,
    473, 495–97, 514, 522, 539, 555–57, 562–68, 573,
    579–80, 585–87, 596, 618, 623, 646
  diagram, 208–14, 219–20
Floyd-Warshall algorithm, 374–79, 646
FM. *See* Frequency modulation or Function management
Forward error correction (FEC), 279–80, 646
Fourier transform 47, 646
FPODA. *See* fixed-priority-oriented demand assignment
Fragmentation, 496, 516–19
  See also Segmentation
Frame, 174–75, 178–79, 208–9, 303
  formats
    ARPANET data packet, 455
    ARPANET DLC frame, 316
    ATM cell, 633
    Bisync frame, 309
    CCITT carrier frame, 178
    DDCMP frame, 317
    DNA end communications data unit, 584
    DNA routing layer header, 462
    DoD IP header, 516
    HDLC frame, 324
    I.430 frame, 175
    I.431 frame, 178
    IEEE 802 LLC frame, 341
    IEEE 802 MAC frames, 338
    ISO IP header, 521
    OSI TPDU headers, 554
    SNA path control FID headers, 468
    SNA transmission control data unit, 591

Frame, formats—Continued
    SONET frames, 634–37
    Start-stop character, 305
    T-1 carrier frame, 178
    TCP header, 552
    TCP/UDP pseudo header, 554
    TDMA frame, 209
    UDP header, 559
    X.25 data packet, 435–36
    X.75 packets, 512
  check sequence (FCS), 222, 323–25, 336, 338–40, 646
  control (FC) field, 338–39, 646
  forwarding, 499, 503
  status (FS) field, 338–39
  synchronization, 262
Framing bits, 62, 174
  overhead, 62
Frequency division multiplexing (FDM), 13, 33–37, 61–62, 646
  hierarchy, 37
  domain, 48
  modulation (FM), 49–50, 62, 646
  shift keying (FSK), 49–50, 646
  spectra, 49–52, 59, 70–72
  translations, 51–52
FS. *See* Frame status field
FSK. *See* Frequency shift keying
Full duplex transmission, 46, 163, 177, 189, 213, 264, 314, 332–33, 362, 646
Fully connected geometry, 111–12
Functional layer, 525–26, 646
  specifications, 118, 156, 165, 646
Function management (FM), 143, 591
Fundamental limits, 18, 31–84

G. See Offered traffic
GAP. *See* Gateway access protocol
Gateway, 489–94, 509–28, 646
  access model, 523
  access protocol (GAP), 523
  -gateway protocol (GGP), 514, 520, 646
General format identifier (GFI), 437, 646
Generating polynomial, 273, 279, 646
Geometry, 106, 109, 205–7, 646
  See also Topology
Geosynchronous communications satellite, 9, 446, 646
GFI. *See* General format identifier
GGP. *See* gateway-gateway protocol
Go-ahead message, 213, 646
Go-back-N ARQ, 284–94, 321, 329–30, 442, 576, 646
Graceful close, 446, 496, 546, 557–59, 587, 646
Guard bands, 62, 208

H, H0, H1, H11 and H12 channels, 179, 613, 646
Half bridge, 492
  duplex transmission, 46, 158, 171, 189, 215, 307, 646
  gateway, 492–93, 589, 646
  router, 492
  session, 588–89
Hamming codes, 269–71, 274
  distance, 268
Handshake, 543–46, 646
Handshaking sequence, 543–46, 550–54, 569–71, 583–85, 646
HDLC. *See* High-Level Data Link Control
Header, 121–22, 144, 303, 310, 315, 435, 454–55, 465–67, 497–98, 516–17, 521, 549, 554, 566–68, 646
HF radio, 39–40
Hierarchical name space, 112, 494
  network structure, 127–28, 139
Higher speed networks, 255–56
High-Level Data Link Control (HDLC), 145, 175, 304, 321–37, 339, 342–44, 347, 410, 419, 432–35, 497–98, 576, 580, 623
Hold down, 385
Hop count, 459, 463–64
  level flow control, 397, 407–10, 439
Host-host layer, 131, 133, 455, 646
Hot potato algorithm, 391
Hub polling, 215–16, 646

I. *See* Indication circuit
I.430, 154, 172–80, 192, 262, 647
I.431, 155, 172, 178–80, 192, 647
IA5. *See* CCITT International Alphabet No. 5
IBL. *See* Input buffer limiting
ICMP. *See* Internet control message protocol
ID#. *See* Connection identifier
Idle periods, 248–50
IDN. *See* Integrated Digital Network
IEEE-802, 106, 145–48, 180–81, 207–24, 304, 322, 337–41, 347, 431, 496, 508, 517, 647
IEEE-802.1, 496, 503
IEEE-802.2, 337, 340–41, 647
IEEE-802.3, 106, 145, 180–81, 338, 647
IEEE-802.4, 106, 145, 180–81, 216, 223–24, 338, 647
IEEE-802.5, 106, 145, 180–81, 216–23, 338, 497–503
IEEE-802.6, 107
  logical link control (LLC) sublayer, 145, 304, 337, 340–41, 648
  medium access control (MAC) sublayer, 145, 304, 337–40
  physical layer, 180–81
I/G bit, 340–41, 499, 647
IGP. *See* Interior gateway protocol
Implicit reservations, 210–11

Incoming Call packet, 435–38
Incoming virtual calls, 433–35
Indication circuit (I), 165–71, 647
   primitive, 124, 647
Information field, 317, 323–24, 338–40, 435–37, 552, 566, 647
   frame, 317–21, 324–37, 647
   sources, 31, 79–81, 358
Initialization, 223, 319, 334
Initial sequence number (ISN), 543, 550–55, 647
Initself, 592–95
Input buffer limiting (IBL), 416–18, 647
Integrated access, 611–12
Integrated Digital Networks (IDN), 606–11, 647
   Services Digital Network (ISDN), 13–17, 146–48, 172–80, 334–35, 347, 431, 473–75, 598, 603–37, 647
     applications, 628–31
     functional groupings, 615–16
     migration towards, 605–6
     reference points R, S, T and U, 615–16
     Reference Model, 615–21, 647
     service classes, 623
     service requirements, 603–5
     trials, 626–27
Interface (between equipment), 9, 153–96, 262, 431, 442–46, 603, 612–16, 637, 647
   (between layers), 119, 123–24, 303, 449, 647
   protocol, 123–24
Interior gateway protocol (IGP), 514
Internal Organization of the Network Layer (IONL), 447–49
International Alphabet No. 5. *See* CCITT International Alphabet No. 5
   Organization for Standardization (ISO), 14, 109, 117–19, 334–35, 431
Internationalbureaucratspeak, 118
Internet control message protocol (ICMP), 514, 519–20, 647
   layer, 131–33, 454, 489, 509–10, 647
   protocol (DoD-IP), 132, 454, 489, 509–20, 548, 647
   protocol (ISO-IP), 489, 520–22, 647
Internetwork, 489, 497
Internetworking, 108, 127–28, 431, 489–533, 541, 647
Interprocess communication, 120, 129–31
Interrupt, 583, 586
   packet, 441, 446, 495, 647
   Request, 583, 586–87
Intersymbol interference, 66–72, 263, 647
Inverse Fourier transform, 47, 647
IONL. *See* Internal Organization of the Network Layer
IP. *See* Internet protocol (DoD or ISO)
Irland's square root rule, 464, 647
Isarithmic flow control, 418, 647
ISDN. *See* Integrated Services Digital Network

ISDN. *See* Integrated Services Digital Network
   user part, 622–24
ISN. *See* Initial sequence number
ISO. *See* International Organization for Standardization
Isochronous transmission, 305, 647

Jacobson algorithms, 547–48
Jamming signal, 231, 253

Labeling algorithm, 365–73, 647
LAN. *See* Local area network
LAPB, 304, 322, 334, 432–35, 647
LAPD, 175–76, 304, 335, 647
Layer choice criteria (ISO), 119
Layered network architectures, 10–12, 117–18, 647
Length field, 338–39, 559, 567–68
   See also Count field
Letter, 555–56
Level I and Level II routes, 462, 647
Line turnaround time, 189–92, 648, 654
Link, 33–47, 109, 116, 648
Little's theorem, 412
LLC. *See* Logical link control sublayer
Loading peaks, 529–30
Local area network (LAN), 1, 106, 205–6, 262, 337–41, 429, 446, 489–90, 496–509, 563, 648
   loop, 26
Locally installed communication facilities, 45–46
Logical
   channel, 316, 432, 456
     number, 432–37, 456, 648
   group number, 436–37
   link, 136, 526
     number, 545, 581–82
     control (LLC) sublayer, 145, 304, 337, 340–41, 648
     Type 1 and Type 2 operation, 341
   unit (LU), 141, 143, 589–95, 648
     Types 0, 1, 2, 3, 4 and 6.1, 590
     Type 6.2, 590, 596–97
Loop geometry. *See* Ring geometry
Looping, 359, 648
Lower speed networks, 255
LU. *See* Logical unit
   services, 592

MAC. *See* Medium access control sublayer
Maintenance frames, 317
MAN. *See* Metropolitan area network
Management, 127–29, 143, 303, 615–21
   plane, 615–21, 637
Manchester encoding, 58–60
Manufacturing Automation Protocol (MAP) 146–48, 180–81, 338, 431, 472, 520–22, 597, 648

Manufacturing Automation Protocol (MAP)—Continued
  logical link control (LLC) sublayer, 145, 304, 337, 340–41, 648
  medium access control (MAC) sublayer, 145, 304, 337–40
  physical layer, 180–81
  network layer, 472, 520–22
  transport layer, 561–81, 597
MAP. *See* Manufacturing Automation Protocol
Master station, 307–12, 321, 648
Maximizing reliability and freedom from errors, 21–22, 153, 261
Maximum flow, 84
M bit, 436
Mechanical characteristics, 155, 165, 648
Medium access control, 205–57, 262, 303, 347, 356, 496–98, 508, 648
  access control (MAC) sublayer, 145, 304, 337–40
Meltdown, 529
Mesh geometry, 109–11
Message switching, 95–98, 648
  synchronization, 262
  transfer part (MTP), 622–23
  transmission time, 104
Metropolitan area network (MAN), 1, 107, 205–6, 563, 648
Microwave radio, 34, 38–40, 429
Miller encoding, 59–60
MILNET. *See* Defense Data Network
Minimal cutsets, 84
Minimizing cost, 23–24, 153
Minimum spanning tree, 84
MLP. *See* Multilink procedure
Modem, 36, 114–15, 154, 163, 189, 648
Modulation, 31, 47–52, 648
Monitor station, 222
MSAT. *See* Ultra small aperture terminal mobile satellite system
MTP. *See* Message transfer part
Multiaccess medium, 205, 210
  techniques, 114–15
Multicast, 356, 494
Multichannel logical interface, 433
Multidrop line, 111, 172–73, 262, 303, 311, 318–19, 322–24, 648
Multiline controller, 188
Multilink procedure (MLP), 511
Multiple token operation, 219–21, 648
Multiplexer capacity, 63–66
Multiplexing, 60–63, 136, 432, 469, 473, 562–63, 581, 648
Multipoint. *See* Multidrop line
Murphy's Law, 336

NAK, 281–85, 308–14, 317–21, 326, 337, 348, 648
Name, 107, 112, 129, 494, 541, 648

Natural operating speed, 32
NAU. *See* Network addressable unit
NCP. *See* Network control point
Negative acknowledge. *See* NAK
Negotiation, 436, 449, 564–66, 593
Network access layer, 130–33, 648
  access level flow control, 397
  addressable unit (NAU), 141, 468, 588–90, 648
  applications layer, 135–36, 648
  control point (NCP), 609–12, 648
  header (NH), 121–22
  layer, 15, 119–22, 125, 147, 345, 429–81, 489–91, 508–9, 520–21, 539, 549, 561–63, 618–23, 648
  management, 24–25, 153, 648
  management layer, 135–37, 648
  service access point (NSAP), 561
  termination types NT1, NT2 and NT12, 616, 648
Network Layer Service Definition, 447–54
Network service types
  Type A, B and C network service, 561–62
NH. *See* Network header
Node, 109–116, 648
Noise, 31, 38, 263
Nomogram, 603–4
Non-data bit, 339, 648
Nonpersistent CSMA, 227–28, 250–51, 648
  CSMA/CD, 231
Normal data packets, 435–36, 439
Normalized average transfer delay ($\hat{T}$), 234, 237–42, 247, 254, 648
  network throughput (S), 233, 245–46, 251–54, 648
  offered traffic (G), 233, 245–46, 251–52, 648
  ring latency (a'), 239, 255
  time delay (a), 239, 251–52, 255
Normal response mode (NRM), 327–32, 649
No sensing, 207
NP. *See* Class NP problems
NP-complete problems, 83
N(R), 324–25, 328–33, 411, 439, 649
NRM. *See* Normal response mode
NRZ encoding, 57–60
N(S), 324–25, 328–33, 439, 649
NSAP. *See* Network service access point
N-SAP, 123
  address, 123
NT1, NT2 or NT12. *See* Network termination types
Null modem, 163, 649
Numbering schemes, 108
N-unitdata, 450, 520
Nyquist intersymbol interference theorem, 68, 649
  pulse, 69–71
  sampling rate, 77, 649
  sampling theorem, 77, 649
  transmission rate, 69, 649

On-off keying (OOK), 49
OOK. *See* On-off keying
OPEN packet, 543–46
Operator protocols. *See* Vendor protocols
Optical communication, 40
Optical fibers. *See* Fiber optics
Optimal packet length, 103–5, 293–95
  routing, 385–90
  routing and flow control, 419–21
Optimizing performance, 22–23, 153
Oscillations, 379–83, 531
OSI Reference Model, 14–15, 117–27, 136, 145– 48, 154,
    165–75, 192, 262, 304, 321–37, 347, 410, 419, 429–
    31, 447–48, 490, 543, 561–81, 597, 649
  application layer, 15, 120–22, 125, 148
  data link layer, 15, 119–26, 148, 175, 304, 321–37, 410,
    419
  network layer, 15, 119–25, 147, 347, 429–447, 520–22
  physical layer, 15, 118–25, 145–47, 154, 165–72, 192,
    262
  presentation layer, 15, 120–25, 147
  session layer, 15, 120–25, 147
  transport layer, 15, 119–25, 147, 543, 561–81, 587
OSI transport layer protocol classes
  Class 0: simple class. *See* TP0
  Class 1: basic error recovery class. *See* TP1
  Class 2: multiplexing class. *See* TP2
  Class 3: error recovery and multiplexing class. *See*
    TP3
  Class 4: error detection and recovery class. *See* TP4
Other Data Acknowledgment, 583
Outgoing virtual calls, 433–35
Overhead, 62–66, 189, 279–80, 306, 337–44
  ARQ versus FEC, 279–80
  Bisync, 342
  DDCMP, 343
  HDLC, 343
  Start-stop, 306

Pacing, 419, 467–72, 596, 649
Packet assembler/disassembler (PAD), 442
  discarding, 418, 444, 495, 559
  layer, 431–42, 445, 497
  level. *See* packet layer
  radio, 205, 446, 563
  switching, 120, 95–99, 429–31, 539, 611–14, 619–21
  switching node (PS), 443–44
  voice, 446
PAD. *See* Packet assembler/disassembler
PAD character, 310, 343, 348
Parallel-to-serial conversion, 183
Parity checks, 264–69, 305–6
Partial response signaling, 71
Passive Open, 550–51, 581, 649

Path, 114, 116, 362–63, 429, 447
  control header (PCH), 143–44, 465
  control layer, 142–44, 464–71, 649
Path Information Unit (PIU), 470–72
PBX. *See* Computerized branch exchange
PCM. *See* Pulse code modulation
PCM multiplexer/demultiplexer, 606–7
PDU. *See* Protocol data unit
Peer-to-peer communication, 116, 121–22, 303, 429, 649
Performance, 189–92, 231–53, 265–66, 268–71, 277–79,
    358
Peripheral node, 139, 465–68, 649
Permanent virtual circuit (PVC), 432–35, 511, 649
Permit, 394–407
  issued after receipt of first packet in window, 401–4
  issued at end of window, 400–401
  window advanced by acknowledgments, 404–7
  See also Credit
P/F. *See* Poll/final bit
PH. *See* Presentation header
Phase modulation (PM), 49–50
  shift keying (PSK), 49–50, 53
Physical constraints, 32–33
  control layer, 141–42
  interface, 114–15
  layer, 15, 118–25, 135–37, 145–47, 153–83, 261–62, 303,
    429–33, 445, 491, 525–29, 618, 622–23
    interface, 153–54
    primitives, 181–83
  unit (PU), 141, 589–92
    Types 1, 2.0, 2.1, 4 and 5, 589–90
Piggy-backed
  acknowledgments, 285–86, 318–24
  credit fields, 555, 580
  sequence counts, 285–86
Pins, 157, 164–65
Pipelining, 98, 103–5, 316
PIU. *See* Path Information Unit
Plain old telephone service (POTS), 603, 626–28
PM. *See* Phase modulation
Point-to-point geometry, 109–11, 172–73, 262, 303, 322–
    23
Poisson distribution, 245
Polar NRZ encoding, 57–60
Poll/final (P/F) bit, 325–36, 649
  pairing, 327–28, 336
Polling, 213–16, 237–42, 254, 312–14, 330–32, 348, 496,
    649
Polynomial algorithms, 83
  code 272–79, 310, 315, 324, 336, 650
Port, 504–7, 526, 550–59, 581
Postal Telephone and Telegraph (PTT), 27
POTS. *See* Plain old telephone service
Power, 22

p-persistent CSMA, 228–29, 650
  CSMA/CD, 231
P(R), 436–42, 650
Preamble, 338–39
Presentation header (PH), 121–22
  layer, 15, 120–25, 147, 430, 539, 618, 650
  services layer, 142–44, 589, 650
Primary logical unit, 592
  rate access, 172, 612
  rate configurations, 179
  station, 322–23, 650
Primitive, 123–24, 181–83, 344–47, 435, 449–54, 467, 473,
    549–54, 560–64, 569–75, 582–83, 591, 650
Priority, 176, 212, 222, 453–54, 456, 464, 470
Procedural specifications, 118, 156, 165, 650
Procedure calls, 467, 591
Process/applications layer, 131–33, 650
Propagation time (τ), 96–99, 102–3, 190–91, 219, 231,
    399–407, 460, 650
Protocol conversion, 114–16, 493
  data unit (PDU), 340, 499, 650
  definition, 117, 650
PS. See Packet switching node
P(S), 436–42, 650
Pseudo code for routing algorithms, 368, 371, 375
  header, 549, 554, 559
PSH bit, 552–53, 556
PSK. See Phase shift keying
PTT. See Postal Telephone and Telegraph
PU. See Physical unit
Public packet network, 107, 431, 443–47
Pulse code modulation (PCM), 78–79, 606, 650
  shaping, 47, 263
Push, 553, 556
PVC. See Permanent virtual circuit

QAM. See Quadrature amplitude modulation
Q bit, 436
QOS. See Quality of service
Quad of STPs, 609–10
Quadrature amplitude modulation (QAM), 53
Quality of service (QOS), 119, 429, 450–54, 520, 564–68,
    650
Quantization, 78, 650
Queueing delay, 97–99, 236, 355, 412

R. See Ring indicator circuit (RS-232C) or Receive circuit
    (X.21)
Radio links, 33, 39–40, 107, 206, 563, 608
Raised cosine spectrum pulse, 69–71, 650
R-ALOHA, 210–11, 651
Random access, 207, 224–31, 242–51, 254, 650
  routing, 391, 650
RD. See Received data circuit

R_e. See Effective transmission rate
Ready for next message (RFNM), 415, 461, 650
Reassembly, 464–65, 495, 586, 650 deadlock.
  See Deadlock
Reassignment, 576, 650
Receive (R) circuit, 165–71, 650
  count, 317–18, 650
  window. See Sliding window flow control
Received data (RD) circuit, 157–62, 193–95, 650
  line signal detector (RLSD) circuit, 157–63, 193–95, 650
Receive Not Ready. See RNR/frame
  Not Ready packet (X.25). See RNR packet (X.25)
  Ready. See RR/frame
  Ready packet (X.25). See RR packet (X.25)
Reed-Solomon codes, 271–72, 279
Reference number, 543–45, 567–69, 581
  points R, S, T and U, 615–16, 650
Reference Model for Open Systems Interconnection. See
    OSI Reference Model
Reject (RJ) TPDU, 566–67, 576, 580
Reject. See REJ/frame
  packet (X.25). See REJ packet
REJ frame, 326–34, 576, 580, 651
REJ packet (X.25), 439–41
Relay, 491, 510–11, 651
Relaying, 429, 445, 473, 618
Reliability, 21, 432, 464, 531–32, 539–40
Repeater, 491–92, 608, 651
REP frame, 318–20, 651
REQALL. See Request for allocation
Request data unit, 591
  for allocation (REQALL), 461
  primitive, 123–24, 651
  response unit (RU), 143–44
  to send (RTS) circuit, 157–63, 171, 193–95, 651
Request/response header (SNA RH) See Transmission
    control header (TCH)
Research computer networks, 109–10, 490
Reservations, 208–13, 254, 496, 651
Reset, 439–42, 450–51, 540–42, 552–53, 562, 651
Residual error rate, 543–44, 561, 565
Resolving incompatibilities of equipment, 18, 153
Resource sharing, 489
Response data unit, 591
  types—definite, exception, and no response, 596
  frame, 143, 322, 328, 332–37
  primitive, 124, 651
  time, 22, 395, 429
Restart, 442, 473, 540–42, 562, 651
Resynchronization, 576, 589, 651
Retransmission, 224, 306, 314–16, 321, 326–33, 394, 562,
    576–78
  timeouts, 547
Reuse of ID#, 545

RFNM. *See* Ready for next message
RH. *See* Transmission control header (TCH)
RI. *See* Routing information field
Ring geometry, 109–11, 206, 216–23, 379–80, 510, 651
    indicator (R) circuit, 157–63, 194–95, 651
    interface, 217–18
    latency (τ'), 239
RJ. *See* Reject TPDU
RLSD. *See* Received line signal detector circuit
RNR frame, 326, 419, 650
RNR packet (X.25), 439–41
Roll call polling, 215, 651
Round trip transmission time estimates, 547–48
Route 112, 357, 651
    designator, 499–501
    matrix, 374–75, 651
Router, 491, 508–22, 651
Routing, 106–8, 114–15, 132–36, 142, 355–93, 419–21, 429,
        445, 454, 456–65, 467–73, 494, 618, 651
    algorithms, 106–8, 355–93, 421, 457–61
    information (RI) field, 499–501
    layer, 135–37, 462–64, 588, 651
    node, 134, 462–63, 651
    tables, 357–61, 421, 458–59, 469, 651
RR frame, 326–31, 650
    packet (X.25), 439–40
RS-232C 9, 154–65, 171, 189, 192–96, 437, 651
RS-449, 164
Rsp(bind), 593–95
Rsp(cdcinit), 593–95
Rsp(cdinit), 592–95
Rsp(cdterm), 594–95
Rsp(unbind), 594–95
RST bit, 552–53
RTS. *See* Request to send circuit
RTS-CTS delay, 160–63, 189
RU. *See* Request response unit
RZ encoding, 57–60

S. *See* Normalized network throughput
S'. *See* Effective network throughput
Sampling, 78, 651
SAP. *See* Service access point
Satellite networks, 34, 42, 107, 205, 207, 256
SCCP. *See* Signaling connection control part
Scheduling, 207–24, 235–42, 651
Scrambler, 188
SD. *See* start delimiter
SDLC, 142, 304, 347
Secondary logical unit, 526, 592
    station, 322–23, 651
Segmentation, 306, 464–65, 473, 495, 520, 585–86, 618,
        651
    See also Fragmentation

Selection, 312–14, 348, 496, 651
Selective Reject. *See* SREJ
    reject ARQ, 284–85, 291–94, 329–32, 651
Self-synchronization, 530
Semantics, 117–20, 651
Send count, 317–18, 651
Sensing before transmission, 207
    both before and during transmission, 207
Sequence number, 315, 324–37, 437, 456, 467, 494, 542,
        568–73, 623
Sequencing, 100–101, 262, 449, 470, 562–63, 581, 586, 618,
        623
Serial-to-parallel conversion, 183
Server model, 523
Service access point (SAP), 123, 651
Session, 142, 467, 588–95
    control layer, 135–37, 651
    header, (SH) 121–22
    layer, 15, 120–25, 147, 430, 539, 618, 651
Sessst, 593–95
SH. *See* Session header
Shannon channel capacity theorem, 7, 73–75, 652
    channel capacity, 74, 651
        transmission at rates above channel capacity, 74–75
Shift register encoders and decoders, 274–77
Shortest-path routing, 84, 362–90, 421, 652
Shortest-path spanning tree, 365, 370, 460
Signal constellation, 31, 52–57, 652
    element timing circuit, 165–66, 171, 193–95, 652
    ground circuit, 157, 193–95
Signaled failures, 561
Signaling channel, 97
    connection control part (SCCP), 622–23
    terminal exchange (STE), 510–11, 652
Signal sets, 47
    transfer point (STP), 608–12, 652
Simplex transmission, 46, 652
Single packet operation, 219–21, 652
    route broadcast, 500–502
    sideband (SSB), 51
    token operation, 219–21, 652
Slave station, 307–12, 321, 652
Sliding window flow control, 295, 398–412, 439, 444–45,
        461, 470, 495, 652
    at receiver, 410–11, 650
    at transmitter, 398–412, 439, 444–45, 461, 470, 495, 654
    See also Permit and credit mechanisms
Slotted-ALOHA, 210–11, 225–26, 243–48, 251–53, 652
Slow circuit switching, 101
S_{max}. *See* Channel capacity
SNA. *See* Systems Network Architecture
    access module, 523–28
    gateway module, 523–28
    interface layer, 525–26, 652

SNAF. *See* Subnetwork Access Functions sublayer
SNDCF. *See* Subnetwork Dependent Convergence
    Functions sublayer
SNICF. *See* Subnetwork Independent Convergence
    Functions sublayer
SOH, 308–9, 317, 348
SONET. *See* Synchronous Optical Network
Source address, 222, 315, 338–41, 455–56, 463–67, 516–
    22, 552–53, 559, 566
   routing, 391, 499, 652
   routing bridge, 497–503, 508–9, 652
   service access point (SSAP) address field, 340–41, 652
Space division multiplexing, 33, 61
Spanning tree bridge, 503–9, 652
SPC. *See* Stored program controlled switching system
Speed matching. 395–96
SREJ frame, 326–31, 411, 651
SSAP. *See* Source service access point address field
SSB. *See* Single sideband
SSCP. *See* System services control point
Stability, 242, 247, 359, 379–85, 421, 458
STACK. *See* Start acknowledgment frame
Star geometry 109, 111, 172–73
Start acknowledgment (STACK) frame, 319, 652
   delimiter (SD), 222, 338–39, 652
   of heading. *See* SOH, 652
   of text. *See* STX, 652
   -stop data link control, 304–7, 341, 347, 652
   (STRT) frame, 319, 652
START bit, 187, 305–6, 652
State diagrams, 158–62, 167–71, 438, 563–65, 652
Static algorithms, 357–58
Station latency, 222, 239
STE. *See* Signaling terminal exchange
Steiner tree, 84
STM. *See* Synchronous transfer mode
Stop-and-wait ARQ, 281–88, 293–94, 310, 314, 336, 652
STOP bit, 305–6, 652
Store-and-forward, 97, 205
   store-and-forward deadlock. *See* Deadlock
Stored program controlled (SPC) switching system, 13,
   603–12, 652
STP. *See* Signal transfer point
STRT. *See* Start frame
Structured buffer pool, 415–16, 652
STS-1, 614, 635–36
STS-N, 614, 636–37
STX, 262, 308–9, 316, 348
Subarea, 141–42, 462–64, 468
   node, 139–40, 464–65, 468, 652
Subnetwork Access Functions (SNAF) sublayer, 448–49,
   653
   Dependent Convergence Functions (SNDCF) sublayer,
    448, 653

Independent Convergence Functions (SNICF) sublayer,
   448, 653
Supervised mode, 311–14, 348, 653
Supervisory frame, 310, 317–21, 324–33, 653
Survivability, 21
SYN, 167–71, 262, 307, 316–18, 348, 653
SYN bit, 551
Synchronization, 19, 115, 183, 208, 215, 222, 261–62, 303,
   309, 319, 347, 530, 618, 653
   of access, 262
   of bits, 261–62
   of bytes, 262
   of content, 262
   of frames or messages, 262
   of waveforms, 261
Synchronous, 46, 222, 307, 341, 653
   communication satellites. *See* Geosynchronous
    communication satellites
   data link control protocols, 304–47, 653
   idle. *See* SYN
   transfer mode (STM), 631, 653
Synchronous Optical Network (SONET), 612, 631, 633–37,
   652
Syntax, 117, 120, 653
System services control point (SSCP), 141, 589–95, 653
Systems Network Architecture (SNA), 12, 107, 138–48,
   180, 419, 431, 653
   data flow control layer, 142–44, 589, 591
   data link control layer, 142–44, 304, 347
   path control layer, 142–44, 464–71
   presentation services layer, 142–44, 589, 591
   transaction services layer, 142–44, 589, 591
   transmission control layer, 142–44, 588–97
systemsnetworkarchitecturecommitteespeak, 139

T. *See* Transmit circuit
$\overline{T}$. *See* Average transfer delay
$\hat{T}$. *See* Normalized average transfer delay
T-1 carrier, 12–13, 37, 178–79, 653
TA. *See* Terminal Adaptor
τ. *See* Propagation delay
τ'. *See* Ring latency
T_c. *See* Average cycle time
TCH. *See* Transmission control header
TCP. *See* Transmission control protocol
TD. *See* Transmitted data circuit
TDM. *See* Time division multiplexing
TDMA. *See* Time division multiple access
TDMA/reservations, 208–13
TE, TE1 and TE2. *See* Terminal Equipment
Technical and Office Products System (TOP), 146–48,
   180–81, 338, 431, 472, 520–22, 597, 653
   layers. *See* Manufacturing Automation Protocol layers
TELENET, 443–44, 477–79, 653

Telephone network, 34–38, 446
user part, 622–23
Teleservices, 146
Teletel, 12
Teletext, 12, 146
Terminal Adaptor (TA), 616, 653
Equipment (TE), 616, 653
Equipment Type 1 (TE1) and Type 2 (TE2), 616, 653
Terminate phase. *See* Disconnection
Termself, 594–95
Terrestrial radio, 107
TH. *See* Transport header
Thermal noise, 38
Three-army problem, 19–20
-way handshake, 543–54, 569–71, 583–85
Throughput, 22, 233, 245–46, 251–54, 359, 394–96, 429–32, 564, 653
Time complexity, 82
constant, 32
division multiple access (TDMA), 208–10, 237, 653
division multiplexing (TDM) 13, 33–37, 61–63, 607, 653
hierarchy, 37, 62, 607
domain, 48
duration, 48
to try reassignment/resynchronization (TTR), 576
to wait for reassignment/resynchronization (TWR), 576
Timestamp, 517–19
Timing diagram, 159–62, 167–70, 653
Timings, 117, 156, 495, 653
Timeouts, 281–84, 318–21, 330–32, 421, 495, 546–48, 556, 562
Token, 218–22
bus, 106, 145–46, 216–24, 339, 503, 508, 653
passing, 106, 145–46, 216–24, 254–55, 339, 653
ring, 106, 145–46, 216–23, 237–42, 255, 339, 499–503, 508, 653
TOP. *See* Technical and Office Products System
Topology, 358, 391, 459
See also Geometry
TP0, 562, 566–74, 580, 642
TP1, 562–63, 566–76, 580, 642–43
TP2, 562–63, 566–74, 580, 643
TP3, 562, 566–76, 580, 643
TP4, 562–63, 566–81, 597, 643
TPDU. *See* Transport protocol data unit
error (ER) TPDU, 566–67
Traffic, 355, 394, 421
Trailer, 121–22, 144, 303, 435, 497–98, 653
Training sequence, 188
Transaction services layer, 142–44, 589, 653
Transfer phase. *See* Data transfer
Transmission control header (TCH), 143–44
control layer, 142–44, 588–97

control protocol (TCP), 132, 454, 514, 543, 548–63, 581, 587, 597, 653
group, 467–71, 654
header (SNA TH). *See* Path control header
time, 399–407, 460
Transmit (T) circuit, 165–67, 654
window. *See* Sliding window flow control
Transmitted data (TD) circuit, 157–62, 193–95, 654
TRANSPAC, 443–44, 478–79, 654
Transparency, 503, 507
Transparent text mode, 307, 311, 315–17, 654
Transport header (TH), 121–22
layer, 15, 119–25, 147, 429–31, 442, 446, 450, 489, 520–22, 539–98, 618, 654
level flow control, 396
protocol data unit (TPDU), 561–63, 566–74
service access point (TSAP), 561, 566
service data unit (TSDU), 561
Traveling salesperson problem, 84
Tree, 206, 362–65, 654
Trellis coding, 37, 54–57, 272, 654
Tributary station, 312–13, 318–19, 654
Triple X, 442, 654
Trunk, 37, 96–97, 603
TSAP. *See* Transport service access point
TSDU. *See* Transport service data unit
TTR. *See* Time to try reassignment/resynchronization
Turnaround time. *See* Line turnaround time
Twisted pairs, 26, 33, 43, 429, 654
Two-dimensional parity check, 266–69, 654
Two-way handshake, 543–44, 569–70
virtual call, 433–35
TWR. *See* Time to wait for reassignment/resynchronization
TYMNET, 375, 443–45, 479–81, 654
Type 1 and Type 2 operation, 341, 654
Type A, Type B and Type C network service, 561–62, 654

UART. *See* Universal synchronous/asynchronous receiver transmitter
UD. *See* Unit data TPDU
UDP. *See* User datagram protocol
UHF transmission, 39–40
U/L bit, 340, 654
Ultra small aperture terminal mobile satellite (MSAT) system, 42
Unbalanced configuration, 322–23, 654
mode transmission, 157, 163–65, 654
Unbind, 594–95
Undersea cables, 34
Unipolar RZ encoding, 57–60
Unitdata (UD) TPDU, 566–67, 571, 654
Universal synchronous/asynchronous receiver transmitter (USART), 184–87, 305, 654

Unnumbered frames, 325–27, 334–37, 654
Update times, 357–58
URG bit, 562–63
USART. *See* Universal synchronous/asynchronous
    receiver transmitter
User commands, 549
   datagram protocol (UDP), 132, 559–60, 654
   layer, 135–37, 654
   plane, 615–21
   process, 432–33
Utilization, 22

V.24, 156–64, 192–96, 654
Variability of parameters, 34, 263, 446
VC. *See* Virtual call
Vendor protocols, 431–33, 443–45
Very small aperture terminal (VSAT) system, 42
VHF transmission, 39–40
Videotex, 12, 146
Virtual call (VC), 432–38, 654
   circuit, 99, 142, 360–61, 383, 432–39, 444–45, 494–97,
     549, 597, 654
   route, 467–69, 654
   Route Change Window Indicator (VR-CWI), 467,
     471
   Route Change Window Reply Indicator (VR-CWRI), 467,
     471
   Route number (VRN), 467–69
   Route Reset Window Indicator (VR-RWI), 467, 471

Voice-grade channel, 36–37, 654
VR-CWI. *See* Virtual Route Change Window Indicator
VR-CWRI. *See* Virtual Route Change Window Reply
    Indicator
VRN. *See* Virtual Route Number
VR-RWI. *See* Virtual Route Reset Window Indicator
VSAT. *See* Very small aperture terminal system
W. See Window size
WACK. *See* Wait acknowledgment
Wait acknowledgment (WACK), 295–96, 326, 419, 654
Walk time, 239
WAN. *See* Wide area network
Waveguide, 34
Weight of code word, 269
Wide area network (WAN), 1, 106–7, 489, 654
Window size (*W*), 398–412, 439–40
Worldnet, 107, 489, 654

X.3, 442–43, 654
X.21, 154, 165–72, 192, 262, 447, 654
X.21bis, 171, 432, 654
X.25, 322, 334, 347, 431–47, 449, 473, 490, 497–98, 509–11,
    540–42, 654
   X.25 based vendor networks, 431, 443–45, 475–481
X.28, 442–43, 654
X.29, 442–43, 654
X.75, 489, 509–12, 654

zero insertion, 323–24, 336, 623, 654